Student Solutic

Jon D. Weerts

Precalculus

Graphing
and Data
Analysis

Second
Edition

Michael Sullivan
Michael Sullivan, III

Prentice
Hall

Upper Saddle River, NJ 07458

Editor in Chief: Sally Yagan
Editorial Assistant: Meisha Welch
Special Projects Manager: Barbara A. Murray
Production Editor: Dawn Murrin
Supplement Cover Manager: Paul Gourhan
Supplement Cover Designer: PM Workshop Inc.
Manufacturing Manager: Trudy Pisciotti

Printed in the United States of America

10 9 8 7 6 5 4 3 2 1

ISBN 0-13-028759-8

Prentice-Hall International (UK) Limited, London
Prentice-Hall of Australia Pty. Limited, Sydney
Prentice-Hall Canada, Inc., Toronto
Prentice-Hall Hispanoamericana, S.A., Mexico
Prentice-Hall of India Private Limited, New Delhi
Pearson Education Asia Pte. Ltd., Singapore
Prentice-Hall of Japan, Inc., Tokyo
Editora Prentice-Hall do Brazil, Ltda., Rio de Janeiro

Contents

Chapter 4   The Zeros of a Polynomial Function

Chapter 5   Exponential and Logarithmic Functions

Chapter 6   Trigonometric Functions

Chapter 7   Analytic Trigonometry

Chapter 8   Applications of Trigonometric Functions

Chapter 9   Polar Coordinates; Vectors

Chapter 10   Analytic Geometry

Chapter 11   Systems of Equations and Inequalities

Chapter 12 Sequences; Induction; The Binomial Theorem

Chapter 13 Counting and Probability

Chapter 14 A Preview of Calculus: The Limit, Derivative, and Integral of a Function

Appendix

Preface

The <u>Student's Solutions Manual</u> to accompany <u>Precalculus: Graphing and Data Analysis, 2nd Edition</u> by Michael Sullivan and Michael Sullivan, III contains detailed solutions to all of the odd numbered problems in the textbook. I have attempted to provide solutions consistent with the procedures introduced in the textbook. TI-83 graphing calculator screens have been included to demonstrate the use of the graphics calculator in solving and in checking solutions to the problems where requested. Every attempt has been made to make this manual as error free as possible. If you have suggestions, error corrections, or comments please feel free to write to me about them.

A number of people need to be recognized for their contributions in the preparation of this manual. Thanks go to Sally Yagan, and Meisha Welch at Prentice Hall for giving me the opportunity to author this manual. Thanks also to Dr. Carole Bauer for thoroughly checking my solutions for errors and providing many useful suggestions to make the solutions more accurate and readable. Thanks also to Mike Sullivan for all his help when I had concerns about the content of the problems.

I wish to express my appreciation to Michael Sullivan and Michael Sullivan, III for providing a thorough Precalculus textbook which assists students in understanding algebraic and trigonometric functions and provides guidance in the use of a graphing utility to assist in that study.

A special word of appreciation goes to my wife, Jan, for her support and understanding during the many hours that went into the preparation of this manual.

Jon D. Weerts
Triton College
2000 Fifth Avenue
River Grove, Il 60171
e-mail: jweerts@triton.cc.il.us

Chapter 1

Graphs and Equations

1.1 Rectangular Coordinates; Graphing Utilities

1. (a) Quadrant II
   (b) Positive x-axis
   (c) Quadrant III
   (d) Quadrant I
   (e) Negative y-axis
   (f) Quadrant IV

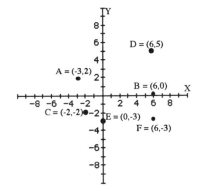

3. The points will be on a vertical line that is two units to the right of the y-axis.

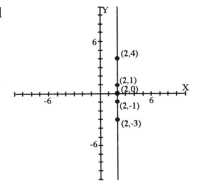

5. $(-1,4)$ Quadrant II

7. $(3,1)$ Quadrant I

9. $X\min = -11$
   $X\max = 5$
   $X\text{scl} = 1$
   $Y\min = -3$
   $Y\max = 6$
   $Y\text{scl} = 1$

11. $X\min = -30$
    $X\max = 50$
    $X\text{scl} = 10$
    $Y\min = -90$
    $Y\max = 50$
    $Y\text{scl} = 10$

13. $X\min = -10$
    $X\max = 110$
    $X\text{scl} = 10$
    $Y\min = -10$
    $Y\max = 160$
    $Y\text{scl} = 10$

1

15. $\begin{array}{l} \text{Xmin} = -6 \\ \text{Xmax} = 6 \\ \text{Xscl} = 2 \\ \text{Ymin} = -4 \\ \text{Ymax} = 4 \\ \text{Yscl} = 2 \end{array}$

17. $\begin{array}{l} \text{Xmin} = -6 \\ \text{Xmax} = 6 \\ \text{Xscl} = 2 \\ \text{Ymin} = -1 \\ \text{Ymax} = 3 \\ \text{Yscl} = 1 \end{array}$

19. $\begin{array}{l} \text{Xmin} = 3 \\ \text{Xmax} = 9 \\ \text{Xscl} = 1 \\ \text{Ymin} = 2 \\ \text{Ymax} = 10 \\ \text{Yscl} = 2 \end{array}$

21. $d(P_1, P_2) = \sqrt{(2-0)^2 + (1-0)^2} = \sqrt{4+1} = \sqrt{5}$

23. $d(P_1, P_2) = \sqrt{(-2-1)^2 + (2-1)^2} = \sqrt{9+1} = \sqrt{10}$

25. $d(P_1, P_2) = \sqrt{(5-3)^2 + (4-(-4))^2} = \sqrt{2^2 + 8^2} = \sqrt{4+64} = \sqrt{68} = 2\sqrt{17}$

27. $d(P_1, P_2) = \sqrt{(6-(-3))^2 + (0-2)^2} = \sqrt{9^2 + (-2)^2} = \sqrt{81+4} = \sqrt{85}$

29. $d(P_1, P_2) = \sqrt{(6-4)^2 + (4-(-3))^2} = \sqrt{2^2 + 7^2} = \sqrt{4+49} = \sqrt{53}$

31. $d(P_1, P_2) = \sqrt{(2.3-(-0.2))^2 + (1.1-0.3)^2} = \sqrt{(2.5)^2 + (0.8)^2}$
$$= \sqrt{6.25+0.64} = \sqrt{6.89} \approx 2.625$$

33. $d(P_1, P_2) = \sqrt{(0-a)^2 + (0-b)^2} = \sqrt{a^2 + b^2}$

35. $P_1 = (1,3), \quad P_2 = (5,15)$
$d(P_1, P_2) = \sqrt{(5-1)^2 + (15-3)^2} = \sqrt{4^2 + 12^2} = \sqrt{16+144} = \sqrt{160} = 4\sqrt{10}$

37. $P_1 = (-4,6), \quad P_2 = (4,-8)$
$d(P_1, P_2) = \sqrt{(4-(-4))^2 + (-8-6)^2} = \sqrt{8^2 + (-14)^2} = \sqrt{64+196} = \sqrt{260} = 2\sqrt{65}$

39. $A = (-2,5), \quad B = (1,3), \quad C = (-1,0)$
$d(A,B) = \sqrt{(1-(-2))^2 + (3-5)^2} = \sqrt{3^2 + (-2)^2}$
$$= \sqrt{9+4} = \sqrt{13}$$
$d(B,C) = \sqrt{(-1-1)^2 + (0-3)^2} = \sqrt{(-2)^2 + (-3)^2}$
$$= \sqrt{4+9} = \sqrt{13}$$
$d(A,C) = \sqrt{(-1-(-2))^2 + (0-5)^2} = \sqrt{1^2 + (-5)^2}$
$$= \sqrt{1+25} = \sqrt{26}$$

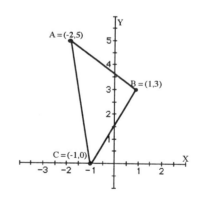

Verifying that $\triangle ABC$ is a right triangle by the Pythagorean Theorem:

$$\left[d(A,B)\right]^2 + \left[d(B,C)\right]^2 = \left[d(A,C)\right]^2$$
$$\left(\sqrt{13}\right)^2 + \left(\sqrt{13}\right)^2 = \left(\sqrt{26}\right)^2$$
$$13 + 13 = 26$$
$$26 = 26$$

The area of a triangle is $A = \frac{1}{2}bh$. In this problem,

$$A = \frac{1}{2}\left[d(A,B)\right] \cdot \left[d(B,C)\right] = \frac{1}{2}\sqrt{13} \cdot \sqrt{13} = \frac{1}{2} \cdot 13 = \frac{13}{2} \text{ square units}$$

41. $A = (-5,3),\ \ B = (6,0),\ \ C = (5,5)$

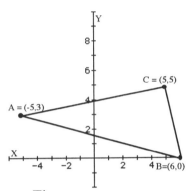

$$d(A,B) = \sqrt{\left(6-(-5)\right)^2 + \left(0-3\right)^2} = \sqrt{11^2 + (-3)^2}$$
$$= \sqrt{121+9} = \sqrt{130}$$
$$d(B,C) = \sqrt{\left(5-6\right)^2 + \left(5-0\right)^2} = \sqrt{(-1)^2 + 5^2}$$
$$= \sqrt{1+25} = \sqrt{26}$$
$$d(A,C) = \sqrt{\left(5-(-5)\right)^2 + \left(5-3\right)^2} = \sqrt{10^2 + 2^2}$$
$$= \sqrt{100+4} = \sqrt{104}$$

Verifying that $\triangle ABC$ is a right triangle by the Pythagorean Theorem:

$$\left[d(A,C)\right]^2 + \left[d(B,C)\right]^2 = \left[d(A,B)\right]^2$$
$$\left(\sqrt{104}\right)^2 + \left(\sqrt{26}\right)^2 = \left(\sqrt{130}\right)^2$$
$$104 + 26 = 130$$
$$130 = 130$$

The area of a triangle is $A = \frac{1}{2}bh$. In this problem,

$$A = \frac{1}{2}\left[d(A,C)\right] \cdot \left[d(B,C)\right] = \frac{1}{2}\sqrt{104} \cdot \sqrt{26} = \frac{1}{2}\sqrt{2704} = \frac{1}{2} \cdot 52 = 26 \text{ square units}$$

43. $A = (4,-3),\ \ B = (0,-3),\ \ C = (4,2)$

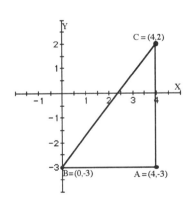

$$d(A,B) = \sqrt{\left(0-4\right)^2 + \left(-3-(-3)\right)^2} = \sqrt{(-4)^2 + 0^2}$$
$$= \sqrt{16+0} = \sqrt{16} = 4$$
$$d(B,C) = \sqrt{\left(4-0\right)^2 + \left(2-(-3)\right)^2} = \sqrt{4^2 + 5^2}$$
$$= \sqrt{16+25} = \sqrt{41}$$
$$d(A,C) = \sqrt{\left(4-4\right)^2 + \left(2-(-3)\right)^2} = \sqrt{0^2 + 5^2}$$
$$= \sqrt{0+25} = \sqrt{25} = 5$$

Verifying that $\triangle ABC$ is a right triangle by the Pythagorean Theorem:

$$\left[d(A,B)\right]^2 + \left[d(A,C)\right]^2 = \left[d(B,C)\right]^2$$
$$4^2 + 5^2 = \left(\sqrt{41}\right)^2$$
$$16 + 25 = 41$$
$$41 = 41$$

The area of a triangle is $A = \frac{1}{2}bh$. In this problem,

$$A = \frac{1}{2}\left[d(A,B)\right] \cdot \left[d(A,C)\right] = \frac{1}{2} \cdot 4 \cdot 5 = 10 \text{ square units}$$

45. All points having an x-coordinate of 2 are of the form (2, y). Those which are 5 units from (−2, −1) are:

$$\sqrt{(2-(-2))^2 + (y-(-1))^2} = 5$$
$$\sqrt{4^2 + (y+1)^2} = 5$$

Squaring both sides: $\quad 4^2 + (y+1)^2 = 25$
$$16 + y^2 + 2y + 1 = 25$$
$$y^2 + 2y - 8 = 0$$
$$(y+4)(y-2) = 0$$
$$y = -4 \text{ or } y = 2$$

Therefore, the points are (2, −4) or (2, 2).

47. All points on the x-axis are of the form (x, 0). Those which are 5 units from (4, −3) are:

$$\sqrt{(x-4)^2 + (0-(-3))^2} = 5$$
$$\sqrt{(x-4)^2 + 3^2} = 5$$

Squaring both sides: $\quad (x-4)^2 + 9 = 25$
$$x^2 - 8x + 16 + 9 = 25$$
$$x^2 - 8x = 0$$
$$x(x-8) = 0$$
$$x = 0 \text{ or } x = 8$$

Therefore, the points are (0, 0) or (8, 0).

49. The coordinates of the midpoint are:

$$(x,y) = \left(\frac{x_1 + x_2}{2}, \frac{y_1 + y_2}{2}\right) = \left(\frac{5+3}{2}, \frac{-4+2}{2}\right) = \left(\frac{8}{2}, \frac{-2}{2}\right) = (4, -1)$$

51. The coordinates of the midpoint are:

$$(x,y) = \left(\frac{x_1 + x_2}{2}, \frac{y_1 + y_2}{2}\right) = \left(\frac{-3+6}{2}, \frac{2+0}{2}\right) = \left(\frac{3}{2}, \frac{2}{2}\right) = \left(\frac{3}{2}, 1\right)$$

53. The coordinates of the midpoint are:

$$(x,y) = \left(\frac{x_1 + x_2}{2}, \frac{y_1 + y_2}{2}\right) = \left(\frac{4+6}{2}, \frac{-3+1}{2}\right) = \left(\frac{10}{2}, \frac{-2}{2}\right) = (5, -1)$$

55. The coordinates of the midpoint are:

$$(x,y) = \left(\frac{x_1 + x_2}{2}, \frac{y_1 + y_2}{2} \right) = \left(\frac{-0.2 + 2.3}{2}, \frac{0.3 + 1.1}{2} \right) = \left(\frac{2.1}{2}, \frac{1.4}{2} \right) = (1.05, 0.7)$$

57. The coordinates of the midpoint are:

$$(x,y) = \left(\frac{x_1 + x_2}{2}, \frac{y_1 + y_2}{2} \right) = \left(\frac{a + 0}{2}, \frac{b + 0}{2} \right) = \left(\frac{a}{2}, \frac{b}{2} \right)$$

59. The midpoint of AB is: $D = \left(\frac{0+0}{2}, \frac{0+6}{2} \right) = (0, 3)$

    The midpoint of AC is: $E = \left(\frac{0+4}{2}, \frac{0+4}{2} \right) = (2, 2)$

    The midpoint of BC is: $F = \left(\frac{0+4}{2}, \frac{6+4}{2} \right) = (2, 5)$

    $$d(C, D) = \sqrt{(0-4)^2 + (3-4)^2} = \sqrt{(-4)^2 + (-1)^2} = \sqrt{16+1} = \sqrt{17}$$

    $$d(B, E) = \sqrt{(2-0)^2 + (2-6)^2} = \sqrt{2^2 + (-4)^2} = \sqrt{4+16} = \sqrt{20} = 2\sqrt{5}$$

    $$d(A, F) = \sqrt{(2-0)^2 + (5-0)^2} = \sqrt{2^2 + 5^2} = \sqrt{4+25} = \sqrt{29}$$

61. $d(P_1, P_2) = \sqrt{(-4-2)^2 + (1-1)^2} = \sqrt{(-6)^2 + 0^2} = \sqrt{36} = 6$

    $d(P_2, P_3) = \sqrt{(-4-(-4))^2 + (-3-1)^2} = \sqrt{0^2 + (-4)^2} = \sqrt{16} = 4$

    $d(P_1, P_3) = \sqrt{(-4-2)^2 + (-3-1)^2} = \sqrt{(-6)^2 + (-4)^2} = \sqrt{36+16} = \sqrt{52} = 2\sqrt{13}$

    Since $\left[d(P_1, P_2) \right]^2 + \left[d(P_2, P_3) \right]^2 = \left[d(P_1, P_3) \right]^2$, the triangle is a right triangle.

63. $d(P_1, P_2) = \sqrt{(0-(-2))^2 + (7-(-1))^2} = \sqrt{2^2 + 8^2} = \sqrt{4+64} = \sqrt{68} = 2\sqrt{17}$

    $d(P_2, P_3) = \sqrt{(3-0)^2 + (2-7)^2} = \sqrt{3^2 + (-5)^2} = \sqrt{9+25} = \sqrt{34}$

    $d(P_1, P_3) = \sqrt{(3-(-2))^2 + (2-(-1))^2} = \sqrt{5^2 + 3^2} = \sqrt{25+9} = \sqrt{34}$

    Since $d(P_2, P_3) = d(P_1, P_3)$, the triangle is isosceles.

    Since $\left[d(P_1, P_3) \right]^2 + \left[d(P_2, P_3) \right]^2 = \left[d(P_1, P_2) \right]^2$, the triangle is also a right triangle.
    Therefore, the triangle is an isosceles right triangle.

65. Plot the vertices of the square at
     $(0, 0)$, $(0, s)$, (s, s), and $(s, 0)$.

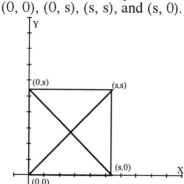

Find the midpoints of the diagonals.

$$M_1 = \left(\frac{0+s}{2}, \frac{0+s}{2}\right) = \left(\frac{s}{2}, \frac{s}{2}\right)$$

$$M_2 = \left(\frac{0+s}{2}, \frac{s+0}{2}\right) = \left(\frac{s}{2}, \frac{s}{2}\right)$$

Since the coordinates of the midpoints
are the same , the diagonals of a square
intersect at their midpoints .

67. Using the Pythagorean Theorem:

$$90^2 + 90^2 = d^2$$
$$8100 + 8100 = d^2$$
$$16200 = d^2$$
$$d = \sqrt{16200}$$
$$d = 90\sqrt{2} \approx 127.28 \text{ feet}$$

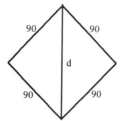

69. (a) First: $(90, 0)$, Second: $(90, 90)$
          Third: $(0, 90)$

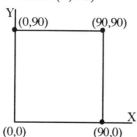

     (b) Using the distance formula:
$$d = \sqrt{(310-90)^2 + (15-90)^2}$$
$$= \sqrt{220^2 + (-75)^2}$$
$$= \sqrt{54025} \approx 232.4 \text{ feet}$$

     (c) Using the distance formula:
$$d = \sqrt{(300-0)^2 + (300-90)^2}$$
$$= \sqrt{300^2 + 210^2}$$
$$= \sqrt{134100} \approx 366.2 \text{ feet}$$

71. The Intrepid heading east moves a distance $30t$ after t
     hours. The truck heading south moves a distance $40t$
     after t hours. Their distance apart after t hours is:

$$d = \sqrt{(30t)^2 + (40t)^2}$$
$$= \sqrt{900t^2 + 1600t^2}$$
$$= \sqrt{2500t^2}$$
$$= 50t$$

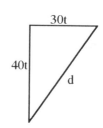

1.2 Graphs of Equations

1. $y = x^4 - \sqrt{x}$

    $0 = 0^4 - \sqrt{0}$ $1 = 1^4 - \sqrt{1}$ $0 = (-1)^4 - \sqrt{-1}$

    $0 = 0$ $1 \neq 0$ $0 \neq 1 - \sqrt{-1}$

    $(0, 0)$ is on the graph of the equation.

3. $y^2 = x^2 + 9$

    $3^2 = 0^2 + 9$ $0^2 = 3^2 + 9$ $0^2 = (-3)^2 + 9$

    $9 = 9$ $0 \neq 18$ $0 \neq 18$

    $(0, 3)$ is on the graph of the equation.

5. $x^2 + y^2 = 4$

    $0^2 + 2^2 = 4$ $(-2)^2 + 2^2 = 4$ $\sqrt{2}^2 + \sqrt{2}^2 = 4$

    $4 = 4$ $8 \neq 4$ $4 = 4$

    $(0, 2)$ and $\left(\sqrt{2}, \sqrt{2}\right)$ are on the graph of the equation.

7. (a)   (b)   (c)   (d)

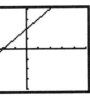 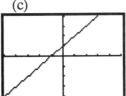 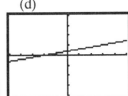

9. (a)   (b)   (c)   (d)

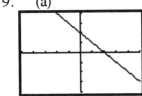 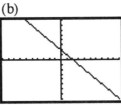 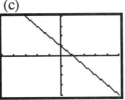 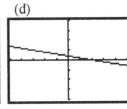

11. (a)   (b)   (c)   (d)

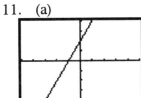 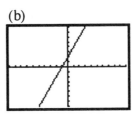 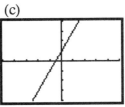 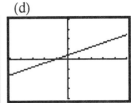

13. (a)   (b)   (c)   (d)

15. (a)   (b)   (c)   (d)

17. (a)   (b)   (c)   (d)

19. (a)   (b)   (c)   (d)

21. (a)   (b)   (c)   (d)

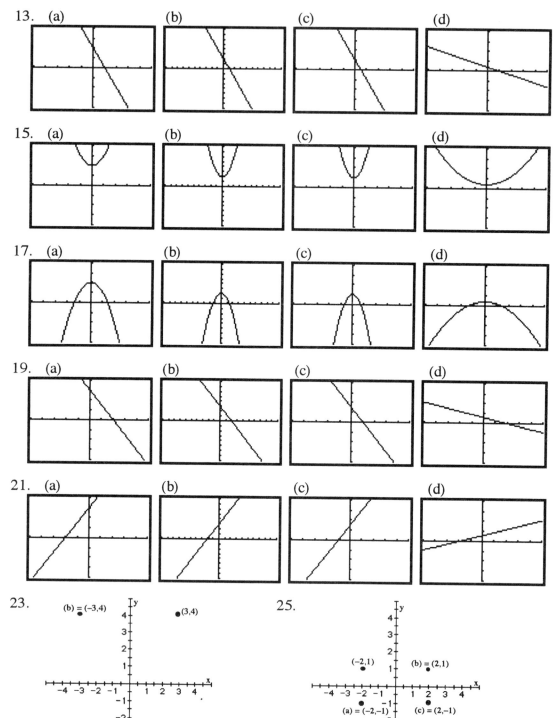

23.

(b) = (-3,4)   (3,4)

(c) = (-3,-4)   (a) = (3,-4)

25.

(-2,1)   (b) = (2,1)

(a) = (-2,-1)   (c) = (2,-1)

27.

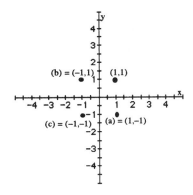

29.

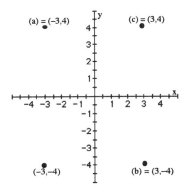

31.

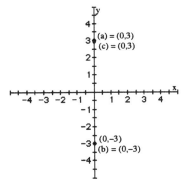

33. (a) $(-1, 0), (1, 0)$ (b) symmetric to the y-axis

35. (a) $\left(\dfrac{-\pi}{2}, 0\right), \left(\dfrac{\pi}{2}, 0\right), (0, 1)$ (b) symmetric to the y-axis

37. (a) $(0, 0)$ (b) symmetric to the x-axis

39. (a) $(1, 0)$ (b) not symmetric to x-axis, y-axis, or origin

41. (a) $(-1.5, 0), (1.5, 0), (0, -2)$ (b) symmetric to the y-axis

43. (a) none   (b) symmetric to the origin

45. $y = 5x + 4$
$2 = 5a + 4$
$5a = -2$
$a = \dfrac{-2}{5}$

47. $2x + 3y = 6$
$2a + 3b = 6$

49.

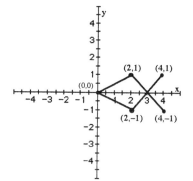

51.

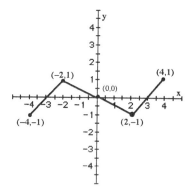

53.

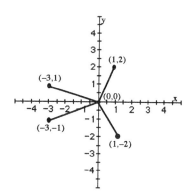

55.

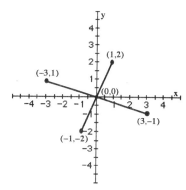

57. $x^2 = y$

    y - intercept : Let $x = 0$, then $y = 0$ (0,0)

    x - intercept : Let $y = 0$, then $x = 0$ (0,0)

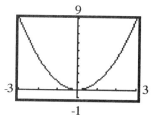

    Test for symmetry:

      x - axis: Replace y by $-y$ so $x^2 = -y$, which is not equivalent to $x^2 = y$.

      y - axis: Replace x by $-x$ so $(-x)^2 = y$ or $x^2 = y$, which is equivalent to $x^2 = y$.

      Origin: Replace x by $-x$ and y by $-y$ so $(-x)^2 = -y$ or $x^2 = -y$,

                 which is not equivalent to $x^2 = y$.

    Therefore, the graph is symmetric with respect to the y - axis .

59. $y = 3x$

    y - intercept : Let $x = 0$, then $y = 0$ (0,0)

    x - intercept : Let $y = 0$, then $x = 0$ (0,0)

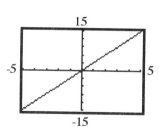

Test for symmetry:

x-axis: Replace y by $-y$ so $-y = 3x$, which is not equivalent to $y = 3x$.

y-axis: Replace x by $-x$ so $y = 3(-x)$ or $y = -3x$,
which is not equivalent to $y = 3x$.

Origin: Replace x by $-x$ and y by $-y$ so $-y = 3(-x)$ or $y = 3x$,
which is equivalent to $y = 3x$.

Therefore, the graph is symmetric with respect to the origin.

61. $x + y - 9 = 0$

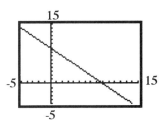

y-intercept: Let $x = 0$, then $0 + y - 9 = 0$
$$y = 9 \qquad (0,9)$$

x-intercept: Let $y = 0$, then $x + 0 - 9 = 0$
$$x = 9 \qquad (9,0)$$

Test for symmetry:

x-axis: Replace y by $-y$ so $x - y - 9 = 0$, which is not
equivalent to $x + y - 9 = 0$.

y-axis: Replace x by $-x$ so $-x + y - 9 = 0$, which is not
equivalent to $x + y - 9 = 0$.

Origin: Replace x by $-x$ and y by $-y$ so $-x - y - 9 = 0$ or $x + y + 9 = 0$,
which is not equivalent to $x + y - 9 = 0$.

Therefore, the graph is not symmetric to the x-axis, the y-axis, or the origin.

63. $9x^2 + 4y = 36$

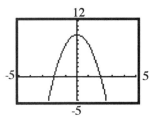

y-intercept: Let $x = 0$, then $0 + 4y = 36$
$$y = 9 \qquad (0,9)$$

x-intercept: Let $y = 0$, then $9x^2 + 0 = 36$
$$x^2 = 4$$
$$x = \pm 2 \quad (2,0), (-2,0)$$

Test for symmetry:

x-axis: Replace y by $-y$ so $9x^2 + 4(-y) = 36$ or $9x^2 - 4y = 36$, which is not
equivalent to $9x^2 + 4y = 36$.

y-axis: Replace x by $-x$ so $9(-x)^2 + 4y = 36$ or $9x^2 + 4y = 36$,
which is equivalent to $9x^2 + 4y = 36$.

Origin: Replace x by $-x$ and y by $-y$ so $9(-x)^2 + 4(-y) = 36$ or
$9x^2 - 4y = 36$, which is not equivalent to $9x^2 + 4y = 36$.

Therefore, the graph is symmetric with respect to the y-axis.

65. $y = x^3 - 27$

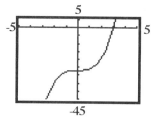

    y - intercept : Let $x = 0$, then $y = 0^3 - 27$

$$y = -27 \qquad (0, -27)$$

    x - intercept : Let $y = 0$, then $0 = x^3 - 27$

$$x^3 = 27$$

$$x = 3 \qquad (3, 0)$$

  Test for symmetry :

    x - axis: Replace y by $-y$ so $-y = x^3 - 27$, which is not

             equivalent to $y = x^3 - 27$.

    y - axis: Replace x by $-x$ so $y = (-x)^3 - 27$ or $y = -x^3 - 27$,

             which is not equivalent to $y = x^3 - 27$.

    Origin: Replace x by $-x$ and y by $-y$ so $-y = (-x)^3 - 27$ or

           $y = x^3 + 27$, which is not equivalent to $y = x^3 - 27$.

  Therefore, the graph is not symmetric to the x-axis, the y-axis, or the origin.

67. $y = x^2 - 3x - 4$

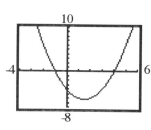

    y - intercept : Let $x = 0$, then $y = 0^2 - 3(0) - 4$

$$y = -4 \qquad (0, -4)$$

    x - intercept : Let $y = 0$, then $0 = x^2 - 3x - 4$

$$(x - 4)(x + 1) = 0$$

$$x = 4 \quad x = -1 \quad (4, 0), (-1, 0)$$

  Test for symmetry :

    x - axis: Replace y by $-y$ so $-y = x^2 - 3x - 4$, which is not

             equivalent to $y = x^2 - 3x - 4$.

    y - axis: Replace x by $-x$ so $y = (-x)^2 - 3(-x) - 4$ or $y = x^2 + 3x - 4$,

             which is not equivalent to $y = x^2 - 3x - 4$.

    Origin: Replace x by $-x$ and y by $-y$ so $-y = (-x)^2 - 3(-x) - 4$ or

           $y = -x^2 - 3x + 4$, which is not equivalent to $y = x^2 - 3x - 4$.

  Therefore, the graph is not symmetric to the x-axis, the y-axis, or the origin.

69. $y = \dfrac{x}{x^2 + 9}$

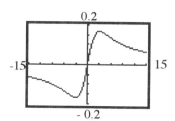

    y - intercept : Let $x = 0$, then $y = \dfrac{0}{0^2 + 9}$

$$y = 0 \qquad (0, 0)$$

    x - intercept : Let $y = 0$, then $0 = \dfrac{x}{x^2 + 9}$

$$x = 0 \qquad (0, 0)$$

Test for symmetry:

x-axis: Replace y by $-y$ so $-y = \dfrac{x}{x^2 + 9}$, which is not

equivalent to $y = \dfrac{x}{x^2 + 9}$.

y-axis: Replace x by $-x$ so $y = \dfrac{-x}{(-x)^2 + 9}$ or $y = \dfrac{-x}{x^2 + 9}$,

which is not equivalent to $y = \dfrac{x}{x^2 + 9}$.

Origin: Replace x by $-x$ and y by $-y$ so $-y = \dfrac{-x}{(-x)^2 + 9}$ or

$y = \dfrac{x}{x^2 + 9}$, which is equivalent to $y = \dfrac{x}{x^2 + 9}$.

Therefore, the graph is symmetric with respect to the origin.

71.   (a)

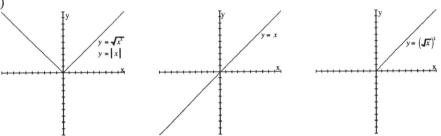

(b)   Since $\sqrt{x^2} = |x|$, then for all x, the graphs of $y = \sqrt{x^2}$ and $y = |x|$ are the same.

(c)   For $y = \left(\sqrt{x}\right)^2$, the domain of the variable x is $x \ge 0$; for $y = x$, the domain of the variable x is all real numbers. Thus, $\left(\sqrt{x}\right)^2 = x$ only for $x \ge 0$.

(d)   For $y = \sqrt{x^2}$, the range of the variable y is $y \ge 0$; for $y = x$, the range of the variable y is all real numbers. Also, $\sqrt{x^2} = x$ only if $x \ge 0$.

1.3   Solving Equations

1.
$$3x + 2 = x + 6$$
$$3x + 2 - 2 = x + 6 - 2$$
$$3x = x + 4$$
$$3x - x = x + 4 - x$$
$$2x = 4$$
$$\frac{2x}{2} = \frac{4}{2}$$
$$x = 2$$

3.
$$2t - 6 = 3 - t$$
$$2t - 6 + 6 = 3 - t + 6$$
$$2t = 9 - t$$
$$2t + t = 9 - t + t$$
$$3t = 9$$
$$\frac{3t}{3} = \frac{9}{3}$$
$$t = 3$$

5.
$$6 - x = 2x + 9$$
$$6 - x - 6 = 2x + 9 - 6$$
$$-x = 2x + 3$$
$$-x - 2x = 2x + 3 - 2x$$
$$-3x = 3$$
$$\frac{-3x}{-3} = \frac{3}{-3}$$
$$x = -1$$

7.
$$3 + 2n = 5n + 7$$
$$3 + 2n - 3 = 5n + 7 - 3$$
$$2n = 5n + 4$$
$$2n - 5n = 5n + 4 - 5n$$
$$-3n = 4$$
$$\frac{-3n}{-3} = \frac{4}{-3}$$
$$n = \frac{-4}{3}$$

9.
$$2(3 + 2x) = 3(x - 4)$$
$$6 + 4x = 3x - 12$$
$$6 + 4x - 6 = 3x - 12 - 6$$
$$4x = 3x - 18$$
$$4x - 3x = 3x - 18 - 3x$$
$$x = -18$$

11.
$$8x - (2x + 1) = 3x - 10$$
$$8x - 2x - 1 = 3x - 10$$
$$6x - 1 = 3x - 10$$
$$6x - 1 + 1 = 3x - 10 + 1$$
$$6x = 3x - 9$$
$$6x - 3x = 3x - 9 - 3x$$
$$3x = -9$$
$$\frac{3x}{3} = \frac{-9}{3}$$
$$x = -3$$

13.
$$\tfrac{2}{3}p = \tfrac{1}{2}p + \tfrac{1}{3}$$
$$6\left(\tfrac{2}{3}p\right) = 6\left(\tfrac{1}{2}p + \tfrac{1}{3}\right)$$
$$4p = 3p + 2$$
$$4p - 3p = 3p + 2 - 3p$$
$$p = 2$$

15.
$$0.9t = 0.4 + 0.1t$$
$$0.9t - 0.1t = 0.4 + 0.1t - 0.1t$$
$$0.8t = 0.4$$
$$\frac{0.8t}{0.8} = \frac{0.4}{0.8}$$
$$t = 0.5$$

17.
$$\frac{x+1}{3} + \frac{x+2}{7} = 5$$
$$21\left(\frac{x+1}{3} + \frac{x+2}{7}\right) = 21(5)$$
$$7(x + 1) + 3(x + 2) = 105$$
$$7x + 7 + 3x + 6 = 105$$
$$10x + 13 = 105$$
$$10x + 13 - 13 = 105 - 13$$
$$10x = 92$$
$$\frac{10x}{10} = \frac{92}{10}$$
$$x = \frac{46}{5}$$

19.
$$\frac{2}{y} + \frac{4}{y} = 3$$
$$y\left(\frac{2}{y} + \frac{4}{y}\right) = y(3)$$
$$2 + 4 = 3y$$
$$6 = 3y$$
$$\frac{6}{3} = \frac{3y}{3}$$
$$y = 2$$

21.
$$x^2 = 9x$$
$$x^2 - 9x = 0$$
$$x(x - 9) = 0$$
$$x = 0 \text{ or } x = 9$$
The solution set is $\{0, 9\}$.

23.
$$x^2 - 25 = 0$$
$$(x + 5)(x - 5) = 0$$
$$x = -5 \text{ or } x = 5$$
The solution set is $\{-5, 5\}$.

25.
$$z^2 + z - 12 = 0$$
$$(z + 4)(z - 3) = 0$$
$$z = -4 \text{ or } z = 3$$
The solution set is $\{-4, 3\}$.

27.
$$2x^2 - 5x - 3 = 0$$
$$(2x + 1)(x - 3) = 0$$
$$x = \frac{-1}{2} \text{ or } x = 3$$
The solution set is $\left\{\frac{-1}{2}, 3\right\}$.

29.
$$3t^2 - 48 = 0$$
$$3(t^2 - 16) = 0$$
$$3(t + 4)(t - 4) = 0$$
$$t = -4 \text{ or } t = 4$$
The solution set is $\{-4, 4\}$.

31.
$$x(x - 7) + 12 = 0$$
$$x^2 - 7x + 12 = 0$$
$$(x - 3)(x - 4) = 0$$
$$x = 3 \text{ or } x = 4$$
The solution set is $\{3, 4\}$.

33.
$$4x^2 + 9 = 12x$$
$$4x^2 - 12x + 9 = 0$$
$$(2x - 3)^2 = 0$$
$$x = \frac{3}{2}$$
The solution set is $\left\{\frac{3}{2}\right\}$.

35.
$$6(p^2 - 1) = 5p$$
$$6p^2 - 6 = 5p$$
$$6p^2 - 5p - 6 = 0$$
$$(3p + 2)(2p - 3) = 0$$
$$p = \frac{-2}{3} \text{ or } p = \frac{3}{2}$$
The solution set is $\left\{\frac{-2}{3}, \frac{3}{2}\right\}$.

37.
$$6x - 5 = \frac{6}{x}$$
$$6x^2 - 5x = 6$$
$$6x^2 - 5x - 6 = 0$$
$$(3x + 2)(2x - 3) = 0$$
$$x = \frac{-2}{3} \text{ or } x = \frac{3}{2}$$
The solution set is $\left\{\frac{-2}{3}, \frac{3}{2}\right\}$.

39.
$$(x + 7)(x - 1) = (x + 1)^2$$
$$x^2 + 6x - 7 = x^2 + 2x + 1$$
$$x^2 + 6x - 7 - x^2 = x^2 + 2x + 1 - x^2$$
$$6x - 7 = 2x + 1$$
$$6x - 7 + 7 = 2x + 1 + 7$$
$$6x = 2x + 8$$
$$6x - 2x = 2x + 8 - 2x$$
$$4x = 8$$
$$\frac{4x}{4} = \frac{8}{4}$$
$$x = 2$$

41.
$$x(2x-3) = (2x+1)(x-4)$$
$$2x^2 - 3x = 2x^2 - 7x - 4$$
$$2x^2 - 3x - 2x^2 = 2x^2 - 7x - 4 - 2x^2$$
$$-3x = -7x - 4$$
$$-3x + 7x = -7x - 4 + 7x$$
$$4x = -4$$
$$\frac{4x}{4} = \frac{-4}{4}$$
$$x = -1$$

43.
$$\sqrt{2t-1} = 1$$
$$\left(\sqrt{2t-1}\right)^2 = 1^2$$
$$2t - 1 = 1$$
$$2t = 2$$
$$t = 1$$
Check: $\sqrt{2(1)-1} = \sqrt{1} = 1$
The solution is $t = 1$.

45.
$$\sqrt{3t+1} = -4$$
$$\left(\sqrt{3t+1}\right)^2 = (-4)^2$$
$$3t + 1 = 16$$
$$3t = 15$$
$$t = 5$$
Check: $\sqrt{3(5)+1} = \sqrt{16} = 4 \neq -4$
The equation has no solution .

47.
$$\sqrt[3]{1-2x} - 3 = 0$$
$$\sqrt[3]{1-2x} = 3$$
$$\left(\sqrt[3]{1-2x}\right)^3 = 3^3$$
$$1 - 2x = 27$$
$$-2x = 26$$
$$x = -13$$
Check: $\sqrt[3]{1-2(-13)} - 3$
$$= \sqrt[3]{27} - 3 = 0$$
The solution is $x = -13$.

49.
$$\sqrt{15-2x} = x$$
$$\left(\sqrt{15-2x}\right)^2 = x^2$$
$$15 - 2x = x^2$$
$$x^2 + 2x - 15 = 0$$
$$(x+5)(x-3) = 0$$
$$x = -5 \text{ or } x = 3$$
Check -5: $\sqrt{15-2(-5)} = \sqrt{25}$
$$= 5 \neq -5$$
Check 3: $\sqrt{15-2(3)} = \sqrt{9} = 3 = 3$
The solution is $x = 3$.

51.
$$x = 2\sqrt{x-1}$$
$$x^2 = \left(2\sqrt{x-1}\right)^2$$
$$x^2 = 4(x-1)$$
$$x^2 = 4x - 4$$
$$x^2 - 4x + 4 = 0$$
$$(x-2)^2 = 0$$
$$x = 2$$
Check: $2 = 2\sqrt{2-1} \rightarrow 2 = 2$
The solution is $x = 2$.

53.
$$3 + \sqrt{3x+1} = x$$
$$\sqrt{3x+1} = x - 3$$
$$\left(\sqrt{3x+1}\right)^2 = (x-3)^2$$
$$3x + 1 = x^2 - 6x + 9$$
$$0 = x^2 - 9x + 8$$
$$(x-1)(x-8) = 0$$
$$x = 1 \text{ or } x = 8$$

Check 1: $3 + \sqrt{3(1)+1}$
$$= 3 + \sqrt{4} = 5 \neq 1$$
Check 8: $3 + \sqrt{3(8)+1}$
$$= 3 + \sqrt{25} = 8 = 8$$
The solution is $x = 8$.

55. $\sqrt{2x+3} - \sqrt{x+1} = 1$

$\qquad \sqrt{2x+3} = 1 + \sqrt{x+1}$

$\qquad \left(\sqrt{2x+3}\right)^2 = \left(1 + \sqrt{x+1}\right)^2$

$\qquad 2x+3 = 1 + 2\sqrt{x+1} + x + 1$

$\qquad x + 1 = 2\sqrt{x+1}$

$\qquad (x+1)^2 = \left(2\sqrt{x+1}\right)^2$

$\qquad x^2 + 2x + 1 = 4(x+1)$

$\qquad x^2 + 2x + 1 = 4x + 4$

$\qquad x^2 - 2x - 3 = 0$

$\qquad (x+1)(x-3) = 0$

$\qquad\qquad x = -1 \text{ or } x = 3$

Check -1: $\sqrt{2(-1)+3} - \sqrt{-1+1}$

$\qquad\qquad = \sqrt{1} - \sqrt{0} = 1 - 0 = 1 = 1$

Check 3: $\sqrt{2(3)+3} - \sqrt{3+1}$

$\qquad\qquad = \sqrt{9} - \sqrt{4} = 3 - 2 = 1 = 1$

The solution is $x = -1$ or $x = 3$.

57. $\sqrt{3x+1} - \sqrt{x-1} = 2$

$\qquad \sqrt{3x+1} = 2 + \sqrt{x-1}$

$\qquad \left(\sqrt{3x+1}\right)^2 = \left(2 + \sqrt{x-1}\right)^2$

$\qquad 3x+1 = 4 + 4\sqrt{x-1} + x - 1$

$\qquad 2x - 2 = 4\sqrt{x-1}$

$\qquad (2x-2)^2 = \left(4\sqrt{x-1}\right)^2$

$\qquad 4x^2 - 8x + 4 = 16(x-1)$

$\qquad x^2 - 2x + 1 = 4x - 4$

$\qquad x^2 - 6x + 5 = 0$

$\qquad (x-1)(x-5) = 0$

$\qquad\qquad x = 1 \text{ or } x = 5$

Check 1: $\sqrt{3(1)+1} - \sqrt{1-1}$

$\qquad\qquad = \sqrt{4} - \sqrt{0} = 2 - 0 = 2 = 2$

Check 5: $\sqrt{3(5)+1} - \sqrt{5-1}$

$\qquad\qquad = \sqrt{16} - \sqrt{4} = 4 - 2 = 2 = 2$

The solution is $x = 1$ or $x = 5$.

59. $|2x| = 8$

$\qquad 2x = 8 \text{ or } 2x = -8$

$\qquad x = 4 \text{ or } \quad x = -4$

The solution set is $\{-4, 4\}$.

61. $|2x+3| = 5$

$\qquad 2x + 3 = 5 \text{ or } 2x + 3 = -5$

$\qquad\qquad 2x = 2 \text{ or } \qquad 2x = -8$

$\qquad\qquad x = 1 \text{ or } \qquad\quad x = -4$

The solution set is $\{-4, 1\}$.

63. $|1 - 4t| = 5$

$\qquad 1 - 4t = 5 \text{ or } 1 - 4t = -5$

$\qquad -4t = 4 \text{ or } \quad -4t = -6$

$\qquad\qquad t = -1 \text{ or } \qquad t = \dfrac{3}{2}$

The solution set is $\left\{-1, \dfrac{3}{2}\right\}$.

65. $|-2x| = 8$

$\qquad -2x = 8 \quad \text{ or } \quad -2x = -8$

$\qquad\quad x = -4 \text{ or } \qquad x = 4$

The solution set is $\{-4, 4\}$.

67. $|-2|x = 4$

$\qquad 2x = 4$

$\qquad\; x = 2$

The solution set is $\{2\}$.

69. $\frac{2}{3}|x| = 8$

$\qquad |x| = 12$

$\qquad\quad x = 12 \text{ or } x = -12$

The solution set is $\{-12, 12\}$.

71. $\left| \dfrac{x}{3} + \dfrac{2}{5} \right| = 2$

$\dfrac{x}{3} + \dfrac{2}{5} = 2$ or $\dfrac{x}{3} + \dfrac{2}{5} = -2$

$5x + 6 = 30$ or $5x + 6 = -30$

$5x = 24$ or $5x = -36$

$x = \dfrac{24}{5}$ or $x = \dfrac{-36}{5}$

The solution set is $\left\{ \dfrac{-36}{5}, \dfrac{24}{5} \right\}$.

73. $ax - b = c$

$ax - b + b = c + b$

$ax = c + b$

$\dfrac{ax}{a} = \dfrac{c + b}{a}$

$x = \dfrac{c + b}{a}$

75. $\dfrac{x}{a} + \dfrac{x}{b} = c$

$ab\left(\dfrac{x}{a} + \dfrac{x}{b} \right) = ab \cdot c$

$bx + ax = abc$

$x(a + b) = abc$

$\dfrac{x(a + b)}{a + b} = \dfrac{abc}{a + b}$

$x = \dfrac{abc}{a + b}$

77. $x^2 - 4x + 2 = 0$

$a = 1, \quad b = -4, \quad c = 2$

$x = \dfrac{-(-4) \pm \sqrt{(-4)^2 - 4(1)(2)}}{2(1)}$

$= \dfrac{4 \pm \sqrt{16 - 8}}{2}$

$= \dfrac{4 \pm 2\sqrt{2}}{2} = 2 \pm \sqrt{2}$

$\left\{ 2 - \sqrt{2}, 2 + \sqrt{2} \right\}$

79. $x^2 - 4x - 1 = 0$

$a = 1, \quad b = -4, \quad c = -1$

$x = \dfrac{-(-4) \pm \sqrt{(-4)^2 - 4(1)(-1)}}{2(1)}$

$= \dfrac{4 \pm \sqrt{16 + 4}}{2}$

$= \dfrac{4 \pm 2\sqrt{5}}{2} = 2 \pm \sqrt{5}$

$\left\{ 2 - \sqrt{5}, 2 + \sqrt{5} \right\}$

81. $2x^2 - 5x + 3 = 0$

$a = 2, \quad b = -5, \quad c = 3$

$x = \dfrac{-(-5) \pm \sqrt{(-5)^2 - 4(2)(3)}}{2(2)}$

$= \dfrac{5 \pm \sqrt{25 - 24}}{4} = \dfrac{5 \pm 1}{4}$

$\left\{ 1, \dfrac{3}{2} \right\}$

83. $4y^2 - y + 2 = 0$

$a = 4, \quad b = -1, \quad c = 2$

$y = \dfrac{-(-1) \pm \sqrt{(-1)^2 - 4(4)(2)}}{2(4)}$

$= \dfrac{1 \pm \sqrt{1 - 32}}{8} = \dfrac{1 \pm \sqrt{-31}}{8}$

No real solution.

85. $4x^2 = 1 - 2x$

$4x^2 + 2x - 1 = 0$

$a = 4, \quad b = 2, \quad c = -1$

$x = \dfrac{-2 \pm \sqrt{2^2 - 4(4)(-1)}}{2(4)}$

$\quad = \dfrac{-2 \pm \sqrt{4 + 16}}{8}$

$\quad = \dfrac{-2 \pm 2\sqrt{5}}{8} = \dfrac{-1 \pm \sqrt{5}}{4}$

$\left\{ \dfrac{-1 - \sqrt{5}}{4}, \dfrac{-1 + \sqrt{5}}{4} \right\}$

87. $9t^2 - 6t + 1 = 0$

$a = 9, \quad b = -6, \quad c = 1$

$t = \dfrac{-(-6) \pm \sqrt{(-6)^2 - 4(9)(1)}}{2(9)}$

$\quad = \dfrac{6 \pm \sqrt{36 - 36}}{18}$

$\quad = \dfrac{6 \pm 0}{18} = \dfrac{1}{3}$

$\left\{ \dfrac{1}{3} \right\}$

89. $x^2 - 4x + 2 = 0$

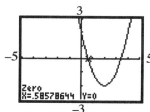 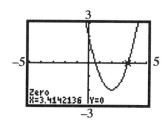

The solution set is: $\{0.59, 3.41\}$.

91. $x^2 + \sqrt{3}\,x = 3$

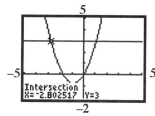 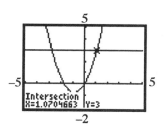

The solution set is: $\{-2.80, 1.07\}$.

93. $\pi x^2 = x + \pi$

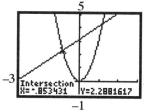 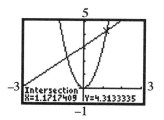

The solution set is: $\{-0.85, 1.17\}$.

95. Solving for R:

$$\frac{1}{R} = \frac{1}{R_1} + \frac{1}{R_2}$$

$$RR_1R_2\left(\frac{1}{R}\right) = RR_1R_2\left(\frac{1}{R_1} + \frac{1}{R_2}\right)$$

$$R_1R_2 = RR_2 + RR_1$$

$$R_1R_2 = R(R_2 + R_1)$$

$$\frac{R_1R_2}{R_2 + R_1} = \frac{R(R_2 + R_1)}{R_2 + R_1}$$

$$\frac{R_1R_2}{R_2 + R_1} = R$$

97. Solving for R:

$$F = \frac{mv^2}{R}$$

$$RF = R\left(\frac{mv^2}{R}\right)$$

$$RF = mv^2$$

$$\frac{RF}{F} = \frac{mv^2}{F}$$

$$R = \frac{mv^2}{F}$$

99. Solving for r:

$$S = \frac{a}{1 - r}$$

$$S(1 - r) = \left(\frac{a}{1 - r}\right)(1 - r)$$

$$S - Sr = a$$

$$S - Sr - S = a - S$$

$$-Sr = a - S$$

$$\frac{-Sr}{-S} = \frac{a - S}{-S}$$

$$r = \frac{S - a}{S}$$

101. Graph the equations and to find the x-coordinate of the points of intersection:

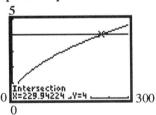

The distance to the water's surface is approximately 229.94 feet.

1.4 Setting Up Equations: Applications

1. Let A represent the area of the circle and r the radius.
   The area of a circle is the product of π times the square of the radius. $\qquad A = \pi r^2$

3. Let A represent the area of the square and s the length of a side.
   The area of the square is the square of the length of a side. $\qquad A = s^2$

5.   Let F represent the force, m the mass, and a the acceleration.
      Force equals the product of the mass times the acceleration.
      $F = ma$

7.   Let W represent the work, F the force, and d the distance.
      Work equals force times distance.
      $W = Fd$

9. C = total variable cost, x = number of dishwashers manufactured.
      $C = 150x$

11.
Amount in Bonds	Amount in CD's	Total
x	$x - 2000$	20,000

$$x + x - 2000 = 20000$$
$$2x - 2000 = 20000$$
$$2x = 22000$$
$$x = 11000$$
$11,000 will be invested in bonds. $9,000 will be invested in CD's.

13.
David	Paige	Dan	Total
x	$\frac{3}{4}x$	$\frac{1}{2}x$	900,000

$$x + \frac{3}{4}x + \frac{1}{2}x = 900,000$$
$$\frac{9}{4}x = 900,000$$
$$x = \frac{4}{9}(900,000)$$
$$x = 400,000$$
David receives $400,000. Paige receives $300,000. Dan receives $200,000.

15. l = length, w = width
$$2l + 2w = 60 \qquad \text{Perimeter } = 2l + 2w$$
$$l = w + 8 \qquad \text{The length is 8 more than the width .}$$
$$2(w + 8) + 2w = 60$$
$$2w + 16 + 2w = 60$$
$$4w + 16 = 60$$
$$4w = 44$$
$$w = 11 \text{ feet}, \quad l = 19 \text{ feet}$$

17. Let x represent the score on the final exam.
     Compute the average score and solve:
     $$\frac{80+83+71+61+95+2x}{7} = 80$$
     $$\frac{390+2x}{7} = 80$$
     $$390+2x = 560$$
     $$2x = 170$$
     $$x = 85$$
     Brooke needs to score an 85 on the final exam to get an average of 80 in the course.

19. Let x represent the original price of the house.
     Then $0.15x$ represents the reduction in the price of the house.
     original price – reduction = new price
     $$x - 0.15x = 125,000$$
     $$0.85x = 125,000$$
     $$x = 147,058.82$$
     The original price of the house was \$147,058.82.
     The amount of the savings is $0.15(\$147,058.82) = \$22,058.82$.

21. Let x represent the price the bookstore pays for the book (publisher price).
     Then $0.25x$ represents the mark up on the book.
     The selling price of the book is \$56.00.
     publisher price + mark up = selling price
     $$x + 0.25x = 56.00$$
     $$1.25x = 56.00$$
     $$x = 44.80$$
     The bookstore pays \$44.80 for the book.

23. Let x represent the amount of money invested in bonds.
     Then $50,000 - x$ represents the amount of money invested in CD's.

	Principle	Rate	Time (yrs)	Interest
Bonds	x	0.15	1	$0.15x$
CD's	$50,000 - x$	0.07	1	$0.07(50,000 - x)$

Since the total interest is to be \$6,000, we have:
$$0.15x + 0.07(50,000 - x) = 6,000$$
$$15x + 7(50,000 - x) = 600,000$$
$$15x + 350,000 - 7x = 600,000$$
$$8x + 350,000 = 600,000$$
$$8x = 250,000$$
$$x = 31,250$$
\$31,250 should be invested in bonds at 15% and \$18,750 should be invested in CD's at 7%.

25.   Let x represent the amount of money loaned at 8%.
      Then $12,000 - x$ represents the amount of money loaned at 18%.

	Principle	Rate	Time (yrs)	Interest
Loan at 8%	x	0.08	1	$0.08x$
Loan at 18%	$12,000 - x$	0.18	1	$0.18(12,000 - x)$

Since the total interest is to be $1,000, we have:

$$0.08x + 0.18(12,000 - x) = 1,000$$
$$8x + 18(12,000 - x) = 100,000$$
$$8x + 216,000 - 18x = 100,000$$
$$-10x + 216,000 = 100,000$$
$$-10x = -116,000$$
$$x = 11,600$$

$11,600 is loaned at 8% and $400 is loaned at 18%.

27.   Let x represent the number of pounds of Earl Gray tea.
      Then $100 - x$ represents the number of pounds of Orange Pekoe tea.

	No. of pounds	Price per pound	Total Value
Earl Gray	x	$5.00	$5x$
Orange Pekoe	$100 - x$	$3.00	$3(100 - x)$
Blend	100	$4.50	$4.50(100)$

$$5x + 3(100 - x) = 4.50(100)$$
$$5x + 300 - 3x = 450$$
$$2x + 300 = 450$$
$$2x = 150$$
$$x = 75$$

75 pounds of Earl Gray tea must be blended with 25 pounds of Orange Pekoe.

29.   Let x represent the number of pounds of cashews.
      Then $x + 60$ represents the number of pounds in the mixture.

	No. of pounds	Price per pound	Total Value
cashews	x	$4.00	$4x$
peanuts	60	$1.50	$1.50(60)$
mixture	$x + 60$	$2.50	$2.50(x + 60)$

$$4x + 1.50(60) = 2.50(x + 60)$$
$$4x + 90 = 2.50x + 150$$
$$1.5x = 60$$
$$x = 40$$

40 pounds of cashews must be added to the 60 pounds of peanuts.

31.   Let x represent the number of ounces of pure water.
      Then $x + 20$ represents the number of ounces in the 30% solution.

	No. of ounces	Concentration of Acid	Pure Acid
water	x	0	0
40% solution	20	0.40	$0.40(20)$
30% solution	$x + 20$	0.30	$0.30(x + 20)$

$$0 + 0.40(20) = 0.30(x + 20)$$
$$8 = 0.3x + 6$$
$$2 = 0.3x$$
$$x = \frac{20}{3} = 6\frac{2}{3}$$

$6\frac{2}{3}$ ounces of pure water should be added.

33. Let r represent the rate of the Metra commuter train.
Then $r + 50$ represents the rate of the Amtrak train.

	Rate	Time	Distance
Metra train	r	3	$3r$
Amtrak train	$r + 50$	1	$r + 50$

Amtrak distance = Metra distance − 10
$$r + 50 = 3r - 10$$
$$60 = 2r$$
$$r = 30$$

The Metra commuter train travels at a rate of 30 miles per hour.
The Amtrak train travels at a rate of 80 miles per hour.

35. Let r represent the speed of the current.

	Rate	Time	Distance
Upstream	$16 - r$	$\frac{20}{60} = \frac{1}{3}$	$\frac{16 - r}{3}$
Downstream	$16 + r$	$\frac{15}{60} = \frac{1}{4}$	$\frac{16 + r}{4}$

Since the distance is the same in each direction:
$$\frac{16 - r}{3} = \frac{16 + r}{4}$$
$$4(16 - r) = 3(16 + r)$$
$$64 - 4r = 48 + 3r$$
$$16 = 7r$$
$$r = \frac{16}{7} \approx 2.286$$

The speed of the current is approximately 2.286 miles per hour.

37. Let t represent the time it takes to do the job together.

	Time to do job	Part of job done in one minute
Trent	30	$\frac{1}{30}$
Lois	20	$\frac{1}{20}$
Together	t	$\frac{1}{t}$

$$\frac{1}{30} + \frac{1}{20} = \frac{1}{t}$$
$$2t + 3t = 60$$
$$5t = 60$$
$$t = 12$$

Working together, the job can be done in 12 minutes.

39. Let w represent the width of window.
Then $l = w + 2$ represents the length of the window.
Since the area is 143 square feet, we have:
$$w(w + 2) = 143$$
$$w^2 + 2w - 143 = 0$$
$$(w + 13)(w - 11) = 0$$
$$w = -13 \text{ or } w = 11$$

The width of the rectangular window is 11 feet and the length is 13 feet.

41. Let l represent the length of the rectangle.
Let w represent the width of the rectangle.
The perimeter is 26 meters and the area is 40 square meters.
$$2l + 2w = 26 \quad \rightarrow \quad l + w = 13 \quad \rightarrow \quad w = 13 - l$$
$$l\,w = 40$$
$$l(13 - l) = 40$$
$$13l - l^2 = 40$$
$$l^2 - 13l + 40 = 0$$
$$(l - 8)(l - 5) = 0$$
$$l = 8 \text{ or } l = 5$$
$$w = 5 \qquad w = 8$$

The dimensions are 5 meters by 8 meters.

43. Let x represent the length of the side of the sheet metal.
$$(x - 2)(x - 2)(1) = 4$$
$$x^2 - 4x + 4 = 4$$
$$x^2 - 4x = 0$$
$$x(x - 4) = 0$$
$$x = 0 \text{ or } x = 4$$

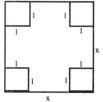

Since the side cannot be 0 feet long, the length of a side of the sheet metal is 4 feet.

45. Let r represent the speed of the current.

	Rate	Time	Distance
Upstream	$15 - r$	$\dfrac{10}{15 - r}$	10
Downstream	$15 + r$	$\dfrac{10}{15 + r}$	10

Since the total time is 1.5 hours, we have:

$$\frac{10}{15-r} + \frac{10}{15+r} = 1.5$$

$$10(15+r) + 10(15-r) = 1.5(15-r)(15+r)$$

$$150 + 10r + 150 - 10r = 1.5(225 - r^2)$$

$$300 = 1.5(225 - r^2)$$

$$200 = 225 - r^2$$

$$r^2 - 25 = 0$$

$$(r-5)(r+5) = 0$$

$$r = 5 \text{ or } r = -5$$

The speed of the current is 5 miles per hour.

47. l = length of the garden
     w = width of the garden

(a)   The length of the garden is to be twice its width. Thus, $l = 2w$.
        The dimensions of the fence are $l + 4$ and $w + 4$.
        The perimeter is 46 feet, so:

$$2(l+4) + 2(w+4) = 46$$

$$2(2w+4) + 2(w+4) = 46$$

$$4w + 8 + 2w + 8 = 46$$

$$6w + 16 = 46$$

$$6w = 30$$

$$w = 5$$

The dimensions of the garden are 5 feet by 10 feet.

(b)   Area = $l \cdot w = 5 \cdot 10 = 50$ square feet.

(c)   If the dimensions of the garden are the same, then the length and width of the
        fence are also the same $(l + 4)$. The perimeter is 46 feet, so:

$$2(l+4) + 2(l+4) = 46$$

$$2l + 8 + 2l + 8 = 46$$

$$4l + 16 = 46$$

$$4l = 30$$

$$l = 7.5$$

The dimensions of the garden are 7.5 feet by 7.5 feet.

(d)   Area = $l \cdot w = 7.5(7.5) = 56.25$ square feet.

49. Let x represent the width of the border measured in feet.
     The radius of the pool is 5 feet.
     Then $x + 5$ represents the radius of the circle, including both the pool and the border.
     The total area of the pool and border is $A_T = \pi(x+5)^2$.

     The area of the pool is $A_P = \pi(5)^2 = 25\pi$.

     The area of the border is $A_B = A_T - A_P = \pi(x+5)^2 - 25\pi$.
     Since the concrete is 3 inches or 0.25 feet thick, the volume of the concrete in the

     border is $0.25 A_B = 0.25\left(\pi(x+5)^2 - 25\pi\right)$

Solving the volume equation:

$$0.25\left(\pi(x+5)^2 - 25\pi\right) = 27$$

$$\pi\left(x^2 + 10x + 25 - 25\right) = 108$$

$$\pi x^2 + 10\pi x - 108 = 0$$

$$x = \frac{-10\pi \pm \sqrt{(10\pi)^2 - 4(\pi)(-108)}}{2(\pi)}$$

$$= \frac{-31.42 \pm \sqrt{2344.1285}}{6.28} = \frac{-31.42 \pm 48.42}{6.28} = 2.71 \text{ or } -12.71$$

The width of the border is approximately 2.71 feet.

51. Let x represent the width of the border measured in feet.
     The total area is $A_T = (6 + 2x)(10 + 2x)$.
     The area of the garden is $A_G = 6 \cdot 10 = 60$.
     The area of the border is $A_B = A_T - A_G = (6 + 2x)(10 + 2x) - 60$.
     Since the concrete is 3 inches or 0.25 feet thick, the volume of the concrete in the
     border is $0.25 A_B = 0.25\left((6 + 2x)(10 + 2x) - 60\right)$
     Solving the volume equation:

$$0.25\left((6 + 2x)(10 + 2x) - 60\right) = 27$$

$$60 + 32x + 4x^2 - 60 = 108$$

$$4x^2 + 32x - 108 = 0$$

$$x^2 + 8x - 27 = 0$$

$$x = \frac{-8 \pm \sqrt{8^2 - 4(1)(-27)}}{2(1)} = \frac{-8 \pm \sqrt{172}}{2} = \frac{-8 \pm 13.11}{2} = 2.56 \text{ or } -10.56$$

The width of the border is approximately 2.56 feet.

53. Let x represent the number of ounces of pure water.
     Then $x + 1$ represents the number of gallons in the 60% solution.

	No. of gallons	Conc. of Antifreeze	Pure Antifreeze
water	x	0	0
100% antifreeze	1	1.00	1(1)
60% antifreeze	$x + 1$	0.60	$0.60(x + 1)$

$$0 + 1(1) = 0.60(x + 1)$$

$$1 = 0.6x + 0.6$$

$$0.4 = 0.6x$$

$$x = \frac{4}{6} = \frac{2}{3}$$

$\frac{2}{3}$ gallon of pure water should be added.

55.   Let x represent the number of pounds of pure cement.
       Then $x + 20$ represents the number of pounds in the 40% mixture.

	No. of pounds	Conc. of Cement	Pure Cement
Pure Cement	x	1.00	x
25% Cement	20	0.25	0.25(20)
40% Cement	$x + 20$	0.40	$0.40(x + 20)$

$$x + 0.25(20) = 0.40(x + 20)$$
$$x + 5 = 0.4x + 8$$
$$0.6x = 3$$
$$x = \frac{30}{6} = 5$$

5 pounds of pure cement should be added.

57.   Let x represent the number of centimeters the length and width should be reduced.
       $12 - x =$ the new length, $7 - x =$ the new width.
       The new volume is 90% of the old volume.
$$(12 - x)(7 - x)(3) = 0.9(12)(7)(3)$$
$$3x^2 - 57x + 252 = 226.8$$
$$3x^2 - 57x + 25.2 = 0$$
$$x^2 - 19x + 8.4 = 0$$
$$x = \frac{-(-19) \pm \sqrt{(-19)^2 - 4(1)(8.4)}}{2(1)} = \frac{19 \pm \sqrt{327.4}}{2}$$
$$= \frac{19 \pm 18.09}{2} = 0.45 \text{ or } 18.55$$

Since 18.55 exceeds the dimensions, it is discarded.
The dimensions of the new chocolate bar are: 11.55 cm by 6.55 cm by 3 cm.

59.   Let x represent the number of grams of pure gold.
       Then $60 - x$ represents the number of grams of 12 karat gold to be used.

	No. of grams	Conc. of gold	Pure gold
Pure gold	x	1.00	x
12 karat gold	$60 - x$	$\frac{1}{2}$	$\frac{1}{2}(60 - x)$
16 karat gold	60	$\frac{2}{3}$	$\frac{2}{3}(60)$

$$x + \tfrac{1}{2}(60 - x) = \tfrac{2}{3}(60)$$
$$x + 30 - 0.5x = 40$$
$$0.5x = 10$$
$$x = 20$$

20 grams of pure gold should be mixed with 40 grams of 12 karat gold.

61. Let t represent the time it takes for Mike to catch up with Dan.

	Time to run mile	Time	Part of mile run in one minute	Distance
Mike	6	t	$\frac{1}{6}$	$\frac{1}{6}t$
Dan	9	$t+1$	$\frac{1}{9}$	$\frac{1}{9}(t+1)$

Since the distances are the same, we have:

$$\frac{1}{6}t = \frac{1}{9}(t+1)$$
$$3t = 2t + 2$$
$$t = 2$$

Mike will pass Dan after 2 minutes, which is a distance of $\frac{1}{3}$ mile.

63. Let t represent the time the auxiliary pump needs to run.

	Time to do job alone	Part of job done in one hour	Time on Job	Part of total job done by each pump
Main Pump	4	$\frac{1}{4}$	3	$\frac{3}{4}$
Auxiliary Pump	9	$\frac{1}{9}$	t	$\frac{1}{9}t$

Since the two pumps are emptying one tanker, we have:

$$\frac{3}{4} + \frac{1}{9}t = 1$$
$$27 + 4t = 36$$
$$4t = 9$$
$$t = \frac{9}{4} = 2.25$$

The auxiliary pump must run for 2.25 hours. It must be started at 9:45 a.m.

65. Let t represent the time for the tub to fill with the faucets on and the stopper removed.

	Time to do job alone	Part of job done in one minute	Time on Job	Part of total job done by each
Faucets open	15	$\dfrac{1}{15}$	t	$\dfrac{t}{15}$
Stopper removed	20	$\dfrac{-1}{20}$	t	$\dfrac{-t}{20}$

Since one tub is being filled, we have:

$$\frac{t}{15} + \frac{-t}{20} = 1$$
$$4t - 3t = 60$$
$$t = 60$$

60 minutes or 1 hour is required to fill the tub.

67. Let x represent the amount of money invested in a CD.
    Then $100,000 - x$ represents the amount of money invested in the bond.

	Principle	Rate	Time (yrs)	Interest
CD	x	0.09	1	$0.09x$
Bond	$100,000 - x$	0.12	1	$0.12(100,000 - x)$

Since the total interest is to be $0.10(100,000) = \$10,000$, we have:

$$0.09x + 0.12(100,000 - x) = 10,000$$
$$9x + 12(100,000 - x) = 1,000,000$$
$$9x + 1,200,000 - 12x = 1,000,000$$
$$-3x + 1,200,000 = 1,000,000$$
$$-3x = -200,000$$
$$x = 66,667$$

The most money that can be invested in the CD is \$66,667 to ensure that the loan payment is made.

69. Let t_1 and t_2 represent the times for the two segments of the trip.

	Rate	Time	Distance
Chicago to Atlanta	45	t_1	$45t_1$
Atlanta to Miami	55	t_2	$55t_2$

Since Atlanta is halfway between Chicago and Miami, the distances are equal.

$$45t_1 = 55t_2 \quad \rightarrow \quad t_1 = \frac{55}{45}t_2 = \frac{11}{9}t_2$$

Computing the average speed:

$$\text{Avg Speed} = \frac{\text{Distance}}{\text{Time}} = \frac{45t_1 + 55t_2}{t_1 + t_2} = \frac{45\left(\frac{11}{9}t_2\right) + 55t_2}{\frac{11}{9}t_2 + t_2}$$

$$= \frac{55t_2 + 55t_2}{\frac{11t_2 + 9t_2}{9}} = \frac{110t_2}{\frac{20t_2}{9}} = \frac{990t_2}{20t_2} = \frac{99}{2} = 49.5 \text{ miles per hour}$$

The average speed for the trip from Chicago to Miami is 49.5 miles per hour.

71. Let x be the original selling price of the shirt.

$$x - 0.40x - 20 = 4$$
$$0.60x = 24$$
$$x = 40$$

The original price should be \$40 to ensure a profit of \$4 after the sale.

If the sale is 50% off, the profit is:

$$40 - 0.50(40) - 20 = 40 - 20 - 20 = 0$$

At 50% off there will be no profit.

73. It is impossible to mix two solutions with a lower concentration and end up with a new solution with a higher concentration.

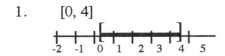

1.   [0, 4]

3.   [4, 6)

5. $[4, \infty)$

7. $(-\infty, -4)$

9. $2 \le x \le 5$

11. $-3 < x < -2$

13. $x \ge 4$

15. $x < -3$

17.   If $x < 5$, then $x - 5 < 0$.

19.   If $x > -4$, then $x + 4 > 0$.

21.   If $x \ge -4$, then $3x \ge -12$.

23.   If $x < 6$, then $-2x > -12$.

25.   If $x \ge 5$, then $-4x \le -20$.

27.   If $2x > 6$, then $x > 3$.

29.   If $\frac{-1}{2} x \le 3$, then $x \ge -6$.

31.   (a)   Graph: $y_1 = x + 1$; $y_2 = 5$.

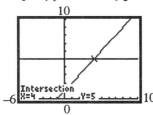

Using INTERSECT, $y_1 = y_2$ at $(4, 5)$. Since $y_1 < y_2$ when $x < 4$, the solution set is $\{x \mid x < 4\}$.

(b)

$$x + 1 < 5$$
$$x + 1 - 1 < 5 - 1$$
$$x < 4$$
$$\{x \mid x < 4\} \text{ or } (-\infty, 4)$$

33. (a) Graph: $y_1 = 1 - 2x$; $y_2 = 3$.

(b)

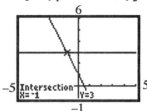

Using INTERSECT, $y_1 = y_2$ at $(-1, 3)$. Since $y_1 \leq y_2$ when $x \geq -1$, the solution set is $\left\{x \mid x \geq -1\right\}$.

$1 - 2x \leq 3$
$-2x \leq 2$
$x \geq -1$
$\left\{x \mid x \geq -1\right\}$ or $[-1, +\infty)$

35. (a) Graph: $y_1 = 3x - 7$; $y_2 = 2$.

(b)

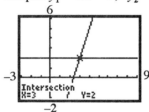

Using INTERSECT, $y_1 = y_2$ at $(3, 2)$. Since $y_1 > y_2$ when $x > 3$, the solution set is $\left\{x \mid x > 3\right\}$.

$3x - 7 > 2$
$3x > 9$
$x > 3$
$\left\{x \mid x > 3\right\}$ or $(3, +\infty)$

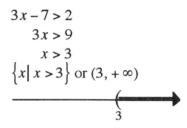

37. (a) Graph: $y_1 = 3x - 1$; $y_2 = 3 + x$.

(b)

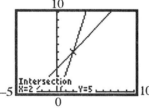

Using INTERSECT, $y_1 = y_2$ at $(2, 5)$. Since $y_1 \geq y_2$ when $x \geq 2$, the solution set is $\left\{x \mid x \geq 2\right\}$.

$3x - 1 \geq 3 + x$
$2x \geq 4$
$x \geq 2$
$\left\{x \mid x \geq 2\right\}$ or $[2, +\infty)$

39. (a)   Graph: $y_1 = -2(x+3)$; $y_2 = 8$.

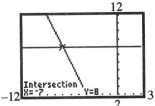

Using INTERSECT, $y_1 = y_2$ at $(-7, 8)$. Since $y_1 < y_2$ when $x > -7$, the solution set is $\left\{x \mid x > -7\right\}$.

(b)
$$-2(x+3) < 8$$
$$-2x - 6 < 8$$
$$-2x < 14$$
$$x > -7$$
$$\left\{x \mid x > -7\right\} \text{ or } (-7, +\infty)$$

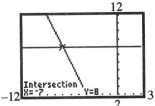

41. (a)   Graph: $y_1 = 4 - 3(1-x)$; $y_2 = 3$.

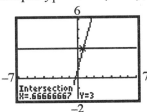

Using INTERSECT, $y_1 = y_2$ at $\left(\frac{2}{3}, 3\right)$. Since $y_1 \leq y_2$ when $x \leq \frac{2}{3}$, the solution set is $\left\{x \mid x \leq \frac{2}{3}\right\}$.

(b)
$$4 - 3(1-x) \leq 3$$
$$4 - 3 + 3x \leq 3$$
$$3x + 1 \leq 3$$
$$3x \leq 2$$
$$x \leq \frac{2}{3}$$
$$\left\{x \mid x \leq \frac{2}{3}\right\} \text{ or } \left(-\infty, \frac{2}{3}\right]$$

43. (a)   Graph: $y_1 = \frac{1}{2}(x-4)$; $y_2 = x + 8$.

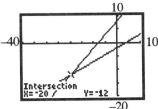

Using INTERSECT, $y_1 = y_2$ at $(-20, -12)$. Since $y_1 > y_2$ when $x < -20$, the solution set is $\left\{x \mid x < -20\right\}$.

(b)
$$\tfrac{1}{2}(x-4) > x + 8$$
$$\tfrac{1}{2}x - 2 > x + 8$$
$$\tfrac{-1}{2}x > 10$$
$$x < -20$$
$$\left\{x \mid x < -20\right\} \text{ or } (-\infty, -20)$$

45. (a) Graph: $y_1 = \dfrac{x}{2}$; $y_2 = 1 - \dfrac{x}{4}$.

(b)

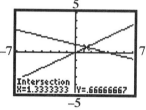

Using INTERSECT, $y_1 = y_2$ at $(1.33, 0.67)$. Since $y_1 \geq y_2$ when , $x \geq 1.33$, the solution set is $\left\{ x \mid x \geq 1.33 \right\}$.

$\dfrac{x}{2} \geq 1 - \dfrac{x}{4}$

$2x \geq 4 - x$

$3x \geq 4$

$x \geq \dfrac{4}{3}$

$\left\{ x \mid x \geq \dfrac{4}{3} \right\}$ or $\left[\dfrac{4}{3}, +\infty \right)$

47. (a) Graph: $y_1 = 0$; $y_2 = 2x - 6$; $y_3 = 4$. (b)

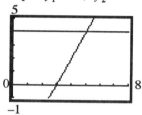

The intersection points are $(3, 0)$ and $(5, 4)$. Since y_2 is between 0 and 4 when $3 \leq x \leq 5$, the solution set is $\left\{ x \mid 3 \leq x \leq 5 \right\}$.

$0 \leq 2x - 6 \leq 4$

$6 \leq 2x \leq 10$

$3 \leq x \leq 5$

$\left\{ x \mid 3 \leq x \leq 5 \right\}$ or $[3, 5]$

49. (a) Graph: $y_1 = -5$; $y_2 = 4 - 3x$; $y_3 = 2$. (b)

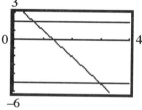

The intersection points are $(3, -5)$ and $(0.67, 2)$. Since y_2 is between -5 and 2 when $0.67 \leq x \leq 3$, the solution set is $\left\{ x \mid 0.67 \leq x \leq 3 \right\}$.

$-5 \leq 4 - 3x \leq 2$

$-9 \leq -3x \leq -2$

$3 \geq x \geq \dfrac{2}{3}$

$\left\{ x \mid \dfrac{2}{3} \leq x \leq 3 \right\}$ or $\left[\dfrac{2}{3}, 3 \right]$

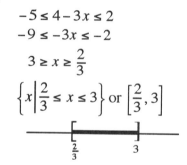

51. (a) Graph: $y_1 = -3$; $y_2 = \dfrac{2x-1}{4}$; $y_3 = 0$.

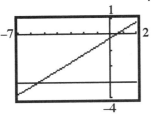

The intersection points are $(-5.5, -3)$ and $(0.5, 0)$. Since y_2 is between -3 and 0 when $-5.5 < x < 0.5$, the solution set is $\{x \mid -5.5 < x < 0.5\}$.

(b)

$$-3 < \dfrac{2x-1}{4} < 0$$
$$-12 < 2x - 1 < 0$$
$$-11 < 2x < 1$$
$$\dfrac{-11}{2} < x < \dfrac{1}{2}$$
$$\left\{x \middle| \dfrac{-11}{2} < x < \dfrac{1}{2}\right\} \text{ or } \left(\dfrac{-11}{2}, \dfrac{1}{2}\right)$$

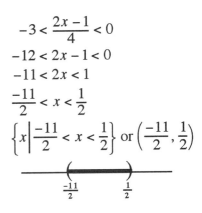

53. (a) Graph: $y_1 = 1$; $y_2 = 1 - \frac{1}{2}x$; $y_3 = 4$.

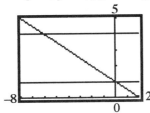

The intersection points are $(-6, 4)$ and $(0, 1)$. Since y_2 is between 1 and 4 when $-6 < x < 0$, the solution set is $\{x \mid -6 < x < 0\}$.

(b)

$$1 < 1 - \tfrac{1}{2}x < 4$$
$$0 < -\tfrac{1}{2}x < 3$$
$$0 > x > -6$$
$$\{x \mid -6 < x < 0\} \text{ or } (-6, 0)$$

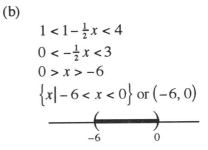

55. (a) Graph: $y_1 = (x+2)(x-3)$; $y_2 = (x-1)(x+1)$.

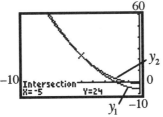

Using INTERSECT, $y_1 = y_2$ at $(-5, 24)$. Since $y_1 > y_2$ when $x < -5$, the solution set is $\{x \mid x < -5\}$.

(b)

$$(x+2)(x-3) > (x-1)(x+1)$$
$$x^2 - x - 6 > x^2 - 1$$
$$-x - 6 > -1$$
$$-x > 5$$
$$x < -5$$
$$\{x \mid x < -5\} \text{ or } (-\infty, -5)$$

57. (a) Graph: $y_1 = x(4x + 3)$; $y_2 = (2x + 1)^2$.

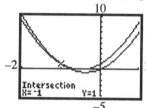

Using INTERSECT, $y_1 = y_2$ at $(-1, 1)$. Since $y_1 \leq y_2$ when $x \geq -1$, the solution set is $\{x \mid x \geq -1\}$.

(b)

$$x(4x + 3) \leq (2x + 1)^2$$
$$4x^2 + 3x \leq 4x^2 + 4x + 1$$
$$3x \leq 4x + 1$$
$$-x \leq 1$$
$$x \geq -1$$
$$\{x \mid x \geq -1\} \text{ or } [-1, +\infty)$$

59. (a) Graph: $y_1 = \dfrac{1}{2}$; $y_2 = \dfrac{x+1}{3}$; $y_3 = \dfrac{3}{4}$.

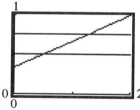

The intersection points are $(0.5, 0.5)$ and $(1.25, 0.75)$. Since y_2 is between 0.5 and 0.75 when $0.5 \leq x < 1.25$, the solution set is $\{x \mid 0.5 \leq x < 1.25\}$.

(b)

$$\frac{1}{2} \leq \frac{x+1}{3} < \frac{3}{4}$$
$$6 \leq 4x + 4 < 9$$
$$2 \leq 4x < 5$$
$$\frac{1}{2} \leq x < \frac{5}{4}$$
$$\left\{ x \mid \frac{1}{2} \leq x < \frac{5}{4} \right\} \text{ or } \left[\frac{1}{2}, \frac{5}{4} \right)$$

61. (a) Graph: $y_1 = |2x|$; $y_2 = 8$.

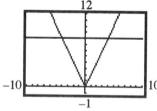

Using INTERSECT, $y_1 = y_2$ at $(-4, 8)$ and $(4, 8)$. Since $y_1 < y_2$ when $-4 < x < 4$, the solution set is $\{x \mid -4 < x < 4\}$.

(b)

$$|2x| < 8$$
$$-8 < 2x < 8$$
$$-4 < x < 4$$
$$\{x \mid -4 < x < 4\} \text{ or } (-4, 4)$$

63. (a)   Graph: $y_1 = |3x|$; $y_2 = 12$.

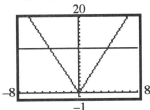

Using INTERSECT, $y_1 = y_2$ at
(−4, 12) and (4, 12). Since $y_1 > y_2$
when $x < −4$ or $x > 4$,the solution
set is $\{x\,|\,x < −4 \text{ or } x > 4\}$.

(b)

$|3x| > 12$
$3x < −12$ or $3x > 12$
$x < −4$ or $x > 4$
$\{x\,|\,x < −4 \text{ or } x > 4\}$ or
$\qquad (−\infty, −4) \cup (4, +\infty)$

65. (a)   Graph: $y_1 = |x − 2|$; $y_2 = 1$.

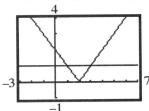

Using INTERSECT, $y_1 = y_2$ at
(1, 1) and (3, 1). Since $y_1 < y_2$
when $1 < x < 3$,the solution set is
$\{x\,|\,1 < x < 3\}$.

(b)

$|x − 2| < 1$
$−1 < x − 2 < 1$
$1 < \;\; x \;\; < 3$
$\{x\,|\,1 < x < 3\}$ or $(1,3)$

67. (a)   Graph: $y_1 = |3x − 2|$; $y_2 = 4$.

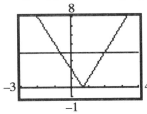

Using INTERSECT, $y_1 = y_2$ at
$\left(\frac{-2}{3}, 4\right)$ and (2, 4). Since $y_1 \le y_2$
when $\frac{-2}{3} \le x \le 2$,the solution set is
$\left\{x\,\middle|\,\frac{-2}{3} \le x \le 2\right\}$.

(b)

$|3t − 2| \le 4$
$−4 \le 3t − 2 \le 4$
$−2 \le \;\; 3t \;\; \le 6$
$\dfrac{-2}{3} \le \;\; t \;\; \le 2$

$\left\{t\,\middle|\,\dfrac{-2}{3} \le t \le 2\right\}$ or $\left[\dfrac{-2}{3}, 2\right]$

69. (a) Graph: $y_1 = |x - 3|$; $y_2 = 2$.

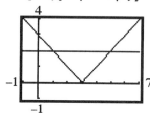

Using INTERSECT, $y_1 = y_2$ at $(1, 2)$ and $(5, 2)$. Since $y_1 \geq y_2$ when $x \leq 1$ or $x \geq 5$, the solution set is $\{x \mid x \leq 1 \text{ or } x \geq 5\}$.

(b)

$|x - 3| \geq 2$

$x - 3 \leq -2$ or $x - 3 \geq 2$

$x \leq 1$ or $x \geq 5$

$\{x \mid x \leq 1 \text{ or } x \geq 5\}$ or

$(-\infty, 1] \cup [5, \infty)$

71. (a) Graph: $y_1 = |1 - 4x|$; $y_2 = 5$.

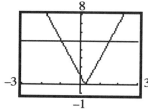

Using INTERSECT, $y_1 = y_2$ at $(-1, 5)$ and $\left(\frac{3}{2}, 5\right)$. Since $y_1 < y_2$ when $-1 < x < \frac{3}{2}$, the solution set is $\left\{x \mid -1 < x < \frac{3}{2}\right\}$.

(b)

$|1 - 4x| < 5$

$-5 < 1 - 4x < 5$

$-6 < -4x < 4$

$\frac{3}{2} > x > -1$

$\left\{x \mid -1 < x < \frac{3}{2}\right\}$ or $\left(-1, \frac{3}{2}\right)$

73. (a) Graph: $y_1 = |1 - 2x|$; $y_2 = 3$.

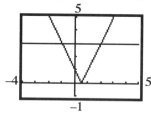

Using INTERSECT, $y_1 = y_2$ at $(-1, 3)$ and $(2, 3)$. Since $y_1 > y_2$ when $x < -1$ or $x > 2$, the solution set is $\{x \mid x < -1 \text{ or } x > 2\}$.

(b)

$|1 - 2x| > 3$

$1 - 2x < -3$ or $1 - 2x > 3$

$-2x < -4$ or $-2x > 2$

$x > 2$ or $x < -1$

$\{x \mid x < -1 \text{ or } x > 2\}$ or

$(-\infty, -1) \cup (2, \infty)$

75. x differs from 2 by less than $\frac{1}{2}$

$$\left| x - 2 \right| < \frac{1}{2}$$
$$\frac{-1}{2} < x - 2 < \frac{1}{2}$$
$$\frac{3}{2} < \quad x \quad < \frac{5}{2}$$
$$\left\{ x \middle| \frac{3}{2} < x < \frac{5}{2} \right\}$$

77. x differs from -3 by more than 2

$$\left| x - (-3) \right| > 2$$
$$x + 3 < -2 \ \text{ or } \ x + 3 > 2$$
$$x < -5 \ \text{ or } \quad x > -1$$
$$\left\{ x \middle| x < -5 \text{ or } x > -1 \right\}$$

79. $21 <$ young adult's age < 30

81. A temperature x that differs from $98.6°$ F by at least $1.5°$

$$\left| x - 98.6° \right| \geq 1.5°$$
$$x - 98.6° \leq -1.5° \ \text{ or } \ x - 98.6° \geq 1.5°$$
$$x \leq 97.1° \ \text{ or } \qquad x \geq 100.1°$$

The temperatures that are considered unhealthy are those that are less than $97.1°$F or greater than $100.1°$F, inclusive.

83. (a) An average 25-year-old male can expect to live at least 48.4 more years. $25 + 48.4 = 73.4$. Therefore, the average age of a 25-year-old male will be ≥ 73.4.

(b) An average 25-year-old female can expect to live at least 54.7 more years. $25 + 54.7 = 79.7$. Therefore, the average age of a 25-year-old female will be ≥ 79.7.

(c) By the given information, a female can expect to live 6.3 years longer.

85. Let P represent the selling price and C represent the commission.
Calculating the commission:
$$C = 45,000 + 0.25(P - 900,000) = 45,000 + 0.25P - 225,000 = 0.25P - 180,000$$
Calculate the commission range, given the price range:
$$900,000 \leq \qquad P \qquad \leq 1,100,000$$
$$0.25(900,000) \leq \quad 0.25P \quad \leq 0.25(1,100,000)$$
$$225,000 \leq \quad 0.25P \quad \leq 275,000$$
$$225,000 - 180,000 \leq 0.25P - 180,000 \leq 275,000 - 180,000$$
$$45,000 \leq \qquad C \qquad \leq 95,000$$
The agent's commission ranges from \$45,000 to \$95,000, inclusive.
$$\frac{45,000}{900,000} = 0.05 = 5\% \ \text{ to } \ \frac{95,000}{1,100,000} = 0.086 = 8.6\%, \text{ inclusive}.$$
As a percent of selling price, the commission ranges from 5% to 8.6%.

87. Let W represent the weekly wage and T represent the withholding tax.
    Calculating the tax:

    $$T = 69.90 + 0.28(W - 517) = 69.90 + 0.28W - 144.76 = 0.28W - 74.86$$

    Calculating the withholding tax range, given the range of weekly wages:

    $$525 \leq \quad W \quad \leq 600$$
    $$0.28(525) \leq \quad 0.28W \quad \leq 0.28(600)$$
    $$147 \leq \quad 0.28W \quad \leq 168$$
    $$147 - 74.86 \leq 0.28W - 74.86 \leq 168 - 74.86$$
    $$72.14 \leq \quad T \quad \leq 93.14$$

    The amount of withholding tax ranges from $72.14 to $93.14, inclusive.

89. Let K represent the monthly usage in kilowatt-hours.
    Let C represent the monthly customer bill.
    Calculating the bill:

    $$C = 0.10494K + 9.36$$

    Calculating the range of kilowatt-hours, given the range of bills:

    $$80.24 \leq \quad C \quad \leq 271.80$$
    $$80.24 \leq 0.10494K + 9.36 \leq 271.80$$
    $$70.88 \leq \quad 0.10494K \quad \leq 262.44$$
    $$675.43 \leq \quad K \quad \leq 2500.86$$

    The range of usage in kilowatt-hours varied from 675.43 to 2500.86.

91. Let C represent the dealer's cost and M represent the markup over dealer's cost.
    If the price is $8800, then $8800 = C + MC = C(1 + M)$

    Solving for C: $C = \dfrac{8800}{1 + M}$

    Calculating the range of dealer costs, given the range of markups:

    $$0.12 \leq M \leq 0.18$$
    $$1.12 \leq 1 + M \leq 1.18$$
    $$\frac{1}{1.12} \geq \frac{1}{1 + M} \geq \frac{1}{1.18}$$
    $$\frac{8800}{1.12} \geq \frac{8800}{1 + M} \geq \frac{8800}{1.18}$$
    $$7857.14 \geq \quad C \quad \geq 7457.63$$

    The dealer's cost ranged from $7457.63 to $7857.14, inclusive.

93. Let T represent the score on the last test and G represent the course grade.
    Calculating the course grade and solving for the last test:

    $$G = \frac{68 + 82 + 87 + 89 + T}{5} = \frac{326 + T}{5}$$

    $$T = 5G - 326$$

    Calculating the range of scores on the last test, given the grade range:

    $$80 \leq \quad G \quad < 90$$
    $$400 \leq \quad 5G \quad < 450$$
    $$74 \leq 5G - 326 < 124$$
    $$74 \leq \quad T \quad < 124$$

    The fifth test must be greater than or equal to 74.

95. Let G represent the number of gallons of gasoline in the tank.
Let D represent the distance traveled.

$$D = 25G \quad \rightarrow \quad G = \frac{D}{25}$$

$$D \geq 300$$

$$\frac{D}{25} \geq 12$$

$$G \geq 12$$

The amount of gasoline at the beginning of the trip ranged from 12 to 20 gallons, inclusive.

97. Since $a < b$

$$\frac{a}{2} < \frac{b}{2} \qquad\qquad \frac{a}{2} < \frac{b}{2}$$

$$\frac{a}{2} + \frac{a}{2} < \frac{a}{2} + \frac{b}{2} \qquad\qquad \frac{a}{2} + \frac{b}{2} < \frac{b}{2} + \frac{b}{2}$$

$$a < \frac{a+b}{2} \qquad\qquad \frac{a+b}{2} < b$$

Thus, $a < \dfrac{a+b}{2} < b$

99. If $0 < a < b$, then $0 < a^2 < ab$ and $0 < ab < b^2$

$$ab - a^2 > 0 \qquad\qquad b^2 - ab > 0$$

$$ab > a^2 > 0 \qquad\qquad b^2 > ab > 0$$

$$\left(\sqrt{ab}\right)^2 > a^2 \qquad\qquad b^2 > \left(\sqrt{ab}\right)^2$$

$$\sqrt{ab} > a \qquad\qquad b > \sqrt{ab}$$

Thus, $a < \sqrt{ab} < b$

101. For $0 < a < b$, $\dfrac{1}{h} = \dfrac{1}{2}\left(\dfrac{1}{a} + \dfrac{1}{b}\right)$

$$h \cdot \frac{1}{h} = \frac{1}{2}\left(\frac{b+a}{ab}\right) \cdot h \quad \rightarrow \quad 1 = \frac{1}{2}\left(\frac{b+a}{ab}\right) \cdot h \quad \rightarrow \quad \frac{2ab}{a+b} = h$$

$$h - a = \frac{2ab}{a+b} - a \qquad\qquad b - h = b - \frac{2ab}{a+b}$$

$$= \frac{2ab - a(a+b)}{a+b} \qquad\qquad = \frac{b(a+b) - 2ab}{a+b}$$

$$= \frac{2ab - a^2 - ab}{a+b} \qquad\qquad = \frac{ab + b^2 - 2ab}{a+b}$$

$$= \frac{ab - a^2}{a+b} \qquad\qquad = \frac{b^2 - ab}{a+b}$$

$$= \frac{a(b-a)}{a+b} > 0 \qquad\qquad = \frac{b(b-a)}{a+b} > 0$$

Therefore, $h > a$. $\qquad\qquad$ Therefore, $h < b$.

Thus, $a < h < b$.

1.6   Lines

1. (a) Slope $= \dfrac{1-0}{2-0} = \dfrac{1}{2}$

   (b) If x increases by 2 units, y will increase by 1 unit.

3. (a) Slope $= \dfrac{1-2}{1-(-2)} = \dfrac{-1}{3}$

   (b) If x increases by 3 units, y will decrease by 1 unit.

5. (x_1, y_1) (x_2, y_2)
   $(2, 3)$ $(4, 0)$

   Slope $= \dfrac{y_2 - y_1}{x_2 - x_1} = \dfrac{0-3}{4-2} = \dfrac{-3}{2}$

   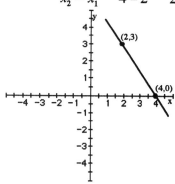

7. (x_1, y_1) (x_2, y_2)
   $(-2, 3)$ $(2, 1)$

   Slope $= \dfrac{y_2 - y_1}{x_2 - x_1} = \dfrac{1-3}{2-(-2)} = \dfrac{-2}{4} = \dfrac{-1}{2}$

9. (x_1, y_1) (x_2, y_2)

   $(-3, -1)$ $(2, -1)$

   Slope $= \dfrac{y_2 - y_1}{x_2 - x_1} = \dfrac{-1-(-1)}{2-(-3)} = \dfrac{0}{5} = 0$

   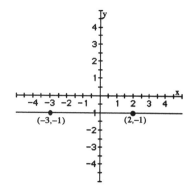

11. (x_1, y_1) (x_2, y_2)
    $(-1, 2)$ $(-1, -2)$

    Slope $= \dfrac{y_2 - y_1}{x_2 - x_1} = \dfrac{-2-2}{-1-(-1)} = \dfrac{-4}{0}$

    Slope is undefined.

    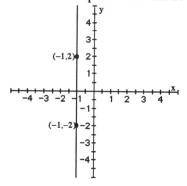

13. (x_1, y_1) (x_2, y_2)
     $(\sqrt{2}, 3)$ $(1, \sqrt{3})$

$$\text{Slope} = \frac{y_2 - y_1}{x_2 - x_1} = \frac{\sqrt{3} - 3}{1 - \sqrt{2}} \approx \frac{1.732 - 3}{1 - 1.414} \approx \frac{-1.268}{-0.414} = 3.063$$

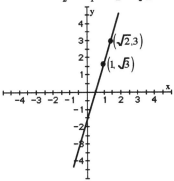

15.

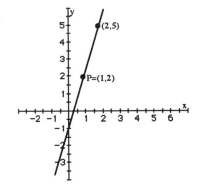

17.

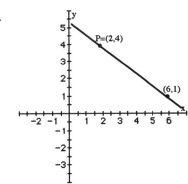

19.

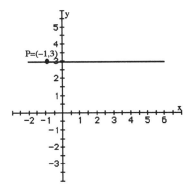

21.

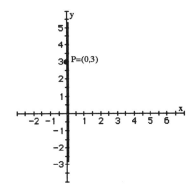

23.   (0,0) and (2,1) are points on the line.
Slope $= \dfrac{1-0}{2-0} = \dfrac{1}{2}$
y-intercept is 0; using $y = mx + b$.
$$y = \frac{1}{2}x + 0$$
$$2y = x$$
$$0 = x - 2y$$
$$x - 2y = 0 \ \text{ or } \ y = \frac{1}{2}x$$

25.   (–1,3) and (1,1) are points on the line.
Slope $= \dfrac{1-3}{1-(-1)} = \dfrac{-2}{2} = -1$
Using $y - y_1 = m(x - x_1)$
$$y - 3 = -1[x - (-1)]$$
$$y - 3 = -(x + 1)$$
$$y - 3 = -x - 1$$
$$y = -x + 2$$
$$x + y = 2 \ \text{ or } \ y = -x + 2$$

27. $y - y_1 = m(x - x_1), \ m = 2$
$$y - 3 = 2(x - 3)$$
$$y - 3 = 2x - 6$$
$$y = 2x - 3$$
$$2x - y = 3 \ \text{ or } \ y = 2x - 3$$

29. $y - y_1 = m(x - x_1), \ m = \frac{-1}{2}$
$$y - 2 = \frac{-1}{2}(x - 1)$$
$$y - 2 = \frac{-1}{2}x + \frac{1}{2}$$
$$y = \frac{-1}{2}x + \frac{5}{2}$$
$$x + 2y = 5 \ \text{ or } \ y = \frac{-1}{2}x + \frac{5}{2}$$

31.   Slope $= 3$; containing (–2,3)
$$y - y_1 = m(x - x_1)$$
$$y - 3 = 3(x - (-2))$$
$$y - 3 = 3x + 6$$
$$y = 3x + 9$$
$$3x - y = -9 \ \text{ or } \ y = 3x + 9$$

33.   Slope $= \frac{-2}{3}$; containing (1,–1)
$$y - y_1 = m(x - x_1)$$
$$y - (-1) = \frac{-2}{3}(x - 1)$$
$$y + 1 = \frac{-2}{3}x + \frac{2}{3}$$
$$y = \frac{-2}{3}x - \frac{1}{3}$$
$$2x + 3y = -1 \ \text{ or } \ y = \frac{-2}{3}x - \frac{1}{3}$$

35.   Containing (1,3) and (–1,2)
$$m = \frac{2-3}{-1-1} = \frac{-1}{-2} = \frac{1}{2}$$
$$y - y_1 = m(x - x_1)$$
$$y - 3 = \frac{1}{2}(x - 1)$$
$$y - 3 = \frac{1}{2}x - \frac{1}{2}$$
$$y = \frac{1}{2}x + \frac{5}{2}$$
$$x - 2y = -5 \ \text{ or } \ y = \frac{1}{2}x + \frac{5}{2}$$

37.   Slope $= -3$; y-intercept $= 3$
$$y = mx + b$$
$$y = -3x + 3$$
$$3x + y = 3 \ \text{ or } \ y = -3x + 3$$

39. x-intercept $= 2$; y-intercept $= -1$
Points are $(2,0)$ and $(0,-1)$
$$m = \frac{-1-0}{0-2} = \frac{-1}{-2} = \frac{1}{2}$$
$$y = mx + b$$
$$y = \frac{1}{2}x - 1$$
$$x - 2y = 2 \text{ or } y = \frac{1}{2}x - 1$$

41. Slope undefined; passing through $(2,4)$
This is a vertical line.
$$x = 2$$
No slope intercept form.

43. Parallel to $y = 2x$; Slope $= 2$
Containing $(-1,2)$
$$y - y_1 = m(x - x_1)$$
$$y - 2 = 2(x - (-1))$$
$$y - 2 = 2x + 2$$
$$y = 2x + 4$$
$$2x - y = -4 \text{ or } y = 2x + 4$$

45. Parallel to $2x - y = -2$; Slope $= 2$
Containing $(0,0)$
$$y - y_1 = m(x - x_1)$$
$$y - 0 = 2(x - 0)$$
$$y = 2x$$
$$2x - y = 0 \text{ or } y = 2x$$

47. Parallel to $x = 5$;
Containing $(4,2)$
This is a vertical line.
$$x = 4$$
No slope intercept form.

49. Perpendicular to $y = \frac{1}{2}x + 4$;
Slope of perpendicular $= -2$
Containing $(1,-2)$
$$y - y_1 = m(x - x_1)$$
$$y - (-2) = -2(x - 1)$$
$$y + 2 = -2x + 2$$
$$y = -2x$$
$$2x + y = 0 \text{ or } y = -2x$$

51. Perpendicular to $2x + y = 2$;
Containing $(-3,0)$
Slope of perpendicular $= \frac{1}{2}$
$$y - y_1 = m(x - x_1)$$
$$y - 0 = \frac{1}{2}(x - (-3))$$
$$y = \frac{1}{2}x + \frac{3}{2}$$
$$x - 2y = -3 \text{ or } y = \frac{1}{2}x + \frac{3}{2}$$

53. Perpendicular to $x = 8$;
Slope of perpendicular $= 0$
Containing $(3,4)$
$$y - y_1 = m(x - x_1)$$
$$y - 4 = 0(x - 3)$$
$$y - 4 = 0$$
$$y = 4$$
$$y = 4 \text{ or } y = 0x + 4$$

55. $y = 2x + 3$
Slope $= 2$
y-intercept $= 3$

57. $\frac{1}{2}y = x - 1$
$y = 2x - 2$
Slope $= 2$
y-intercept $= -2$

59. $y = \frac{1}{2}x + 2$
Slope $= \frac{1}{2}$
y-intercept $= 2$

61. $x + 2y = 4$
$2y = -x + 4$
$y = \frac{-1}{2}x + 2$
Slope $= \frac{-1}{2}$
y-intercept $= 2$

63. $2x - 3y = 6$
$-3y = -2x + 6$
$y = \frac{2}{3}x - 2$
Slope $= \frac{2}{3}$
y-intercept $= -2$

65. $x + y = 1$
$y = -x + 1$
Slope $= -1$
y-intercept $= 1$

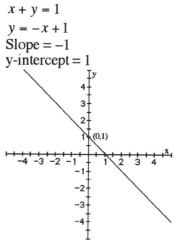

67. $x = -4$
Slope is undefined
y-intercept - none

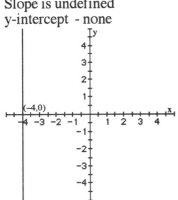

69. $y = 5$
Slope $= 0$
y-intercept $= 5$

71. $y - x = 0$
$y = x$
Slope $= 1$
y-intercept $= 0$

73. $2y - 3x = 0$
$2y = 3x$
$y = \frac{3}{2}x$
Slope $= \frac{3}{2}$
y-intercept $= 0$

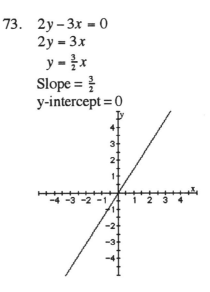

75. The equation of the x-axis is $y = 0$. (The slope is 0 and the y-intercept is 0.)

77. (b)   79. (d)

81. Slope $= 1$; y-intercept $= 2$
$y = x + 2$ or $x - y = -2$

83. Slope $= \frac{-1}{3}$; y-intercept $= 1$
$y = \frac{-1}{3}x + 1$ or $x + 3y = 3$

85. $(^{\circ}C, ^{\circ}F) = (0, 32);$ $(^{\circ}C, ^{\circ}F) = (100, 212)$

$$\text{slope } = \frac{212 - 32}{100 - 0} = \frac{180}{100} = \frac{9}{5}$$

$$^{\circ}F - 32 = \frac{9}{5}(^{\circ}C - 0)$$

$$^{\circ}F - 32 = \frac{9}{5}(^{\circ}C)$$

$$^{\circ}C = \frac{5}{9}(^{\circ}F - 32)$$

If $^{\circ}F = 70$, then

$$^{\circ}C = \frac{5}{9}(70 - 32) = \frac{5}{9}(38)$$

$$^{\circ}C \approx 21^{\circ}$$

87. (a) Since there is only a profit of
$0.50 per copy and the expense
of $100 must be deducted, the
profit is:
$$P = 0.50x - 100$$
    (b) $P = 0.50(1000) - 100$
    $$= 500 - 100$$
    $$= \$400$$
    (c) $P = 0.50(5000) - 100$
    $$= 2500 - 100$$
    $$= \$2400$$

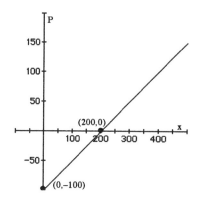

89. $C = 0.06543x + 5.65$
For 300 kWh,
$$C = 0.06543(300) + 5.65$$
$$= \$25.28$$
For 750 kWh,
$$C = 0.06543(750) + 5.65$$
$$= \$54.72$$

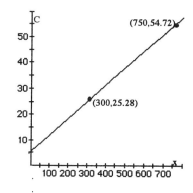

91. $2x - y = C$
Graph the lines:
$$2x - y = -2$$
$$2x - y = 0$$
$$2x - y = 4$$
All the lines have the same slope, 2.
The lines are parallel.

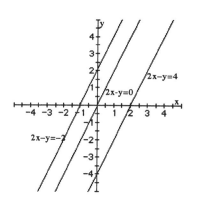

1.7 Scatter Diagrams; Linear Curve Fitting

1. Linear, $m > 0$

3. Linear, $m < 0$

5. Nonlinear

7.   (a)

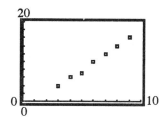

(b)   Answers will vary. We select (4,6) and (8,14). The slope of the line containing these points is:
$$m = \frac{14-6}{8-4} = \frac{8}{4} = 2$$
The equation of the line is:
$$y - y_1 = m(x - x_1)$$
$$y - 6 = 2(x - 4)$$
$$y - 6 = 2x - 8$$
$$y = 2x - 2$$

(c)

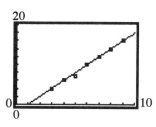

(d)   Using the LINear REGresssion program, the line of best fit is:
$$y = 2.0357x - 2.3571$$

(e)

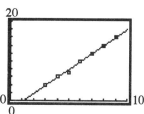

9.   (a)

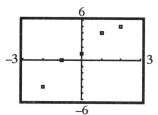

(b)   Answers will vary. We select (–2,–4) and (1,4). The slope of the line containing these points is:
$$m = \frac{4-(-4)}{1-(-2)} = \frac{8}{3}$$
The equation of the line is:
$$y - y_1 = m(x - x_1)$$
$$y - (-4) = \frac{8}{3}(x - (-2))$$
$$y + 4 = \frac{8}{3}x + \frac{16}{3}$$
$$y = \frac{8}{3}x + \frac{4}{3}$$

(c)

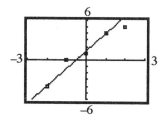

(d)   Using the LINear REGresssion program, the line of best fit is:
$$y = 2.2x + 1.2$$

(e)

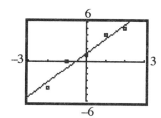

11. (a)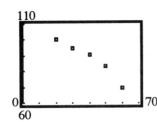

(b) Answers will vary. We select (30,95) and (60,70). The slope of the line containing these points is:
$$m = \frac{70 - 95}{60 - 30} = \frac{-25}{30} = \frac{-5}{6}$$
The equation of the line is:
$$y - y_1 = m(x - x_1)$$
$$y - 95 = \frac{-5}{6}(x - 30)$$
$$y - 95 = \frac{-5}{6}x + 25$$
$$y = \frac{-5}{6}x + 120$$

(c)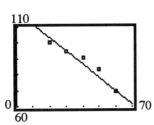

(d) Using the LINear REGresssion program, the line of best fit is:
$$y = -0.72x + 116.6$$

(e)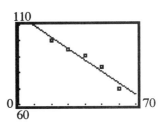

13. (a)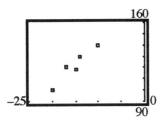

(b) Answers will vary. We select (−20,100) and (−15,118). The slope of the line containing these points is:
$$m = \frac{118 - 100}{-15 - (-20)} = \frac{18}{5}$$
The equation of the line is:
$$y - y_1 = m(x - x_1)$$
$$y - 100 = \frac{18}{5}(x - (-20))$$
$$y - 100 = \frac{18}{5}x + 72$$
$$y = \frac{18}{5}x + 172$$

(c)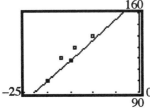

(d) Using the LINear REGresssion program, the line of best fit is:
$$y = 3.8613x + 180.2920$$

(e)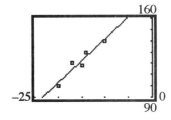

15. (a)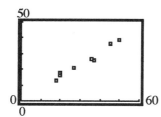

(b) $C = 0.7549I + 0.6266$
(c) As disposable income increases by $1, consumption increases by $0.7549.
(d) $C = 0.7549(42) + 0.6266$
$= 32.332$
A family with disposable income of $42,000 consumes about $32,332.

17. (a) (Data used in graphs is in thousands.)

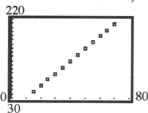

(b) $L = 2.9814I - 0.0761$
(c)

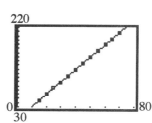

(d) As annual income increases by $1, the loan amount increases by $2.9814.

(e) $L = 2.9814(42) - 0.0761 = 125.143$
A person with an annual income of $42,000 would qualify for a loan of about $125,143.

19. (a)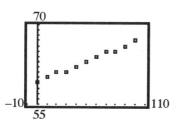

(b) $T = 0.0782h + 59.0909$
(c)

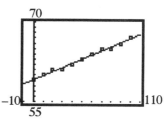

(d) As relative humidity increases by 1%, the apparent temperature increases by $0.0782°$.

(e) $T = 0.0782(75) + 59.0909 = 64.96$
A relative humidity of 75% would give an apparent temperature of $65°$.

21. (a)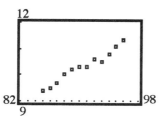

(b) $M = 0.1633x - 4.4691$
(c)

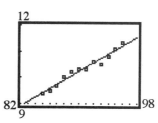

(d) As the year increases by 1, the average miles per car (in thousands) increases by 0.1633.

(e) $M = 0.1633(97) - 4.4691 = 11.371$
In 1997, the average number of miles driven per car is 11,371.

23. (a)

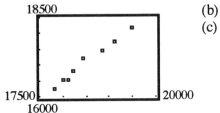

(b) $y = 1.1228x - 3687.5376$

(c)

18500

17500 ⌊ 20000
  16000

(d) As the per capita income increases by 1, the per capita consumption expenditure increases by 1.1228.

(e) $y = 1.1228(19200) - 3687.5376$
$\quad = 17870.2224$
If the per capita income is \$19,200, the per capita consumption expenditure is \$17,870.

1.8   Circles

1. Center = (2, 1)
Radius = distance from (0,1) to (2,1)
$$= \sqrt{(2-0)^2 + (1-1)^2}$$
$$= \sqrt{4} = 2$$
$$(x-2)^2 + (y-1)^2 = 4$$

3. Center = midpoint of (1,2) and (4,2)
$$= \left(\frac{1+4}{2}, \frac{2+2}{2}\right)$$
$$= \left(\frac{5}{2}, 2\right)$$
Radius = distance from $\left(\frac{5}{2},2\right)$ to (4,2)
$$= \sqrt{\left(4 - \frac{5}{2}\right)^2 + (2-2)^2}$$
$$= \sqrt{\frac{9}{4}} = \frac{3}{2}$$
$$\left(x - \frac{5}{2}\right)^2 + (y-2)^2 = \frac{9}{4}$$

5. $(x-h)^2 + (y-k)^2 = r^2$
$(x-1)^2 + (y-(-1))^2 = 1^2$
$(x-1)^2 + (y+1)^2 = 1$
General form:
$x^2 - 2x + 1 + y^2 + 2y + 1 = 1$
$x^2 + y^2 - 2x + 2y + 1 = 0$

(1,−1)

7. $(x - h)^2 + (y - k)^2 = r^2$

    $(x - 0)^2 + (y - 2)^2 = 2^2$

       $x^2 + (y - 2)^2 = 4$

General form:

   $x^2 + y^2 - 4y + 4 = 4$

     $x^2 + y^2 - 4y = 0$

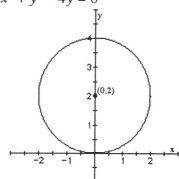

9. $(x - h)^2 + (y - k)^2 = r^2$

    $(x - 4)^2 + (y - (-3))^2 = 5^2$

      $(x - 4)^2 + (y + 3)^2 = 25$

General form:

  $x^2 - 8x + 16 + y^2 + 6y + 9 = 25$

     $x^2 + y^2 - 8x + 6y = 0$

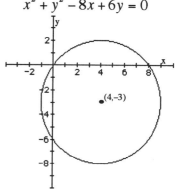

11. $(x - h)^2 + (y - k)^2 = r^2$

     $(x - 0)^2 + (y - 0)^2 = 2^2$

         $x^2 + y^2 = 4$

General form:

   $x^2 + y^2 - 4 = 0$

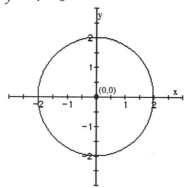

13. $(x - h)^2 + (y - k)^2 = r^2$

    $\left(x - \dfrac{1}{2}\right)^2 + (y - 0)^2 = \left(\dfrac{1}{2}\right)^2$

    $\left(x - \dfrac{1}{2}\right)^2 + y^2 = \dfrac{1}{4}$

General form:

   $x^2 - x + \dfrac{1}{4} + y^2 = \dfrac{1}{4}$

     $x^2 + y^2 - x = 0$

$\left(x - \tfrac{1}{2}\right)^2 + y^2 - \tfrac{1}{4} = 0$

$x^2 - x + \tfrac{1}{4} + y^2 - \tfrac{1}{4} = 0$

$x^2 + y^2 - x = 0$

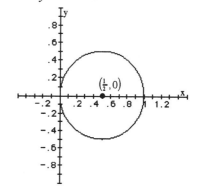

15. $x^2 + y^2 = 4$

     $x^2 + y^2 = 2^2$

     Center: $(0,0)$

     Radius $= 2$

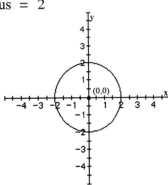

17. $(x - 3)^2 + y^2 = 4$

     $(x - 3)^2 + y^2 = 2^2$

     Center: $(3, 0)$

     Radius $= 2$

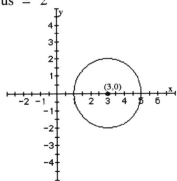

19. $x^2 + y^2 + 4x - 4y - 1 = 0$

$$x^2 + 4x + y^2 - 4y = 1$$

$$(x^2 + 4x + 4) + (y^2 - 4y + 4) = 1 + 4 + 4$$

$$(x + 2)^2 + (y - 2)^2 = 3^2$$

     Center: $(-2, 2)$

     Radius $= 3$

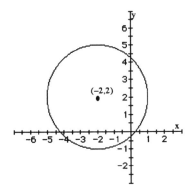

21. $x^2 + y^2 - x + 2y + 1 = 0$

$$x^2 - x + y^2 + 2y = -1$$

$$\left(x^2 - x + \frac{1}{4}\right) + (y^2 + 2y + 1) = -1 + \frac{1}{4} + 1$$

$$\left(x - \frac{1}{2}\right)^2 + (y + 1)^2 = \left(\frac{1}{2}\right)^2$$

     Center: $\left(\frac{1}{2}, -1\right)$

     Radius $= \frac{1}{2}$

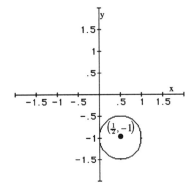

23. $2x^2 + 2y^2 - 12x + 8y - 24 = 0$
$$x^2 + y^2 - 6x + 4y = 12$$
$$x^2 - 6x + y^2 + 4y = 12$$
$$(x^2 - 6x + 9) + (y^2 + 4y + 4) = 12 + 9 + 4$$
$$(x - 3)^2 + (y + 2)^2 = 5^2$$
Center: (3,–2)
Radius = 5

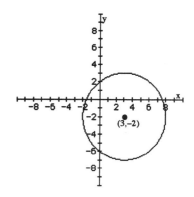

25. Center at (0,0); containing the point (–2,3).
$$r = d(C,P) = \sqrt{(-2-0)^2 + (3-0)^2} = \sqrt{4+9} = \sqrt{13}$$
Equation:
$$(x-0)^2 + (y-0)^2 = \left(\sqrt{13}\right)^2$$
$$x^2 + y^2 = 13$$
$$x^2 + y^2 - 13 = 0$$

27. Center at (2,3); tangent to the x-axis.
$$r = 3$$
Equation:
$$(x-2)^2 + (y-3)^2 = 3^2$$
$$x^2 - 4x + 4 + y^2 - 6y + 9 = 9$$
$$x^2 + y^2 - 4x - 6y + 4 = 0$$

29. Endpoints of a diameter are (1,4) and (–3,2).
The center is at the midpoint of the diameter:
$$\text{Center: } \left(\frac{1+(-3)}{2}, \frac{4+2}{2}\right) = (-1,3)$$
$$\text{Radius: } r = \sqrt{(1-(-1))^2 + (4-3)^2} = \sqrt{4+1} = \sqrt{5}$$
Equation:
$$(x-(-1))^2 + (y-3)^2 = \left(\sqrt{5}\right)^2$$
$$x^2 + 2x + 1 + y^2 - 6y + 9 = 5$$
$$x^2 + y^2 + 2x - 6y + 5 = 0$$

31. (c)   33. (b)

35. $(x+3)^2 + (y-1)^2 = 16$ 37. $(x-2)^2 + (y-2)^2 = 9$

39. (a)
$$x^2 + (mx + b)^2 = r^2$$
$$x^2 + m^2 x^2 + 2bmx + b^2 = r^2$$
$$(1 + m^2)x^2 + 2bmx + b^2 - r^2 = 0$$
There is one solution if and only if the discriminant is zero.
$$(2bm)^2 - 4(1 + m^2)(b^2 - r^2) = 0$$
$$4b^2 m^2 - 4b^2 + 4r^2 - 4b^2 m^2 + 4m^2 r^2 = 0$$
$$-4b^2 + 4r^2 + 4m^2 r^2 = 0$$
$$-b^2 + r^2 + m^2 r^2 = 0$$
$$r^2(1 + m^2) = b^2$$

(b) Using the quadratic formula, knowing that the discriminant is zero:
$$x = \frac{-2bm}{2(1 + m^2)} = \frac{-bm}{\dfrac{b^2}{r^2}} = \frac{-bmr^2}{b^2} = \frac{-mr^2}{b}$$

$$y = m\left(\frac{-mr^2}{b}\right) + b = \frac{-m^2 r^2}{b} + b = \frac{-m^2 r^2 + b^2}{b} = \frac{r^2}{b}$$

(c) The slope of the tangent line is m.
The slope of the line joining the point of tangency and the center is:
$$\frac{\dfrac{r^2}{b} - 0}{\dfrac{-mr^2}{b} - 0} = \frac{r^2}{b} \cdot \frac{b}{-mr^2} = -\frac{1}{m}$$

41. $x^2 + y^2 - 4x + 6y + 4 = 0$
$$(x^2 - 4x + 4) + (y^2 + 6y + 9) = -4 + 4 + 9$$
$$(x - 2)^2 + (y + 3)^2 = 9$$
Center: $(2, -3)$

Slope from center to $\left(3, 2\sqrt{2} - 3\right)$ is $\dfrac{2\sqrt{2} - 3 - (-3)}{3 - 2} = \dfrac{2\sqrt{2}}{1} = 2\sqrt{2}$

Slope of the tangent line is: $\dfrac{-1}{2\sqrt{2}} = \dfrac{-\sqrt{2}}{4}$

Equation of the tangent line:
$$y - \left(2\sqrt{2} - 3\right) = \frac{-\sqrt{2}}{4}(x - 3)$$
$$y - 2\sqrt{2} + 3 = \frac{-\sqrt{2}}{4}x + \frac{3\sqrt{2}}{4}$$
$$4y - 8\sqrt{2} + 12 = -\sqrt{2}x + 3\sqrt{2}$$
$$\sqrt{2}x + 4y = 11\sqrt{2} - 12$$

43. Find the centers of the two circles:
$$x^2 + y^2 - 4x + 6y + 4 = 0$$
$$(x^2 - 4x + 4) + (y^2 + 6y + 9) = -4 + 4 + 9$$
$$(x - 2)^2 + (y + 3)^2 = 9 \qquad \text{Center: } (2, -3)$$
$$x^2 + y^2 + 6x + 4y + 9 = 0$$
$$(x^2 + 6x + 9) + (y^2 + 4y + 4) = -9 + 9 + 4$$
$$(x + 3)^2 + (y + 2)^2 = 4 \qquad \text{Center: } (-3, -2)$$

Find the slope of the line containing the centers:
$$m = \frac{-2 - (-3)}{-3 - 2} = \frac{1}{-5}$$

Find the equation of the line containing the centers:
$$y + 3 = \frac{-1}{5}(x - 2)$$
$$5y + 15 = -x + 2$$
$$x + 5y = -13$$

1   Chapter Review

1. (a) Graph: $y_1 = 2 - \dfrac{x}{3}$; $y_2 = 6$.   (b)

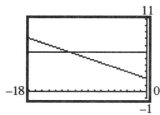

Using INTERSECT, $y_1 = y_2$ at $(-12, 6)$. The solution set is $\{-12\}$.

$$2 - \frac{x}{3} = 6$$
$$6 - x = 18$$
$$-x = 12$$
$$x = -12$$

3. (a) Graph: $y_1 = -2(5 - 3x) + 8$;
$$y_2 = 4 + 5x.$$
(b)

Using INTERSECT, $y_1 = y_2$ at $(6, 34)$. The solution set is $\{6\}$.

$$-2(5 - 3x) + 8 = 4 + 5x$$
$$-10 + 6x + 8 = 4 + 5x$$
$$6x - 2 = 4 + 5x$$
$$x = 6$$

5.   (a)   Graph: $y_1 = \dfrac{3x}{4} - \dfrac{x}{3}$; $y_2 = \dfrac{1}{12}$.   (b)

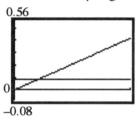

0.56

0

1

−0.08

Using INTERSECT, $y_1 = y_2$ at (0.20, 0.083). The solution set is {0.20}.

$$\frac{3x}{4} - \frac{x}{3} = \frac{1}{12}$$
$$9x - 4x = 1$$
$$5x = 1$$
$$x = \frac{1}{5}$$

7.   (a)   Graph: $y_1 = \dfrac{x}{x-1}$; $y_2 = \dfrac{5}{6}$.   (b)

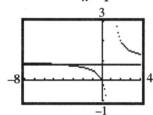

3

−8

4

−1

Using INTERSECT, $y_1 = y_2$ at (−5, 0.83). The solution set is {−5}.

$$\frac{x}{x-1} = \frac{5}{6}$$
$$6x = 5(x-1)$$
$$6x = 5x - 5$$
$$x = -5$$

9.   (a)   Graph: $y_1 = x(1-x)$; $y_2 = -6$.   (b)

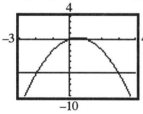

4

−3

4

−10

Using INTERSECT, $y_1 = y_2$ at (−2, −6) and (3, −6). The solution set is {−2, 3}.

$$x(1-x) = -6$$
$$x - x^2 = -6$$
$$x^2 - x - 6 = 0$$
$$(x+2)(x-3) = 0$$
$$x = -2 \ \text{ or } \ x = 3$$

11. (a)
Graph: $y_1 = \frac{1}{2}\left(x - \frac{1}{3}\right)$; $y_2 = \frac{3}{4} - \frac{x}{6}$.

(b)

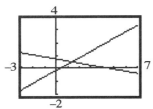

Using INTERSECT, $y_1 = y_2$ at (1.375, 0.52). The solution set is {1.375}.

$$\frac{1}{2}\left(x - \frac{1}{3}\right) = \frac{3}{4} - \frac{x}{6}$$
$$\frac{x}{2} - \frac{1}{6} = \frac{3}{4} - \frac{x}{6}$$
$$6x - 2 = 9 - 2x$$
$$8x = 11$$
$$x = \frac{11}{8}$$

13. (a) Graph: $y_1 = (x - 1)(2x + 3)$; $y_2 = 3$.

(b)

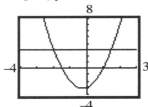

Using INTERSECT, $y_1 = y_2$ at (–2, 3) and (1.5, 3). The solution set is {–2, 1.5}.

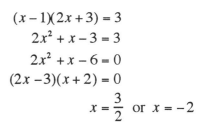

$$(x - 1)(2x + 3) = 3$$
$$2x^2 + x - 3 = 3$$
$$2x^2 + x - 6 = 0$$
$$(2x - 3)(x + 2) = 0$$
$$x = \frac{3}{2} \text{ or } x = -2$$

15. (a) Graph: $y_1 = 4x + 3$; $y_2 = 4x^2$.

(b)

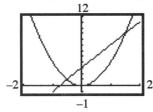

Using INTERSECT, $y_1 = y_2$ at (–0.5, 1) and (1.5, 9). The solution set is {–0.5, 1.5}.

$$4x + 3 = 4x^2$$
$$4x^2 - 4x - 3 = 0$$
$$(2x - 3)(2x + 1) = 0$$
$$x = \frac{3}{2} \text{ or } x = \frac{-1}{2}$$

17. (a) Graph: $y_1 = \sqrt[3]{x - 1}$; $y_2 = 2$.

(b)

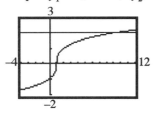

Using INTERSECT, $y_1 = y_2$ at (9, 2). The solution set is {9}.

$$\sqrt[3]{x - 1} = 2$$
$$\left(\sqrt[3]{x - 1}\right)^3 = 2^3$$
$$x - 1 = 8$$
$$x = 9$$

Check: $\sqrt[3]{9 - 1} = \sqrt[3]{8} = 2$
The solution is $x = 9$.

19.   (a)   Graph: $y_1 = x(x+1) - 2$; $y_2 = 0$.   (b)

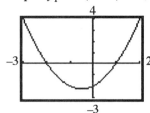

$$x(x+1) - 2 = 0$$
$$x^2 + x - 2 = 0$$
$$(x+2)(x-1) = 0$$
$$x = -2 \text{ or } x = 1$$

Using INTERSECT, $y_1 = y_2$ at $(-2, 0)$ and $(1, 0)$. The solution set is $\{-2, 1\}$.

21.   (a)   Graph: $y_1 = \sqrt{2x-3} + x$; $y_2 = 3$.   (b)

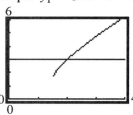

$$\sqrt{2x-3} + x = 3$$
$$\sqrt{2x-3} = 3 - x$$
$$2x - 3 = 9 - 6x + x^2$$
$$x^2 - 8x + 12 = 0$$
$$(x-2)(x-6) = 0$$
$$x = 2 \text{ or } x = 6$$

Using INTERSECT, $y_1 = y_2$ at $(2, 3)$. The solution set is $\{2\}$.

Check 2: $\sqrt{2(2) - 3} + 2 = \sqrt{1} + 2 = 3$

Check 6: $\sqrt{2(6) - 3} + 6 = \sqrt{9} + 6$
$$= 9 \neq 3$$

The solution is $x = 2$.

23.   (a)   Graph: $y_1 = \sqrt{x+1} + \sqrt{x-1}$;
$$y_2 = \sqrt{2x+1}.$$

(b)

$$\sqrt{x+1} + \sqrt{x-1} = \sqrt{2x+1}$$
$$\left(\sqrt{x+1} + \sqrt{x-1}\right)^2 = \left(\sqrt{2x+1}\right)^2$$
$$x + 1 + 2\sqrt{x+1}\sqrt{x-1} + x - 1 = 2x + 1$$
$$2\sqrt{x+1}\sqrt{x-1} = 1$$
$$\left(2\sqrt{x+1}\sqrt{x-1}\right)^2 = 1^2$$
$$4(x+1)(x-1) = 1$$
$$4x^2 - 4 = 1$$
$$4x^2 = 5$$
$$x^2 = \frac{5}{4}$$
$$x = \frac{\pm\sqrt{5}}{2}$$

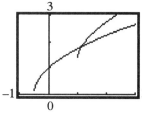

Using INTERSECT, $y_1 = y_2$ at $(1.12, 1.80)$. The solution set is $\{1.12\}$.

Check $\dfrac{-\sqrt{5}}{2}$: $\sqrt{\dfrac{-\sqrt{5}}{2} + 1} + \sqrt{\dfrac{-\sqrt{5}}{2} - 1} = \sqrt{2\left(\dfrac{-\sqrt{5}}{2}\right) + 1}$

$$\sqrt{-0.118} + \sqrt{-2.118} = \sqrt{-1.236}$$

This is not defined.

Check $\dfrac{\sqrt{5}}{2}$: $\sqrt{\dfrac{\sqrt{5}}{2}+1} + \sqrt{\dfrac{\sqrt{5}}{2}-1} = \sqrt{2\left(\dfrac{\sqrt{5}}{2}\right)+1}$

$$\sqrt{2.118} + \sqrt{0.118} = \sqrt{3.236}$$
$$1.455 + 0.344 = 1.799$$
$$1.799 = 1.799$$

The solution is $x = \dfrac{\sqrt{5}}{2}$.

25. (a) Graph: $y_1 = |2x+3|$; $y_2 = 7$.

    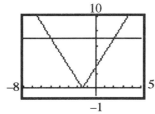

    Using INTERSECT, $y_1 = y_2$ at $(-5, 7)$ and $(2, 7)$. The solution set is $\{-5, 2\}$.

    (b)
    $$|2x+3| = 7$$
    $$2x + 3 = 7 \ \text{ or } \ 2x + 3 = -7$$
    $$2x = 4 \ \text{ or } \ \ \ \ 2x = -10$$
    $$x = 2 \ \text{ or } \ \ \ \ \ \ x = -5$$
    The solution set is $\{-5, 2\}$.

27. (a) Graph: $y_1 = |2-3x|$; $y_2 = 7$.

    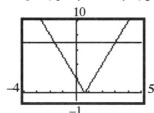

    Using INTERSECT, $y_1 = y_2$ at $(-1.67, 7)$ and $(3, 7)$. The solution set is $\{-1.67, 3\}$.

    (b)
    $$|2-3x| = 7$$
    $$2 - 3x = 7 \ \ \text{ or } \ 2 - 3x = -7$$
    $$-3x = 5 \ \ \text{ or } \ \ -3x = -9$$
    $$x = \dfrac{-5}{3} \ \ \text{ or } \ \ \ \ \ x = 3$$
    The solution set is $\left\{\dfrac{-5}{3}, 3\right\}$.

29. (a) Graph: $y_1 = \dfrac{2x-3}{5}+2$; $y_2 = \dfrac{x}{2}$.

    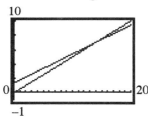

    Using INTERSECT, $y_1 \le y_2$ at $(14, 7)$. The solution set is $\{x \mid x \ge 14\}$.

    (b)
    $$\dfrac{2x-3}{5} + 2 \le \dfrac{x}{2}$$
    $$2(2x-3) + 10(2) \le 5x$$
    $$4x - 6 + 20 \le 5x$$
    $$14 \le x$$
    $$x \ge 14$$
    $$\{x \mid x \ge 14\} \ \text{ or } \ [14, +\infty)$$

31. (a) Graph: $y_1 = -9$; $y_2 = \dfrac{2x+3}{-4}$; $y_3 = 7$. (b)

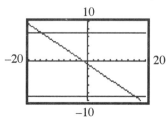

The intersection points are $(-15.5, 7)$ and $(16.5, -9)$. Since y_2 is between -9 and 7 when $-15.5 \le x \le 16.5$, the solution set is $\{x | -15.5 \le x \le 16.5\}$.

$$-9 \le \frac{2x+3}{-4} \le 7$$
$$36 \ge 2x+3 \ge -28$$
$$33 \ge 2x \ge -31$$
$$\frac{33}{2} \ge x \ge \frac{-31}{2}$$
$$\frac{-31}{2} \le x \le \frac{33}{2}$$
$$\left\{x \left| \frac{-31}{2} \le x \le \frac{33}{2} \right.\right\} \text{ or } \left[\frac{-31}{2}, \frac{33}{2}\right]$$

33. (a) Graph: $y_1 = 6$; $y_2 = \dfrac{3-3x}{12}$; $y_3 = 2$. (b)

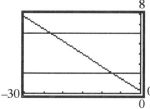

The intersection points are $(-23, 6)$ and $(-7, 2)$. Since y_2 is between 2 and 6 when $-23 < x < -7$, the solution set is $\{x | -23 < x < -7\}$.

$$6 > \frac{3-3x}{12} > 2$$
$$72 > 3-3x > 24$$
$$69 > -3x > 21$$
$$-23 < x < -7$$
$$\{x | -23 < x < -7\} \text{ or } (-23, -7)$$

35. (a) Graph: $y_1 = |3x+4|$; $y_2 = \frac{1}{2}$. (b)

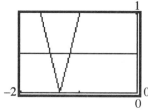

Using INTERSECT, $y_1 < y_2$ when $-1.50 < x < -1.17$. The solution set is $\{x | -1.50 < x < -1.17\}$.

$$|3x+4| < \frac{1}{2}$$
$$\frac{-1}{2} < 3x+4 < \frac{1}{2}$$
$$\frac{-9}{2} < 3x < \frac{-7}{2}$$
$$\frac{-3}{2} < x < \frac{-7}{6}$$
$$\left\{x \left| \frac{-3}{2} < x < \frac{-7}{6} \right.\right\} \text{ or } \left(\frac{-3}{2}, \frac{-7}{6}\right)$$

37. (a) Graph: $y_1 = |2x - 5|$; $y_2 = 9$.

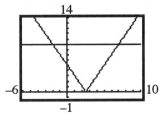

Using INTERSECT, $y_1 \geq y_2$ when $x \leq -2$ or $x \geq 7$. The solution set is $\{x \mid x \leq -2 \text{ or } x \geq 7\}$.

(b) $|2x - 5| \geq 9$

$2x - 5 \leq -9$ or $2x - 5 \geq 9$

$2x \leq -4$ or $\quad 2x \geq 14$

$x \leq -2$ or $\quad\quad x \geq 7$

$\{x \mid x \leq -2 \text{ or } x \geq 7\}$

$\quad$ or $(-\infty, -2] \cup [7, +\infty)$

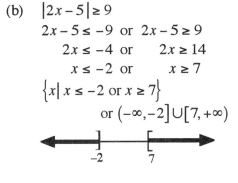

39. Slope $= -2$; containing $(3, -1)$

$y - y_1 = m(x - x_1)$

$y - (-1) = -2(x - 3)$

$y + 1 = -2x + 6$

$y = -2x + 5$

$2x + y = 5$ or $y = -2x + 5$

41. Slope undefined; containing $(-3, 4)$
This is a vertical line.

$x = -3$

$\quad$ No slope intercept form.

43. y-intercept $= -2$; containing $(5, -3)$
Points are $(5, -3)$ and $(0, -2)$

$m = \dfrac{-2 - (-3)}{0 - 5} = \dfrac{1}{-5} = \dfrac{-1}{5}$

$y = mx + b$

$y = \dfrac{-1}{5}x - 2$

$x + 5y = -10$ or $y = \dfrac{-1}{5}x - 2$

45. Parallel to $2x - 3y + 4 = 0$;

$\quad$ Slope $= \frac{2}{3}$; containing $(-5, 3)$

$y - y_1 = m(x - x_1)$

$y - 3 = \dfrac{2}{3}(x - (-5))$

$y - 3 = \dfrac{2}{3}x + \dfrac{10}{3}$

$y = \dfrac{2}{3}x + \dfrac{19}{3}$

$2x - 3y = -19$ or $y = \dfrac{2}{3}x + \dfrac{19}{3}$

47. Perpendicular to $x + y - 2 = 0$;
Containing $(4, -3)$
Slope of perpendicular $= 1$

$y - y_1 = m(x - x_1)$

$y - (-3) = 1(x - 4)$

$y + 3 = x - 4$

$y = x - 7$

$x - y = 7$ or $y = x - 7$

49. $4x - 5y + 20 = 0$
x-intercept $= -5$
y-intercept $= 4$

51. $\dfrac{1}{2}x - \dfrac{1}{3}y + \dfrac{1}{6} = 0$

    x-intercept $= \dfrac{-1}{3}$

    y-intercept $= \dfrac{1}{2}$

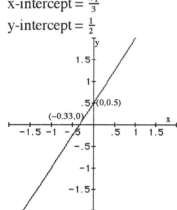

53. $\sqrt{2}x + \sqrt{3}y = \sqrt{6}$

    x-intercept $= \sqrt{3}$

    y-intercept $= \sqrt{2}$

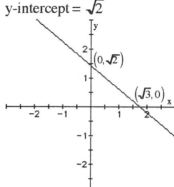

55. Intercepts: $(0,0)$

    Test for symmetry:

    x-axis: Replace y by $-y$ so $2x = 3(-y)^2$ or $2x = 3y^2$, which is
    equivalent to $2x = 3y^2$.

    y-axis: Replace x by $-x$ so $2(-x) = 3y^2$ or $-2x = 3y^2$,
    which is not equivalent to $2x = 3y^2$.

    Origin: Replace x by $-x$ and y by $-y$ so $2(-x) = 3(-y)^2$ or
    $-2x = 3y^2$, which is not equivalent to $2x = 3y^2$.

    Therefore, the graph is symmetric with respect to the x-axis.

57. Intercepts: $(0,1)$, $(0,-1)$, $\left(\dfrac{1}{2},0\right)$, and $\left(\dfrac{-1}{2},0\right)$

    Test for symmetry:

    x-axis: Replace y by $-y$ so $4x^2 + (-y)^2 = 1$ or $4x^2 + y^2 = 1$, which is
    equivalent to $4x^2 + y^2 = 1$.

    y-axis: Replace x by $-x$ so $4(-x)^2 + y^2 = 1$ or $4x^2 + y^2 = 1$,
    which is equivalent to $4x^2 + y^2 = 1$.

    Origin: Replace x by $-x$ and y by $-y$ so $4(-x)^2 + (-y)^2 = 1$ or
    $4x^2 + y^2 = 1$, which is equivalent to $4x^2 + y^2 = 1$.

    Therefore, the graph is symmetric with respect to the x-axis, y-axis, and origin.

59. Intercepts: $(0,1)$
    Test for symmetry:

    x-axis: Replace y by $-y$ so $-y = x^4 + 2x^2 + 1$, which is not
    equivalent to $y = x^4 + 2x^2 + 1$.

    y-axis: Replace x by $-x$ so $y = (-x)^4 + 2(-x)^2 + 1$ or $y = x^4 + 2x^2 + 1$,
    which is equivalent to $y = x^4 + 2x^2 + 1$.

    Origin: Replace x by $-x$ and y by $-y$ so $-y = (-x)^4 + 2(-x)^2 + 1$ or
    $-y = x^4 + 2x^2 + 1$, which is not equivalent to $y = x^4 + 2x^2 + 1$.

    Therefore, the graph is symmetric with respect to the y-axis.

61. Intercepts: $(0,0)$, $(0,-2)$, $(-1,0)$
    Test for symmetry:

    x-axis: Replace y by $-y$ so $x^2 + x + (-y)^2 + 2(-y) = 0$ or $x^2 + x + y^2 - 2y = 0$,
    which is not equivalent to $x^2 + x + y^2 + 2y = 0$.

    y-axis: Replace x by $-x$ so $(-x)^2 + (-x) + y^2 + 2y = 0$ or $x^2 - x + y^2 + 2y = 0$,
    which is not equivalent to $x^2 + x + y^2 + 2y = 0$.

    Origin: Replace x by $-x$ and y by $-y$ so $(-x)^2 + (-x) + (-y)^2 + 2(-y) = 0$ or
    $x^2 - x + y^2 - 2y = 0$, which is not equivalent to
    $x^2 + x + y^2 + 2y = 0$.

    Therefore, the graph is not symmetric to the x-axis, the y-axis, or the origin.

63. $x^2 + y^2 - 2x + 4y - 4 = 0$
    $$x^2 - 2x + y^2 + 4y = 4$$
    $$(x^2 - 2x + 1) + (y^2 + 4y + 4) = 4 + 1 + 4$$
    $$(x - 1)^2 + (y + 2)^2 = 3^2$$
    Center: $(1,-2)$
    Radius $= 3$

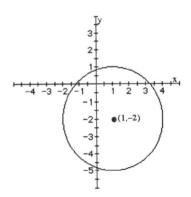

65. $3x^2 + 3y^2 - 6x + 12y = 0$
    $$x^2 + y^2 - 2x + 4y = 0$$
    $$x^2 - 2x + y^2 + 4y = 0$$
    $$(x^2 - 2x + 1) + (y^2 + 4y + 4) = 1 + 4$$
    $$(x - 1)^2 + (y + 2)^2 = \left(\sqrt{5}\right)^2$$
    Center: $(1,-2)$
    Radius $= \sqrt{5}$

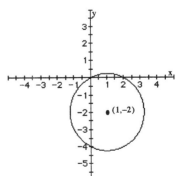

67. Given the points (7,4) and (–3,2).

Slope: $m = \dfrac{2-4}{-3-7} = \dfrac{-2}{-10} = \dfrac{1}{5}$

Distance: $d = \sqrt{(-3-7)^2 + (2-4)^2} = \sqrt{100+4} = \sqrt{104} = 2\sqrt{26}$

Midpoint: $\left(\dfrac{7+(-3)}{2}, \dfrac{4+2}{2}\right) = (2,3)$

69. Find the distance between each pair of points.

$d_{A,B} = \sqrt{(1-3)^2 + (1-4)^2} = \sqrt{4+9} = \sqrt{13}$

$d_{B,C} = \sqrt{(-2-1)^2 + (3-1)^2} = \sqrt{9+4} = \sqrt{13}$

$d_{A,C} = \sqrt{(-2-3)^2 + (3-4)^2} = \sqrt{25+1} = \sqrt{26}$

Since $AB = BC$, triangle ABC is isosceles.

71. Find the slope between each pair of points.

$m_{A,B} = \dfrac{1-5}{6-2} = \dfrac{-4}{4} = -1$

$m_{A,C} = \dfrac{-1-5}{8-2} = \dfrac{-6}{6} = -1$

$m_{B,C} = \dfrac{-1-1}{8-6} = \dfrac{-2}{2} = -1$

Since the slopes are all the same, the points all lie on the same line.

73. Endpoints of the diameter are (–3, 2) and (5,–6).
The center is at the midpoint of the diameter:

Center: $\left(\dfrac{-3+5}{2}, \dfrac{2+(-6)}{2}\right) = (1,-2)$

Radius: $r = \sqrt{(1-(-3))^2 + (-2-2)^2} = \sqrt{16+16} = \sqrt{32} = 4\sqrt{2}$

Equation:

$$(x-1)^2 + (y+2)^2 = \left(4\sqrt{2}\right)^2$$
$$x^2 - 2x + 1 + y^2 + 4y + 4 = 32$$
$$x^2 + y^2 - 2x + 4y - 27 = 0$$

75. (a)

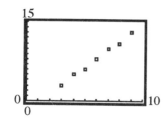

(b) $y = 1.6429x - 1.8571$

(c) As x increases by 1, y increases by 1.6429.

77. (a)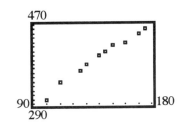

    (b) $y = 2.0183x + 112.4326$
    (c) As x increases by 1, y increases by 2.0183.

79. Using $s = vt$, we have $t = 3$ and $v = 1100$. Finding the distance s in feet:

    $s = 1100(3)$

    $s = 3300$

    The storm is 3300 feet away.

81. Let s represent the distance the plane can travel.

	Rate	Time	Distance
With wind	250+30=280	$\dfrac{s/2}{280}$	$\dfrac{s}{2}$
Against wind	250–30=220	$\dfrac{s/2}{220}$	$\dfrac{s}{2}$

    Since the total time is at most 5 hours, we have:

    $$\frac{s/2}{280} + \frac{s/2}{220} \le 5$$

    $$\frac{s}{560} + \frac{s}{440} \le 5$$

    $$11s + 14s \le 5(6160)$$

    $$25s \le 30800$$

    $$s \le 1232$$

    The plane can travel at most 1232 miles or 616 miles one way and return 616 miles.

83. Let t represent the time it takes the helicopter to reach the raft.

	Rate	Time	Distance
Raft	5	t	$5t$
Helicopter	90	t	$90t$

    Since the total distance is 150 miles, we have:

    $$5t + 90t = 150$$

    $$95t = 150$$

    $$t = 1.58 \text{ hours} = 1 \text{ hour and } 35 \text{ minutes}$$

    The helicopter will reach the raft in 1 hour and 35 minutes.

85. Let t represent the time it takes Clarissa to complete the job by herself.

	Time to do job alone	Part of job done in one day	Time on Job	Part of total job done by each person
Clarissa	t	$\dfrac{1}{t}$	6	$\dfrac{6}{t}$
Shawna	$t+5$	$\dfrac{1}{t+5}$	6	$\dfrac{6}{t+5}$

Since the two people paint one house, we have:

$$\frac{6}{t} + \frac{6}{t+5} = 1$$

$$6(t+5) + 6t = t(t+5)$$

$$6t + 30 + 6t = t^2 + 5t$$

$$t^2 - 7t - 30 = 0$$

$$(t-10)(t+3) = 0$$

$$t = 10 \ \text{ or } \ t = -3$$

It takes Clarissa 10 days to paint the house when working by herself.

87. Let x represent the number of cubic centimeters of 15% HCl.
Then $100 - x$ represents the number of cubic centimeters of 5% HCl.

	No. of cu.cm.	Conc. of salt solution	Pure salt
15% HCl	x	0.15	$0.15x$
5% HCl	$100 - x$	0.05	$0.05(100 - x)$
8% HCl	100	0.08	$0.08(100)$

$$0.15x + 0.05(100 - x) = 0.08(100)$$

$$0.15x + 5 - 0.05x = 8$$

$$0.10x = 3$$

$$x = 30$$

30 cubic centimeters of the 15% HCl solution are mixed with 70 cubic centimeters of the 5% HCl solution.

89. Let x represent the number of adult tickets sold.
Then $5200 - x$ represents the number of children's tickets sold.

	No. of tickets	Price per ticket	Total Value
Adult	x	\$4.75	$4.75x$
Children	$5200 - x$	\$2.50	$2.50(5200 - x)$
Total	5200		20,335

$$4.75x + 2.50(5200 - x) = 20335$$

$$4.75x + 13000 - 2.50x = 20335$$

$$2.25x = 7335$$

$$x = 3260$$

3,260 adult tickets and 1,940 children's tickets were sold.

91. The effective speed of the train (i.e., relative to the man) is $30 - 4 = 26$ miles per hour. The time is 5 seconds $= \dfrac{5}{60}$ minutes $= \dfrac{5}{3600}$ hours $= \dfrac{1}{720}$ hours.

$$s = vt = 26\left(\frac{1}{720}\right) = \frac{26}{720} \text{ miles} = \frac{26}{720} \cdot 5280 = 190.67 \text{ feet}$$

The freight train is 190.67 feet long.

93.   Let x represent the number of passengers over 20.
Then $20 + x$ represents the total number of passengers.
$15 - 0.1x$ represents the fare for each passenger.
Solving the equation for total cost ($482.40), we have:
$$(20 + x)(15 - 0.1x) = 482.40$$
$$300 + 13x - 0.1x^2 = 482.40$$
$$-0.1x^2 + 13x - 182.40 = 0$$
$$x^2 - 130x + 1824 = 0$$
$$(x - 114)(x - 16) = 0$$
$$x = 114 \text{ or } x = 16$$
Since the capacity of the bus is 44, we discard the 114. The total number of passengers is $20 + 16 = 36$, and the ticket price per passenger is $15 - 0.1(16) = \$13.40$.
36 people went on the trip; each person paid \$13.40.

95.   (a)

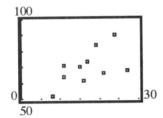

(b) $C = 1.0184A + 52.6898$
(c)   As algebra scores increase by 1, calculus scores increase by 1.0184.

(d) $C = 1.0184(20) + 52.6898 = 73.06$
A student with a score of 20 on the algebra test will score 73 on the calculus test.

97.   (a)

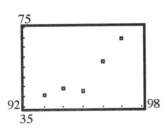

(b)   As time increases, the value increases.
(c)

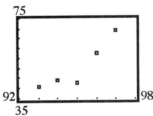

(d) $m = \dfrac{42.97 - 40.97}{1995 - 1993} = \dfrac{2}{2} = 1$

(f) $m = \dfrac{69.17 - 42.97}{1997 - 1995} = \dfrac{26.2}{2} = 13.1$

(e)   Each year the value of the portfolio increases by \$1.

(g)   Each year the value of the portfolio increases by \$13.10.

99.   (a) $x = 0$ is the y-axis
       (b) $y = 0$ is the x-axis
       (c) $x + y = 0$ is a line through the origin with slope -1
       (d) $xy = 0$ is the x-axis or the y-axis or both
       (e) $x^2 + y^2 = 0$ is the origin

Functions

2.1 Functions

1. Function
   Domain: {20, 25, 30, 40}
   Range: {$200, $300, $350, $425}

3. Not a function

5. Function
   Domain: {2, –3, 4, 1}
   Range: {6, 9, 10}

7. Function
   Domain: {1, 2, 3, 4}
   Range: {3}

9. Not a function

11. Function
    Domain: {–2, –1, 0, 1}
    Range: {4, 1, 0}

13. $f(x) = 3x^2 + 2x - 4$

   (a) $f(0) = 3(0)^2 + 2(0) - 4 = -4$

   (b) $f(1) = 3(1)^2 + 2(1) - 4 = 3 + 2 - 4 = 1$

   (c) $f(-1) = 3(-1)^2 + 2(-1) - 4 = 3 - 2 - 4 = -3$

   (d) $f(-x) = 3(-x)^2 + 2(-x) - 4 = 3x^2 - 2x - 4$

   (e) $-f(x) = -(3x^2 + 2x - 4) = -3x^2 - 2x + 4$

   (f) $f(x + 1) = 3(x + 1)^2 + 2(x + 1) - 4 = 3(x^2 + 2x + 1) + 2x + 2 - 4$
   $$= 3x^2 + 6x + 3 + 2x + 2 - 4 = 3x^2 + 8x + 1$$

   (g) $f(2x) = 3(2x)^2 + 2(2x) - 4 = 12x^2 + 4x - 4$

   (h) $f(x + h) = 3(x + h)^2 + 2(x + h) - 4 = 3(x^2 + 2xh + h^2) + 2x + 2h - 4$
   $$= 3x^2 + 6xh + 3h^2 + 2x + 2h - 4$$

15. $f(x) = \dfrac{x}{x^2 + 1}$

   (a) $f(0) = \dfrac{0}{0^2 + 1} = \dfrac{0}{1} = 0$

   (b) $f(1) = \dfrac{1}{1^2 + 1} = \dfrac{1}{2}$

(c) $f(-1) = \dfrac{-1}{(-1)^2 + 1} = \dfrac{-1}{1+1} = \dfrac{-1}{2}$

(d) $f(-x) = \dfrac{-x}{(-x)^2 + 1} = \dfrac{-x}{x^2 + 1}$

(e) $-f(x) = -\dfrac{x}{x^2 + 1} = \dfrac{-x}{x^2 + 1}$

(f) $f(x+1) = \dfrac{x+1}{(x+1)^2 + 1} = \dfrac{x+1}{x^2 + 2x + 1 + 1} = \dfrac{x+1}{x^2 + 2x + 2}$

(g) $f(2x) = \dfrac{2x}{(2x)^2 + 1} = \dfrac{2x}{4x^2 + 1}$

(h) $f(x+h) = \dfrac{x+h}{(x+h)^2 + 1} = \dfrac{x+h}{x^2 + 2xh + h^2 + 1}$

17. $f(x) = |x| + 4$

    (a) $f(0) = |0| + 4 = 0 + 4 = 4$

    (b) $f(1) = |1| + 4 = 1 + 4 = 5$

    (c) $f(-1) = |-1| + 4 = 1 + 4 = 5$

    (d) $f(-x) = |-x| + 4 = |x| + 4$

    (e) $-f(x) = -(|x| + 4) = -|x| - 4$

    (f) $f(x+1) = |x+1| + 4$

    (g) $f(2x) = |2x| + 4 = 2|x| + 4$

    (h) $f(x+h) = |x+h| + 4$

19. $f(x) = \dfrac{2x+1}{3x-5}$

    (a) $f(0) = \dfrac{2(0)+1}{3(0)-5} = \dfrac{0+1}{0-5} = \dfrac{-1}{5}$

    (b) $f(1) = \dfrac{2(1)+1}{3(1)-5} = \dfrac{2+1}{3-5} = \dfrac{3}{-2} = \dfrac{-3}{2}$

    (c) $f(-1) = \dfrac{2(-1)+1}{3(-1)-5} = \dfrac{-2+1}{-3-5} = \dfrac{-1}{-8} = \dfrac{1}{8}$

    (d) $f(-x) = \dfrac{2(-x)+1}{3(-x)-5} = \dfrac{-2x+1}{-3x-5} = \dfrac{2x-1}{3x+5}$

    (e) $-f(x) = -\dfrac{2x+1}{3x-5} = \dfrac{-2x-1}{3x-5}$

    (f) $f(x+1) = \dfrac{2(x+1)+1}{3(x+1)-5} = \dfrac{2x+2+1}{3x+3-5} = \dfrac{2x+3}{3x-2}$

    (g) $f(2x) = \dfrac{2(2x)+1}{3(2x)-5} = \dfrac{4x+1}{6x-5}$

    (h) $f(x+h) = \dfrac{2(x+h)+1}{3(x+h)-5} = \dfrac{2x+2h+1}{3x+3h-5}$

21. $f(0) = 3$ since $(0,3)$ is on the graph.
      $f(-6) = -3$ since $(-6,-3)$ is on the graph.

23. $f(2)$ is positive since $f(2) = 4$.

25. $f(x) = 0$ when $x = -3$, $x = 6$, and $x = 10$.

27.   The domain of f is $\{x|-6 \le x \le 11\}$ or $[-6, 11]$

29.   The x-intercepts are $(-3, 0)$, $(6, 0)$, and $(11, 0)$.

31.   The line $y = \frac{1}{2}$ intersects the graph 3 times.

33. $f(x) = 3$ when $x = 0$ and $x = 4$.

35. $f(x) = \dfrac{x+2}{x-6}$

   (a) $f(3) = \dfrac{3+2}{3-6} = \dfrac{5}{-3} \ne 14$ $(3,14)$ is not on the graph of f.

   (b) $f(4) = \dfrac{4+2}{4-6} = \dfrac{6}{-2} = -3$ $(4,-3)$ is the point on the graph of f.

   (c)   Solve for x:
   $$2 = \frac{x+2}{x-6}$$
   $$2x - 12 = x + 2 \qquad (14, 2) \text{ is a point on the graph of } f.$$
   $$x = 14$$

   (d)   The domain of f is $\{x| x \ne 6\}$.

37. $f(x) = \dfrac{2x^2}{x^4 + 1}$

   (a) $f(-1) = \dfrac{2(-1)^2}{(-1)^4 + 1} = \dfrac{2}{2} = 1$ $(-1,1)$ is a point on the graph of f.

   (b) $f(2) = \dfrac{2(2)^2}{(2)^4 + 1} = \dfrac{8}{17}$ $\left(2, \dfrac{8}{17}\right)$ is a point on the graph of f.

   (c)   Solve for x:
   $$1 = \frac{2x^2}{x^4 + 1}$$
   $$x^4 + 1 = 2x^2$$
   $$x^4 - 2x^2 + 1 = 0$$
   $$(x^2 - 1)^2 = 0$$
   $$x^2 - 1 = 0$$
   $$x^2 = 1$$
   $$x = \pm 1$$
   $(1,1)$ and $(-1,1)$ are points on the graph of f.

(d) The domain of f is: $\{$Real Numbers$\}$.

39. Not a function since vertical lines will intersect the graph in more than one point.

41. Function   (a) Domain: $\{x|-\pi \le x \le \pi\}$; Range: $\{y|-1 \le y \le 1\}$

   (b) $\left(\dfrac{-\pi}{2},0\right)$, $\left(\dfrac{\pi}{2},0\right)$, $(0,1)$

   (c) y-axis

43. Not a function since vertical lines will intersect the graph in more than one point.

45. Function   (a) Domain: $\{x|x > 0\}$; Range: $\{y|y \in \text{Real Numbers}\}$

   (b) $(1,0)$

   (c) No symmetry to the x-axis, y-axis, or origin.

47. Function   (a) Domain: $\{x|x \in \text{Real Numbers}\}$; Range: $\{y|y \le 2\}$

   (b) $(-3,0)$, $(3,0)$, $(0,2)$

   (c) y-axis

49. Function   (a) Domain: $\{x|x \in \text{Real Numbers}\}$; Range: $\{y|y \ge -3\}$

   (b) $(1,0)$, $(3,0)$, $(0,9)$

   (c) No symmetry to the x-axis, y-axis, or origin.

51. $f(x) = -5x + 4$

   Domain: $\{$Real Numbers$\}$

53. $f(x) = \dfrac{x}{x^2 + 1}$

   Domain: $\{$Real Numbers$\}$

55. $g(x) = \dfrac{x}{x^2 - 16}$

   $x^2 - 16 \ne 0$

   $x^2 \ne 16$

   $x \ne \pm 4$

   Domain: $\{x|x \ne -4, x \ne 4\}$

57. $F(x) = \dfrac{x-2}{x^3 + x}$

   $x^3 + x \ne 0$

   $x(x^2 + 1) \ne 0$

   $x \ne 0$, $x^2 \ne -1$

   Domain: $\{x|x \ne 0\}$

59. $h(x) = \sqrt{3x - 12}$

   $3x - 12 \ge 0$

   $3x \ge 12$

   $x \ge 4$

   Domain: $\{x|x \ge 4\}$

61. $f(x) = \dfrac{4}{\sqrt{x-9}}$

   $x - 9 > 0$

   $x > 9$

   Domain: $\{x|x > 9\}$

63. $p(x) = \sqrt{\dfrac{2}{x-1}}$

$\dfrac{2}{x-1} > 0 \quad \rightarrow \quad x - 1 > 0 \quad \rightarrow \quad x > 1$

Domain: $\{x \mid x > 1\}$

65. $y = x^2 + 2x$

Graph $y_1 = x^2 + 2x$. The graph passes the vertical line test. Thus, the equation represents a function.

67. $y = \dfrac{2}{x}$

Graph $y_1 = \dfrac{2}{x}$. The graph passes the vertical line test. Thus, the equation represents a function.

69. $y^2 = 1 - x^2$

Solve for y: $y = \pm\sqrt{1 - x^2}$

For $x = 0$, $y = \pm 1$. Thus, $(0,1)$ and $(0,-1)$ are on the graph. This is not a function, since a distinct x corresponds to two different y's.

71. $x^2 + y = 1$

Graph $y_1 = 1 - x^2$. The graph passes the vertical line test. Thus, the equation represents a function.

73. (a)   III   (b)   IV   (c)   I   (d)   V   (e)   II

75.

77. (a)   Two times
     (b)   Kevin's distance from home increased steadily at a rate of 1.5 miles per hour.
     (c)   Kevin's distance from home did not change.
     (d)   Kevin's distance from home decreased steadily at a rate of 10 miles per hour.
     (e)   Kevin stayed at home for 0.2 hours.

(f)   Kevin's distance from home increased rapidly at the beginning and then tapered off to a very slow change in distance from home.

(g)   Kevin's distance from home did not change.

(h)   Kevin's distance from home decreased rapidly at the beginning of the interval and then tapered off as he got closer to home.

(i)   The furthest distance Kevin is from home is 3 miles.

79.   Solving for A:

$f(x) = 2x^3 + Ax^2 + 4x - 5$ and $f(2) = 5$

$f(2) = 2(2)^3 + A(2)^2 + 4(2) - 5$

$5 = 16 + 4A + 8 - 5$

$5 = 4A + 19$

$-14 = 4A$

$A = \dfrac{-7}{2}$

81.   Solving for A:

$f(x) = \dfrac{3x + 8}{2x - A}$ and $f(0) = 2$

$f(0) = \dfrac{3(0) + 8}{2(0) - A}$

$2 = \dfrac{8}{-A}$

$-2A = 8$

$A = -4$

83.   Solving for A:

$f(x) = \dfrac{2x - A}{x - 3}$ and $f(4) = 0$

$f(4) = \dfrac{2(4) - A}{4 - 3}$

$0 = \dfrac{8 - A}{1}$

$0 = 8 - A$

$A = 8$

f is undefined when $x = 3$.

85.   (a)   The relation is not a function because 23 is paired with both 56 and 53.

(b)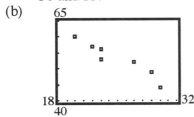

(c) $D = -1.3355p + 86.1974$

(d)   As the price of the jeans increases by \$1, the demand for the jeans decreases by 1.3355.

(e) $D(p) = -1.3355p + 86.1974$

(f)   Domain: $\{p \mid p > 0\}$

(g) $D(28) = -1.3355(28) + 86.1974$

$= 48.80 \approx 49$

87.   (a)   The relation is a function. Each time value is paired with exactly one distance value.

(b)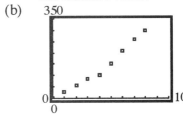

(c) $s = 37.7833t - 19.1333$

(d)   As the time increases by 1 hour, the distances increases by 37.7833 miles.

(e) $s(t) = 37.7833t - 19.1333$

(f)   Domain: $\{t \mid t \geq 0\}$

(g) $s(11) = 37.7833(11) - 19.1333$

$= 396.483 \approx 396$ miles

89. (a) $H(x) = 20 - 4.9x^2$

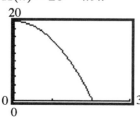

(b) $H(1) = 20 - 4.9(1)^2 = 20 - 4.9 = 15.1$ meters

$H(1.1) = 20 - 4.9(1.1)^2 = 20 - 4.9(1.21) = 20 - 5.929 = 14.071$ meters

$H(1.2) = 20 - 4.9(1.2)^2 = 20 - 4.9(1.44) = 20 - 7.056 = 12.944$ meters

$H(1.3) = 20 - 4.9(1.3)^2 = 20 - 4.9(1.69) = 20 - 8.281 = 11.719$ meters

(c)

$H(x) = 15$	$H(x) = 10$	$H(x) = 5$
$15 = 20 - 4.9x^2$	$10 = 20 - 4.9x^2$	$5 = 20 - 4.9x^2$
$-5 = -4.9x^2$	$-10 = -4.9x^2$	$-15 = -4.9x^2$
$x^2 = 1.0204$	$x^2 = 2.0408$	$x^2 = 3.0612$
$x = 1.01$ seconds	$x = 1.43$ seconds	$x = 1.75$ seconds

(d) $H(x) = 0$

$0 = 20 - 4.9x^2$

$-20 = -4.9x^2$

$x^2 = 4.0816$

$x = 2.02$ seconds

91. Let x represent the length of the rectangle.

Then $\dfrac{x}{2}$ represents the width of the rectangle, since the length is twice the width.

The function for the area is: $A(x) = x \cdot \dfrac{x}{2} = \dfrac{x^2}{2} = \dfrac{1}{2}x^2$

93. Let x represent the number of hours worked.
The function for the gross salary is: $G(x) = 10x$

95. $h(x) = \dfrac{-32x^2}{130^2} + x$

(a) $h(100) = \dfrac{-32(100)^2}{130^2} + 100 = \dfrac{-320000}{16900} + 100 = -18.93 + 100 = 81.07$ feet

(b) $h(300) = \dfrac{-32(300)^2}{130^2} + 300 = \dfrac{-2880000}{16900} + 300 = -170.41 + 300 = 129.59$ feet

(c) $h(500) = \dfrac{-32(500)^2}{130^2} + 500 = \dfrac{-8000000}{16900} + 500 = -473.37 + 500 = 26.63$ feet

(d) Graphing:

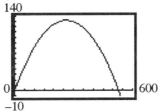

(e) Find x when $h(x)$ is 90 feet:

$$90 = \frac{-32x^2}{130^2} + x$$

$$\frac{32x^2}{16900} - x + 90 = 0$$

$$32x^2 - 16900x + 1521000 = 0$$

$$x = \frac{16900 \pm \sqrt{16900^2 - 4(32)(1521000)}}{2(32)}$$

$$x = \frac{16900 \pm \sqrt{90922000}}{64}$$

$$x = \frac{16900 \pm 9535.3}{64}$$

$$x \approx 115.07 \text{ feet} \quad \text{or} \quad x \approx 413.05 \text{ feet}$$

(f) Creating a TABLE:

X	Y₁
150	107.4
175	117.01
200	124.26
225	129.14
250	131.66
275	131.8
300	129.59

X=275

The ball will travel approximately 275 feet before it reaches maximum height. The maximum height is approximately 131.8 feet.

(g) Refining the TABLE:

X	Y₁
261	132.01
262	132.02
263	132.03
264	132.03
265	132.03
266	132.02
267	132.01

X=264

The ball will travel approximately 264 feet when it reaches a maximum height of 132.03 feet.

(h) The domain of h is: $\{x \mid 0 \le x \le 528.125\}$.

97. $C(x) = 100 + \dfrac{x}{10} + \dfrac{36000}{x}$

(a) $C(500) = 100 + \dfrac{500}{10} + \dfrac{36000}{500} = 100 + 50 + 72 = \$222$

(b) $C(450) = 100 + \dfrac{450}{10} + \dfrac{36000}{450} = 100 + 45 + 80 = \$225$

(c) $C(600) = 100 + \dfrac{600}{10} + \dfrac{36000}{600} = 100 + 60 + 60 = \$220$

(d) $C(400) = 100 + \dfrac{400}{10} + \dfrac{36000}{400} = 100 + 40 + 90 = \$230$

(e)   Graphing:

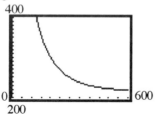

(f)   Creating a TABLE:

X	Y1
450	225
500	222
550	220.45
600	220
650	220.38
700	221.43
750	223

X=600

A ground speed of 600 miles per hour produces a minimum cost of $220.

99. (a) $h(x) = 2x$

$h(a + b) = 2(a + b) = 2a + 2b = h(a) + h(b)$

$h(x) = 2x$ has the property.

(b) $g(x) = x^2$

$g(a + b) = (a + b)^2 = a^2 + 2ab + b^2 \neq a^2 + b^2 = h(a) + h(b)$

$g(x) = x^2$ does not have the property.

(c) $F(x) = 5x - 2$

$F(a + b) = 5(a + b) - 2 = 5a + 5b - 2 \neq 5a - 2 + 5b - 2 = h(a) + h(b)$

$F(x) = 5x - 2$ does not have the property.

(d) $G(x) = \dfrac{1}{x}$

$G(a + b) = \dfrac{1}{a + b} \neq \dfrac{1}{a} + \dfrac{1}{b} = h(a) + h(b)$

$G(x) = \dfrac{1}{x}$ does not have the property.

101. No, $x = -1$ is not in the domain of g, but it is in the domain of f.

103. A function may have any number of x-intercepts; it can have only one y-intercept.

105. A function cannot be symmetric to the x-axis, because it would then fail the vertical line test.

2.2   Characteristics of Functions

1.   Yes

3.   No It only increases on (5, 10).

5. f is increasing on the intervals: (–8, –2), (0, 2), (5, 10).

7.   Yes. The local maximum at $x = 2$ is 10.

9. f has local maxima at $x = -2$ and $x = 2$. The local maxima are 6 and 10, respectively.

11.   (a)   Intercepts: (–2,0), (2,0), and (0,3).
      (b)   Domain: $\{x|-4 \le x \le 4\}$; Range: $\{y|0 \le y \le 3\}$.
      (c)   Interval notation: Increasing: (–2, 0) and (2, 4); Decreasing: (–4, –2) and (0, 2).
            Inequality notation: Increasing: $-2 < x < 0$ and $2 < x < 4$
                        Decreasing: $-4 < x < -2$ and $0 < x < 2$
      (d)   Since the graph is symmetric to the y-axis, the function is <u>even</u>.

13.   (a)   Intercepts: (0,1).
      (b)   Domain: { Real Numbers }; Range: $\{y|y > 0\}$.
      (c)   Interval notation: Increasing: $(-\infty, +\infty)$; Decreasing: never.
            Inequality notation: Increasing: $-\infty < x < +\infty$
                        Decreasing: never
      (d)   Since the graph is not symmetric to the y-axis or the origin, the function is <u>neither</u> even nor odd.

15.   (a)   Intercepts: $(-\pi,0)$, (0,0), and $(\pi,0)$.
      (b)   Domain: $\{x|-\pi \le x \le \pi\}$; Range: $\{y|-1 \le y \le 1\}$.
      (c)   Interval notation: Increasing: $\left(\frac{-\pi}{2}, \frac{\pi}{2}\right)$; Decreasing: $\left(-\pi, \frac{-\pi}{2}\right)$ and $\left(\frac{\pi}{2}, \pi\right)$.

            Inequality notation: Increasing: $\frac{-\pi}{2} < x < \frac{\pi}{2}$

                        Decreasing: $-\pi < x < \frac{-\pi}{2}$ and $\frac{\pi}{2} < x < \pi$
      (d)   Since the graph is symmetric to the origin, the function is <u>odd</u>.

17.   (a)   Intercepts: $\left(0, \frac{1}{2}\right), \left(\frac{1}{3}, 0\right),$ and $\left(\frac{5}{2}, 0\right)$.
      (b)   Domain: $\{x|-3 \le x \le 3\}$; Range: $\{y|-1 \le y \le 2\}$.

(c)   Interval notation: Increasing: $(2, 3)$; Decreasing: $(-1, 1)$;

Constant: $(-3, -1)$ and $(1, 2)$.

Inequality notation: Increasing: $2 < x < 3$; Decreasing: $-1 < x < 1$;

Constant: $-3 < x < -1$ and $1 < x < 2$.

(d)   Since the graph is not symmetric to the y-axis or the origin, the function is <u>neither</u> even nor odd.

19. (a)   Intercepts: $(0, 2), (-2, 0)$, and $(2, 0)$.

    (b)   Domain: $\{x \mid -4 \le x \le 4\}$; Range: $\{y \mid 0 \le y \le 2\}$.

    (c)   Interval notation: Increasing: $(-2, 0)$ and $(2, 4)$;

Decreasing: $(-4, -2)$ and $(0, 2)$.

Inequality notation: Increasing: $-2 < x < 0$ and $2 < x < 4$;

Decreasing: $-4 < x < -2$ and $0 < x < 2$.

    (d)   Since the graph is symmetric to the y-axis, the function is <u>even</u>.

21. (a) f has a local maximum of 3 at $x = 0$.

    (b) f has a local minimum of 0 at both $x = -2$ and $x = 2$.

23. (a) f has a local maximum of 1 at $x = \frac{\pi}{2}$.

    (b) f has a local minimum of -1 at $x = \frac{-\pi}{2}$.

25. $f(x) = 5x$

    (a) $\dfrac{f(x) - f(1)}{x - 1} = \dfrac{5x - 5}{x - 1} = \dfrac{5(x - 1)}{x - 1} = 5$

    (b) $\dfrac{f(2) - f(1)}{2 - 1} = \dfrac{10 - 5}{2 - 1} = \dfrac{5}{1} = 5$

    (c)   Slope = 5; Containing $(1, 5)$:

$$y - 5 = 5(x - 1)$$
$$y - 5 = 5x - 5$$
$$y = 5x$$

    d)   Graphing:

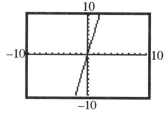

27. $f(x) = 1 - 3x$

    (a) $\dfrac{f(x) - f(1)}{x - 1} = \dfrac{1 - 3x - (-2)}{x - 1}$

$$= \dfrac{-3x + 3}{x - 1} = \dfrac{-3(x - 1)}{x - 1} = -3$$

    (b) $\dfrac{f(2) - f(1)}{2 - 1} = \dfrac{1 - 3(2) - (-2)}{2 - 1} = \dfrac{-3}{1} = -3$

    (c)   Slope = -3; Containing $(1, -2)$:

$$y - (-2) = -3(x - 1)$$
$$y + 2 = -3x + 3$$
$$y = -3x + 1$$

    d)   Graphing:

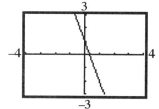

29.

$$f(x) = x^2 - 2x$$

(a) $\dfrac{f(x) - f(1)}{x - 1} = \dfrac{x^2 - 2x - (-1)}{x - 1}$

$\qquad = \dfrac{x^2 - 2x + 1}{x - 1} = \dfrac{(x-1)^2}{x-1} = x - 1$

(b) $\dfrac{f(2) - f(1)}{2 - 1} = \dfrac{2^2 - 2(2) - (-1)}{2 - 1} = \dfrac{1}{1} = 1$

(c) Slope = 1; Containing $(1, -1)$:

$\qquad y - (-1) = 1(x - 1)$

$\qquad\qquad y + 1 = 1x - 1$

$\qquad\qquad\quad y = x - 2$

d) Graphing:

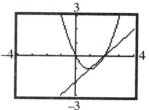

31.

$$f(x) = x^3 - x$$

(a) $\dfrac{f(x) - f(1)}{x - 1} = \dfrac{x^3 - x - 0}{x - 1} = \dfrac{x^3 - x}{x - 1}$

$\qquad = \dfrac{x(x - 1)(x + 1)}{x - 1} = x^2 + x$

(b) $\dfrac{f(2) - f(1)}{2 - 1} = \dfrac{2^3 - 2 - 0}{2 - 1} = \dfrac{6}{1} = 6$

(c) Slope = 6; Containing $(1, 0)$:

$\qquad y - 0 = 6(x - 1)$

$\qquad\quad y = 6x - 6$

d) Graphing:

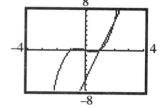

33.

$$f(x) = \dfrac{2}{x + 1}$$

(a) $\dfrac{f(x) - f(1)}{x - 1} = \dfrac{\dfrac{2}{x + 1} - 1}{x - 1} = \dfrac{\dfrac{2 - x - 1}{x + 1}}{x - 1}$

$\qquad = \dfrac{1 - x}{(x - 1)(x + 1)} = \dfrac{-1}{x + 1}$

(b) $\dfrac{f(2) - f(1)}{2 - 1} = \dfrac{\dfrac{2}{2 + 1} - 1}{2 - 1} = \dfrac{\dfrac{-1}{3}}{1} = \dfrac{-1}{3}$

(c) Slope = $\frac{-1}{3}$; Containing $(1, 1)$:

$\qquad y - 1 = \dfrac{-1}{3}(x - 1)$

$\qquad y - 1 = \dfrac{-1}{3}x + \dfrac{1}{3}$

$\qquad\quad y = \dfrac{-1}{3}x + \dfrac{4}{3}$

d) Graphing:

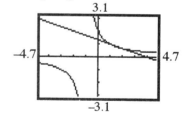

35. $f(x) = \sqrt{x}$

(a) $\dfrac{f(x) - f(1)}{x - 1} = \dfrac{\sqrt{x} - 1}{x - 1}$

d)   Graphing:

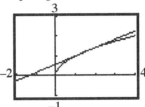

(b) $\dfrac{f(2) - f(1)}{2 - 1} = \dfrac{\sqrt{2} - 1}{1} = \sqrt{2} - 1$

(c)   Slope $= \sqrt{2} - 1$; Containing $(1, 1)$:
$$y - 1 = \left(\sqrt{2} - 1\right)(x - 1)$$
$$y - 1 = \left(\sqrt{2} - 1\right)x - \left(\sqrt{2} - 1\right)$$
$$y = \left(\sqrt{2} - 1\right)x - \sqrt{2} + 2$$

37. $f(x) = 4x^3$
$$f(-x) = 4(-x)^3 = -4x^3$$
f is odd.

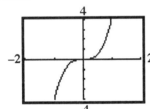

39. $g(x) = -3x^2 - 5$
$$g(-x) = -3(-x)^2 - 5 = -3x^2 - 5$$
g is even.

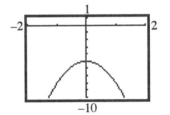

41. $F(x) = \sqrt[3]{x}$
$$F(-x) = \sqrt[3]{-x} = -\sqrt[3]{x}$$
F is odd.

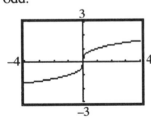

43. $f(x) = x + |x|$
$$f(-x) = -x + |-x| = -x + |x|$$
f is neither even nor odd.

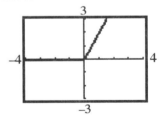

45. $g(x) = \dfrac{1}{x^2}$
$$g(-x) = \dfrac{1}{(-x)^2} = \dfrac{1}{x^2}$$
g is even.

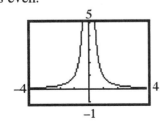

47. $h(x) = \dfrac{-x^3}{3x^2 - 9}$
$$h(-x) = \dfrac{-(-x)^3}{3(-x)^2 - 9} = \dfrac{x^3}{3x^2 - 9}$$
h is odd.

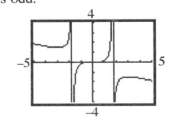

49. One at most because if f is increasing it could only cross the x-axis at most one time. It could not "turn" and cross it again or it would start to decrease.

51. $f(x) = x^3 - 3x + 2$

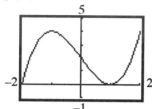

Local maximum: $(-1, 4)$
Local minimum: $(1, 0)$
Increasing: $(-2, -1), (1, 2)$
Decreasing: $(-1, 1)$

53. $f(x) = x^5 - x^3$

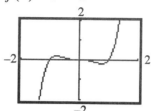

Local maximum: $(-0.77, 0.19)$
Local minimum: $(0.77, -0.19)$
Increasing: $(-2, -0.77), (0.77, 2)$
Decreasing: $(-0.77, 0.77)$

55. $f(x) = -0.2x^3 - 0.6x^2 + 4x - 6$

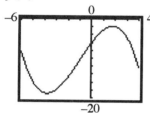

Local maximum: $(1.77, -1.91)$
Local minimum: $(-3.77, -18.89)$
Increasing: $(-3.77, 1.77)$
Decreasing: $(-6, -3.77), (1.77, 4)$

57. $f(x) = 0.25x^4 + 0.3x^3 - 0.9x^2 + 3$

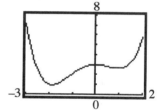

Local maximum: $(0, 3)$
Local minima: $(-1.87, 0.95),$
                      $(0.97, 2.65)$
Increasing: $(-1.87, 0), (0.97, 2)$
Decreasing: $(-3, -1.87), (0, 0.97)$

59. $f(x) = 2x + 5$

(a) $m_{sec} = \dfrac{f(x+h) - f(x)}{h} = \dfrac{2(x+h) + 5 - 2x - 5}{h} = \dfrac{2h}{h} = 2$

(b) $h = 0.5, x = 1$

$m_{sec} = 2$

$h = 0.1, x = 1$

$m_{sec} = 2$

$h = 0.01, x = 1$

$m_{sec} = 2$ m_{sec} approaches 2 as h approaches 0.

(c) $y - 7 = 2(x - 1)$

$y - 7 = 2x - 2$

$y = 2x + 5$

(d)

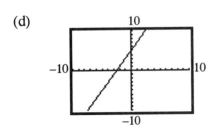

61. $f(x) = x^2 + 2x$

(a) $m_{sec} = \dfrac{f(x+h) - f(x)}{h} = \dfrac{(x+h)^2 + 2(x+h) - (x^2 + 2x)}{h}$

$= \dfrac{x^2 + 2xh + h^2 + 2x + 2h - x^2 - 2x}{h} = \dfrac{2xh + h^2 + 2h}{h} = 2x + h + 2$

(b) $h = 0.5,\ x = 1$

$m_{sec} = 2(1) + 0.5 + 2 = 4.5$

$h = 0.1,\ x = 1$

$m_{sec} = 2(1) + 0.1 + 2 = 4.1$

$h = 0.01,\ x = 1$

$m_{sec} = 2(1) + 0.01 + 2 = 4.01$ m_{sec} approaches 4 as h approaches 0.

(c) $y - 3 = 4.01(x - 1)$

$y - 3 = 4.01x - 4.01$

$y = 4.01x - 1.01$

(d)

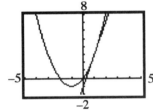

63. $f(x) = \dfrac{1}{x + 1}$

(a)

$m_{sec} = \dfrac{f(x+h) - f(x)}{h} = \dfrac{\dfrac{1}{x+h+1} - \dfrac{1}{x+1}}{h}$

$= \dfrac{\dfrac{x+1 - (x+h+1)}{(x+h+1)(x+1)}}{h} = \dfrac{\dfrac{-h}{(x+h+1)(x+1)}}{h} = \dfrac{-1}{(x+h+1)(x+1)}$

(b) $h = 0.5, x = 1$

$$m_{sec} = \frac{-1}{(1 + 0.5 + 1)(1 + 1)} = \frac{-1}{2.5(2)} = \frac{-1}{5}$$

$h = 0.1, x = 1$

$$m_{sec} = \frac{-1}{(1 + 0.1 + 1)(1 + 1)} = \frac{-1}{2.1(2)} = \frac{-1}{4.2}$$

$h = 0.01, x = 1$

$$m_{sec} = \frac{-1}{(1 + 0.01 + 1)(1 + 1)} = \frac{-1}{2.01(2)} = \frac{-1}{4.02}$$

m_{sec} approaches $-\frac{1}{4}$ as h approaches 0.

(c)
$$y - \frac{1}{2} = \frac{-1}{4.02}(x - 1)$$

$$y - \frac{1}{2} = -0.2488x + 0.2488$$

$$y = -0.2488x + 0.7488$$

(d)

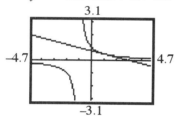

65.   (a), (b), (e)

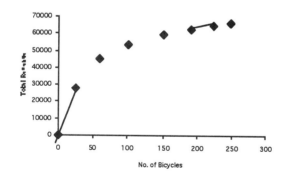

(c)   Average rate of change $= \dfrac{28000 - 0}{25 - 0} = \dfrac{28000}{25} = 1120$

(d)   For each additional bicycle sold between 0 and 25, the total revenue increases by \$1120.

(f)   Average rate of change $= \dfrac{64835 - 62360}{223 - 190} = \dfrac{2475}{33} = 75$

(g)   For each additional bicycle sold between 190 and 223, the total revenue increases by \$75.

67. (a), (b), (e)

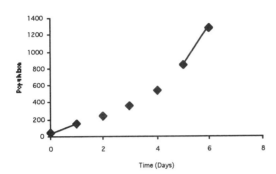

(c) Average rate of change $= \dfrac{153 - 50}{1 - 0} = \dfrac{103}{1} = 103$

(d) The population is increasing at a rate of 103 per day between day 0 and day 1.

(f) Average rate of change $= \dfrac{1280 - 839}{6 - 5} = \dfrac{441}{1} = 441$

(g) The population is increasing at a rate of 441 per day between day 5 and day 6.

(h) As time passes, the average rate of change of the population is increasing.

69. Graphing $V(x) = x(24 - 2x)^2$:

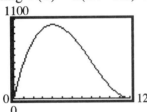

The volume is largest when $x = 4$ inches.

71. (a) Graphing $s(t) = -16t^2 + 80t + 6$

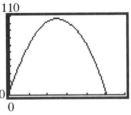

(b) The height is a maximum at 2.5 seconds.

(c) The maximum height is 106 feet.

2.3   Library of Functions; Piecewise Defined Functions

1. C   3. E   5. B   7. F

9.

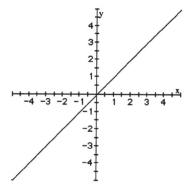

11.

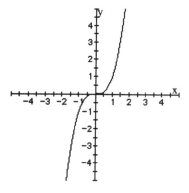

13.

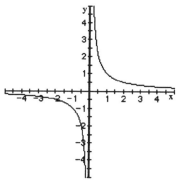

15. (a) $f(-2) = (-2)^2 = 4$
     (b) $f(0) = 2$
     (c) $f(2) = 2(2) + 1 = 5$

17. (a) $f(1.2) = \text{int}(2(1.2)) = \text{int}(2.4) = 2$
     (b) $f(1.6) = \text{int}(2(1.6)) = \text{int}(3.2) = 3$
     (c) $f(-1.8) = \text{int}(2(-1.8)) = \text{int}(-3.6) = -4$

19. $f(x) = \begin{cases} 2x & \text{if } x \neq 0 \\ 1 & \text{if } x = 0 \end{cases}$
     (a)   Domain: {Real Numbers}
     (b)   x-intercept: none
           y-intercept: (0,1)
     (c)

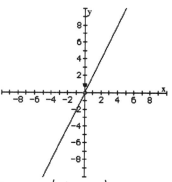

     (d)   Range: $\{y \mid y \neq 0\}$

21. $f(x) = \begin{cases} -2x + 3 & \text{if } x < 1 \\ 3x - 2 & \text{if } x \geq 1 \end{cases}$
     (a)   Domain: {Real Numbers}
     (b)   x-intercept: none
           y-intercept: (0,3)
     (c)

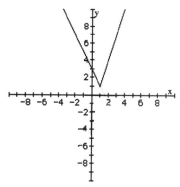

     (d)   Range: $\{y \mid y \geq 1\}$

23. $f(x) = \begin{cases} x+3 & \text{if } -2 \le x < 1 \\ 5 & \text{if } x = 1 \\ -x+2 & \text{if } x > 1 \end{cases}$

    (a) Domain: $\{x \mid x \ge -2\}$

    (b) x-intercept: $(2, 0)$
         y-intercept: $(0, 3)$

    (c)

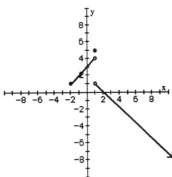

    (d) Range: $\{y \mid y < 4\} \cup \{5\}$

25. $f(x) = \begin{cases} 1+x & \text{if } x < 0 \\ x^2 & \text{if } x \ge 0 \end{cases}$

    (a) Domain: {Real Numbers}

    (b) x-intercept: $(-1,0)$, $(0,0)$
         y-intercept: $(0,0)$

    (c)

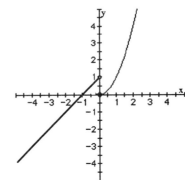

    (d) Range: {Real Numbers}

27. $f(x) = \begin{cases} |x| & \text{if } -2 \le x < 0 \\ 1 & \text{if } x = 0 \\ x^3 & \text{if } x > 0 \end{cases}$

    (a) Domain: $\{x \mid x \ge -2\}$

    (b) x-intercept: none
         y-intercept: $(0, 1)$

    (c)

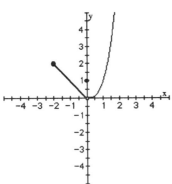

    (d) Range: $\{y \mid y > 0\}$

29. $h(x) = 2\text{int}(x)$

    (a) Domain: {Real Numbers}

    (b) x-intercept: all ordered pairs
         $(x,0)$ when $0 \le x < 1$.
         y-intercept: $(0,0)$

    (c)

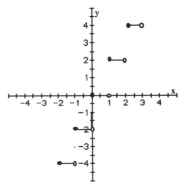

    (d) Range: {Even Integers}

31. $f(x) = \begin{cases} -x & \text{if } -1 \le x \le 0 \\ \frac{1}{2}x & \text{if } 0 < x \le 2 \end{cases}$

33. $f(x) = \begin{cases} -x & \text{if } x \le 0 \\ -x+2 & \text{if } 0 < x \le 2 \end{cases}$

35. (a)   Charge for 50 therms: $C = 9.45 + 0.36375(50) + 0.3128(50) = \$43.28$
   (b)   Charge for 500 therms:
$$C = 9.45 + 0.36375(50) + 0.11445(450) + 0.3128(500) = \$235.54$$
   (c)   The monthly charge function:
$$C = \begin{cases} 9.45 + 0.36375x + 0.3128x & \text{for } 0 \le x \le 50 \\ 9.45 + 0.36375(50) + 0.11445(x - 50) + 0.3128x & \text{for } x > 50 \end{cases}$$
$$= \begin{cases} 9.45 + 0.67655x & \text{for } 0 \le x \le 50 \\ 9.45 + 18.1875 + 0.11445x - 5.7225 + 0.3128x & \text{for } x > 50 \end{cases}$$
$$= \begin{cases} 9.45 + 0.67655x & \text{for } 0 \le x \le 50 \\ 21.915 + 0.42725x & \text{for } x > 50 \end{cases}$$

   (d)

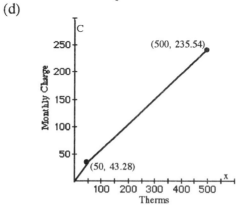

37. (a) $W = 10° C$

   (b) $W = 33 - \dfrac{(10.45 + 10\sqrt{5} - 5)(33 - 10)}{22.04} = 3.98° C$

   (c) $W = 33 - \dfrac{(10.45 + 10\sqrt{15} - 15)(33 - 10)}{22.04} = -2.67° C$

   (d) $W = 33 - 1.5958(33 - 10) = -3.7° C$

   (e)   When $0 \le v < 1.79$, the wind speed is so small that there is no effect on the temperature.

   (f)   For each drop of $1°$ in temperature, the wind chill factor drops approximately 1.6°C. When the wind speed exceeds 20, there is a constant drop in temperature.

39.   Each graph is that of $y = x^2$, but shifted vertically. If $y = x^2 + k$, $k > 0$, the shift is up k units; if $y = x^2 + k$, $k < 0$, the shift is down $|k|$ units. The graph of $y = x^2 - 4$ is the same as the graph of $y = x^2$, but shifted down 4 units. The graph of $y = x^2 + 5$ is the graph of $y = x^2$, but shifted up 5 units.

41. Each graph is that of $y = |x|$, but either compressed or stretched. If $y = k|x|$ and $k > 1$, the graph is stretched; if $y = k|x|$ and $0 < k < 1$, the graph is compressed. The graph of $y = \frac{1}{4}|x|$ is the same as the graph of $y = |x|$, but compressed. The graph of $y = 5|x|$ is the same as the graph of $y = |x|$, but stretched.

43. The graph of $y = \sqrt{-x}$ is the reflection about the y-axis of the graph of $y = \sqrt{x}$. The same type of reflection occurs when graphing $y = 2x + 1$ and $y = 2(-x) + 1$. The conclusion is that the graph of $y = f(-x)$ is the reflection about the y-axis of the graph of $y = f(x)$.

45. For the graph of $y = x^n$, n a positive even integer, as n increases, the graph of the function is narrower for $|x| > 1$ and flatter for $|x| < 1$.

2.4 Graphing Techniques: Transformations

1. B	3. H	5. I	7. L
9. F	11. G	13. C	15. B

17. $y = (x - 4)^3$ 19. $y = x^3 + 4$ 21. $y = -x^3$ 23. $y = 4x^3$

25. (1) $y = \sqrt{x} + 2$
    (2) $y = -\left(\sqrt{x} + 2\right)$
    (3) $y = -\left(\sqrt{-x} + 2\right)$

27. (1) $y = -\sqrt{x}$
    (2) $y = -\sqrt{x} + 2$
    (3) $y = -\sqrt{x + 3} + 2$

29. C

31. C

33. $f(x) = x^2 - 1$
Using the graph of $y = x^2$, vertically shift downward 1 unit.

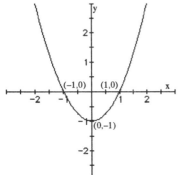

35. $g(x) = x^3 + 1$
Using the graph of $y = x^3$, vertically shift upward 1 unit.

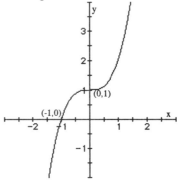

37. $h(x) = \sqrt{x-2}$

Using the graph of $y = \sqrt{x}$,
horizontally shift to the right 2 units.

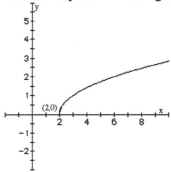

39. $f(x) = (x-1)^3 + 2$

Using the graph of $y = x^3$,
horizontally shift to the right 1 unit,
then vertically shift up 2 units.

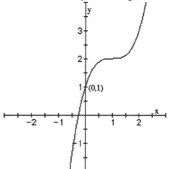

41. $g(x) = 4\sqrt{x}$

Using the graph of $y = \sqrt{x}$, vertically
stretch by a factor of 4.

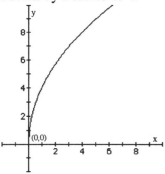

43. $h(x) = \dfrac{1}{2x}$

Using the graph of $y = \dfrac{1}{x}$, vertically

compress by a factor of $\frac{1}{2}$.

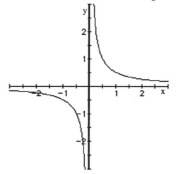

45. $f(x) = -|x|$

Reflect the graph of $y = |x|$, about
the x-axis.

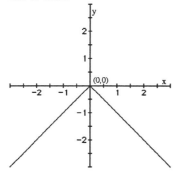

47. $g(x) = -\sqrt{-x}$

Reflect the graph of $y = \sqrt{x}$, about
the y-axis; and reflect about the x-
axis.

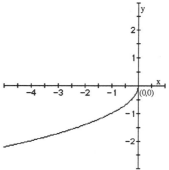

49. $h(x) = \text{int}(-x)$

Reflect the graph of $y = \text{int}(x)$, about the y-axis.

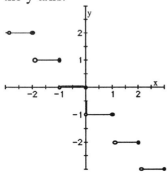

51. $f(x) = 2(x+1)^2 - 3$

Using the graph of $y = x^2$, horizontally shift to the left 1 unit, vertically stretch by a factor of 2, and vertically shift downward 3 units.

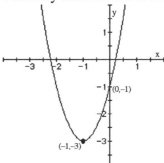

53. $g(x) = \sqrt{x-2} + 1$

Using the graph of $y = \sqrt{x}$, horizontally shift to the right 2 units and vertically shift upward 1 unit.

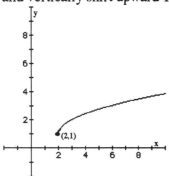

55. $h(x) = \sqrt{-x} - 2$

Reflect the graph of $y = \sqrt{x}$, about the y-axis and vertically shift downward 2 units.

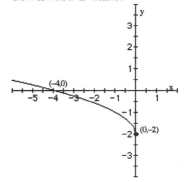

57. $f(x) = -(x+1)^3 - 1$

Using the graph of $y = x^3$, horizontally shift to the left 1 units, reflect the graph on the x-axis, and vertically shift downward 1 unit.

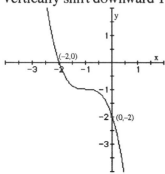

59. $g(x) = 2|1-x| = 2|x-1|$

Using the graph of $y = |x|$, horizontally shift to the right 1 unit, and vertically stretch by a factor or 2.

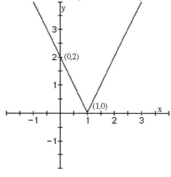

61. $h(x) = 2\text{int}(x - 1)$

Using the graph of $y = \text{int}(x)$, horizontally shift to the right 1 unit, and vertically stretch by a factor of 2.

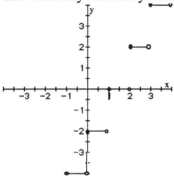

63. (a) $F(x) = f(x) + 3$
Shift up 3 units.

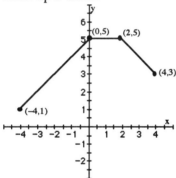

(b) $G(x) = f(x + 2)$
Shift left 2 units.

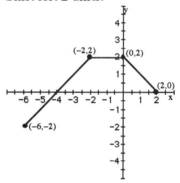

(c) $P(x) = -f(x)$
Reflect about the x-axis.

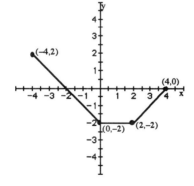

(d) $H(x) = f(x + 1) - 2$
Shift left 1 unit and shift down 2 units.

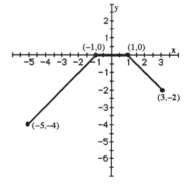

(e) $Q(x) = \frac{1}{2}f(x)$
Compress vertically by a factor
of $\frac{1}{2}$.

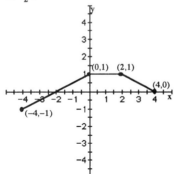

(f) $g(x) = f(-x)$
Reflect about y-axis.

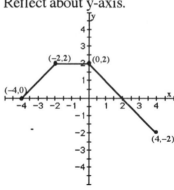

(g) $h(x) = f(2x)$
Compress horizontally by a
factor of $\frac{1}{2}$.

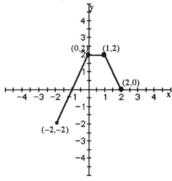

65.   (a) $F(x) = f(x) + 3$
Shift up 3 units.

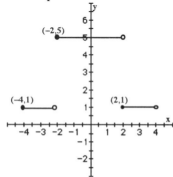

(b) $G(x) = f(x + 2)$
Shift left 2 units.

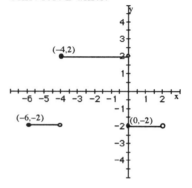

(c) $P(x) = -f(x)$
Reflect about the x-axis.

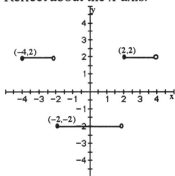

(d) $H(x) = f(x+1) - 2$
Shift left 1 unit and shift down 2 units.

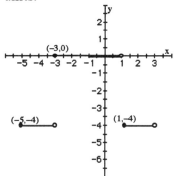

(e) $Q(x) = \frac{1}{2} f(x)$
Compress vertically by a factor of $\frac{1}{2}$.

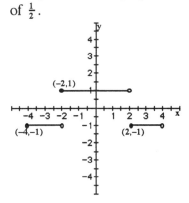

(f) $g(x) = f(-x)$
Reflect about y-axis.

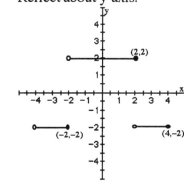

(g) $h(x) = f(2x)$
Compress horizontally by a factor of $\frac{1}{2}$.

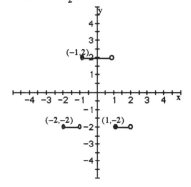

67.   (a) $F(x) = f(x) + 3$
      Shift up 3 units.

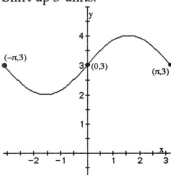

(b) $G(x) = f(x + 2)$
      Shift left 2 units.

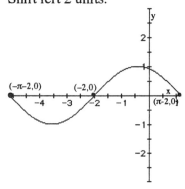

(c) $P(x) = -f(x)$
      Reflect about the x-axis.

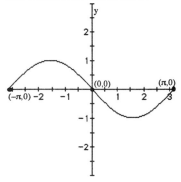

(d) $H(x) = f(x + 1) - 2$
      Shift left 1 unit and shift down 2 units.

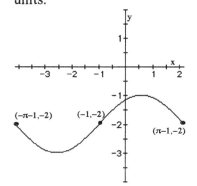

(e) $Q(x) = \frac{1}{2} f(x)$
      Compress vertically by a factor of $\frac{1}{2}$.

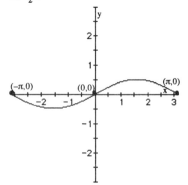

(f) $g(x) = f(-x)$
      Reflect about y-axis.

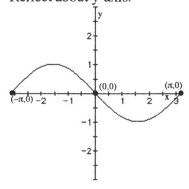

(g) $h(x) = f(2x)$
Compress horizontally by a
factor of $\frac{1}{2}$.

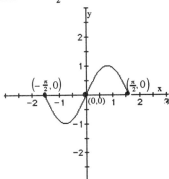

69. (a)
$y = |x + 1|$

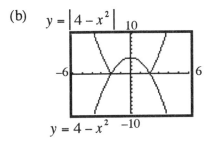

(b) $y = |4 - x^2|$

(c) $y = |x^3 + x|$

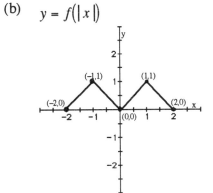

(d)   Any part of the graph of $y = f(x)$
that lies below the x-axis is
reflected about the x-axis to obtain
the graph of $y = |f(x)|$.

71. (a) $y = |f(x)|$

(b) $y = f(|x|)$

73. $f(x) = x^2 + 2x$
$f(x) = (x^2 + 2x + 1) - 1$
$f(x) = (x + 1)^2 - 1$
Using $f(x) = x^2$, shift left 1 unit and shift down 1 unit.

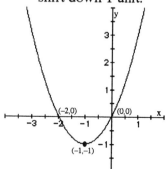

75. $f(x) = x^2 - 8x + 1$
$f(x) = (x^2 - 8x + 16) + 1 - 16$
$f(x) = (x - 4)^2 - 15$
Using $f(x) = x^2$, shift right 4 units and shift down 15 units.

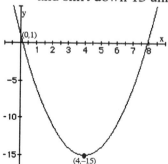

77. $f(x) = x^2 + x + 1$
$f(x) = \left(x^2 + x + \frac{1}{4}\right) + 1 - \frac{1}{4}$
$f(x) = \left(x + \frac{1}{2}\right)^2 + \frac{3}{4}$
Using $f(x) = x^2$, shift left $\frac{1}{2}$ unit and shift up $\frac{3}{4}$ unit.

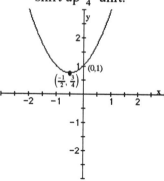

79. $y = (x - c)^2$
If $c = 0$, $y = x^2$.
If $c = 3$, $y = (x - 3)^2$; shift right 3 units.
If $c = -2$, $y = (x + 2)^2$; shift left 2 units.

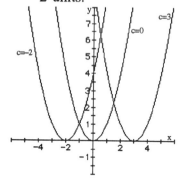

81. $F = \frac{9}{5}C + 32$

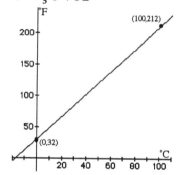

$F = \frac{9}{5}(K - 273) + 32$

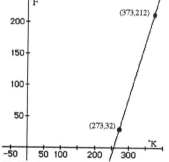

Shift the graph 273 units to the right.

83.   (a)

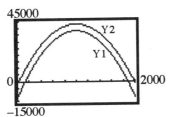

(b)   Select the 10% tax since the profits are higher.

(c)   The graph of Y1 is obtained by shifting the graph of $p(x)$ vertically down 10,000. The graph of Y2 is obtained by multiplying the y-coordinate of the graph of $p(x)$ by 0.9. Thus, Y2 is the graph of $p(x)$ vertically compressed by a factor of 0.9.

(d)   Select the 10% tax since the graph of

$Y1 = 0.9p(x) \geq Y2 = -0.05x^2 + 100x - 6800$ for all x in the domain.

2.5   Operations on Functions; Composite Functions

1. $f(x) = 3x + 4$ $g(x) = 2x - 3$

(a) $(f + g)(x) = 3x + 4 + 2x - 3 = 5x + 1$
The domain is all real numbers.

(b) $(f - g)(x) = (3x + 4) - (2x - 3) = 3x + 4 - 2x + 3 = x + 7$
The domain is all real numbers.

(c) $(f \cdot g)(x) = (3x + 4)(2x - 3) = 6x^2 - 9x + 8x - 12 = 6x^2 - x - 12$
The domain is all real numbers.

(d) $\left(\dfrac{f}{g}\right)(x) = \dfrac{3x + 4}{2x - 3}$

The domain is all real numbers except $\frac{3}{2}$.

3. $f(x) = x - 1$ $g(x) = 2x^2$

(a) $(f + g)(x) = x - 1 + 2x^2 = 2x^2 + x - 1$
The domain is all real numbers.

(b) $(f - g)(x) = (x - 1) - (2x^2) = x - 1 - 2x^2 = -2x^2 + x - 1$
The domain is all real numbers.

(c) $(f \cdot g)(x) = (x - 1)(2x^2) = 2x^3 - 2x^2$
The domain is all real numbers.

(d) $\left(\dfrac{f}{g}\right)(x) = \dfrac{x - 1}{2x^2}$

The domain is all real numbers except 0.

5. $f(x) = \sqrt{x}$ $g(x) = 3x - 5$

    (a) $(f + g)(x) = \sqrt{x} + 3x - 5$

    The domain is $\{x \mid x \geq 0\}$.

    (b) $(f - g)(x) = \sqrt{x} - (3x - 5) = \sqrt{x} - 3x + 5$

    The domain is $\{x \mid x \geq 0\}$.

    (c) $(f \cdot g)(x) = \sqrt{x}(3x - 5) = 3x\sqrt{x} - 5\sqrt{x}$

    The domain is $\{x \mid x \geq 0\}$.

    (d) $\left(\dfrac{f}{g}\right)(x) = \dfrac{\sqrt{x}}{3x - 5}$

    The domain is $\left\{x \mid x \geq 0 \text{ and } x \neq \dfrac{5}{3}\right\}$.

7. $f(x) = 1 + \dfrac{1}{x}$ $g(x) = \dfrac{1}{x}$

    (a) $(f + g)(x) = 1 + \dfrac{1}{x} + \dfrac{1}{x} = 1 + \dfrac{2}{x}$

    The domain is $\{x \mid x \neq 0\}$.

    (b) $(f - g)(x) = 1 + \dfrac{1}{x} - \dfrac{1}{x} = 1$

    The domain is $\{x \mid x \neq 0\}$.

    (c) $(f \cdot g)(x) = \left(1 + \dfrac{1}{x}\right)\dfrac{1}{x} = \dfrac{1}{x} + \dfrac{1}{x^2}$

    The domain is $\{x \mid x \neq 0\}$.

    (d) $\left(\dfrac{f}{g}\right)(x) = \dfrac{1 + \dfrac{1}{x}}{\dfrac{1}{x}} = \dfrac{\dfrac{x+1}{x}}{\dfrac{1}{x}} = \dfrac{x+1}{x} \cdot \dfrac{x}{1} = x + 1$

    The domain is $\{x \mid x \neq 0\}$.

9. $f(x) = \dfrac{2x + 3}{3x - 2}$ $g(x) = \dfrac{4x}{3x - 2}$

    (a) $(f + g)(x) = \dfrac{2x + 3}{3x - 2} + \dfrac{4x}{3x - 2} = \dfrac{2x + 3 + 4x}{3x - 2} = \dfrac{6x + 3}{3x - 2}$

    The domain is $\left\{x \mid x \neq \dfrac{2}{3}\right\}$.

    (b) $(f - g)(x) = \dfrac{2x + 3}{3x - 2} - \dfrac{4x}{3x - 2} = \dfrac{2x + 3 - 4x}{3x - 2} = \dfrac{-2x + 3}{3x - 2}$

    The domain is $\left\{x \mid x \neq \dfrac{2}{3}\right\}$.

(c) $(f \cdot g)(x) = \left(\dfrac{2x+3}{3x-2}\right)\left(\dfrac{4x}{3x-2}\right) = \dfrac{8x^2+12x}{(3x-2)^2}$

The domain is $\left\{x \,\middle|\, x \neq \dfrac{2}{3}\right\}$.

(d) $\left(\dfrac{f}{g}\right)(x) = \dfrac{\dfrac{2x+3}{3x-2}}{\dfrac{4x}{3x-2}} = \dfrac{2x+3}{3x-2} \cdot \dfrac{3x-2}{4x} = \dfrac{2x+3}{4x}$

The domain is $\left\{x \,\middle|\, x \neq \dfrac{2}{3} \text{ and } x \neq 0\right\}$.

11. $f(x) = 3x + 1 \qquad (f+g)(x) = 6 - \tfrac{1}{2}x$
$6 - \tfrac{1}{2}x = 3x + 1 + g(x)$
$5 - \tfrac{7}{2}x = g(x)$
$g(x) = 5 - \tfrac{7}{2}x$

13. $f(x) = 2x \qquad g(x) = 3x^2 + 1$
   (a) $(f \circ g)(4) = f(g(4)) = f\left(3(4)^2 + 1\right) = f(49) = 2(49) = 98$
   (b) $(g \circ f)(2) = g(f(2)) = g(2 \cdot 2) = g(4) = 3(4)^2 + 1 = 48 + 1 = 49$
   (c) $(f \circ f)(1) = f(f(1)) = f(2(1)) = f(2) = 2(2) = 4$
   (d) $(g \circ g)(0) = g(g(0)) = g\left(3(0)^2 + 1\right) = g(1) = 3(1)^2 + 1 = 4$

15. $f(x) = 4x^2 - 3 \qquad g(x) = 3 - \tfrac{1}{2}x^2$
   (a) $(f \circ g)(4) = f(g(4)) = f\left(3 - \tfrac{1}{2}(4)^2\right) = f(-5) = 4(-5)^2 - 3 = 97$
   (b) $(g \circ f)(2) = g(f(2)) = g(4(2)^2 - 3) = g(13) = 3 - \tfrac{1}{2}(13)^2 = 3 - \dfrac{169}{2} = \dfrac{-163}{2}$
   (c) $(f \circ f)(1) = f(f(1)) = f(4(1)^2 - 3) = f(1) = 4(1)^2 - 3 = 1$
   (d) $(g \circ g)(0) = g(g(0)) = g\left(3 - \tfrac{1}{2}(0)^2\right) = g(3) = 3 - \tfrac{1}{2}(3)^2 = 3 - \dfrac{9}{2} = \dfrac{-3}{2}$

17. $f(x) = \sqrt{x} \qquad g(x) = 2x$
   (a) $(f \circ g)(4) = f(g(4)) = f(2(4)) = f(8) = \sqrt{8} = 2\sqrt{2}$
   (b) $(g \circ f)(2) = g(f(2)) = g\left(\sqrt{2}\right) = 2\sqrt{2}$
   (c) $(f \circ f)(1) = f(f(1)) = f\left(\sqrt{1}\right) = f(1) = \sqrt{1} = 1$
   (d) $(g \circ g)(0) = g(g(0)) = g(2(0)) = g(0) = 2(0) = 0$

19. $f(x) = |x|$ $g(x) = \dfrac{1}{x^2 + 1}$

    (a) $(f \circ g)(4) = f(g(4)) = f\!\left(\dfrac{1}{4^2 + 1}\right) = f\!\left(\dfrac{1}{17}\right) = \left|\dfrac{1}{17}\right| = \dfrac{1}{17}$

    (b) $(g \circ f)(2) = g(f(2)) = g(|2|) = g(2) = \dfrac{1}{2^2 + 1} = \dfrac{1}{5}$

    (c) $(f \circ f)(1) = f(f(1)) = f(|1|) = f(1) = |1| = 1$

    (d) $(g \circ g)(0) = g(g(0)) = g\!\left(\dfrac{1}{0^2 + 1}\right) = g(1) = \dfrac{1}{1^2 + 1} = \dfrac{1}{2}$

21. $f(x) = \dfrac{3}{x^2 + 1}$ $g(x) = \sqrt{x}$

    (a) $(f \circ g)(4) = f(g(4)) = f(\sqrt{4}) = f(2) = \dfrac{3}{2^2 + 1} = \dfrac{3}{5}$

    (b) $(g \circ f)(2) = g(f(2)) = g\!\left(\dfrac{3}{2^2 + 1}\right) = g\!\left(\dfrac{3}{5}\right) = \sqrt{\dfrac{3}{5}} = \dfrac{\sqrt{15}}{5}$

    (c) $(f \circ f)(1) = f(f(1)) = f\!\left(\dfrac{3}{1^2 + 1}\right) = f\!\left(\dfrac{3}{2}\right) = \dfrac{3}{\left(\frac{3}{2}\right)^2 + 1} = \dfrac{3}{\frac{13}{4}} = \dfrac{12}{13}$

    (d) $(g \circ g)(0) = g(g(0)) = g(\sqrt{0}) = g(0) = \sqrt{0} = 0$

23. The domain of g is $\{x \mid x \ne 0\}$. The domain of f is $\{x \mid x \ne 1\}$.
    Thus, $g(x) \ne 1$, so we solve:

$$g(x) = 1$$
$$\dfrac{2}{x} = 1$$
$$x = 2$$

    Thus, $x \ne 2$; so the domain of $f \circ g$ is $\{x \mid x \ne 0,\ x \ne 2\}$.

25. The domain of g is $\{x \mid x \ne 0\}$. The domain of f is $\{x \mid x \ne 1\}$.
    Thus, $g(x) \ne 1$, so we solve:

$$g(x) = 1$$
$$\dfrac{-4}{x} = 1$$
$$x = -4$$

    Thus, $x \ne -4$; so the domain of $f \circ g$ is $\{x \mid x \ne -4,\ x \ne 0\}$.

27. The domain of g is $\{\text{Real Numbers}\}$. The domain of f is $\{x \mid x \geq 0\}$.

Thus, $g(x) \geq 0$, so we solve:

$$g(x) \geq 0$$
$$2x + 3 \geq 0$$
$$x \geq \frac{-3}{2}$$

Thus, the domain of $f \circ g$ is $\left\{ x \mid x \geq \frac{-3}{2} \right\}$.

29. The domain of g is $\{x \mid x \neq 1\}$. The domain of f is $\{x \mid x \geq -1\}$.

Thus, $g(x) \geq -1$, so we solve:

$$g(x) \geq -1$$
$$\frac{2}{x-1} \geq -1$$
$$\frac{2}{x-1} + 1 \geq 0$$
$$\frac{2 + x - 1}{x-1} \geq 0$$
$$\frac{x+1}{x-1} \geq 0$$

$$(x + 1 \geq 0 \text{ and } x - 1 > 0) \quad \text{or} \quad (x + 1 \leq 0 \text{ and } x - 1 < 0)$$
$$(x \geq -1 \text{ and } x > 1) \quad \text{or} \quad (x \leq -1 \text{ and } x < 1)$$
$$x > 1 \quad \text{or} \quad x \leq -1$$

Thus, the domain of $f \circ g$ is $\{x \mid x \leq -1 \text{ or } x > 1\}$.

31. $f(x) = 2x + 3 \qquad g(x) = 3x$

The domain of f is all real numbers. The domain of g is all real numbers.

  (a) $(f \circ g)(x) = f(g(x)) = f(3x) = 2(3x) + 3 = 6x + 3$ Domain: All real numbers.

  (b) $(g \circ f)(x) = g(f(x)) = g(2x + 3) = 3(2x + 3) = 6x + 9$
        Domain: All real numbers.

  (c) $(f \circ f)(x) = f(f(x)) = f(2x + 3) = 2(2x + 3) + 3 = 4x + 6 + 3 = 4x + 9$
        Domain: All real numbers.

  (d) $(g \circ g)(x) = g(g(x)) = g(3x) = 3(3x) = 9x$ Domain: All real numbers.

33. $f(x) = 3x + 1 \qquad g(x) = x^2$

The domain of f is all real numbers. The domain of g is all real numbers.

  (a) $(f \circ g)(x) = f(g(x)) = f(x^2) = 3x^2 + 1$ Domain: All real numbers.

  (b) $(g \circ f)(x) = g(f(x)) = g(3x + 1) = (3x + 1)^2 = 9x^2 + 6x + 1$
        Domain: All real numbers.

  (c) $(f \circ f)(x) = f(f(x)) = f(3x + 1) = 3(3x + 1) + 1 = 9x + 3 + 1 = 9x + 4$
        Domain: All real numbers.

  (d) $(g \circ g)(x) = g(g(x)) = g(x^2) = (x^2)^2 = x^4$ Domain: All real numbers.

35. $f(x) = x^2$ $g(x) = x^2 + 4$

The domain of f is all real numbers. The domain of g is all real numbers.

(a) $(f \circ g)(x) = f(g(x)) = f(x^2 + 4) = (x^2 + 4)^2 = x^4 + 8x^2 + 16$
Domain: All real numbers.

(b) $(g \circ f)(x) = g(f(x)) = g(x^2) = (x^2)^2 + 4 = x^4 + 4$ Domain: All real numbers.

(c) $(f \circ f)(x) = f(f(x)) = f(x^2) = (x^2)^2 = x^4$ Domain: All real numbers.

(d) $(g \circ g)(x) = g(g(x)) = g(x^2 + 4) = (x^2 + 4)^2 + 4 = x^4 + 8x^2 + 16 + 4$
$= x^4 + 8x^2 + 20$ Domain: All real numbers.

37. $f(x) = \dfrac{3}{x-1}$ $g(x) = \dfrac{2}{x}$

The domain of f is $\{x \mid x \neq 1\}$. The domain of g is $\{x \mid x \neq 0\}$.

(a) $(f \circ g)(x) = f(g(x)) = f\left(\dfrac{2}{x}\right) = \dfrac{3}{\dfrac{2}{x} - 1} = \dfrac{3}{\dfrac{2-x}{x}} = \dfrac{3x}{2-x}$

Domain of $f \circ g$ is $\{x \mid x \neq 0,\ x \neq 2\}$.

(b) $(g \circ f)(x) = g(f(x)) = g\left(\dfrac{3}{x-1}\right) = \dfrac{2}{\dfrac{3}{x-1}} = \dfrac{2(x-1)}{3}$

Domain of $g \circ f$ is $\{x \mid x \neq 1\}$.

(c) $(f \circ f)(x) = f(f(x)) = f\left(\dfrac{3}{x-1}\right) = \dfrac{3}{\dfrac{3}{x-1} - 1} = \dfrac{3}{\dfrac{3-(x-1)}{x-1}} = \dfrac{3(x-1)}{4-x}$

Domain of $f \circ f$ is $\{x \mid x \neq 1,\ x \neq 4\}$.

(d) $(g \circ g)(x) = g(g(x)) = g\left(\dfrac{2}{x}\right) = \dfrac{2}{\dfrac{2}{x}} = \dfrac{2x}{2} = x$ Domain of $g \circ g$ is $\{x \mid x \neq 0\}$.

39. $f(x) = \dfrac{x}{x-1}$ $g(x) = \dfrac{-4}{x}$

The domain of f is $\{x \mid x \neq 1\}$. The domain of g is $\{x \mid x \neq 0\}$.

(a) $(f \circ g)(x) = f(g(x)) = f\left(\dfrac{-4}{x}\right) = \dfrac{\dfrac{-4}{x}}{\dfrac{-4}{x} - 1} = \dfrac{\dfrac{-4}{x}}{\dfrac{-4-x}{x}} = \dfrac{-4}{-4-x}$

Domain of $f \circ g$ is $\{x \mid x \neq -4,\ x \neq 0\}$.

(b) $(g \circ f)(x) = g(f(x)) = g\left(\dfrac{x}{x-1}\right) = \dfrac{-4}{\dfrac{x}{x-1}} = \dfrac{-4(x-1)}{x}$

Domain of $g \circ f$ is $\{x \mid x \neq 0,\ x \neq 1\}$.

(c) $(f \circ f)(x) = f(f(x)) = f\left(\dfrac{x}{x-1}\right) = \dfrac{\dfrac{x}{x-1}}{\dfrac{x}{x-1}-1} = \dfrac{\dfrac{x}{x-1}}{\dfrac{x-(x-1)}{x-1}} = \dfrac{x}{1} = x$

Domain of $f \circ f$ is $\{x \mid x \neq 1\}$.

(d) $(g \circ g)(x) = g(g(x)) = g\left(\dfrac{-4}{x}\right) = \dfrac{-4}{\dfrac{-4}{x}} = \dfrac{-4x}{-4} = x$

Domain of $g \circ g$ is $\{x \mid x \neq 0\}$.

41. $f(x) = \sqrt{x}$ $\qquad g(x) = 2x + 3$

The domain of f is $\{x \mid x \geq 0\}$. The domain of g is $\{\text{Real Numbers}\}$.

(a) $(f \circ g)(x) = f(g(x)) = f(2x+3) = \sqrt{2x+3}$ Domain of $f \circ g$ is $\left\{x \mid x \geq \dfrac{-3}{2}\right\}$.

(b) $(g \circ f)(x) = g(f(x)) = g(\sqrt{x}) = 2\sqrt{x} + 3$ Domain of $g \circ f$ is $\{x \mid x \geq 0\}$.

(c) $(f \circ f)(x) = f(f(x)) = f(\sqrt{x}) = \sqrt{\sqrt{x}} = x^{\frac{1}{4}} = \sqrt[4]{x}$

Domain of $f \circ f$ is $\{x \mid x \geq 0\}$.

(d) $(g \circ g)(x) = g(g(x)) = g(2x+3) = 2(2x+3)+3 = 4x+6+3 = 4x+9$

Domain of $g \circ g$ is $\{\text{Real Numbers}\}$.

43. $f(x) = \sqrt{x+1}$ $\qquad g(x) = \dfrac{2}{x-1}$

The domain of f is $\{x \mid x \geq -1\}$. The domain of g is $\{x \mid x \neq 1\}$.

(a) $(f \circ g)(x) = f(g(x)) = f\left(\dfrac{2}{x-1}\right) = \sqrt{\dfrac{2}{x-1}+1} = \sqrt{\dfrac{2+x-1}{x-1}} = \sqrt{\dfrac{x+1}{x-1}}$

Domain of $f \circ g$ is $\{x \mid x \leq -1 \text{ or } x > 1\}$. (See Problem 29.)

(b) $(g \circ f)(x) = g(f(x)) = g(\sqrt{x+1}) = \dfrac{2}{\sqrt{x+1}-1}$

Domain of $g \circ f$ is $\{x \mid x \geq -1, x \neq 0\}$.

(c) $(f \circ f)(x) = f(f(x)) = f(\sqrt{x+1}) = \sqrt{\sqrt{x+1}+1}$

Domain of $f \circ f$ is $\{x \mid x \geq -1\}$.

(d) $(g \circ g)(x) = g(g(x)) = g\left(\dfrac{2}{x-1}\right) = \dfrac{2}{\dfrac{2}{x-1}-1} = \dfrac{2}{\dfrac{2-(x-1)}{x-1}} = \dfrac{2(x-1)}{3-x}$

Domain of $g \circ g$ is $\{x \mid x \neq 1, x \neq 3\}$.

45. $f(x) = ax + b$ $g(x) = cx + d$ The domain of f is {Real Numbers}.
The domain of g is {Real Numbers}.

(a) $(f \circ g)(x) = f(g(x)) = f(cx + d) = a(cx + d) + b = acx + ad + b$
Domain of $f \circ g$ is {Real Numbers}.

(b) $(g \circ f)(x) = g(f(x)) = g(ax + b) = c(ax + b) + d = acx + bc + d$
Domain of $g \circ f$ is {Real Numbers}.

(c) $(f \circ f)(x) = f(f(x)) = f(ax + b) = a(ax + b) + b = a^2 x + ab + b$
Domain of $f \circ f$ is {Real Numbers}.

(d) $(g \circ g)(x) = g(g(x)) = g(cx + d) = c(cx + d) + d = c^2 x + cd + d$
Domain of $g \circ g$ is {Real Numbers}.

47. $(f \circ g)(x) = f(g(x)) = f\left(\frac{1}{2}x\right) = 2\left(\frac{1}{2}x\right) = x$
$(g \circ f)(x) = g(f(x)) = g(2x) = \frac{1}{2}(2x) = x$

49. $(f \circ g)(x) = f(g(x)) = f\left(\sqrt[3]{x}\right) = \left(\sqrt[3]{x}\right)^3 = x$
$(g \circ f)(x) = g(f(x)) = g\left(x^3\right) = \sqrt[3]{x^3} = x$

51. $(f \circ g)(x) = f(g(x)) = f\left(\frac{1}{2}(x + 6)\right) = 2\left(\frac{1}{2}(x + 6)\right) - 6 = x + 6 - 6 = x$
$(g \circ f)(x) = g(f(x)) = g(2x - 6) = \frac{1}{2}((2x - 6) + 6) = \frac{1}{2}(2x) = x$

53. $(f \circ g)(x) = f(g(x)) = f\left(\frac{1}{a}(x - b)\right) = a\left(\frac{1}{a}(x - b)\right) + b = x - b + b = x$
$(g \circ f)(x) = g(f(x)) = g(ax + b) = \frac{1}{a}((ax + b) - b) = \frac{1}{a}(ax) = x$

55. $H(x) = (2x + 3)^4$
$f(x) = x^4, \quad g(x) = 2x + 3$

57. $H(x) = \sqrt{x^2 + 1}$
$f(x) = \sqrt{x}, \quad g(x) = x^2 + 1$

59. $H(x) = |2x + 1|$
$f(x) = |x|, \quad g(x) = 2x + 1$

61. $f(x) = 2x^3 - 3x^2 + 4x - 1$ $g(x) = 2$
$(f \circ g)(x) = f(g(x)) = f(2) = 2(2)^3 - 3(2)^2 + 4(2) - 1 = 16 - 12 + 8 - 1 = 11$
$(g \circ f)(x) = g(f(x)) = g\left(2x^3 - 3x^2 + 4x - 1\right) = 2$

63. $f(x) = 2x^2 + 5 \qquad g(x) = 3x + a$
$(f \circ g)(x) = f(g(x)) = f(3x + a) = 2(3x + a)^2 + 5$
When $x = 0$, $(f \circ g)(0) = 23$
Solving:
$$2(3 \cdot 0 + a)^2 + 5 = 23$$
$$2a^2 + 5 = 23$$
$$2a^2 = 18$$
$$a^2 = 9$$
$$a = -3 \text{ or } 3$$

65. $S(r) = 4\pi r^2 \qquad r(t) = \frac{2}{3}t^3,\ t \geq 0$
$S(r(t)) = S\left(\frac{2}{3}t^3\right) = 4\pi\left(\frac{2}{3}t^3\right)^2 = 4\pi\left(\frac{4}{9}t^6\right) = \frac{16}{9}\pi t^6$

67. $N(t) = 100t - 5t^2,\ 0 \leq t \leq 10 \qquad C(N) = 15000 + 8000N$
$C(N(t)) = C\left(100t - 5t^2\right) = 15000 + 8000\left(100t - 5t^2\right)$
$\qquad\qquad = 15,000 + 800,000t - 40,000t^2$

69. $p = \frac{-1}{4}x + 100 \qquad 0 \leq x \leq 400$
$\frac{1}{4}x = 100 - p$
$x = 4(100 - p)$
$$C = \frac{\sqrt{x}}{25} + 600 = \frac{\sqrt{4(100 - p)}}{25} + 600$$
$$= \frac{2\sqrt{100 - p}}{25} + 600$$

71. $V = \pi r^2 h \qquad h = 2r$
$V(r) = \pi r^2 (2r) = 2\pi r^3$

2.6 Mathematical Models: Constructing Functions

1.   If $V = \pi r^2 h$ and $h = 2r$, then $V(r) = \pi r^2(2r) = 2\pi r^3$.

3.   (a)   If $p = \frac{-1}{6}x + 100$ and $R = xp$, then $R(x) = x\left(\frac{-1}{6}x + 100\right) = \frac{-1}{6}x^2 + 100x$.
     (b) $R(200) = \frac{-1}{6}(200)^2 + 100(200) = \$13,333$

(c) Graphing:

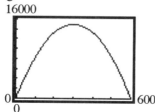

(d) 300; $15,000

(e) $p = \frac{-1}{6}(300) + 100 = -50 + 100 = \$50$

5. (a) If $x = -5p + 100$ and $R = xp$, then $p = \frac{100 - x}{5}$ and

$$R(x) = x\left(\frac{100 - x}{5}\right) = \frac{-1}{5}x^2 + 20x.$$

(b) $R(15) = \frac{-1}{5}(15)^2 + 20(15) = \$255$

(c) Graphing:

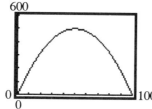

(d) 50; $500

(e) $p = \frac{100 - 50}{5} = \frac{50}{5} = \$10$

7. (a) Let x be the width of the rectangle and let y be the length of the rectangle.
Then, the perimeter is: $P = 2y + 2x = 400$.

Solving for y: $y = \frac{400 - 2x}{2} = 200 - x$.

The area function is: $A(x) = y(x) = (200 - x)x = -x^2 + 200x$.

(b) The domain is: $\{x \mid 0 < x < 200\}$

(c) Graphing:

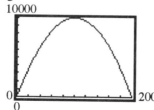

The area is largest when $x = 100$ yards.

9.   (a)   The distance d from P to the origin is $d = \sqrt{x^2 + y^2}$. Since P is a point on the graph of $y = x^2 - 8$, we have:

$$d(x) = \sqrt{x^2 + (x^2 - 8)^2} = \sqrt{x^4 - 15x^2 + 64}$$

   (b) $d(0) = \sqrt{0^4 - 15(0)^2 + 64} = \sqrt{64} = 8$

   (c) $d(1) = \sqrt{(1)^4 - 15(1)^2 + 64} = \sqrt{1 - 15 + 64} = \sqrt{50} = 5\sqrt{2} \approx 7.07$

   (d)   Graphing:

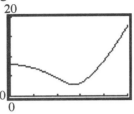

   (e) d is smallest when x is 2.74.

11.   (a)   The distance d from P to the point $(1, 0)$ is $d = \sqrt{(x-1)^2 + y^2}$. Since P is a point on the graph of $y = \sqrt{x}$, we have:

$$d(x) = \sqrt{(x-1)^2 + \left(\sqrt{x}\right)^2} = \sqrt{x^2 - x + 1}$$

   (b)   Graphing:

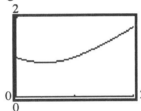

   (e) d is smallest when x is 0.50.

13.   By definition, a triangle has area $A = \frac{1}{2}bh$, $b = $ base, $h = $ height. Because a vertex of the triangle is at the origin, we know that $b = x$ and $h = y$. Expressing the area of the triangle as a function of x, we have: $A(x) = \frac{1}{2}xy = \frac{1}{2}x(x^3) = \frac{1}{2}x^4$.

15.   (a) $A(x) = xy = x(16 - x^2) = -x^3 + 16x$

   (b)   Domain: $\{x \mid 0 < x < 4\}$

   (c)   Graphing:

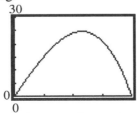

The area is largest when x is approximately 2.31.

17. (a) $A(x) = (2x)(2y) = 4x(4-x^2)^{\frac{1}{2}}$

(b) $p(x) = 2(2x) + 2(2y) = 4x + 4(4-x^2)^{\frac{1}{2}}$

(c) Graphing: (d) Graphing:

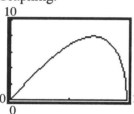

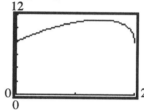

The area is largest when x is   The perimeter is largest when x is
approximately 1.41.   approximately 1.41.

19. (a) C = circumference , A = area, r = radius, x = side of square

$$C = 2\pi r = 10 - 4x \quad \rightarrow \quad r = \frac{5-2x}{\pi}$$

$$A(x) = x^2 + \pi r^2 = x^2 + \pi\left(\frac{5-2x}{\pi}\right)^2 = x^2 + \frac{25-20x+4x^2}{\pi}$$

(b) Since the lengths must be positive, we have:
$$10 - 4x > 0 \quad \text{and } x > 0$$
$$-4x > -10$$
$$x < 2.5 \text{ and } x > 0$$
Domain: $\{x \mid 0 < x < 2.5\}$

(c) Graphing:

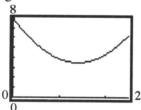

The area is smallest when x is approximately 1.40 meters.

21. (a) Since the wire of length x is bent into a circle, the circumference is x.
Therefore, $C(x) = x$.

(b) Since $C = x = 2\pi r$, $r = \frac{x}{2\pi}$.
$$A(x) = \pi r^2 = \pi\left(\frac{x}{2\pi}\right)^2 = \frac{x^2}{4\pi}.$$

23. (a) A = area, r = radius; diameter = $2r$ (b) p = perimeter
$A(r) = (2r)(r) = 2r^2$ $p(r) = 2(2r) + 2r = 6r$

25. Area of the equilateral triangle $= \dfrac{1}{2}x \cdot \dfrac{\sqrt{3}}{2}x = \dfrac{\sqrt{3}}{4}x^2$

Area of $\frac{1}{3}$ of the equilateral triangle $= \dfrac{1}{2}x\sqrt{r^2 - \left(\dfrac{x}{2}\right)^2} = \dfrac{1}{2}x\sqrt{r^2 - \dfrac{x^2}{4}} = \dfrac{1}{3} \cdot \dfrac{\sqrt{3}}{4}x^2$

Solving for r^2:

$$\dfrac{1}{2}x\sqrt{r^2 - \dfrac{x^2}{4}} = \dfrac{1}{3} \cdot \dfrac{\sqrt{3}}{4}x^2$$

$$\sqrt{r^2 - \dfrac{x^2}{4}} = \dfrac{2}{x} \cdot \dfrac{\sqrt{3}}{12}x^2$$

$$\sqrt{r^2 - \dfrac{x^2}{4}} = \dfrac{\sqrt{3}}{6}x$$

$$r^2 - \dfrac{x^2}{4} = \dfrac{3}{36}x^2$$

$$r^2 = \dfrac{x^2}{3}$$

Area inside the circle, but outside the triangle:

$$A(x) = \pi r^2 - \dfrac{\sqrt{3}}{4}x^2 = \pi\dfrac{x^2}{3} - \dfrac{\sqrt{3}}{4}x^2 = \left(\dfrac{\pi}{3} - \dfrac{\sqrt{3}}{4}\right)x^2$$

27.
$$C = \begin{cases} 95 & \text{if } x = 7 \\ 119 & \text{if } 7 < x \le 8 \\ 143 & \text{if } 8 < x \le 9 \\ 167 & \text{if } 9 < x \le 10 \\ 190 & \text{if } 10 < x \le 14 \end{cases}$$

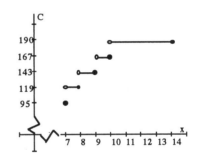

29.
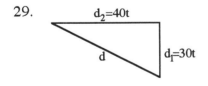

$d^2 = d_1^{\,2} + d_2^{\,2}$

$d^2 = (30t)^2 + (40t)^2$

$d(t) = \sqrt{900t^2 + 1600t^2}$

$d(t) = \sqrt{2500t^2} = 50t$

31. (a) length $= 24 - 2x$
    width $= 24 - 2x$
    height $= x$
    $V(x) = x(24 - 2x)(24 - 2x) = x(24 - 2x)^2$

    (b) $V(3) = 3(24 - 2(3))^2 = 3(18)^2 = 3(324) = 972$ cu. in.

    (c) $V(10) = 3(24 - 2(10))^2 = 3(4)^2 = 3(16) = 48$ cu. in.

(d)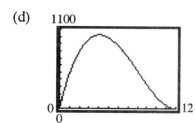

The volume is largest when
$x = 4$ inches.

33. r = radius of cylinder , h = height of cylinder , V = volume of cylinder

$$r^2 + \left(\frac{h}{2}\right)^2 = R^2$$

$$r^2 + \frac{h^2}{4} = R^2$$

$$r^2 = R^2 - \frac{h^2}{4}$$

$$r^2 = \frac{4R^2 - h^2}{4}$$

$$V = \pi r^2 h$$

$$V(h) = \pi\left(\frac{4R^2 - h^2}{4}\right)h = \frac{\pi}{4}\left(4R^2 h - h^3\right)$$

35.   (a)   The total cost of installing the cable along the road is
$10x$. If cable is installed x miles along the road, there
are $5 - x$ miles left from the road to the house and
where the cable ends.

$$d = \sqrt{(5-x)^2 + 2^2} = \sqrt{25 - 10x + x^2 + 4}$$
$$= \sqrt{x^2 - 10x + 29}$$

The total cost of installing the cable is:

$$C(x) = 10x + 14\sqrt{x^2 - 10x + 29}$$

Domain: $\left\{x \mid 0 < x < 5\right\}$

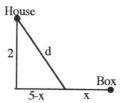

(b) $C(1) = 10(1) + 14\sqrt{1^2 - 10(1) + 29} = 10 + 14\sqrt{20} \approx 10 + 62.61 = \$72.61$

(c) $C(3) = 10(3) + 14\sqrt{3^2 - 10(3) + 29} = 30 + 14\sqrt{8} \approx 30 + 39.60 = \$69.60$

(d)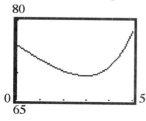

(e)

X	Y₁
1.5	71.436
2	70.478
2.5	69.822
3	69.598
3.5	70
4	71.305
4.5	73.862

X=4.5

The table indicates that $x = 3$
results in the least cost.

(f)   Using MINIMUM, the graph indicates that $x = 2.96$ results in the least cost.

37. For schedule X:

$$f(x) = \begin{cases} 0.15x & \text{if } 0 < x \le 25{,}750 \\ 3{,}862.50 + 0.28(x - 25{,}750) & \text{if } 25{,}750 < x \le 62{,}450 \\ 14{,}138.50 + 0.31(x - 62{,}450) & \text{if } 62{,}450 < x \le 130{,}250 \\ 35{,}156.50 + 0.36(x - 130{,}250) & \text{if } 130{,}250 < x \le 283{,}150 \\ 90{,}200.50 + 0.396(x - 283{,}150) & \text{if } x > 283{,}150 \end{cases}$$

For Schedule Y-1:

$$f(x) = \begin{cases} 0.15x & \text{if } 0 < x \le 43{,}050 \\ 6{,}457.50 + 0.28(x - 43{,}050) & \text{if } 43{,}050 < x \le 104{,}050 \\ 23{,}537.50 + 0.31(x - 104{,}050) & \text{if } 104{,}050 < x \le 158{,}550 \\ 40{,}432.50 + 0.36(x - 158{,}550) & \text{if } 158{,}550 < x \le 283{,}150 \\ 85{,}288.50 + 0.396(x - 283{,}150) & \text{if } x > 283{,}150 \end{cases}$$

2   Chapter Review

1. $f(4) = -5$ gives the ordered pair $(4, -5)$. $f(0) = 3$ gives $(0, 3)$.

   Finding the slope: $m = \dfrac{3 - (-5)}{0 - 4} = \dfrac{8}{-4} = -2$

   Using slope-intercept form: $f(x) = -2x + 3$

3. $f(x) = \dfrac{Ax + 5}{6x - 2}$ and $f(1) = 4$

   Solving:

   $$\frac{A(1) + 5}{6(1) - 2} = 4$$

   $$\frac{A + 5}{4} = 4$$

   $$A + 5 = 16$$

   $$A = 11$$

5. (b), (c), and (d) pass the vertical line test and therefore are functions.

7. $f(x) = \dfrac{3x}{x^2 - 4}$

   (a) $f(-x) = \dfrac{3(-x)}{(-x)^2 - 4} = \dfrac{-3x}{x^2 - 4}$

   (b) $-f(x) = -\left(\dfrac{3x}{x^2 - 4}\right) = \dfrac{-3x}{x^2 - 4}$

   (c) $f(x + 2) = \dfrac{3(x + 2)}{(x + 2)^2 - 4} = \dfrac{3x + 6}{x^2 + 4x + 4 - 4} = \dfrac{3x + 6}{x^2 + 4x}$

   (d) $f(x - 2) = \dfrac{3(x - 2)}{(x - 2)^2 - 4} = \dfrac{3x - 6}{x^2 - 4x + 4 - 4} = \dfrac{3x - 6}{x^2 - 4x}$

(e) $f(2x) = \dfrac{3(2x)}{(2x)^2 - 4} = \dfrac{6x}{4x^2 - 4} = \dfrac{3x}{2x^2 - 2}$

9. $f(x) = \sqrt{x^2 - 4}$

(a) $f(-x) = \sqrt{(-x)^2 - 4} = \sqrt{x^2 - 4}$

(b) $-f(x) = -\sqrt{x^2 - 4}$

(c) $f(x + 2) = \sqrt{(x+2)^2 - 4} = \sqrt{x^2 + 4x + 4 - 4} = \sqrt{x^2 + 4x}$

(d) $f(x - 2) = \sqrt{(x-2)^2 - 4} = \sqrt{x^2 - 4x + 4 - 4} = \sqrt{x^2 - 4x}$

(e) $f(2x) = \sqrt{(2x)^2 - 4} = \sqrt{4x^2 - 4} = 2\sqrt{x^2 - 1}$

11. $f(x) = \dfrac{x^2 - 4}{x^2}$

(a) $f(-x) = \dfrac{(-x)^2 - 4}{(-x)^2} = \dfrac{x^2 - 4}{x^2}$

(b) $-f(x) = -\left(\dfrac{x^2 - 4}{x^2}\right) = \dfrac{4 - x^2}{x^2}$

(c) $f(x + 2) = \dfrac{(x+2)^2 - 4}{(x+2)^2} = \dfrac{x^2 + 4x + 4 - 4}{x^2 + 4x + 4} = \dfrac{x^2 + 4x}{x^2 + 4x + 4}$

(d) $f(x - 2) = \dfrac{(x-2)^2 - 4}{(x-2)^2} = \dfrac{x^2 - 4x + 4 - 4}{x^2 - 4x + 4} = \dfrac{x^2 - 4x}{x^2 - 4x + 4}$

(e) $f(2x) = \dfrac{(2x)^2 - 4}{(2x)^2} = \dfrac{4x^2 - 4}{4x^2} = \dfrac{x^2 - 1}{x^2}$

13. $f(x) = \dfrac{x}{x^2 - 9}$

The denominator cannot be zero:

$$x^2 - 9 \neq 0$$
$$(x + 3)(x - 3) \neq 0$$
$$x \neq -3 \text{ or } 3$$

Domain: $\left\{x \mid x \neq -3, x \neq 3\right\}$

15. $f(x) = \sqrt{2 - x}$

The radicand must be positive:

$$2 - x \geq 0$$
$$x \leq 2$$

Domain: $\left\{x \mid x \leq 2\right\}$ or $(-\infty, 2]$

17. $f(x) = \dfrac{\sqrt{x}}{|x|}$

The radicand must be positive and the denominator cannot be zero:

$$x > 0$$

Domain: $\left\{x \mid x > 0\right\}$ or $(0, +\infty)$

19. $f(x) = \dfrac{x}{x^2 + 2x - 3}$

The denominator cannot be zero:

$$x^2 + 2x - 3 \neq 0$$
$$(x + 3)(x - 1) \neq 0$$
$$x \neq -3 \text{ or } 1$$

Domain: $\left\{x \mid x \neq -3, x \neq 1\right\}$

21. $f(x) = \begin{cases} 3x - 2 & \text{if } x \le 1 \\ x + 1 & \text{if } x > 1 \end{cases}$

   (a)   Domain: {Real Numbers}

   (b)   x-intercept: $\left(\frac{2}{3}, 0\right)$

         y-intercept: $(0, -2)$

   (c)

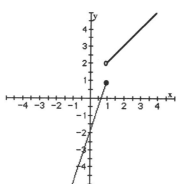

   (d)   Range: $\{y > 2\} \cup \{y \le 1\}$

23. $f(x) = \begin{cases} x & \text{if } -4 \le x < 0 \\ 1 & \text{if } x = 0 \\ 3x & \text{if } x > 0 \end{cases}$

   (a)   Domain: $\{x \mid x \ge -4\}$

   (b)   x-intercept: none

         y-intercept: $(0, 1)$

   (c)

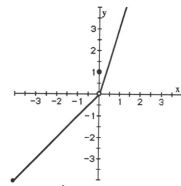

   (d)   Range: $\{y \mid y \ge -4,\ y \ne 0\}$

25. $f(x) = 2 - 5x$

$$\frac{f(x) - f(2)}{x - 2} = \frac{2 - 5x - (-8)}{x - 2} = \frac{-5x + 10}{x - 2} = \frac{-5(x - 2)}{x - 2} = -5$$

27. $f(x) = 3x - 4x^2$

$$\frac{f(x) - f(2)}{x - 2} = \frac{3x - 4x^2 - (-10)}{x - 2} = \frac{-4x^2 + 3x + 10}{x - 2}$$

$$= \frac{-(4x^2 - 3x - 10)}{x - 2} = \frac{-(4x + 5)(x - 2)}{x - 2} = -4x - 5$$

29. $f(x) = x^3 - 4x$

$$f(-x) = (-x)^3 - 4(-x) = -x^3 + 4x = -\left(x^3 - 4x\right) = -f(x)$$

f is odd.

31. $h(x) = \dfrac{1}{x^4} + \dfrac{1}{x^2} + 1$

$$h(-x) = \frac{1}{(-x)^4} + \frac{1}{(-x)^2} + 1 = \frac{1}{x^4} + \frac{1}{x^2} + 1 = h(x)$$

h is even.

33. $G(x) = 1 - x + x^3$

$$G(-x) = 1 - (-x) + (-x)^3 = 1 + x - x^3 \ne -G(x) \ne G(x)$$

G is neither even nor odd.

35. $f(x) = \dfrac{x}{1+x^2}$

$$f(-x) = \frac{-x}{1+(-x)^2} = \frac{-x}{1+x^2} = -f(x)$$

f is odd.

37. $F(x) = |x| - 4$

Using the graph of $y = |x|$, vertically shift the graph downward 4 units.

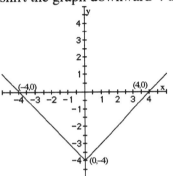

Intercepts: $(-4,0), (4,0), (0,-4)$
Domain: {Real Numbers}
Range: $\{y \mid y \geq -4\}$

39. $g(x) = -2|x|$

Reflect the graph of $y = |x|$ about the x-axis and vertically stretch the graph by a factor of 2.

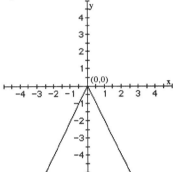

Intercepts: $(0,0)$
Domain: {Real Numbers}
Range: $\{y \mid y \leq 0\}$

41. $h(x) = \sqrt{x-1}$

Using the graph of $y = \sqrt{x}$, horizontally shift the graph to the right 1 unit.

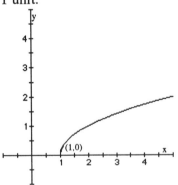

Intercepts: $(1,0)$
Domain: $\{x \mid x \geq 1\}$
Range: $\{y \mid y \geq 0\}$

43. $f(x) = \sqrt{1-x} = \sqrt{-1(x-1)}$

Reflect the graph of $y = \sqrt{x}$ about the y-axis and horizontally shift the graph to the right 1 unit..

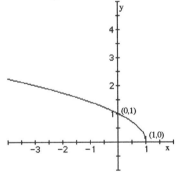

Intercepts: $(1,0), (0,1)$
Domain: $\{x \mid x \leq 1\}$
Range: $\{y \mid y \geq 0\}$

45. $h(x) = (x - 1)^2 + 2$

Using the graph of $y = x^2$, horizontally shift the graph to the right 1 unit and vertically shift the graph up 2 units.

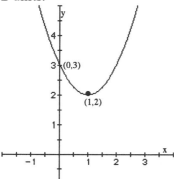

Intercepts: (0,3)
Domain: {Real Numbers}
Range: $\left\{ y \mid y \geq 2 \right\}$

47. $g(x) = 3(x - 1)^3 + 1$

Using the graph of $y = x^3$, horizontally shift the graph to the right 1 unit, vertically stretch the graph by a factor of 3, and vertically shift the graph up 1 unit.

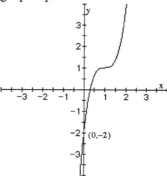

Intercepts: $(0,-2), \left(1 - \dfrac{\sqrt[3]{3}}{3}, 0 \right)$

Domain: {Real Numbers}
Range: {Real Numbers}

49. $f(x) = 2x^3 - 5x + 1$

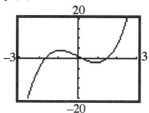

Local maximum: (−0.91, 4.04)
Local minimum: (0.91, −2.04)
Increasing: (−3, −0.91), (0.91, 3)
Decreasing: (−0.91, 0.91)

51. $f(x) = 2x^4 - 5x^3 + 2x + 1$

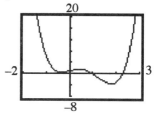

Local maximum: (0.41, 1.53)
Local minima: (−0.34, 0.54),
    (1.80, −3.56)
Increasing: (−0.34, 0.41), (1.80, 3)
Decreasing: (−2,−0.34), (0.41, 1.80)

53. $f(x) = 3x - 5 \qquad g(x) = 1 - 2x^2$

(a) $(f \circ g)(2) = f(g(2)) = f\left(1 - 2(2)^2\right) = f(-7) = 3(-7) - 5 = -26$

(b) $(g \circ f)(-2) = g(f(-2)) = g(3(-2) - 5) = g(-11) = 1 - 2(-11)^2 = -241$

(c) $(f \circ f)(4) = f(f(4)) = f(3(4) - 5) = f(7) = 3(7) - 5 = 16$

(d) $(g \circ g)(-1) = g(g(-1)) = g\left(1 - 2(-1)^2\right) = g(-1) = 1 - 2(-1)^2 = -1$

55. $f(x) = \sqrt{x+2}$ $g(x) = 2x^2 + 1$

(a) $(f \circ g)(2) = f(g(2)) = f(2(2)^2 + 1) = f(9) = \sqrt{9+2} = \sqrt{11}$

(b) $(g \circ f)(-2) = g(f(-2)) = g(\sqrt{-2+2}) = g(0) = 2(0)^2 + 1 = 1$

(c) $(f \circ f)(4) = f(f(4)) = f(\sqrt{4+2}) = f(\sqrt{6}) = \sqrt{\sqrt{6}+2}$

(d) $(g \circ g)(-1) = g(g(-1)) = g(2(-1)^2 + 1) = g(3) = 2(3)^2 + 1 = 19$

57. $f(x) = \dfrac{1}{x^2 + 4}$ $g(x) = 3x - 2$

(a) $(f \circ g)(2) = f(g(2)) = f(3(2) - 2) = f(4) = \dfrac{1}{4^2 + 4} = \dfrac{1}{20}$

(b) $(g \circ f)(-2) = g(f(-2)) = g\left(\dfrac{1}{(-2)^2 + 4}\right) = g\left(\dfrac{1}{8}\right) = 3\left(\dfrac{1}{8}\right) - 2 = \dfrac{-13}{8}$

(c) $(f \circ f)(4) = f(f(4)) = f\left(\dfrac{1}{4^2 + 4}\right) = f\left(\dfrac{1}{20}\right) = \dfrac{1}{\left(\dfrac{1}{20}\right)^2 + 4} = \dfrac{1}{\dfrac{1601}{400}} = \dfrac{400}{1601}$

(d) $(g \circ g)(-1) = g(g(-1)) = g(3(-1) - 2) = g(-5) = 3(-5) - 2 = -17$

59. $f(x) = 2 - x$ $g(x) = 3x + 1$
The domain of f is all real numbers. The domain of g is all real numbers.

(a) $(f \circ g)(x) = f(g(x)) = f(3x + 1) = 2 - (3x + 1) = 2 - 3x - 1 = 1 - 3x$
Domain: All real numbers.

(b) $(g \circ f)(x) = g(f(x)) = g(2 - x) = 3(2 - x) + 1 = 6 - 3x + 1 = 7 - 3x$
Domain: All real numbers.

(c) $(f \circ f)(x) = f(f(x)) = f(2 - x) = 2 - (2 - x) = 2 - 2 + x = x$
Domain: All real numbers.

(d) $(g \circ g)(x) = g(g(x)) = g(3x + 1) = 3(3x + 1) + 1 = 9x + 3 + 1 = 9x + 4$
Domain: All real numbers.

61. $f(x) = 3x^2 + x + 1$ $g(x) = |3x|$
The domain of f is all real numbers. The domain of g is all real numbers.

(a) $(f \circ g)(x) = f(g(x)) = f(|3x|) = 3(|3x|)^2 + (|3x|) + 1 = 27x^2 + 3|x| + 1$
Domain: All real numbers.

(b) $(g \circ f)(x) = g(f(x)) = g(3x^2 + x + 1) = |3(3x^2 + x + 1)| = |9x^2 + 3x + 3|$
Domain: All real numbers.

(c) $(f \circ f)(x) = f(f(x)) = f(3x^2 + x + 1) = 3(3x^2 + x + 1)^2 + (3x^2 + x + 1) + 1$

$= 3(9x^4 + 6x^3 + 7x^2 + 2x + 1) + 3x^2 + x + 1 + 1$

$= 27x^4 + 18x^3 + 24x^2 + 7x + 5$
Domain: All real numbers.

(d) $(g \circ g)(x) = g(g(x)) = g(|3x|) = |3|3x|| = 9|x|$
Domain: All real numbers.

63. $f(x) = \dfrac{x+1}{x-1}$ $\qquad g(x) = \dfrac{1}{x}$

The domain of f is $\{x\,|\,x \neq 1\}$. The domain of g is $\{x\,|\,x \neq 0\}$.

(a) $(f \circ g)(x) = f(g(x)) = f\!\left(\dfrac{1}{x}\right) = \dfrac{\dfrac{1}{x}+1}{\dfrac{1}{x}-1} = \dfrac{\dfrac{1+x}{x}}{\dfrac{1-x}{x}} = \dfrac{1+x}{1-x}$

Domain of $f \circ g$ is $\{x\,|\,x \neq 0,\ x \neq 1\}$.

(b) $(g \circ f)(x) = g(f(x)) = g\!\left(\dfrac{x+1}{x-1}\right) = \dfrac{1}{\dfrac{x+1}{x-1}} = \dfrac{x-1}{x+1}$

Domain of $g \circ f$ is $\{x\,|\,x \neq -1,\ x \neq 1\}$.

(c) $(f \circ f)(x) = f(f(x)) = f\!\left(\dfrac{x+1}{x-1}\right) = \dfrac{\dfrac{x+1}{x-1}+1}{\dfrac{x+1}{x-1}-1} = \dfrac{\dfrac{x+1+x-1}{x-1}}{\dfrac{x+1-(x-1)}{x-1}} = \dfrac{2x}{2} = x$

Domain of $f \circ f$ is $\{x\,|\,x \neq 1\}$.

(d) $(g \circ g)(x) = g(g(x)) = g\!\left(\dfrac{1}{x}\right) = \dfrac{1}{\dfrac{1}{x}} = x$ $\quad$ Domain of $g \circ g$ is $\{x\,|\,x \neq 0\}$.

65. (a) $y = f(-x)$
Reflect about the y-axis.

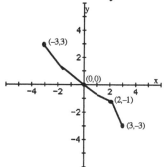

(b) $y = -f(x)$
Reflect about the x-axis.

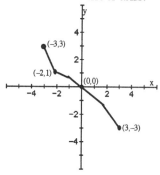

(c) $y = f(x+2)$
Horizontally shift left 2 units.

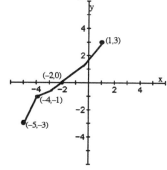

(d) $y = f(x) + 2$
Vertically shift up 2 units.

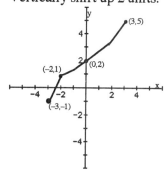

(e) $y = 2f(x)$
Vertical stretch by a factor of 2.

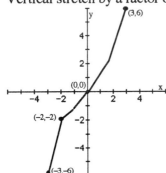

(f) $y = f(3x)$
Horizontal compression by $\frac{1}{3}$.

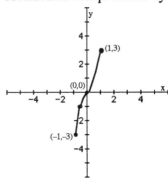

67. Let (x,y) be any point on the line $y = x$. Then the distance from (x,y) to $(3,1)$ is given by the distance formula:

$$d = \sqrt{(x-3)^2 + (y-1)^2} = \sqrt{x^2 - 6x + 9 + y^2 - 2y + 1}$$

Since $y = x$, we have:

$$d = \sqrt{x^2 - 6x + 9 + x^2 - 2x + 1} = \sqrt{2x^2 - 8x + 10}$$

Graphing the function and finding the minimum:

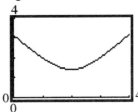

The minimum occurs when $x = 2$.
Since $y = x$ and $x = 2$, the closest point on the line to $(3, 1)$ is $(2, 2)$.

69. Graph:

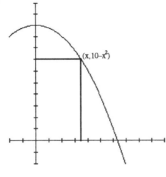

$A(x) = x(10 - x^2) = 10x - x^3$
Graphing the function:

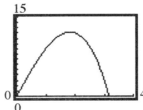

The maximum area is 12.17 when $x = 1.83$.

71. $S = 4\pi r^2 \rightarrow r = \sqrt{\dfrac{S}{4\pi}}$ $V(S) = \dfrac{4}{3}\pi r^3 = \dfrac{4\pi}{3}\left(\sqrt{\dfrac{S}{4\pi}}\right)^3 = \dfrac{4\pi}{3} \cdot \dfrac{S}{4\pi}\sqrt{\dfrac{S}{4\pi}} = \dfrac{S}{6}\sqrt{\dfrac{S}{\pi}}$

$V(2S) = \dfrac{2S}{6}\sqrt{\dfrac{2S}{\pi}} = 2\sqrt{2}\left(\dfrac{S}{6}\sqrt{\dfrac{S}{\pi}}\right)$ The volume is $2\sqrt{2}$ times as large.

Chapter 3

Polynomial and Rational Functions

3.1 Quadratic Functions; Curve Fitting

1. D   3. A   5. B   7. E   9. D   11. B

13. $f(x) = \frac{1}{4}x^2$

Using the function $y = x^2$, compress vertically by a factor of $\frac{1}{4}$.

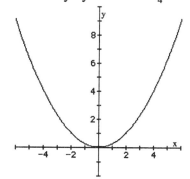

15. $f(x) = \frac{1}{4}x^2 - 2$

Using the function $y = x^2$, compress vertically by a factor of $\frac{1}{4}$, and shift downward 2 units.

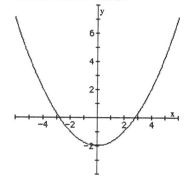

17. $f(x) = \frac{1}{4}x^2 + 2$

Using the function $y = x^2$, compress vertically by a factor of $\frac{1}{4}$, and shift upward 2 units.

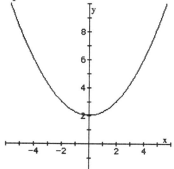

19. $f(x) = \frac{1}{4}x^2 + 1$

Using the function $y = x^2$, compress vertically by a factor of $\frac{1}{4}$, and shift upward 1 unit.

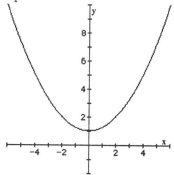

21. $f(x) = x^2 + 4x + 2$
Completing the square:
$f(x) = (x^2 + 4x + 4) + 2 - 4$
$\qquad = (x+2)^2 - 2$

Using the function $y = x^2$, shift the graph to the left 2 units, and shift downward 2 units.

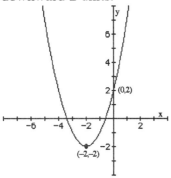

23. $f(x) = 2x^2 - 4x + 1$
Completing the square:
$f(x) = 2(x^2 - 2x + 1) + 1 - 2$
$\qquad = 2(x-1)^2 - 1$

Using the function $y = x^2$, shift the graph to the right 1 unit, stretch the graph vertically by a factor of 2, and shift downward 1 unit.

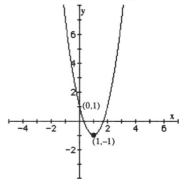

25. $f(x) = -x^2 - 2x$
Completing the square:
$f(x) = -(x^2 + 2x + 1) + 1$
$\qquad = -(x+1)^2 + 1$

Using the function $y = x^2$, shift the graph to the left 1 unit, reflect the graph on the x-axis, and shift upward 1 unit.

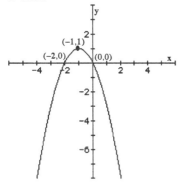

27. $f(x) = \frac{1}{2}x^2 + x - 1$
Completing the square:
$f(x) = \frac{1}{2}(x^2 + 2x + 1) - 1 - \frac{1}{2}$
$\qquad = \frac{1}{2}(x+1)^2 - \frac{3}{2}$

Using the function $y = x^2$, shift the graph to the left 1 unit, compress the graph vertically by a factor of $\frac{1}{2}$, and shift downward $\frac{3}{2}$ units.

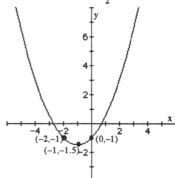

29. $f(x) = -x^2 - 6x$

$a = -1, b = -6, c = 0$. Since $a = -1 < 0$, the graph opens down.

The x-coordinate of the vertex is $x = \dfrac{-b}{2a} = \dfrac{-(-6)}{2(-1)} = \dfrac{6}{-2} = -3$.

The y-coordinate of the vertex is $f\left(\dfrac{-b}{2a}\right) = f(-3) = -(-3)^2 - 6(-3) = -9 + 18 = 9$.

Thus, the vertex is $(-3, 9)$.
The axis of symmetry is the line $x = -3$.
The discriminant is:

$b^2 - 4ac = (-6)^2 - 4(-1)(0) = 36 > 0$,

so the graph has two x-intercepts.
The x-intercepts are found by solving:

$-x^2 - 6x = 0$

$-x(x + 6) = 0$

$x = 0$ or $x = -6$

The x-intercepts are –6 and 0.
The y-intercept is $f(0) = 0$.

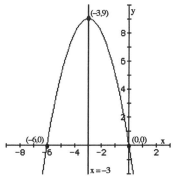

31. $f(x) = 2x^2 - 8x$

$a = 2, b = -8, c = 0$. Since $a = 2 > 0$, the graph opens up.

The x-coordinate of the vertex is $x = \dfrac{-b}{2a} = \dfrac{-(-8)}{2(2)} = \dfrac{8}{4} = 2$.

The y-coordinate of the vertex is $f\left(\dfrac{-b}{2a}\right) = f(2) = 2(2)^2 - 8(2) = 8 - 16 = -8$.

Thus, the vertex is $(2, -8)$.
The axis of symmetry is the line $x = 2$.
The discriminant is:

$b^2 - 4ac = (-8)^2 - 4(2)(0) = 64 > 0$,

so the graph has two x-intercepts.
The x-intercepts are found by solving:

$2x^2 - 8x = 0$

$2x(x - 4) = 0$

$x = 0$ or $x = 4$

The x-intercepts are 0 and 4.
The y-intercept is $f(0) = 0$.

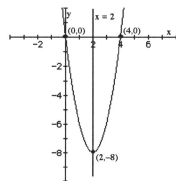

33. $f(x) = x^2 + 2x - 8$

$a = 1, b = 2, c = -8$. Since $a = 1 > 0$, the graph opens up.

The x-coordinate of the vertex is $x = \dfrac{-b}{2a} = \dfrac{-2}{2(1)} = \dfrac{-2}{2} = -1$.

The y-coordinate of the vertex is $f\left(\dfrac{-b}{2a}\right) = f(-1) = (-1)^2 + 2(-1) - 8 = 1 - 2 - 8 = -9$.

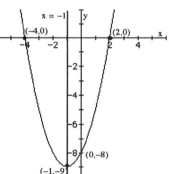

Thus, the vertex is $(-1, -9)$.
The axis of symmetry is the line $x = -1$.
The discriminant is:

$b^2 - 4ac = 2^2 - 4(1)(-8) = 4 + 32 = 36 > 0$,

so the graph has two x-intercepts.
The x-intercepts are found by solving:

$x^2 + 2x - 8 = 0$

$(x + 4)(x - 2) = 0$

$x = -4$ or $x = 2$

The x-intercepts are –4 and 2.
The y-intercept is $f(0) = -8$.

35. $f(x) = x^2 + 2x + 1$

$a = 1, b = 2, c = 1$. Since $a = 1 > 0$, the graph opens up.

The x-coordinate of the vertex is $x = \dfrac{-b}{2a} = \dfrac{-2}{2(1)} = \dfrac{-2}{2} = -1$.

The y-coordinate of the vertex is $f\left(\dfrac{-b}{2a}\right) = f(-1) = (-1)^2 + 2(-1) + 1 = 1 - 2 + 1 = 0$.

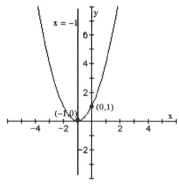

Thus, the vertex is $(-1, 0)$.
The axis of symmetry is the line $x = -1$.
The discriminant is:

$b^2 - 4ac = 2^2 - 4(1)(1) = 4 - 4 = 0$,

so the graph has one x-intercept.
The x-intercept is found by solving:

$x^2 + 2x + 1 = 0$

$(x + 1)^2 = 0$

$x = -1$

The x-intercept is –1.
The y-intercept is $f(0) = 1$.

37. $f(x) = 2x^2 - x + 2$

$a = 2, b = -1, c = 2.$ Since $a = 2 > 0,$ the graph opens up.

The x-coordinate of the vertex is $x = \dfrac{-b}{2a} = \dfrac{-(-1)}{2(2)} = \dfrac{1}{4}.$

The y-coordinate of the vertex is $f\left(\dfrac{-b}{2a}\right) = f\left(\dfrac{1}{4}\right) = 2\left(\dfrac{1}{4}\right)^2 - \dfrac{1}{4} + 2 = \dfrac{1}{8} - \dfrac{1}{4} + 2 = \dfrac{15}{8}.$

Thus, the vertex is $\left(\dfrac{1}{4}, \dfrac{15}{8}\right).$

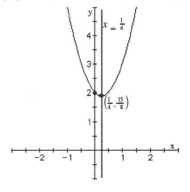

The axis of symmetry is the line $x = \dfrac{1}{4}.$

The discriminant is:

$\qquad b^2 - 4ac = (-1)^2 - 4(2)(2) = 1 - 16 = -15,$

so the graph has no x-intercepts.

The y-intercept is $f(0) = 2.$

39. $f(x) = -2x^2 + 2x - 3$

$a = -2, b = 2, c = -3.$ Since $a = -2 < 0,$ the graph opens down.

The x-coordinate of the vertex is $x = \dfrac{-b}{2a} = \dfrac{-(2)}{2(-2)} = \dfrac{-2}{-4} = \dfrac{1}{2}.$ The y-coordinate of

the vertex is $f\left(\dfrac{-b}{2a}\right) = f\left(\dfrac{1}{2}\right) = -2\left(\dfrac{1}{2}\right)^2 + 2\left(\dfrac{1}{2}\right) - 3 = \dfrac{-1}{2} + 1 - 3 = \dfrac{-5}{2}.$

Thus, the vertex is $\left(\dfrac{1}{2}, \dfrac{-5}{2}\right).$

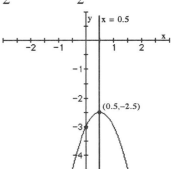

The axis of symmetry is the line $x = \dfrac{1}{2}.$

The discriminant is:

$\qquad b^2 - 4ac = 2^2 - 4(-2)(-3) = 4 - 24 = -20,$

so the graph has no x-intercepts.

The y-intercept is $f(0) = -3.$

41. $f(x) = 3x^2 + 6x + 2$

$a = 3,\ b = 6,\ c = 2$. Since $a = 3 > 0$, the graph opens up.

The x-coordinate of the vertex is $x = \dfrac{-b}{2a} = \dfrac{-6}{2(3)} = \dfrac{-6}{6} = -1$. The y-coordinate of the

vertex is $f\!\left(\dfrac{-b}{2a}\right) = f(-1) = 3(-1)^2 + 6(-1) + 2 = 3 - 6 + 2 = -1$.

Thus, the vertex is $(-1, -1)$.
The axis of symmetry is the line $x = -1$.
The discriminant is:

$\quad b^2 - 4ac = 6^2 - 4(3)(2) = 36 - 24 = 12$,

so the graph has two x-intercepts.
The x-intercepts are found by solving:

$$x = \frac{-b \pm \sqrt{b^2 - 4ac}}{2a} = \frac{-6 \pm \sqrt{12}}{2(3)}$$

$$= \frac{-6 \pm 2\sqrt{3}}{6} = \frac{-3 \pm \sqrt{3}}{3} = \frac{-3 \pm 1.732}{3}$$

The x-intercepts are approximately -0.42 and -1.58.
The y-intercept is $f(0) = 2$.

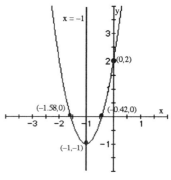

43. $f(x) = -4x^2 - 6x + 2$

$a = -4,\ b = -6,\ c = 2$. Since $a = -4 < 0$, the graph opens down.

The x-coordinate of the vertex is $x = \dfrac{-b}{2a} = \dfrac{-(-6)}{2(-4)} = \dfrac{6}{-8} = \dfrac{-3}{4}$. The y-coordinate of

the vertex is $f\!\left(\dfrac{-b}{2a}\right) = f\!\left(\dfrac{-3}{4}\right) = -4\!\left(\dfrac{-3}{4}\right)^2 - 6\!\left(\dfrac{-3}{4}\right) + 2 = \dfrac{-9}{4} + \dfrac{9}{2} + 2 = \dfrac{17}{4}$.

Thus, the vertex is $\left(\dfrac{-3}{4}, \dfrac{17}{4}\right)$.

The axis of symmetry is the line $x = \dfrac{-3}{4}$.

The discriminant is:

$\quad b^2 - 4ac = (-6)^2 - 4(-4)(2) = 36 + 32 = 68$,

so the graph has two x-intercepts.
The x-intercepts are found by solving:

$$x = \frac{-b \pm \sqrt{b^2 - 4ac}}{2a} = \frac{-(-6) \pm \sqrt{68}}{2(-4)}$$

$$= \frac{6 \pm 2\sqrt{17}}{-8} = \frac{-3 \pm \sqrt{17}}{4} = \frac{-3 \pm 4.123}{4}$$

The x-intercepts are approximately -1.78 and 0.28.
The y-intercept is $f(0) = 2$.

45. $f(x) = 2x^2 + 12x$

$a = 2, b = 12, c = 0$. Since $a = 2 > 0$, the graph opens up, so the vertex is a minimum point. The minimum occurs at $x = \dfrac{-b}{2a} = \dfrac{-12}{2(2)} = \dfrac{-12}{4} = -3$. The minimum value is $f\left(\dfrac{-b}{2a}\right) = f(-3) = 2(-3)^2 + 12(-3) = 18 - 36 = -18$.

47. $f(x) = 2x^2 + 12x - 3$

$a = 2, b = 12, c = -3$. Since $a = 2 > 0$, the graph opens up, so the vertex is a minimum point. The minimum occurs at $x = \dfrac{-b}{2a} = \dfrac{-12}{2(2)} = \dfrac{-12}{4} = -3$. The minimum value is $f\left(\dfrac{-b}{2a}\right) = f(-3) = 2(-3)^2 + 12(-3) - 3 = 18 - 36 - 3 = -21$.

49. $f(x) = -x^2 + 10x - 4$

$a = -1, b = 10, c = -4$. Since $a = -1 < 0$, the graph opens down, so the vertex is a maximum point. The maximum occurs at $x = \dfrac{-b}{2a} = \dfrac{-10}{2(-1)} = \dfrac{-10}{-2} = 5$. The maximum value is $f\left(\dfrac{-b}{2a}\right) = f(5) = -(5)^2 + 10(5) - 4 = -25 + 50 - 4 = 21$.

51. $f(x) = -3x^2 + 12x + 1$

$a = -3, b = 12, c = 1$. Since $a = -3 < 0$, the graph opens down, so the vertex is a maximum point. The maximum occurs at $x = \dfrac{-b}{2a} = \dfrac{-12}{2(-3)} = \dfrac{-12}{-6} = 2$. The maximum value is $f\left(\dfrac{-b}{2a}\right) = f(2) = -3(2)^2 + 12(2) + 1 = -12 + 24 + 1 = 13$.

53. (a) $f(x) = 1(x - (-3))(x - 1) = 1(x + 3)(x - 1) = 1\left(x^2 + 2x - 3\right) = x^2 + 2x - 3$

$f(x) = 2(x - (-3))(x - 1) = 2(x + 3)(x - 1) = 2\left(x^2 + 2x - 3\right) = 2x^2 + 4x - 6$

$f(x) = -2(x - (-3))(x - 1) = -2(x + 3)(x - 1)$

$\qquad\qquad\qquad\qquad = -2\left(x^2 + 2x - 3\right) = -2x^2 - 4x + 6$

$f(x) = 5(x - (-3))(x - 1) = 5(x + 3)(x - 1) = 5\left(x^2 + 2x - 3\right) = 5x^2 + 10x - 15$

(b)   The value of a multiplies the value of the y-intercept by the value of a. The values of the x-intercepts are not changed.

(c)   The axis of symmetry is unaffected by the value of a.

(d)   The y-coordinate of the vertex is multiplied by the value of a.

(e)   The x-coordinate of the vertex is the midpoint of the x-intercepts.

55. $R(p) = -4p^2 + 4000p$

$a = -4, b = 4000, c = 0$. Since $a = -4 < 0$, the graph is a parabola that opens down, so the vertex is a maximum point. The maximum occurs at

$$p = \frac{-b}{2a} = \frac{-4000}{2(-4)} = \frac{-4000}{-8} = 500.$$

$R(500) = -4(500)^2 + 4000(500) = -1000000 + 2000000 = 1,000,000$.

Thus, the unit price should be \$500 for maximum revenue. The maximum revenue is \$1,000,000.

57.   (a) $R(x) = x\left(\frac{-1}{6}x + 100\right) = \frac{-1}{6}x^2 + 100x$

(b) $R(200) = \frac{-1}{6}(200)^2 + 100(200) = \frac{-20000}{3} + 20000 = \frac{40000}{3} \approx \$13,333$

(c) $x = \frac{-b}{2a} = \frac{-100}{2\left(\frac{-1}{6}\right)} = \frac{-100}{\frac{-1}{3}} = \frac{300}{1} = 300$

$R(300) = \frac{-1}{6}(300)^2 + 100(300) = -15000 + 30000 = \$15,000$

(d) $p = \frac{-1}{6}(300) + 100 = -50 + 100 = \$50$

59.   (a)   If $x = -5p + 100$, then $p = \frac{100 - x}{5}$. $R(x) = x\left(\frac{100 - x}{5}\right) = \frac{-1}{5}x^2 + 20x$

(b) $R(15) = \frac{-1}{5}(15)^2 + 20(15) = -45 + 300 = \$255$

(c) $x = \frac{-b}{2a} = \frac{-20}{2\left(\frac{-1}{5}\right)} = \frac{-20}{\frac{-2}{5}} = \frac{100}{2} = 50$

$R(50) = \frac{-1}{5}(50)^2 + 20(50) = -500 + 1000 = \$500$

(d) $p = \frac{100 - 50}{5} = \frac{50}{5} = \$10$

61.   (a)   Let $x =$ width and $y =$ length of the rectangular area.

$P = 2x + 2y = 400$

$y = \frac{400 - 2x}{2} = 200 - x$

Then $A(x) = (200 - x)x = 200x - x^2 = -x^2 + 200x$

(b) $x = \frac{-b}{2a} = \frac{-200}{2(-1)} = \frac{-200}{-2} = 100$ yards

(c) $A(100) = -100^2 + 200(100) = -10000 + 20000 = 10,000$ sq yds.

63. Let x = width and y = length of the rectangular area.

$$2x + y = 4000 \quad \rightarrow \quad y = 4000 - 2x$$

Then $A(x) = (4000 - 2x)x = 4000x - 2x^2 = -2x^2 + 4000x$

$$x = \frac{-b}{2a} = \frac{-4000}{2(-2)} = \frac{-4000}{-4} = 1000$$

$$A(1000) = -2(1000)^2 + 4000(1000) = -2000000 + 4000000 = 2,000,000$$

The largest area that can be enclosed is 2,000,000 square meters.

65. (a) $a = \frac{-32}{2500}$, $b = 1$, $c = 200$. The maximum height occurs when

$$x = \frac{-b}{2a} = \frac{-1}{2\left(\frac{-32}{2500}\right)} = \frac{2500}{64} = 39.0625 \text{ feet from base of the cliff.}$$

(b) The maximum height is

$$h(39.0625) = \frac{-32(39.0625)^2}{2500} + 39.0625 + 200 = 219.53 \text{ feet}.$$

(c) Solving when $h(x) = 0$:

$$\frac{-32}{2500}x^2 + x + 200 = 0$$

$$x = \frac{-1 \pm \sqrt{1^2 - 4\left(\frac{-32}{2500}\right)(200)}}{2\left(\frac{-32}{2500}\right)} = \frac{-1 \pm \sqrt{11.24}}{-0.0256}$$

$$x = -91.90 \text{ or } x = 170.02$$

Since the distance cannot be negative, the projectile strikes the water 170.02 feet from the base of the cliff.

(d) Solving when $h(x) = 100$:

$$\frac{-32}{2500}x^2 + x + 200 = 100$$

$$\frac{-32}{2500}x^2 + x + 100 = 0$$

$$x = \frac{-1 \pm \sqrt{1^2 - 4\left(\frac{-32}{2500}\right)(100)}}{2\left(\frac{-32}{2500}\right)} = \frac{-1 \pm \sqrt{6.12}}{-0.0256}$$

$$x = -57.57 \text{ or } x = 135.70$$

Since the distance cannot be negative, the projectile is 100 feet above the water 135.70 feet from the base of the cliff.

(e) Graphing:

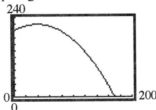

67. Locate the origin at the point where the cable touches the road. Then the equation of the parabola is of the form: $y = ax^2$, where $a > 0$. Since the point (200, 75) is on the parabola, we can find the constant a:

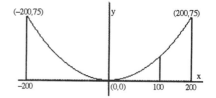

$$75 = a(200)^2 \quad \rightarrow \quad a = \frac{75}{200^2} = 0.001875$$

When $x = 100$, we have:
$$y = 0.001875(100)^2 = 18.75 \text{ meters} .$$

69. Let x = the depth of the gutter and y = the width of the gutter.
Then $A = xy$ is the cross-sectional area of the gutter.
Since the aluminum sheets for the gutter are 12 inches wide, we have
$$2x + y = 12 \text{ or } y = 12 - 2x.$$

The area is to be maximized, so: $A = xy = x(12 - 2x) = -2x^2 + 12x$.
This equation is a parabola opening down; thus, it has a maximum when
$$x = \frac{-b}{2a} = \frac{-12}{2(-2)} = \frac{-12}{-4} = 3.$$

Thus, a depth of 3 inches produces a maximum cross-sectional area.

71. Let x = the width of the rectangle or the diameter of the semicircle.
Let y = the length of the rectangle.

The perimeter of each semicircle is $\dfrac{\pi x}{2}$.

The perimeter of the track is given by: $\dfrac{\pi x}{2} + \dfrac{\pi x}{2} + y + y = 1500$.
Solving for x:
$$\frac{\pi x}{2} + \frac{\pi x}{2} + y + y = 1500$$
$$\pi x + 2y = 1500$$
$$\pi x = 1500 - 2y$$
$$x = \frac{1500 - 2y}{\pi}$$

The area of the rectangle is: $A = xy = \left(\dfrac{1500 - 2y}{\pi}\right)y = \dfrac{-2}{\pi}y^2 + \dfrac{1500}{\pi}y$

This equation is a parabola opening down; thus, it has a maximum when
$$y = \frac{-b}{2a} = \frac{\dfrac{-1500}{\pi}}{2\left(\dfrac{-2}{\pi}\right)} = \frac{-1500}{-4} = 375. \quad \text{Thus, } x = \frac{1500 - 2(375)}{\pi} = \frac{750}{\pi} \approx 238.73.$$

The dimensions for the rectangle with maximum area are $\dfrac{750}{\pi} \approx 238.73$ meters by 375 meters.

Problems 73 – 77. The equations for the curves that are graphed on the screens use many more decimal places in order to get the desired accuracy.

73. (a) Graphing:

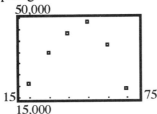

The data appear to be quadratic with $a < 0$.

(b) $I(x) = -42.588x^2 + 3805.527x - 38526.004$

(c) Graphing the quadratic function of best fit:

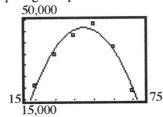

(d) $x = \dfrac{-b}{2a} = \dfrac{-3805.527}{2(-42.588)} = 44.678$

An individual will earn the most income at an age of 44.7 years.

(e) The maximum income will be:

$I(44.7) = -42.588(44.7)^2 + 3805.527(44.7) - 38526.004 = \$46,486$

(f) The data show that the age is a bit greater and the maximum income is also greater than what the quadratic function of best fit indicates.

75. (a) Graphing:

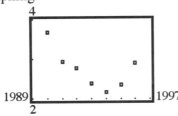

The data appears to be quadratic with $a > 0$.

(b) $p(x) = 0.102x^2 - 406.340x + 405067.721$

(c) Graphing the quadratic function of best fit:

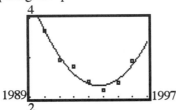

(d)   The predicted price in 1997 (using the equation to many more decimal places)
is: $p(1997) = 3.37$

(e) $x = \dfrac{-b}{2a} = \dfrac{406.340}{2(0.102)} = 1993.8$ (using the equation to many more decimal places)

The price was lowest in 1994.

(f)   This result from the quadratic function of best fit is close to the actual data.

77. (a)   Graphing:

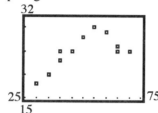

The data appears to be quadratic with $a < 0$.

(b) $M(s) = -0.017s^2 + 1.935s - 25.341$

(c)   Graphing the quadratic function of best fit:

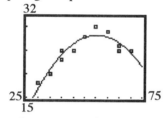

(d) $s = \dfrac{-b}{2a} = \dfrac{-1.935}{2(-0.017)} = 56.9$ (using the equation to many more decimal places)

The speed that maximizes miles per gallon is about 57 miles per hour.

(e)   This result from the quadratic function of best fit is close to the actual data.

(f)   The predicted miles per gallon when the speed is 63 miles per hour is:
$M(63) = 27.2$ miles per gallon . (using the equation to many more decimal places)

79.   We are given: $V(x) = kx(a - x) = -kx^2 + akx$

The reaction rate is a maximum when: $x = \dfrac{-b}{2a} = \dfrac{-ak}{2(-k)} = \dfrac{ak}{2k} = \dfrac{a}{2}$

81.   We have:
$$a(-h)^2 + b(-h) + c = ah^2 - bh + c = y_0$$
$$a(0)^2 + b(0) + c = c = y_1$$
$$a(h)^2 + b(h) + c = ah^2 + bh + c = y_2$$

Equating the two equations for the area, we have:
$$y_0 + 4y_1 + y_2 = ah^2 - bh + c + 4c + ah^2 + bh + c = 2ah^2 + 6c$$

Therefore, Area $= \dfrac{h}{3}\left(2ah^2 + 6c\right) = \dfrac{h}{3}\left(y_0 + 4y_1 + y_2\right)$.

83. $f(x) = 2x^2 + 8, \quad h = 2$

   Area $= \frac{2}{3}\left(2(2)(2)^2 + 6(8)\right) = \frac{2}{3}(16 + 48) = \frac{2}{3}(64) = \frac{128}{3}$

85. $f(x) = -x^2 + x + 4, \quad h = 1$

   Area $= \frac{1}{3}\left(2(-1)(1)^2 + 6(4)\right) = \frac{1}{3}(-2 + 24) = \frac{1}{3}(22) = \frac{22}{3}$

3.2   Power Functions; Curve Fitting

1. $f(x) = (x+1)^4$

   Using the graph of $y = x^4$, shift the graph horizontally, 1 unit to the left.

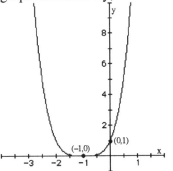

3. $f(x) = x^5 - 3$

   Using the graph of $y = x^5$, shift the graph vertically, 3 units down.

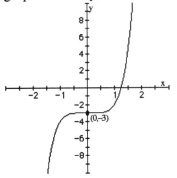

5. $f(x) = \frac{1}{2}x^4$

   Using the graph of $y = x^4$, compress the graph vertically by a factor of $\frac{1}{2}$.

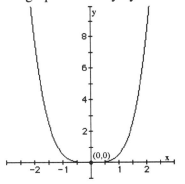

7. $f(x) = -x^5$

   Using the graph of $y = x^5$, reflect the graph about the x-axis.

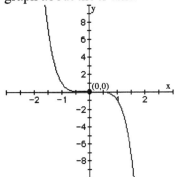

9. $f(x) = (x-1)^5 + 2$

Using the graph of $y = x^5$, shift the graph horizontally, 1 unit to the right, and shift the graph vertically 2 units up.

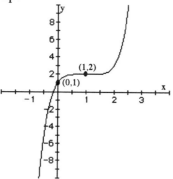

11. $f(x) = 2(x+1)^4 + 1$

Using the graph of $y = x^4$, shift the graph horizontally, 1 unit to the left, stretch vertically by a factor of 2, and shift vertically 1 unit up.

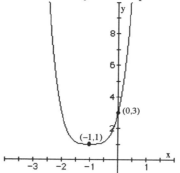

13. $f(x) = 4 - (x-2)^5 = -(x-2)^5 + 4$

Using the graph of $y = x^5$, shift the graph horizontally, 2 units to the right, reflect about the x-axis, and shift vertically 4 units up.

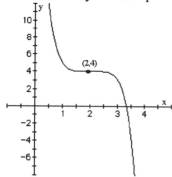

15. $f(x) = \frac{-1}{2}(x-2)^4 - 1$

Using the graph of $y = x^4$, shift the graph horizontally, 2 units to the right, compress vertically by a factor of $\frac{1}{2}$, reflect about the x-axis, and shift vertically 1 unit down.

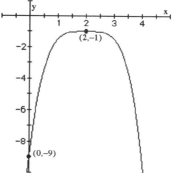

17. (a) Graphing:

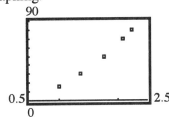

(b) $s(t) = 15.973\,t^{2.002}$

(c) Graphing the power function of best fit:

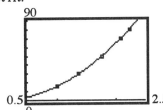

(d) Graphing $s = 100$ and using INTERSECT:

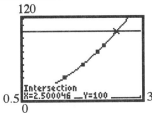

The object will fall 100 feet in approximately 2.5 seconds.

(e) Rewriting the equation:
$$s(t) = \tfrac{1}{2}(31.946)\,t^2$$
David's estimate of g is 31.946 ft/sec^2.

19. (a) Graphing:

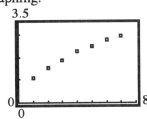

(b) $T(l) = 1.095\,l^{0.510}$

(c) Graphing the power function of best fit:

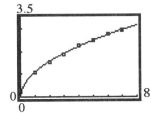

(d) $T(2.3) = 1.095(2.3)^{0.510}$
≈ 1.67 seconds
The period is approximately 1.67 seconds when the length is 2.3 feet.

3.3 Polynomial Functions; Curve Fitting

1. $f(x) = 4x + x^3$ is a polynomial function of degree 3.

3. $g(x) = \dfrac{1 - x^2}{2} = \dfrac{1}{2} - \dfrac{1}{2}x^2$ is a polynomial function of degree 2.

5. $f(x) = 1 - \dfrac{1}{x} = 1 - x^{-1}$ is not a polynomial function because it contains a negative exponent.

7. $g(x) = x^{\frac{3}{2}} - x^2 + 2$ is not a polynomial function because it contains a fractional exponent.

9. $F(x) = 5x^4 - \pi x^3 + \frac{1}{2}$ is a polynomial function of degree 4.

11. $f(x) = a(x - (-1))(x - 1)(x - 3)$
     For $a = 1$: $f(x) = (x + 1)(x - 1)(x - 3)$
     $$f(x) = \left(x^2 - 1\right)(x - 3)$$
     $$f(x) = x^3 - 3x^2 - x + 3$$

13. $f(x) = a(x - (-3))(x - 0)(x - 4)$
     For $a = 1$: $f(x) = (x + 3)(x)(x - 4)$
     $$f(x) = \left(x^2 + 3x\right)(x - 4)$$
     $$f(x) = x^3 - 4x^2 + 3x^2 - 12x$$
     $$f(x) = x^3 - x^2 - 12x$$

15. $f(x) = a(x - (-4))(x - (-1))(x - 2)(x - 3)$
     For $a = 1$: $f(x) = (x + 4)(x + 1)(x - 2)(x - 3)$
     $$f(x) = \left(x^2 + 5x + 4\right)\left(x^2 - 5x + 6\right)$$
     $$f(x) = x^4 - 5x^3 + 6x^2 + 5x^3 - 25x^2 + 30x + 4x^2 - 20x + 24$$
     $$f(x) = x^4 - 15x^2 + 10x + 24$$

17. The real zeros of $f(x) = 3(x - 7)(x + 3)^2$ are: 7, with multiplicity one; and –3, with multiplicity two. The graph crosses the x-axis at 7 and touches it at –3.

19. The real zeros of $f(x) = 4\left(x^2 + 1\right)(x - 2)^3$ are: 2, with multiplicity three. $x^2 + 1 = 0$ has no real solution. The graph crosses the x-axis at 2.

21. The real zeros of $f(x) = -2\left(x + \frac{1}{2}\right)^2\left(x^2 + 4\right)^2$ are: $\frac{-1}{2}$, with multiplicity two.

     $x^2 + 4 = 0$ has no real solution. The graph touches the x-axis at $\frac{-1}{2}$.

23. The real zeros of $f(x) = (x - 5)^3(x + 4)^2$ are: 5, with multiplicity three; and –4, with multiplicity two. The graph crosses the x-axis at 5 and touches it at –4.

25. $f(x) = 3\left(x^2 + 8\right)\left(x^2 + 9\right)^2$ has no real zeros. $x^2 + 8 = 0$ and $x^2 + 9 = 0$ have no real solutions. The graph neither touches nor crosses the x-axis.

27. (a) $f(x) = (x-1)^2$

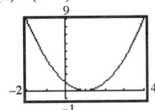

(b) x-intercept: 1; y-intercept: 1
(c) Even
(d) $y = x^2$
(e) 1
(f) Local minimum: $(1, 0)$

29. (a) $f(x) = x^2(x-3)$

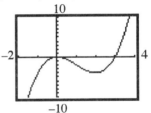

(b) x-intercepts: 0, 3; y-intercept: 0
(c) Even: 0; Odd: 3
(d) $y = x^3$
(e) 2
(f) Local maximum: $(0, 0)$
Local minimum: $(2, -4)$

31. (a) $f(x) = 6x^3(x+4)$

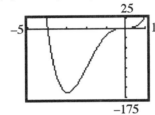

(b) x-intercepts: -4, 0; y-intercept: 0
(c) Odd: -4, 0
(d) $y = 6x^4$
(e) 1
(f) Local minimum: $(-3, -162)$

33. (a) $f(x) = -4x^2(x+2)$

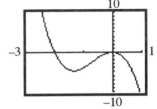

(b) x-intercepts: -2, 0; y-intercept: 0
(c) Even: 0; Odd: -2
(d) $y = -4x^3$
(e) 2
(f) Local maximum: $(0, 0)$
Local minimum: $(-1.33, -4.74)$

35. (a) $f(x) = x(x-2)(x+4)$

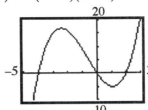

(b) x-intercepts: -4, 0, 2;
y-intercept: 0
(c) Odd: -4, 0, 2
(d) $y = x^3$
(e) 2
(f) Local maximum: $(-2.43, 16.90)$
Local minimum: $(1.10, -5.05)$

37. (a) $f(x) = 4x - x^3$

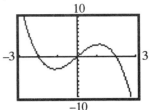

(b) x-intercepts: -2, 0, 2;
y-intercept: 0
(c) Odd: -2, 0, 2
(d) $y = -x^3$
(e) 2
(f) Local maximum: $(1.15, 3.08)$
Local minimum: $(-1.15, -3.08)$

39. (a) $f(x) = x^2(x-2)(x+2)$

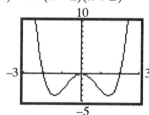

   (b) x-intercepts: $-2, 0, 2$;
       y-intercept: 0
   (c) Even: 0; Odd: $-2, 2$
   (d) $y = x^4$
   (e) 3
   (f) Local maximum: $(0, 0)$
       Local minima: $(-1.41, -4)$,
       $\qquad\qquad\quad (1.41, -4)$

41. (a) $f(x) = x^2(x-2)^2$

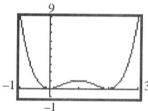

   (b) x-intercepts: $0, 2$; y-intercept: 0
   (c) Even: $0, 2$
   (d) $y = x^4$
   (e) 3
   (f) Local maximum: $(1, 1)$
       Local minima: $(0, 0), (2, 0)$

43. (a) $f(x) = x^2(x-3)(x+1)$

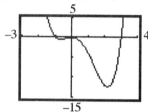

   (b) x-intercepts: $-1, 0, 3$;
       y-intercept: 0
   (c) Even: 0; Odd: $-1, 3$
   (d) $y = x^4$
   (e) 3
   (f) Local maximum: $(0, 0)$
       Local minima: $(-0.69, -0.54)$,
       $\qquad\qquad\quad (2.19, -12.39)$

45. (a) $f(x) = x(x+2)(x-4)(x-6)$

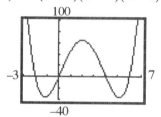

   (b) x-intercepts: $-2, 0, 4, 6$;
       y-intercept: 0
   (c) Odd: $-2, 0, 4, 6$
   (d) $y = x^4$
   (e) 3
   (f) Local maximum: $(2, 64)$
       Local minima: $(-1.16, -36)$,
       $\qquad\qquad\quad (5.16, -36)$

47. (a) $f(x) = x^2(x-2)(x^2+3)$

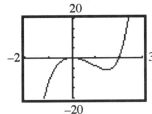

(b) x-intercepts: 0, 2; y-intercept: 0
(c) Even: 0; Odd: 2
(d) $y = x^5$
(e) 2
(f) Local maximum: (0, 0)
     Local minimum: (1.48, –5.91)

49. (a)
     $f(x) = x^3 + 0.2x^2 - 1.5876x - 0.31752$

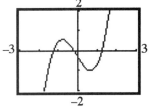

(b) x-intercepts: –1.26, –0.2, 1.26;
     y-intercept: –0.31752
(c) Odd: –1.26, –0.2, 1.26
(d) $y = x^3$
(e) 2
(f) Local maximum: (–0.80, 0.57)
     Local minimum: (0.66, –0.99)

51. (a)
     $f(x) = x^3 + 2.56x^2 - 3.31x + 0.89$

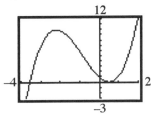

(b) x-intercepts: –3.56, 0.50;
     y-intercept: 0.89
(c) Even: 0.50; Odd: –3.56
(d) $y = x^3$
(e) 2
(f) Local maximum: (–2.21, 9.91)
     Local minimum: (0.50, 0)

53. (a) $f(x) = x^4 - 2.5x^2 + 0.5625$

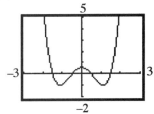

(b) x-intercepts: –1.50, –0.50,
     0.50, 1.50;
     y-intercept: 0.5625
(c) Odd: –1.50, –0.50, 0.50, 1.50
(d) $y = x^4$
(e) 3
(f) Local maximum: (0, 0.5625)
     Local minima: (–1.12, –1),
     (1.12, –1)

55. (a) $f(x) = x^4 + 0.65x^3 - 16.6319x^2$
    $+ 14.209335x - 3.1264785$

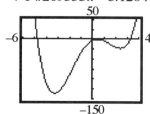

(b) x-intercepts: −4.78, 0.45, 3.23;
y-intercept: −3.1264785
(c) Even: 0.45; Odd: −4.78, 3.23
(d) $y = x^4$
(e) 3
(f) Local maximum: (0.45, 0)
Local minima: (−3.32, −135.92)
(2.38, −22.67)

57. (a) $f(x) = \pi x^3 + \sqrt{2}x^2 - x - 2$

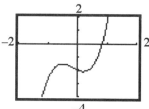

(b) x-intercept: 0.84; y-intercept: −2
(c) Odd: 0.84
(d) $y = \pi x^3$
(e) 2
(f) Local maximum: (−0.51, −1.54)
Local minimum: (0.21, −2.12)

59. (a) $f(x) = 2x^4 - \pi x^3 + \sqrt{5}x - 4$

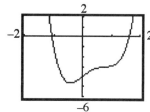

(b) x-intercepts: −1.07, 1.62;
y-intercept: −4
(c) Odd: −1.07, 1.62
(d) $y = 2x^4$
(e) 1
(f) Local minimum: (−0.42, −4.64)

61. (a) $f(x) = -2x^5 - \sqrt{2}x^2 - x - \sqrt{2}$

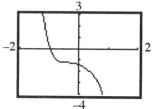

(b) x-intercept: −0.98;
y-intercept: −1.41
(c) Odd: −0.98
(d) $y = -2x^5$
(e) None
(f) None

63. Answers will vary.
    (a) $f(x) = (x - 0)(x - 1)(x - 2) = x^3 - 3x^2 + 2x$
    (b) $f(x) = (x - 0)(x - 1)^2(x - 2) = x^4 - 4x^3 + 5x^2 - 2x$
    (c) $f(x) = \frac{-1}{2}(x + 1)(x - 1)(x - 2) = \frac{-1}{2}(x^2 - 1)(x - 2) = \frac{-1}{2}x^3 + x^2 + \frac{1}{2}x - 1$
    (d) $f(x) = -\frac{1}{2}(x + 1)(x - 1)^2(x - 2) = -\frac{1}{2}x^4 + \frac{3}{2}x^3 - \frac{1}{2}x^2 - \frac{3}{2}x + 1$

65. (a) Graphing:

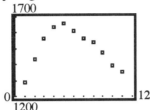

The graph may be a cubic relation.

(b) $M(x) = 1.524x^3 - 39.809x^2 + 282.288x + 1035.5$

(c) Graphing the cubic function of best fit:

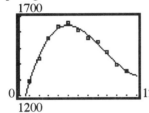

(d) $M(12) = 1.524(12)^3 - 39.809(12)^2 + 282.288(12) + 1035.5 = 1324$

According to the function there would be approximately 1,324,000 motor vehicle thefts in 1998.

67. (a) Graphing:

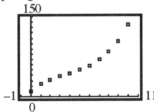

The graph may be a cubic relation.

(b) Average rate of change $= \dfrac{50 - 43}{5 - 4} = \dfrac{7}{1} = 7$

(c) Average rate of change $= \dfrac{105 - 85}{9 - 8} = \dfrac{20}{1} = 20$

(d) $C(x) = 0.216x^3 - 2.347x^2 + 14.328x + 10.224$

(e) Graphing the cubic function of best fit:

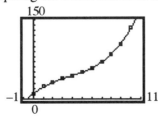

(f) $C(11) = 0.216(11)^3 - 2.347(11)^2 + 14.328(11) + 10.224 \approx 171$

The cost of manufacturing 11 Cavaliers in 1 hour would be approximately $171,000.

(g) The y-intercept would indicate the fixed costs before any cars are made.

3.4 Rational Functions I

1. In $R(x) = \dfrac{4x}{x-3}$, the denominator, $q(x) = x-3$, has a zero at 3. Thus, the domain of $R(x)$ is all real numbers except 3.

3. In $H(x) = \dfrac{-4x^2}{(x-2)(x+4)}$, the denominator, $q(x) = (x-2)(x+4)$, has zeros at 2 and -4. Thus, the domain of $H(x)$ is all real numbers except 2 and -4.

5. In $F(x) = \dfrac{3x(x-1)}{2x^2-5x-3}$, the denominator, $q(x) = 2x^2-5x-3 = (2x+1)(x-3)$, has zeros at $\frac{-1}{2}$ and 3. Thus, the domain of $F(x)$ is all real numbers except $\frac{-1}{2}$ and 3.

7. In $R(x) = \dfrac{x}{x^3-8}$, the denominator, $q(x) = x^3-8 = (x-2)(x^2+2x+4)$, has a zero at 2. ($x^2+2x+4$ has no real zeros.) Thus, the domain of $R(x)$ is all real numbers except 2.

9. In $H(x) = \dfrac{3x^2+x}{x^2+4}$, the denominator, $q(x) = x^2+4$, has no real zeros. Thus, the domain of $H(x)$ is all real numbers.

11. In $R(x) = \dfrac{3(x^2-x-6)}{4(x^2-9)}$, the denominator, $q(x) = 4(x^2-9) = 4(x-3)(x+3)$, has zeros at 3 and -3. Thus, the domain of $R(x)$ is all real numbers except 3 and -3.

13. (a) Domain: $\left\{x \mid x \neq 2\right\}$; Range: $\left\{y \mid y \neq 1\right\}$
    (b) Intercept: $(0,0)$ (c) Horizontal Asymptote: $y = 1$
    (d) Vertical Asymptote: $x = 2$ (e) Oblique Asymptote: none

15. (a) Domain: $\left\{x \mid x \neq 0\right\}$; Range: all real numbers
    (b) Intercepts: $(-1,0),(1,0)$ (c) Horizontal Asymptote: none
    (d) Vertical Asymptote: $x = 0$ (e) Oblique Asymptote: $y = 2x$

17. (a) Domain: $\left\{x \mid x \neq -2, x \neq 2\right\}$; Range: $\left\{y \mid y \leq 0 \text{ or } y > 1\right\}$
    (b) Intercept: $(0,0)$ (c) Horizontal Asymptote: $y = 1$
    (d) Vertical Asymptotes: $x = -2, x = 2$ (e) Oblique Asymptote: none

19. (a) Domain: $\{x \mid x \neq -1\}$; Range: $\{y \mid y \neq 2\}$
    (b) Intercepts: $(-1.5, 0), (0, 3)$ (c) Horizontal Asymptote: $y = 2$
    (d) Vertical Asymptote: $x = -1$ (e) Oblique Asymptote: none

21. (a) Domain: $\{x \mid x \neq -4, x \neq 3\}$; Range: all real numbers
    (b) Intercept: $(0, 0)$ (c) Horizontal Asymptote: $y = 0$
    (d) Vertical Asymptotes: $x = -4, x = 3$ (e) Oblique Asymptote: none

23. $R(x) = \dfrac{1}{(x-1)^2}$

Using the function $y = \dfrac{1}{x^2}$, shift the graph horizontally 1 unit to the right.

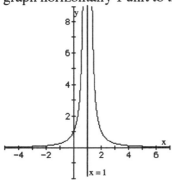

25. $H(x) = \dfrac{-2}{x+1}$

Using the function $y = \dfrac{1}{x}$, shift the graph horizontally 1 unit to the left, reflect about the x-axis, and stretch vertically by a factor of 2.

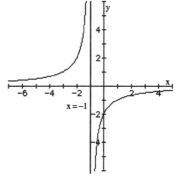

27. $R(x) = \dfrac{1}{x^2 + 4x + 4} = \dfrac{1}{(x+2)^2}$

Using the function $y = \dfrac{1}{x^2}$, shift the graph horizontally 2 units to the left.

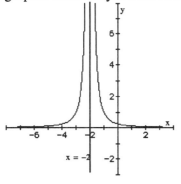

29. $F(x) = 1 - \dfrac{1}{x} = \dfrac{-1}{x} + 1$

Using the function $y = \dfrac{1}{x}$, reflect about the x-axis, and shift vertically 1 unit up.

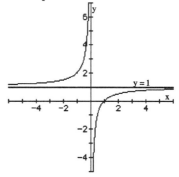

31. $R(x) = \dfrac{x^2 - 4}{x^2} = \dfrac{-4}{x^2} + 1$

Using the function $y = \dfrac{1}{x^2}$, reflect the graph about the x-axis, stretch vertically by a factor of 4, and shift vertically 1 unit up.

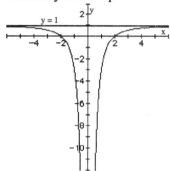

33. $F(x) = 1 + \dfrac{2}{(x-3)^2} = \dfrac{2}{(x-3)^2} + 1$

Using the function $y = \dfrac{1}{x^2}$, shift the graph 3 units right, stretch vertically by a factor of 2, and shift vertically 1 unit up.

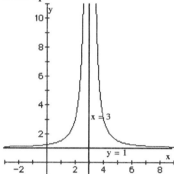

35. $R(x) = \dfrac{3x}{x+4}$

The degree of the numerator, $p(x) = 3x$, is $n = 1$. The degree of the denominator, $q(x) = x + 4$, is $m = 1$. Since $n = m$, the line $y = \dfrac{3}{1} = 3$ is a horizontal asymptote. The denominator is zero at $x = -4$, so $x = -4$ is a vertical asymptote.

37. $H(x) = \dfrac{x^4 + 2x^2 + 1}{x^2 - x + 1}$

The degree of the numerator, $p(x) = x^4 + 2x^2 + 1$, is $n = 4$. The degree of the denominator, $q(x) = x^2 - x + 1$, is $m = 2$. Since $n > m + 1$, there is no horizontal asymptote or oblique asymptote. The denominator has no real zeros, so there is no vertical asymptote.

39. $T(x) = \dfrac{x^3}{x^4 - 1}$

The degree of the numerator, $p(x) = x^3$, is $n = 3$. The degree of the denominator, $q(x) = x^4 - 1$ is $m = 4$. Since $n < m$, the line $y = 0$ is a horizontal asymptote. The denominator is zero at $x = -1$ and $x = 1$, so $x = -1$ and $x = 1$ are vertical asymptotes.

41. $Q(x) = \dfrac{5 - x^2}{3x^4}$

The degree of the numerator, $p(x) = 5 - x^2$, is $n = 2$. The degree of the denominator, $q(x) = 3x^4$ is $m = 4$. Since $n < m$, the line $y = 0$ is a horizontal asymptote. The denominator is zero at $x = 0$, so $x = 0$ is a vertical asymptote.

43. $R(x) = \dfrac{3x^4 + 4}{x^3 + 3x}$

The degree of the numerator, $p(x) = 3x^4 + 4$, is $n = 4$. The degree of the denominator, $q(x) = x^3 + 3x$ is $m = 3$. Since $n = m + 1$, there is an oblique asymptote. Dividing:

$$
\begin{array}{r}
3x \\
x^3 + 3x \overline{)3x^4 + 0x^3 + 0x^2 + 0x + 4} \\
\underline{3x^4 \qquad + 9x^2} \\
-9x^2 + 0x + 4
\end{array}
\qquad
R(x) = 3x + \dfrac{-9x^2 + 4}{x^3 + 3x}
$$

Thus, the oblique asymptote is $y = 3x$.
The denominator is zero at $x = 0$, so $x = 0$ is a vertical asymptote.

45. $G(x) = \dfrac{x^3 - 1}{x - x^2}$, $x \neq 1$

The degree of the numerator, $p(x) = x^3 - 1$, is $n = 3$. The degree of the denominator, $q(x) = x - x^2$ is $m = 2$. Since $n = m + 1$, there is an oblique asymptote. Dividing:

$$
\begin{array}{r}
-x - 1 \\
-x^2 + x \overline{)x^3 + 0x^2 + 0x - 1} \\
\underline{x^3 - x^2} \\
x^2 + 0x \\
\underline{x^2 - x} \\
x - 1
\end{array}
\qquad
G(x) = -x - 1 + \dfrac{x - 1}{x - x^2} = -x - 1 - \dfrac{1}{x}, \; x \neq 1
$$

Thus, the oblique asymptote is $y = -x - 1$.
$G(x)$ must be in lowest terms to find the vertical asymptote:

$$
G(x) = \frac{x^3 - 1}{x - x^2} = \frac{(x - 1)(x^2 + x + 1)}{-x(x - 1)} = \frac{x^2 + x + 1}{-x}
$$

The denominator is zero at $x = 0$, so $x = 0$ is a vertical asymptotes.

3.5 Rational Functions II

In problems 1-43, we will use the terminology: $R(x) = \dfrac{p(x)}{q(x)}$, where the degree of $p(x) = n$ and the degree of $q(x) = m$. The graphs in Step 6 are in dot mode.

1. $R(x) = \dfrac{x + 1}{x(x + 4)}$ $p(x) = x + 1$; $q(x) = x(x + 4) = x^2 + 4x$; $n = 1$; $m = 2$

Step 1: Domain: $\left\{ x \mid x \neq -4, \; x \neq 0 \right\}$

Step 2:   (a)   The x-intercept is the zero of $p(x)$: -1

(b)   There is no y-intercept; $R(0)$ is not defined, since $q(0) = 0$.

Step 3: $R(-x) = \dfrac{-x+1}{-x(-x+4)} = \dfrac{-x+1}{x^2-4x}$; this is neither $R(x)$ nor $-R(x)$, so there is no symmetry.

Step 4:   The vertical asymptotes are the zeros of $q(x)$: $x = -4$ and $x = 0$

Step 5:   Since $n < m$, the line $y = 0$ is the horizontal asymptote.

$R(x)$ intersects $y = 0$ at $(-1, 0)$.

Step 6:   Graphing:

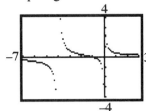

Step 7:   Graphing by hand:

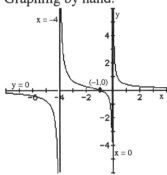

3. $R(x) = \dfrac{3x+3}{2x+4}$ $p(x) = 3x+3$; $q(x) = 2x+4$; $n = 1$; $m = 1$

Step 1:   Domain: $\{x \mid x \neq -2\}$

Step 2:   (a)   The x-intercept is the zero of $p(x)$: -1

(b)   The y-intercept is $R(0) = \dfrac{3(0)+3}{2(0)+4} = \dfrac{3}{4}$.

Step 3: $R(-x) = \dfrac{3(-x)+3}{2(-x)+4} = \dfrac{-3x+3}{-2x+4} = \dfrac{3x-3}{2x-4}$; this is neither $R(x)$ nor $-R(x)$, so there is no symmetry.

Step 4:   The vertical asymptote is the zero of $q(x)$: $x = -2$

Step 5:   Since $n = m$, the line $y = \dfrac{3}{2}$ is the horizontal asymptote.

$R(x)$ does not intersect $y = \dfrac{3}{2}$.

Step 6:   Graphing:

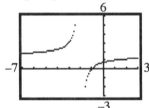

Step 7:   Graphing by hand:

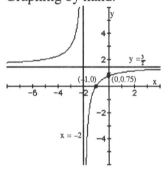

5. $R(x) = \dfrac{3}{x^2 - 4}$ $p(x) = 3$; $q(x) = x^2 - 4$; $n = 0$; $m = 2$

Step 1:   Domain: $\left\{x \mid x \neq -2, \, x \neq 2\right\}$

Step 2:   (a)   There is no x-intercept.

(b)   The y-intercept is $R(0) = \dfrac{3}{0^2 - 4} = \dfrac{3}{-4} = \dfrac{-3}{4}$.

Step 3: $R(-x) = \dfrac{3}{(-x)^2 - 4} = \dfrac{3}{x^2 - 4} = R(x)$; $R(x)$ is symmetric to the y-axis.

Step 4:   The vertical asymptotes are the zeros of $q(x)$: $x = -2$ and $x = 2$

Step 5:   Since $n < m$, the line $y = 0$ is the horizontal asymptote.

$R(x)$ does not intersect $y = 0$.

Step 6:   Graphing: Step 7:   Graphing by hand:

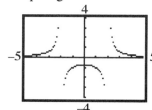   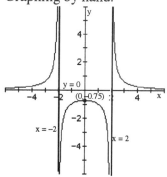

7. $P(x) = \dfrac{x^4 + x^2 + 1}{x^2 - 1}$ $p(x) = x^4 + x^2 + 1$; $q(x) = x^2 - 1$; $n = 4$; $m = 2$

Step 1:   Domain: $\left\{x \mid x \neq -1, \, x \neq 1\right\}$

Step 2:   (a)   There is no x-intercept.

(b)   The y-intercept is $P(0) = \dfrac{0^4 + 0^2 + 1}{0^2 - 1} = \dfrac{1}{-1} = -1$.

Step 3: $P(-x) = \dfrac{(-x)^4 + (-x)^2 + 1}{(-x)^2 - 1} = \dfrac{x^4 + x^2 + 1}{x^2 - 1} = P(x)$; $P(x)$ is symmetric to the y-axis.

Step 4:   The vertical asymptotes are the zeros of $q(x)$: $x = -1$ and $x = 1$

Step 5:   Since $n > m + 1$, there is no horizontal asymptote and no oblique asymptote.

Step 6:   Graphing: Step 7:   Graphing by hand:

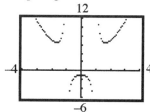

9. $H(x) = \dfrac{x^3 - 1}{x^2 - 9}$ $p(x) = x^3 - 1$; $q(x) = x^2 - 9$; $n = 3$; $m = 2$

Step 1:   Domain: $\left\{x \mid x \neq -3, x \neq 3\right\}$

Step 2:   (a)   The x-intercept is the zero of $p(x)$: 1.

          (b)   The y-intercept is $H(0) = \dfrac{0^3 - 1}{0^2 - 9} = \dfrac{-1}{-9} = \dfrac{1}{9}$.

Step 3: $H(-x) = \dfrac{(-x)^3 - 1}{(-x)^2 - 9} = \dfrac{-x^3 - 1}{x^2 - 9}$; this is neither $H(x)$ nor $-H(x)$, so there is no symmetry.

Step 4:   The vertical asymptotes are the zeros of $q(x)$: $x = -3$ and $x = 3$

Step 5:   Since $n = m + 1$, there is an oblique asymptote. Dividing:

$$x^2 - 9 \,\overline{\smash{\big)}\, x^3 + 0x^2 + 0x - 1} \qquad \begin{array}{c} x \\ \end{array}$$

$$\underline{x^3 - 9x}$$

$$9x - 1$$

$$H(x) = x + \dfrac{9x - 1}{x^2 - 9}$$

The oblique asymptote is $y = x$.

Solve to find intersection points:

$$\dfrac{x^3 - 1}{x^2 - 9} = x$$

$$x^3 - 1 = x^3 - 9x$$

$$-1 = -9x$$

$$x = \dfrac{1}{9}$$

The oblique asymptote intersects $H(x)$ at $\left(\dfrac{1}{9}, \dfrac{1}{9}\right)$.

Step 6:   Graphing:

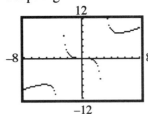

Step 7:   Graphing by hand:

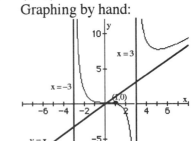

11. $R(x) = \dfrac{x^2}{x^2 + x - 6} = \dfrac{x^2}{(x+3)(x-2)}$ $p(x) = x^2$; $q(x) = x^2 + x - 6$; $n = 2$; $m = 2$

Step 1: Domain: $\{x \mid x \neq -3, x \neq 2\}$

Step 2: (a) The x-intercept is the zero of $p(x)$: 0

   (b) The y-intercept is $R(0) = \dfrac{0^2}{0^2 + 0 - 6} = \dfrac{0}{-6} = 0$.

Step 3: $R(-x) = \dfrac{(-x)^2}{(-x)^2 + (-x) - 6} = \dfrac{x^2}{x^2 - x - 6}$; this is neither $R(x)$ nor $-R(x)$, so there is no symmetry.

Step 4: The vertical asymptotes are the zeros of $q(x)$: $x = -3$ and $x = 2$

Step 5: Since $n = m$, the line $y = 1$ is the horizontal asymptote.

   $R(x)$ intersects $y = 1$ at (6, 1), since:

$$\frac{x^2}{x^2 + x - 6} = 1$$
$$x^2 = x^2 + x - 6$$
$$0 = x - 6$$
$$x = 6$$

Step 6: Graphing:

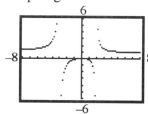

Step 7: Graphing by hand:

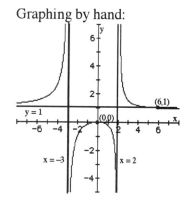

13. $G(x) = \dfrac{x}{x^2 - 4} = \dfrac{x}{(x+2)(x-2)}$ $p(x) = x$; $q(x) = x^2 - 4$; $n = 1$; $m = 2$

Step 1: Domain: $\{x \mid x \neq -2, x \neq 2\}$

Step 2: (a) The x-intercept is the zero of $p(x)$: 0

   (b) The y-intercept is $G(0) = \dfrac{0}{0^2 - 4} = \dfrac{0}{-4} = 0$.

Step 3: $G(-x) = \dfrac{-x}{(-x)^2 - 4} = \dfrac{-x}{x^2 - 4} = -G(x)$; $G(x)$ is symmetric to the origin.

Step 4: The vertical asymptotes are the zeros of $q(x)$: $x = -2$ and $x = 2$

Step 5: Since $n < m$, the line $y = 0$ is the horizontal asymptote.

   $G(x)$ intersects $y = 0$ at (0, 0).

Step 6: Graphing:

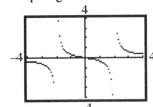

Step 7: Graphing by hand:

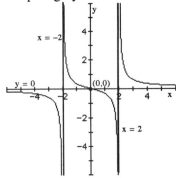

15. $R(x) = \dfrac{3}{(x-1)(x^2-4)} = \dfrac{3}{(x-1)(x+2)(x-2)}$ $p(x) = 3;\ q(x) = (x-1)(x^2-4);$

$n = 0;\ m = 3$

Step 1: Domain: $\left\{x \mid x \neq -2,\ x \neq 1,\ x \neq 2\right\}$

Step 2: (a) There is no x-intercept.

(b) The y-intercept is $R(0) = \dfrac{3}{(0-1)(0^2-4)} = \dfrac{3}{4}$.

Step 3: $R(-x) = \dfrac{3}{(-x-1)\big((-x)^2-4\big)} = \dfrac{3}{(-x-1)(x^2-4)}$; this is neither $R(x)$ nor

$-R(x)$, so there is no symmetry.

Step 4: The vertical asymptotes are the zeros of $q(x)$: $x = -2$, $x = 1$, and $x = 2$

Step 5: Since $n < m$, the line $y = 0$ is the horizontal asymptote.

$R(x)$ does not intersect $y = 0$.

Step 6: Graphing:

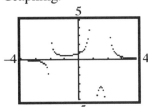

Step 7: Graphing by hand:

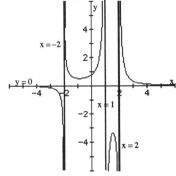

17. $H(x) = \dfrac{4(x^2 - 1)}{x^4 - 16} = \dfrac{4(x-1)(x+1)}{(x^2+4)(x+2)(x-2)}$ $\quad p(x) = 4(x^2 - 1); \quad q(x) = x^4 - 16;$

$n = 2; \quad m = 4$

Step 1: Domain: $\{x \mid x \neq -2, \, x \neq 2\}$

Step 2: (a) The x-intercepts are the zeros of $p(x)$: -1 and 1

(b) The y-intercept is $H(0) = \dfrac{4(0^2 - 1)}{0^4 - 16} = \dfrac{-4}{-16} = \dfrac{1}{4}$.

Step 3: $H(-x) = \dfrac{4((-x)^2 - 1)}{(-x)^4 - 16} = \dfrac{4(x^2 - 1)}{x^4 - 16} = H(x)$; $H(x)$ is symmetric to the y-axis.

Step 4: The vertical asymptotes are the zeros of $q(x)$: $x = -2$, and $x = 2$

Step 5: Since $n < m$, the line $y = 0$ is the horizontal asymptote.

$H(x)$ intersects $y = 0$ at $(-1, 0)$ and $(1, 0)$.

Step 6: Graphing: Step 7: Graphing by hand:

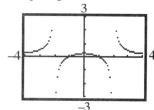 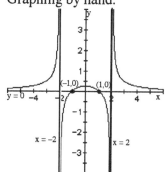

19. $F(x) = \dfrac{x^2 - 3x - 4}{x + 2} = \dfrac{(x+1)(x-4)}{x+2}$ $\quad p(x) = x^2 - 3x - 4; \quad q(x) = x + 2; \quad n = 2; \quad m = 1$

Step 1: Domain: $\{x \mid x \neq -2\}$

Step 2: (a) The x-intercepts are the zeros of $p(x)$: -1 and 4.

(b) The y-intercept is $F(0) = \dfrac{0^2 - 3(0) - 4}{0 + 2} = \dfrac{-4}{2} = -2$.

Step 3: $F(-x) = \dfrac{(-x)^2 - 3(-x) - 4}{-x + 2} = \dfrac{x^2 + 3x - 4}{-x + 2}$; this is neither $F(x)$ nor $-F(x)$,

so there is no symmetry.

Step 4: The vertical asymptote is the zero of $q(x)$: $x = -2$

Step 5: Since $n = m + 1$, there is an oblique asymptote. Dividing:

$$\begin{array}{r} x - 5 \\ x + 2 \overline{) x^2 - 3x - 4} \\ \underline{x^2 + 2x} \\ -5x - 4 \\ \underline{-5x - 10} \\ 6 \end{array} \qquad F(x) = x - 5 + \dfrac{6}{x+2}$$

The oblique asymptote is $y = x - 5$.

Solve to find intersection points:

$$\frac{x^2 - 3x - 4}{x + 2} = x - 5$$

$$x^2 - 3x - 4 = x^2 - 3x - 10$$

$$-4 = -10$$

Since there is no solution, the oblique asymptote does not intersect $F(x)$.

Step 6:   Graphing: Step 7:   Graphing by hand:

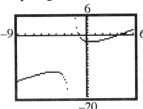
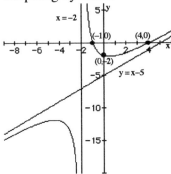

21. $R(x) = \dfrac{x^2 + x - 12}{x - 4} = \dfrac{(x + 4)(x - 3)}{x - 4}$ $p(x) = x^2 + x - 12$; $q(x) = x - 4$; $n = 2$; $m = 1$

Step 1:   Domain: $\left\{ x \mid x \ne 4 \right\}$

Step 2:   (a)   The x-intercepts are the zeros of $p(x)$: −4 and 3.

(b)   The y-intercept is $R(0) = \dfrac{0^2 + 0 - 12}{0 - 4} = \dfrac{-12}{-4} = 3$.

Step 3: $R(-x) = \dfrac{(-x)^2 + (-x) - 12}{-x - 4} = \dfrac{x^2 - x - 12}{-x - 4}$; this is neither $R(x)$ nor $-R(x)$,

so there is no symmetry.

Step 4:   The vertical asymptote is the zero of $q(x)$: $x = 4$

Step 5:   Since $n = m + 1$, there is an oblique asymptote. Dividing:

$$\begin{array}{r} x + 5 \\ x - 4 \overline{)\, x^2 + x - 12} \\ \underline{x^2 - 4x } \\ 5x - 12 \\ \underline{5x - 20} \\ 8 \end{array}$$ $R(x) = x + 5 + \dfrac{8}{x - 4}$

The oblique asymptote is $y = x + 5$.

Solve to find intersection points:

$$\frac{x^2 + x - 12}{x - 4} = x + 5$$

$$x^2 + x - 12 = x^2 + x - 20$$

$$-12 = -20$$

Since there is no solution, the oblique asymptote does not intersect $R(x)$.

Step 6: Graphing:

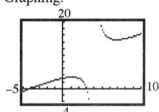

Step 7: Graphing by hand:

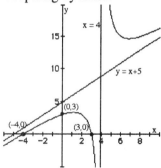

23. $F(x) = \dfrac{x^2 + x - 12}{x + 2} = \dfrac{(x+4)(x-3)}{x+2}$ $p(x) = x^2 + x - 12;\; q(x) = x + 2;\; n = 2;\; m = 1$

Step 1: Domain: $\{x \mid x \neq -2\}$

Step 2: (a) The x-intercepts are the zeros of $p(x)$: -4 and 3.

(b) The y-intercept is $F(0) = \dfrac{0^2 + 0 - 12}{0 + 2} = \dfrac{-12}{2} = -6$.

Step 3: $F(-x) = \dfrac{(-x)^2 + (-x) - 12}{-x + 2} = \dfrac{x^2 - x - 12}{-x + 2}$; this is neither $F(x)$ nor $-F(x)$, so there is no symmetry.

Step 4: The vertical asymptote is the zero of $q(x)$: $x = -2$

Step 5: Since $n = m + 1$, there is an oblique asymptote. Dividing:

$$
\begin{array}{r}
x - 1 \\
x + 2 \overline{)\, x^2 + x - 12} \\
\underline{x^2 + 2x} \\
-x - 12 \\
\underline{-x - 2} \\
-10
\end{array}
\qquad F(x) = x - 1 + \dfrac{-10}{x + 2}
$$

The oblique asymptote is $y = x - 1$.
Solve to find intersection points:

$$\dfrac{x^2 + x - 12}{x + 2} = x - 1$$
$$x^2 + x - 12 = x^2 + x - 2$$
$$-12 = -2$$

Since there is no solution, the oblique asymptote does not intersect $F(x)$.

Step 6:   Graphing: Step 7:   Graphing by hand:

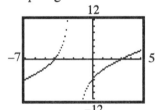 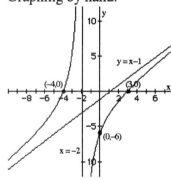

25. $R(x) = \dfrac{x(x-1)^2}{(x+3)^3}$ $p(x) = x(x-1)^2$; $q(x) = (x+3)^3$; $n = 3$; $m = 3$

Step 1:   Domain: $\{x \mid x \neq -3\}$

Step 2:   (a)   The x-intercepts are the zeros of $p(x)$: 0 and 1

         (b)   The y-intercept is $R(0) = \dfrac{0(0-1)^2}{(0+3)^3} = \dfrac{0}{27} = 0$.

Step 3: $R(-x) = \dfrac{-x(-x-1)^2}{(-x+3)^3}$; this is neither $R(x)$ nor $-R(x)$, so there is no symmetry.

Step 4:   The vertical asymptote is the zero of $q(x)$: $x = -3$

Step 5:   Since $n = m$, the line $y = 1$ is the horizontal asymptote.
Solve to find intersection points:
$$\frac{x(x-1)^2}{(x+3)^3} = 1$$
$$x^3 - 2x^2 + x = x^3 + 9x^2 + 27x + 27$$
$$0 = 11x^2 + 26x + 27$$
Since there is no real solution, $R(x)$ does not intersect $y = 1$.

Step 6:   Graphing: Step 7:   Graphing by hand:

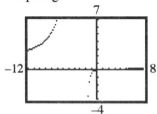 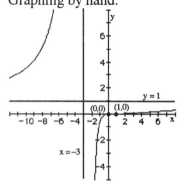

27. $R(x) = \dfrac{x^2 + x - 12}{x^2 - x - 6} = \dfrac{(x+4)(x-3)}{(x-3)(x+2)} = \dfrac{x+4}{x+2}$ $p(x) = x^2 + x - 12;\ q(x) = x^2 - x - 6;$

$n = 2;\ m = 2$

Step 1: Domain: $\left\{x \mid x \ne -2,\ x \ne 3\right\}$

Step 2: (a) The x-intercept is the zero of $p(x)$: -4 (3 is not a zero because reduced form must be used to find the zeros.)

(b) The y-intercept is $R(0) = \dfrac{0^2 + 0 - 12}{0^2 - 0 - 6} = \dfrac{-12}{-6} = 2$.

Step 3: $R(-x) = \dfrac{(-x)^2 + (-x) - 12}{(-x)^2 - (-x) - 6} = \dfrac{x^2 - x - 12}{x^2 + x - 6}$; this is neither $R(x)$ nor $-R(x)$, so there is no symmetry.

Step 4: The vertical asymptote is the zero of $q(x)$: $x = -2$ ($x = 3$ is not a vertical asymptote because reduced form must be used to find the them.)

Step 5: Since $n = m$, the line $y = 1$ is the horizontal asymptote.

$R(x)$ does not intersect $y = 1$ because $R(x)$ is not defined at $x = 3$.

$$\frac{x^2 + x - 12}{x^2 - x - 6} = 1$$
$$x^2 + x - 12 = x^2 - x - 6$$
$$2x = 6$$
$$x = 3$$

Step 6: Graphing:

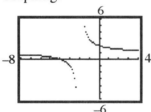

Step 7: Graphing by hand:

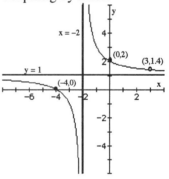

29. $R(x) = \dfrac{6x^2 - 7x - 3}{2x^2 - 7x + 6} = \dfrac{(3x+1)(2x-3)}{(2x-3)(x-2)} = \dfrac{3x+1}{x-2}$ $p(x) = 6x^2 - 7x - 3;$

$q(x) = 2x^2 - 7x + 6;\ n = 2;\ m = 2$

Step 1: Domain: $\left\{x \mid x \ne \dfrac{3}{2},\ x \ne 2\right\}$

Step 2: (a) The x-intercept is the zero of $p(x)$: $\dfrac{-1}{3}$ ($\frac{3}{2}$ is not a zero because reduced form must be used to find the zeros.)

(b) The y-intercept is $R(0) = \dfrac{6(0)^2 - 7(0) - 3}{2(0)^2 - 7(0) + 6} = \dfrac{-3}{6} = \dfrac{-1}{2}$.

Step 3: $R(-x) = \dfrac{6(-x)^2 - 7(-x) - 3}{2(-x)^2 - 7(-x) + 6} = \dfrac{6x^2 + 7x - 3}{2x^2 + 7x + 6}$; this is neither $R(x)$ nor $-R(x)$, so there is no symmetry.

Step 4:   The vertical asymptote is the zero of $q(x)$: $x = 2$ ($x = \frac{3}{2}$ is not a vertical asymptote because reduced form must be used to find the them.)

Step 5:   Since $n = m$, the line $y = 3$ is the horizontal asymptote.

$R(x)$ does not intersect $y = 3$ because $R(x)$ is not defined at $x = \dfrac{3}{2}$.

$$\frac{6x^2 - 7x - 3}{2x^2 - 7x + 6} = 3$$
$$6x^2 - 7x - 3 = 6x^2 - 21x + 18$$
$$14x = 21$$
$$x = \frac{3}{2}$$

Step 6:   Graphing:

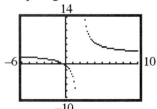

Step 7:   Graphing by hand:

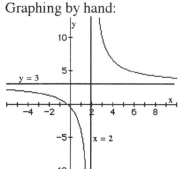

31. $R(x) = \dfrac{x^2 + 5x + 6}{x + 3} = \dfrac{(x + 2)(x + 3)}{x + 3} = x + 2$ $p(x) = x^2 + 5x + 6;$ $q(x) = x + 3;$
$n = 2;$ $m = 1$

Step 1:   Domain: $\{x \mid x \neq -3\}$

Step 2:   (a)   The x-intercept is the zero of $p(x)$: -2 (-3 is not a zero because reduced form must be used to find the zeros.)

(b)   The y-intercept is $R(0) = \dfrac{0^2 + 5(0) + 6}{0 + 3} = \dfrac{6}{3} = 2$.

Step 3: $R(-x) = \dfrac{(-x)^2 + 5(-x) + 6}{-x + 3} = \dfrac{x^2 - 5x + 6}{-x + 3}$; this is neither $R(x)$ nor $-R(x)$, so there is no symmetry.

Step 4:   There are no vertical asymptotes. ($x = -3$ is not a vertical asymptote because reduced form must be used to find the them.)

Step 5:   Since $n = m + 1$ there is a oblique asymptote. The line $y = x + 2$ is the oblique asymptote.

The oblique asymptote does not intersect $R(x)$.

Step 6:   Graphing:

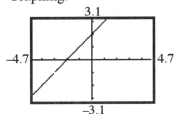

Step 7:   Graphing by hand:

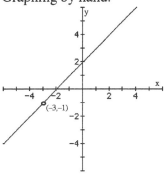

33. $f(x) = x + \dfrac{1}{x} = \dfrac{x^2 + 1}{x}$ $p(x) = x^2 + 1;\ q(x) = x;\ n = 2;\ m = 1$

Step 1:   Domain: $\left\{ x \mid x \neq 0 \right\}$

Step 2:   (a)   There are no x-intercepts.
         (b)   There is no y-intercept because 0 is not in the domain.

Step 3: $f(-x) = \dfrac{(-x)^2 + 1}{-x} = \dfrac{x^2 + 1}{-x} = -f(x)$; The graph of $f(x)$ is symmetric to the origin.

Step 4:   The vertical asymptote is the zero of $q(x)$: $x = 0$

Step 5:   Since $n = m + 1$, there is an oblique asymptote. Dividing:

$$x\overline{\smash{\big)}\,x^2 + 1} \quad \begin{array}{c} x \\ \end{array}$$

$$\underline{x^2}$$
$$1$$

$$f(x) = x + \dfrac{1}{x}$$

The oblique asymptote is $y = x$.
Solve to find intersection points:

$$\dfrac{x^2 + 1}{x} = x$$
$$x^2 + 1 = x^2$$
$$1 = 0$$

Since there is no solution, the oblique asymptote does not intersect $f(x)$.

Step 6:   Graphing:

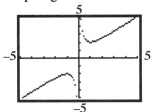

Step 7:   Graphing by hand:

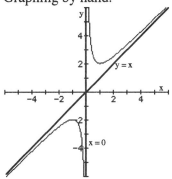

35. $f(x) = x^2 + \dfrac{1}{x} = \dfrac{x^3 + 1}{x}$ $p(x) = x^3 + 1;$ $q(x) = x;$ $n = 3;$ $m = 1$

Step 1: Domain: $\{x \mid x \neq 0\}$

Step 2: (a)   The x-intercept is the zero of $p(x)$: -1
 (b)   There is no y-intercept because 0 is not in the domain.

Step 3: $f(-x) = \dfrac{(-x)^3 + 1}{-x} = \dfrac{-x^3 + 1}{-x}$; this is neither $f(x)$ nor $-f(x)$, so there is
 no symmetry.

Step 4: The vertical asymptote is the zero of $q(x)$: $x = 0$

Step 5: Since $n > m + 1$, there is no horizontal or oblique asymptote.

Step 6: Graphing: Step 7:   Graphing by hand:

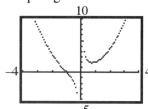
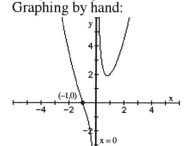

37. $f(x) = x + \dfrac{1}{x^3} = \dfrac{x^4 + 1}{x^3}$ $p(x) = x^4 + 1;$ $q(x) = x^3;$ $n = 4;$ $m = 3$

Step 1: Domain: $\{x \mid x \neq 0\}$

Step 2: (a)   There are no x-intercepts.
 (b)   There is no y-intercept because 0 is not in the domain.

Step 3: $f(-x) = \dfrac{(-x)^4 + 1}{(-x)^3} = \dfrac{x^4 + 1}{-x^3} = -f(x)$; The graph of $f(x)$ is symmetric to
 the origin.

Step 4: The vertical asymptote is the zero of $q(x)$: $x = 0$

Step 5: Since $n = m + 1$, there is an oblique asymptote. Dividing:

$$f(x) = x + \frac{1}{x^3}$$

The oblique asymptote is $y = x$.
Solve to find intersection points:

$$\frac{x^4 + 1}{x^3} = x$$
$$x^4 + 1 = x^4$$
$$1 = 0$$

Since there is no solution, the oblique asymptote does not intersect $f(x)$.

Step 6: Graphing:

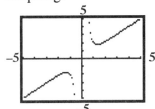

Step 7: Graphing by hand:

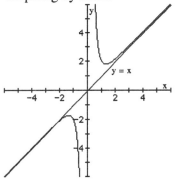

39. (a) $f(x) = \dfrac{x^2}{x^2 - 4}$

(b) $f(x) = \dfrac{-3x}{x^2 - 1}$

(c) $f(x) = \dfrac{(x-1)^3(x-3)}{(x+1)^2(x-2)^2}$

(d) $f(x) = \dfrac{3(x+2)(x-1)^2}{(x+3)(x-4)^2}$

41. $f(x) = \dfrac{3(x-2)(x+1)^2}{(x+5)^2(x-6)}$ Answers will vary.

43. (a) $g(0) = \dfrac{3.99 \times 10^{14}}{\left(6.374 \times 10^6 + 0\right)^2} = \dfrac{3.99 \times 10^{14}}{40.628 \times 10^{12}} = 0.0982 \times 10^2 = 9.821$

(b) $g(443) = \dfrac{3.99 \times 10^{14}}{\left(6.374 \times 10^6 + 443\right)^2} = \dfrac{3.99 \times 10^{14}}{\left(6.3744 \times 10^6\right)^2} = \dfrac{3.99 \times 10^{14}}{40.633 \times 10^{12}} = 9.819$

(c) $g(8848) = \dfrac{3.99 \times 10^{14}}{\left(6.374 \times 10^6 + 8848\right)^2} = \dfrac{3.99 \times 10^{14}}{\left(6.3828 \times 10^6\right)^2} = \dfrac{3.99 \times 10^{14}}{40.743 \times 10^{12}} = 9.794$

(d) Since the degree of the numerator is 0 and the degree of the denominator is 2, the horizontal asymptote is $y = 0$.

(e) Graphing:

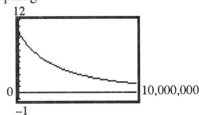

(f) $g(h) = 0$ has no solution. Thus, it is never possible to escape the pull of earth's gravity.

45. (a) Graphing:

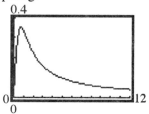

(b) Using MAXIMUM, the concentration is highest when $t = 0.71$ hours.
(c) The degree of the numerator is 1 and the degree of the denominator is 2. Thus, the horizontal asymptote is $C(t) = 0$. The concentration of the drug decreases to 0 as time increases.

47. (a) The average cost function is: $\overline{C}(x) = \dfrac{0.2x^3 - 2.3x^2 + 14.3x + 10.2}{x}$

(b) $\overline{C}(6) = \dfrac{0.2(6)^3 - 2.3(6)^2 + 14.3(6) + 10.2}{6} = \dfrac{56.4}{6} = 9.4$

The average cost of producing 6 Cavaliers per hour is $9400.

(c) $\overline{C}(9) = \dfrac{0.2(9)^3 - 2.3(9)^2 + 14.3(9) + 10.2}{9} = \dfrac{98.4}{9} = 10.933$

The average cost of producing 9 Cavaliers per hour is $10,933.

(d) Graphing:

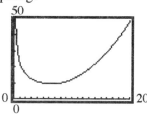

(e) Using MINIMUM, the number of Cavaliers that should be produced per hour to minimize cost is 6.38.
(f) The minimum average cost is $9,366.

49. (a) The surface area is the sum of the areas of the six sides.
$$S = xy + xy + xy + xy + x^2 + x^2 = 4xy + 2x^2$$

The volume is $x \cdot x \cdot y = x^2 y = 10,000 \quad \rightarrow \quad y = \dfrac{10000}{x^2}$

Thus, $S(x) = 4x\left(\dfrac{10000}{x^2}\right) + 2x^2 = 2x^2 + \dfrac{40000}{x} = \dfrac{2x^3 + 40000}{x}$

(b) Graphing:

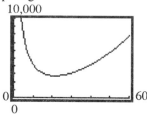

(c)   The minimum surface area (amount of cardboard) is 2,785 square inches.

(d)   The surface area is a minimum when $x = 21.544$.

$$y = \frac{10000}{21.544^2} = 21.545$$

The dimensions of the box are: 21.544 in. by 21.544 in. by 21.545 in.

51.   (a) $500 = \pi r^2 h \quad \rightarrow \quad h = \frac{500}{\pi r^2}$

$$C(r) = 6(2\pi r^2) + 4(2\pi rh) = 12\pi r^2 + 8\pi r\left(\frac{500}{\pi r^2}\right) = 12\pi r^2 + \frac{4000}{r}$$

(b)   Graphing:

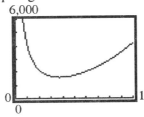

The cost is least for $r = 3.76$ cm.

3   Chapter Review

1. $f(x) = \frac{1}{4}x^2 - 16$

$a = \frac{1}{4}, b = 0, c = -16$. Since $a = \frac{1}{4} > 0$, the graph opens up.

The x-coordinate of the vertex is $x = \frac{-b}{2a} = \frac{-0}{2(\frac{1}{4})} = \frac{0}{\frac{1}{2}} = 0$.

The y-coordinate of the vertex is $f\left(\frac{-b}{2a}\right) = f(0) = \frac{1}{4}(0)^2 - 16 = -16$.

Thus, the vertex is $(0, -16)$.

The axis of symmetry is the line $x = 0$.

The discriminant is:

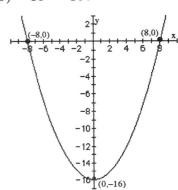

$$b^2 - 4ac = (0)^2 - 4\left(\frac{1}{4}\right)(-16) = 16 > 0,$$

so the graph has two x-intercepts.

The x-intercepts are found by solving:

$$\frac{1}{4}x^2 - 16 = 0$$
$$x^2 - 64 = 0$$
$$x^2 = 64$$
$$x = 8 \text{ or } x = -8$$

The x-intercepts are –8 and 8.

The y-intercept is $f(0) = -16$.

3. $f(x) = -4x^2 + 4x$

$a = -4,\ b = 4,\ c = 0$. Since $a = -4 < 0$, the graph opens down.

The x-coordinate of the vertex is $x = \dfrac{-b}{2a} = \dfrac{-4}{2(-4)} = \dfrac{-4}{-8} = \dfrac{1}{2}$.

The y-coordinate of the vertex is $f\left(\dfrac{-b}{2a}\right) = f\left(\dfrac{1}{2}\right) = -4\left(\dfrac{1}{2}\right)^2 + 4\left(\dfrac{1}{2}\right) = -1 + 2 = 1$.

Thus, the vertex is $\left(\dfrac{1}{2}, 1\right)$.

The axis of symmetry is the line $x = \dfrac{1}{2}$.

The discriminant is:

$b^2 - 4ac = 4^2 - 4(-4)(0) = 16 > 0$,

so the graph has two x-intercepts.

The x-intercepts are found by solving:

$-4x^2 + 4x = 0$

$-4x(x - 1) = 0$

$x = 0\ $ or $\ x = 1$

The x-intercepts are 0 and 1.

The y-intercept is $f(0) = -4(0)^2 + 4(0) = 0$.

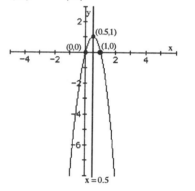

5. $f(x) = \dfrac{9}{2}x^2 + 3x + 1$

$a = \dfrac{9}{2},\ b = 3,\ c = 1$. Since $a = \dfrac{9}{2} > 0$, the graph opens up.

The x-coordinate of the vertex is $x = \dfrac{-b}{2a} = \dfrac{-3}{2\left(\frac{9}{2}\right)} = \dfrac{-3}{9} = \dfrac{-1}{3}$.

The y-coordinate of the vertex is

$$f\left(\dfrac{-b}{2a}\right) = f\left(\dfrac{-1}{3}\right) = \dfrac{9}{2}\left(\dfrac{-1}{3}\right)^2 + 3\left(\dfrac{-1}{3}\right) + 1 = \dfrac{1}{2} - 1 + 1 = \dfrac{1}{2}.$$

Thus, the vertex is $\left(\dfrac{-1}{3}, \dfrac{1}{2}\right)$.

The axis of symmetry is the line $x = \dfrac{-1}{3}$.

The discriminant is:

$b^2 - 4ac = 3^2 - 4\left(\dfrac{9}{2}\right)(1) = 9 - 18 = -9 < 0$,

so the graph has no x-intercepts.

The y-intercept is $f(0) = \dfrac{9}{2}(0)^2 + 3(0) + 1 = 1$.

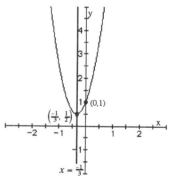

7. $f(x) = 3x^2 - 4x - 1$

$a = 3, b = -4, c = -1.$ Since $a = 3 > 0$, the graph opens up.

The x-coordinate of the vertex is $x = \dfrac{-b}{2a} = \dfrac{-(-4)}{2(3)} = \dfrac{4}{6} = \dfrac{2}{3}$.

The y-coordinate of the vertex is

$$f\left(\dfrac{-b}{2a}\right) = f\left(\dfrac{2}{3}\right) = 3\left(\dfrac{2}{3}\right)^2 - 4\left(\dfrac{2}{3}\right) - 1 = \dfrac{4}{3} - \dfrac{8}{3} - 1 = \dfrac{-7}{3}.$$

Thus, the vertex is $\left(\dfrac{2}{3}, \dfrac{-7}{3}\right)$.

The axis of symmetry is the line $x = \dfrac{2}{3}$.

The discriminant is:

$$b^2 - 4ac = (-4)^2 - 4(3)(-1) = 16 + 12 = 28 > 0,$$

so the graph has two x-intercepts.

The x-intercepts are found by solving:

$$x = \dfrac{-b \pm \sqrt{b^2 - 4ac}}{2a} = \dfrac{-(-4) \pm \sqrt{28}}{2(3)}$$

$$= \dfrac{4 \pm 2\sqrt{7}}{6} = \dfrac{2 \pm \sqrt{7}}{3} = \dfrac{-3 \pm 2.646}{3}$$

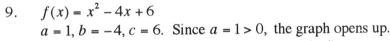

The x-intercepts are –0.22 and 1.55.

The y-intercept is $f(0) = 3(0)^2 - 4(0) - 1 = -1$.

9. $f(x) = x^2 - 4x + 6$

$a = 1, b = -4, c = 6.$ Since $a = 1 > 0$, the graph opens up.

The x-coordinate of the vertex is $x = \dfrac{-b}{2a} = \dfrac{-(-4)}{2(1)} = \dfrac{4}{2} = 2$.

The y-coordinate of the vertex is $f\left(\dfrac{-b}{2a}\right) = f(2) = (2)^2 - 4(2) + 6 = 4 - 8 + 6 = 2$.

Thus, the vertex is $(2, 2)$.

The axis of symmetry is the line $x = 2$.

The discriminant is:

$$b^2 - 4ac = (-4)^2 - 4(1)(6) = 16 - 24 = -8 < 0,$$

so the graph has no x-intercepts.

The y-intercept is $f(0) = (0)^2 - 4(0) + 6 = 6$.

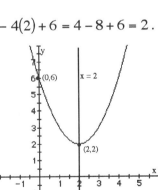

11. $f(x) = (x+2)^3$

Using the graph of $y = x^3$, shift the graph horizontally, 2 units to the left.

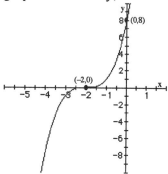

13. $f(x) = -(x-1)^4$

Using the graph of $y = x^4$, shift the graph horizontally, 1 unit right, and reflect about the x-axis.

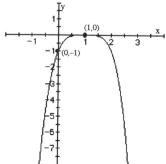

15. $f(x) = (x-1)^4 + 2$

Using the graph of $y = x^4$, shift the graph horizontally, 1 unit to the right, and shift vertically 2 units up.

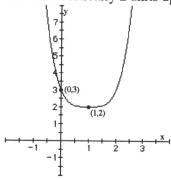

17. $f(x) = 3x^2 - 6x + 4$

$a = 3, b = -6, c = 4$. Since $a = 3 > 0$, the graph opens up, so the vertex is a minimum point. The minimum occurs at $x = \dfrac{-b}{2a} = \dfrac{-(-6)}{2(3)} = \dfrac{6}{6} = 1$. The minimum value is $f\left(\dfrac{-b}{2a}\right) = f(1) = 3(1)^2 - 6(1) + 4 = 3 - 6 + 4 = 1$.

19. $f(x) = -x^2 + 8x - 4$

$a = -1, b = 8, c = -4$. Since $a = -1 < 0$, the graph opens down, so the vertex is a maximum point. The maximum occurs at $x = \dfrac{-b}{2a} = \dfrac{-8}{2(-1)} = \dfrac{-8}{-2} = 4$. The maximum value is $f\left(\dfrac{-b}{2a}\right) = f(4) = -(4)^2 + 8(4) - 4 = -16 + 32 - 4 = 12$.

21. $f(x) = -3x^2 + 12x + 4$

    $a = -3, b = 12, c = 4$. Since $a = -3 < 0$, the graph opens down, so the vertex is a maximum point. The maximum occurs at $x = \dfrac{-b}{2a} = \dfrac{-12}{2(-3)} = \dfrac{-12}{-6} = 2$. The maximum value is $f\left(\dfrac{-b}{2a}\right) = f(2) = -3(2)^2 + 12(2) + 4 = -12 + 24 + 4 = 16$.

23. (a) $f(x) = x(x+2)(x+4)$

    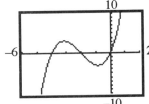

    (b) x-intercepts: –4, –2, 0;
    y-intercept: 0
    (c) Odd: –4, –2, 0
    (d) $y = x^3$
    (e) 2
    (f) Local maximum: (–3.15, 3.08)
    Local minimum: (–0.85, –3.08)

25. (a) $f(x) = (x-2)^2(x+4)$

    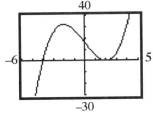

    (b) x-intercepts: –4, 2;
    y-intercept: 16
    (c) Odd: –4; Even: 2
    (d) $y = x^3$
    (e) 2
    (f) Local maximum: (–2, 32)
    Local minimum: (2, 0)

27. (a) $f(x) = x^3 - 4x^2 = x^2(x-4)$

    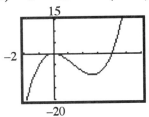

    (b) x-intercepts: 4, 0;
    y-intercept: 0
    (c) Odd: 4; Even: 0
    (d) $y = x^3$
    (e) 2
    (f) Local maximum: (0, 0)
    Local minimum: (2.67, –9.48)

29. (a) $f(x) = (x-1)^2(x+3)(x+1)$

    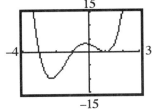

    (b) x-intercepts: –3, –1, 1;
    y-intercept: 3
    (c) Odd: –3, –1; Even: 1
    (d) $y = x^4$
    (e) 3
    (f) Local maximum: (–0.22, 3.23)
    Local minima: (–2.28, –9.91),
    (1, 0)

31. $R(x) = \dfrac{2x-6}{x}$ $p(x) = 2x-6;$ $q(x) = x;$ $n = 1;$ $m = 1$

Step 1: Domain: $\{x \mid x \neq 0\}$

Step 2: (a) The x-intercept is the zero of $p(x)$: 3
        (b) There is no y-intercept because 0 is not in the domain.

Step 3: $R(-x) = \dfrac{2(-x)-6}{-x} = \dfrac{-2x-6}{-x} = \dfrac{2x+6}{x}$; this is neither $R(x)$ nor $-R(x)$,
        so there is no symmetry.

Step 4: The vertical asymptote is the zero of $q(x)$: $x = 0$

Step 5: Since $n = m$, the line $y = 2$ is the horizontal asymptote.
        $R(x)$ does not intersect $y = 2$.

Step 6: Graphing: Step 7:   Graphing by hand:

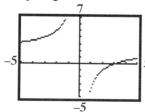
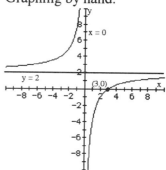

33. $H(x) = \dfrac{x+2}{x(x-2)}$ $p(x) = x+2;$ $q(x) = x(x-2) = x^2 - 2x;$ $n = 1;$ $m = 2$

Step 1: Domain: $\{x \mid x \neq 0,\ x \neq 2\}$

Step 2: (a) The x-intercept is the zero of $p(x)$: -2
        (b) There is no y-intercept because 0 is not in the domain.

Step 3: $H(-x) = \dfrac{-x+2}{-x(-x-2)} = \dfrac{-x+2}{x^2+2x}$; this is neither $H(x)$ nor $-H(x)$, so there
        is no symmetry.

Step 4: The vertical asymptotes are the zeros of $q(x)$: $x = 0$ and $x = 2$

Step 5: Since $n < m$, the line $y = 0$ is the horizontal asymptote.
        $R(x)$ intersects $y = 0$ at $(-2, 0)$.

Step 6: Graphing: Step 7:   Graphing by hand:

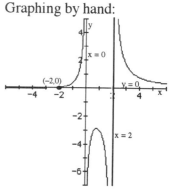

35. $R(x) = \dfrac{x^2 + x - 6}{x^2 - x - 6} = \dfrac{(x+3)(x-2)}{(x-3)(x+2)}$ $\quad p(x) = x^2 + x - 6; \quad q(x) = x^2 - x - 6;$

$$n = 2; \quad m = 2$$

Step 1: Domain: $\left\{x \mid x \ne -2, \, x \ne 3\right\}$

Step 2: (a) The x-intercepts are the zeros of $p(x)$: -3 and 2

(b) The y-intercept is $R(0) = \dfrac{0^2 + 0 - 6}{0^2 - 0 - 6} = \dfrac{-6}{-6} = 1$.

Step 3: $R(-x) = \dfrac{(-x)^2 + (-x) - 6}{(-x)^2 - (-x) - 6} = \dfrac{x^2 - x - 6}{x^2 + x - 6}$; this is neither $R(x)$ nor $-R(x)$, so there is no symmetry.

Step 4: The vertical asymptotes are the zeros of $q(x)$: $x = -2$ and $x = 3$

Step 5: Since $n = m$, the line $y = 1$ is the horizontal asymptote.

$R(x)$ intersects $y = 1$ at $(0, 1)$, since:

$$\frac{x^2 + x - 6}{x^2 - x - 6} = 1$$
$$x^2 + x - 6 = x^2 - x - 6$$
$$2x = 0$$
$$x = 0$$

Step 6: Graphing:

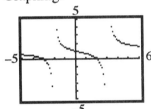

Step 7: Graphing by hand:

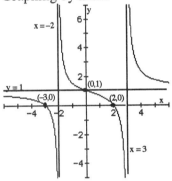

37. $F(x) = \dfrac{x^3}{x^2 - 4}$ $\quad p(x) = x^3; \quad q(x) = x^2 - 4; \quad n = 3; \quad m = 2$

Step 1: Domain: $\left\{x \mid x \ne -2, \, x \ne 2\right\}$

Step 2: (a) The x-intercept is the zero of $p(x)$: 0.

(b) The y-intercept is $F(0) = \dfrac{0^3}{0^2 - 4} = \dfrac{0}{-4} = 0$.

Step 3: $F(-x) = \dfrac{(-x)^3}{(-x)^2 - 4} = \dfrac{-x^3}{x^2 - 4} = -F(x)$; $F(x)$ is symmetric to the origin.

Step 4: The vertical asymptotes are the zeros of $q(x)$: $x = -2$ and $x = 2$

Step 5:   Since $n = m + 1$, there is an oblique asymptote. Dividing:

$$
\begin{array}{r}
x \\
x^2 - 4\overline{)x^3 + 0x^2 + 0x + 0} \\
\underline{x^3 - 4x} \\
4x
\end{array}
\qquad F(x) = x + \frac{4x}{x^2 - 4}
$$

The oblique asymptote is $y = x$.
Solve to find intersection points:

$$\frac{x^3}{x^2 - 4} = x$$

$$x^3 = x^3 - 4x$$

$$0 = -4x$$

$$x = 0$$

The oblique asymptote intersects $F(x)$ at $(0, 0)$.

Step 6:   Graphing: Step 7:   Graphing by hand:

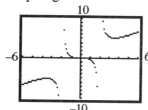   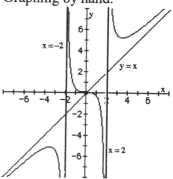

39. $R(x) = \dfrac{2x^4}{(x - 1)^2}$ $p(x) = 2x^4$; $q(x) = (x - 1)^2$; $n = 4$; $m = 2$

Step 1:   Domain: $\left\{x \mid x \neq 1\right\}$

Step 2:   (a)   The x-intercept is the zero of $p(x)$: 0

(b)   The y-intercept is $R(0) = \dfrac{2(0)^4}{(0 - 1)^2} = \dfrac{0}{1} = 0$.

Step 3: $R(-x) = \dfrac{2(-x)^4}{(-x - 1)^2} = \dfrac{2x^4}{(x + 1)^2}$; this is neither $R(x)$ nor $-R(x)$, so there is no symmetry.

Step 4:   The vertical asymptote is the zero of $q(x)$: $x = 1$

Step 5:   Since $n > m + 1$, there is no horizontal asymptote and no oblique asymptote.

Step 6:   Graphing:

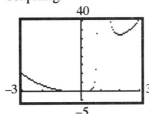

Step 7:   Graphing by hand:

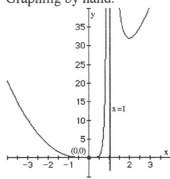

41. $G(x) = \dfrac{x^2 - 4}{x^2 - x - 2} = \dfrac{(x+2)(x-2)}{(x-2)(x+1)} = \dfrac{x+2}{x+1}$ $p(x) = x^2 - 4;$ $q(x) = x^2 - x - 2;$

$n = 2;\ m = 2$

Step 1:   Domain: $\left\{x \mid x \neq -1,\ x \neq 2\right\}$

Step 2:   (a)   The x-intercept is the zero of $p(x)$: -2 (2 is not a zero because reduced form must be used to find the zeros.)

(b)   The y-intercept is $G(0) = \dfrac{0^2 - 4}{0^2 - 0 - 2} = \dfrac{-4}{-2} = 2$.

Step 3: $G(-x) = \dfrac{(-x)^2 - 4}{(-x)^2 - (-x) - 2} = \dfrac{x^2 - 4}{x^2 + x - 2}$; this is neither $G(x)$ nor $-G(x)$, so there is no symmetry.

Step 4:   The vertical asymptote is the zero of $q(x)$: $x = -1$ ($x = 2$ is not a vertical asymptote because reduced form must be used to find the them.)

Step 5:   Since $n = m$, the line $y = 1$ is the horizontal asymptote.

$G(x)$ does not intersect $y = 1$ because $G(x)$ is not defined at $x = 2$.

$$\frac{x^2 - 4}{x^2 - x - 2} = 1$$
$$x^2 - 4 = x^2 - x - 2$$
$$-2 = -x$$
$$x = 2$$

Step 6:   Graphing:

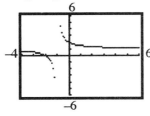

Step 7:   Graphing by hand:

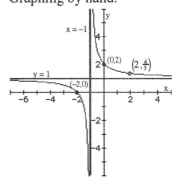

43. Since there are 200 feet of border, we know that $2x + 2y = 200$.
     The area is to be maximized, so $A = x \cdot y$.
     Solving the perimeter formula for y:
$$2x + 2y = 200$$
$$2y = 200 - 2x$$
$$y = 100 - x$$
     The area function is: $A(x) = x(100 - x) = -x^2 + 100x$
     The maximum value occurs at the vertex:
$$x = \frac{-b}{2a} = \frac{-100}{2(-1)} = \frac{-100}{-2} = 50$$
     The pond should be 50 feet by 50 feet for maximum area.

45. The area function is: $A(x) = x(10 - x) = -x^2 + 10x$
     The maximum value occurs at the vertex:
$$x = \frac{-b}{2a} = \frac{-10}{2(-1)} = \frac{-10}{-2} = 5$$
     The maximum area is:
$$A(5) = -(5)^2 + 10(5) = -25 + 50 = 25 \text{ square units.}$$

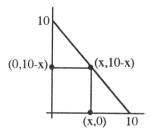

47. $C(x) = 0.003x^2 - 30x + 111{,}800$
     (a)   The vertex will be a minimum:
$$x = \frac{-b}{2a} = \frac{-(-30)}{2(0.003)} = \frac{30}{0.006} = 5000$$
           5000 textbooks should be manufactured for minimum cost.
     (b) $C(5000) = 0.003(5000)^2 - 30(5000) + 111{,}800 = \$36{,}800$ (total cost)
           $\text{Cost per book} = \dfrac{36{,}800}{5{,}000} = \$7.36$
     (c) $C(1) = 0.003(1)^2 - 30(1) + 111{,}800 = \$111{,}770$

49. (a)   Graphing:

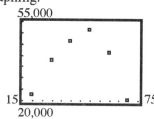

           This appears to be a quadratic function.
     (b) $I(x) = -45.669x^2 + 4081.54x - 41021.213$

(c)   Graphing the quadratic function of best fit:

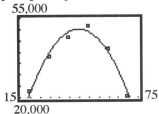

(d)   Using MAXIMUM, an individual will earn the most income at age 44.7 years.
(e)   The peak income will be $50,173.
(f)   These values are a bit less than the data values of approximately 49 years with an income of about $51,875.
(g)   The age for maximum income remains about the same, but the maximum income has increased steadily.

51.   (a)   Graphing:

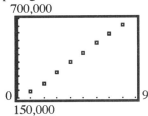

(b) $A(t) = -212.008t^3 + 2429.132t^2 + 59568.854t + 130003.143$
(c)   Graphing the cubic function of best fit:

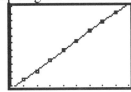

(d) $A(11) = -212.008(11)^3 + 2429.132(11)^2 + 59568.854(11) + 130003.143$
        $= 797,003$
There will be approximately 797,003 cumulative cases of AIDS reported in the United States in 2000.
(e)   The cumulative number of cases reaches 850,000 in year 13 which is the year 2002.

Chapter 4

The Zeros of a Polynomial Function

4.1   The Real Zeros of a Polynomial Function

1. $f(x) = 4x^3 - 3x^2 - 8x + 4; \quad c = 3$
$f(3) = 4(3)^3 - 3(3)^2 - 8(3) + 4 = 108 - 27 - 24 + 4 = 61 \neq 0$
Thus, 3 is not a zero of f.

3. $f(x) = 3x^4 - 6x^3 - 5x + 10; \quad c = 1$
$f(1) = 3(1)^4 - 6(1)^3 - 5(1) + 10 = 3 - 6 - 5 + 10 = 2 \neq 0$
Thus, 1 is not a zero of f.

5. $f(x) = 3x^6 + 2x^3 - 176; \quad c = -2$
$f(-2) = 3(-2)^6 + 2(-2)^3 - 176 = 192 - 16 - 176 = 0$
Thus, –2 is a zero of f.   Use synthetic division to find the factors.

$$
\begin{array}{r|rrrrrrr}
-2 & 3 & 0 & 0 & 2 & 0 & 0 & -176 \\
   &   & -6 & 12 & -24 & 44 & -88 & 176 \\
\hline
   & 3 & -6 & 12 & -22 & 44 & -88 & 0
\end{array}
$$

The factored form is: $f(x) = (x + 2)\left(3x^5 - 6x^4 + 12x^3 - 22x^2 + 44x - 88\right)$.

7. $f(x) = 4x^6 - 64x^4 + x^2 - 16; \quad c = 4$
Use synthetic division to determine whether 4 is a zero.

$$
\begin{array}{r|rrrrrrr}
4 & 4 & 0 & -64 & 0 & 1 & 0 & -16 \\
  &   & 16 & 64 & 0 & 0 & 4 & 16 \\
\hline
  & 4 & 16 & 0 & 0 & 1 & 4 & 0
\end{array}
$$

Thus, 4 is a zero of f. The factored form is: $f(x) = (x - 4)\left(4x^5 + 16x^4 + x + 4\right)$.

9. $f(x) = 2x^4 - x^3 + 2x - 1; \quad c = \frac{-1}{2}$
$f\left(\frac{-1}{2}\right) = 2\left(\frac{-1}{2}\right)^4 - \left(\frac{-1}{2}\right)^3 + 2\left(\frac{-1}{2}\right) - 1 = \frac{1}{8} + \frac{1}{8} - 1 - 1 = \frac{-7}{4} \neq 0$
Thus, $\frac{-1}{2}$ is not a zero of f.

11. $f(x) = -4x^7 + x^3 - x^2 + 2$
The maximum number of zeros is the degree of the polynomial which is 7.
Examining $f(x) = -4x^7 + x^3 - x^2 + 2$, there are 3 variations in sign; thus, there are 3 or 1 positive real zeros.
Examining $f(-x) = -4(-x)^7 + (-x)^3 - (-x)^2 + 2 = 4x^7 - x^3 - x^2 + 2$, there are 2 variations in sign; thus, there are 2 or 0 negative real zeros.

13. $f(x) = 2x^6 - 3x^2 - x + 1$
The maximum number of zeros is the degree of the polynomial which is 6.
Examining $f(x) = 2x^6 - 3x^2 - x + 1$, there are 2 variations in sign; thus, there are 2 or 0 positive real zeros.
Examining $f(-x) = 2(-x)^6 - 3(-x)^2 - (-x) + 1 = 2x^6 - 3x^2 + x + 1$, there are 2 variations in sign; thus, there are 2 or 0 negative real zeros.

15. $f(x) = 3x^3 - 2x^2 + x + 2$
The maximum number of zeros is the degree of the polynomial which is 3.
Examining $f(x) = 3x^3 - 2x^2 + x + 2$, there are 2 variations in sign; thus, there are 2 or 0 positive real zeros.
Examining $f(-x) = 3(-x)^3 - 2(-x)^2 + (-x) + 2 = -3x^3 - 2x^2 - x + 2$, there is 1 variation in sign; thus, there is 1 negative real zero.

17. $f(x) = -x^4 + x^2 - 1$
The maximum number of zeros is the degree of the polynomial which is 4.
Examining $f(x) = -x^4 + x^2 - 1$, there are 2 variations in sign; thus, there are 2 or 0 positive real zeros.
Examining $f(-x) = -(-x)^4 + (-x)^2 - 1 = -x^4 + x^2 - 1$, there are 2 variations in sign; thus, there are 2 or 0 negative real zeros.

19. $f(x) = x^5 + x^4 + x^2 + x + 1$
The maximum number of zeros is the degree of the polynomial which is 5.
Examining $f(x) = x^5 + x^4 + x^2 + x + 1$, there are no variations in sign; thus, there are 0 positive real zeros.
Examining $f(-x) = (-x)^5 + (-x)^4 + (-x)^2 + (-x) + 1 = -x^5 + x^4 + x^2 - x + 1$, there are 3 variations in sign; thus, there are 3 or 1 negative real zeros.

21. $f(x) = x^6 - 1$
The maximum number of zeros is the degree of the polynomial which is 6.
Examining $f(x) = x^6 - 1$, there is 1 variation in sign; thus, there is 1 positive real zero.
Examining $f(-x) = (-x)^6 - 1 = x^6 - 1$, there is 1 variation in sign; thus, there is 1 negative real zero.

23. $f(x) = 3x^4 - 3x^3 + x^2 - x + 1$

p must be a factor of 1: $p = \pm 1$

q must be a factor of 3: $q = \pm 1, \pm 3$

The possible rational zeros are: $\dfrac{p}{q} = \pm 1, \pm \dfrac{1}{3}$

25. $f(x) = x^5 - 6x^2 + 9x - 3$

p must be a factor of -3: $p = \pm 1, \pm 3$

q must be a factor of 1: $q = \pm 1$

The possible rational zeros are: $\dfrac{p}{q} = \pm 1, \pm 3$

27. $f(x) = -4x^3 - x^2 + x + 2$

p must be a factor of 2: $p = \pm 1, \pm 2$

q must be a factor of -4: $q = \pm 1, \pm 2, \pm 4$

The possible rational zeros are: $\dfrac{p}{q} = \pm 1, \pm 2, \pm \dfrac{1}{2}, \pm \dfrac{1}{4}$

29. $f(x) = 3x^4 - x^2 + 2$

p must be a factor of 2: $p = \pm 1, \pm 2$

q must be a factor of 3: $q = \pm 1, \pm 3$

The possible rational zeros are: $\dfrac{p}{q} = \pm 1, \pm 2, \pm \dfrac{1}{3}, \pm \dfrac{2}{3}$

31. $f(x) = 2x^5 - x^3 + 2x^2 + 4$

p must be a factor of 4: $p = \pm 1, \pm 2, \pm 4$

q must be a factor of 2: $q = \pm 1, \pm 2$

The possible rational zeros are: $\dfrac{p}{q} = \pm 1, \pm 2, \pm 4, \pm \dfrac{1}{2}$

33. $f(x) = 6x^4 + 2x^3 - x^2 + 2$

p must be a factor of 2: $p = \pm 1, \pm 2$

q must be a factor of 6: $q = \pm 1, \pm 2, \pm 3, \pm 6$

The possible rational zeros are: $\dfrac{p}{q} = \pm 1, \pm 2, \pm \dfrac{1}{2}, \pm \dfrac{1}{3}, \pm \dfrac{2}{3}, \pm \dfrac{1}{6}$

35. $f(x) = 2x^3 + x^2 - 1 = 2\left(x^3 + \frac{1}{2}x^2 + 0x - \frac{1}{2}\right)$

$a_2 = \frac{1}{2}$, $a_1 = 0$, $a_0 = \frac{-1}{2}$

Max $\left\{1, \left|\frac{-1}{2}\right| + |0| + \left|\frac{1}{2}\right|\right\} = $ Max $\{1, 1\} = 1$

$1 + $ Max $\left\{\left|\frac{-1}{2}\right|, |0|, \left|\frac{1}{2}\right|\right\} = 1 + \frac{1}{2} = \frac{3}{2}$

The smaller of the two numbers is 1. Thus, every zero of f lies between -1 and 1.

Graphing using the bounds and ZOOM-FIT:

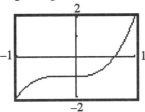

37. $f(x) = x^3 - 5x^2 - 11x + 11$

$a_2 = -5, \quad a_1 = -11, \quad a_0 = 11$

Max $\left\{1, |11| + |-11| + |-5|\right\}$ = Max $\left\{1, 27\right\} = 27$

$1 + $ Max $\left\{|11|, |-11|, |-5|\right\} = 1 + 11 = 12$

The smaller of the two numbers is 12. Thus, every zero of f lies between -12 and 12.

Graphing using the bounds and ZOOM-FIT: (Second graph has a better window.)

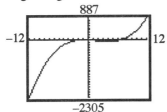 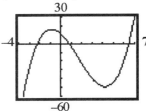

39. $f(x) = x^4 + 3x^3 - 5x^2 + 9 = x^4 + 3x^3 - 5x^2 + 0x + 9$

$a_3 = 3, \quad a_2 = -5, \quad a_1 = 0, \quad a_0 = 9$

Max $\left\{1, |9| + |0| + |-5| + |3|\right\}$ = Max $\left\{1, 17\right\} = 17$

$1 + $ Max $\left\{|9|, |0|, |-5|, |3|\right\} = 1 + 9 = 10$

The smaller of the two numbers is 10. Thus, every zero of f lies between -10 and 10.

Graphing using the bounds and ZOOM-FIT: (Second graph has a better window.)

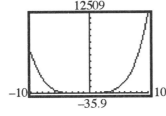 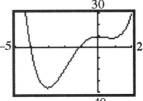

41. $f(x) = x^3 + 2x^2 - 5x - 6$

Step 1: $f(x)$ has at most 3 real zeros.

Step 2: By Descartes Rule of Signs, there is 1 positive real zero.

Also because $f(-x) = (-x)^3 + 2(-x)^2 - 5(-x) - 6 = -x^3 + 2x^2 + 5x - 6$, there are 2 or 0 negative real zeros.

Step 3:   Possible rational zeros:

$$p = \pm 1, \pm 2, \pm 3, \pm 6; \quad q = \pm 1; \quad \frac{p}{q} = \pm 1, \pm 2, \pm 3, \pm 6$$

Step 4:   Using the Bounds on Zeros Theorem:

$$a_2 = 2, \quad a_1 = -5, \quad a_0 = -6$$

$$\text{Max}\left\{1, |-6| + |-5| + |2|\right\} = \text{Max}\left\{1, 13\right\} = 13$$

$$1 + \text{Max}\left\{|-6|, |-5|, |2|\right\} = 1 + 6 = 7$$

The smaller of the two numbers is 7. Thus, every zero of f lies between –7 and 7.

Graphing using the bounds and ZOOM-FIT: (Second has a better window.)

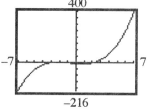   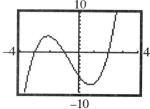

Step 5:   (a)   From the graph it appears that there are x-intercepts at –3, –1, and 2.
          (b)   Using synthetic division:

$$-3\overline{)\begin{array}{rrrr} 1 & 2 & -5 & -6 \\ & -3 & 3 & 6 \\ \hline 1 & -1 & -2 & 0 \end{array}}$$

Since the remainder is 0, $x - (-3) = x + 3$ is a factor. The other factor is the quotient: $x^2 - x - 2$.

          (c)   Thus, $f(x) = (x + 3)\left(x^2 - x - 2\right) = (x + 3)(x + 1)(x - 2)$.
The zeros are –3, -1, and 2.

43. $f(x) = 2x^3 - 13x^2 + 24x - 9$

Step 1: $f(x)$ has at most 3 real zeros.

Step 2:   By Descartes Rule of Signs, there are 3 or 1 positive real zeros.

$f(-x) = 2(-x)^3 - 13(-x)^2 + 24(-x) - 9 = -2x^3 - 13x^2 - 24x - 9$; thus, there are no negative real zeros.

Step 3:   Possible rational zeros:

$$p = \pm 1, \pm 3, \pm 9; \quad q = \pm 1, \pm 2; \quad \frac{p}{q} = \pm 1, \pm 3, \pm 9, \pm \frac{1}{2}, \pm \frac{3}{2}, \pm \frac{9}{2}$$

Step 4:   Using the Bounds on Zeros Theorem:

$$f(x) = 2\left(x^3 - \frac{13}{2}x^2 + 12x - \frac{9}{2}\right)$$

$$a_2 = \frac{-13}{2}, \quad a_1 = 12, \quad a_0 = \frac{-9}{2}$$

$$\text{Max}\left\{1, \left|\frac{-9}{2}\right| + |12| + \left|\frac{-13}{2}\right|\right\} = \text{Max}\left\{1, 23\right\} = 23$$

$$1 + \text{Max}\left\{\left|\frac{-9}{2}\right|, |12|, \left|\frac{-13}{2}\right|\right\} = 1 + 12 = 13$$

The smaller of the two numbers is 13. Thus, every zero of f lies between –13 and 13.

Graphing using the bounds and ZOOM-FIT: (Second has a better window.)

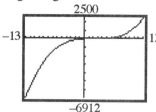 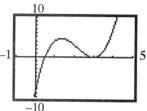

Step 5: (a) From the graph it appears that there are x-intercepts at $\frac{1}{2}$ and 3.

(b) Using synthetic division:

$$3\overline{)2 \quad -13 \quad 24 \quad -9}$$
$$\phantom{3\overline{)2}} \quad 6 \quad -21 \quad 9$$
$$\overline{ 2 \quad -7 \quad 3 \quad 0}$$

Since the remainder is 0, $x - 3$ is a factor. The other factor is the quotient: $2x^2 - 7x + 3$.

(c) Thus, $f(x) = (x - 3)(2x^2 - 7x + 3) = (x - 3)(2x - 1)(x - 3)$.

The zeros are $\frac{1}{2}$ and 3 (multiplicity 2).

45. $f(x) = 3x^3 + 4x^2 + 4x + 1$

Step 1: $f(x)$ has at most 3 real zeros.

Step 2: By Descartes Rule of Signs, there are no positive real zeros.

$f(-x) = 3(-x)^3 + 4(-x)^2 + 4(-x) + 1 = -3x^3 + 4x^2 - 4x + 1$; thus, there are 3 or 1 negative real zeros.

Step 3: Possible rational zeros:

$$p = \pm 1; \quad q = \pm 1, \pm 3; \quad \frac{p}{q} = \pm 1, \pm \frac{1}{3}$$

Step 4: Using the Bounds on Zeros Theorem:

$$f(x) = 3\left(x^3 + \tfrac{4}{3}x^2 + \tfrac{4}{3}x + \tfrac{1}{3}\right)$$

$$a_2 = \tfrac{4}{3}, \quad a_1 = \tfrac{4}{3}, \quad a_0 = \tfrac{1}{3}$$

$$\text{Max}\left\{1, \left|\tfrac{1}{3}\right| + \left|\tfrac{4}{3}\right| + \left|\tfrac{4}{3}\right|\right\} = \text{Max}\left\{1, 3\right\} = 3$$

$$1 + \text{Max}\left\{\left|\tfrac{1}{3}\right|, \left|\tfrac{4}{3}\right|, \left|\tfrac{4}{3}\right|\right\} = 1 + \tfrac{4}{3} = \tfrac{7}{3}$$

The smaller of the two numbers is $\frac{7}{3}$. Thus, every zero of f lies between $\frac{-7}{3}$ and $\frac{7}{3}$.

Graphing using the bounds and ZOOM-FIT: (Second has a better window.)

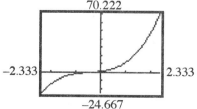 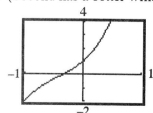

Step 5:   (a)   From the graph it appears that there is an x-intercepts at $\frac{-1}{3}$.
          (b)   Using synthetic division:

$$\frac{-1}{3}\overline{)\begin{array}{cccc} 3 & 4 & 4 & 1 \\ & -1 & -1 & -1 \end{array}}$$
$$\begin{array}{cccc} 3 & 3 & 3 & 0 \end{array}$$

Since the remainder is 0, $x - \left(\frac{-1}{3}\right) = x + \frac{1}{3}$ is a factor. The other factor

is the quotient: $3x^2 + 3x + 3$.

          (c)   Thus, $f(x) = \left(x + \frac{1}{3}\right)\left(3x^2 + 3x + 3\right) = 3\left(x + \frac{1}{3}\right)\left(x^2 + x + 1\right)$.

$x^2 + x + 1 = 0$ has no real solution.
The zero is $\frac{-1}{3}$.

47. $f(x) = x^3 - 8x^2 + 17x - 6$

Step 1: $f(x)$ has at most 3 real zeros.
Step 2:   By Descartes Rule of Signs, there are 3 or 1 positive real zeros.
Also because $f(-x) = (-x)^3 - 8(-x)^2 + 17(-x) - 6 = -x^3 - 8x^2 - 17x - 6$,
there are no negative real zeros.
Step 3:   Possible rational zeros:

$$p = \pm 1, \pm 2, \pm 3, \pm 6; \quad q = \pm 1; \quad \frac{p}{q} = \pm 1, \pm 2, \pm 3, \pm 6$$

Step 4:   Using the Bounds on Zeros Theorem:

$a_2 = -8, \quad a_1 = 17, \quad a_0 = -6$

$\text{Max}\left\{1, |-6| + |17| + |-8|\right\} = \text{Max}\left\{1, 31\right\} = 31$

$1 + \text{Max}\left\{|-6|, |17|, |-8|\right\} = 1 + 17 = 18$

The smaller of the two numbers is 18. Thus, every zero of f lies between
−18 and 18.
Graphing using the bounds and ZOOM-FIT: (Second has a better window.)

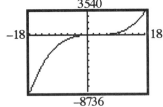
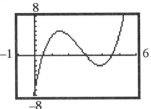

Step 5:   (a)   From the graph it appears that there are x-intercepts at 0.5, 3, and 4.5.
          (b)   Using synthetic division:

$$3\overline{)\begin{array}{cccc} 1 & -8 & 17 & -6 \\ & 3 & -15 & 6 \end{array}}$$
$$\begin{array}{cccc} 1 & -5 & 2 & 0 \end{array}$$

Since the remainder is 0, $x - 3$ is a factor. The other factor is the
quotient: $x^2 - 5x + 2$.

(c)   Thus, $f(x) = (x-3)(x^2 - 5x + 2)$. Using the quadratic formula to

find the solutions of the depressed equation $x^2 - 5x + 2 = 0$:

$$x = \frac{-(-5) \pm \sqrt{(-5)^2 - 4(1)(2)}}{2(1)} = \frac{5 \pm \sqrt{17}}{2}$$

Thus, $f(x) = (x-3)\left(x - \left(\frac{5 + \sqrt{17}}{2}\right)\right)\left(x - \left(\frac{5 - \sqrt{17}}{2}\right)\right)$.

The zeros are 3, $\dfrac{5 + \sqrt{17}}{2}$, and $\dfrac{5 - \sqrt{17}}{2}$ or 3, 4.56, and 0.44.

49. $f(x) = x^4 + x^3 - 3x^2 - x + 2$

Step 1: $f(x)$ has at most 4 real zeros.

Step 2:   By Descartes Rule of Signs, there are 2 or 0 positive real zeros.

$f(-x) = (-x)^4 + (-x)^3 - 3(-x)^2 - (-x) + 2 = x^4 - x^3 - 3x^2 + x + 2$; thus, there are 2 or 0 negative real zeros.

Step 3:   Possible rational zeros:

$$p = \pm 1, \pm 2; \quad q = \pm 1; \quad \frac{p}{q} = \pm 1, \pm 2$$

Step 4:   Using the Bounds on Zeros Theorem:

$a_3 = 1, \ a_2 = -3, \ a_1 = -1, \ a_0 = 2$

$\text{Max}\left\{1, |2| + |-1| + |-3| + |1|\right\} = \text{Max}\left\{1, 7\right\} = 7$

$1 + \text{Max}\left\{|2|, |-1|, |-3|, |1|\right\} = 1 + 3 = 4$

The smaller of the two numbers is 4. Thus, every zero of f lies between –4 and 4.

Graphing using the bounds and ZOOM-FIT: (Second has a better window.)

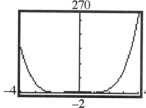 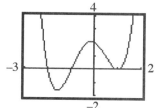

Step 5:   (a)   From the graph it appears that there are x-intercepts at –2, –1, and 1.

(b)   Using synthetic division:

$$
\begin{array}{r|rrrr}
-2 & 1 & 1 & -3 & -1 & 2 \\
   &   & -2 & 2 & 2 & -2 \\
\hline
   & 1 & -1 & -1 & 1 & 0
\end{array}
\qquad
\begin{array}{r|rrrr}
-1 & 1 & -1 & -1 & 1 \\
   &   & -1 & 2 & -1 \\
\hline
   & 1 & -2 & 1 & 0
\end{array}
$$

Since the remainder is 0, $x + 2$ and $x + 1$ are factors. The other factor is the quotient: $x^2 - 2x + 1$.

(c)   Thus, $f(x) = (x + 2)(x + 1)(x - 1)^2$.

The zeros are –2, –1, and 1 (multiplicity 2).

51. $f(x) = 2x^4 + 17x^3 + 35x^2 - 9x - 45$

   Step 1: $f(x)$ has at most 4 real zeros.

   Step 2:   By Descartes Rule of Signs, there is 1 positive real zero.

$$f(-x) = 2(-x)^4 + 17(-x)^3 + 35(-x)^2 - 9(-x) - 45$$
$$= 2x^4 - 17x^3 + 35x^2 + 9x - 45$$

   thus, there are 3 or 1 negative real zeros.

   Step 3:   Possible rational zeros:

$$p = \pm 1, \pm 3, \pm 5, \pm 9, \pm 15, \pm 45; \quad q = \pm 1, \pm 2;$$

$$\frac{p}{q} = \pm 1, \pm 3, \pm 5, \pm 9, \pm 15, \pm 45, \pm \frac{1}{2}, \pm \frac{3}{2}, \pm \frac{5}{2}, \pm \frac{9}{2}, \pm \frac{15}{2}, \pm \frac{45}{2}$$

   Step 4:   Using the Bounds on Zeros Theorem:

$$f(x) = 2\left(x^4 + \frac{17}{2}x^3 + \frac{35}{2}x^2 - \frac{9}{2}x - \frac{45}{2} \right)$$

$$a_3 = \frac{17}{2}, \quad a_2 = \frac{35}{2}, \quad a_1 = \frac{-9}{2}, \quad a_0 = \frac{-45}{2}$$

$$\text{Max}\left\{ 1, \left|\frac{-45}{2}\right| + \left|\frac{-9}{2}\right| + \left|\frac{35}{2}\right| + \left|\frac{17}{2}\right| \right\} = \text{Max}\left\{ 1, 53 \right\} = 53$$

$$1 + \text{Max}\left\{ \left|\frac{-45}{2}\right|, \left|\frac{-9}{2}\right|, \left|\frac{35}{2}\right|, \left|\frac{17}{2}\right| \right\} = 1 + \frac{45}{2} = 23.5$$

   The smaller of the two numbers is 23.5. Thus, every zero of f lies
         between -23.5 and 23.5.

   Graphing using the bounds and ZOOM-FIT: (Second has a better window.)

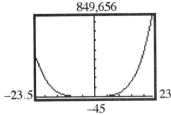

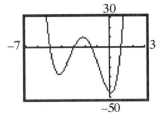

   Step 5:   (a)   From the graph it appears that there are x-intercepts at -5, -3, -1.5,
         and 1.

      (b)   Using synthetic division:

$$
\begin{array}{r|rrrrr}
-5 & 2 & 17 & 35 & -9 & -45 \\
   &   & -10 & -35 & 0 & 45 \\
\hline
   & 2 & 7 & 0 & -9 & 0
\end{array}
\qquad
\begin{array}{r|rrrr}
-3 & 2 & 7 & 0 & -9 \\
   &   & -6 & -3 & 9 \\
\hline
   & 2 & 1 & -3 & 0
\end{array}
$$

      Since the remainder is 0, $x + 5$ and $x + 3$ are factors. The other factor
      is the quotient: $2x^2 + x - 3$.

      (c)   Thus, $f(x) = (x + 5)(x + 3)(2x + 3)(x - 1)$.
         The zeros are -5, -3, -1.5 and 1.

53. $f(x) = 2x^4 - 3x^3 - 21x^2 - 2x + 24$

Step 1: $f(x)$ has at most 4 real zeros.

Step 2: By Descartes Rule of Signs, there are 2 or 0 positive real zeros.

$$f(-x) = 2(-x)^4 - 3(-x)^3 - 21(-x)^2 - 2(-x) + 24$$

$$= 2x^4 + 3x^3 - 21x^2 + 2x + 24$$

thus, there are 2 or 0 negative real zeros.

Step 3: Possible rational zeros:

$$p = \pm 1, \pm 2, \pm 3, \pm 4, \pm 6, \pm 8, \pm 12, \pm 24; \quad q = \pm 1, \pm 2;$$

$$\frac{p}{q} = \pm 1, \pm 2, \pm 3, \pm 4, \pm 6, \pm 8, \pm 12, \pm 24, \pm \frac{1}{2}, \pm \frac{3}{2}$$

Step 4: Using the Bounds on Zeros Theorem:

$$f(x) = 2\left(x^4 - \frac{3}{2}x^3 - \frac{21}{2}x^2 - x + 12\right)$$

$$a_3 = \frac{-3}{2}, \quad a_2 = \frac{-21}{2}, \quad a_1 = -1, \quad a_0 = 12$$

$$\text{Max}\left\{1, |12| + |-1| + \left|\frac{-21}{2}\right| + \left|\frac{-3}{2}\right|\right\} = \text{Max}\left\{1, 25\right\} = 25$$

$$1 + \text{Max}\left\{|12|, |-1|, \left|\frac{-21}{2}\right|, \left|\frac{-3}{2}\right|\right\} = 1 + 12 = 13$$

The smaller of the two numbers is 13. Thus, every zero of f lies between -13 and 13.

Graphing using the bounds and ZOOM-FIT: (Second has a better window.)

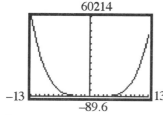
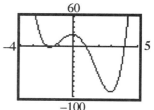

Step 5: (a) From the graph it appears that there are x-intercepts at -2, -1.5, 1, and 4.

(b) Using synthetic division:

$$-2\overline{)2 \quad -3 \quad -21 \quad -2 \quad 24}$$
$$\underline{-4 \quad 14 \quad 14 \quad -24}$$
$$2 \quad -7 \quad -7 \quad 12 \quad 0$$

$$4\overline{)2 \quad -7 \quad -7 \quad 12}$$
$$\underline{8 \quad 4 \quad -12}$$
$$2 \quad 1 \quad -3 \quad 0$$

Since the remainder is 0, $x + 2$ and $x - 4$ are factors. The other factor is the quotient: $2x^2 + x - 3$.

(c) Thus, $f(x) = (x + 2)(2x + 3)(x - 1)(x - 4)$.

The zeros are -2, -1.5, 1 and 4.

55. $f(x) = 4x^4 + 7x^2 - 2$

  Step 1: $f(x)$ has at most 4 real zeros.

  Step 2: By Descartes Rule of Signs, there is 1 positive real zero.

  $f(-x) = 4(-x)^4 + 7(-x)^2 - 2 = 4x^4 + 7x^2 - 2$;

  thus, there is 1 negative real zero.

  Step 3: Possible rational zeros:

  $$p = \pm 1, \pm 2; \quad q = \pm 1, \pm 2, \pm 4; \quad \frac{p}{q} = \pm 1, \pm 2, \pm \frac{1}{2}, \pm \frac{1}{4}$$

  Step 4: Using the Bounds on Zeros Theorem:

  $$f(x) = 4\left(x^4 + \frac{7}{4}x^2 - \frac{1}{2} \right)$$

  $$a_3 = 0, \; a_2 = \frac{7}{4}, \; a_1 = 0, \; a_0 = \frac{-1}{2}$$

  $$\text{Max}\left\{ 1, \left| \frac{-1}{2} \right| + |0| + \left| \frac{7}{4} \right| + |0| \right\} = \text{Max}\left\{ 1, 2.25 \right\} = 2.25$$

  $$1 + \text{Max}\left\{ \left| \frac{-1}{2} \right|, |0|, \left| \frac{7}{4} \right|, |0| \right\} = 1 + 1.75 = 2.75$$

  The smaller of the two numbers is 2.25. Thus, every zero of f lies
    between –2.25 and 2.25.

  Graphing using the bounds and ZOOM-FIT: (Second has a better window.)

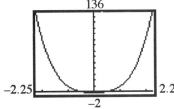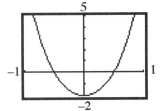

  Step 5: (a) From the graph it appears that there are x-intercepts at –0.5, and 0.5.

  (b) Using synthetic division:

$$-0.5 \overline{)\begin{array}{rrrrr} 4 & 0 & 7 & 0 & -2 \\ & -2 & 1 & -4 & 2 \\ \hline 4 & -2 & 8 & -4 & 0 \end{array}} \qquad 0.5 \overline{)\begin{array}{rrrr} 4 & -2 & 8 & -4 \\ & 2 & 0 & 4 \\ \hline 4 & 0 & 8 & 0 \end{array}}$$

  Since the remainder is 0, $x + \frac{1}{2}$ and $x - \frac{1}{2}$ are factors. The other factor
    is the quotient: $4x^2 + 8$.

  (c) Thus, $f(x) = 4\left(x + \frac{1}{2} \right)\left(x - \frac{1}{2} \right)\left(x^2 + 2 \right) = (2x + 1)(2x - 1)\left(x^2 + 2 \right)$. The
    depressed equation has no real zeros.
    The zeros are –0.5, and 0.5.

57. $f(x) = 4x^5 - 8x^4 - x + 2$

Step 1: $f(x)$ has at most 5 real zeros.

Step 2: By Descartes Rule of Signs, there are 2 or 0 positive real zeros.
$f(-x) = 4(-x)^5 - 8(-x)^4 - (-x) + 2 = -4x^5 - 8x^4 + x + 2$;
thus, there is 1 negative real zero.

Step 3: Possible rational zeros:

$$p = \pm 1, \pm 2; \quad q = \pm 1, \pm 2, \pm 4; \quad \frac{p}{q} = \pm 1, \pm 2, \pm \frac{1}{2}, \pm \frac{1}{4}$$

Step 4: Using the Bounds on Zeros Theorem:

$$f(x) = 4\left(x^5 - 2x^4 - \frac{1}{4}x + \frac{1}{2} \right)$$

$$a_4 = -2, \ a_3 = 0, \ a_2 = 0, \ a_1 = \frac{-1}{4}, \ a_0 = \frac{1}{2}$$

$$\text{Max}\left\{ 1, \left|\frac{1}{2}\right| + \left|\frac{-1}{4}\right| + |0| + |0| + |-2| \right\} = \text{Max}\left\{ 1, 2.75 \right\} = 2.75$$

$$1 + \text{Max}\left\{ \left|\frac{1}{2}\right|, \left|\frac{-1}{4}\right|, |0|, |0|, |-2| \right\} = 1 + 2 = 3$$

The smaller of the two numbers is 2.75. Thus, every zero of f lies between –2.75 and 2.75.

Graphing using the bounds and ZOOM-FIT: (Second has a better window.)

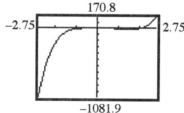 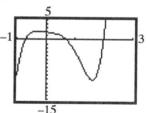

Step 5: (a) From the graph it appears that there are x-intercepts at –0.7, 0.7 and 2.

(b) Using synthetic division:

$$2\overline{)\begin{array}{rrrrrr} 4 & -8 & 0 & 0 & -1 & 2 \\ & 8 & 0 & 0 & 0 & -2 \\ \hline 4 & 0 & 0 & 0 & -1 & 0 \end{array}}$$

Since the remainder is 0, $x - 2$ is a factor. The other factor is the quotient: $4x^4 - 1$.

(c) Factoring,

$$f(x) = (x - 2)\left(4x^4 - 1\right) = (x - 2)(2x^2 - 1)(2x^2 + 1)$$

$$= (x - 2)\left(\sqrt{2}x - 1\right)\left(\sqrt{2}x + 1\right)\left(2x^2 + 1\right)$$

The zeros are $\dfrac{-\sqrt{2}}{2}, \dfrac{\sqrt{2}}{2}$, and 2 or –0.71, –0.71, and 2.

59. $f(x) = x^3 + 3.2x^2 - 16.83x - 5.31$

$f(x)$ has at most 3 real zeros.
Solving by graphing (using ZERO):

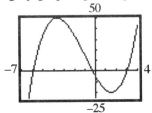

The zeros are approximately –5.9, –0.3, and 3.

61. $f(x) = x^4 - 1.4x^3 - 33.71x^2 + 23.94x + 292.41$

$f(x)$ has at most 4 real zeros.
Solving by graphing (using ZERO):

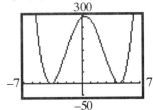

The zeros are approximately –3.80 and 4.50. These zeros are each of multiplicity 2.

63. $f(x) = x^3 + 19.5x^2 - 1021x + 1000.5$

$f(x)$ has at most 3 real zeros.
Solving by graphing (using ZERO):

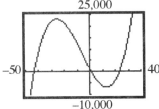

The zeros are approximately –43.5, 1, and 23.

65. $x^4 - x^3 + 2x^2 - 4x - 8 = 0$

The solutions of the equation are the zeros of $f(x) = x^4 - x^3 + 2x^2 - 4x - 8$.

Step 1: $f(x)$ has at most 4 real zeros.

Step 2: By Descartes Rule of Signs, there are 3 or 1 positive real zeros.

$f(-x) = (-x)^4 - (-x)^3 + 2(-x)^2 - 4(-x) - 8 = x^4 + x^3 + 2x^2 + 4x - 8$;
thus, there is 1 negative real zero.

Step 3: Possible rational zeros:

$$p = \pm 1, \pm 2, \pm 4, \pm 8; \quad q = \pm 1; \quad \frac{p}{q} = \pm 1, \pm 2, \pm 4, \pm 8$$

Step 4: Using the Bounds on Zeros Theorem:

$$a_3 = -1, \quad a_2 = 2, \quad a_1 = -4, \quad a_0 = -8$$

$$\text{Max}\left\{1, |-8| + |-4| + |2| + |-1|\right\} = \text{Max}\left\{1, 15\right\} = 15$$

$$1 + \text{Max}\left\{|-8|, |-4|, |2|, |-1|\right\} = 1 + 8 = 9$$

The smaller of the two numbers is 9. Thus, every zero of f lies between –9 and 9.

Graphing using the bounds and ZOOM-FIT: (Second has a better window.)

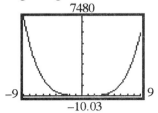
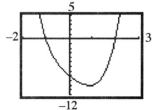

Step 5: (a) From the graph it appears that there are x-intercepts at –1 and 2.

(b) Using synthetic division:

$$-1\overline{)\begin{array}{rrrrr} 1 & -1 & 2 & -4 & -8 \\ & -1 & 2 & -4 & 8 \\ \hline 1 & -2 & 4 & -8 & 0 \end{array}}$$

$$2\overline{)\begin{array}{rrrr} 1 & -2 & 4 & -8 \\ & 2 & 0 & 8 \\ \hline 1 & 0 & 4 & 0 \end{array}}$$

Since the remainder is 0, $x + 1$ and $x - 2$ are factors. The other factor is the quotient: $x^2 + 4$.

(c) The zeros are –1 and 2. ($x^2 + 4 = 0$ has no real solutions.)

67. $3x^3 + 4x^2 - 7x + 2 = 0$

The solutions of the equation are the zeros of $f(x) = 3x^3 + 4x^2 - 7x + 2$.

Step 1: $f(x)$ has at most 3 real zeros.

Step 2: By Descartes Rule of Signs, there are 2 or 0 positive real zeros.

$$f(-x) = 3(-x)^3 + 4(-x)^2 - 7(-x) + 2 = -3x^3 + 4x^2 + 7x + 2;$$

thus, there is 1 negative real zero.

Step 3: Possible rational zeros:

$$p = \pm 1, \pm 2; \quad q = \pm 1, \pm 3; \quad \frac{p}{q} = \pm 1, \pm 2, \pm \frac{1}{3}, \pm \frac{2}{3}$$

Step 4: Using the Bounds on Zeros Theorem:

$$f(x) = 3\left(x^3 + \frac{4}{3}x^2 - \frac{7}{3}x + \frac{2}{3}\right)$$

$$a_2 = \frac{4}{3}, \quad a_1 = \frac{-7}{3}, \quad a_0 = \frac{2}{3}$$

$$\text{Max}\left\{1, \left|\frac{2}{3}\right| + \left|\frac{-7}{3}\right| + \left|\frac{4}{3}\right|\right\} = \text{Max}\left\{1, \frac{13}{3}\right\} = \frac{13}{3}$$

$$1 + \text{Max}\left\{\left|\frac{2}{3}\right|, \left|\frac{-7}{3}\right|, \left|\frac{4}{3}\right|\right\} = 1 + \frac{7}{3} = \frac{10}{3}$$

The smaller of the two numbers is 3.33. Thus, every zero of f lies between –3.33 and 3.33.

Graphing using the bounds and ZOOM-FIT: (Second has a better window.)

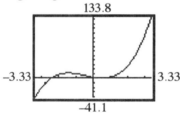

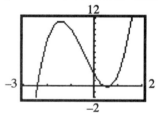

Step 5: (a) From the graph it appears that there are x-intercepts at $\frac{1}{3}$, $\frac{2}{3}$, and -2.4.

(b) Using synthetic division:

$$\frac{2}{3} \overline{) \begin{array}{rrrr} 3 & 4 & -7 & 2 \\ & 2 & 4 & -2 \\ \hline 3 & 6 & -3 & 0 \end{array}}$$

Since the remainder is 0, $x - \frac{2}{3}$ is a factor. The other factor is the quotient: $3x^2 + 6x - 3$.

$$f(x) = \left(x - \tfrac{2}{3}\right)\left(3x^2 + 6x - 3\right) = 3\left(x - \tfrac{2}{3}\right)\left(x^2 + 2x - 1\right)$$

Using the quadratic formula to solve $x^2 + 2x - 1 = 0$:

$$x = \frac{-2 \pm \sqrt{4 - 4(1)(-1)}}{2(1)} = \frac{-2 \pm \sqrt{8}}{2} = \frac{-2 \pm 2\sqrt{2}}{2} = -1 \pm \sqrt{2}$$

(c) The zeros are $\frac{2}{3}$, $-1 + \sqrt{2}$, and $-1 - \sqrt{2}$ or 0.67, 0.41, and -2.41.

69. $3x^3 - x^2 - 15x + 5 = 0$

Solving by factoring:

$$x^2(3x - 1) - 5(3x - 1) = 0$$

$$(3x - 1)\left(x^2 - 5\right) = 0$$

$$(3x - 1)\left(x - \sqrt{5}\right)\left(x + \sqrt{5}\right) = 0$$

The solutions of the equation are $\frac{1}{3}$, $\sqrt{5}$, and $-\sqrt{5}$ or 0.33, 2.24, and -2.24.

71. $x^4 + 4x^3 + 2x^2 - x + 6 = 0$

The solutions of the equation are the zeros of $f(x) = x^4 + 4x^3 + 2x^2 - x + 6$.

Step 1: $f(x)$ has at most 4 real zeros.

Step 2: By Descartes Rule of Signs, there are 2 or 0 positive real zeros.

$$f(-x) = (-x)^4 + 4(-x)^3 + 2(-x)^2 - (-x) + 6 = x^4 - 4x^3 + 2x^2 + x + 6;$$

thus, there are 2 or 0 negative real zeros.

Step 3: Possible rational zeros:

$$p = \pm 1, \pm 2, \pm 3, \pm 6; \quad q = \pm 1; \quad \frac{p}{q} = \pm 1, \pm 2, \pm 3, \pm 6$$

Step 4:   Using the Bounds on Zeros Theorem:
$$a_3 = 4, \; a_2 = 2, \; a_1 = -1, \; a_0 = 6$$
$$\text{Max} \left\{ 1, |6| + |-1| + |2| + |4| \right\} = \text{Max} \left\{ 1, 13 \right\} = 13$$
$$1 + \text{Max} \left\{ |6|, |-1|, |2|, |4| \right\} = 1 + 6 = 7$$

The smaller of the two numbers is 7. Thus, every zero of f lies between
-7 and 7.
Graphing using the bounds and ZOOM-FIT: (Second has a better window.)

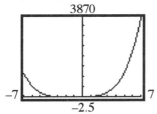
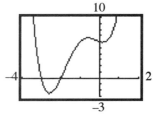

Step 5:   (a)   From the graph it appears that there are x-intercepts at -3 and -2.
           (b)   Using synthetic division:

$$-3\overline{)\begin{array}{ccccc} 1 & 4 & 2 & -1 & 6 \\ & -3 & -3 & 3 & -6 \\ \hline 1 & 1 & -1 & 2 & 0 \end{array}} \qquad -2\overline{)\begin{array}{cccc} 1 & 1 & -1 & 2 \\ & -2 & 2 & -2 \\ \hline 1 & -1 & 1 & 0 \end{array}}$$

Since the remainder is 0, $x + 3$ and $x + 2$ are factors. The other factor
is the quotient: $x^2 - x + 1$.

           (c)   The zeros are -3 and -2. ($x^2 - x + 1 = 0$ has no real solutions.)

73. $x^3 - \frac{2}{3} x^2 + \frac{8}{3} x + 1 = 0$

The solutions of the equation are the zeros of $f(x) = x^3 - \frac{2}{3} x^2 + \frac{8}{3} x + 1$.

Step 1: $f(x)$ has at most 3 real zeros.
Step 2:   By Descartes Rule of Signs, there are 2 or 0 positive real zeros.
$$f(-x) = (-x)^3 - \frac{2}{3}(-x)^2 + \frac{8}{3}(-x) + 1 = -x^3 - \frac{2}{3}x^2 - \frac{8}{3}x + 1;$$
thus, there is 1 negative real zero.
Step 3:   Use the equivalent equation $3x^3 - 2x^2 + 8x + 3 = 0$ to find the possible
rational zeros:
$$p = \pm 1, \pm 3; \quad q = \pm 1, \pm 3; \quad \frac{p}{q} = \pm 1, \pm 3, \pm \frac{1}{3}$$
Step 4:   Using the Bounds on Zeros Theorem:
$$a_2 = \frac{-2}{3}, \quad a_1 = \frac{8}{3}, \quad a_0 = 1$$
$$\text{Max} \left\{ 1, |1| + \left| \frac{8}{3} \right| + \left| \frac{-2}{3} \right| \right\} = \text{Max} \left\{ 1, \frac{13}{3} \right\} = \frac{13}{3}$$
$$1 + \text{Max} \left\{ |1|, \left| \frac{8}{3} \right|, \left| \frac{-2}{3} \right| \right\} = 1 + \frac{8}{3} = \frac{11}{3}$$

The smaller of the two numbers is 3.67. Thus, every zero of f lies
between -3.67 and 3.67.

Graphing using the bounds and ZOOM-FIT: (Second has a better window.)

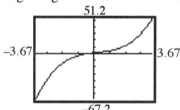

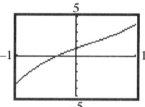

Step 5: (a) From the graph it appears that there is an x-intercepts at $\frac{-1}{3}$.

(b) Using synthetic division:

$$\frac{-1}{3}\overline{)\,1 \quad \frac{-2}{3} \quad \frac{8}{3} \quad 1}$$
$$\phantom{\frac{-1}{3}\overline{)\,1}}\quad \frac{-1}{3} \quad \frac{1}{3} \quad -1$$
$$\overline{\phantom{\frac{-1}{3})}\;1 \quad -1 \quad 3 \quad 0}$$

Since the remainder is 0, $x + \frac{1}{3}$ is a factor. The other factor is the quotient: $x^2 - x + 3$.

(c) The real zero is $\frac{-1}{3}$. ($x^2 - x + 3 = 0$ has no real solutions.)

75. Using the TABLE feature to show that there is a zero in the interval:

$f(x) = 8x^4 - 2x^2 + 5x - 1; \quad [0, 1]$

X	Y1
-1	0
0	-1
1	10
2	129
3	644
4	2035
5	4974

X=-1

$f(0) = -1 < 0$ and $f(1) = 10 > 0$
Since one is positive and one is negative , there is a zero in the interval .

Using the TABLE feature to approximate the zero to two decimal places:

X	Y1
.213	-.0093
.214	-.0048
.215	-4E-4
.216	.0041
.217	.00856
.218	.01302
	.01748

X=.219

The zero is approximately 0.22.

77. Using the TABLE feature to show that there is a zero in the interval:

$f(x) = 2x^3 + 6x^2 - 8x + 2; \quad [-5, -4]$

X	Y1
-8	-574
-7	-334
-6	-166
-5	-58
-4	2
-3	26
-2	26

Y1=2X^3+6X²-8X+2

$f(-5) = -58 < 0$ and $f(-4) = 2 > 0$
Since one is positive and one is negative , there is a zero in the interval .

Using the TABLE feature to approximate the zero to two decimal places:

X	Y1
-4.052	-.129
-4.051	-.0871
-4.05	-.0453
-4.049	-.0035
-4.048	.03831
-4.047	.08003
-4.046	.12172

Y1=2X^3+6X²-8X+2

The zero is approximately -4.05.

79. Using the TABLE feature to show that there is a zero in the interval:

$$f(x) = x^5 - x^4 + 7x^3 - 7x^2 - 18x + 18; \quad [1.4, 1.5]$$

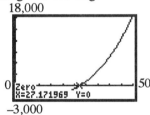

$f(1.4) = -0.1754 < 0$ and $f(1.5) = 1.4063 > 0$

Since one is positive and one is negative ,

there is a zero in the interval .

Using the TABLE feature to approximate the zero to two decimal places:

X	Y1
1.411	-.041
1.412	-.0283
1.413	-.0156
1.414	-.0028
1.415	.01018
1.416	.02315
1.417	.03621

Y1◻X^5-X^4+7X^3...

The zero is approximately 1.41.

81. $C(x) = 3000$

$$0.216x^3 - 2.347x^2 + 14.328x + 10.224 = 3000$$

$$0.216x^3 - 2.347x^2 + 14.328x - 2989.776 = 0$$

There are at most 3 solutions.

Graphing and solving:

18,000

0 ⌐Zero   50
 X=27.171969 Y=0

–3,000

About 27 Cavaliers can be manufactured.

83. $x - 2$ is a factor of $f(x) = x^3 - kx^2 + kx + 2$ only if the remainder that results when $f(x)$ is divided by $x - 2$ is 0. Dividing, we have:

$$
\begin{array}{r|rrrr}
2) & 1 & -k & k & 2 \\
  &   & 2 & -2k+4 & -2k+8 \\
\hline
  & 1 & -k+2 & -k+4 & -2k+10 \\
\end{array}
$$

Since we want the remainder to equal 0, set the remainder equal to zero and solve:

$$-2k + 10 = 0$$
$$-2k = -10$$
$$k = 5$$

85. By the Remainder Theorem we know that the remainder from synthetic division by c is equal to $f(c)$. Thus the easiest way to find the remainder is to evaluate:

$$f(1) = 2(1)^{20} - 8(1)^{10} + 1 - 2 = 2 - 8 + 1 - 2 = -7$$

The remainder is –7.

87. $f(x) = 2x^3 + 3x^2 - 6x + 7$
     By the Rational Zero Theorem, the only possible rational zeros are:
     $$\frac{p}{q} = \pm 1, \pm 7, \pm \frac{1}{2}, \pm \frac{7}{2}$$
     Since $\frac{1}{3}$ is not in the list of possible rational zeros, it is not a zero of $f(x)$.

89. $f(x) = 2x^6 - 5x^4 + x^3 - x + 1$
     By the Rational Zero Theorem, the only possible rational zeros are:
     $$\frac{p}{q} = \pm 1, \pm \frac{1}{2}$$
     Since $\frac{3}{5}$ is not in the list of possible rational zeros, it is not a zero of $f(x)$.

91. Let x be the length of a side of the original cube.
     After removing the 1 inch slice, one dimension will be $x - 1$.
     The volume of the new solid will be:
     $$(x - 1) \cdot x \cdot x = 294$$
     $$x^3 - x^2 = 294$$
     $$x^3 - x^2 - 294 = 0$$
     By Descartes Rule of Signs, we know that there is one positive real solution.
     The possible rational zeros are:
     $$p = \pm 1, \pm 2, \pm 3, \pm 6, \pm 7, \pm 14, \pm 21, \pm 42, \pm 49, \pm 98, \pm 147, \pm 294; \quad q = \pm 1$$
     The rational zeros are the same as the values for p.
     Using synthetic division:

     $$
     \begin{array}{r|rrrr}
     7 & 1 & -1 & 0 & -294 \\
       &   & 7 & 42 & 294 \\
     \hline
       & 1 & 6 & 42 & 0 \\
     \end{array}
     $$

     7 is a zero, so the length of the original edge of the cube was 7 inches.

99. $x^3 + 3x - 14 = 0$
     $$x = \sqrt[3]{\frac{-(-14)}{2} + \sqrt{\frac{(-14)^2}{4} + \frac{3^3}{27}}} + \sqrt[3]{\frac{-(-14)}{2} - \sqrt{\frac{(-14)^2}{4} + \frac{3^3}{27}}}$$
     $$= \sqrt[3]{7 + \sqrt{49 + 1}} + \sqrt[3]{7 - \sqrt{49 + 1}}$$
     $$= \sqrt[3]{7 + \sqrt{50}} + \sqrt[3]{7 - \sqrt{50}}$$
     $$= \sqrt[3]{7 + 5\sqrt{2}} + \sqrt[3]{7 - 5\sqrt{2}}$$
     $$= 2$$

101. From the stated property, we can make the following statement about zero:
     If $2 + a$ is a zero, then $2 - a$ is also a zero. If $2 + b$ is a zero, then $2 - b$ is also a
     zero. These are four distinct zeros. If these are the only zeros of the function, then
     the sum of the zeros is: $2 + a + 2 - a + 2 + b + 2 - b = 8$. The sum of the zeros is 8.

4.2 Complex Numbers; Quadratic Equations with a Negative Discriminant

1. $(2-3i)+(6+8i) = (2+6)+(-3+8)i = 8+5i$

3. $(-3+2i)-(4-4i) = (-3-4)+(2-(-4))i = -7+6i$

5. $(2-5i)-(8+6i) = (2-8)+(-5-6)i = -6-11i$

7. $3(2-6i) = 6-18i$

9. $2i(2-3i) = 4i-6i^2 = 4i-6(-1) = 6+4i$

11. $(3-4i)(2+i) = 6+3i-8i-4i^2 = 6-5i-4(-1) = 10-5i$

13. $(-6+i)(-6-i) = 36+6i-6i-i^2 = 36-(-1) = 37$

15. $\dfrac{10}{3-4i} = \dfrac{10}{3-4i} \cdot \dfrac{3+4i}{3+4i} = \dfrac{30+40i}{9+12i-12i-16i^2} = \dfrac{30+40i}{9-16(-1)} = \dfrac{30+40i}{25}$

    $= \dfrac{30}{25}+\dfrac{40}{25}i = \dfrac{6}{5}+\dfrac{8}{5}i$

17. $\dfrac{2+i}{i} = \dfrac{2+i}{i} \cdot \dfrac{-i}{-i} = \dfrac{-2i-i^2}{-i^2} = \dfrac{-2i-(-1)}{-(-1)} = \dfrac{1-2i}{1} = 1-2i$

19. $\dfrac{6-i}{1+i} = \dfrac{6-i}{1+i} \cdot \dfrac{1-i}{1-i} = \dfrac{6-6i-i+i^2}{1-i+i-i^2} = \dfrac{6-7i+(-1)}{1-(-1)} = \dfrac{5-7i}{2} = \dfrac{5}{2}-\dfrac{7}{2}i$

21. $\left(\dfrac{1}{2}+\dfrac{\sqrt{3}}{2}i\right)^2 = \dfrac{1}{4}+2\left(\dfrac{1}{2}\right)\left(\dfrac{\sqrt{3}}{2}i\right)+\dfrac{3}{4}i^2 = \dfrac{1}{4}+\dfrac{\sqrt{3}}{2}i+\dfrac{3}{4}(-1) = \dfrac{-1}{2}+\dfrac{\sqrt{3}}{2}i$

23. $(1+i)^2 = 1+2i+i^2 = 1+2i+(-1) = 2i$

25. $i^{23} = i^{22+1} = i^{22} \cdot i = \left(i^2\right)^{11} \cdot i = (-1)^{11}i = -i$

27. $i^{-15} = \dfrac{1}{i^{15}} = \dfrac{1}{i^{14+1}} = \dfrac{1}{i^{14} \cdot i} = \dfrac{1}{(i^2)^7 \cdot i} = \dfrac{1}{(-1)^7 i} = \dfrac{1}{-i} = \dfrac{1}{-i} \cdot \dfrac{i}{i} = \dfrac{i}{-i^2} = \dfrac{i}{-(-1)} = i$

29. $i^6-5 = \left(i^2\right)^3-5 = (-1)^3-5 = -1-5 = -6$

31. $6i^3-4i^5 = i^3(6-4i^2) = i^2 \cdot i(6-4(-1)) = -1 \cdot i(10) = -10i$

33. $(1+i)^3 = (1+i)(1+i)(1+i) = (1+2i+i^2)(1+i) = (1+2i-1)(1+i) = 2i(1+i)$
$$= 2i + 2i^2 = 2i + 2(-1) = -2 + 2i$$

35. $i^7(1+i^2) = i^7(1+(-1)) = i^7(0) = 0$

37. $i^6 + i^4 + i^2 + 1 = (i^2)^3 + (i^2)^2 + i^2 + 1 = (-1)^3 + (-1)^2 + (-1) + 1 = -1 + 1 - 1 + 1 = 0$

39. $\sqrt{-4} = 2i$

41. $\sqrt{-25} = 5i$

43. $\sqrt{(3+4i)(4i-3)} = \sqrt{12i - 9 + 16i^2 - 12i} = \sqrt{-9 + 16(-1)} = \sqrt{-25} = 5i$

45. $x^2 + 4 = 0$
$a = 1, b = 0, c = 4, \quad b^2 - 4ac = 0^2 - 4(1)(4) = -16$
$$x = \frac{-0 \pm \sqrt{-16}}{2(1)} = \frac{\pm 4i}{2} = \pm 2i$$
The solution set is $\{-2i, 2i\}$.

47. $x^2 - 16 = 0$
$a = 1, b = 0, c = -16, \quad b^2 - 4ac = 0^2 - 4(1)(-16) = 64$
$$x = \frac{-0 \pm \sqrt{64}}{2(1)} = \frac{\pm 8}{2} = \pm 4$$
The solution set is $\{-4, 4\}$.

49. $x^2 - 6x + 13 = 0$
$a = 1, b = -6, c = 13, \quad b^2 - 4ac = (-6)^2 - 4(1)(13) = 36 - 52 = -16$
$$x = \frac{-(-6) \pm \sqrt{-16}}{2(1)} = \frac{6 \pm 4i}{2} = 3 \pm 2i$$
The solution set is $\{3 - 2i, 3 + 2i\}$.

51. $x^2 - 6x + 10 = 0$
$a = 1, b = -6, c = 10, \quad b^2 - 4ac = (-6)^2 - 4(1)(10) = 36 - 40 = -4$
$$x = \frac{-(-6) \pm \sqrt{-4}}{2(1)} = \frac{6 \pm 2i}{2} = 3 \pm i$$
The solution set is $\{3 - i, 3 + i\}$.

53. $8x^2 - 4x + 1 = 0$

    $a = 8, b = -4, c = 1, \quad b^2 - 4ac = (-4)^2 - 4(8)(1) = 16 - 32 = -16$

    $x = \dfrac{-(-4) \pm \sqrt{-16}}{2(8)} = \dfrac{4 \pm 4i}{16} = \dfrac{1}{4} \pm \dfrac{1}{4}i$

    The solution set is $\left\{\dfrac{1}{4} - \dfrac{1}{4}i, \ \dfrac{1}{4} + \dfrac{1}{4}i\right\}$.

55. $5x^2 + 2x + 1 = 0$

    $a = 5, b = 2, c = 1, \quad b^2 - 4ac = 2^2 - 4(5)(1) = 4 - 20 = -16$

    $x = \dfrac{-2 \pm \sqrt{-16}}{2(5)} = \dfrac{-2 \pm 4i}{10} = \dfrac{-1}{5} \pm \dfrac{2}{5}i$

    The solution set is $\left\{\dfrac{-1}{5} - \dfrac{2}{5}i, \ \dfrac{-1}{5} + \dfrac{2}{5}i\right\}$.

57. $x^2 + x + 1 = 0$

    $a = 1, b = 1, c = 1, \quad b^2 - 4ac = 1^2 - 4(1)(1) = 1 - 4 = -3$

    $x = \dfrac{-1 \pm \sqrt{-3}}{2(1)} = \dfrac{-1 \pm \sqrt{3}\,i}{2} = \dfrac{-1}{2} \pm \dfrac{\sqrt{3}}{2}i$

    The solution set is $\left\{\dfrac{-1}{2} - \dfrac{\sqrt{3}}{2}i, \ \dfrac{-1}{2} + \dfrac{\sqrt{3}}{2}i\right\}$.

59. $x^3 - 8 = 0$

    $(x - 2)(x^2 + 2x + 4) = 0$

    $x - 2 = 0 \ \text{ or } \ x^2 + 2x + 4 = 0$

    $x = 2$

    $a = 1, b = 2, c = 4, \quad b^2 - 4ac = 2^2 - 4(1)(4) = 4 - 16 = -12$

    $x = \dfrac{-2 \pm \sqrt{-12}}{2(1)} = \dfrac{-2 \pm 2\sqrt{3}\,i}{2} = -1 \pm \sqrt{3}i$

    The solution set is $\left\{2, \ -1 - \sqrt{3}i, \ -1 + \sqrt{3}i\right\}$.

61. $x^4 - 16 = 0$

    $(x^2 - 4)(x^2 + 4) = 0$

    $(x - 2)(x + 2)(x^2 + 4) = 0$

    $x - 2 = 0 \ \text{ or } \ x + 2 = 0 \ \text{ or } \ x^2 + 4 = 0$

    $x = 2 \ \text{ or } \quad x = -2$

    $a = 1, b = 0, c = 4, \quad b^2 - 4ac = 0^2 - 4(1)(4) = 0 - 16 = -16$

    $x = \dfrac{-0 \pm \sqrt{-16}}{2(1)} = \dfrac{\pm 4i}{2} = \pm 2i$

    The solution set is $\left\{-2, \ 2, \ -2i, \ 2i\right\}$.

63. $x^4 + 13x^2 + 36 = 0$

$(x^2 + 9)(x^2 + 4) = 0$

$x^2 + 9 = 0$ or $x^2 + 4 = 0$

$a = 1, b = 0, c = 9,$ $b^2 - 4ac = 0^2 - 4(1)(9) = 0 - 36 = -36$

$x = \dfrac{-0 \pm \sqrt{-36}}{2(1)} = \dfrac{\pm 6i}{2} = \pm 3i$

$a = 1, b = 0, c = 4,$ $b^2 - 4ac = 0^2 - 4(1)(4) = 0 - 16 = -16$

$x = \dfrac{-0 \pm \sqrt{-16}}{2(1)} = \dfrac{\pm 4i}{2} = \pm 2i$

The solution set is $\{-3i,\ 3i,\ -2i,\ 2i\}$.

65. $3x^2 - 3x + 4 = 0$

$a = 3, b = -3, c = 4,$ $b^2 - 4ac = (-3)^2 - 4(3)(4) = 9 - 48 = -39$

The equation has two complex conjugate solutions.

67. $2x^2 + 3x - 4 = 0$

$a = 2, b = 3, c = -4,$ $b^2 - 4ac = 3^2 - 4(2)(-4) = 9 + 32 = 41$

The equation has two unequal real solutions.

69. $9x^2 - 12x + 4 = 0$

$a = 9, b = -12, c = 4,$ $b^2 - 4ac = (-12)^2 - 4(9)(4) = 144 - 144 = 0$

The equation has a repeated real solution.

71. The other solution is the conjugate of $2 + 3i$, or $2 - 3i$.

73. $z + \bar{z} = 3 - 4i + \overline{3 - 4i} = 3 - 4i + 3 + 4i = 6$

75. $z \cdot \bar{z} = (3 - 4i)(\overline{3 - 4i}) = (3 - 4i)(3 + 4i) = 9 + 12i - 12i - 16i^2 = 9 - 16(-1) = 25$

77. $z + \bar{z} = a + bi + \overline{a + bi} = a + bi + a - bi = 2a$

$z - \bar{z} = a + bi - (\overline{a + bi}) = a + bi - (a - bi) = a + bi - a + bi = 2bi$

79. $\overline{z + w} = \overline{(a + bi) + (c + di)} = \overline{(a + c) + (b + d)i} = (a + c) - (b + d)i$

$= (a - bi) + (c - di) = \overline{a + bi} + \overline{c + di} = \bar{z} + \bar{w}$

4.3   Complex Zeros; Fundamental Theorem of Algebra

1. Since complex zeros appear in conjugate pairs, $4 + i$, the conjugate of $4 - i$, is the remaining zero of f.

3. Since complex zeros appear in conjugate pairs, $-i$, the conjugate of i, and $1 - i$, the conjugate of $1 + i$, are the remaining zeros of f.

5. Since complex zeros appear in conjugate pairs, $-i$, the conjugate of i, and $-2i$, the conjugate of $2i$, are the remaining zeros of f.

7. Since complex zeros appear in conjugate pairs, $-i$, the conjugate of i, is the remaining zero of f.

9. Since complex zeros appear in conjugate pairs, $2 - i$, the conjugate of $2 + i$, and $-3 + i$, the conjugate of $-3 - i$, are the remaining zeros of f.

11. Since $3 + 2i$ is a zero, its conjugate $3 - 2i$ is also a zero of f. Finding the function:
$$\begin{aligned} f(x) &= (x - 4)(x - 4)(x - (3 + 2i))(x - (3 - 2i)) \\ &= \left(x^2 - 8x + 16\right)((x - 3) - 2i)((x - 3) + 2i) \\ &= \left(x^2 - 8x + 16\right)\left(x^2 - 6x + 9 - 4i^2\right) \\ &= \left(x^2 - 8x + 16\right)\left(x^2 - 6x + 13\right) \\ &= x^4 - 6x^3 + 13x^2 - 8x^3 + 48x^2 - 104x + 16x^2 - 96x + 208 \\ &= x^4 - 14x^3 + 77x^2 - 200x + 208 \end{aligned}$$

13. Since $-i$ is a zero, its conjugate i is also a zero, and since $1 + i$ is a zero, its conjugate $1 - i$ is also a zero of f. Finding the function:
$$\begin{aligned} f(x) &= (x - 2)(x + i)(x - i)(x - (1 + i))(x - (1 - i)) \\ &= (x - 2)\left(x^2 - i^2\right)((x - 1) - i)((x - 1) + i) \\ &= (x - 2)\left(x^2 + 1\right)\left(x^2 - 2x + 1 - i^2\right) \\ &= \left(x^3 - 2x^2 + x - 2\right)\left(x^2 - 2x + 2\right) \\ &= x^5 - 2x^4 + 2x^3 - 2x^4 + 4x^3 - 4x^2 + x^3 - 2x^2 + 2x - 2x^2 + 4x - 4 \\ &= x^5 - 4x^4 + 7x^3 - 8x^2 + 6x - 4 \end{aligned}$$

15. Since $-i$ is a zero, its conjugate i is also a zero of f. Finding the function:
$$\begin{aligned} f(x) &= (x - 3)(x - 3)(x + i)(x - i) \\ &= \left(x^2 - 6x + 9\right)\left(x^2 - i^2\right) \\ &= \left(x^2 - 6x + 9\right)\left(x^2 + 1\right) \\ &= x^4 + x^2 - 6x^3 - 6x + 9x^2 + 9 \\ &= x^4 - 6x^3 + 10x^2 - 6x + 9 \end{aligned}$$

17. Since $2i$ is a zero, its conjugate $-2i$ is also a zero of f. $x - 2i$ and $x + 2i$ are factors of f. Thus, $(x - 2i)(x + 2i) = x^2 + 4$ is a factor of f. Using division to find the other factor:

$$
\begin{array}{r}
x - 4 \\
x^2 + 4 \overline{\smash{)}\, x^3 - 4x^2 + 4x - 16} \\
\underline{x^3 \qquad\ + 4x} \\
-4x^2 \qquad\ -16 \\
\underline{-4x^2 \qquad\ -16}
\end{array}
$$

$x - 4$ is a factor and the remaining zero is 4. The zeros of f are $4, 2i, -2i$.

19. Since $-2i$ is a zero, its conjugate $2i$ is also a zero of f. $x - 2i$ and $x + 2i$ are factors of f. Thus, $(x - 2i)(x + 2i) = x^2 + 4$ is a factor of f. Using division to find the other factor:

$$
\begin{array}{r}
2x^2 + 5x - 3 \\
x^2 + 4 \overline{\smash{)}\, 2x^4 + 5x^3 + 5x^2 + 20x - 12} \\
\underline{2x^4 \qquad\quad +8x^2} \\
5x^3 - 3x^2 + 20x \\
\underline{5x^3 \qquad\ + 20x} \\
-3x^2 \qquad\ -12 \\
\underline{-3x^2 \qquad\ -12}
\end{array}
$$

$2x^2 + 5x - 3 = (2x - 1)(x + 3)$ are factors and the remaining zeros are $\frac{1}{2}$ and -3. The zeros of f are $2i, -2i, -3, \frac{1}{2}$.

21. Since $3 - 2i$ is a zero, its conjugate $3 + 2i$ is also a zero of h.
     $x - (3 - 2i)$ and $x - (3 + 2i)$ are factors of h. Thus,
     $(x - (3 - 2i))(x - (3 + 2i)) = ((x - 3) + 2i)((x - 3) - 2i) = x^2 - 6x + 9 - 4i^2 = x^2 - 6x + 13$
     is a factor of h. Using division to find the other factor:

$$
\begin{array}{r}
x^2 - 3x - 10 \\
x^2 - 6x + 13 \overline{\smash{)}\, x^4 - 9x^3 + 21x^2 + 21x - 130} \\
\underline{x^4 - 6x^3 + 13x^2} \\
-3x^3 + 8x^2 + 21x \\
\underline{-3x^3 + 18x^2 - 39x} \\
-10x^2 + 60x - 130 \\
\underline{-10x^2 + 60x - 130}
\end{array}
$$

$x^2 - 3x - 10 = (x + 2)(x - 5)$ are factors and the remaining zeros are -2 and 5. The zeros of h are $3 - 2i, 3 + 2i, -2, 5$.

23. Since $-4i$ is a zero, its conjugate $4i$ is also a zero of h. $x - 4i$ and $x + 4i$ are factors of h. Thus, $(x - 4i)(x + 4i) = x^2 + 16$ is a factor of h. Using division to find the other factor:

$$
\begin{array}{r}
3x^3 + 2x^2 - 33x - 22 \\
x^2 + 16 \overline{)3x^5 + 2x^4 + 15x^3 + 10x^2 - 528x - 352} \\
\underline{3x^5 \qquad\quad + 48x^3} \\
2x^4 - 33x^3 + 10x^2 \\
\underline{2x^4 \qquad\quad + 32x^2} \\
-33x^3 - 22x^2 - 528x \\
\underline{-33x^3 \qquad\quad - 528x} \\
-22x^2 \qquad\quad - 352 \\
\underline{-22x^2 \qquad\quad - 352}
\end{array}
$$

$3x^3 + 2x^2 - 33x - 22 = x^2(3x + 2) - 11(3x + 2) = (3x + 2)(x^2 - 11)$

$= (3x + 2)\left(x - \sqrt{11}\right)\left(x + \sqrt{11}\right)$ are factors and the remaining zeros are

$\frac{-2}{3}, \sqrt{11}$, and $-\sqrt{11}$. The zeros of h are $4i, -4i, -\sqrt{11}, \sqrt{11}, \frac{-2}{3}$.

25. $f(x) = x^3 - 1 = (x - 1)\left(x^2 + x + 1\right)$ The zeros of $x^2 + x + 1 = 0$ are:

$$x = \frac{-1 \pm \sqrt{1^2 - 4(1)(1)}}{2(1)} = \frac{-1 \pm \sqrt{-3}}{2} = \frac{-1}{2} + \frac{\sqrt{3}}{2}i \text{ or } \frac{-1}{2} - \frac{\sqrt{3}}{2}i$$

The zeros are: $1, \ \frac{-1}{2} + \frac{\sqrt{3}}{2}i, \ \frac{-1}{2} - \frac{\sqrt{3}}{2}i$.

27. $f(x) = x^3 - 8x^2 + 25x - 26$

Step 1: $f(x)$ has 3 complex zeros.

Step 2: By Descartes Rule of Signs, there are 3 or 1 positive real zeros.

$f(-x) = (-x)^3 - 8(-x)^2 + 25(-x) - 26 = -x^3 - 8x^2 - 25x - 26$; thus, there are no negative real zeros.

Step 3: Possible rational zeros:

$$p = \pm 1, \pm 2, \pm 13, \pm 26; \quad q = \pm 1; \quad \frac{p}{q} = \pm 1, \pm 2, \pm 13, \pm 26$$

Step 4: Graphing the function:

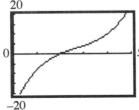

From the graph it appears that there is an x-intercept at 2.

Step 5: Using synthetic division:

$$2)\overline{)1 \quad -8 \quad 25 \quad -26}$$
$$\underline{\quad\quad 2 \quad -12 \quad 26}$$
$$1 \quad -6 \quad 13 \quad 0$$

Since the remainder is 0, $x - 2$ is a factor. The other factor is the quotient: $x^2 - 6x + 13$.

Using the quadratic formula to find the zeros of $x^2 - 6x + 13 = 0$:

$$x = \frac{-(-6) \pm \sqrt{(-6)^2 - 4(1)(13)}}{2(1)} = \frac{6 \pm \sqrt{-16}}{2} = \frac{6 \pm 4i}{2} = 3 \pm 2i.$$

The complex zeros are 2, $3 - 2i$, $3 + 2i$.

29. $f(x) = x^4 + 5x^2 + 4 = \left(x^2 + 4\right)\left(x^2 + 1\right) = (x + 2i)(x - 2i)(x + i)(x - i)$
The zeros are: $-2i$, $-i$, i, $2i$.

31. $f(x) = x^4 + 2x^3 + 22x^2 + 50x - 75$
Step 1: $f(x)$ has 4 complex zeros.
Step 2:   By Descartes Rule of Signs, there is 1 positive real zero.
$$f(-x) = (-x)^4 + 2(-x)^3 + 22(-x)^2 + 50(-x) - 75$$
$$= x^4 - 2x^3 + 22x^2 - 50x - 75$$
thus, there are 3 or 1 negative real zeros.
Step 3:   Possible rational zeros:
$$p = \pm 1, \pm 3, \pm 5, \pm 15, \pm 25, \pm 75; \quad q = \pm 1;$$
$$\frac{p}{q} = \pm 1, \pm 3, \pm 5, \pm 15, \pm 25, \pm 75$$
Step 4:   Graphing the function:

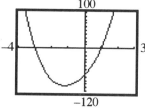

From the graph it appears that there are x-intercepts at -3 and 1.
Step 5:   Using synthetic division:

$$-3)\overline{)1 \quad 2 \quad 22 \quad 50 \quad -75}$$
$$\underline{\quad\quad -3 \quad 3 \quad -75 \quad 75}$$
$$1 \quad -1 \quad 25 \quad -25 \quad 0$$

Since the remainder is 0, $x + 3$ is a factor. The other factor is the quotient:
$$x^3 - x^2 + 25x - 25 = x^2(x - 1) + 25(x - 1) = (x - 1)\left(x^2 + 25\right)$$
$$= (x - 1)(x + 5i)(x - 5i)$$
The complex zeros are -3, 1, $-5i$, $5i$.

33. $f(x) = 3x^4 - x^3 - 9x^2 + 159x - 52$

Step 1: $f(x)$ has 4 complex zeros.

Step 2:   By Descartes Rule of Signs, there are 3 or 1 positive real zeros.
$$f(-x) = 3(-x)^4 - (-x)^3 - 9(-x)^2 + 159(-x) - 52$$
$$= 3x^4 + x^3 - 9x^2 - 159x - 52$$
thus, there is 1 negative real zero.

Step 3:   Possible rational zeros:
$$p = \pm 1, \pm 2, \pm 4, \pm 13, \pm 26, \pm 52; \quad q = \pm 1, \pm 3;$$
$$\frac{p}{q} = \pm 1, \pm 2, \pm 4, \pm 13, \pm 26, \pm 52, \pm \frac{1}{3}, \pm \frac{2}{3}, \pm \frac{4}{3}, \pm \frac{13}{3}, \pm \frac{26}{3}, \pm \frac{52}{3}$$

Step 4:   Graphing the function:

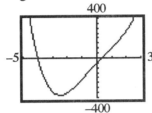

From the graph it appears that there are x-intercepts at -4 and $\frac{1}{3}$.

Step 5:   Using synthetic division:

$$-4\overline{)\,3 \quad -1 \quad -9 \quad 159 \quad -52\,}$$
$$\underline{\quad -12 \quad 52 \quad -172 \quad 52}$$
$$3 \quad -13 \quad 43 \quad -13 \quad 0$$

$$\tfrac{1}{3}\overline{)\,3 \quad -13 \quad 43 \quad -13\,}$$
$$\underline{\qquad 1 \quad -4 \quad 13}$$
$$3 \quad -12 \quad 39 \quad 0$$

Since the remainder is 0, $x + 4$ and $x - \frac{1}{3}$ are factors. The other factor is the quotient: $3x^2 - 12x + 39 = 3\left(x^2 - 4x + 13\right)$.

Using the quadratic formula to find the zeros of $x^2 - 4x + 13 = 0$:
$$x = \frac{-(-4) \pm \sqrt{(-4)^2 - 4(1)(13)}}{2(1)} = \frac{4 \pm \sqrt{-36}}{2} = \frac{4 \pm 6i}{2} = 2 \pm 3i.$$

The complex zeros are -4, $\frac{1}{3}$, $2 - 3i$, $2 + 3i$.

35.   If the coefficients are real numbers and $2 + i$ is a zero, then $2 - i$ would also be a zero. This would then require a polynomial of degree 4.

37.   If the coefficients are real numbers, then complex zeros must appear in conjugate pairs. We have a conjugate pair and one real zero. Thus, there is only one remaining zero and it must be real because a complex zero would require a pair and the polynomial would then have to be of degree 5.

4.4 Polynomial and Rational Inequalities

1. (a) $(x-5)(x+2) < 0$

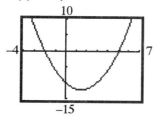

Graph $f(x) = (x-5)(x+2)$.
The x-intercepts are $x = -2$ and $x = 5$.
The graph of f is below the x-axis for
$-2 < x < 5$. Thus, the solution set is
$\left\{ x \mid -2 < x < 5 \right\}$.

(b) $(x-5)(x+2) < 0$ $f(x) = (x-5)(x+2)$

$x = 5, x = -2$ are the zeros.

Interval	Test Number	$f(x)$	Positive/Negative
$-\infty < x < -2$	-3	8	Positive
$-2 < x < 5$	0	-10	Negative
$5 < x < \infty$	6	8	Positive

The solution set is $\left\{ x \mid -2 < x < 5 \right\}$.

3. (a) $x^2 - 4x > 0$

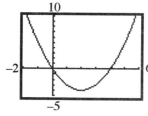

Graph $f(x) = x^2 - 4x$.
The x-intercepts are $x = 0$ and $x = 4$.
The graph of f is above the x-axis for
$x < 0$ or $x > 4$. Thus, the solution set is
$\left\{ x \mid x < 0 \text{ or } x > 4 \right\}$.

(b) $x^2 - 4x > 0$ $f(x) = x^2 - 4x$

$x(x-4) > 0$

$x = 0, x = 4$ are the zeros.

Interval	Test Number	$f(x)$	Positive/Negative
$-\infty < x < 0$	-1	5	Positive
$0 < x < 4$	1	-3	Negative
$4 < x < \infty$	5	5	Positive

The solution set is $\left\{ x \mid x < 0 \text{ or } x > 4 \right\}$.

5. (a) $x^2 - 9 < 0$

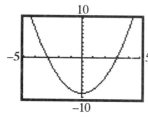

Graph $f(x) = x^2 - 9$.
The x-intercepts are $x = -3$ and $x = 3$.
The graph of f is below the x-axis for
$-3 < x < 3$. Thus, the solution set is
$\left\{ x \mid -3 < x < 3 \right\}$.

(b) $x^2 - 9 < 0$ $f(x) = x^2 - 9$

$(x + 3)(x - 3) < 0$

$x = -3$, $x = 3$ are the zeros.

Interval	Test Number	$f(x)$	Positive/Negative
$-\infty < x < -3$	-4	7	Positive
$-3 < x < 3$	0	-9	Negative
$3 < x < \infty$	4	7	Positive

The solution set is $\left\{ x \mid -3 < x < 3 \right\}$.

7.   (a) $x^2 + x > 12$

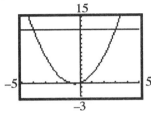

Graph $y_1 = x^2 + x$, $y_2 = 12$.
y_1 intersects y_2 at $x = -4$ and $x = 3$.
$y_1 > y_2$ for $x < -4$ or $x > 3$. Thus, the
solution set is $\left\{ x \mid x < -4 \text{ or } x > 3 \right\}$.

(b) $x^2 + x > 12$ $f(x) = x^2 + x - 12$

$x^2 + x - 12 > 0$

$(x + 4)(x - 3) > 0$

$x = -4$, $x = 3$ are the zeros.

Interval	Test Number	$f(x)$	Positive/Negative
$-\infty < x < -4$	-5	8	Positive
$-4 < x < 3$	0	-12	Negative
$3 < x < \infty$	4	8	Positive

The solution set is $\left\{ x \mid x < -4 \text{ or } x > 3 \right\}$.

9.   (a) $2x^2 < 5x + 3$

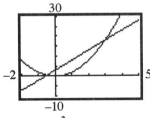

Graph $y_1 = 2x^2$, $y_2 = 5x + 3$.
y_1 intersects y_2 at $x = -0.50$ and $x = 3$.
$y_1 < y_2$ for $-0.50 < x < 3$. Thus, the
solution set is $\left\{ x \mid -0.50 < x < 3 \right\}$.

(b) $2x^2 < 5x + 3$ $f(x) = 2x^2 - 5x - 3$

$2x^2 - 5x - 3 < 0$

$(2x + 1)(x - 3) < 0$

$x = \frac{-1}{2}$, $x = 3$ are the zeros.

Interval	Test Number	$f(x)$	Positive/Negative
$-\infty < x < -1/2$	-1	4	Positive
$-1/2 < x < 3$	0	-3	Negative
$3 < x < \infty$	4	9	Positive

The solution set is $\left\{ x \mid \frac{-1}{2} < x < 3 \right\}$.

11. (a) $x(x-7) > 8$

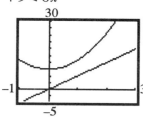

Graph $y_1 = x(x-7)$, $y_2 = 8$.
y_1 intersects y_2 at $x = -1$ and $x = 8$.
$y_1 > y_2$ for $x < -1$ or $x > 8$. Thus, the solution set is $\left\{ x \mid x < -1 \text{ or } x > 8 \right\}$.

(b) $x(x-7) > 8$ $f(x) = x^2 - 7x - 8$

$x^2 - 7x > 8$

$x^2 - 7x - 8 > 0$

$(x+1)(x-8) > 0$

$x = -1$, $x = 8$ are the zeros.

Interval	Test Number	$f(x)$	Positive/Negative
$-\infty < x < -1$	-2	10	Positive
$-1 < x < 8$	0	-8	Negative
$8 < x < \infty$	9	10	Positive

The solution set is $\left\{ x \mid x < -1 \text{ or } x > 8 \right\}$.

13. (a) $4x^2 + 9 < 6x$

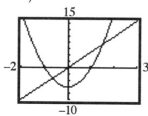

Graph $y_1 = 4x^2 + 9$, $y_2 = 6x$.
y_1 is always greater than y_2. Thus, there is no solution.

(b) $4x^2 + 9 < 6x$ $f(x) = 4x^2 - 6x + 9$

$4x^2 - 6x + 9 < 0$

$b^2 - 4ac = (-6)^2 - 4(4)(9) = 36 - 144 = -108$

Since the discriminant is negative, there are no real zeros.
There is only one interval, the entire number line; choose any value and test.
For $x = 0$, $4x^2 - 6x + 9 = 9 > 0$. Thus, there are no real zeros.

15. (a) $6(x^2 - 1) > 5x$

Graph $y_1 = 6(x^2 - 1)$, $y_2 = 5x$.
y_1 intersects y_2 at $x = -0.67$ and $x = 1.5$.
$y_1 > y_2$ for $x < -0.67$ or $x > 1.5$. Thus, the solution set is
$\left\{ x \mid x < -0.67 \text{ or } x > 1.5 \right\}$.

(b) $\qquad 6(x^2 - 1) > 5x \qquad\qquad f(x) = 6x^2 - 5x - 6$

$$6x^2 - 6 > 5x$$
$$6x^2 - 5x - 6 > 0$$
$$(3x + 2)(2x - 3) > 0$$

$x = \frac{-2}{3}$, $x = \frac{3}{2}$ are the zeros.

Interval	Test Number	$f(x)$	Positive/Negative
$-\infty < x < -2/3$	-1	5	Positive
$-2/3 < x < 3/2$	0	-6	Negative
$3/2 < x < \infty$	2	8	Positive

The solution set is $\left\{ x \,\middle|\, x < \frac{-2}{3} \text{ or } x > \frac{3}{2} \right\}$.

17. (a) $(x - 1)(x^2 + x + 1) > 0$

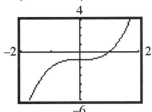

Graph $f(x) = (x - 1)(x^2 + x + 1)$.
The x-intercept is $x = 1$. The graph of f is above the x-axis for $x > 1$. Thus, the solution set is $\left\{ x \,\middle|\, x > 1 \right\}$.

(b) $(x - 1)(x^2 + x + 1) > 0 \qquad f(x) = (x - 1)(x^2 + x + 1)$

$x = 1$ is the zero. $x^2 + x + 1 = 0$ has no real zeros.

Interval	Test Number	$f(x)$	Positive/Negative
$-\infty < x < 1$	0	-1	Negative
$1 < x < \infty$	2	7	Positive

The solution set is $\left\{ x \,\middle|\, x > 1 \right\}$.

19. (a) $(x - 1)(x - 2)(x - 3) < 0$

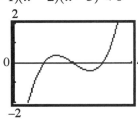

Graph $f(x) = (x - 1)(x - 2)(x - 3)$.
The x-intercepts are $x = 1$, $x = 2$, and $x = 3$. The graph of f is below the x-axis for $x < 1$ or $2 < x < 3$. Thus, the solution set is $\left\{ x \,\middle|\, x < 1 \text{ or } 2 < x < 3 \right\}$.

(b) $(x - 1)(x - 2)(x - 3) < 0 \qquad f(x) = (x - 1)(x - 2)(x - 3)$

$x = 1$, $x = 2$, $x = 3$ are the zeros.

Interval	Test Number	$f(x)$	Positive/Negative
$-\infty < x < 1$	0	-6	Negative
$1 < x < 2$	1.5	0.375	Positive
$2 < x < 3$	2.5	-0.375	Negative
$3 < x < \infty$	4	6	Positive

The solution set is $\left\{ x \,\middle|\, x < 1 \text{ or } 2 < x < 3 \right\}$.

21. (a) $x^3 - 2x^2 - 3x > 0$

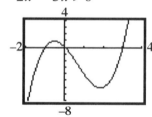

Graph $f(x) = x^3 - 2x^2 - 3x$.
The x-intercepts are $x = -1$, $x = 0$,
and $x = 3$. The graph of f is above the
x-axis for $-1 < x < 0$ or $x > 3$. Thus,
the solution set is
$\left\{ x \middle| -1 < x < 0 \ \text{ or } \ x > 3 \right\}$.

(b) $x^3 - 2x^2 - 3x > 0$ $f(x) = x^3 - 2x^2 - 3x$

$x\left(x^2 - 2x - 3\right) > 0$

$x(x + 1)(x - 3) > 0$

$x = -1$, $x = 0$, $x = 3$ are the zeros.

Interval	Test Number	$f(x)$	Positive/Negative
$-\infty < x < -1$	-2	-10	Negative
$-1 < x < 0$	-0.5	0.875	Positive
$0 < x < 3$	1	-4	Negative
$3 < x < \infty$	4	20	Positive

The solution set is $\left\{ x \middle| -1 < x < 0 \ \text{ or } \ x > 3 \right\}$.

23. (a) $x^4 > x^2$

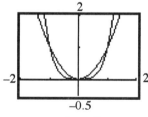

Graph $y_1 = x^4$, $y_2 = x^2$.
y_1 intersects y_2 at $x = -1$, $x = 0$,
and $x = 1$. $y_1 > y_2$ for $x < -1$ or $x > 1$.
Thus, the solution set is
$\left\{ x \middle| x < -1 \ \text{ or } \ x > 1 \right\}$.

(b) $x^4 > x^2$ $f(x) = x^4 - x^2$

$x^4 - x^2 > 0$

$x^2\left(x^2 - 1\right) > 0$

$x^2(x + 1)(x - 1) > 0$

$x = -1$, $x = 0$, $x = 1$ are the zeros.

Interval	Test Number	$f(x)$	Positive/Negative
$-\infty < x < -1$	-2	12	Positive
$-1 < x < 0$	-0.5	-0.1875	Negative
$0 < x < 1$	0.5	-0.1875	Negative
$1 < x < \infty$	2	12	Positive

The solution set is $\left\{ x \middle| x < -1 \ \text{ or } \ x > 1 \right\}$.

25. (a) $x^3 > x^2$

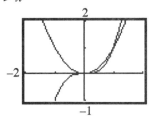

Graph $y_1 = x^3$, $y_2 = x^2$.
y_1 intersects y_2 at $x = 0$ and $x = 1$.
$y_1 > y_2$ for $x > 1$. Thus, the solution set is $\left\{ x \mid x > 1 \right\}$.

(b) $x^3 > x^2$ $f(x) = x^3 - x^2$
$x^3 - x^2 > 0$
$x^2(x - 1) > 0$
$x = 0$, $x = 1$ are the zeros.

Interval	Test Number	$f(x)$	Positive/Negative
$-\infty < x < 0$	-1	-2	Negative
$0 < x < 1$	0.5	-0.125	Negative
$1 < x < \infty$	2	4	Positive

The solution set is $\left\{ x \mid x > 1 \right\}$.

27. (a) $x^4 > 1$

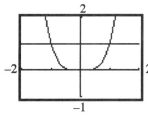

Graph $y_1 = x^4$, $y_2 = 1$.
y_1 intersects y_2 at $x = -1$ and $x = 1$.
$y_1 > y_2$ for $x < -1$ or $x > 1$. Thus, the solution set is $\left\{ x \mid x < -1 \text{ or } x > 1 \right\}$.

(b) $x^4 > 1$ $f(x) = x^4 - 1$
$x^4 - 1 > 0$
$\left(x^2 + 1 \right)\left(x^2 - 1 \right) > 0$
$\left(x^2 + 1 \right)(x + 1)(x - 1) > 0$
$x = -1$, $x = 1$ are the zeros.

Interval	Test Number	$f(x)$	Positive/Negative
$-\infty < x < -1$	-2	15	Positive
$-1 < x < 1$	0	-1	Negative
$1 < x < \infty$	2	15	Positive

The solution set is $\left\{ x \mid x < -1 \text{ or } x > 1 \right\}$.

29. (a) $x^2 - 7x - 8 < 0$

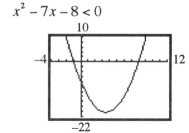

Graph $f(x) = x^2 - 7x - 8$.
The x-intercepts are $x = -1$ and $x = 8$.
The graph of f is below the x-axis for $-1 < x < 8$. Thus, the solution set is $\left\{ x \mid -1 < x < 8 \right\}$.

(b) $x^2 - 7x - 8 < 0$ $f(x) = x^2 - 7x - 8$

$(x + 1)(x - 8) < 0$

$x = -1, \; x = 8$ are the zeros.

Interval	Test Number	$f(x)$	Positive/Negative
$-\infty < x < -1$	-2	10	Positive
$-1 < x < 8$	0	-8	Negative
$8 < x < \infty$	9	10	Positive

The solution set is $\left\{ x \mid -1 < x < 8 \right\}$.

31. (a) $x^3 + x - 12 \geq 0$

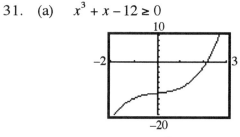

Graph $f(x) = x^3 + x - 12$.
The x-intercept is $x = 2.14$. The graph of f is above the x-axis for $x > 2.14$.
Thus, the solution set is $\left\{ x \mid x \geq 2.14 \right\}$.

(b) Graph $f(x) = x^3 + x - 12$ and use ZERO to find the zero of the function.

$x = 2.14$ is the zero (rounded to two decimal places).

Interval	Test Number	$f(x)$	Positive/Negative
$-\infty < x < 2.14$	0	-12	Negative
$2.14 < x < \infty$	3	18	Positive

The solution set is $\left\{ x \mid x \geq 2.14 \right\}$.

33. (a) $x^4 - 3x^2 - 4 > 0$

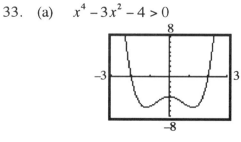

Graph $f(x) = x^4 - 3x^2 - 4$.
The x-intercepts are $x = -2$, and $x = 2$.
The graph of f is above the x-axis for $x < -2$ or $x > 2$. Thus, the solution set is $\left\{ x \mid x < -2 \; \text{or} \; x > 2 \right\}$.

(b) $x^4 - 3x^2 - 4 > 0$ $f(x) = x^4 - 3x^2 - 4$

$\left(x^2 - 4\right)\left(x^2 + 1\right) > 0$

$(x + 2)(x - 2)\left(x^2 + 1\right) > 0$

$x = -2$, $x = 2$ are the zeros.

Interval	Test Number	$f(x)$	Positive/Negative
$-\infty < x < -2$	-3	50	Positive
$-2 < x < 2$	0	-4	Negative
$2 < x < \infty$	3	50	Positive

The solution set is $\left\{ x \mid x < -2 \text{ or } x > 2 \right\}$.

35. (a) $x^3 - 4 \geq 3x^2 + 5x - 3$

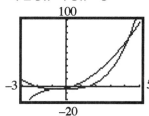

Graph $y_1 = x^3 - 4$, $y_2 = 3x^2 + 5x - 3$.
y_1 intersects y_2 at $x = -1$, $x = -0.24$,
and $x = 4.24$. $y_1 \geq y_2$ for
$-1 \leq x \leq -0.24$ or $x \geq 4.24$.
Thus, the solution set is
$\left\{ x \mid -1 \leq x \leq -0.24 \text{ or } x \geq 4.24 \right\}$.

(b) $x^3 - 4 \geq 3x^2 + 5x - 3$ $f(x) = x^3 - 3x^2 - 5x - 1$

$x^3 - 3x^2 - 5x - 1 \geq 0$

Do synthetic division by -1:

$$-1\overline{)\begin{array}{cccc} 1 & -3 & -5 & -1 \\ & -1 & 4 & 1 \\ \hline 1 & -4 & -1 & 0 \end{array}}$$

$x = -1$ is a solution. Solving the remaining factor $x^2 - 4x - 1 = 0$:

$$x = \frac{-(-4) \pm \sqrt{(-4)^2 - 4(1)(-1)}}{2(1)} = \frac{4 \pm \sqrt{20}}{2} = \frac{4 \pm 2\sqrt{5}}{2} = 2 \pm \sqrt{5}$$

$x = -1$, $x = 2 - \sqrt{5}$, $x = 2 + \sqrt{5}$ or $x = -1$, $x = -0.24$, $x = 4.24$ are the zeros.

Interval	Test Number	$f(x)$	Positive/Negative
$-\infty < x < -1$	-2	-11	Negative
$-1 < x < 2 - \sqrt{5}$	-0.5	0.625	Positive
$2 - \sqrt{5} < x < 2 + \sqrt{5}$	0	-1	Negative
$2 + \sqrt{5} < x < \infty$	5	24	Positive

The solution set is $\left\{ x \mid -1 \leq x \leq 2 - \sqrt{5} \text{ or } x \geq 2 + \sqrt{5} \right\}$.

37.   (a) $\dfrac{x+1}{x-1} > 0$

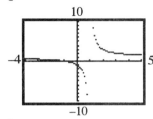

Graph $f(x) = \dfrac{x+1}{x-1}$.

The x-intercept is $x = -1$. The expression is undefined at $x = 1$. The graph of f is above the x-axis for $x < -1$ or $x > 1$. Thus, the solution set is $\left\{ x \mid x < -1 \text{ or } x > 1 \right\}$.

  (b) $\dfrac{x+1}{x-1} > 0$ $f(x) = \dfrac{x+1}{x-1}$

The zeros and values where the expression is undefined are $x = -1$, and $x = 1$.

Interval	Test Number	$f(x)$	Positive/Negative
$-\infty < x < -1$	-2	$1/3$	Positive
$-1 < x < 1$	0	-1	Negative
$1 < x < \infty$	2	3	Positive

The solution set is $\left\{ x \mid x < -1 \text{ or } x > 1 \right\}$.

39.   (a) $\dfrac{(x-1)(x+1)}{x} < 0$

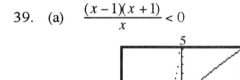

Graph $f(x) = \dfrac{(x-1)(x+1)}{x}$.

The x-intercepts are $x = -1$ and $x = 1$. The expression is undefined at $x = 0$. The graph of f is below the x-axis for $x < -1$ or $0 < x < 1$. Thus, the solution set is $\left\{ x \mid x < -1 \text{ or } 0 < x < 1 \right\}$.

  (b) $\dfrac{(x-1)(x+1)}{x} < 0$ $f(x) = \dfrac{(x-1)(x+1)}{x}$

The zeros and values where the expression is undefined are
$x = -1$, $x = 0$, and $x = 1$.

Interval	Test Number	$f(x)$	Positive/Negative
$-\infty < x < -1$	-2	-1.5	Negative
$-1 < x < 0$	-0.5	1.5	Positive
$0 < x < 1$	0.5	-1.5	Negative
$1 < x < \infty$	2	1.5	Positive

The solution set is $\left\{ x \mid x < -1 \text{ or } 0 < x < 1 \right\}$.

41.   (a) $\dfrac{(x-2)^2}{x^2 - 1} \geq 0$

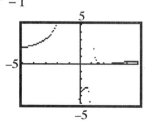

Graph $f(x) = \dfrac{(x-2)^2}{x^2 - 1}$.

The x-intercept is $x = 2$. The expression is undefined at $x = -1$ and $x = 1$. The graph of f is above the x-axis for $x < -1$ or $x > 1$. Thus, the solution set is $\left\{ x \mid x < -1 \text{ or } x > 1 \right\}$.

(b) $\dfrac{(x-2)^2}{x^2-1} \geq 0$ $f(x) = \dfrac{(x-2)^2}{x^2-1}$

$\dfrac{(x-2)^2}{(x+1)(x-1)} \geq 0$

The zeros and values where the expression is undefined are

$x = -1$, $x = 1$, and $x = 2$.

Interval	Test Number	$f(x)$	Positive/Negative
$-\infty < x < -1$	-2	$16/3$	Positive
$-1 < x < 1$	0	-4	Negative
$1 < x < 2$	1.5	0.2	Positive
$2 < x < \infty$	3	0.125	Positive

The solution set is $\left\{ x \mid x < -1 \text{ or } x > 1 \right\}$.

43. (a) $6x - 5 < \dfrac{6}{x}$

                         Graph $y_1 = 6x - 5$, $y_2 = \dfrac{6}{x}$.

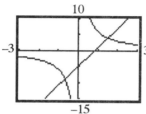

y_1 intersects y_2 at $x = -0.67$ and $x = 1.5$. y_2 is undefined at $x = 0$. $y_1 < y_2$ for $x < 0.67$ or $0 < x < 1.5$. Thus, the solution set is

$\left\{ x \mid x < 0.67 \text{ or } 0 < x < 1.5 \right\}$.

(b) $6x - 5 < \dfrac{6}{x}$ $f(x) = 6x - 5 - \dfrac{6}{x}$

$6x - 5 - \dfrac{6}{x} < 0$

$\dfrac{6x^2 - 5x - 6}{x} < 0$

$\dfrac{(2x-3)(3x+2)}{x} < 0$

The zeros and values where the expression is undefined are

$x = \frac{-2}{3}$, $x = 0$, and $x = \frac{3}{2}$.

Interval	Test Number	$f(x)$	Positive/Negative
$-\infty < x < -2/3$	-1	-5	Negative
$-2/3 < x < 0$	-0.5	4	Positive
$0 < x < 3/2$	1	-5	Negative
$3/2 < x < \infty$	2	4	Positive

The solution set is $\left\{ x \mid x < \frac{-2}{3} \text{ or } 0 < x < \frac{3}{2} \right\}$.

45. (a) $\dfrac{x+4}{x-2} \le 1$

Graph $y_1 = \dfrac{x+4}{x-2},\ y_2 = 1.$

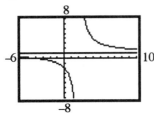

y_1 is not defined at $x = 2$. $y_1 < y_2$ for $x < 2$. Thus, the solution set is $\left\{ x \mid x < 2 \right\}.$

(b) $\dfrac{x+4}{x-2} \le 1$

$f(x) = \dfrac{x+4}{x-2} - 1$

$\dfrac{x+4}{x-2} - 1 \le 0$

$\dfrac{x+4-(x-2)}{x-2} \le 0$

$\dfrac{6}{x-2} \le 0$

The value where the expression is undefined is $x = 2$.

Interval	Test Number	$f(x)$	Positive/Negative
$-\infty < x < 2$	0	-3	Negative
$2 < x < \infty$	3	6	Positive

The solution set is $\left\{ x \mid x < 2 \right\}.$

47. (a) $\dfrac{3x-5}{x+2} \le 2$

Graph $y_1 = \dfrac{3x-5}{x+2},\ y_2 = 2.$

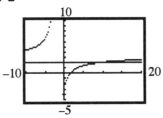

y_1 is not defined at $x = -2$.
y_1 intersects y_2 at $x = 9$. $y_1 < y_2$ for $-2 < x < 9$. Thus, the solution set is $\left\{ x \mid -2 < x \le 9 \right\}.$

(b) $\dfrac{3x-5}{x+2} \le 2$

$f(x) = \dfrac{3x-5}{x+2} - 2$

$\dfrac{3x-5}{x+2} - 2 \le 0$

$\dfrac{3x-5-2(x+2)}{x+2} \le 0$

$\dfrac{x-9}{x+2} \le 0$

The zeros and values where the expression is undefined are $x = -2$, and $x = 9$.

Interval	Test Number	$f(x)$	Positive/Negative
$-\infty < x < -2$	-3	12	Positive
$-2 < x < 9$	0	-4.5	Negative
$9 < x < \infty$	10	$1/12$	Positive

The solution set is $\left\{ x \mid -2 < x \le 9 \right\}.$

49. (a) $\dfrac{1}{x-2} < \dfrac{2}{3x-9}$

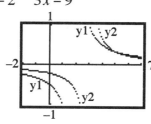

Graph $y_1 = \dfrac{1}{x-2}$, $y_2 = \dfrac{2}{3x-9}$.

The undefined values are at $x = 2$, $x = 3$.

y_1 intersects y_2 at $x = 5$. $y_1 < y_2$ for $x < 2$ or $3 < x < 5$. Thus, the solution set is $\left\{ x \mid x < 2 \text{ or } 3 < x < 5 \right\}$.

(b) $\dfrac{1}{x-2} < \dfrac{2}{3x-9}$

$f(x) = \dfrac{1}{x+2} - \dfrac{2}{3x-9}$

$\dfrac{1}{x-2} - \dfrac{2}{3x-9} < 0$

$\dfrac{3x-9-2(x-2)}{(x-2)(3x-9)} < 0$

$\dfrac{x-5}{(x-2)(3x-9)} < 0$

The zeros and values where the expression is undefined are
$x = 2$, $x = 3$, and $x = 5$.

Interval	Test Number	$f(x)$	Positive/Negative
$-\infty < x < 2$	0	$-5/18$	Negative
$2 < x < 3$	2.5	$10/3$	Positive
$3 < x < 5$	4	$-1/6$	Negative
$5 < x < \infty$	6	$1/36$	Positive

The solution set is $\left\{ x \mid x < 2 \text{ or } 3 < x < 5 \right\}$.

51. (a) $\dfrac{2x+5}{x+1} > \dfrac{x+1}{x-1}$

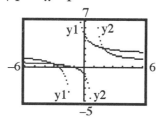

Graph $y_1 = \dfrac{2x+5}{x+1}$, $y_2 = \dfrac{x+1}{x-1}$.

The undefined values are at
$x = -1$, $x = 1$. y_1 intersects y_2 at
$x = -3$ and $x = 2$. $y_1 > y_2$ for
$x < -3$, $-1 < x < 1$, or $x > 2$. Thus, the solution set is
$\left\{ x \mid x < -3, \ -1 < x < 1, \ x > 2 \right\}$.

(b)
$$\frac{2x+5}{x+1} > \frac{x+1}{x-1} \qquad\qquad f(x) = \frac{2x+5}{x+1} - \frac{x+1}{x-1}$$

$$\frac{2x+5}{x+1} - \frac{x+1}{x-1} > 0$$

$$\frac{(2x+5)(x-1)-(x+1)(x+1)}{(x+1)(x-1)} > 0$$

$$\frac{2x^2+3x-5-\left(x^2+2x+1\right)}{(x+1)(x-1)} > 0$$

$$\frac{x^2+x-6}{(x+1)(x-1)} > 0$$

$$\frac{(x+3)(x-2)}{(x+1)(x-1)} > 0$$

The zeros and values where the expression is undefined are

$x = -3,\ x = -1,\ x = 1,$ and $x = 2$.

Interval	Test Number	$f(x)$	Positive/Negative
$-\infty < x < -3$	-4	$2/5$	Positive
$-3 < x < -1$	-2	$-4/3$	Negative
$-1 < x < 1$	0	6	Positive
$1 < x < 2$	1.5	$-9/5$	Negative
$2 < x < \infty$	3	$3/4$	Positive

The solution set is $\left\{ x \mid x < -3,\ -1 < x < 1,\ x > 2 \right\}$.

53. (a)
$$\frac{x^2(3+x)(x+4)}{(x+5)(x-1)} > 0$$

Graph $f(x) = \dfrac{x^2(3+x)(x+4)}{(x+5)(x-1)}$.

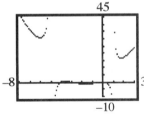

The x-intercepts are $x = -4,\ x = -3,$ and $x = 0$. The expression is **undefined** at $x = -5$ and $x = 1$. The graph of f is above the x-axis for

$x < -5,\ -4 < x < -3,$ or $x > 1$. Thus, the solution set is

$\left\{ x \mid x < -5,\ -4 < x < -3,\ x > 1 \right\}$.

(b)
$$\frac{x^2(3+x)(x+4)}{(x+5)(x-1)} > 0 \qquad\qquad f(x) = \frac{x^2(3+x)(x+4)}{(x+5)(x-1)}$$

The zeros and values where the expression is undefined are

$x = -5,\ x = -4,\ x = -3,\ x = 0$ and $x = 1$.

Interval	Test Number	$f(x)$	Positive/Negative
$-\infty < x < -5$	-6	$216/7$	Positive
$-5 < x < -4$	-4.5	$-243/44$	Negative
$-4 < x < -3$	-3.5	$49/108$	Positive
$-3 < x < 0$	-1	$-3/4$	Negative
$0 < x < 1$	0.5	$-63/44$	Negative
$1 < x < \infty$	2	$120/7$	Positive

The solution set is $\left\{ x \mid x < -5,\ -4 < x < -3,\ x > 1 \right\}$.

55. (a) $\dfrac{2x^2 - x - 1}{x - 4} \le 0$

Graph $f(x) = \dfrac{2x^2 - x - 1}{x - 4}$.

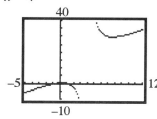

The x-intercepts are $x = -0.5$, and $x = 1$. The expression is undefined at $x = 4$. The graph of f is below the x-axis for $x < -0.5$, or $1 < x < 4$. Thus, the solution set is

$\left\{ x \mid x \le -0.5, \text{ or } 1 \le x < 4 \right\}$.

(b) $\dfrac{2x^2 - x - 1}{x - 4} \le 0$

$f(x) = \dfrac{2x^2 - x - 1}{x - 4}$

$\dfrac{(2x + 1)(x - 1)}{x - 4} \le 0$

The zeros and values where the expression is undefined are

$x = \frac{-1}{2}, \; x = 1, \text{ and } x = 4.$

Interval	Test Number	$f(x)$	Positive/Negative
$-\infty < x < -1/2$	-1	$-2/5$	Negative
$-1/2 < x < 1$	0	$1/4$	Positive
$1 < x < 4$	2	$-5/2$	Negative
$4 < x < \infty$	5	44	Positive

The solution set is $\left\{ x \mid x \le \frac{-1}{2}, \text{ or } 1 \le x < 4 \right\}$.

57. (a) $\dfrac{x^2 + 3x - 1}{x + 3} > 0$

Graph $f(x) = \dfrac{x^2 + 3x - 1}{x + 3}$.

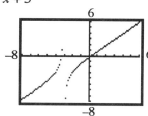

The x-intercepts are $x = -3.30$ and $x = 0.30$. The expression is undefined at $x = -3$. The graph of f is above the x-axis for $-3.30 < x < -3$, or $x > 0.30$. Thus, the solution set is

$\left\{ x \mid -3.30 < x < -3, \text{ or } x > 0.30 \right\}$.

(b) $\dfrac{x^2 + 3x - 1}{x + 3} > 0$ $\qquad$ $f(x) = \dfrac{x^2 + 3x - 1}{x + 3}$

Solving the numerator:

$$x = \dfrac{-3 \pm \sqrt{3^2 - 4(1)(-1)}}{2(1)} = \dfrac{-3 \pm \sqrt{13}}{2}$$

The zeros and values where the expression is undefined are

$$x = \dfrac{-3 - \sqrt{13}}{2}, x = \dfrac{-3 + \sqrt{13}}{2}, \text{ and } x = -3 \text{ or } x = -3.30, x = 0.30, x = -3.$$

Interval	Test Number	$f(x)$	Positive/Negative
$-\infty < x < \dfrac{-3 - \sqrt{13}}{2}$	-4	-3	Negative
$\dfrac{-3 - \sqrt{13}}{2} < x < -3$	-3.2	$9/5$	Positive
$-3 < x < \dfrac{-3 + \sqrt{13}}{2}$	0	$-1/3$	Negative
$\dfrac{-3 + \sqrt{13}}{2} < x < \infty$	1	$3/4$	Positive

The solution set is $\left\{ x \,\middle|\, \dfrac{-3 - \sqrt{13}}{2} < x < -3 \text{ or } x > \dfrac{-3 + \sqrt{13}}{2} \right\}$.

59. Let x be the positive number. Then

$$x^3 > 4x^2$$
$$x^3 - 4x^2 > 0$$
$$x^2(x - 4) > 0$$

The zeros are $x = 0$ and $x = 4$. $\qquad$ $f(x) = x^3 - 4x^2$

Interval	Test Number	$f(x)$	Positive/Negative
$-\infty < x < 0$	-1	-5	Negative
$0 < x < 4$	1	-3	Negative
$4 < x < \infty$	5	25	Positive

The solution set is $\left\{ x \mid x > 4 \right\}$. All real number larger than 4 satisfy the condition.

61. The domain of the expression includes all values for which

$$x^2 - 16 \geq 0$$
$$(x + 4)(x - 4) \geq 0$$

The zeros are $x = -4$ and $x = 4$. $\qquad$ $f(x) = x^2 - 16$

Interval	Test Number	$f(x)$	Positive/Negative
$-\infty < x < -4$	-5	9	Positive
$-4 < x < 4$	0	-16	Negative
$4 < x < \infty$	5	9	Positive

The solution or domain is $\left\{ x \mid x \leq -4 \text{ or } x \geq 4 \right\}$.

63. The domain of the expression includes all values for which
$$\frac{x-2}{x+4} \geq 0$$
The zeros and values where the expression is undefined are $x = -4$ and $x = 2$.

$$f(x) = \frac{x-2}{x+4}$$

Interval	Test Number	$f(x)$	Positive/Negative
$-\infty < x < -4$	-5	7	Positive
$-4 < x < 2$	0	$-1/2$	Negative
$2 < x < \infty$	3	$1/7$	Positive

The solution or domain is $\left\{ x \mid x < -4 \text{ or } x \geq 2 \right\}$.

65. (a) Find the values of t for which
$$80t - 16t^2 > 96$$
$$-16t^2 + 80t - 96 > 0$$
$$16t^2 - 80t + 96 < 0$$
$$16(t^2 - 5t + 6) < 0$$
$$16(t - 2)(t - 3) < 0$$

The zeros are $t = 2$ and $t = 3$. $\qquad s(t) = 16t^2 - 80t + 96$

Interval	Test Number	$s(t)$	Positive/Negative
$-\infty < t < 2$	1	32	Positive
$2 < t < 3$	2.5	-4	Negative
$3 < t < \infty$	4	32	Positive

The solution set is $\left\{ t \mid 2 < t < 3 \right\}$. The ball is more than 96 feet above the ground for times between 2 and 3 seconds.

(b) Graphing: $\left(s = 80t - 16t^2 \right)$

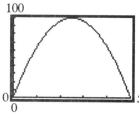

(c) Using MAXIMUM, the maximum height is 100 feet.
(d) The maximum height occurs at 2.5 seconds.

67. (a) Profit = Revenue – Cost
$$x(40 - 0.2x) - 32x \geq 50$$
$$40x - 0.2x^2 - 32x \geq 50$$
$$-0.2x^2 + 8x - 50 \geq 0$$
$$2x^2 - 80x + 500 \leq 0$$
$$x^2 - 40x + 250 \leq 0$$

The zeros are approximately $x = 7.75$ and $x = 32.25$.

$$f(x) = x^2 - 40x + 250$$

Interval	Test Number	$f(x)$	Positive/Negative
$0 < x < 7.75$	7	19	Positive
$7.75 < x < 32.25$	10	−50	Negative
$32.25 < x < \infty$	40	250	Positive

The profit is at least \$50 when at least 8 and no more than 32 watches are sold.

(b)   Graphing the revenue function: $(R(x) = x(40 - 0.2x))$

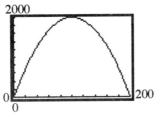

(c)   Using MAXIMUM, the maximum revenue is \$2,000.
(d)   Using MAXIMUM, the company should sell 100 wristwatches to maximize revenue.

(e)   Graphing the profit function: $(P(x) = x(40 - 0.2x) - 32x)$

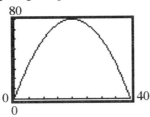

(f)   Using MAXIMUM, the maximum profit is \$80.
(g)   Using MAXIMUM, the company should sell 20 watches for maximum profit.

69.   The cost of manufacturing x Chevy Cavaliers in a day was found to be:

$$C(x) = 0.216x^3 - 2.347x^2 + 14.328x + 10.224$$

Since the budget constraints require that the cost must be less than or equal to \$97,000, we need to solve: $C(x) \leq 97$ or

$$0.216x^3 - 2.347x^2 + 14.328x + 10.224 \leq 97$$

Graphing $y_1 = 0.216x^3 - 2.347x^2 + 14.328x + 10.224$ and $y_2 = 97$:

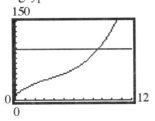

y_1 intersects y_2 at $x = 8.59$. $y_1 < y_2$ when $x < 8.59$. Chevy can produce at most eight Cavaliers in a day, assuming cars cannot be partially completed.

71. (a)   Graphing:

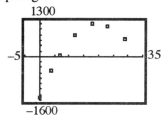

(b)   The quadratic function of best fit is $p(x) = -6.776x^2 + 270.668x - 1500.202$.

(c)   Graphing $y_1 = -6.776x^2 + 270.668x - 1500.202$ and $y_2 = 1000$:

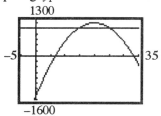

y_1 intersects y_2 at $x = 14.50$ and $x = 25.44$. $y_1 > y_2$ when $14.50 < x < 25.44$. Since the number of computers must be a positive integer, Marissa must sell between 15 and 25 computers per week for the profit to exceed $1000 per week.

(d)   Using MAXIMUM, the profits are maximized when Marissa sells about 20 computers.

(e)   Finding the maximum value:
$$p(20) = -6.776(20)^2 + 270.668(20) - 1500.202 = \$1202.76$$

73. Prove that if a, b are real numbers and $a \geq 0$, $b \geq 0$, then $a \leq b$ is equivalent to $\sqrt{a} \leq \sqrt{b}$.

$$a \leq b \Rightarrow b - a \geq 0$$
$$\left(\sqrt{b} - \sqrt{a}\right)\left(\sqrt{b} + \sqrt{a}\right) \geq 0$$

Either   (1) $\sqrt{b} - \sqrt{a} \geq 0$ and $\sqrt{b} + \sqrt{a} \geq 0$

  or   (2) $\sqrt{b} - \sqrt{a} \leq 0$ and $\sqrt{b} + \sqrt{a} \leq 0$

In (1) $\sqrt{b} \geq \sqrt{a}$ and $\sqrt{b} \geq -\sqrt{a}$

$\sqrt{a} \leq \sqrt{b}$ and $-\sqrt{a} \leq \sqrt{b}$

In (2) $\sqrt{b} \leq \sqrt{a}$ and $\sqrt{b} \leq -\sqrt{a}$

Case (2) is impossible since $\sqrt{a} \geq 0$ and $\sqrt{b} \geq 0$ which means

$\sqrt{b} \leq -\sqrt{a}$ is impossible.

Case (1) is true and if $a \leq b$, then $\sqrt{a} \leq \sqrt{b}$

So $a \leq b$ is equivalent to $\sqrt{a} \leq \sqrt{b}$.

4   Chapter Review

1. $f(x) = 12x^8 - x^7 + 8x^4 - 2x^3 + x + 3$

   Examining $f(x)$, there are 4 variations in sign; thus, there are 4 or 2 or 0 positive real zeros.

   Examining $f(-x) = 12(-x)^8 - (-x)^7 + 8(-x)^4 - 2(-x)^3 + (-x) + 3$

   $= 12x^8 + x^7 + 8x^4 + 2x^3 - x + 3$, there are 2 variations in sign; thus, there are 2 or 0 negative real zeros.

3. $f(x) = 12x^8 - x^7 + 6x^4 - x^3 + x - 3$

   p must be a factor of -3: $p = \pm 1, \pm 3$

   q must be a factor of 12: $q = \pm 1, \pm 2, \pm 3, \pm 4, \pm 6, \pm 12$

   The possible rational zeros are: $\dfrac{p}{q} = \pm 1, \pm 3, \pm \dfrac{1}{2}, \pm \dfrac{3}{2}, \pm \dfrac{1}{3}, \pm \dfrac{1}{4}, \pm \dfrac{3}{4}, \pm \dfrac{1}{6}, \pm \dfrac{1}{12}$

5. $f(x) = x^3 - 3x^2 - 6x + 8$

   Step 1: $f(x)$ has at most 3 real zeros.

   Step 2: By Descartes Rule of Signs, there are 2 or 0 positive real zeros.

   Also because $f(-x) = (-x)^3 - 3(-x)^2 - 6(-x) + 8 = -x^3 - 3x^2 + 6x + 8$, there is 1 negative real zero.

   Step 3: Possible rational zeros:

   $$p = \pm 1, \pm 2, \pm 4, \pm 8; \quad q = \pm 1; \quad \dfrac{p}{q} = \pm 1, \pm 2, \pm 4, \pm 8$$

   Step 4: Using the Bounds on Zeros Theorem:

   $$a_2 = -3, \quad a_1 = -6, \quad a_0 = 8$$

   $$\text{Max}\left\{1, |8| + |-6| + |-3|\right\} = \text{Max}\left\{1, 17\right\} = 17$$

   $$1 + \text{Max}\left\{|8|, |-6|, |-3|\right\} = 1 + 8 = 9$$

   The smaller of the two numbers is 9. Thus, every zero of f lies between -9 and 9.

   Graphing using the bounds and ZOOM-FIT: (Second has a better window.)

   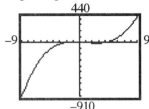   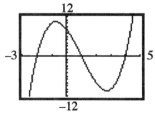

Step 5:   (a)   From the graph it appears that there are x-intercepts at –2, 1, and 4.
          (b)   Using synthetic division:

$$-2\overline{)\begin{array}{cccc} 1 & -3 & -6 & 8 \\ & -2 & 10 & -8 \end{array}}$$
$$\begin{array}{cccc} 1 & -5 & 4 & 0 \end{array}$$

Since the remainder is 0, $x - (-2) = x + 2$ is a factor. The other
factor is the quotient: $x^2 - 5x + 4$.

          (c)   Thus, $f(x) = (x + 2)(x^2 - 5x + 4) = (x + 2)(x - 1)(x - 4)$.
          The zeros are –2, 1, and 4.

7. $f(x) = 4x^3 + 4x^2 - 7x + 2$

Step 1: $f(x)$ has at most 3 real zeros.
Step 2:   By Descartes Rule of Signs, there are 2 or 0 positive real zeros.
          $f(-x) = 4(-x)^3 + 4(-x)^2 - 7(-x) + 2 = -4x^3 + 4x^2 + 7x + 2$; thus, there
          is 1 negative real zero.
Step 3:   Possible rational zeros:

$$p = \pm 1, \pm 2; \quad q = \pm 1, \pm 2, \pm 4; \quad \frac{p}{q} = \pm 1, \pm 2, \pm \frac{1}{2}, \pm \frac{1}{4}$$

Step 4:   Using the Bounds on Zeros Theorem:

$$f(x) = 4\left(x^3 + x^2 - \tfrac{7}{4}x + \tfrac{1}{2}\right)$$
$$a_2 = 1, \quad a_1 = \tfrac{-7}{4}, \quad a_0 = \tfrac{1}{2}$$
$$\text{Max}\left\{1, \left|\tfrac{1}{2}\right| + \left|\tfrac{-7}{4}\right| + |1|\right\} = \text{Max}\left\{1, \tfrac{13}{4}\right\} = 3.25$$
$$1 + \text{Max}\left\{\left|\tfrac{1}{2}\right|, \left|\tfrac{-7}{4}\right|, |1|\right\} = 1 + \tfrac{7}{4} = \tfrac{11}{4} = 2.75$$

The smaller of the two numbers is 2.75. Thus, every zero of f lies
          between –2.75 and 2.75.
Graphing using the bounds and ZOOM-FIT: (Second has a better window.)

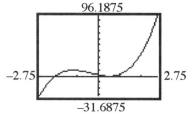
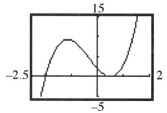

Step 5:   (a)   From the graph it appears that there are x-intercepts at –2 and $\tfrac{1}{2}$.
          (b)   Using synthetic division:

$$-2\overline{)\begin{array}{cccc} 4 & 4 & -7 & 2 \\ & -8 & 8 & -2 \end{array}}$$
$$\begin{array}{cccc} 4 & -4 & 1 & 0 \end{array}$$

Since the remainder is 0, $x - (-2) = x + 2$ is a factor. The other
factor is the quotient: $4x^2 - 4x + 1$.

          (c)   Thus, $f(x) = (x + 2)(4x^2 - 4x + 1) = (x + 2)(2x - 1)(2x - 1)$.
          The zeros are –2 and $\tfrac{1}{2}$ (multiplicity 2).

9. $f(x) = x^4 - 4x^3 + 9x^2 - 20x + 20$

Step 1: $f(x)$ has at most 4 real zeros.

Step 2:   By Descartes Rule of Signs, there are 4 or 2 or 0 positive real zeros.

$$f(-x) = (-x)^4 - 4(-x)^3 + 9(-x)^2 - 20(-x) + 20$$
$$= x^4 + 4x^3 + 9x^2 + 20x + 20$$

thus, there are no negative real zeros.

Step 3:   Possible rational zeros:

$$p = \pm 1, \pm 2, \pm 4, \pm 5, \pm 10, \pm 20; \quad q = \pm 1;$$

$$\frac{p}{q} = \pm 1, \pm 2, \pm 4, \pm 5, \pm 10, \pm 20$$

Step 4:   Using the Bounds on Zeros Theorem:

$$a_3 = -4, \quad a_2 = 9, \quad a_1 = -20, \quad a_0 = 20$$

$$\text{Max}\left\{1, |20| + |-20| + |9| + |-4|\right\} = \text{Max}\left\{1, 53\right\} = 53$$

$$1 + \text{Max}\left\{|20|, |-20|, |9|, |-4|\right\} = 1 + 20 = 21$$

The smaller of the two numbers is 21. Thus, every zero of f lies between −21 and 21.

Graphing using the bounds and ZOOM-FIT: (Second has a better window.)

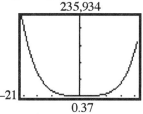 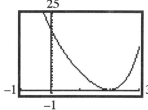

Step 5:   (a)   From the graph it appears that there is an x-intercept at 2.

(b)   Using synthetic division:

$$\begin{array}{r|rrrrr} 2 & 1 & -4 & 9 & -20 & 20 \\ & & 2 & -4 & 10 & -20 \\ \hline & 1 & -2 & 5 & -10 & 0 \end{array} \qquad \begin{array}{r|rrrr} 2 & 1 & -2 & 5 & -10 \\ & & 2 & 0 & 10 \\ \hline & 1 & 0 & 5 & 0 \end{array}$$

Since the remainder is 0, $x - 2$ is a factor twice. The other factor is the quotient: $x^2 + 5$.

(c)   Thus, $f(x) = (x - 2)(x - 2)\left(x^2 + 5\right) = (x - 2)^2\left(x^2 + 5\right)$.

The zero is 2 (multiplicity 2). ($x^2 + 5 = 0$ has no real solutions.)

11. $f(x) = 2x^3 - 11.84x^2 - 9.116x + 82.46$

$f(x)$ has at most 3 real zeros.

Solving by graphing (using ZERO):

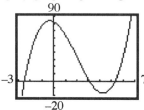

The zeros are approximately −2.5, 3.1, and 5.32.

13. $g(x) = 15x^4 - 21.5x^3 - 1718.3x^2 + 5308x + 3796.8$

$g(x)$ has at most 4 real zeros.
Solving by graphing (using ZERO):

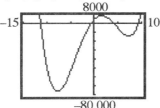

The zeros are approximately –11.3, –0.6, 4, and 9.33.

15. $f(x) = 3x^3 + 18.02x^2 + 11.0467x - 53.8756$

$f(x)$ has at most 3 real zeros.
Solving by graphing (using ZERO):

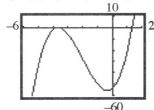

The zeros are approximately –3.67 and 1.33.

17. $2x^4 + 2x^3 - 11x^2 + x - 6 = 0$

The solutions of the equation are the zeros of $f(x) = 2x^4 + 2x^3 - 11x^2 + x - 6$.

Step 1: $f(x)$ has at most 4 real zeros.

Step 2: By Descartes Rule of Signs, there are 3 or 1 positive real zeros.
$f(-x) = 2(-x)^4 + 2(-x)^3 - 11(-x)^2 + (-x) - 6 = 2x^4 - 2x^3 - 11x^2 - x - 6$;
thus, there is 1 negative real zero.

Step 3: Possible rational zeros:

$$p = \pm 1, \pm 2, \pm 3, \pm 6; \quad q = \pm 1, \pm 2; \quad \frac{p}{q} = \pm 1, \pm 2, \pm 3, \pm 6, \pm \frac{1}{2}, \pm \frac{3}{2}$$

Step 4: Using the Bounds on Zeros Theorem:

$$f(x) = 2\left(x^4 + x^3 - \tfrac{11}{2}x^2 + \tfrac{1}{2}x - 3\right)$$

$$a_3 = 1, \ a_2 = -\tfrac{11}{2}, \ a_1 = \tfrac{1}{2}, \ a_0 = -3$$

$$\text{Max}\left\{1, |-3| + \left|\tfrac{1}{2}\right| + \left|-\tfrac{11}{2}\right| + |1|\right\} = \text{Max}\left\{1, 10\right\} = 10$$

$$1 + \text{Max}\left\{|-3|, \left|\tfrac{1}{2}\right|, \left|-\tfrac{11}{2}\right|, |1|\right\} = 1 + \tfrac{11}{2} = \tfrac{13}{2} = 6.5$$

The smaller of the two numbers is 6.5. Thus, every zero of f lies between –6.5 and 6.5.

Graphing using the bounds and ZOOM-FIT: (Second has a better window.)

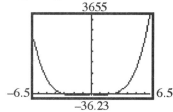

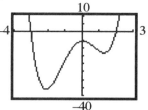

Step 5:   (a)   From the graph it appears that there are x-intercepts at –3 and 2.
          (b)   Using synthetic division:

$$
\begin{array}{r|rrrrr}
-3 & 2 & 2 & -11 & 1 & -6 \\
   &   & -6 & 12 & -3 & 6 \\
\hline
   & 2 & -4 & 1 & -2 & 0
\end{array}
\qquad
\begin{array}{r|rrrr}
2 & 2 & -4 & 1 & -2 \\
  &   & 4 & 0 & 2 \\
\hline
  & 2 & 0 & 1 & 0
\end{array}
$$

Since the remainder is 0, $x + 3$ and $x - 2$ are factors. The other factor
is the quotient: $2x^2 + 1$.

          (c)   The zeros are –3 and 2. ($2x^2 + 1 = 0$ has no real solutions.)

19. $2x^4 + 7x^3 + x^2 - 7x - 3 = 0$

The solutions of the equation are the zeros of $f(x) = 2x^4 + 7x^3 + x^2 - 7x - 3$.

Step 1: $f(x)$ has at most 4 real zeros.

Step 2:   By Descartes Rule of Signs, there is 1 positive real zero.
$f(-x) = 2(-x)^4 + 7(-x)^3 + (-x)^2 - 7(-x) - 3 = 2x^4 - 7x^3 + x^2 + 7x - 3$;
thus, there are 3 or 1 negative real zeros.

Step 3:   Possible rational zeros:

$$p = \pm 1, \pm 3; \quad q = \pm 1, \pm 2; \quad \frac{p}{q} = \pm 1, \pm 3, \pm \frac{1}{2}, \pm \frac{3}{2}$$

Step 4:   Using the Bounds on Zeros Theorem:

$$f(x) = 2\left(x^4 + \tfrac{7}{2}x^3 + \tfrac{1}{2}x^2 - \tfrac{7}{2}x - \tfrac{3}{2}\right)$$

$$a_3 = \tfrac{7}{2}, \ a_2 = \tfrac{1}{2}, \ a_1 = -\tfrac{7}{2}, \ a_0 = -\tfrac{3}{2}$$

$$\text{Max}\left\{1, \left|-\tfrac{3}{2}\right| + \left|-\tfrac{7}{2}\right| + \left|\tfrac{1}{2}\right| + \left|\tfrac{7}{2}\right|\right\} = \text{Max}\left\{1, 9\right\} = 9$$

$$1 + \text{Max}\left\{\left|-\tfrac{3}{2}\right|, \left|-\tfrac{7}{2}\right|, \left|\tfrac{1}{2}\right|, \left|\tfrac{7}{2}\right|\right\} = 1 + \tfrac{7}{2} = \tfrac{9}{2} = 4.5$$

The smaller of the two numbers is 4.5. Thus, every zero of f lies between
–4.5 and 4.5.

Graphing using the bounds and ZOOM-FIT: (Second has a better window.)

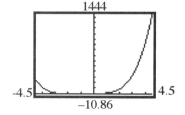

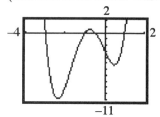

Step 5: (a) From the graph it appears that there are x-intercepts at $-3, -1, \frac{-1}{2}$, and 1.
    (b) Using synthetic division:

$$-3\overline{)\,2 \quad 7 \quad 1 \quad -7 \quad -3\,}$$
$$\underline{\quad\quad -6 \quad -3 \quad 6 \quad 3\,}$$
$$2 \quad 1 \quad -2 \quad -1 \quad 0$$

$$-1\overline{)\,2 \quad 1 \quad -2 \quad -1\,}$$
$$\underline{\quad\quad -2 \quad 1 \quad 1\,}$$
$$2 \quad -1 \quad -1 \quad 0$$

Since the remainder is 0, $x + 3$ and $x + 1$ are factors. The other factor is the quotient: $2x^2 - x - 1$.

    (c) Thus,

$$f(x) = (x+3)(x+1)\left(2x^2 - x - 1\right) = (x+3)(x+1)(2x+1)(x-1).$$

The zeros are $-3, -1, \frac{-1}{2}$, and 1.

21. $f(x) = x^3 - x^2 - 4x + 2$

$a_2 = -1, \quad a_1 = -4, \quad a_0 = 2$

$\text{Max}\left\{1, |2| + |-4| + |-1|\right\} = \text{Max}\left\{1, 7\right\} = 7$

$1 + \text{Max}\left\{|2|, |-4|, |-1|\right\} = 1 + 4 = 5$

The smaller of the two numbers is 5. Thus, every zero of f lies between -5 and 5.
Graphing using the bounds and ZOOM-FIT: (Second graph has a better window.)

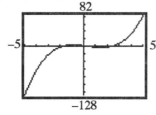
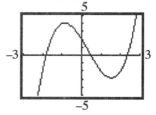

23. $f(x) = 2x^3 - 7x^2 - 10x + 35 = 2\left(x^3 - \frac{7}{2}x^2 - 5x + \frac{35}{2}\right)$

$a_2 = -\frac{7}{2}, \quad a_1 = -5, \quad a_0 = \frac{35}{2}$

$\text{Max}\left\{1, \left|\frac{35}{2}\right| + |-5| + \left|-\frac{7}{2}\right|\right\} = \text{Max}\left\{1, 26\right\} = 26$

$1 + \text{Max}\left\{\left|\frac{35}{2}\right|, |-5|, \left|-\frac{7}{2}\right|\right\} = 1 + \frac{35}{2} = \frac{37}{2} = 18.5$

The smaller of the two numbers is 18.5. Thus, every zero of f lies between -18.5 and 18.5.
Graphing using the bounds and ZOOM-FIT: (Second graph has a better window.)

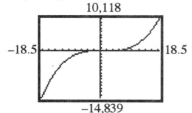
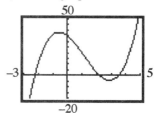

25. Using the TABLE feature to show that there is a zero in the interval:
$$f(x) = 3x^3 - x - 1; \quad [0, 1]$$

X	Y1
-2	-23
-1	-3
0	-1
1	1
2	21
3	77
4	187

Y1☐3X^3-X-1

$f(0) = -1 < 0$ and $f(1) = 1 > 0$
Since one is positive and one is negative, there is a zero in the interval.

Using the TABLE feature to approximate the zero to two decimal places:

X	Y1
.82	-.1659
.83	-.1146
.84	-.0619
.85	-.0076
.86	.04817
.87	.10551
.88	.16442

Y1☐3X^3-X-1

The zero is approximately 0.85.

27. Using the TABLE feature to show that there is a zero in the interval:
$$f(x) = 8x^4 - 4x^3 - 2x - 1; \quad [0, 1]$$

X	Y1
-2	163
-1	13
0	-1
1	1
2	91
3	533
4	1783

Y1☐8X^4-4X^3-2X...

$f(0) = -1 < 0$ and $f(1) = 1 > 0$
Since one is positive and one is negative, there is a zero in the interval.

Using the TABLE feature to approximate the zero to two decimal places:

X	Y1
.9	-.4672
.91	-.3483
.92	-.2236
.93	-.093
.94	.04366
.95	.18655
.96	.33583

Y1☐8X^4-4X^3-2X...

The zero is approximately 0.94.

29. $(6 + 3i) - (2 - 4i) = (6 - 2) + (3 - (-4))i = 4 + 7i$

31. $4(3 - i) + 3(-5 + 2i) = 12 - 4i - 15 + 6i = -3 + 2i$

33. $\dfrac{3}{3+i} = \dfrac{3}{3+i} \cdot \dfrac{3-i}{3-i} = \dfrac{9 - 3i}{9 - 3i + 3i - i^2} = \dfrac{9 - 3i}{10} = \dfrac{9}{10} - \dfrac{3}{10}i$

35. $i^{50} = i^{48} \cdot i^2 = (i^4)^{12} \cdot i^2 = 1^{12}(-1) = -1$

37. $(2 + 3i)^3 = (2 + 3i)^2(2 + 3i) = (4 + 12i + 9i^2)(2 + 3i) = (-5 + 12i)(2 + 3i)$
$$= -10 - 15i + 24i + 36i^2 = -46 + 9i$$

39. Since complex zeros appear in conjugate pairs, $4 - i$, the conjugate of $4 + i$, is the remaining zero of f.

41. Since complex zeros appear in conjugate pairs, $-i$, the conjugate of i, and $1 - i$, the conjugate of $1 + i$, are the remaining zeros of f.

43. $x^2 + x + 1 = 0$

$a = 1, b = 1, c = 1, \quad b^2 - 4ac = 1^2 - 4(1)(1) = 1 - 4 = -3$

$x = \dfrac{-1 \pm \sqrt{-3}}{2(1)} = \dfrac{-1 \pm \sqrt{3}\,i}{2} = \dfrac{-1}{2} \pm \dfrac{\sqrt{3}}{2}\,i$

The solution set is $\left\{ \dfrac{-1}{2} - \dfrac{\sqrt{3}}{2}i, \ \dfrac{-1}{2} + \dfrac{\sqrt{3}}{2}i \right\}$.

45. $2x^2 + x - 2 = 0$

$a = 2, b = 1, c = -2, \quad b^2 - 4ac = 1^2 - 4(2)(-2) = 1 + 16 = 17$

$x = \dfrac{-1 \pm \sqrt{17}}{2(2)} = \dfrac{-1 \pm \sqrt{17}}{4}$

The solution set is $\left\{ \dfrac{-1 - \sqrt{17}}{4}, \ \dfrac{-1 + \sqrt{17}}{4} \right\}$.

47. $x^2 + 3 = x$

$x^2 - x + 3 = 0$

$a = 1, b = -1, c = 3, \quad b^2 - 4ac = (-1)^2 - 4(1)(3) = 1 - 12 = -11$

$x = \dfrac{-(-1) \pm \sqrt{-11}}{2(1)} = \dfrac{1 \pm \sqrt{11}\,i}{2} = \dfrac{1}{2} \pm \dfrac{\sqrt{11}}{2}\,i$

The solution set is $\left\{ \dfrac{1}{2} - \dfrac{\sqrt{11}}{2}i, \ \dfrac{1}{2} + \dfrac{\sqrt{11}}{2}i \right\}$.

49. $x(1 - x) = 6$

$-x^2 + x - 6 = 0$

$a = -1, b = 1, c = -6, \quad b^2 - 4ac = 1^2 - 4(-1)(-6) = 1 - 24 = -23$

$x = \dfrac{-1 \pm \sqrt{-23}}{2(-1)} = \dfrac{-1 \pm \sqrt{23}\,i}{-2} = \dfrac{1}{2} \pm \dfrac{\sqrt{23}}{2}\,i$

The solution set is $\left\{ \dfrac{1}{2} - \dfrac{\sqrt{23}}{2}i, \ \dfrac{1}{2} + \dfrac{\sqrt{23}}{2}i \right\}$.

51. $x^4 + 2x^2 - 8 = 0$

$\left(x^2 + 4\right)\left(x^2 - 2\right) = 0$

$x^2 + 4 = 0 \ \text{ or } \ x^2 - 2 = 0$

$x^2 = -4 \ \text{ or } \ x^2 = 2$

$x = \pm 2i \ \text{ or } \ x = \pm\sqrt{2}$

The solution set is $\left\{ -2i, \ 2i, \ -\sqrt{2}, \ \sqrt{2} \right\}$.

53. $x^3 - x^2 - 8x + 12 = 0$

The solutions of the equation are the zeros of the function $f(x) = x^3 - x^2 - 8x + 12$.

Step 1: $f(x)$ has 3 complex zeros.

Step 2:   By Descartes Rule of Signs, there are 2 or 0 positive real zeros.

$f(-x) = (-x)^3 - (-x)^2 - 8(-x) + 12 = -x^3 - x^2 + 8x + 12$; thus, there is 1 negative real zero.

Step 3:   Possible rational zeros:

$$p = \pm 1, \pm 2, \pm 3, \pm 4, \pm 6, \pm 12; \quad q = \pm 1; \quad \frac{p}{q} = \pm 1, \pm 2, \pm 3, \pm 4, \pm 6, \pm 12$$

Step 4:   Graphing the function:

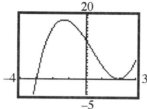

From the graph it appears that there are x-intercepts at –3 and 2.

Step 5:   Using synthetic division:

$$2\overline{)1 \quad -1 \quad -8 \quad 12}$$
$$2 \quad 2 \quad -12$$
$$\overline{1 \quad 1 \quad -6 \quad 0}$$

Since the remainder is 0, $x - 2$ is a factor. The other factor is the quotient:

$x^2 + x - 6 = (x + 3)(x - 2)$.

The complex zeros are –3, 2 (multiplicity 2).

55. $3x^4 - 4x^3 + 4x^2 - 4x + 1 = 0$

The solutions of the equation are the zeros of the function

$$f(x) = 3x^4 - 4x^3 + 4x^2 - 4x + 1$$

Step 1: $f(x)$ has 4 complex zeros.

Step 2:   By Descartes Rule of Signs, there are 4 or 2 or 0 positive real zeros.

$$f(-x) = 3(-x)^4 - 4(-x)^3 + 4(-x)^2 - 4(-x) + 1$$
$$= 3x^4 + 4x^3 + 4x^2 + 4x + 1$$

thus, there are no negative real zeros.

Step 3:   Possible rational zeros:

$$p = \pm 1; \quad q = \pm 1, \pm 3; \quad \frac{p}{q} = \pm 1, \pm \frac{1}{3}$$

Step 4:   Graphing the function:

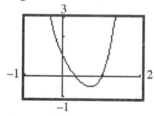

From the graph it appears that there are x-intercepts at $\frac{1}{3}$ and 1.

Step 5:   Using synthetic division:

$$1)\overline{\begin{array}{rrrrr} 3 & -4 & 4 & -4 & 1 \\ & 3 & -1 & 3 & -1 \\ \hline 3 & -1 & 3 & -1 & 0 \end{array}} \qquad \tfrac{1}{3})\overline{\begin{array}{rrrr} 3 & -1 & 3 & -1 \\ & 1 & 0 & 1 \\ \hline 3 & 0 & 3 & 0 \end{array}}$$

Since the remainder is 0, $x - 1$ and $x - \tfrac{1}{3}$ are factors. The other factor is the quotient: $3x^2 + 3 = 3(x^2 + 1)$.

Solving $x^2 + 1 = 0$:
$$x^2 = -1$$
$$x = \pm i$$

The complex zeros are 1, $\tfrac{1}{3}$, $-i$, i.

57.   (a) $2x^2 + 5x - 12 < 0$

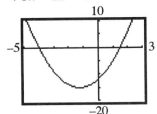

Graph $f(x) = 2x^2 + 5x - 12$.
The x-intercepts are $x = -4$ and $x = 1.5$.
The graph of f is below the x-axis for
$-4 < x < 1.5$. Thus, the solution set is
$\left\{ x \mid -4 < x < 1.5 \right\}$.

(b) $2x^2 + 5x - 12 < 0$ $f(x) = 2x^2 + 5x - 12$
$(x + 4)(2x - 3) < 0$
$x = -4, x = \tfrac{3}{2}$ are the zeros.

Interval	Test Number	$f(x)$	Positive/Negative
$-\infty < x < -4$	-5	13	Positive
$-4 < x < 3/2$	0	-12	Negative
$3/2 < x < \infty$	2	6	Positive

The solution set is $\left\{ x \mid -4 < x < \dfrac{3}{2} \right\}$.

59.   (a) $\dfrac{6}{x + 3} \geq 1$

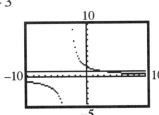

Graph $y_1 = \dfrac{6}{x + 3}$, $y_2 = 1$.
y_1 is undefined at $x = -3$. y_1 intersects y_2
at $x = 3$. $y_1 > y_2$ for $-3 < x < 3$. Thus,
the solution set is $\left\{ x \mid -3 < x \leq 3 \right\}$.

(b)
$$\frac{6}{x+3} \geq 1 \qquad\qquad f(x) = \frac{6}{x+3} - 1$$

$$\frac{6}{x+3} - 1 \geq 0$$

$$\frac{6 - 1(x+3)}{x+3} \geq 0$$

$$\frac{-x+3}{x+3} \geq 0$$

The zeros and values where the expression is undefined are $x = -3$, and $x = 3$.

Interval	Test Number	$f(x)$	Positive/Negative
$-\infty < x < -3$	-4	-7	Negative
$-3 < x < 3$	0	1	Positive
$3 < x < \infty$	4	$-1/7$	Negative

The solution set is $\left\{ x \mid -3 < x \leq 3 \right\}$.

61. (a)
$$\frac{2x-6}{1-x} < 2$$

Graph $y_1 = \frac{2x-6}{1-x}$, $y_2 = 2$.

y_1 is undefined at $x = 1$. y_1 intersects y_2 at $x = 2$. $y_1 < y_2$ for $x < 1$ or $x > 2$. Thus, the solution set is $\left\{ x \mid x < 1 \text{ or } x > 2 \right\}$.

(b)
$$\frac{2x-6}{1-x} < 2 \qquad\qquad f(x) = \frac{2x-6}{1-x} - 2$$

$$\frac{2x-6}{1-x} - 2 < 0$$

$$\frac{2x-6-2(1-x)}{1-x} < 0$$

$$\frac{4x-8}{1-x} < 0$$

The zeros and values where the expression is undefined are $x = 1$, and $x = 2$.

Interval	Test Number	$f(x)$	Positive/Negative
$-\infty < x < 1$	0	-8	Negative
$1 < x < 2$	1.5	4	Positive
$2 < x < \infty$	3	-2	Negative

The solution set is $\left\{ x \mid x < 1 \text{ or } x > 2 \right\}$.

63. (a) $\dfrac{(x-2)(x-1)}{x-3} > 0$

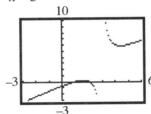

Graph $f(x) = \dfrac{(x-2)(x-1)}{x-3}$.

The x-intercepts are $x = 2$ and $x = 1$.
The expression is undefined at $x = 3$.
The graph of f is above the x-axis for
$1 < x < 2$ or $x > 3$. Thus, the solution
set is $\left\{ x \mid 1 < x < 2 \text{ or } x > 3 \right\}$.

(b) $\dfrac{(x-2)(x-1)}{x-3} > 0$

$f(x) = \dfrac{(x-2)(x-1)}{x-3}$

The zeros and values where the expression is undefined are

$x = 1$, $x = 2$, and $x = 3$.

Interval	Test Number	$f(x)$	Positive/Negative
$-\infty < x < 1$	0	$-2/3$	Negative
$1 < x < 2$	1.5	$1/6$	Positive
$2 < x < 3$	2.5	$-3/2$	Negative
$3 < x < \infty$	4	6	Positive

The solution set is $\left\{ x \mid 1 < x < 2 \text{ or } x > 3 \right\}$.

65. (a) $\dfrac{x^2 - 8x + 12}{x^2 - 16} > 0$

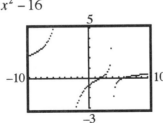

Graph $f(x) = \dfrac{x^2 - 8x + 12}{x^2 - 16}$.

The x-intercepts are $x = 2$, and $x = 6$.
The expression is undefined at
$x = -4$ and $x = 4$. The graph of f is
above the x-axis for
$x < -4, 2 < x < 4$, or $x > 6$. Thus, the
solution set is
$\left\{ x \mid x < -4, 2 < x < 4, x > 6 \right\}$.

(b) $\dfrac{x^2 - 8x + 12}{x^2 - 16} > 0$

$f(x) = \dfrac{x^2 - 8x + 12}{x^2 - 16}$

$\dfrac{(x-2)(x-6)}{(x+4)(x-4)} > 0$

The zeros and values where the expression is undefined are

$x = -4$, $x = 2$, $x = 4$, and $x = 6$.

Interval	Test Number	$f(x)$	Positive/Negative
$-\infty < x < -4$	-5	$77/9$	Positive
$-4 < x < 2$	0	$-3/4$	Negative
$2 < x < 4$	3	$3/7$	Positive
$4 < x < 6$	5	$-1/3$	Negative
$6 < x < \infty$	7	$5/33$	Positive

The solution set is $\left\{ x \mid x < -4, 2 < x < 4, x > 6 \right\}$.

Chapter 5

Exponential and Logarithmic Functions

5.1 One-to-One Functions; Inverse Functions

1. (a)
Domain	Range
$200	20 hours
$300	25 hours
$350	30 hours
$425	40 hours

    (b)   Inverse is a function.

3. (a)
Domain	Range
	20 hours
$200	25 hours
$350	30 hours
$425	40 hours

    (b)   Inverse is not a function since $200 corresponds to two elements in the range.

5. (a) $\{(6,2), (6,-3), (9,4), (10,1)\}$

    (b)   Inverse is not a function since 6 corresponds to 2 and –3.

7. (a) $\{(0,0), (1,1), (16,2), (81,3)\}$

    (b)   Inverse is a function.

9.   Every horizontal line intersects the graph of f at exactly one point. One-to-One.

11. There are horizontal lines that intersect the graph of f at more than one point. Not One-to-One.

13. Every horizontal line intersects the graph of f at exactly one point. One-to-One.

15. Graphing the inverse:

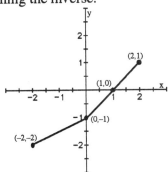

17. Graphing the inverse:

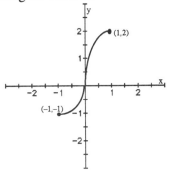

19. Graphing the inverse:

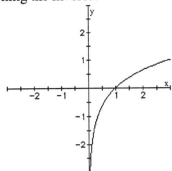

21. $f(x) = 3x + 4, \qquad g(x) = \frac{1}{3}(x - 4)$

$f(g(x)) = f\left(\frac{1}{3}(x-4)\right) = 3\left(\frac{1}{3}(x-4)\right) + 4$

$\qquad\qquad = (x - 4) + 4 = x$

$g(f(x)) = g(3x + 4) = \frac{1}{3}\left((3x + 4) - 4\right)$

$\qquad\qquad = \frac{1}{3}(3x) = x$

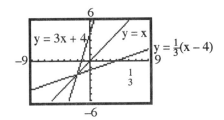

23. $f(x) = 4x - 8, \qquad g(x) = \frac{x}{4} + 2$

$f(g(x)) = f\left(\frac{x}{4} + 2\right) = 4\left(\frac{x}{4} + 2\right) - 8$

$\qquad\qquad = (x + 8) - 8 = x$

$g(f(x)) = g(4x - 8) = \frac{4x - 8}{4} + 2$

$\qquad\qquad = x - 2 + 2 = x$

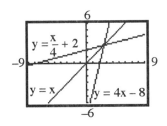

25. $f(x) = x^3 - 8, \qquad g(x) = \sqrt[3]{x + 8}$

$f(g(x)) = f\left(\sqrt[3]{x + 8}\right) = \left(\sqrt[3]{x + 8}\right)^3 - 8$

$\qquad\qquad = (x + 8) - 8 = x$

$g(f(x)) = g(x^3 - 8) = \sqrt[3]{(x^3 - 8) + 8}$

$\qquad\qquad = \sqrt[3]{x^3} = x$

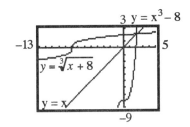

27. $f(x) = \dfrac{1}{x}, \qquad g(x) = \dfrac{1}{x}$

$f(g(x)) = f\left(\dfrac{1}{x}\right) = \dfrac{1}{\dfrac{1}{x}} = x$

$g(f(x)) = g\left(\dfrac{1}{x}\right) = \dfrac{1}{\dfrac{1}{x}} = x$

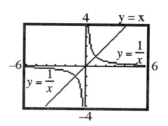

29. $f(x) = \dfrac{2x+3}{x+4}, \qquad g(x) = \dfrac{4x-3}{2-x}$

$f(g(x)) = f\left(\dfrac{4x-3}{2-x}\right) = \dfrac{2\left(\dfrac{4x-3}{2-x}\right)+3}{\left(\dfrac{4x-3}{2-x}\right)+4} = \dfrac{\dfrac{8x-6+6-3x}{2-x}}{\dfrac{4x-3+8-4x}{2-x}}$

$= \dfrac{\dfrac{5x}{2-x}}{\dfrac{5}{2-x}} = \dfrac{5x}{2-x} \cdot \dfrac{2-x}{5} = x$

$g(f(x)) = g\left(\dfrac{2x+3}{x+4}\right) = \dfrac{4\left(\dfrac{2x+3}{x+4}\right)-3}{2-\left(\dfrac{2x+3}{x+4}\right)} = \dfrac{\dfrac{8x+12-3x-12}{x+4}}{\dfrac{2x+8-2x-3}{x+4}}$

$= \dfrac{\dfrac{5x}{x+4}}{\dfrac{5}{x+4}} = \dfrac{5x}{x+4} \cdot \dfrac{x+4}{5} = x$

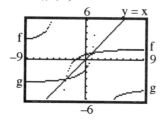

31. $f(x) = 3x$

$y = 3x$

$x = 3y \qquad$ Inverse

$y = \dfrac{x}{3}$

$f^{-1}(x) = \dfrac{x}{3}$

Verify: $f\left(f^{-1}(x)\right) = f\left(\dfrac{x}{3}\right) = 3\left(\dfrac{x}{3}\right) = x$

$f^{-1}\left(f(x)\right) = f^{-1}(3x) = \dfrac{3x}{3} = x$

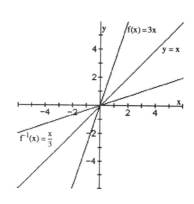

Domain of f = range of f^{-1} = $(-\infty, \infty)$
Range of f = domain of f^{-1} = $(-\infty, \infty)$

33. $f(x) = 4x + 2$
      $y = 4x + 2$
      $x = 4y + 2$ Inverse
      $4y = x - 2$
      $y = \dfrac{x - 2}{4}$
      $f^{-1}(x) = \dfrac{x - 2}{4}$

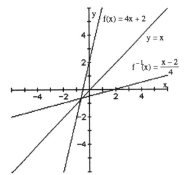

Verify: $f\left(f^{-1}(x)\right) = f\left(\dfrac{x-2}{4}\right) = 4\left(\dfrac{x-2}{4}\right) + 2 = x - 2 + 2 = x$

$f^{-1}(f(x)) = f^{-1}(4x + 2) = \dfrac{(4x + 2) - 2}{4} = \dfrac{4x}{4} = x$

Domain of f = range of f^{-1} = $(-\infty, \infty)$
Range of f = domain of f^{-1} = $(-\infty, \infty)$

35. $f(x) = x^3 - 1$
      $y = x^3 - 1$
      $x = y^3 - 1$ Inverse
      $y^3 = x + 1$
      $y = \sqrt[3]{x + 1}$
      $f^{-1}(x) = \sqrt[3]{x + 1}$

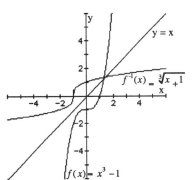

Verify: $f\left(f^{-1}(x)\right) = f\left(\sqrt[3]{x+1}\right) = \left(\sqrt[3]{x+1}\right)^3 - 1 = x + 1 - 1 = x$

$f^{-1}(f(x)) = f^{-1}(x^3 - 1) = \sqrt[3]{(x^3 - 1) + 1} = \sqrt[3]{x^3} = x$

Domain of f = range of f^{-1} = $(-\infty, \infty)$
Range of f = domain of f^{-1} = $(-\infty, \infty)$

37. $f(x) = x^2 + 4, \ x \geq 0$

   $y = x^2 + 4 \ \ x \geq 0$

   $x = y^2 + 4 \ \ y \geq 0$ Inverse

   $y^2 = x - 4 \ \ y \geq 0$

   $y = \sqrt{x - 4}$

   $f^{-1}(x) = \sqrt{x - 4}$

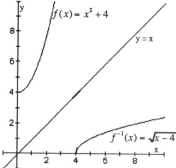

Verify: $f\left(f^{-1}(x)\right) = f\left(\sqrt{x-4}\right) = \left(\sqrt{x-4}\right)^2 + 4 = x - 4 + 4 = x$

   $f^{-1}(f(x)) = f^{-1}(x^2 + 4) = \sqrt{(x^2 + 4) - 4} = \sqrt{x^2} = |x| = x, \ x \geq 0$

Domain of f = range of $f^{-1} = [0, \infty)$

Range of f = domain of $f^{-1} = [4, \infty)$

39. $f(x) = \dfrac{4}{x}$

   $y = \dfrac{4}{x}$

   $x = \dfrac{4}{y}$ Inverse

   $xy = 4$

   $y = \dfrac{4}{x}$

   $f^{-1}(x) = \dfrac{4}{x}$

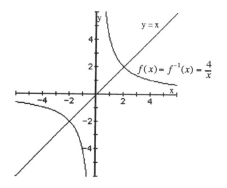

Verify: $f\left(f^{-1}(x)\right) = f\left(\dfrac{4}{x}\right) = \dfrac{4}{\dfrac{4}{x}} = 4 \cdot \dfrac{x}{4} = x$

   $f^{-1}(f(x)) = f^{-1}\left(\dfrac{4}{x}\right) = \dfrac{4}{\dfrac{4}{x}} = 4 \cdot \dfrac{x}{4} = x$

Domain of f = range of f^{-1} = all real numbers except 0

Range of f = domain of f^{-1} = all real numbers except 0

41. $f(x) = \dfrac{1}{x-2}$

$\qquad y = \dfrac{1}{x-2}$

$\qquad x = \dfrac{1}{y-2}$ Inverse

$\qquad x(y-2) = 1$

$\qquad xy - 2x = 1$

$\qquad\quad xy = 2x + 1$

$\qquad\quad y = \dfrac{2x+1}{x}$

$\quad f^{-1}(x) = \dfrac{2x+1}{x}$

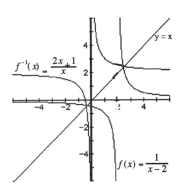

Verify: $f\left(f^{-1}(x)\right) = f\left(\dfrac{2x+1}{x}\right) = \dfrac{1}{\dfrac{2x+1}{x} - 2} = \dfrac{1}{\dfrac{2x+1-2x}{x}} = \dfrac{1}{\dfrac{1}{x}} = x$

$f^{-1}(f(x)) = f^{-1}\left(\dfrac{1}{x-2}\right) = \dfrac{2\left(\dfrac{1}{x-2}\right)+1}{\left(\dfrac{1}{x-2}\right)} = \dfrac{\dfrac{2+x-2}{x-2}}{\dfrac{1}{x-2}} = \dfrac{x}{x-2} \cdot \dfrac{x-2}{1} = x$

Domain of f = range of f^{-1} = all real numbers except 2
Range of f = domain of f^{-1} = all real numbers except 0

43. $f(x) = \dfrac{2}{3+x}$

$\qquad y = \dfrac{2}{3+x}$

$\qquad x = \dfrac{2}{3+y}$ Inverse

$\qquad x(3+y) = 2$

$\qquad 3x + xy = 2$

$\qquad\quad xy = 2 - 3x$

$\qquad\quad y = \dfrac{2-3x}{x}$

$\quad f^{-1}(x) = \dfrac{2-3x}{x}$

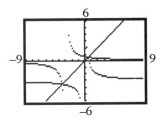

Verify: $f\left(f^{-1}(x)\right) = f\left(\dfrac{2-3x}{x}\right) = \dfrac{2}{3 + \dfrac{2-3x}{x}} = \dfrac{2}{\dfrac{3x+2-3x}{x}} = \dfrac{2}{\dfrac{2}{x}} = 2 \cdot \dfrac{x}{2} = x$

$f^{-1}(f(x)) = f^{-1}\left(\dfrac{2}{3+x}\right) = \dfrac{2 - 3\left(\dfrac{2}{3+x}\right)}{\left(\dfrac{2}{3+x}\right)} = \dfrac{\dfrac{6+2x-6}{3+x}}{\dfrac{2}{3+x}} = \dfrac{2x}{3+x} \cdot \dfrac{3+x}{2} = x$

Domain of f = range of f^{-1} = all real numbers except -3
Range of f = domain of f^{-1} = all real numbers except 0

45. $f(x) = (x+2)^2, \quad x \geq -2$

$\qquad y = (x+2)^2 \quad x \geq -2$

$\qquad x = (y+2)^2 \quad y \geq -2 \quad$ Inverse

$\qquad \sqrt{x} = y + 2, \qquad x \geq 0$

$\qquad y = \sqrt{x} - 2, \quad x \geq 0$

$\quad f^{-1}(x) = \sqrt{x} - 2, \quad x \geq 0$

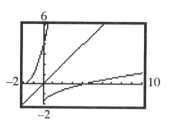

Verify: $f\left(f^{-1}(x)\right) = f\left(\sqrt{x} - 2\right) = \left(\sqrt{x} - 2 + 2\right)^2 = \left(\sqrt{x}\right)^2 = x$

$\qquad f^{-1}(f(x)) = f^{-1}\left((x+2)^2\right) = \sqrt{(x+2)^2} - 2 = x + 2 - 2 = x, \quad x \geq -2$

Domain of f = range of f^{-1} = $[-2, \infty)$

Range of f = domain of f^{-1} = $[0, \infty)$

47. $f(x) = \dfrac{2x}{x-1}$

$\qquad y = \dfrac{2x}{x-1}$

$\qquad x = \dfrac{2y}{y-1} \quad$ Inverse

$\qquad x(y-1) = 2y$

$\qquad xy - x = 2y$

$\qquad xy - 2y = x$

$\qquad y(x-2) = x$

$\qquad y = \dfrac{x}{x-2}$

$\quad f^{-1}(x) = \dfrac{x}{x-2}$

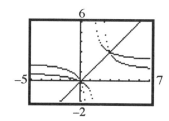

Verify: $f\left(f^{-1}(x)\right) = f\left(\dfrac{x}{x-2}\right) = \dfrac{2\left(\dfrac{x}{x-2}\right)}{\left(\dfrac{x}{x-2}\right) - 1} = \dfrac{\dfrac{2x}{x-2}}{\dfrac{x - x + 2}{x-2}} = \dfrac{\dfrac{2x}{x-2}}{\dfrac{2}{x-2}} = \dfrac{2x}{x-2} \cdot \dfrac{x-2}{2} = x$

$\qquad f^{-1}(f(x)) = f^{-1}\left(\dfrac{2x}{x-1}\right) = \dfrac{\left(\dfrac{2x}{x-1}\right)}{\left(\dfrac{2x}{x-1}\right) - 2} = \dfrac{\dfrac{2x}{x-1}}{\dfrac{2x - 2x + 2}{x-1}} = \dfrac{2x}{x-1} \cdot \dfrac{x-1}{2} = x$

Domain of f = range of f^{-1} = all real numbers except 1

Range of f = domain of f^{-1} = all real numbers except 2

49. $f(x) = \dfrac{3x+4}{2x-3}$

$y = \dfrac{3x+4}{2x-3}$

$x = \dfrac{3y+4}{2y-3}$ Inverse

$x(2y-3) = 3y+4$

$2xy - 3x = 3y + 4$

$2xy - 3y = 3x + 4$

$y(2x-3) = 3x+4$

$y = \dfrac{3x+4}{2x-3}$

$f^{-1}(x) = \dfrac{3x+4}{2x-3}$

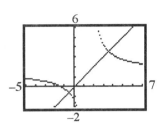

Verify:

$f\left(f^{-1}(x)\right) = f\left(\dfrac{3x+4}{2x-3}\right) = \dfrac{3\left(\dfrac{3x+4}{2x-3}\right)+4}{2\left(\dfrac{3x+4}{2x-3}\right)-3} = \dfrac{\dfrac{9x+12+8x-12}{2x-3}}{\dfrac{6x+8-6x+9}{2x-3}} = \dfrac{\dfrac{17x}{2x-3}}{\dfrac{17}{2x-3}}$

$= \dfrac{17x}{2x-3} \cdot \dfrac{2x-3}{17} = x$

$f^{-1}\left(f(x)\right) = f^{-1}\left(\dfrac{3x+4}{2x-3}\right) = \dfrac{3\left(\dfrac{3x+4}{2x-3}\right)+4}{2\left(\dfrac{3x+4}{2x-3}\right)-3} = \dfrac{\dfrac{9x+12+8x-12}{2x-3}}{\dfrac{6x+8-6x+9}{2x-3}} = \dfrac{\dfrac{17x}{2x-3}}{\dfrac{17}{2x-3}}$

$= \dfrac{17x}{2x-3} \cdot \dfrac{2x-3}{17} = x$

Domain of f = range of f^{-1} = all real numbers except $\frac{3}{2}$

Range of f = domain of f^{-1} = all real numbers except $\frac{3}{2}$

51. $f(x) = \dfrac{2x+3}{x+2}$

$y = \dfrac{2x+3}{x+2}$

$x = \dfrac{2y+3}{y+2}$ Inverse

$x(y+2) = 2y+3$

$xy + 2x = 2y + 3$

$xy - 2y = -2x + 3$

$y(x-2) = -2x+3$

$y = \dfrac{-2x+3}{x-2}$

$f^{-1}(x) = \dfrac{-2x+3}{x-2}$

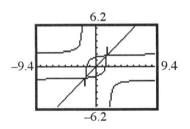

Verify: $f\left(f^{-1}(x)\right) = f\left(\dfrac{-2x+3}{x-2}\right) = \dfrac{2\left(\dfrac{-2x+3}{x-2}\right)+3}{\left(\dfrac{-2x+3}{x-2}\right)+2} = \dfrac{\dfrac{-4x+6+3x-6}{x-2}}{\dfrac{-2x+3+2x-4}{x-2}} = \dfrac{\dfrac{-x}{x-2}}{\dfrac{-1}{x-2}}$

$$= \dfrac{-x}{x-2} \cdot \dfrac{x-2}{-1} = x$$

$f^{-1}(f(x)) = f^{-1}\left(\dfrac{2x+3}{x+2}\right) = \dfrac{-2\left(\dfrac{2x+3}{x+2}\right)+3}{\left(\dfrac{2x+3}{x+2}\right)-2} = \dfrac{\dfrac{-4x-6+3x+6}{x+2}}{\dfrac{2x+3-2x-4}{x+2}} = \dfrac{\dfrac{-x}{x+2}}{\dfrac{-1}{x+2}}$

$$= \dfrac{-x}{x+2} \cdot \dfrac{x+2}{-1} = x$$

Domain of f = range of f^{-1} = all real numbers except –2
Range of f = domain of f^{-1} = all real numbers except 2

53. $f(x) = 2\sqrt[3]{x}$
$\quad\quad y = 2\sqrt[3]{x}$
$\quad\quad x = 2\sqrt[3]{y}$ Inverse
$\quad\quad x^3 = 8y$
$\quad\quad y = \dfrac{x^3}{8}$
$\quad f^{-1}(x) = \dfrac{x^3}{8}$

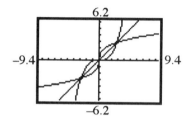

Verify: $f\left(f^{-1}(x)\right) = f\left(\dfrac{x^3}{8}\right) = 2\sqrt[3]{\dfrac{x^3}{8}} = 2 \cdot \dfrac{x}{2} = x$

$$f^{-1}(f(x)) = f^{-1}\left(2\sqrt[3]{x}\right) = \dfrac{\left(2\sqrt[3]{x}\right)^3}{8} = \dfrac{8x}{8} = x$$

Domain of f = range of f^{-1} = $(-\infty, \infty)$
Range of f = domain of f^{-1} = $(-\infty, \infty)$

55. $f(x) = mx + b, \quad m \neq 0$
$\quad\quad y = mx + b$
$\quad\quad x = my + b$ Inverse
$\quad\quad x - b = my$
$\quad\quad y = \dfrac{x-b}{m}$
$\quad f^{-1}(x) = \dfrac{x-b}{m}, \quad m \neq 0$

57. An even function cannot be one-to-one. When a function is even, $f(-x) = f(x)$.
Thus, both x and $-x$ produce the same y value.

59. f^{-1} lies in quadrant I. Whenever (a,b) is on f, then (b,a) is on f^{-1}. Since both coordinates of (a,b) are positive, both coordinates of (b,a) are positive and it is in quadrant I.

61. $f(x) = |x|$, $x \geq 0$ is one-to-one. Thus, $f(x) = x$, $x \geq 0$ and $f^{-1}(x) = x$, $x \geq 0$.

63. $f(x) = \frac{9}{5}x + 32$ $g(x) = \frac{5}{9}(x - 32)$

$f(g(x)) = f\left(\frac{5}{9}(x - 32)\right) = \frac{9}{5}\left(\frac{5}{9}(x - 32)\right) + 32 = x - 32 + 32 = x$

$g(f(x)) = g\left(\frac{9}{5}x + 32\right) = \frac{5}{9}\left(\frac{9}{5}x + 32 - 32\right) = \frac{5}{9}\left(\frac{9}{5}x\right) = x$

65. $T(l) = 2\pi\sqrt{\dfrac{l}{g}}$, $g \approx 32.2$

$T = 2\pi\sqrt{\dfrac{l}{g}}$

$\dfrac{T}{2\pi} = \sqrt{\dfrac{l}{g}}$

$\dfrac{T^2}{4\pi^2} = \dfrac{l}{g}$

$l = \dfrac{gT^2}{4\pi^2}$

$l(T) = \dfrac{gT^2}{4\pi^2}$

67. $f(x) = \dfrac{ax + b}{cx + d}$

$y = \dfrac{ax + b}{cx + d}$

$x = \dfrac{ay + b}{cy + d}$ Inverse

$x(cy + d) = ay + b$

$cxy + dx = ay + b$

$cxy - ay = b - dx$

$y(cx - a) = b - dx$

$y = \dfrac{b - dx}{cx - a}$

$f^{-1}(x) = \dfrac{-dx + b}{cx - a}$

$f = f^{-1}$ if $\dfrac{ax + b}{cx + d} = \dfrac{-dx + b}{cx - a}$

This is true if $a = -d$.

5.2 Exponential Functions

1. (a) $3^{2.2} = 11.212$ (b) $3^{2.23} = 11.587$ (c) $3^{2.236} = 11.664$ (d) $3^{\sqrt{5}} = 11.665$

3. (a) $2^{3.14} = 8.815$ (b) $2^{3.141} = 8.821$ (c) $2^{3.1415} = 8.824$ (d) $2^{\pi} = 8.825$

5. (a) $3.1^{2.7} = 21.217$ (b) $3.14^{2.71} = 22.217$
    (c) $3.141^{2.718} = 22.440$ (d) $\pi^{e} = 22.459$

7. $e^{1.2} = 3.320$ 9. $e^{-0.85} = 0.427$

11. B   13. D   15. A   17. E   19. A   21. E   23. B

25. $f(x) = 2^x + 1$

Using the graph of $y = 2^x$, shift the graph up 1 unit.
Domain: $(-\infty, \infty)$
Range: $(1, \infty)$
Horizontal Asymptote: $y = 1$

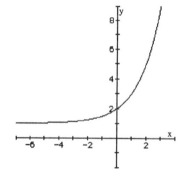

27. $f(x) = 3^{-x} - 2$

Using the graph of $y = 3^x$, reflect the graph about the y-axis, and shift down 2 units.
Domain: $(-\infty, \infty)$
Range: $(-2, \infty)$
Horizontal Asymptote: $y = -2$

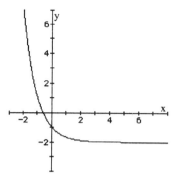

29. $f(x) = 2 + 3(4^x)$

Using the graph of $y = 4^x$, stretch the graph vertically by a factor of 3, and shift up 2 units.
Domain: $(-\infty, \infty)$
Range: $(2, \infty)$
Horizontal Asymptote: $y = 2$

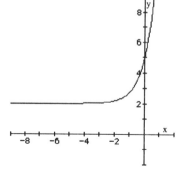

31. $f(x) = 2 + 3^{\frac{x}{2}}$

Using the graph of $y = 3^x$, stretch the graph horizontally by a factor of 2, and shift up 2 units.
Domain: $(-\infty, \infty)$
Range: $(2, \infty)$
Horizontal Asymptote: $y = 2$

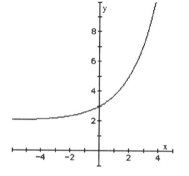

33. $f(x) = 5 - 2\left(3^{-(x+1)}\right)$

Using the graph of $y = 3^x$, reflect on the y-axis, shift the graph one unit to the left, stretch vertically by a factor of 2, reflect on the x-axis, and shift up 5 units.

Domain: $(-\infty, \infty)$

Range: $(-\infty, 5)$

Horizontal Asymptote: $y = 5$

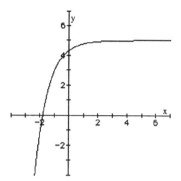

35. $f(x) = e^{-x}$

Using the graph of $y = e^x$, reflect the graph about the y-axis.

Domain: $(-\infty, \infty)$

Range: $(0, \infty)$

Horizontal Asymptote: $y = 0$

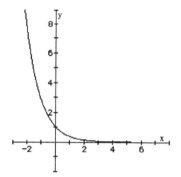

37. $f(x) = e^{x+2}$

Using the graph of $y = e^x$, shift the graph 2 units to the left.

Domain: $(-\infty, \infty)$

Range: $(0, \infty)$

Horizontal Asymptote: $y = 0$

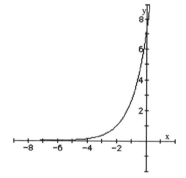

39. $f(x) = 5 - e^{-x}$

Using the graph of $y = e^x$, reflect the graph about the y-axis, reflect about the x-axis, and shift up 5 units.

Domain: $(-\infty, \infty)$

Range: $(-\infty, 5)$

Horizontal Asymptote: $y = 5$

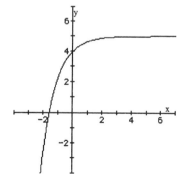

41. $f(x) = 2 - e^{\frac{-x}{2}}$

Using the graph of $y = e^x$, reflect the graph about the y-axis, stretch horizontally by a factor of 2, reflect about the x-axis, and shift up 2 units.

Domain: $(-\infty, \infty)$

Range: $(-\infty, 2)$

Horizontal Asymptote: $y = 2$

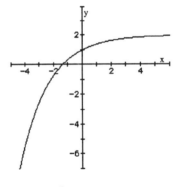

43. $2^{2x+1} = 4$

$2^{2x+1} = 2^2$

$2x + 1 = 2$

$2x = 1$

$x = \dfrac{1}{2}$

The solution is $\left\{\dfrac{1}{2}\right\}$.

45. $3^{x^3} = 9^x$

$3^{x^3} = \left(3^2\right)^x$

$3^{x^3} = 3^{2x}$

$x^3 = 2x$

$x^3 - 2x = 0$

$x(x^2 - 2) = 0$

$x = 0$ or $x^2 = 2$

$x = 0$ or $x = \pm\sqrt{2}$

The solution is $\left\{-\sqrt{2},\ 0,\ \sqrt{2}\right\}$.

47. $8^{x^2-2x} = \frac{1}{2}$

$\left(2^3\right)^{x^2-2x} = 2^{-1}$

$2^{3x^2-6x} = 2^{-1}$

$3x^2 - 6x = -1$

$3x^2 - 6x + 1 = 0$

$x = \dfrac{-(-6) \pm \sqrt{(-6)^2 - 4(3)(1)}}{2(3)} = \dfrac{6 \pm \sqrt{24}}{6} = \dfrac{6 \pm 2\sqrt{6}}{6} = \dfrac{3 \pm \sqrt{6}}{3}$

The solution is $\left\{\dfrac{3 - \sqrt{6}}{3},\ \dfrac{3 + \sqrt{6}}{3}\right\}$.

49. $2^x \cdot 8^{-x} = 4^x$

$2^x \cdot \left(2^3\right)^{-x} = \left(2^2\right)^x$

$2^x \cdot 2^{-3x} = 2^{2x}$

$2^{-2x} = 2^{2x}$

$-2x = 2x$

$-4x = 0$

$x = 0$

The solution is $\{0\}$.

51. $\left(\frac{1}{5}\right)^{2-x} = 25$

$\left(5^{-1}\right)^{2-x} = 5^2$

$5^{x-2} = 5^2$

$x - 2 = 2$

$x = 4$

The solution is $\{4\}$.

53. $4^x = 8$

$(2^2)^x = 2^3$

$2^{2x} = 2^3$

$2x = 3$

$x = \dfrac{3}{2}$

The solution is $\left\{\dfrac{3}{2}\right\}$.

55. $e^{x^2} = e^{3x} \cdot \dfrac{1}{e^2}$

$e^{x^2} = e^{3x-2}$

$x^2 = 3x - 2$

$x^2 - 3x + 2 = 0$

$(x-1)(x-2) = 0$

$x = 1 \text{ or } x = 2$

The solution is $\{1, 2\}$.

57. $4^x = 7$

$(4^x)^{-2} = 7^{-2}$

$4^{-2x} = \dfrac{1}{7^2}$

$4^{-2x} = \dfrac{1}{49}$

59. $3^{-x} = 2$

$(3^{-x})^{-2} = 2^{-2}$

$3^{2x} = \dfrac{1}{2^2}$

$3^{2x} = \dfrac{1}{4}$

61. $p = 100e^{-0.03n}$

    (a) $p = 100e^{-0.03(10)} = 100e^{-0.3} = 100(0.741) = 74.1\%$ of light

    (b) $p = 100e^{-0.03(25)} = 100e^{-0.75} = 100(0.472) = 47.2\%$ of light

63. $w(d) = 50e^{-0.004d}$

    (a) $w(30) = 50e^{-0.004(30)} = 50e^{-0.12} = 50(0.887) = 44.35$ watts

    (b) $w(365) = 50e^{-0.004(365)} = 50e^{-1.46} = 50(0.232) = 11.61$ watts

65. $D(h) = 5e^{-0.4h}$

    $D(1) = 5e^{-0.4(1)} = 5e^{-0.4} = 5(0.670) = 3.35$ milligrams

    $D(6) = 5e^{-0.4(6)} = 5e^{-2.4} = 5(0.091) = 0.45$ milligrams

67. $F(t) = 1 - e^{-0.1t}$

    (a) $F(10) = 1 - e^{-0.1(10)} = 1 - e^{-1} = 1 - 0.368 = 0.632 = 63.2\%$

    (b) $F(40) = 1 - e^{-0.1(40)} = 1 - e^{-4} = 1 - 0.018 = 0.982 = 98.2\%$

    (c) Graphing the function:

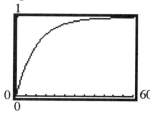

    (d) As $t \to \infty$, $F(t) \to 1$.

69. $P(x) = \dfrac{20^x e^{-20}}{x!}$

    (a) $P(15) = \dfrac{20^{15} e^{-20}}{15!} = 0.0516 = 5.16\%$ The probability that 15 cars will arrive between 5:00 p.m. and 6:00 p.m. is 5.16%.

    (b) $P(20) = \dfrac{20^{20} e^{-20}}{20!} = 0.0888 = 8.88\%$ The probability that 20 cars will arrive between 5:00 p.m. and 6:00 p.m. is 8.88%.

71. $R = 10^{\left(\frac{2345}{T} - \frac{2345}{D} + 2\right)}$

    (a) $R = 10^{\left(\frac{2345}{283} - \frac{2345}{278} + 2\right)} = 10^{1.851} = 70.96\%$

    (b) $R = 10^{\left(\frac{2345}{293} - \frac{2345}{288} + 2\right)} = 10^{1.861} = 72.61\%$

    (c) $R = 10^{\left(\frac{2345}{x} - \frac{2345}{x} + 2\right)} = 10^2 = 100\%$

73. $I = \dfrac{E}{R}\left[1 - e^{-\left(\frac{R}{L}\right)t}\right]$

    (a) $I = \dfrac{120}{10}\left[1 - e^{-\left(\frac{10}{5}\right)0.3}\right] = 12\left[1 - e^{-0.6}\right] = 5.414$ amperes after 0.3 second

    $I = \dfrac{120}{10}\left[1 - e^{-\left(\frac{10}{5}\right)0.5}\right] = 12\left[1 - e^{-1}\right] = 7.585$ amperes after 0.5 second

    $I = \dfrac{120}{10}\left[1 - e^{-\left(\frac{10}{5}\right)1}\right] = 12\left[1 - e^{-2}\right] = 10.376$ amperes after 1 second

    (b) As $t \to \infty$, $e^{-\left(\frac{10}{5}\right)t} \to 0$. Therefore, the maximum current is 12 amperes.

    (c) Graphing the function:

    (d) $I = \dfrac{120}{5}\left[1 - e^{-\left(\frac{5}{10}\right)0.3}\right] = 24\left[1 - e^{-0.15}\right] = 3.343$ amperes after 0.3 second

    $I = \dfrac{120}{5}\left[1 - e^{-\left(\frac{5}{10}\right)0.5}\right] = 24\left[1 - e^{-0.25}\right] = 5.309$ amperes after 0.5 second

    $I = \dfrac{120}{5}\left[1 - e^{-\left(\frac{5}{10}\right)1}\right] = 24\left[1 - e^{-0.5}\right] = 9.443$ amperes after 1 second

(e)   As $t \to \infty$, $e^{-\left(\frac{5}{10}\right)t} \to 0$. Therefore, the maximum current is 24 amperes.

(f)   Graphing the function:

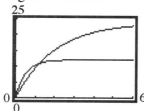

75. $2 + \dfrac{1}{2!} + \dfrac{1}{3!} + \dfrac{1}{4!} + \dots + \dfrac{1}{n!}$

$n = 4;\quad 2 + \dfrac{1}{2!} + \dfrac{1}{3!} + \dfrac{1}{4!} = 2.7083$

$n = 6;\quad 2 + \dfrac{1}{2!} + \dfrac{1}{3!} + \dfrac{1}{4!} + \dfrac{1}{5!} + \dfrac{1}{6!} = 2.7181$

$n = 8;\quad 2 + \dfrac{1}{2!} + \dfrac{1}{3!} + \dfrac{1}{4!} + \dfrac{1}{5!} + \dfrac{1}{6!} + \dfrac{1}{7!} + \dfrac{1}{8!} = 2.7182788$

$n = 10;\quad 2 + \dfrac{1}{2!} + \dfrac{1}{3!} + \dfrac{1}{4!} + \dfrac{1}{5!} + \dfrac{1}{6!} + \dfrac{1}{7!} + \dfrac{1}{8!} + \dfrac{1}{9!} + \dfrac{1}{10!} = 2.7182818$

$e = 2.718281828$

77. $f(x) = a^x$

$\dfrac{f(x+h) - f(x)}{h} = \dfrac{a^{x+h} - a^x}{h} = \dfrac{a^x a^h - a^x}{h} = \dfrac{a^x\left(a^h - 1\right)}{h} = a^x\left(\dfrac{a^h - 1}{h}\right)$

79. $f(x) = a^x$

$f(-x) = a^{-x} = \dfrac{1}{a^x} = \dfrac{1}{f(x)}$

81. $\sinh x = \dfrac{1}{2}\left(e^x - e^{-x}\right)$

(a) $f(-x) = \sinh(-x) = \dfrac{1}{2}\left(e^{-x} - e^x\right) = -\dfrac{1}{2}\left(e^x - e^{-x}\right) = -\sinh x = -f(x)$

Therefore, $f(x) = \sinh x$ is an odd function.

(b)   Graphing:

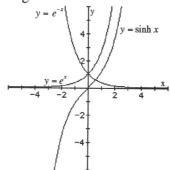

83. $f(x) = 2^{(2^x)} + 1$

$f(1) = 2^{(2^1)} + 1 = 2^2 + 1 = 4 + 1 = 5$

$f(2) = 2^{(2^2)} + 1 = 2^4 + 1 = 16 + 1 = 17$

$f(3) = 2^{(2^3)} + 1 = 2^8 + 1 = 256 + 1 = 257$

$f(4) = 2^{(2^4)} + 1 = 2^{16} + 1 = 65536 + 1 = 65537$

$f(5) = 2^{(2^5)} + 1 = 2^{32} + 1 = 4,294,967,296 + 1 = 4,294,967,297$

$\qquad 4,294,967,297 = 641 \times 6,700,417$

5.3   Logarithmic Functions

1. $9 = 3^2$ is equivalent to $2 = \log_3 9$

3. $a^2 = 1.6$ is equivalent to $2 = \log_a 1.6$

5. $1.1^2 = M$ is equivalent to $2 = \log_{1.1} M$

7. $2^x = 7.2$ is equivalent to $x = \log_2 7.2$

9. $x^{\sqrt{2}} = \pi$ is equivalent to $\sqrt{2} = \log_x \pi$

11. $e^x = 8$ is equivalent to $x = \ln 8$

13. $\log_2 8 = 3$ is equivalent to $2^3 = 8$

15. $\log_a 3 = 6$ is equivalent to $a^6 = 3$

17. $\log_3 2 = x$ is equivalent to $3^x = 2$

19. $\log_2 M = 1.3$ is equivalent to $2^{1.3} = M$

21. $\log_{\sqrt{2}} \pi = x$ is equivalent to $\left(\sqrt{2}\right)^x = \pi$

23. $\ln 4 = x$ is equivalent to $e^x = 4$

25. $\log_2 1 = 0$ since $2^0 = 1$

27. $\log_5 25 = 2$ since $5^2 = 25$

29. $\log_{\frac{1}{2}} 16 = -4$ since $\left(\frac{1}{2}\right)^{-4} = 2^4 = 16$

31. $\log_{10} \sqrt{10} = \frac{1}{2}$ since $10^{\frac{1}{2}} = \sqrt{10}$

33. $\log_{\sqrt{2}} 4 = 4$ since $\left(\sqrt{2}\right)^4 = 4$

35. $\ln \sqrt{e} = \frac{1}{2}$ since $e^{\frac{1}{2}} = \sqrt{e}$

37. The domain of $f(x) = \ln(x - 3)$ is:

$\qquad x - 3 > 0$

$\qquad\quad x > 3$

$\qquad \{x \mid x > 3\}$

39. The domain of $F(x) = \log_2 x^2$ is:

$\qquad x^2 > 0$

$\qquad \{x \mid x \neq 0\}$

41. The domain of
    $$h(x) = \log_{\frac{1}{2}}\left(x^2 - 2x + 1\right) \text{ is:}$$
    $$x^2 - 2x + 1 > 0$$
    $$(x-1)^2 > 0$$
    $$\{x \mid x \ne 1\}$$

43. The domain of $f(x) = \ln\left(\dfrac{1}{x+1}\right)$ is:
    $$\frac{1}{x+1} > 0$$
    $$x + 1 > 0$$
    $$x > -1$$
    $$\{x \mid x > -1\}$$

45. The domain of $g(x) = \log_5\left(\dfrac{x+1}{x}\right)$ requires that $\dfrac{x+1}{x} > 0$.

    The expression is zero or undefined when $x = -1$ or $x = 0$.

Interval	Test Number	$f(x) = \dfrac{x+1}{x}$	Positive/Negative
$-\infty < x < -1$	-2	$1/2$	Positive
$-1 < x < 0$	-0.5	-1	Negative
$0 < x < \infty$	1	2	Positive

    The domain is $\{x \mid x < -1 \text{ or } x > 0\}$.

47. $\ln \dfrac{5}{3} = 0.511$

49. $\dfrac{\ln(10/3)}{0.04} = 30.099$

51. For $f(x) = \log_a x$, find a so that $f(2) = \log_a 2 = 2$ or $a^2 = 2$ or $a = \sqrt{2}$.
    (The base a must be positive by definition.)

53. B   55. D   57. A   59. E

61. $f(x) = \ln(x + 4)$
    Using the graph of $y = \ln x$, shift the
    graph 4 units to the left.
    Domain: $(-4, \infty)$
    Range: $(-\infty, \infty)$
    Vertical Asymptote: $x = -4$

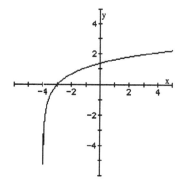

63. $f(x) = \ln(-x)$
Using the graph of $y = \ln x$, reflect the graph about the y-axis.
Domain: $(-\infty, 0)$
Range: $(-\infty, \infty)$
Vertical Asymptote: $x = 0$

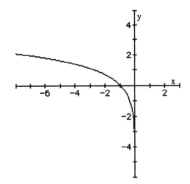

65. $g(x) = \ln(2x)$
Using the graph of $y = \ln x$, compress the graph horizontally by a factor of $\frac{1}{2}$.
Domain: $(0, \infty)$
Range: $(-\infty, \infty)$
Vertical Asymptote: $x = 0$

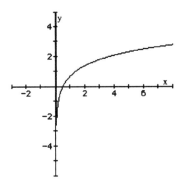

67. $f(x) = 3\ln x$
Using the graph of $y = \ln x$, stretch the graph vertically by a factor of 3.
Domain: $(0, \infty)$
Range: $(-\infty, \infty)$
Vertical Asymptote: $x = 0$

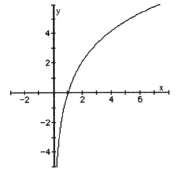

69. $g(x) = \ln(3 - x) = \ln(-(x-3))$
Using the graph of $y = \ln x$, reflect the graph about the y-axis, and shift 3 units to the right.
Domain: $(-\infty, 3)$
Range: $(-\infty, \infty)$
Vertical Asymptote: $x = 3$

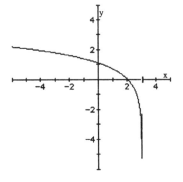

71. $f(x) = -\ln(x-1)$
    Using the graph of $y = \ln x$, shift the graph 1 unit to the right, and reflect about the x-axis.
    Domain: $(1, \infty)$
    Range: $(-\infty, \infty)$
    Vertical Asymptote: $x = 1$

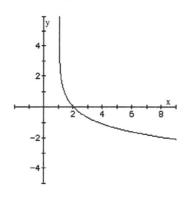

73. $\log_3 x = 2$
    $$x = 3^2$$
    $$x = 9$$

75. $\log_2 (2x + 1) = 3$
    $$2x + 1 = 2^3$$
    $$2x + 1 = 8$$
    $$2x = 7$$
    $$x = \frac{7}{2}$$

77. $\log_x 4 = 2$
    $$x^2 = 4$$
    $$x = 2 \quad (x \neq -2, \text{ base is positive })$$

79. $\ln e^x = 5$
    $$e^x = e^5$$
    $$x = 5$$

81. $\log_4 64 = x$
    $$4^x = 64$$
    $$4^x = 4^3$$
    $$x = 3$$

83. $\log_3 243 = 2x + 1$
    $$3^{2x+1} = 243$$
    $$3^{2x+1} = 3^5$$
    $$2x + 1 = 5$$
    $$2x = 4$$
    $$x = 2$$

85. $e^{3x} = 10$
    $$3x = \ln 10$$
    $$x = \frac{\ln 10}{3}$$

87. $e^{2x+5} = 8$
    $$2x + 5 = \ln 8$$
    $$2x = -5 + \ln 8$$
    $$x = \frac{-5 + \ln 8}{2}$$

89. $\log_3 \left(x^2 + 1\right) = 2$
    $$x^2 + 1 = 3^2$$
    $$x^2 + 1 = 9$$
    $$x^2 = 8$$
    $$x = \pm\sqrt{8} = \pm 2\sqrt{2}$$

91. $\log_2 8^x = -3$
    $$8^x = 2^{-3}$$
    $$8^x = \tfrac{1}{8}$$
    $$8^x = 8^{-1}$$
    $$x = -1$$

93. (a)   Graphing $f(x) = 2^x$:

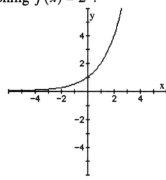

Domain: $(-\infty, \infty)$
Range: $(0, \infty)$
Horizontal asymptote: $y = 0$

(b)   Finding the inverse:
$$f(x) = 2^x$$
$$y = 2^x$$
$$x = 2^y \quad \text{Inverse}$$
$$y = \log_2 x$$
$$f^{-1}(x) = \log_2 x$$

(c)   Graphing the inverse:

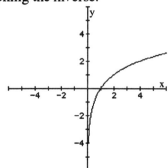

Domain: $(0, \infty)$
Range: $(-\infty, \infty)$
Vertical asymptote: $x = 0$

95. (a)   Graphing $f(x) = 2^{x+3}$:

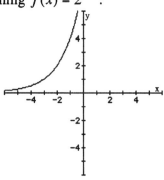

Domain: $(-\infty, \infty)$
Range: $(0, \infty)$
Horizontal asymptote: $y = 0$

(b)   Finding the inverse:
$$f(x) = 2^{x+3}$$
$$y = 2^{x+3}$$
$$x = 2^{y+3} \quad \text{Inverse}$$
$$y + 3 = \log_2 x$$
$$y = -3 + \log_2 x$$
$$f^{-1}(x) = -3 + \log_2 x$$

(c)   Graphing the inverse:

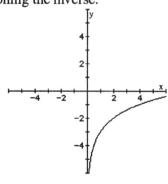

Domain: $(0, \infty)$
Range: $(-\infty, \infty)$
Vertical asymptote: $x = 0$

97. (a)   Graphing $f(x) = 1 + 2^{x-3}$:

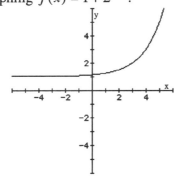

Domain: $(-\infty, \infty)$
Range: $(1, \infty)$
Horizontal asymptote: $y = 1$

(b)   Finding the inverse:
$$f(x) = 1 + 2^{x-3}$$
$$y = 1 + 2^{x-3}$$
$$x = 1 + 2^{y-3} \qquad \text{Inverse}$$
$$x - 1 = 2^{y-3}$$
$$y - 3 = \log_2(x - 1)$$
$$y = 3 + \log_2(x - 1)$$
$$f^{-1}(x) = 3 + \log_2(x - 1)$$

(c)   Graphing the inverse:

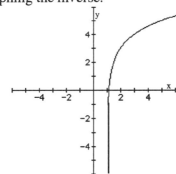

Domain: $(1, \infty)$
Range: $(-\infty, \infty)$
Vertical asymptote: $x = 1$

99. $P = 100e^{-0.1n}$

(a)
$$50 = 100e^{-0.1n}$$
$$0.5 = e^{-0.1n}$$
$$\ln 0.5 = -0.1n$$
$$n = \frac{\ln 0.5}{-0.1}$$
$$n \approx 6.93$$
7 panes of glass are needed.

(b)
$$25 = 100e^{-0.1n}$$
$$0.25 = e^{-0.1n}$$
$$\ln 0.25 = -0.1n$$
$$n = \frac{\ln 0.25}{-0.1}$$
$$n \approx 13.86$$
14 panes of glass are needed.

101. $w = 50e^{-0.004d}$

(a)
$$30 = 50e^{-0.004d}$$
$$0.6 = e^{-0.004d}$$
$$\ln 0.6 = -0.004d$$
$$d = \frac{\ln 0.6}{-0.004}$$
$$d \approx 127.7$$
Approximately 128 days.

(b)
$$5 = 50e^{-0.004d}$$
$$0.1 = e^{-0.004d}$$
$$\ln 0.1 = -0.004d$$
$$d = \frac{\ln 0.1}{-0.004}$$
$$d \approx 575.6$$
Approximately 576 days.

103. $F(t) = 1 - e^{-0.1t}$

(a)
$$0.5 = 1 - e^{-0.1t}$$
$$-0.5 = -e^{-0.1t}$$
$$0.5 = e^{-0.1t}$$
$$\ln 0.5 = -0.1t$$
$$t = \frac{\ln 0.5}{-0.1}$$
$$n \approx 6.93$$
Approximately 7 minutes.

(b)
$$0.8 = 1 - e^{-0.1t}$$
$$-0.2 = -e^{-0.1t}$$
$$0.2 = e^{-0.1t}$$
$$\ln 0.2 = -0.1t$$
$$t = \frac{\ln 0.2}{-0.1}$$
$$n \approx 16.09$$
Approximately 16 minutes.

(c) It is impossible for the probability to reach 100% because $e^{-0.1t}$ will never equal zero.

105. $D = 5e^{-0.4h}$
$$2 = 5e^{-0.4h}$$
$$0.4 = e^{-0.4h}$$
$$\ln 0.4 = -0.4h$$
$$h = \frac{\ln 0.4}{-0.4}$$
$$h \approx 2.29 \text{ hours}$$

107. $I = \frac{E}{R}\left[1 - e^{-\left(\frac{R}{L}\right)t}\right]$

0.5 ampere:

$$0.5 = \frac{12}{10}\left[1 - e^{-\left(\frac{10}{5}\right)t}\right]$$
$$0.4167 = 1 - e^{-2t}$$
$$e^{-2t} = 0.5833$$
$$-2t = \ln 0.5833$$
$$t = \frac{\ln 0.5833}{-2}$$
$$t = 0.2695 \text{ seconds}$$

1.0 ampere:

$$1.0 = \frac{12}{10}\left[1 - e^{-\left(\frac{10}{5}\right)t}\right]$$
$$0.8333 = 1 - e^{-2t}$$
$$e^{-2t} = 0.1667$$
$$-2t = \ln 0.1667$$
$$t = \frac{\ln 0.1667}{-2}$$
$$t = 0.8958 \text{ seconds}$$

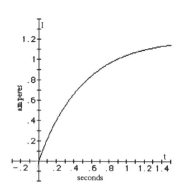

109. $L(10^{-7}) = 10 \log \frac{10^{-7}}{10^{-12}} = 10 \log 10^5 = 10 \cdot 5 = 50$ decibels

111. $L(10^{-1}) = 10 \log \frac{10^{-1}}{10^{-12}} = 10 \log 10^{11} = 10 \cdot 11 = 110$ decibels

113. $M(125{,}892) = \log \frac{125{,}892}{10^{-3}} = 8.1$

115. $R = 3e^{kx}$

    (a) $10 = 3e^{k(0.06)}$

    $3.3333 = e^{0.06k}$

    $\ln 3.3333 = 0.06k$

    $k = \dfrac{\ln 3.3333}{0.06}$

    $k = 20.066$

    (b) $R = 3e^{20.066(0.17)}$

    $R = 3e^{3.41122}$

    $R = 90.9\%$

    (c) $100 = 3e^{20.066x}$

    $33.3333 = e^{20.066x}$

    $\ln 33.3333 = 20.066x$

    $x = \dfrac{\ln 33.3333}{20.066}$

    $x = 0.175$

    (d) $15 = 3e^{20.066x}$

    $5 = e^{20.066x}$

    $\ln 5 = 20.066x$

    $x = \dfrac{\ln 5}{20.066}$

    $x = 0.08$

5.4 Properties of Logarithms

1. $\ln 6 = \ln(3 \cdot 2) = \ln 3 + \ln 2 = b + a$

3. $\ln 1.5 = \ln \dfrac{3}{2} = \ln 3 - \ln 2 = b - a$

5. $\ln(2e) = \ln 2 + \ln e = a + 1$

7. $\ln 12 = \ln(3 \cdot 4) = \ln 3 + \ln 4 = \ln 3 + \ln 2^2 = \ln 3 + 2\ln 2 = b + 2a$

9. $\ln \sqrt[5]{18} = \ln 18^{\frac{1}{5}} = \frac{1}{5}\ln 18 = \frac{1}{5}\ln\left(2 \cdot 3^2\right) = \frac{1}{5}\left(\ln 2 + \ln 3^2\right)$
      $= \frac{1}{5}\left(\ln 2 + 2\ln 3\right) = \frac{1}{5}(a + 2b)$

11. $\log_2 3 = \dfrac{\ln 3}{\ln 2} = \dfrac{b}{a}$

13. $\log_a\left(u^2 v^3\right) = \log_a u^2 + \log_a v^3 = 2\log_a u + 3\log_a v$

15. $\log \dfrac{1}{M^3} = \log M^{-3} = -3\log M$

17. $\log_5 \sqrt{\dfrac{a^3}{b}} = \log_5\left(\dfrac{a^3}{b}\right)^{\frac{1}{2}} = \log_5 \dfrac{a^{\frac{3}{2}}}{b^{\frac{1}{2}}} = \log_5 a^{\frac{3}{2}} - \log_5 b^{\frac{1}{2}} = \frac{3}{2}\log_5 a - \frac{1}{2}\log_5 b$

19. $\ln\left(x^2\sqrt{1-x}\right) = \ln x^2 + \ln\sqrt{1-x} = \ln x^2 + \ln(1-x)^{\frac{1}{2}} = 2\ln x + \frac{1}{2}\ln(1-x)$

21. $\log_2\left(\dfrac{x^3}{x-3}\right) = \log_2 x^3 - \log_2(x-3) = 3\log_2 x - \log_2(x-3)$

23. $\log\left[\dfrac{x(x+2)}{(x+3)^2}\right] = \log x(x+2) - \log(x+3)^2 = \log x + \log(x+2) - 2\log(x+3)$

25. $\ln\left[\dfrac{x^2-x-2}{(x+4)^2}\right]^{\frac{1}{3}} = \frac{1}{3}\ln\left[\dfrac{(x-2)(x+1)}{(x+4)^2}\right] = \frac{1}{3}\left[\ln(x-2)(x+1) - \ln(x+4)^2\right]$
$$= \frac{1}{3}\left[\ln(x-2) + \ln(x+1) - 2\ln(x+4)\right] = \frac{1}{3}\ln(x-2) + \frac{1}{3}\ln(x+1) - \frac{2}{3}\ln(x+4)$$

27. $\ln\dfrac{5x\sqrt{1-3x}}{(x-4)^3} = \ln 5x\sqrt{1-3x} - \ln(x-4)^3 = \ln 5 + \ln x + \ln\sqrt{1-3x} - 3\ln(x-4)$
$$= \ln 5 + \ln x + \ln(1-3x)^{\frac{1}{2}} - 3\ln(x-4) = \ln 5 + \ln x + \frac{1}{2}\ln(1-3x) - 3\ln(x-4)$$

29. $3\log_5 u + 4\log_5 v = \log_5 u^3 + \log_5 v^4 = \log_5(u^3 v^4)$

31. $\log_{\frac{1}{2}}\sqrt{x} - \log_{\frac{1}{2}} x^3 = \log_{\frac{1}{2}}\dfrac{\sqrt{x}}{x^3} = \log_{\frac{1}{2}}\dfrac{x^{\frac{1}{2}}}{x^3} = \log_{\frac{1}{2}} x^{\frac{-5}{2}} = \frac{-5}{2}\log_{\frac{1}{2}} x$

33. $\ln\dfrac{x}{x-1} + \ln\dfrac{x+1}{x} - \ln\left(x^2-1\right) = \ln\left[\dfrac{x}{x-1}\cdot\dfrac{x+1}{x}\right] - \ln\left(x^2-1\right) = \ln\left[\dfrac{x+1}{x-1} \div \left(x^2-1\right)\right]$
$$= \ln\left[\dfrac{x+1}{(x-1)(x-1)(x+1)}\right] = \ln\dfrac{1}{(x-1)^2} = \ln(x-1)^{-2} = -2\ln(x-1)$$

35. $8\log_2\sqrt{3x-2} - \log_2\dfrac{4}{x} + \log_2 4 = \log_2\left(\sqrt{3x-2}\right)^8 - \left(\log_2 4 - \log_2 x\right) + \log_2 4$
$$= \log_2(3x-2)^4 - \log_2 4 + \log_2 x + \log_2 4 = \log_2\left[x(3x-2)^4\right]$$

37. $2\log_a 5x^3 - \frac{1}{2}\log_a(2x+3) = \log_a\left(5x^3\right)^2 - \log_a(2x-3)^{\frac{1}{2}} = \log_a\left[\dfrac{25x^6}{(2x-3)^{\frac{1}{2}}}\right]$

39. $\quad y = ab^x$
$\ln y = \ln\left(ab^x\right)$
$\ln y = \ln a + \ln b^x$
$\ln y = \ln a + x\ln b$
$\ln y = \ln a + (\ln b)x$

41. $3^{\log_3 5 - \log_3 4} = 3^{\log_3\left(\frac{5}{4}\right)} = \frac{5}{4}$

43. $e^{\log_{e^2} 16}$
     Simplify the exponent:

$$\text{Let} \qquad a = \log_{e^2} 16$$
$$\left(e^2\right)^a = 16$$
$$e^{2a} = 16 = 4^2$$
$$e^a = 4$$
$$a = \ln 4$$

Thus, $e^{\log_{e^2} 16} = e^{\ln 4} = 4$

45. $\log_3 21 = \dfrac{\log 21}{\log 3} = \dfrac{1.32222}{0.47712} = 2.771$

47. $\log_{\frac{1}{3}} 71 = \dfrac{\log 71}{\log \frac{1}{3}} = \dfrac{\log 71}{-\log 3} = \dfrac{1.85126}{-0.47712} = -3.880$

49. $\log_{\sqrt{2}} 7 = \dfrac{\log 7}{\log \sqrt{2}} = \dfrac{\log 7}{\log 2^{\frac{1}{2}}} = \dfrac{\log 7}{\frac{1}{2}\log 2} = \dfrac{0.84510}{0.5(0.30103)} = 5.615$

51. $\log_\pi e = \dfrac{\ln e}{\ln \pi} = \dfrac{1}{1.14473} = 0.874$

53. $y = \log_4 x = \dfrac{\ln x}{\ln 4}$ or $y = \dfrac{\log x}{\log 4}$

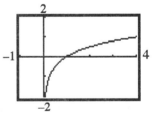

55. $y = \log_2(x+2) = \dfrac{\ln(x+2)}{\ln 2}$

or $y = \dfrac{\log(x+2)}{\log 2}$

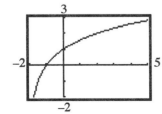

57. $y = \log_{x-1}(x+1) = \dfrac{\ln(x+1)}{\ln(x-1)}$

or $y = \dfrac{\log(x+1)}{\log(x-1)}$

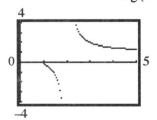

59. $\ln y = \ln x + \ln C$
$\ln y = \ln(xC)$
$y = Cx$

61. $\ln y = \ln x + \ln(x+1) + \ln C$
$\ln y = \ln(x(x+1)C)$
$y = Cx(x+1)$

63. $\ln y = 3x + \ln C$
$\ln y = \ln e^{3x} + \ln C$
$\ln y = \ln(Ce^{3x})$
$y = Ce^{3x}$

65. $\ln(y-3) = -4x + \ln C$
$\ln(y-3) = \ln e^{-4x} + \ln C$
$\ln(y-3) = \ln(Ce^{-4x})$
$y - 3 = Ce^{-4x}$
$y = Ce^{-4x} + 3$

67. $3\ln y = \frac{1}{2}\ln(2x+1) - \frac{1}{3}\ln(x+4) + \ln C$

$\ln y^3 = \ln(2x+1)^{\frac{1}{2}} - \ln(x+4)^{\frac{1}{3}} + \ln C$

$\ln y^3 = \ln\left[\dfrac{C(2x+1)^{\frac{1}{2}}}{(x+4)^{\frac{1}{3}}}\right]$

$y^3 = \dfrac{C(2x+1)^{\frac{1}{2}}}{(x+4)^{\frac{1}{3}}}$

$y = \left[\dfrac{C(2x+1)^{\frac{1}{2}}}{(x+4)^{\frac{1}{3}}}\right]^{\frac{1}{3}}$

$y = \dfrac{\sqrt[3]{C}(2x+1)^{\frac{1}{6}}}{(x+4)^{\frac{1}{9}}}$

69. $\log_2 3 \cdot \log_3 4 \cdot \log_4 5 \cdot \log_5 6 \cdot \log_6 7 \cdot \log_7 8$

$= \dfrac{\log 3}{\log 2} \cdot \dfrac{\log 4}{\log 3} \cdot \dfrac{\log 5}{\log 4} \cdot \dfrac{\log 6}{\log 5} \cdot \dfrac{\log 7}{\log 6} \cdot \dfrac{\log 8}{\log 7} = \dfrac{\log 8}{\log 2} = \dfrac{\log 2^3}{\log 2} = \dfrac{3\log 2}{\log 2} = 3$

71. $\log_2 3 \cdot \log_3 4 \cdot \ldots \cdot \log_n(n+1) \cdot \log_{n+1} 2$

$= \dfrac{\log 3}{\log 2} \cdot \dfrac{\log 4}{\log 3} \cdot \ldots \cdot \dfrac{\log(n+1)}{\log n} \cdot \dfrac{\log 2}{\log(n+1)} = \dfrac{\log 2}{\log 2} = 1$

73. Verifying:
$$\log_a\left(x + \sqrt{x^2-1}\right) + \log_a\left(x - \sqrt{x^2-1}\right) = \log_a\left[\left(x + \sqrt{x^2-1}\right)\left(x - \sqrt{x^2-1}\right)\right]$$
$$= \log_a\left[x^2 - \left(x^2-1\right)\right] = \log_a\left[x^2 - x^2 + 1\right] = \log_a 1 = 0$$

75. Verifying:
$$2x + \ln\left(1 + e^{-2x}\right) = \ln e^{2x} + \ln\left(1 + e^{-2x}\right) = \ln\left(e^{2x}\left(1 + e^{-2x}\right)\right) = \ln\left(e^{2x} + e^0\right) = \ln\left(e^{2x} + 1\right)$$

77. $f(x) = \log_a x$

$$x = a^{f(x)}$$

$$x^{-1} = a^{-f(x)} = \left(a^{-1}\right)^{f(x)} = \left(\frac{1}{a}\right)^{f(x)}$$

$$\log_{\frac{1}{a}} x^{-1} = f(x)$$

$$-\log_{\frac{1}{a}} x = f(x)$$

$$-f(x) = \log_{\frac{1}{a}} x$$

79. $f(x) = \log_a x$

$$a^{f(x)} = x$$

$$\frac{1}{a^{f(x)}} = \frac{1}{x}$$

$$a^{-f(x)} = \frac{1}{x}$$

$$-f(x) = \log_a \frac{1}{x} = f\left(\frac{1}{x}\right)$$

81. If $A = \log_a M$ and $B = \log_a N$, then $a^A = M$ and $a^B = N$.

$$\log_a\left(\frac{M}{N}\right) = \log_a\left(\frac{a^A}{a^B}\right) = \log_a a^{A-B} = A - B = \log_a M - \log_a N$$

5.5   Logarithmic and Exponential Equations

1. $\log_4(x+2) = \log_4 8$

$$x + 2 = 8$$

$$x = 6$$

3. $\frac{1}{2}\log_3 x = 2\log_3 2$

$$\log_3 x^{\frac{1}{2}} = \log_3 2^2$$

$$x^{\frac{1}{2}} = 4$$

$$x = 16$$

5. $2\log_5 x = 3\log_5 4$

$$\log_5 x^2 = \log_5 4^3$$

$$x^2 = 64$$

$$x = \pm 8$$

Since $\log_5(-8)$ is undefined, the only solution is $x = 8$.

7. $3\log_2(x-1) + \log_2 4 = 5$

$$\log_2(x-1)^3 + \log_2 4 = 5$$

$$\log_2 4(x-1)^3 = 5$$

$$4(x-1)^3 = 2^5$$

$$(x-1)^3 = \frac{32}{4}$$

$$(x-1)^3 = 8$$

$$x - 1 = 2$$

$$x = 3$$

9. $\log x + \log(x+15) = 2$

$$\log x(x+15) = 2$$

$$x(x+15) = 10^2$$

$$x^2 + 15x - 100 = 0$$

$$(x+20)(x-5) = 0$$

$$x = -20 \text{ or } x = 5$$

Since $\log(-20)$ is undefined, the only solution is $x = 5$.

11. $\ln x + \ln(x+2) = 4$

$\quad \ln x(x+2) = 4$

$\quad\quad x(x+2) = e^4$

$\quad x^2 + 2x - e^4 = 0$

$$x = \frac{-2 \pm \sqrt{2^2 - 4(1)(-e^4)}}{2(1)} = \frac{-2 \pm \sqrt{4 + 4e^4}}{2}$$

$$= \frac{-2 \pm 2\sqrt{1+e^4}}{2} = -1 \pm \sqrt{1+e^4}$$

Since $\ln\left(-1 - \sqrt{1+e^4}\right)$ is undefined, the only solution is $x = -1 + \sqrt{1+e^4} \approx 6.456$.

13. $\quad 2^{2x} + 2^x - 12 = 0$

$\quad \left(2^x\right)^2 + 2^x - 12 = 0$

$\quad \left(2^x - 3\right)\left(2^x + 4\right) = 0$

$\quad\quad 2^x - 3 = 0 \quad\quad$ or $\;2^x + 4 = 0$

$\quad\quad\quad 2^x = 3 \quad\quad$ or $\quad\quad 2^x = -4$

$\quad\quad\quad\quad x = \log_2 3 \quad\quad$ No solution

$\quad\quad\quad\quad x \approx 1.585$

15. $\quad 3^{2x} + 3^{x+1} - 4 = 0$

$\quad \left(3^x\right)^2 + 3 \cdot 3^x - 4 = 0$

$\quad \left(3^x - 1\right)\left(3^x + 4\right) = 0$

$\quad\quad 3^x - 1 = 0 \;$ or $\; 3^x + 4 = 0$

$\quad\quad\quad 3^x = 1 \;$ or $\quad\quad 3^x = -4$

$\quad\quad\quad\quad x = 0 \quad\quad$ No solution

17. $\quad\quad 2^x = 10$

$\quad \log(2^x) = \log 10$

$\quad\quad x \log 2 = 1$

$\quad\quad\quad x = \dfrac{1}{\log 2} \approx 3.322$

19. $\quad\quad 8^{-x} = 1.2$

$\quad \log(8^{-x}) = \log 1.2$

$\quad -x \log 8 = \log 1.2$

$\quad\quad\quad x = \dfrac{\log 1.2}{-\log 8} \approx -0.088$

21. $\quad\quad\quad 3^{1-2x} = 4^x$

$\quad \log(3^{1-2x}) = \log(4^x)$

$\quad (1 - 2x)\log 3 = x \log 4$

$\quad \log 3 - 2x \log 3 = x \log 4$

$\quad\quad\quad \log 3 = x \log 4 + 2x \log 3$

$\quad\quad\quad \log 3 = x(\log 4 + 2\log 3)$

$\quad\quad\quad\quad x = \dfrac{\log 3}{\log 4 + 2\log 3} \approx 0.307$

23.
$$\left(\tfrac{3}{5}\right)^x = 7^{1-x}$$
$$\log\left(\left(\tfrac{3}{5}\right)^x\right) = \log\left(7^{1-x}\right)$$
$$x\log\tfrac{3}{5} = (1-x)\log 7$$
$$x(\log 3 - \log 5) = \log 7 - x\log 7$$
$$x\log 3 - x\log 5 + x\log 7 = \log 7$$
$$x(\log 3 - \log 5 + \log 7) = \log 7$$
$$x = \frac{\log 7}{\log 3 - \log 5 + \log 7} \approx 1.356$$

25.
$$1.2^x = (0.5)^{-x}$$
$$\log 1.2^x = \log(0.5)^{-x}$$
$$x\log 1.2 = -x\log 0.5$$
$$x\log 1.2 + x\log 0.5 = 0$$
$$x(\log 1.2 + \log 0.5) = 0$$
$$x = 0$$

27.
$$\pi^{1-x} = e^x$$
$$\ln \pi^{1-x} = \ln e^x$$
$$(1-x)\ln \pi = x$$
$$\ln \pi - x\ln \pi = x$$
$$\ln \pi = x + x\ln \pi$$
$$\ln \pi = x(1 + \ln \pi)$$
$$x = \frac{\ln \pi}{1 + \ln \pi} \approx 0.534$$

29.
$$5\left(2^{3x}\right) = 8$$
$$2^{3x} = \tfrac{8}{3}$$
$$\log 2^{3x} = \log\left(\tfrac{8}{3}\right)$$
$$3x\log 2 = \log 8 - \log 5$$
$$x = \frac{\log 8 - \log 5}{3\log 2} \approx 0.226$$

31.
$$\log_a(x-1) - \log_a(x+6) = \log_a(x-2) - \log_a(x+3)$$
$$\log_a\frac{x-1}{x+6} = \log_a\frac{x-2}{x+3}$$
$$\frac{x-1}{x+6} = \frac{x-2}{x+3}$$
$$(x-1)(x+3) = (x-2)(x+6)$$
$$x^2 + 2x - 3 = x^2 + 4x - 12$$
$$-2x = -9$$
$$x = \frac{9}{2}$$

33. $\log_{\frac{1}{3}}(x^2 + x) - \log_{\frac{1}{3}}(x^2 - x) = -1$

$$\log_{\frac{1}{3}}\frac{x^2 + x}{x^2 - x} = -1$$

$$\frac{x^2 + x}{x^2 - x} = \left(\frac{1}{3}\right)^{-1}$$

$$\frac{x(x+1)}{x(x-1)} = 3$$

$$x + 1 = 3(x - 1)$$

$$x + 1 = 3x - 3$$

$$-2x = -4$$

$$x = 2$$

35. $\log_2(x + 1) - \log_4 x = 1$

$$\log_2(x + 1) - \frac{\log_2 x}{\log_2 4} = 1$$

$$\log_2(x + 1) - \frac{\log_2 x}{2} = 1$$

$$2\log_2(x + 1) - \log_2 x = 2$$

$$\log_2(x + 1)^2 - \log_2 x = 2$$

$$\log_2\frac{(x + 1)^2}{x} = 2$$

$$\frac{(x + 1)^2}{x} = 2^2$$

$$x^2 + 2x + 1 = 4x$$

$$x^2 - 2x + 1 = 0$$

$$(x - 1)^2 = 0$$

$$x - 1 = 0$$

$$x = 1$$

37. $\log_{16} x + \log_4 x + \log_2 x = 7$

$$\frac{\log_2 x}{\log_2 16} + \frac{\log_2 x}{\log_2 4} + \log_2 x = 7$$

$$\frac{\log_2 x}{4} + \frac{\log_2 x}{2} + \log_2 x = 7$$

$$\log_2 x + 2\log_2 x + 4\log_2 x = 28$$

$$7\log_2 x = 28$$

$$\log_2 x = 4$$

$$x = 2^4 = 16$$

39. $\left(\sqrt[3]{2}\right)^{2-x} = 2^{x^2}$

$$\left(2^{\frac{1}{3}}\right)^{2-x} = 2^{x^2}$$

$$2^{\frac{1}{3}(2-x)} = 2^{x^2}$$

$$\frac{1}{3}(2 - x) = x^2$$

$$2 - x = 3x^2$$

$$3x^2 + x - 2 = 0$$

$$(3x - 2)(x + 1) = 0$$

$$x = \frac{2}{3} \text{ or } x = -1$$

41. $\dfrac{e^x + e^{-x}}{2} = 1$

$$e^x + e^{-x} = 2$$

$$e^x\left(e^x + e^{-x}\right) = 2e^x$$

$$e^{2x} + 1 = 2e^x$$

$$(e^x)^2 - 2e^x + 1 = 0$$

$$\left(e^x - 1\right)^2 = 0$$

$$e^x - 1 = 0$$

$$e^x = 1$$

$$x = 0$$

43.
$$\frac{e^x - e^{-x}}{2} = 2$$

$$e^x - e^{-x} = 4$$

$$e^x\left(e^x - e^{-x}\right) = 4e^x$$

$$e^{2x} - 1 = 4e^x$$

$$(e^x)^2 - 4e^x - 1 = 0$$

$$e^x = \frac{-(-4) \pm \sqrt{(-4)^2 - 4(1)(-1)}}{2(1)} = \frac{4 \pm \sqrt{20}}{2} = \frac{4 \pm 2\sqrt{5}}{2} = 2 \pm \sqrt{5}$$

$$x = \ln\left(2 + \sqrt{5}\right)$$

$\ln\left(2 - \sqrt{5}\right)$ is undefined; it is not a solution.

45. Using INTERSECT to solve:
$$y_1 = \ln(x) / \ln(5) + \ln(x) / \ln(3)$$
$$y_2 = 1$$

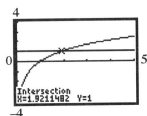

The solution is 1.92.

47. Using INTERSECT to solve:
$$y_1 = \ln(x + 1) / \ln(5) - \ln(x - 2) / \ln(4)$$
$$y_2 = 1$$

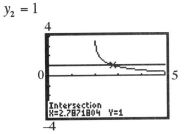

The solution is 2.79.

49. Using INTERSECT to solve:
$$y_1 = e^x; \ y_2 = -x$$

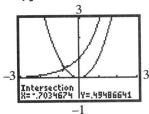

The solution is –0.57.

51. Using INTERSECT to solve:
$$y_1 = e^x; \ y_2 = x^2$$

The solution is –0.70.

53. Using INTERSECT to solve:
$y_1 = \ln x; \quad y_2 = -x$

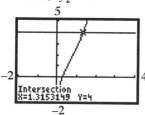

The solution is 0.57.

55. Using INTERSECT to solve:
$y_1 = \ln x; \quad y_2 = x^3 - 1$

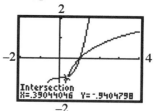

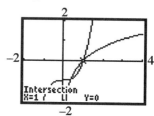

The solutions are 0.39, 1.00.

57. Using INTERSECT to solve:
$y_1 = e^x + \ln x; \quad y_2 = 4$

The solution is 1.32.

59. Using INTERSECT to solve:
$y_1 = e^{-x}; \quad y_2 = \ln x$

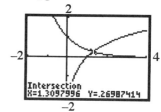

The solution is 1.31.

5.6 Compound Interest

1. $P = \$100, \ r = 0.04, \ n = 4, \ t = 2$
$$A = P\left[1 + \frac{r}{n}\right]^{nt} = 100\left[1 + \frac{0.04}{4}\right]^{(4)(2)} = \$108.29$$

3. $P = \$500, \ r = 0.08, \ n = 4, \ t = 2.5$
$$A = P\left[1 + \frac{r}{n}\right]^{nt} = 500\left[1 + \frac{0.08}{4}\right]^{(4)(2.5)} = \$609.50$$

5. $P = \$600, \ r = 0.05, \ n = 365, \ t = 3$
$$A = P\left[1 + \frac{r}{n}\right]^{nt} = 600\left[1 + \frac{0.05}{365}\right]^{(365)(3)} = \$697.09$$

7. $P = \$10,\ r = 0.11,\ t = 2$
    $$A = Pe^{rt} = 10e^{(0.11)(2)} = \$12.46$$

9. $P = \$100,\ r = 0.10,\ t = 2.25$
    $$A = Pe^{rt} = 100e^{(0.10)(2.25)} = \$125.23$$

11. $A = \$100,\ r = 0.06,\ n = 12,\ t = 2$
    $$P = A\left[1 + \frac{r}{n}\right]^{-nt} = 100\left[1 + \frac{0.06}{12}\right]^{(-12)(2)} = \$88.72$$

13. $A = \$1000,\ r = 0.06,\ n = 365,\ t = 2.5$
    $$P = A\left[1 + \frac{r}{n}\right]^{-nt} = 1000\left[1 + \frac{0.06}{365}\right]^{(-365)(2.5)} = \$860.72$$

15. $A = \$600,\ r = 0.04,\ n = 4,\ t = 2$
    $$P = A\left[1 + \frac{r}{n}\right]^{-nt} = 600\left[1 + \frac{0.04}{4}\right]^{(-4)(2)} = \$554.09$$

17. $A = \$80,\ r = 0.09,\ t = 3.25$
    $$P = Ae^{-rt} = 80e^{(-0.09)(3.25)} = \$59.71$$

19. $A = \$400,\ r = 0.10,\ t = 1$
    $$P = Ae^{-rt} = 400e^{(-0.10)(1)} = \$361.93$$

21. $r_e = \left[1 + \dfrac{r}{n}\right]^{n} - 1 = \left[1 + \dfrac{0.0525}{4}\right]^{4} - 1 = 1.0535 - 1 = 0.0535 = 5.35\%$

23. $2P = P(1 + r)^3$
    $2 = (1 + r)^3$
    $\sqrt[3]{2} = 1 + r$
    $r = \sqrt[3]{2} - 1 \approx 1.26 - 1 = 0.26 = 26\%$

25. 6% compounded quarterly:
    $$A = 10,000\left[1 + \frac{0.06}{4}\right]^{(4)(1)} = \$10,613.64$$
    $6\frac{1}{4}\%$ compounded annually:
    $$A = 10,000[1 + 0.0625]^{1} = \$10,625$$
    $6\frac{1}{4}\%$ compounded annually yields the larger amount.

27. 9% compounded monthly:

$$A = 10,000\left[1+\frac{0.09}{12}\right]^{(12)(1)} = \$10,938.07$$

8.8% compounded daily:

$$A = 10,000\left[1+\frac{0.088}{365}\right]^{365} = \$10,919.77$$

9% compounded monthly yields the larger amount.

29. Compounded monthly:

$$2P = P\left[1+\frac{0.08}{12}\right]^{12t}$$

$$2 = (1.00667)^{12t}$$

$$\ln 2 = 12t \ln(1.00667)$$

$$t = \frac{\ln 2}{12 \ln(1.00667)} \approx 8.69 \text{ years}$$

Compounded continuously:

$$2P = Pe^{0.08t}$$

$$2 = e^{0.08t}$$

$$\ln 2 = 0.08t$$

$$t = \frac{\ln 2}{0.08} \approx 8.66 \text{ years}$$

31. Compounded monthly:

$$150 = 100\left[1+\frac{0.08}{12}\right]^{12t}$$

$$1.5 = (1.00667)^{12t}$$

$$\ln 1.5 = 12t \ln(1.00667)$$

$$t = \frac{\ln 1.5}{12 \ln(1.00667)} \approx 5.083 \text{ years}$$

Compounded continuously:

$$150 = 100e^{0.08t}$$

$$1.5 = e^{0.08t}$$

$$\ln 1.5 = 0.08t$$

$$t = \frac{\ln 1.5}{0.08} \approx 5.068 \text{ years}$$

33. $$25,000 = 10,000e^{0.06t}$$

$$2.5 = e^{0.06t}$$

$$\ln 2.5 = 0.06t$$

$$t = \frac{\ln 2.5}{0.06} \approx 15.27 \text{ years}$$

35. $$A = 90,000(1+0.03)^5 = \$104,335$$

37. $$P = 15,000e^{(-0.05)(3)} = \$12,910.62$$

39. $$A = 1500(1+0.15)^5 = 1500(1.15)^5 = \$3017$$

41. $$850,000 = 650,000(1+r)^3$$

$$1.3077 = (1+r)^3$$

$$\sqrt[3]{1.3077} = 1+r$$

$$r = \sqrt[3]{1.3077} - 1 \approx 0.0935 = 9.35\%$$

43. 5.6% compounded continuously:
$$A = 1000e^{(0.056)(1)} = \$1057.60$$
Jim does not have enough money to buy the computer.
5.9% compounded monthly:
$$A = 1000\left[1 + \frac{0.059}{12}\right]^{12} = \$1060.62$$
The second bank offers the better deal.

45. Will - 9% compounded semiannually:
$$A = 2000\left[1 + \frac{0.09}{2}\right]^{(2)(20)} = \$11,632.73$$
Henry - 8.5% compounded continuously:
$$A = 2000e^{(0.085)(20)} = \$10,947.89$$
Will has more money after 20 years.

47. $P = 50,000;\ t = 5$
    (a) Simple interest at 12% per annum:
    $$A = 50,000 + 50,000(0.12)(5) = \$80,000$$
    (b) 11.5% compounded monthly:
    $$A = 50,000\left[1 + \frac{0.115}{12}\right]^{(12)(5)} = \$88,613.59$$
    (c) 11.25% compounded continuously:
    $$A = 50,000e^{(0.1125)(5)} = \$87,752.73$$
    Subtract $50,000 from each to get the amount of interest:
    (a) $30,000
    (b) $38,613.59
    (c) $37.752.73
    Option (a) results in the least interest.

49. (a) $A = \$10,000,\ r = 0.10,\ n = 12,\ t = 20$ (compounded monthly)
    $$P = 10,000\left[1 + \frac{0.10}{12}\right]^{(-12)(20)} = \$1364.62$$
    (b) $A = \$10,000,\ r = 0.10,\ t = 20$ (compounded continuously)
    $$P = 10,000e^{(-0.10)(20)} = \$1353.35$$

51. $A = \$10,000,\ r = 0.08,\ n = 1,\ t = 10$ (compounded annually)
    $$P = 10,000\left[1 + \frac{0.08}{1}\right]^{(-1)(10)} = \$4631.93$$

55. (a) $y = \dfrac{\ln 2}{1 \cdot \ln\left(1 + \dfrac{0.12}{1}\right)} = \dfrac{\ln 2}{\ln 1.12} = 6.12$ years

(b) $y = \dfrac{\ln 3}{4 \cdot \ln\left(1 + \dfrac{0.06}{4}\right)} = \dfrac{\ln 3}{4\ln 1.015} = 18.45$ years

(c) $mP = P\left[1 + \dfrac{r}{n}\right]^{nt}$

$m = \left[1 + \dfrac{r}{n}\right]^{nt}$

$\ln m = nt \cdot \ln\left[1 + \dfrac{r}{n}\right]$

$t = \dfrac{\ln m}{n \cdot \ln\left[1 + \dfrac{r}{n}\right]}$

5.7   Growth and Decay

1. $P(t) = 500e^{0.02t}$

(a)   Find t when $P = 1000$:

$1000 = 500e^{0.02t}$

$2 = e^{0.02t}$

$\ln 2 = 0.02t$

$t = \dfrac{\ln 2}{0.02} \approx 34.7$ days

Find t when $P = 2000$:

$2000 = 500e^{0.02t}$

$4 = e^{0.02t}$

$\ln 4 = 0.02t$

$t = \dfrac{\ln 4}{0.02} \approx 69.3$ days

(b)   Graphing:

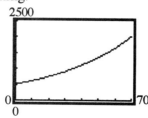

3.   Find t when $A(t) = \frac{1}{2}A_0$:

$\frac{1}{2}A_0 = A_0 e^{-0.0244t}$

$\frac{1}{2} = e^{-0.0244t}$

$\ln \frac{1}{2} = -0.0244t$

$t = \dfrac{\ln \frac{1}{2}}{-0.0244} \approx 28.4$ years

5. (a) $A = 10$; $A_0 = 100$ (b)
$$10 = 100e^{-0.0244t}$$
$$0.1 = e^{-0.0244t}$$
$$\ln 0.1 = -0.0244t$$
$$t = \frac{\ln 0.1}{-0.0244} \approx 94.4 \text{ years}$$

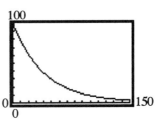

7. Use $N(t) = N_0 e^{kt}$ and solve for k:
$$1800 = 1000e^{k(1)}$$
$$1.8 = e^k$$
$$k = \ln 1.8 \approx 0.5878$$
When $t = 3$:
$$N(3) = 1000e^{0.5878(3)} = 5832 \text{ mosquitos}$$
Find t when $N(t) = 10,000$:
$$10,000 = 1000e^{0.5878t}$$
$$10 = e^{0.5878t}$$
$$\ln 10 = 0.5878t$$
$$t = \frac{\ln 10}{0.5878} \approx 3.9 \text{ days}$$

9. Use $P(t) = P_0 e^{kt}$ and solve for k:
$$2P_0 = P_0 e^{k(1.5)}$$
$$2 = e^{1.5k}$$
$$\ln 2 = 1.5k$$
$$k = \frac{\ln 2}{1.5} \approx 0.4621$$
When $t = 2$:
$$P(2) = 10,000e^{0.4621(2)} = 25,199 \text{ is the population 2 years from now.}$$

11. Use $A = A_0 e^{kt}$ and solve for k:
$$\tfrac{1}{2} A_0 = A_0 e^{k(1690)}$$
$$\tfrac{1}{2} = e^{1690k}$$
$$\ln \tfrac{1}{2} = 1690k$$
$$k = \frac{\ln 0.5}{1690} \approx -0.00041$$
When $A_0 = 10$ and $t = 50$:
$$A = 10e^{-0.00041(50)} = 9.797 \text{ grams}$$

13. (a) Use $A = A_0 e^{kt}$ and solve for k:

$$\frac{1}{2}A_0 = A_0 e^{k(5600)}$$

$$\frac{1}{2} = e^{5600k}$$

$$\ln\frac{1}{2} = 5600k$$

$$k = \frac{\ln 0.5}{5600} \approx -0.000124$$

Solve for t when $A = 0.3A_0$:

$$0.3A_0 = A_0 e^{-0.000124t}$$

$$0.3 = e^{-0.000124t}$$

$$\ln 0.3 = -0.000124t$$

$$t = \frac{\ln 0.3}{-0.000124}$$

$$\approx 9709 \text{ years ago}$$

(b) Graphing:

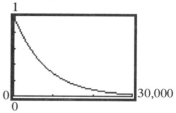

(c) Using INTERSECT to find the time until half remains:

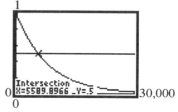

5590 years elapse.

15. (a) Using $u = T + (u_0 - T)e^{kt}$ where $t = 5$,
         $T = 70$, $u_0 = 450$, $u = 300$:

$$300 = 70 + (450 - 70)e^{k(5)}$$

$$230 = 380e^{5k}$$

$$0.6053 = e^{5k}$$

$$5k = \ln 0.6053$$

$$k = \frac{\ln 0.6053}{5} \approx -0.1004$$

$T = 70$, $u_0 = 450$, $u = 135$:

$$135 = 70 + (450 - 70)e^{-0.1004t}$$

$$65 = 380e^{-0.1004t}$$

$$0.17105 = e^{-0.1004t}$$

$$-0.1004t = \ln 0.17105$$

$$k = \frac{\ln 0.17105}{-0.1004} \approx 17.6 \text{ minutes}$$

The pizza will be cool enough to eat at 5:18 p.m.

(b) Graphing:

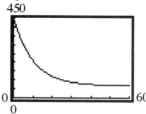

(c) Using INTERSECT:

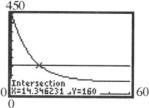

The pizza will be 160°F after about 14.3 minutes.

(d) As time passes the temperature gets closer to 70°F.

17. (a)   Using $u = T + (u_0 - T)e^{kt}$ where $t = 3$,   (b)   Graphing:
$T = 35$, $u_0 = 8$, $u = 15$:

$$15 = 35 + (8 - 35)e^{k(3)}$$
$$-20 = -27e^{3k}$$
$$0.74074 = e^{3k}$$
$$3k = \ln 0.74074$$
$$k = \frac{\ln 0.74074}{3} \approx -0.100035$$

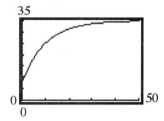

At $t = 5$:
$$u = 35 + (8 - 35)e^{-0.100035(5)} = 18.63°\,C$$
At $t = 10$:
$$u = 35 + (8 - 35)e^{-0.100035(10)} = 25.1°\,C$$

19. Use $A = A_0 e^{kt}$ and solve for k:
$$15 = 25e^{k(10)}$$
$$0.6 = e^{10k}$$
$$\ln 0.6 = 10k$$
$$k = \frac{\ln 0.6}{10} \approx -0.0511$$
When $A_0 = 25$ and $t = 24$:
$$A = 25e^{-0.0511(24)} = 7.33 \text{ kilograms}$$
Find t when $A = \frac{1}{2}A_0$:
$$0.5 = 25e^{-0.0511t}$$
$$0.02 = e^{-0.0511t}$$
$$\ln 0.02 = -0.0511t$$
$$t = \frac{\ln 0.02}{-0.0511} \approx 76.6 \text{ hours}$$

21. Use $A = A_0 e^{kt}$ and solve for k: Find t when $A = 0.1A_0$:
$$\frac{1}{2}A_0 = A_0 e^{k(8)} \qquad\qquad 0.1A_0 = A_0 e^{-0.0866t}$$
$$0.5 = e^{8k} \qquad\qquad\qquad 0.1 = e^{-0.0866t}$$
$$\ln 0.5 = 8k \qquad\qquad\qquad \ln 0.1 = -0.0866t$$
$$k = \frac{\ln 0.5}{8} \approx -0.0866 \qquad\qquad t = \frac{\ln 0.1}{-0.0866} \approx 26.6 \text{ days}$$
The farmers need to wait about 27 days before using the hay.

23. (a) $P(0) = \dfrac{0.9}{1 + 6e^{-0.32(0)}} = \dfrac{0.9}{1 + 6 \cdot 1} = \dfrac{0.9}{7} = 0.1286$

    (b) The maximum proportion is the carrying capacity, 0.9.

    (c) Graphing:

    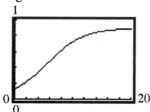

    (d)
    $$0.8 = \dfrac{0.9}{1 + 6e^{-0.32t}}$$
    $$0.8\left(1 + 6e^{-0.32t}\right) = 0.9$$
    $$1 + 6e^{-0.32t} = 1.125$$
    $$6e^{-0.32t} = 0.125$$
    $$e^{-0.32t} = 0.020833$$
    $$-0.32t = \ln(0.020833)$$
    $$t = \dfrac{\ln(0.020833)}{-0.32} \approx 12.1$$

    80% of households will own VCR's in 1996 (t = 12).

25. (a) Graphing:

    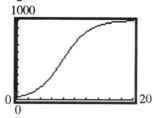

    (b) As $t \to \infty$, $e^{-0.439t} \to 0$. Thus, $P(t) \to 1000$. The carrying capacity is 1000.

    (c) $P(0) = \dfrac{1000}{1 + 32.33e^{-0.439(0)}} = \dfrac{1000}{33.33} = 30$

    (d)
    $$800 = \dfrac{1000}{1 + 32.33e^{-0.439t}}$$
    $$800\left(1 + 32.33e^{-0.439t}\right) = 1000$$
    $$1 + 32.33e^{-0.439t} = 1.25$$
    $$32.33e^{-0.439t} = 0.25$$
    $$e^{-0.439t} = 0.007733$$
    $$-0.439t = \ln(0.007733)$$
    $$t = \dfrac{\ln(0.007733)}{-0.439} \approx 11.076 \text{ hours}$$

27. (a) $y = \dfrac{6}{1 + e^{-(5.085 - 0.1156(100))}} = 0.0092$ O - rings

(b) $y = \dfrac{6}{1 + e^{-(5.085 - 0.1156(60))}} = 0.8145$ O - rings

(c) $y = \dfrac{6}{1 + e^{-(5.085 - 0.1156(30))}} = 5.0063$ O - rings

(d) Graphing:

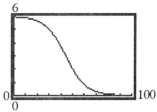

At 58°F, there would be 1 leaky O-ring.
At 44°F, there would be 3 leaky O-rings.
At 30°F, there would be 5 leaky O-rings.

5.8   Exponential, Logarithmic, and Logistic Curve Fitting

All functions graphed on the scatter plots use values with many more decimal places than shown in the stated equations.

1. (a) Graphing:

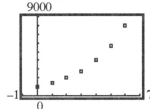

(b) $P = 1000.188(1.414)^t$

(c) $1000.187781(1.41414215)^t = A_0 e^{kt}$

$$A_0 = 1000.187781$$
$$e^k = 1.41414215$$
$$k = \ln 1.41414215$$
$$= 0.3465$$
$$P = 1000.188e^{0.3465t}$$

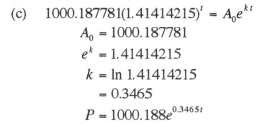

(d) Graphing:

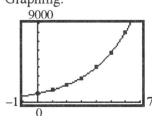

(e) $P = 1000.188e^{0.3465(7)} = 11312$

3. (a) Graphing:

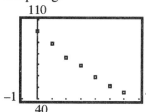

(b) $y = 100.326(0.877)^t$

(c) $100.3262508(0.876865102)^t = A_0 e^{kt}$

$A_0 = 100.3262508$

$e^k = 0.8768651017$

$k = \ln 0.8768651017$

$= -0.1314$

$A = 100.326 e^{-0.1314t}$

(d) Graphing:

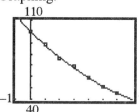

(e) $\frac{1}{2}(100.326) = 100.326 e^{-0.1314t}$

$\frac{1}{2} = e^{-0.1314t}$

$\ln 0.5 = -0.1314\, t$

$t = \dfrac{\ln 0.5}{-0.1314} \approx 5.3$ weeks

(f) $A = 100.326 e^{-0.1314(50)} = 0.14$ grams

5. (a) (Let $t = 0$ represent 1990)
Graphing :

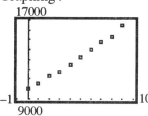

(b) $V = 10014.4963(1.05655)^t$

(c) $A = P(1 + r)^t$ where r is the annual rate of return. Thus,
$r = 0.05655$ or 5.655%.

(d) $V = 10014.4963(1.05655)^{30}$
$= \$52,166.20$

7. (a) (Let $t = 0$ represent 1989)
Graphing:

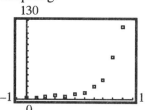

(b) $P = 1.2450(1.52556)^t$

(c) $1.245013118(1.525561679)^t = A_0 e^{kt}$

$A_0 = 1.245013118$

$e^k = 1.525561679$

$k = \ln 1.525561679$

$= 0.42236$

$P = 1.2450 e^{0.42236t}$

(d) Graphing:

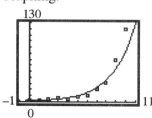

(e) $P = 1.2450 e^{0.42236(11)} = \$129.69$

9. (a) Graphing:

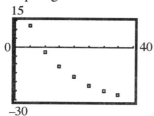

    (b) $C = 44.1977 - 20.3314 \ln S$

    (c) Graphing:

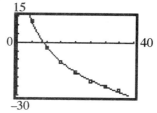

    (d) $C = 44.1977 - 20.3314 \ln 23$
         $= -19.6°F$

11. (a) Graphing:

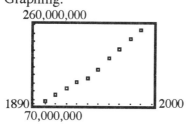

    (b) $P = \dfrac{695,129,657}{1 + (4.1136 \times 10^{14})e^{-0.01662t}}$

    (c) Graphing:

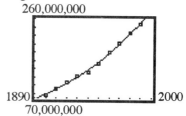

    (d) Carrying capacity = 695,129,657

    (e) $P = \dfrac{695,129,657}{1 + (4.1136 \times 10^{14})e^{-0.01662(2000)}}$
         $= 277,214,000$

13. (a) Graphing:

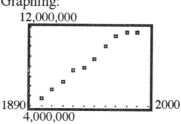

    (b) $P = \dfrac{14,168,687}{1 + (1.5076 \times 10^{21})e^{-0.02531t}}$

    (c) Graphing:

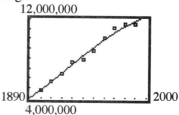

    (d) Carrying capacity = 14,168,687

    (e) $P = \dfrac{14,168,687}{1 + (1.5076 \times 10^{21})e^{-0.02531(2000)}}$
         $= 12,252,170$

15. (a)   Graphing the scatter diagram:

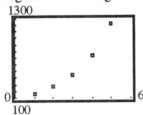

   (b)   Finding a model:

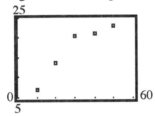

   The exponential model is the best model.

17. (a)   Graphing the scatter diagram:

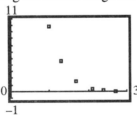

   (b)   Finding a model:

ExpReg	LnReg	PwrReg	LinReg
y=a*b^x	y=a+blnx	y=a*x^b	y=ax+b
a=6.727330625	a=-16.55688464	a=1.310727352	a=.3947
b=1.028644888	b=10.31473128	b=-.7615955413	b=5.229
r²=.8068914138	r²=.9724599362	r²=.9479118718	r²=.8814265605
r=.8982713476	r=.9861338328	r=.973607658	r=.9388432033

   The logarithmic model is the best model.

19. (a)   Graphing the scatter diagram:

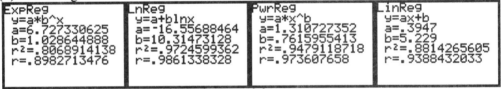

   (b)   Finding a model:

ExpReg	LnReg	PwrReg	LinReg
y=a*b^x	y=a+blnx	y=a*x^b	y=ax+b
a=346.113717	a=8.193177195	a=16.35402705	a=-5.26921875
b=.0352268374	b=-9.6182215	b=-5.62949942	b=12.572875
r²=.9923916306	r²=.8752814613	r²=.9521398312	r²=.775058396
r=-.9961885517	r=-.935564782	r=-.9757765273	r=-.8803740091

   The exponential model is the best model.

21. (a)   Graphing the scatter diagram: (Let $t = 1$ represent 1988)

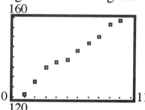

(b)   Finding a model:

ExpReg	LnReg	PwrReg	LinReg
y=a*b^x	y=a+blnx	y=a*x^b	y=ax+b
a=120.9703266	a=117.6533438	a=118.7461836	a=3.936363636
b=1.028414036	b=15.84745921	b=.1143077374	b=119.94
r²=.9758175319	r²=.9335503629	r²=.9520838952	r²=.982616606
r=.9878347695	r=.9662041	r=.9757478645	r=.9912701983

The linear model is the best model.

(c) $CPI(11) = 3.93636(11) + 119.94 = 163.24$

23. (a)   Graphing the scatter diagram: (Let $t = 1$ represent 1992)

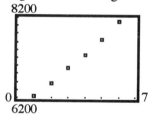

(b)   Finding a model:

ExpReg	LnReg	PwrReg	LinReg
y=a*b^x	y=a+blnx	y=a*x^b	y=ax+b
a=5922.055268	a=6034.907615	a=6081.189915	a=370.4542857
b=1.053364042	b=1000.486849	b=.1417586535	b=5835.393333
r²=.9991726755	r²=.9121944209	r²=.9319881948	r²=.9968781053
r=.9995862522	r=.9550886979	r=.9653953567	r=.9984378325

The exponential model is the best model.

(c) $GDP(7) = 5922.055(1.0534)^7 = 8521.6$

5   Chapter Review

1. $f(x) = \dfrac{2x+3}{5x-2}$

$\quad\quad y = \dfrac{2x+3}{5x-2}$

$\quad\quad x = \dfrac{2y+3}{5y-2}$ Inverse

$\quad x(5y-2) = 2y+3$

$\quad 5xy - 2x = 2y+3$

$\quad 5xy - 2y = 2x+3$

$\quad y(5x-2) = 2x+3$

$\quad\quad y = \dfrac{2x+3}{5x-2}$

$\quad f^{-1}(x) = \dfrac{2x+3}{5x-2}$

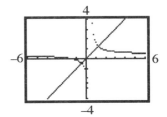

Verify: $f\left(f^{-1}(x)\right) = f\left(\dfrac{2x+3}{5x-2}\right) = \dfrac{2\left(\dfrac{2x+3}{5x-2}\right)+3}{5\left(\dfrac{2x+3}{5x-2}\right)-2} = \dfrac{\dfrac{4x+6+15x-6}{5x-2}}{\dfrac{10x+15-10x+4}{5x-2}} = \dfrac{\dfrac{19x}{5x-2}}{\dfrac{19}{5x-2}}$

$\quad\quad\quad\quad = \dfrac{19x}{5x-2}\cdot\dfrac{5x-2}{19} = x$

$\quad f^{-1}\left(f(x)\right) = f^{-1}\left(\dfrac{2x+3}{5x-2}\right) = \dfrac{2\left(\dfrac{2x+3}{5x-2}\right)+3}{5\left(\dfrac{2x+3}{5x-2}\right)-2} = \dfrac{\dfrac{4x+6+15x-6}{5x-2}}{\dfrac{10x+15-10x+4}{5x-2}} = \dfrac{\dfrac{19x}{5x-2}}{\dfrac{19}{5x-2}}$

$\quad\quad\quad\quad = \dfrac{19x}{5x-2}\cdot\dfrac{5x-2}{19} = x$

Domain of f = range of f^{-1} = all real numbers except $\frac{2}{5}$

Range of f = domain of f^{-1} = all real numbers except $\frac{2}{5}$

3. $f(x) = \dfrac{1}{x-1}$

$\quad\quad y = \dfrac{1}{x-1}$

$\quad\quad x = \dfrac{1}{y-1}$ Inverse

$\quad x(y-1) = 1$

$\quad xy - x = 1$

$\quad\quad xy = x+1$

$\quad\quad y = \dfrac{x+1}{x}$

$\quad f^{-1}(x) = \dfrac{x+1}{x}$

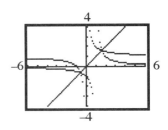

Verify: $f\left(f^{-1}(x)\right) = f\left(\dfrac{x+1}{x}\right) = \dfrac{1}{\dfrac{x+1}{x}-1} = \dfrac{1}{\dfrac{x+1-x}{x}} = \dfrac{1}{\dfrac{1}{x}} = 1\cdot\dfrac{x}{1} = x$

$f^{-1}\left(f(x)\right) = f^{-1}\left(\dfrac{1}{x-1}\right) = \dfrac{\dfrac{1}{x-1}+1}{\dfrac{1}{x-1}} = \dfrac{\dfrac{1+x-1}{x-1}}{\dfrac{1}{x-1}} = \dfrac{x}{x-1}\cdot\dfrac{x-1}{1} = x$

Domain of f = range of f^{-1} = all real numbers except 1
Range of f = domain of f^{-1} = all real numbers except 0

5. $f(x) = \dfrac{3}{x^{\frac{1}{3}}}$

$y = \dfrac{3}{x^{\frac{1}{3}}}$

$x = \dfrac{3}{y^{\frac{1}{3}}}$ Inverse

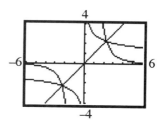

$xy^{\frac{1}{3}} = 3$

$y^{\frac{1}{3}} = \dfrac{3}{x}$

$y = \dfrac{27}{x^3}$

$f^{-1}(x) = \dfrac{27}{x^3}$

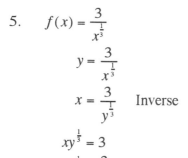

Verify: $f\left(f^{-1}(x)\right) = f\left(\dfrac{27}{x^3}\right) = \dfrac{3}{\left(\dfrac{27}{x^3}\right)^{\frac{1}{3}}} = \dfrac{3}{\dfrac{3}{x}} = 3\cdot\dfrac{x}{3} = x$

$f^{-1}\left(f(x)\right) = f^{-1}\left(\dfrac{3}{x^{\frac{1}{3}}}\right) = \dfrac{27}{\left(\dfrac{3}{x^{\frac{1}{3}}}\right)^3} = \dfrac{27}{\dfrac{27}{x}} = 27\cdot\dfrac{x}{27} = x$

Domain of f = range of f^{-1} = all real numbers except 0
Range of f = domain of f^{-1} = all real numbers except 0

7. $\log_2\left(\tfrac{1}{8}\right) = \log_2 2^{-3} = -3\log_2 2 = -3$

9. $\ln e^{\sqrt{2}} = \sqrt{2}$

11. $2^{\log_2 0.4} = 0.4$

13. $\log_3\left(\dfrac{uv^2}{w}\right) = \log_3 uv^2 - \log_3 w = \log_3 u + \log_3 v^2 - \log_3 w = \log_3 u + 2\log_3 v - \log_3 w$

15. $\log\left(x^2\sqrt{x^3+1}\right) = \log x^2 + \log\left(x^3+1\right)^{\frac{1}{2}} = 2\log x + \frac{1}{2}\log\left(x^3+1\right)$

17. $\ln\left(\dfrac{x\sqrt[3]{x^2+1}}{x-3}\right) = \ln\left(x\sqrt[3]{x^2+1}\right) - \ln(x-3) = \ln x + \ln\left(x^2+1\right)^{\frac{1}{3}} - \ln(x-3)$

$\qquad\qquad\qquad = \ln x + \frac{1}{3}\ln\left(x^2+1\right) - \ln(x-3)$

19. $3\log_4 x^2 + \frac{1}{2}\log_4\sqrt{x} = \log_4\left(x^2\right)^3 + \log_4\left(x^{\frac{1}{2}}\right)^{\frac{1}{2}} = \log_4 x^6 + \log_4 x^{\frac{1}{4}} = \log_4 x^6 \cdot x^{\frac{1}{4}}$

$\qquad\qquad\qquad = \log_4 x^{\frac{25}{4}} = \frac{25}{4}\log_4 x$

21. $\ln\left(\dfrac{x-1}{x}\right) + \ln\left(\dfrac{x}{x+1}\right) - \ln\left(x^2-1\right) = \ln\dfrac{x-1}{x}\cdot\dfrac{x}{x+1} - \ln\left(x^2-1\right) = \ln\dfrac{\frac{x-1}{x+1}}{x^2-1}$

$\qquad\qquad = \ln\left(\dfrac{x-1}{x+1}\cdot\dfrac{1}{(x-1)(x+1)}\right) = \ln\dfrac{1}{(x+1)^2} = \ln(x+1)^{-2} = -2\ln(x+1)$

23. $2\log 2 + 3\log x - \frac{1}{2}\left[\log(x+3)+\log(x-2)\right] = \log 2^2 + \log x^3 - \frac{1}{2}\log(x+3)(x-2)$

$\qquad\qquad = \log 4x^3 - \log((x+3)(x-2))^{\frac{1}{2}} = \log\dfrac{4x^3}{((x+3)(x-2))^{\frac{1}{2}}}$

25. $\log_4 19 = \dfrac{\log 19}{\log 4} = 2.124$

27. $\ln y = 2x^2 + \ln C$

$\ln y = \ln e^{2x^2} + \ln C$

$\ln y = \ln\left(Ce^{2x^2}\right)$

$y = Ce^{2x^2}$

29. $\ln(y-3) + \ln(y+3) = x + C$

$\ln(y-3)(y+3) = x + C$

$(y-3)(y+3) = e^{x+C}$

$y^2 - 9 = e^{x+C}$

$y^2 = 9 + e^{x+C}$

$y = \sqrt{9 + e^{x+C}}$

31. $e^{y+C} = x^2 + 4$

$\ln e^{y+C} = \ln(x^2+4)$

$y + C = \ln(x^2+4)$

$y = \ln(x^2+4) - C$

33. $f(x) = 2^{x-3}$

Using the graph of $y = 2^x$, shift the graph 3 units to the right.
Domain: $(-\infty, \infty)$
Range: $(0, \infty)$
Horizontal Asymptote: $y = 0$

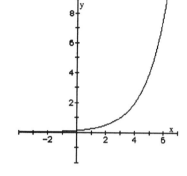

35. $f(x) = \frac{1}{2} \cdot 3^{-x}$

Using the graph of $y = 3^x$, reflect the graph about the y-axis, and shrink vertically by a factor of $\frac{1}{2}$.
Domain: $(-\infty, \infty)$
Range: $(0, \infty)$
Horizontal Asymptote: $y = 0$

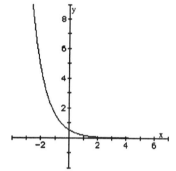

37. $f(x) = 1 - e^x$

Using the graph of $y = e^x$, reflect about the x-axis, and shift up 1 unit.
Domain: $(-\infty, \infty)$
Range: $(-\infty, 1)$
Horizontal Asymptote: $y = 1$

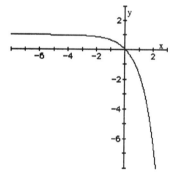

39. $f(x) = 3e^x$

Using the graph of $y = e^x$, stretch vertically by a factor of 3.
Domain: $(-\infty, \infty)$
Range: $(0, \infty)$
Horizontal Asymptote: $y = 0$

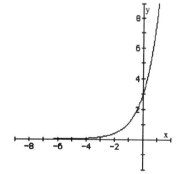

41. $f(x) = 3 - e^{-x}$

Using the graph of $y = e^x$, reflect the graph about the y-axis, reflect about the x-axis, and shift up 3 units.

Domain: $(-\infty, \infty)$

Range: $(-\infty, 3)$

Horizontal Asymptote: $y = 3$

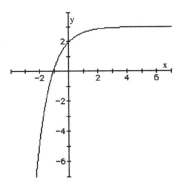

43. $4^{1-2x} = 2$

$\left(2^2\right)^{1-2x} = 2$

$2^{2-4x} = 2^1$

$2 - 4x = 1$

$-4x = -1$

$x = \frac{1}{4}$

45. $3^{x^2+x} = \sqrt{3}$

$3^{x^2+x} = 3^{\frac{1}{2}}$

$x^2 + x = \frac{1}{2}$

$2x^2 + 2x - 1 = 0$

$x = \dfrac{-2 \pm \sqrt{4 - 4(2)(-1)}}{2(2)} = \dfrac{-2 \pm \sqrt{12}}{4} = \dfrac{-2 \pm 2\sqrt{3}}{4} = \dfrac{-1 \pm \sqrt{3}}{2}$

$x = \dfrac{-1 - \sqrt{3}}{2}$ or $x = \dfrac{-1 + \sqrt{3}}{2}$

47. $\log_x 64 = -3$

$x^{-3} = 64$

$\left(x^{-3}\right)^{\frac{-1}{3}} = 64^{\frac{-1}{3}}$

$x = \dfrac{1}{\sqrt[3]{64}} = \dfrac{1}{4}$

49. $5^x = 3^{x+2}$

$\log\left(5^x\right) = \log\left(3^{x+2}\right)$

$x \log 5 = (x + 2)\log 3$

$x \log 5 = x \log 3 + 2\log 3$

$x \log 5 - x \log 3 = 2\log 3$

$x(\log 5 - \log 3) = 2\log 3$

$x = \dfrac{2\log 3}{\log 5 - \log 3} \approx 4.301$

51.
$$9^{2x} = 27^{3x-4}$$
$$\left(3^2\right)^{2x} = \left(3^3\right)^{3x-4}$$
$$3^{4x} = 3^{9x-12}$$
$$4x = 9x - 12$$
$$-5x = -12$$
$$x = \frac{12}{5}$$

53.
$$\log_3 \sqrt{x-2} = 2$$
$$\sqrt{x-2} = 3^2$$
$$x - 2 = 9^2$$
$$x - 2 = 81$$
$$x = 83$$

55.
$$8 = 4^{x^2} \cdot 2^{5x}$$
$$2^3 = \left(2^2\right)^{x^2} \cdot 2^{5x}$$
$$2^3 = 2^{2x^2+5x}$$
$$3 = 2x^2 + 5x$$
$$0 = 2x^2 + 5x - 3$$
$$0 = (2x-1)(x+3)$$
$$x = \tfrac{1}{2} \text{ or } x = -3$$

57.
$$\log_6(x+3) + \log_6(x+4) = 1$$
$$\log_6(x+3)(x+4) = 1$$
$$(x+3)(x+4) = 6^1$$
$$x^2 + 7x + 12 = 6$$
$$x^2 + 7x + 6 = 0$$
$$(x+6)(x+1) = 0$$
$$x = -6 \text{ or } x = -1$$
The logarithms are undefined when $x = -6$, so $x = -1$ is the only solution.

59.
$$e^{1-x} = 5$$
$$1 - x = \ln 5$$
$$-x = -1 + \ln 5$$
$$x = 1 - \ln 5 \approx -0.609$$

61.
$$2^{3x} = 3^{2x+1}$$
$$\ln 2^{3x} = \ln 3^{2x+1}$$
$$3x \ln 2 = (2x+1) \ln 3$$
$$3x \ln 2 = 2x \ln 3 + \ln 3$$
$$3x \ln 2 - 2x \ln 3 = \ln 3$$
$$x(3 \ln 2 - 2 \ln 3) = \ln 3$$
$$x = \frac{\ln 3}{3 \ln 2 - 2 \ln 3}$$
$$x \approx -9.327$$

63. $h(300) = (30(0) + 8000) \log\left(\frac{760}{300}\right) = 8000 \log 2.53333 = 3229.5$ meters

65. $P = 25e^{0.1d}$

(a)
$$P = 25e^{0.1(4)}$$
$$= 25e^{0.4}$$
$$= 37.3 \text{ watts}$$

(b)
$$50 = 25e^{0.1d}$$
$$2 = e^{0.1d}$$
$$\ln 2 = 0.1d$$
$$d = \frac{\ln 2}{0.1} = 6.9 \text{ decibels}$$

67.

(a) $n = \dfrac{\log 10000 - \log 90000}{\log(1 - 0.20)} = 9.85$ years

(b) $n = \dfrac{\log 0.5i - \log i}{\log(1 - 0.15)} = \dfrac{\log \dfrac{0.5i}{i}}{\log 0.85} = \dfrac{\log 0.5}{\log 0.85} = 4.27$ years

69. $P = A\left(1 + \dfrac{r}{n}\right)^{-nt} = 85000\left(1 + \dfrac{0.04}{2}\right)^{-2(18)} = \$41,668.97$

71. $A = A_0 e^{kt}$

$\frac{1}{2}A_0 = A_0 e^{k(5600)}$ $0.05A_0 = A_0 e^{-0.000124t}$

$0.5 = e^{5600k}$ $0.05 = e^{-0.000124t}$

$\ln 0.5 = 5600k$ $\ln 0.05 = -0.000124t$

$k = \dfrac{\ln 0.5}{5600} \approx -0.000124$ $t = \dfrac{\ln 0.05}{-0.000124} \approx 24{,}159$ years ago

73. $P = P_0 e^{kt} = 5{,}840{,}445{,}216 e^{0.0133(3)} = 6{,}078{,}190{,}457$

75. (a) $P(0) = \dfrac{0.8}{1 + 1.67 e^{-0.16(0)}} = \dfrac{0.8}{1 + 1.67} = 0.2996$

(b) 0.8

(c) Graphing: (d) Using INTERSECT we have:

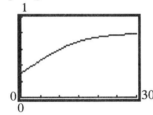   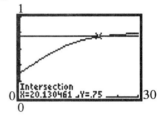

75% use Windows 98 in 2018.

77. (a) Graphing the scatter diagram:

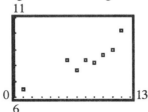

(b) Finding a model:

ExpReg	LnReg	PwrReg	LinReg
y=a*b^x	y=a+blnx	y=a*x^b	y=ax+b
a=6.27760714	a=6.256679556	a=6.342730419	a=.280952381
b=1.035566181	b=1.108346766	b=.141721551	b=6.114880952
r²=.8784257077	r²=.7046522173	r²=.7732420293	r²=.8458588666
r=.9372436757	r=.8394356541	r=.8793418159	r=.9197058587

The exponential model is the best model.

(c) $C(13) = (6.2776)(1.035566)^{13} = 9.9$

Chapter 6

Trigonometric Functions

6.1 Angles and Their Measure

1.

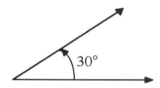

3.

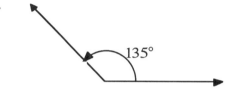

5.

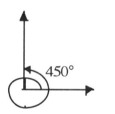

7.

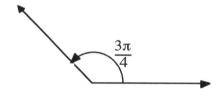

9.

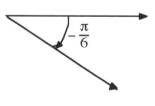

11.

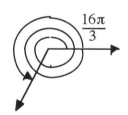

13. $30° = 30 \cdot \dfrac{\pi}{180}$ radian $= \dfrac{\pi}{6}$ radians

15. $240° = 240 \cdot \dfrac{\pi}{180}$ radian $= \dfrac{4\pi}{3}$ radians

17. $-60° = -60 \cdot \dfrac{\pi}{180}$ radian $= -\dfrac{\pi}{3}$ radians

19. $180° = 180 \cdot \dfrac{\pi}{180}$ radian $= \pi$ radians

21. $-135° = -135 \cdot \dfrac{\pi}{180}$ radian $= -\dfrac{3\pi}{4}$ radians

23. $-90° = -90 \cdot \dfrac{\pi}{180}$ radian $= -\dfrac{\pi}{2}$ radians

25. $\dfrac{\pi}{3} = \dfrac{\pi}{3} \cdot \dfrac{180}{\pi}$ degrees $= 60°$

27. $-\dfrac{5\pi}{4} = -\dfrac{5\pi}{4} \cdot \dfrac{180}{\pi}$ degrees $= -225°$

29. $\dfrac{\pi}{2} = \dfrac{\pi}{2} \cdot \dfrac{180}{\pi}$ degrees $= 90°$

31. $\dfrac{\pi}{12} = \dfrac{\pi}{12} \cdot \dfrac{180}{\pi}$ degrees $= 15°$

33. $-\dfrac{\pi}{2} = -\dfrac{\pi}{2} \cdot \dfrac{180}{\pi}$ degrees $= -90°$

35. $-\dfrac{\pi}{6} = -\dfrac{\pi}{6} \cdot \dfrac{180}{\pi}$ degrees $= -30°$

37. $r = 10$ meters; $\theta = \frac{1}{2}$ radian; $s = r\theta = 10 \cdot \frac{1}{2} = 5$ meters

39. $\theta = \frac{1}{3}$ radian; $s = 2$ feet; $s = r\theta$ or $r = \dfrac{s}{\theta} = \dfrac{2}{\frac{1}{3}} = 6$ feet

41. $r = 5$ miles; $s = 3$ miles; $s = r\theta$ or $\theta = \dfrac{s}{r} = \dfrac{3}{5} = 0.6$ radians

43. $r = 2$ inches; $\theta = 30°$; Convert to radians: $30° = 30 \cdot \dfrac{\pi}{180} = \dfrac{\pi}{6}$ radians

  $s = r\theta = 2 \cdot \dfrac{\pi}{6} = \dfrac{\pi}{3}$ inches

45. $17° = 17 \cdot \dfrac{\pi}{180}$ radian $= \dfrac{17\pi}{180}$ radians $= 0.30$

47. $-40° = -40 \cdot \dfrac{\pi}{180}$ radian $= -\dfrac{2\pi}{9}$ radians $= -0.70$

49. $125° = 125 \cdot \dfrac{\pi}{180}$ radian $= \dfrac{25\pi}{36}$ radians $= 2.18$

51. $340° = 340 \cdot \dfrac{\pi}{180}$ radian $= \dfrac{17\pi}{9}$ radians $= 5.93$

53. 3.14 radians = $3.14 \cdot \dfrac{180}{\pi}$ degrees = 179.91°

55. 10.25 radians = $10.25 \cdot \dfrac{180}{\pi}$ degrees = 587.28°

57. 2 radians = $2 \cdot \dfrac{180}{\pi}$ degrees = 114.59°

59. 6.32 radians = $6.32 \cdot \dfrac{180}{\pi}$ degrees = 362.11°

61. $40°10'25'' = \left(40 + 10 \cdot \dfrac{1}{60} + 25 \cdot \dfrac{1}{60} \cdot \dfrac{1}{60}\right)^{\circ} = (40 + 0.1667 + 0.00694)^{\circ} = 40.17°$

63. $1°2'3'' = \left(1 + 2 \cdot \dfrac{1}{60} + 3 \cdot \dfrac{1}{60} \cdot \dfrac{1}{60}\right)^{\circ} = (1 + 0.0333 + 0.00083)^{\circ} = 1.03°$

65. $9°9'9'' = \left(9 + 9 \cdot \dfrac{1}{60} + 9 \cdot \dfrac{1}{60} \cdot \dfrac{1}{60}\right)^{\circ} = (9 + 0.15 + 0.0025)^{\circ} = 9.15°$

67. 40.32° = ?
     $0.32° = 0.32(1°) = 0.32(60') = 19.2'$
     $0.2' = 0.2(1') = 0.2(60'') = 12''$
     $40.32° = 40° + 0.32° = 40° + 19.2' = 40° + 19' + 0.2' = 40° + 19' + 12'' = 40°19'12''$

69. 18.255° = ?
     $0.255° = 0.255(1°) = 0.255(60') = 15.3'$
     $0.3' = 0.3(1') = 0.3(60'') = 18''$
     $18.255° = 18° + 0.255° = 18° + 15.3' = 18° + 15' + 0.3' = 18° + 15' + 18'' = 18°15'18''$

71. 19.99° = ?
     $0.99° = 0.99(1°) = 0.99(60') = 59.4'$
     $0.4' = 0.4(1') = 0.4(60'') = 24''$
     $19.99° = 19° + 0.99° = 19° + 59.4' = 19° + 59' + 0.4' = 19° + 59' + 24'' = 19°59'24''$

73. $r = 6$ inches; $\theta = 90° = \dfrac{\pi}{2}$ radians

     $s = r\theta = 6 \cdot \dfrac{\pi}{2} = 3\pi$ inches ≈ 9.42 inches

     $r = 6$ inches; $\theta = \dfrac{25}{60}$ rev $= \dfrac{5}{12} \cdot 360° = 150° = \dfrac{5\pi}{6}$ radians

     $s = r\theta = 6 \cdot \dfrac{5\pi}{6} = 5\pi$ inches ≈ 15.71 inches

75. $r = 5$ cm.; $t = 20$ seconds; $\theta = \frac{1}{3}$ radian

$$\omega = \frac{\theta}{t} = \frac{\frac{1}{3}}{20} = \frac{1}{3} \cdot \frac{1}{20} = \frac{1}{60} \text{ radian / sec}$$

$$v = \frac{s}{t} = \frac{r\theta}{t} = \frac{5 \cdot \frac{1}{3}}{20} = \frac{5}{3} \cdot \frac{1}{20} = \frac{1}{12} \text{ cm / sec}$$

77. $d = 26$ inches; $r = 13$ inches; $v = 35$ mi / hr

$$v = \frac{35 \text{ mi}}{\text{hr}} \cdot \frac{5280 \text{ ft}}{\text{mi}} \cdot \frac{12 \text{ in}}{\text{ft}} \cdot \frac{1 \text{ hr}}{60 \text{ min}} = 36960 \text{ in / min}$$

$$\omega = \frac{v}{r} = \frac{36960 \text{ in / min}}{13 \text{ in}} = 2843.08 \text{ radians / min}$$

$$= \frac{2843.08 \text{ rad}}{\text{min}} \cdot \frac{1 \text{ rev}}{2\pi \text{ rad}} = 452.5 \text{ rev / min}$$

79. $r = 3960$ miles; $\theta = 35°9' - 29°57' = 5°12' = 5.2° = 5.2 \cdot \frac{\pi}{180} = 0.09076$ radian

$$s = r\theta = 3960 \cdot 0.09076 \approx 359.4 \text{ miles}$$

81. $r = 3429.5$ miles; $\omega = 1$ rev / day $= 2\pi$ radians / day $= \frac{\pi}{12}$ radians / hr

$$v = r\omega = 3429.5 \cdot \frac{\pi}{12} = 897.8 \text{ miles / hr}$$

83. $r = 2.39 \times 10^5$ miles;

$$\omega = 1 \text{ rev} / 27.3 \text{ days} = 2\pi \text{ radians} / 27.3 \text{ day} = \frac{\pi}{12 \cdot 27.3} \text{ radians / hr}$$

$$v = r\omega = \left(2.39 \times 10^5\right) \cdot \frac{\pi}{327.6} = 2292 \text{ miles / hr}$$

85. $r_1 = 2$ inches; $r_2 = 8$ inches; $\omega_1 = 3$ rev / min $= 6\pi$ radians / min
Find ω_2:

$$v_1 = v_2$$
$$r_1\omega_1 = r_2\omega_2$$
$$2(6\pi) = 8\omega_2$$
$$\omega_2 = \frac{12\pi}{8} = 1.5\pi \text{ radians / min } = \frac{1.5\pi}{2\pi} \text{ rev / min} = \frac{3}{4} \text{ rev / min}$$

87. $r = 4$ feet; $\omega = 10$ rev / min $= 20\pi$ radians / min

$$v = r\omega = 4 \cdot 20\pi = 80\pi \frac{\text{ft}}{\text{min}} = \frac{80\pi \text{ ft}}{\text{min}} \cdot \frac{1 \text{ mi}}{5280 \text{ ft}} \cdot \frac{60 \text{ min}}{\text{hr}} = 2.86 \text{ mi / hr}$$

89. $d = 8.5$ feet; $r = 4.25$ feet; $v = 9.55$ mi / hr

$$\omega = \frac{v}{r} = \frac{9.55 \text{ mi / hr}}{4.25 \text{ ft}} = \frac{9.55 \text{ mi}}{\text{hr}} \cdot \frac{1}{4.25 \text{ ft}} \cdot \frac{5280 \text{ ft}}{\text{mi}} \cdot \frac{1 \text{ hr}}{60 \text{ min}} \cdot \frac{1 \text{ rev}}{2\pi} = 31.47 \text{ rev / min}$$

91. The earth makes one full rotation in 24 hours. The distance traveled in 24 hours is the circumference of the earth. At the equator the circumference is $2\pi(3960)$ miles. Therefore, the linear velocity a person must travel to keep up with the sun is:

$$v = \frac{s}{t} = \frac{2\pi(3960)}{24} \approx 1037 \text{ miles / hr}$$

93. r_1 rotates at ω_1 rev / min; r_2 rotates at ω_2 rev / min;

$v = r_1\omega_1 = r_2\omega_2$

So, $\dfrac{r_1}{r_2} = \dfrac{\omega_2}{\omega_1}$

6.2 Trigonometric Functions: Unit Circle Approach

1. $\left(\dfrac{1}{4}, \dfrac{\sqrt{15}}{4}\right)$

$\sin t = \dfrac{\sqrt{15}}{4}$

$\csc t = \dfrac{1}{\dfrac{\sqrt{15}}{4}} = 1 \cdot \dfrac{4}{\sqrt{15}} \cdot \dfrac{\sqrt{15}}{\sqrt{15}} = \dfrac{4\sqrt{15}}{15}$

$\cos t = \dfrac{1}{4}$

$\sec t = \dfrac{1}{\dfrac{1}{4}} = 1 \cdot \dfrac{4}{1} = 4$

$\tan t = \dfrac{\dfrac{\sqrt{15}}{4}}{\dfrac{1}{4}} = \dfrac{\sqrt{15}}{4} \cdot \dfrac{4}{1} = \sqrt{15}$

$\cot t = \dfrac{\dfrac{1}{4}}{\dfrac{\sqrt{15}}{4}} = \dfrac{1}{4} \cdot \dfrac{4}{\sqrt{15}} = \dfrac{1}{\sqrt{15}} \cdot \dfrac{\sqrt{15}}{\sqrt{15}} = \dfrac{\sqrt{15}}{15}$

3. $\left(-\dfrac{2}{5}, \dfrac{\sqrt{21}}{5}\right)$

$\sin t = \dfrac{\sqrt{21}}{5}$

$\csc t = \dfrac{1}{\dfrac{\sqrt{21}}{5}} = 1 \cdot \dfrac{5}{\sqrt{21}} \cdot \dfrac{\sqrt{21}}{\sqrt{21}} = \dfrac{5\sqrt{21}}{21}$

$\cos t = -\dfrac{2}{5}$

$\sec t = \dfrac{1}{-\dfrac{2}{5}} = 1 \cdot -\dfrac{5}{2} = -\dfrac{5}{2}$

$\tan t = \dfrac{\dfrac{\sqrt{21}}{5}}{-\dfrac{2}{5}} = \dfrac{\sqrt{21}}{5} \cdot -\dfrac{5}{2} = -\dfrac{\sqrt{21}}{2}$

$\cot t = \dfrac{-\dfrac{2}{5}}{\dfrac{\sqrt{21}}{5}} = -\dfrac{2}{5} \cdot \dfrac{5}{\sqrt{21}} \cdot \dfrac{\sqrt{21}}{\sqrt{21}} = -\dfrac{2\sqrt{21}}{21}$

5. $\left(-\dfrac{\sqrt{35}}{6}, -\dfrac{1}{6}\right)$

$$\sin t = -\dfrac{1}{6} \qquad\qquad \cos t = -\dfrac{\sqrt{35}}{6}$$

$$\tan t = \dfrac{-\dfrac{1}{6}}{-\dfrac{\sqrt{35}}{6}} = -\dfrac{1}{6} \cdot -\dfrac{6}{\sqrt{35}} \cdot \dfrac{\sqrt{35}}{\sqrt{35}} = \dfrac{\sqrt{35}}{35}$$

$$\csc t = \dfrac{1}{-\dfrac{1}{6}} = 1 \cdot -\dfrac{6}{1} = -6$$

$$\sec t = \dfrac{1}{-\dfrac{\sqrt{35}}{6}} = 1 \cdot -\dfrac{6}{\sqrt{35}} \cdot \dfrac{\sqrt{35}}{\sqrt{35}} = -\dfrac{6\sqrt{35}}{35}$$

$$\cot t = \dfrac{-\dfrac{\sqrt{35}}{6}}{-\dfrac{1}{6}} = -\dfrac{\sqrt{35}}{6} \cdot -\dfrac{6}{1} = \sqrt{35}$$

7. $\left(\dfrac{2\sqrt{2}}{3}, -\dfrac{1}{3}\right)$

$$\sin t = -\dfrac{1}{3} \qquad\qquad \csc t = \dfrac{1}{-\dfrac{1}{3}} = 1 \cdot -\dfrac{3}{1} = -3$$

$$\cos t = \dfrac{2\sqrt{2}}{3} \qquad\qquad \sec t = \dfrac{1}{\dfrac{2\sqrt{2}}{3}} = 1 \cdot \dfrac{3}{2\sqrt{2}} \cdot \dfrac{\sqrt{2}}{\sqrt{2}} = \dfrac{3\sqrt{2}}{4}$$

$$\tan t = \dfrac{-\dfrac{1}{3}}{\dfrac{2\sqrt{2}}{3}} = -\dfrac{1}{3} \cdot \dfrac{3}{2\sqrt{2}} \cdot \dfrac{\sqrt{2}}{\sqrt{2}} = -\dfrac{\sqrt{2}}{4} \qquad \cot t = \dfrac{\dfrac{2\sqrt{2}}{3}}{-\dfrac{1}{3}} = \dfrac{2\sqrt{2}}{3} \cdot -\dfrac{3}{1} = -2\sqrt{2}$$

9. $\left(-\dfrac{3\sqrt{5}}{7}, -\dfrac{2}{7}\right)$

$$\sin t = -\dfrac{2}{7} \qquad\qquad \csc t = \dfrac{1}{-\dfrac{2}{7}} = 1 \cdot -\dfrac{7}{2} = -\dfrac{7}{2}$$

$$\cos t = -\dfrac{3\sqrt{5}}{7} \qquad\qquad \sec t = \dfrac{1}{-\dfrac{3\sqrt{5}}{7}} = 1 \cdot -\dfrac{7}{3\sqrt{5}} \cdot \dfrac{\sqrt{5}}{\sqrt{5}} = -\dfrac{7\sqrt{5}}{15}$$

$$\tan t = \dfrac{-\dfrac{2}{7}}{-\dfrac{3\sqrt{5}}{7}} = -\dfrac{2}{7} \cdot -\dfrac{7}{3\sqrt{5}} \cdot \dfrac{\sqrt{5}}{\sqrt{5}} = \dfrac{2\sqrt{5}}{15} \qquad \cot t = \dfrac{-\dfrac{3\sqrt{5}}{7}}{-\dfrac{2}{7}} = -\dfrac{3\sqrt{5}}{7} \cdot -\dfrac{7}{2} = \dfrac{3\sqrt{5}}{2}$$

11. $\sin 45° + \cos 60° = \dfrac{\sqrt{2}}{2} + \dfrac{1}{2} = \dfrac{1+\sqrt{2}}{2}$ 13. $\sin 90° + \tan 45° = 1+1 = 2$

15. $\sin 45° \cos 45° = \dfrac{\sqrt{2}}{2} \cdot \dfrac{\sqrt{2}}{2} = \dfrac{2}{4} = \dfrac{1}{2}$ 17. $\csc 45° \tan 60° = \sqrt{2} \cdot \sqrt{3} = \sqrt{6}$

19. $4\sin 90° - 3\tan 180° = 4 \cdot 1 - 3 \cdot 0 = 4$

21. $2\sin \dfrac{\pi}{3} - 3\tan \dfrac{\pi}{6} = 2 \cdot \dfrac{\sqrt{3}}{2} - 3 \cdot \dfrac{\sqrt{3}}{3} = \sqrt{3} - \sqrt{3} = 0$

23. $\sin \dfrac{\pi}{4} - \cos \dfrac{\pi}{4} = \dfrac{\sqrt{2}}{2} - \dfrac{\sqrt{2}}{2} = 0$

25. $2\sec \dfrac{\pi}{4} + 4\cot \dfrac{\pi}{3} = 2 \cdot \sqrt{2} + 4 \cdot \dfrac{\sqrt{3}}{3} = 2\sqrt{2} + \dfrac{4\sqrt{3}}{3}$

27. $\tan \pi - \cos 0 = 0 - 1 = -1$ 29. $\csc \dfrac{\pi}{2} + \cot \dfrac{\pi}{2} = 1 + 0 = 1$

31. The point on the unit circle that corresponds to $\theta = \dfrac{2\pi}{3} = 120°$ is $\left(-\dfrac{1}{2}, \dfrac{\sqrt{3}}{2} \right)$.

$\sin t = \dfrac{\sqrt{3}}{2}$ $\csc t = \dfrac{1}{\dfrac{\sqrt{3}}{2}} = 1 \cdot \dfrac{2}{\sqrt{3}} \cdot \dfrac{\sqrt{3}}{\sqrt{3}} = \dfrac{2\sqrt{3}}{3}$

$\cos t = -\dfrac{1}{2}$ $\sec t = \dfrac{1}{-\dfrac{1}{2}} = 1 \cdot -\dfrac{2}{1} = -2$

$\tan t = \dfrac{\dfrac{\sqrt{3}}{2}}{-\dfrac{1}{2}} = \dfrac{\sqrt{3}}{2} \cdot -\dfrac{2}{1} = -\sqrt{3}$ $\cot t = \dfrac{-\dfrac{1}{2}}{\dfrac{\sqrt{3}}{2}} = -\dfrac{1}{2} \cdot \dfrac{2}{\sqrt{3}} \cdot \dfrac{\sqrt{3}}{\sqrt{3}} = -\dfrac{\sqrt{3}}{3}$

33. The point on the unit circle that corresponds to $\theta = 210° = \dfrac{7\pi}{6}$ is $\left(-\dfrac{\sqrt{3}}{2}, -\dfrac{1}{2} \right)$.

$\sin t = -\dfrac{1}{2}$ $\csc t = \dfrac{1}{-\dfrac{1}{2}} = 1 \cdot -\dfrac{2}{1} = -2$

$\cos t = -\dfrac{\sqrt{3}}{2}$ $\sec t = \dfrac{1}{-\dfrac{\sqrt{3}}{2}} = 1 \cdot -\dfrac{2}{\sqrt{3}} \cdot \dfrac{\sqrt{3}}{\sqrt{3}} = -\dfrac{2\sqrt{3}}{3}$

$\tan t = \dfrac{-\dfrac{1}{2}}{-\dfrac{\sqrt{3}}{2}} = -\dfrac{1}{2} \cdot -\dfrac{2}{\sqrt{3}} \cdot \dfrac{\sqrt{3}}{\sqrt{3}} = \dfrac{\sqrt{3}}{3}$ $\cot t = \dfrac{-\dfrac{\sqrt{3}}{2}}{-\dfrac{1}{2}} = -\dfrac{\sqrt{3}}{2} \cdot -\dfrac{2}{1} = \sqrt{3}$

35.   The point on the unit circle that corresponds to $\theta = \dfrac{5\pi}{3} = 300°$ is $\left(\dfrac{1}{2}, -\dfrac{\sqrt{3}}{2}\right)$.

$$\sin t = -\dfrac{\sqrt{3}}{2}$$

$$\csc t = \dfrac{1}{-\dfrac{\sqrt{3}}{2}} = 1 \cdot -\dfrac{2}{\sqrt{3}} \cdot \dfrac{\sqrt{3}}{\sqrt{3}} = -\dfrac{2\sqrt{3}}{3}$$

$$\cos t = \dfrac{1}{2}$$

$$\sec t = \dfrac{1}{\dfrac{1}{2}} = 1 \cdot \dfrac{2}{1} = 2$$

$$\tan t = \dfrac{-\dfrac{\sqrt{3}}{2}}{\dfrac{1}{2}} = -\dfrac{\sqrt{3}}{2} \cdot \dfrac{2}{1} = -\sqrt{3}$$

$$\cot t = \dfrac{\dfrac{1}{2}}{-\dfrac{\sqrt{3}}{2}} = \dfrac{1}{2} \cdot -\dfrac{2}{\sqrt{3}} \cdot \dfrac{\sqrt{3}}{\sqrt{3}} = -\dfrac{\sqrt{3}}{3}$$

37.   The point on the unit circle that corresponds to $\theta = \dfrac{7\pi}{3} = \dfrac{\pi}{3} = 60°$ is $\left(\dfrac{1}{2}, \dfrac{\sqrt{3}}{2}\right)$.

$$\sin t = \dfrac{\sqrt{3}}{2}$$

$$\csc t = \dfrac{1}{\dfrac{\sqrt{3}}{2}} = 1 \cdot \dfrac{2}{\sqrt{3}} \cdot \dfrac{\sqrt{3}}{\sqrt{3}} = \dfrac{2\sqrt{3}}{3}$$

$$\cos t = \dfrac{1}{2}$$

$$\sec t = \dfrac{1}{\dfrac{1}{2}} = 1 \cdot \dfrac{2}{1} = 2$$

$$\tan t = \dfrac{\dfrac{\sqrt{3}}{2}}{\dfrac{1}{2}} = \dfrac{\sqrt{3}}{2} \cdot \dfrac{2}{1} = \sqrt{3}$$

$$\cot t = \dfrac{\dfrac{1}{2}}{\dfrac{\sqrt{3}}{2}} = \dfrac{1}{2} \cdot \dfrac{2}{\sqrt{3}} \cdot \dfrac{\sqrt{3}}{\sqrt{3}} = \dfrac{\sqrt{3}}{3}$$

39.   The point on the unit circle that corresponds to $\theta = 405° = 45° = \dfrac{\pi}{4}$ is $\left(\dfrac{\sqrt{2}}{2}, \dfrac{\sqrt{2}}{2}\right)$.

$$\sin t = \dfrac{\sqrt{2}}{2}$$

$$\csc t = \dfrac{1}{\dfrac{\sqrt{2}}{2}} = 1 \cdot \dfrac{2}{\sqrt{2}} \cdot \dfrac{\sqrt{2}}{\sqrt{2}} = \sqrt{2}$$

$$\cos t = \dfrac{\sqrt{2}}{2}$$

$$\sec t = \dfrac{1}{\dfrac{\sqrt{2}}{2}} = 1 \cdot \dfrac{2}{\sqrt{2}} \cdot \dfrac{\sqrt{2}}{\sqrt{2}} = \sqrt{2}$$

$$\tan t = \dfrac{\dfrac{\sqrt{2}}{2}}{\dfrac{\sqrt{2}}{2}} = \dfrac{\sqrt{2}}{2} \cdot \dfrac{2}{\sqrt{2}} = 1$$

$$\cot t = \dfrac{\dfrac{\sqrt{2}}{2}}{\dfrac{\sqrt{2}}{2}} = 1$$

41. The point on the unit circle that corresponds to $\theta = -\frac{\pi}{6} = \frac{11\pi}{6} = 330°$ is $\left(\frac{\sqrt{3}}{2}, -\frac{1}{2}\right)$.

$$\sin t = -\frac{1}{2}$$

$$\csc t = \frac{1}{-\frac{1}{2}} = 1 \cdot -\frac{2}{1} = -2$$

$$\cos t = \frac{\sqrt{3}}{2}$$

$$\sec t = \frac{1}{\frac{\sqrt{3}}{2}} = 1 \cdot \frac{2}{\sqrt{3}} \cdot \frac{\sqrt{3}}{\sqrt{3}} = \frac{2\sqrt{3}}{3}$$

$$\tan t = \frac{-\frac{1}{2}}{\frac{\sqrt{3}}{2}} = -\frac{1}{2} \cdot \frac{2}{\sqrt{3}} \cdot \frac{\sqrt{3}}{\sqrt{3}} = -\frac{\sqrt{3}}{3}$$

$$\cot t = \frac{\frac{\sqrt{3}}{2}}{-\frac{1}{2}} = \frac{\sqrt{3}}{2} \cdot -\frac{2}{1} = -\sqrt{3}$$

43. The point on the unit circle that corresponds to $\theta = -45° = 315° = \frac{7\pi}{4}$ is $\left(\frac{\sqrt{2}}{2}, -\frac{\sqrt{2}}{2}\right)$.

$$\sin t = -\frac{\sqrt{2}}{2}$$

$$\csc t = \frac{1}{-\frac{\sqrt{2}}{2}} = 1 \cdot -\frac{2}{\sqrt{2}} \cdot \frac{\sqrt{2}}{\sqrt{2}} = -\sqrt{2}$$

$$\cos t = \frac{\sqrt{2}}{2}$$

$$\sec t = \frac{1}{\frac{\sqrt{2}}{2}} = 1 \cdot \frac{2}{\sqrt{2}} \cdot \frac{\sqrt{2}}{\sqrt{2}} = \sqrt{2}$$

$$\tan t = \frac{-\frac{\sqrt{2}}{2}}{\frac{\sqrt{2}}{2}} = -\frac{\sqrt{2}}{2} \cdot \frac{2}{\sqrt{2}} = -1$$

$$\cot t = \frac{\frac{\sqrt{2}}{2}}{-\frac{\sqrt{2}}{2}} = -1$$

45. The point on the unit circle that corresponds to $\theta = \frac{5\pi}{2} = \frac{\pi}{2} = 90°$ is $(0, 1)$.

$$\sin t = 1 \qquad \csc t = \frac{1}{1} = 1$$

$$\cos t = 0 \qquad \sec t = \frac{1}{0} = \text{not defined}$$

$$\tan t = \frac{1}{0} = \text{not defined} \qquad \cot t = \frac{0}{1} = 0$$

47. The point on the unit circle that corresponds to $\theta = -180° = 180° = \pi$ is $(-1, 0)$.

$$\sin t = 0 \qquad \csc t = \frac{1}{0} = \text{not defined}$$

$$\cos t = -1 \qquad \sec t = \frac{1}{-1} = -1$$

$$\tan t = \frac{0}{-1} = 0 \qquad \cot t = \frac{-1}{0} = \text{not defined}$$

49.   The point on the unit circle that corresponds to $\theta = -\dfrac{\pi}{2} = \dfrac{3\pi}{2} = 270°$ is $(0, -1)$.

$$\sin t = -1 \qquad\qquad\qquad \csc t = \dfrac{1}{-1} = -1$$

$$\cos t = 0 \qquad\qquad\qquad \sec t = \dfrac{1}{0} = \text{not defined}$$

$$\tan t = \dfrac{-1}{0} = \text{not defined} \qquad\qquad \cot t = \dfrac{0}{-1} = 0$$

51.   The point on the unit circle that corresponds to $\theta = 480° = 120° = \dfrac{2\pi}{3}$ is $\left(-\dfrac{1}{2}, \dfrac{\sqrt{3}}{2}\right)$.

$$\sin t = \dfrac{\sqrt{3}}{2} \qquad\qquad\qquad \csc t = \dfrac{1}{\dfrac{\sqrt{3}}{2}} = 1 \cdot \dfrac{2}{\sqrt{3}} \cdot \dfrac{\sqrt{3}}{\sqrt{3}} = \dfrac{2\sqrt{3}}{3}$$

$$\cos t = -\dfrac{1}{2} \qquad\qquad\qquad \sec t = \dfrac{1}{-\dfrac{1}{2}} = 1 \cdot -\dfrac{2}{1} = -2$$

$$\tan t = \dfrac{\dfrac{\sqrt{3}}{2}}{-\dfrac{1}{2}} = \dfrac{\sqrt{3}}{2} \cdot -\dfrac{2}{1} = -\sqrt{3} \qquad \cot t = \dfrac{-\dfrac{1}{2}}{\dfrac{\sqrt{3}}{2}} = -\dfrac{1}{2} \cdot \dfrac{2}{\sqrt{3}} \cdot \dfrac{\sqrt{3}}{\sqrt{3}} = -\dfrac{\sqrt{3}}{3}$$

53.   Using your calculator, $\sin 28° = 0.47$.

55.   Using your calculator, $\tan 21° = 0.38$.

57.   Using your calculator, $\sec 41° = \dfrac{1}{\cos 41°} = 1.33$.

59.   Using your calculator, $\cot 70° = \dfrac{1}{\tan 70°} = 0.36$.

61.   Set the calculator to radian mode: $\sin \dfrac{\pi}{10} = 0.31$.

63.   Set the calculator to radian mode: $\tan \dfrac{5\pi}{12} = 3.73$.

65.   Set the calculator to radian mode: $\sec \dfrac{\pi}{12} = \dfrac{1}{\cos \dfrac{\pi}{12}} = 1.04$.

67.   Set the calculator to radian mode: $\sin 1 = 0.84$.

69.   Set the calculator to degree mode: $\sin 1° = 0.02$

71. $\sin 60° = \dfrac{\sqrt{3}}{2}$

73. $\sin\dfrac{60°}{2} = \sin 30° = \dfrac{1}{2}$

75. $(\sin 60°)^2 = \left(\dfrac{\sqrt{3}}{2}\right)^2 = \dfrac{3}{4}$

77. $\sin(2 \cdot 60°) = \sin 120° = \dfrac{\sqrt{3}}{2}$

79. $2\sin 60° = 2 \cdot \dfrac{\sqrt{3}}{2} = \sqrt{3}$

81. $\sin(-60°) = \sin 300° = -\dfrac{\sqrt{3}}{2}$

83. Complete the table:

θ	0.5	0.4	0.2	0.1	0.01	0.001	0.0001	0.00001
$\sin\theta$	0.4794	0.3894	0.1987	0.0998	0.0100	0.0010	0.0001	0.00001
$\dfrac{\sin\theta}{\theta}$	0.9589	0.9735	0.9933	0.9983	1.0000	1.0000	1.0000	1.0000

The ratio $\dfrac{\sin\theta}{\theta}$ as θ approaches 0 is 1.

85. (a) $t = 1$ $(0.5, 0.8)$

$\sin 1 = 0.8;$ $\cos 1 = 0.5;$ $\tan 1 = \dfrac{0.8}{0.5} = 1.6;$

$\csc 1 = \dfrac{1}{0.8} = 1.3;$ $\sec 1 = \dfrac{1}{0.5} = 2;$ $\cot 1 = \dfrac{0.5}{0.8} = 0.6$

Calculator values:

$\sin 1 = 0.8415;$ $\cos 1 = 0.5403;$ $\tan 1 = 1.5574;$

$\csc 1 = 1.1884;$ $\sec 1 = 1.8508;$ $\cot 1 = 0.6421$

(b) $t = 5.1$ $(0.4, -0.9)$

$\sin 5.1 = -0.9;$ $\cos 5.1 = 0.4;$ $\tan 5.1 = \dfrac{-0.9}{0.4} = -2.3;$

$\csc 5.1 = \dfrac{1}{-0.9} = -1.1;$ $\sec 5.1 = \dfrac{1}{0.4} = 2.5;$ $\cot 5.1 = \dfrac{0.4}{-0.9} = -0.4$

Calculator values:

$\sin 5.1 = -0.9258;$ $\cos 5.1 = 0.3780;$ $\tan 5.1 = -2.4494;$

$\csc 5.1 = -1.0801;$ $\sec 5.1 = 2.6457;$ $\cot 5.1 = -0.4083$

(c) $t = 2.4$ $(-0.8, 0.6)$

$\sin 2.4 = 0.6;$ $\cos 2.4 = -0.8;$ $\tan 2.4 = \dfrac{0.6}{-0.8} = -0.8;$

$\csc 2.4 = \dfrac{1}{0.6} = 1.7;$ $\sec 2.4 = \dfrac{1}{-0.8} = -1.3;$ $\cot 2.4 = \dfrac{-0.8}{0.6} = -1.3$

Calculator values:

$\sin 2.4 = 0.6755;$ $\cos 2.4 = -0.7374;$ $\tan 2.4 = -0.9160;$

$\csc 2.4 = 1.4805;$ $\sec 2.4 = -1.3561;$ $\cot 2.4 = -1.0917$

87. (a) $t = 1.5$ (0, 1)

$\sin 1.5 = 1;$ $\cos 1.5 = 0;$ $\tan 1.5 = \dfrac{1}{0} = $ undefined;

$\csc 1.5 = \dfrac{1}{1} = 1;$ $\sec 1.5 = \dfrac{1}{0} = $ undefined; $\cot 1.5 = \dfrac{0}{1} = 0$

Calculator values:

$\sin 1.5 = 0.9975;$ $\cos 1.5 = 0.0707;$ $\tan 1.5 = 14.1014;$

$\csc 1.5 = 1.0025;$ $\sec 1.5 = 14.1368;$ $\cot 1.5 = 0.0709$

(b) $t = 4.3$ (−0.3, −0.9)

$\sin 4.3 = -0.9;$ $\cos 4.3 = -0.3;$ $\tan 4.3 = \dfrac{-0.9}{-0.3} = 3;$

$\csc 4.3 = \dfrac{1}{-0.9} = -1.1;$ $\sec 4.3 = \dfrac{1}{-0.3} = -3.3;$ $\cot 4.3 = \dfrac{-0.3}{-0.9} = 0.3$

Calculator values:

$\sin 4.3 = -0.9162;$ $\cos 4.3 = -0.4008;$ $\tan 4.3 = 2.2858;$

$\csc 4.3 = -1.0915;$ $\sec 4.3 = -2.4950;$ $\cot 4.3 = 0.4375$

(c) $t = 5.3$ (0.6, −0.8)

$\sin 5.3 = -0.8;$ $\cos 5.3 = 0.6;$ $\tan 5.3 = \dfrac{-0.8}{0.6} = -1.3;$

$\csc 5.3 = \dfrac{1}{-0.8} = -1.3;$ $\sec 5.3 = \dfrac{1}{0.6} = 1.7;$ $\cot 5.3 = \dfrac{0.6}{-0.8} = -0.8$

Calculator values:

$\sin 5.3 = -0.8323;$ $\cos 5.3 = 0.5544;$ $\tan 5.3 = -1.5013;$

$\csc 5.3 = -1.2015;$ $\sec 5.3 = 1.8038;$ $\cot 5.3 = -0.6661$

89. For the point (−3, 4), $x = -3,$ $y = 4,$ $r = \sqrt{x^2 + y^2} = \sqrt{9 + 16} = \sqrt{25} = 5$

$\sin \theta = \dfrac{4}{5}$ $\cos \theta = \dfrac{-3}{5}$ $\tan \theta = \dfrac{4}{-3}$

$\csc \theta = \dfrac{5}{4}$ $\sec \theta = \dfrac{5}{-3}$ $\cot \theta = \dfrac{-3}{4}$

91. For the point (2, −3), $x = 2,$ $y = -3,$ $r = \sqrt{x^2 + y^2} = \sqrt{4 + 9} = \sqrt{13}$

$\sin \theta = \dfrac{-3}{\sqrt{13}} \cdot \dfrac{\sqrt{13}}{\sqrt{13}} = \dfrac{-3\sqrt{13}}{13}$ $\cos \theta = \dfrac{2}{\sqrt{13}} \cdot \dfrac{\sqrt{13}}{\sqrt{13}} = \dfrac{2\sqrt{13}}{13}$ $\tan \theta = \dfrac{-3}{2}$

$\csc \theta = -\dfrac{\sqrt{13}}{3}$ $\sec \theta = \dfrac{\sqrt{13}}{2}$ $\cot \theta = -\dfrac{2}{3}$

93. For the point (−2, −2), $x = -2,$ $y = -2,$ $r = \sqrt{x^2 + y^2} = \sqrt{4 + 4} = \sqrt{8} = 2\sqrt{2}$

$\sin \theta = \dfrac{-2}{2\sqrt{2}} \cdot \dfrac{\sqrt{2}}{\sqrt{2}} = \dfrac{-\sqrt{2}}{2}$ $\cos \theta = \dfrac{-2}{2\sqrt{2}} \cdot \dfrac{\sqrt{2}}{\sqrt{2}} = \dfrac{-\sqrt{2}}{2}$ $\tan \theta = \dfrac{-2}{-2} = 1$

$\csc \theta = \dfrac{2\sqrt{2}}{-2} = -\sqrt{2}$ $\sec \theta = \dfrac{2\sqrt{2}}{-2} = -\sqrt{2}$ $\cot \theta = \dfrac{-2}{-2} = 1$

95. For the point $(-3, -2)$, $x = -3$, $y = -2$, $r = \sqrt{x^2 + y^2} = \sqrt{9 + 4} = \sqrt{13}$

$$\sin\theta = \frac{-2}{\sqrt{13}} \cdot \frac{\sqrt{13}}{\sqrt{13}} = \frac{-2\sqrt{13}}{13} \qquad \cos\theta = \frac{-3}{\sqrt{13}} \cdot \frac{\sqrt{13}}{\sqrt{13}} = \frac{-3\sqrt{13}}{13} \qquad \tan\theta = \frac{-2}{-3} = \frac{2}{3}$$

$$\csc\theta = \frac{\sqrt{13}}{-2} = -\frac{\sqrt{13}}{2} \qquad\qquad \sec\theta = \frac{\sqrt{13}}{-3} = -\frac{\sqrt{13}}{3} \qquad\qquad \cot\theta = \frac{-3}{-2} = \frac{3}{2}$$

97. For the point $\left(\frac{1}{3}, -\frac{1}{4}\right)$, $x = \frac{1}{3}$, $y = -\frac{1}{4}$, $r = \sqrt{x^2 + y^2} = \sqrt{\frac{1}{9} + \frac{1}{16}} = \sqrt{\frac{25}{144}} = \frac{5}{12}$

$$\sin\theta = \frac{-\frac{1}{4}}{\frac{5}{12}} = -\frac{1}{4} \cdot \frac{12}{5} = -\frac{3}{5} \qquad\qquad \csc\theta = \frac{\frac{5}{12}}{-\frac{1}{4}} = \frac{5}{12} \cdot -\frac{4}{1} = -\frac{5}{3}$$

$$\cos\theta = \frac{\frac{1}{3}}{\frac{5}{12}} = \frac{1}{3} \cdot \frac{12}{5} = \frac{4}{5} \qquad\qquad \sec\theta = \frac{\frac{5}{12}}{\frac{1}{3}} = \frac{5}{12} \cdot \frac{3}{1} = \frac{5}{4}$$

$$\tan\theta = \frac{-\frac{1}{4}}{\frac{1}{3}} = -\frac{1}{4} \cdot \frac{3}{1} = -\frac{3}{4} \qquad\qquad \cot\theta = \frac{\frac{1}{3}}{-\frac{1}{4}} = \frac{1}{3} \cdot -\frac{4}{1} = -\frac{4}{3}$$

99. $\sin 45° + \sin 135° + \sin 225° + \sin 315° = \dfrac{\sqrt{2}}{2} + \dfrac{\sqrt{2}}{2} - \dfrac{\sqrt{2}}{2} - \dfrac{\sqrt{2}}{2} = 0$

101. If $\sin\theta = 0.1$, then $\sin(\theta + \pi) = -0.1$

103. If $\tan\theta = 3$, then $\tan(\theta + \pi) = 3$

105. If $\sin\theta = \frac{1}{5}$, then $\csc\theta = \dfrac{1}{\frac{1}{5}} = 1 \cdot \dfrac{5}{1} = 5$

107. Use the formula $R = \dfrac{{v_0}^2 \sin(2\theta)}{g}$ with $g = 32.2\text{ft} / \sec^2$; $\theta = 45°$; $v_0 = 100\text{ ft} / \sec$:

$$R = \frac{100^2 \sin(2(45°))}{32.2} = \frac{10000\sin 90°}{32.2} = \frac{10000}{32.2} \approx 310.56\text{ feet}$$

Use the formula $H = \dfrac{{v_0}^2 \sin^2\theta}{2g}$ with $g = 32.2\text{ft} / \sec^2$; $\theta = 45°$; $v_0 = 100\text{ ft} / \sec$:

$$H = \frac{100^2 \sin^2 45°}{2(32.2)} = \frac{10000(0.7071)^2}{64.4} \approx 77.64\text{ feet}$$

109. Use the formula $R = \dfrac{{v_0}^2 \sin(2\theta)}{g}$ with $g = 9.8\text{ m} / \sec^2$; $\theta = 25°$; $v_0 = 500\text{ m} / \sec$:

$$R = \frac{500^2 \sin(2(25°))}{9.8} = \frac{250{,}000\sin 50°}{9.8} \approx 19{,}542\text{ meters}$$

Use the formula $H = \dfrac{{v_0}^2 \sin^2\theta}{2g}$ with $g = 9.8\text{ m} / \sec^2$; $\theta = 25°$; $v_0 = 500\text{ m} / \sec$:

$$H = \frac{500^2 \sin^2 25°}{2(9.8)} = \frac{250{,}000(0.4226)^2}{19.6} \approx 2278\text{ meters}$$

111. Use the formula $t = \sqrt{\dfrac{2a}{g\sin\theta\cos\theta}}$ with $g = 32$ ft / sec^2 and $a = 10$ feet :

(a) $t = \sqrt{\dfrac{2(10)}{32\sin 30°\cos 30°}} = \sqrt{\dfrac{20}{32\cdot\frac{1}{2}\cdot\frac{\sqrt{3}}{2}}} = \sqrt{\dfrac{20}{8\sqrt{3}}} = \sqrt{\dfrac{5}{2\sqrt{3}}} \approx 1.20$ seconds

(b) $t = \sqrt{\dfrac{2(10)}{32\sin 45°\cos 45°}} = \sqrt{\dfrac{20}{32\cdot\frac{\sqrt{2}}{2}\cdot\frac{\sqrt{2}}{2}}} = \sqrt{\dfrac{20}{16}} = \sqrt{\dfrac{5}{4}} \approx 1.12$ seconds

(c) $t = \sqrt{\dfrac{2(10)}{32\sin 60°\cos 60°}} = \sqrt{\dfrac{20}{32\cdot\frac{\sqrt{3}}{2}\cdot\frac{1}{2}}} = \sqrt{\dfrac{20}{8\sqrt{3}}} = \sqrt{\dfrac{5}{2\sqrt{3}}} \approx 1.20$ seconds

113. (a) $T(30°) = 1 + \dfrac{2}{3\sin 30°} - \dfrac{1}{4\tan 30°} = 1 + \dfrac{2}{3\cdot\frac{1}{2}} - \dfrac{1}{4\cdot\frac{1}{\sqrt{3}}} = 1 + \dfrac{4}{3} - \dfrac{\sqrt{3}}{4} \approx 1.9$ hrs

$\dfrac{1}{x} = \tan\theta \;\rightarrow\; x = \dfrac{1}{\tan\theta}$

Distance traveled on road is : $8 - 2x = 8 - \dfrac{2}{\tan\theta}$

Time on road $= \dfrac{\text{distance on road}}{\text{rate on road}} = \dfrac{8 - \frac{2}{\tan\theta}}{8}$

Time on road $= \dfrac{8 - \frac{2}{\tan 30°}}{8} \approx 0.57$ hours

(b) $T(45°) = 1 + \dfrac{2}{3\sin 45°} - \dfrac{1}{4\tan 45°} = 1 + \dfrac{2}{3\cdot\frac{1}{\sqrt{2}}} - \dfrac{1}{4\cdot 1} = 1 + \dfrac{2\sqrt{2}}{3} - \dfrac{1}{4} \approx 1.69$ hrs

Time on road $= \dfrac{8 - \frac{2}{\tan 45°}}{8} \approx 0.75$ hours

(c) $T(60°) = 1 + \dfrac{2}{3\sin 60°} - \dfrac{1}{4\tan 60°} = 1 + \dfrac{2}{3\cdot\frac{\sqrt{3}}{2}} - \dfrac{1}{4\cdot\sqrt{3}}$

$= 1 + \dfrac{4}{3\sqrt{3}} - \dfrac{1}{4\sqrt{3}} \approx 1.63$ hrs

Time on road $= \dfrac{8 - \frac{2}{\tan 60°}}{8} \approx 0.86$ hours

(d) $T(90°) = 1 + \dfrac{2}{3\sin 90°} - \dfrac{1}{4\tan 90°} =$ ($\tan 90°$ is undefined)

The distance would be 2 miles in the sand and 8 miles on the road. The total time would be: $\frac{2}{3} + 1 = \frac{5}{3} \approx 1.67$ hours.

115. (a) $R = \dfrac{32^2\sqrt{2}}{32}\left[\sin(2(60°)) - \cos(2(60°)) - 1\right] = 32\sqrt{2}\left(0.866 - (-0.5) - 1\right) \approx 16.6 \text{ ft}$

(b) Graph:

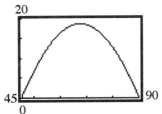

(c) Using MAXIMUM, R is largest when $\theta = 67.5°$.

6.3 Properties of the Trigonometric Functions

1. $\sin 405° = \sin(360° + 45°) = \sin 45° = \dfrac{\sqrt{2}}{2}$

3. $\tan 405° = \tan(180° + 180° + 45°) = \tan 45° = 1$

5. $\csc 450° = \csc(360° + 90°) = \csc 90° = 1$

7. $\cot 390° = \cot(180° + 180° + 30°) = \cot 30° = \sqrt{3}$

9. $\cos\dfrac{33\pi}{4} = \cos\left(\dfrac{\pi}{4} + \dfrac{32\pi}{4}\right) = \cos\left(\dfrac{\pi}{4} + 8\pi\right) = \cos\left(\dfrac{\pi}{4} + 4\cdot 2\pi\right) = \cos\dfrac{\pi}{4} = \dfrac{\sqrt{2}}{2}$

11. $\tan 21\pi = \tan(0 + 21\pi) = \tan 0 = 0$

13. $\sec\dfrac{17\pi}{4} = \sec\left(\dfrac{\pi}{4} + \dfrac{16\pi}{4}\right) = \sec\left(\dfrac{\pi}{4} + 4\pi\right) = \sec\left(\dfrac{\pi}{4} + 2\cdot 2\pi\right) = \sec\dfrac{\pi}{4} = \sqrt{2}$

15. $\tan\dfrac{19\pi}{6} = \tan\left(\dfrac{\pi}{6} + \dfrac{18\pi}{6}\right) = \tan\left(\dfrac{\pi}{6} + 3\pi\right) = \tan\dfrac{\pi}{6} = \dfrac{\sqrt{3}}{3}$

17. Since $\sin\theta > 0$ for points in quadrants I and II, and $\cos\theta < 0$ for points in quadrants II and III, the angle θ lies in quadrant II.

19. Since $\sin\theta < 0$ for points in quadrants III and IV, and $\tan\theta < 0$ for points in quadrants II and IV, the angle θ lies in quadrant IV.

21. Since $\cos\theta > 0$ for points in quadrants I and IV, and $\tan\theta < 0$ for points in quadrants II and IV, the angle θ lies in quadrant IV.

23. Since $\sec\theta < 0$ for points in quadrants II and III, and $\sin\theta > 0$ for points in quadrants I and II, the angle θ lies in quadrant II.

25. $\sin\theta = \dfrac{2\sqrt{5}}{5}, \quad \cos\theta = \dfrac{\sqrt{5}}{5}$

$\tan\theta = \dfrac{\sin\theta}{\cos\theta} = \dfrac{\frac{2\sqrt{5}}{5}}{\frac{\sqrt{5}}{5}} = \dfrac{2\sqrt{5}}{5}\cdot\dfrac{5}{\sqrt{5}} = 2$ 　　　$\sec\theta = \dfrac{1}{\cos\theta} = \dfrac{1}{\frac{\sqrt{5}}{5}} = \dfrac{5}{\sqrt{5}}\cdot\dfrac{\sqrt{5}}{\sqrt{5}} = \sqrt{5}$

$\csc\theta = \dfrac{1}{\sin\theta} = \dfrac{1}{\frac{2\sqrt{5}}{5}} = 1\cdot\dfrac{5}{2\sqrt{5}}\cdot\dfrac{\sqrt{5}}{\sqrt{5}} = \dfrac{\sqrt{5}}{2}$ 　　　$\cot\theta = \dfrac{1}{\tan\theta} = \dfrac{1}{2}$

27. $\sin\theta = \dfrac{1}{2}, \quad \cos\theta = \dfrac{\sqrt{3}}{2}$

$\tan\theta = \dfrac{\sin\theta}{\cos\theta} = \dfrac{\frac{1}{2}}{\frac{\sqrt{3}}{2}} = \dfrac{1}{2}\cdot\dfrac{2}{\sqrt{3}}\cdot\dfrac{\sqrt{3}}{\sqrt{3}} = \dfrac{\sqrt{3}}{3}$ 　　　$\sec\theta = \dfrac{1}{\cos\theta} = \dfrac{1}{\frac{\sqrt{3}}{2}} = \dfrac{2}{\sqrt{3}}\cdot\dfrac{\sqrt{3}}{\sqrt{3}} = \dfrac{2\sqrt{3}}{3}$

$\csc\theta = \dfrac{1}{\sin\theta} = \dfrac{1}{\frac{1}{2}} = 1\cdot\dfrac{2}{1} = 2$ 　　　$\cot\theta = \dfrac{1}{\tan\theta} = \dfrac{1}{\frac{\sqrt{3}}{3}} = \dfrac{3}{\sqrt{3}}\cdot\dfrac{\sqrt{3}}{\sqrt{3}} = \sqrt{3}$

29. $\sin\theta = -\dfrac{1}{3}, \quad \cos\theta = \dfrac{2\sqrt{2}}{3}$

$\tan\theta = \dfrac{\sin\theta}{\cos\theta} = \dfrac{-\frac{1}{3}}{\frac{2\sqrt{2}}{3}} = -\dfrac{1}{3}\cdot\dfrac{3}{2\sqrt{2}}\cdot\dfrac{\sqrt{2}}{\sqrt{2}} = -\dfrac{\sqrt{2}}{4}$

$\csc\theta = \dfrac{1}{\sin\theta} = \dfrac{1}{-\frac{1}{3}} = 1\cdot-\dfrac{3}{1} = -3$

$\sec\theta = \dfrac{1}{\cos\theta} = \dfrac{1}{\frac{2\sqrt{2}}{3}} = \dfrac{3}{2\sqrt{2}}\cdot\dfrac{\sqrt{2}}{\sqrt{2}} = \dfrac{3\sqrt{2}}{4}$

$\cot\theta = \dfrac{1}{\tan\theta} = \dfrac{1}{-\frac{\sqrt{2}}{4}} = -\dfrac{4}{\sqrt{2}}\cdot\dfrac{\sqrt{2}}{\sqrt{2}} = -2\sqrt{2}$

31. $\sin\theta = 0.2588, \quad \cos\theta = 0.9659$

$\tan\theta = \dfrac{\sin\theta}{\cos\theta} = \dfrac{0.2588}{0.9659} \approx 0.2679$ 　　　$\sec\theta = \dfrac{1}{\cos\theta} = \dfrac{1}{0.9659} \approx 1.0353$

$\csc\theta = \dfrac{1}{\sin\theta} = \dfrac{1}{0.2588} \approx 3.8640$ 　　　$\cot\theta = \dfrac{\cos\theta}{\sin\theta} = \dfrac{0.9659}{0.2588} \approx 3.7322$

33. $\sin\theta = \frac{12}{13}$, θ in quadrant II

Solve for $\cos\theta$:

$$\sin^2\theta + \cos^2\theta = 1$$
$$\cos^2\theta = 1 - \sin^2\theta$$
$$\cos\theta = \pm\sqrt{1 - \sin^2\theta}$$

Since θ is in quadrant II, $\cos\theta < 0$.

$$\cos\theta = -\sqrt{1-\sin^2\theta} = -\sqrt{1-\left(\frac{12}{13}\right)^2} = -\sqrt{1-\frac{144}{169}} = -\sqrt{\frac{25}{169}} = -\frac{5}{13}$$

$$\tan\theta = \frac{\sin\theta}{\cos\theta} = \frac{\frac{12}{13}}{-\frac{5}{13}} = \frac{12}{13}\cdot-\frac{13}{5} = -\frac{12}{5} \qquad \sec\theta = \frac{1}{\cos\theta} = \frac{1}{-\frac{5}{13}} = -\frac{13}{5}$$

$$\csc\theta = \frac{1}{\sin\theta} = \frac{1}{\frac{12}{13}} = \frac{13}{12} \qquad\qquad \cot\theta = \frac{1}{\tan\theta} = \frac{1}{-\frac{12}{5}} = -\frac{5}{12}$$

35. $\cos\theta = -\frac{4}{5}$, θ in quadrant III

Solve for $\sin\theta$:

$$\sin^2\theta + \cos^2\theta = 1$$
$$\sin^2\theta = 1 - \cos^2\theta$$
$$\sin\theta = \pm\sqrt{1 - \cos^2\theta}$$

Since θ is in quadrant III, $\sin\theta < 0$.

$$\sin\theta = -\sqrt{1-\cos^2\theta} = -\sqrt{1-\left(-\frac{4}{5}\right)^2} = -\sqrt{1-\frac{16}{25}} = -\sqrt{\frac{9}{25}} = -\frac{3}{5}$$

$$\tan\theta = \frac{\sin\theta}{\cos\theta} = \frac{-\frac{3}{5}}{-\frac{4}{5}} = -\frac{3}{5}\cdot-\frac{5}{4} = \frac{3}{4} \qquad \sec\theta = \frac{1}{\cos\theta} = \frac{1}{-\frac{4}{5}} = -\frac{5}{4}$$

$$\csc\theta = \frac{1}{\sin\theta} = \frac{1}{-\frac{3}{5}} = -\frac{5}{3} \qquad\qquad \cot\theta = \frac{1}{\tan\theta} = \frac{1}{\frac{3}{4}} = \frac{4}{3}$$

37. $\sin\theta = \frac{5}{13}$, $90° < \theta < 180°$, θ in quadrant II

Solve for $\cos\theta$:

$$\sin^2\theta + \cos^2\theta = 1$$
$$\cos^2\theta = 1 - \sin^2\theta$$
$$\cos\theta = \pm\sqrt{1 - \sin^2\theta}$$

Since θ is in quadrant II, $\cos\theta < 0$.

$$\cos\theta = -\sqrt{1-\sin^2\theta} = -\sqrt{1-\left(\frac{5}{13}\right)^2} = -\sqrt{1-\frac{25}{169}} = -\sqrt{\frac{144}{169}} = -\frac{12}{13}$$

$$\tan\theta = \frac{\sin\theta}{\cos\theta} = \frac{\frac{5}{13}}{-\frac{12}{13}} = \frac{5}{13}\cdot-\frac{13}{12} = -\frac{5}{12} \qquad \sec\theta = \frac{1}{\cos\theta} = \frac{1}{-\frac{12}{13}} = -\frac{13}{12}$$

$$\csc\theta = \frac{1}{\sin\theta} = \frac{1}{\frac{5}{13}} = \frac{13}{5} \qquad\qquad \cot\theta = \frac{1}{\tan\theta} = \frac{1}{-\frac{5}{12}} = -\frac{12}{5}$$

39. $\cos\theta = -\frac{1}{3}$, $\frac{\pi}{2} < \theta < \pi$, θ in quadrant II

Solve for $\sin\theta$:

$$\sin^2\theta + \cos^2\theta = 1$$
$$\sin^2\theta = 1 - \cos^2\theta$$
$$\sin\theta = \pm\sqrt{1 - \cos^2\theta}$$

Since θ is in quadrant II, $\sin\theta > 0$.

$$\sin\theta = \sqrt{1 - \cos^2\theta} = \sqrt{1 - \left(-\frac{1}{3}\right)^2} = \sqrt{1 - \frac{1}{9}} = \sqrt{\frac{8}{9}} = \frac{2\sqrt{2}}{3}$$

$$\tan\theta = \frac{\sin\theta}{\cos\theta} = \frac{\frac{2\sqrt{2}}{3}}{-\frac{1}{3}} = \frac{2\sqrt{2}}{3} \cdot -\frac{3}{1} = -2\sqrt{2} \qquad \sec\theta = \frac{1}{\cos\theta} = \frac{1}{-\frac{1}{3}} = -3$$

$$\csc\theta = \frac{1}{\sin\theta} = \frac{1}{\frac{2\sqrt{2}}{3}} = \frac{3}{2\sqrt{2}} \cdot \frac{\sqrt{2}}{\sqrt{2}} = \frac{3\sqrt{2}}{4} \qquad \cot\theta = \frac{1}{\tan\theta} = \frac{1}{-2\sqrt{2}} \cdot \frac{\sqrt{2}}{\sqrt{2}} = -\frac{\sqrt{2}}{4}$$

41. $\sin\theta = \frac{2}{3}$, $\tan\theta < 0$, $\Rightarrow$ θ in quadrant II

Solve for $\cos\theta$:

$$\sin^2\theta + \cos^2\theta = 1$$
$$\cos^2\theta = 1 - \sin^2\theta$$
$$\cos\theta = \pm\sqrt{1 - \sin^2\theta}$$

Since θ is in quadrant II, $\cos\theta < 0$.

$$\cos\theta = -\sqrt{1 - \sin^2\theta} = -\sqrt{1 - \left(\frac{2}{3}\right)^2} = -\sqrt{1 - \frac{4}{9}} = -\sqrt{\frac{5}{9}} = -\frac{\sqrt{5}}{3}$$

$$\tan\theta = \frac{\sin\theta}{\cos\theta} = \frac{\frac{2}{3}}{-\frac{\sqrt{5}}{3}} = \frac{2}{3} \cdot -\frac{3}{\sqrt{5}} = -\frac{2\sqrt{5}}{5} \qquad \sec\theta = \frac{1}{\cos\theta} = \frac{1}{-\frac{\sqrt{5}}{3}} = -\frac{3}{\sqrt{5}} = -\frac{3\sqrt{5}}{5}$$

$$\csc\theta = \frac{1}{\sin\theta} = \frac{1}{\frac{2}{3}} = \frac{3}{2} \qquad\qquad \cot\theta = \frac{1}{\tan\theta} = \frac{1}{-\frac{2\sqrt{5}}{5}} = -\frac{5}{2\sqrt{5}} = -\frac{\sqrt{5}}{2}$$

43. $\sec\theta = 2$, $\sin\theta < 0$, $\Rightarrow$ θ in quadrant IV

Solve for $\cos\theta$:

$$\cos\theta = \frac{1}{\sec\theta} = \frac{1}{2}$$

Solve for $\sin\theta$:

$$\sin^2\theta + \cos^2\theta = 1$$
$$\sin^2\theta = 1 - \cos^2\theta$$
$$\sin\theta = \pm\sqrt{1 - \cos^2\theta}$$

Since θ is in quadrant IV, $\sin\theta < 0$.

$$\sin\theta = -\sqrt{1 - \cos^2\theta} = -\sqrt{1 - \left(\frac{1}{2}\right)^2} = -\sqrt{1 - \frac{1}{4}} = -\sqrt{\frac{3}{4}} = -\frac{\sqrt{3}}{2}$$

$$\tan\theta = \frac{\sin\theta}{\cos\theta} = \frac{-\frac{\sqrt{3}}{2}}{\frac{1}{2}} = -\frac{\sqrt{3}}{2} \cdot \frac{2}{1} = -\sqrt{3} \qquad \cot\theta = \frac{1}{\tan\theta} = \frac{1}{-\sqrt{3}} = -\frac{\sqrt{3}}{3}$$

$$\csc\theta = \frac{1}{\sin\theta} = \frac{1}{-\frac{\sqrt{3}}{2}} = -\frac{2}{\sqrt{3}} = -\frac{2\sqrt{3}}{3}$$

45. $\tan\theta = \frac{3}{4}$, $\sin\theta < 0$, $\Rightarrow$ θ in quadrant III

Solve for $\sec\theta$:

$$\sec^2\theta = 1 + \tan^2\theta$$

$$\sec\theta = \pm\sqrt{1 + \tan^2\theta}$$

Since θ is in quadrant III, $\sec\theta < 0$.

$$\sec\theta = -\sqrt{1 + \tan^2\theta} = -\sqrt{1 + \left(\frac{3}{4}\right)^2} = -\sqrt{1 + \frac{9}{16}} = -\sqrt{\frac{25}{16}} = -\frac{5}{4}$$

$$\cos\theta = \frac{1}{\sec\theta} = -\frac{4}{5}$$

$$\sin\theta = -\sqrt{1 - \cos^2\theta} = -\sqrt{1 - \left(-\frac{4}{5}\right)^2} = -\sqrt{1 - \frac{16}{25}} = -\sqrt{\frac{9}{25}} = -\frac{3}{5}$$

$$\csc\theta = \frac{1}{\sin\theta} = \frac{1}{-\frac{3}{5}} = -\frac{5}{3} \qquad\qquad \cot\theta = \frac{1}{\tan\theta} = \frac{1}{\frac{3}{4}} = \frac{4}{3}$$

47. $\tan\theta = -\frac{1}{3}$, $\sin\theta > 0$, $\Rightarrow$ θ in quadrant II

Solve for $\sec\theta$:

$$\sec^2\theta = 1 + \tan^2\theta$$

$$\sec\theta = \pm\sqrt{1 + \tan^2\theta}$$

Since θ is in quadrant II, $\sec\theta < 0$.

$$\sec\theta = -\sqrt{1 + \tan^2\theta} = -\sqrt{1 + \left(-\frac{1}{3}\right)^2} = -\sqrt{1 + \frac{1}{9}} = -\sqrt{\frac{10}{9}} = -\frac{\sqrt{10}}{3}$$

$$\cos\theta = \frac{1}{\sec\theta} = \frac{1}{-\frac{\sqrt{10}}{3}} = -\frac{3}{\sqrt{10}} = -\frac{3\sqrt{10}}{10}$$

$$\sin\theta = \sqrt{1 - \cos^2\theta} = \sqrt{1 - \left(-\frac{3\sqrt{10}}{10}\right)^2} = \sqrt{1 - \frac{90}{100}} = \sqrt{\frac{10}{100}} = \frac{\sqrt{10}}{10}$$

$$\csc\theta = \frac{1}{\sin\theta} = \frac{1}{\frac{\sqrt{10}}{10}} = \sqrt{10} \qquad\qquad \cot\theta = \frac{1}{\tan\theta} = \frac{1}{-\frac{1}{3}} = -3$$

49. $\sin(-60°) = -\sin 60° = -\dfrac{\sqrt{3}}{2}$

51. $\tan(-30°) = -\tan 30° = -\dfrac{\sqrt{3}}{3}$

53. $\sec(-60°) = \sec 60° = 2$

55. $\sin(-90°) = -\sin 90° = -1$

57. $\tan\left(-\dfrac{\pi}{4}\right) = -\tan\dfrac{\pi}{4} = -1$

59. $\cos\left(-\dfrac{\pi}{4}\right) = \cos\dfrac{\pi}{4} = \dfrac{\sqrt{2}}{2}$

61. $\tan(-\pi) = -\tan\pi = 0$

63. $\csc\left(-\dfrac{\pi}{4}\right) = -\csc\dfrac{\pi}{4} = -\sqrt{2}$

65. $\sec\left(-\dfrac{\pi}{6}\right) = \sec\dfrac{\pi}{6} = \dfrac{2\sqrt{3}}{3}$

67. $\sin(-\pi) + \cos 5\pi = -\sin\pi + \cos(\pi + 4\pi) = 0 + \cos\pi = -1$

69. $\sec(-\pi) + \csc\left(-\frac{\pi}{2}\right) = \sec \pi - \csc \frac{\pi}{2} = -1 - 1 = -2$

71. $\sin\left(-\frac{9\pi}{4}\right) - \tan\left(-\frac{9\pi}{4}\right) = -\sin\frac{9\pi}{4} + \tan\frac{9\pi}{4} = -\sin\left(\frac{\pi}{4} + \frac{8\pi}{4}\right) + \tan\left(\frac{\pi}{4} + \frac{8\pi}{4}\right)$

$$= -\sin\frac{\pi}{4} + \tan\frac{\pi}{4} = -\frac{\sqrt{2}}{2} + 1$$

73. $\sin^2 40° + \cos^2 40° = 1$

75. $\sin 80° \csc 80° = \sin 80° \cdot \frac{1}{\sin 80°} = 1$

77. $\tan 40° - \frac{\sin 40°}{\cos 40°} = \tan 40° - \tan 40° = 0$

79. Given: $\sin \theta = 0.3$
    $\sin \theta + \sin(\theta + 2\pi) + \sin(\theta + 4\pi) = \sin \theta + \sin \theta + \sin \theta = 0.3 + 0.3 + 0.3 = 0.9$

81. Given: $\tan \theta = 3$
    $\tan \theta + \tan(\theta + \pi) + \tan(\theta + 2\pi) = \tan \theta + \tan \theta + \tan \theta = 3 + 3 + 3 = 9$

83. Find the value:
    $\sin 1° + \sin 2° + \sin 3° + ... + \sin 357° + \sin 358° + \sin 359°$
    $= \sin 1° + \sin 2° + \sin 3° + ... + \sin(360° - 3°) + \sin(360° - 2°) + \sin(360° - 1°)$
    $= \sin 1° + \sin 2° + \sin 3° + ... + \sin(-3°) + \sin(-2°) + \sin(-1°)$
    $= \sin 1° + \sin 2° + \sin 3° + ... - \sin 3° - \sin 2° - \sin 1°$
    $= \sin 180°$
    $= 0$

85. The domain of the sine function is the set of all real numbers.

87. $f(\theta) = \tan \theta$ is not defined for numbers that are odd multiples of $\frac{\pi}{2}$.

89. $f(\theta) = \sec \theta$ is not defined for numbers that are odd multiples of $\frac{\pi}{2}$.

91. The range of the sine function is the set of all real numbers between -1 and 1, inclusive.

93. The range of the tangent function is the set of all real numbers.

95. The range of the secant function is the set of all real number greater than or equal to 1 and all real numbers less than or equal to -1.

97. The sine function is odd because $\sin(-\theta) = -\sin\theta$. Its graph is symmetric to the origin.

99. The tangent function is odd because $\tan(-\theta) = -\tan\theta$. Its graph is symmetric to the origin.

101. The secant function is even because $\sec(-\theta) = \sec\theta$. Its graph is symmetric to the y-axis.

103. (a) $f(-a) = -f(a) = -\dfrac{1}{3}$

   (b) $f(a) + f(a + 2\pi) + f(a + 4\pi) = f(a) + f(a) + f(a) = \dfrac{1}{3} + \dfrac{1}{3} + \dfrac{1}{3} = 1$

105. (a) $f(-a) = -f(a) = -2$
   (b) $f(a) + f(a + \pi) + f(a + 2\pi) = f(a) + f(a) + f(a) = 2 + 2 + 2 = 6$

107. (a) $f(-a) = f(a) = -4$
   (b) $f(a) + f(a + 2\pi) + f(a + 4\pi) = f(a) + f(a) + f(a) = (-4) + (-4) + (-4) = -12$

109. Since $\tan\theta = \dfrac{500}{1500} = \dfrac{1}{3}$, then $\sin\theta = \dfrac{1}{\sqrt{1+9}} = \dfrac{1}{\sqrt{10}}$.

   $T = 5 - \dfrac{5}{3 \cdot \dfrac{1}{3}} + \dfrac{5}{\dfrac{1}{\sqrt{10}}} = 5 - 5 + 5\sqrt{10} = 5\sqrt{10} \approx 15.8$ minutes

111. Let $P = (x, y)$ be the point on the unit circle that corresponds to an angle θ. Consider the equation $\tan\theta = \dfrac{y}{x} = a$. Then $y = ax$. Now $x^2 + y^2 = 1$, so $x^2 + a^2 x^2 = 1$. Thus, $x = \pm\dfrac{1}{\sqrt{1+a^2}}$ and $y = \pm\dfrac{a}{\sqrt{1+a^2}}$; that is, for any real number a, there is a point $P = (x, y)$ on the unit circle for which $\tan\theta = a$. In other words, $-\infty < \tan\theta < +\infty$, and the range of the tangent function is the set of all real numbers.

113. Suppose there is a number p, $0 < p < 2\pi$, for which $\sin(\theta + p) = \sin\theta$ for all θ. If $\theta = 0$, then $\sin(0 + p) = \sin p = \sin 0 = 0$; so that $p = \pi$. If $\theta = \dfrac{\pi}{2}$, then $\sin\left(\dfrac{\pi}{2} + p\right) = \sin\dfrac{\pi}{2}$. But $p = \pi$. Thus, $\sin\dfrac{3\pi}{2} = -1 = \sin\dfrac{\pi}{2} = 1$, or $-1 = 1$. This is impossible. The smallest positive number p for which $\sin(\theta + p) = \sin\theta$ for all θ is therefore $p = 2\pi$.

115. $\sec\theta = \dfrac{1}{\cos\theta}$: since $\cos\theta$ has period 2π, so does $\sec\theta$.

117. If $P = (a, b)$ is the point on the unit circle corresponding to θ, then $Q = (-a, -b)$ is the point on the unit circle corresponding to $\theta + \pi$.

Thus, $\tan(\theta + \pi) = \dfrac{-b}{-a} = \dfrac{b}{a} = \tan \theta$; that is, the period of the tangent function is π.

119. Let $P = (a, b)$ be the point on the unit circle corresponding to θ.

Then $\csc \theta = \dfrac{1}{b} = \dfrac{1}{\sin \theta}$; $\sec \theta = \dfrac{1}{a} = \dfrac{1}{\cos \theta}$; $\cot \theta = \dfrac{a}{b} = \dfrac{1}{\dfrac{b}{a}} = \dfrac{1}{\tan \theta}$

121. $(\sin \theta \cos \phi)^2 + (\sin \theta \sin \phi)^2 + \cos^2 \theta = \sin^2 \theta \cos^2 \phi + \sin^2 \theta \sin^2 \phi + \cos^2 \theta$
$$= \sin^2 \theta (\cos^2 \phi + \sin^2 \phi) + \cos^2 \theta = \sin^2 \theta + \cos^2 \theta = 1$$

6.4 Graphs of the Sine and Cosine Functions

1. 0

3. The graph of $y = \sin x$ is increasing for $-\dfrac{\pi}{2} < x < \dfrac{\pi}{2}$.

5. The largest value of $y = \sin x$ is 1.

7. $\sin x = 0$ when $x = 0, \pi, 2\pi$

9. $\sin x = 1$ when $x = -\dfrac{3\pi}{2}, \dfrac{\pi}{2}$; $\sin x = -1$ when $x = -\dfrac{\pi}{2}, \dfrac{3\pi}{2}$

11. B, C, F   13. C   15. D

17. $y = 3\sin x$

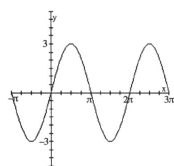

19. $y = \cos\left(x + \dfrac{\pi}{4}\right)$

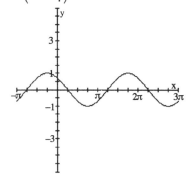

21. $y = \sin x - 1$

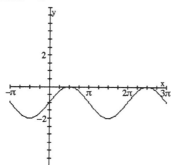

23. $y = -2\sin x$

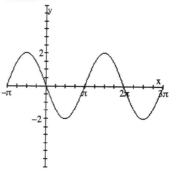

25. $y = \sin(\pi x)$

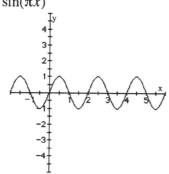

27. $y = 2\sin x + 2$

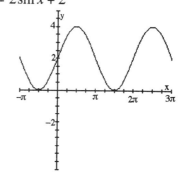

29. $y = -2\cos\left(x - \dfrac{\pi}{2}\right)$

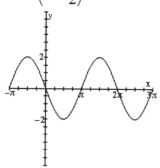

31. $y = 3\sin(\pi - x)$

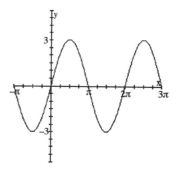

33.   Graph:

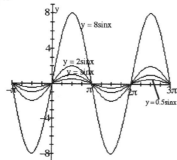

The value of A stretches the graph vertically by a factor of A.

35. Graph:

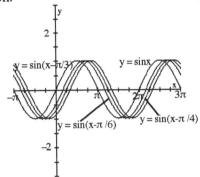

Subtracting ϕ from x shifts the graph ϕ units to the right.

37. Graph:

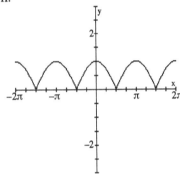

6.5 Graphs of the Tangent, Cotangent, Cosecant, and Secant Functions

1. y-intercept: 0

3. y-intercept: 1

5. $\sec x = 1$ for $x = -2\pi, 0, 2\pi$; $\sec x = -1$ for $x = -\pi, \pi$

7. $y = \sec x$ has vertical asymptotes for $x = -\dfrac{3\pi}{2}, -\dfrac{\pi}{2}, \dfrac{\pi}{2}, \dfrac{3\pi}{2}$

9. $y = \tan x$ has vertical asymptotes for $x = -\dfrac{3\pi}{2}, -\dfrac{\pi}{2}, \dfrac{\pi}{2}, \dfrac{3\pi}{2}$

11. B

13. A

15. $y = -\sec x$

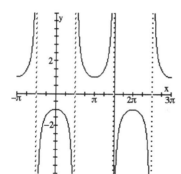

17. $y = \sec\left(x - \dfrac{\pi}{2}\right)$

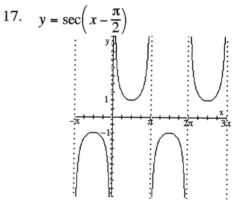

19. $y = \tan(x - \pi)$

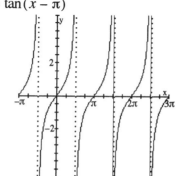

21. $y = 3\tan(2x)$

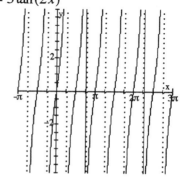

23. $y = \sec(2x)$

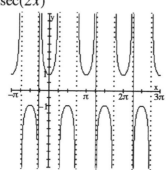

25. $y = \cot(\pi x)$

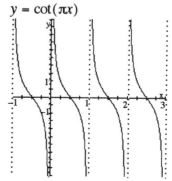

27. $y = -3\tan(4x)$

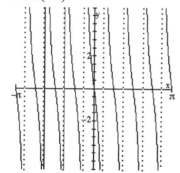

29. $y = 2\sec\left(\tfrac{1}{2}x\right)$

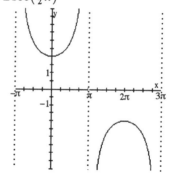

31. $y = -3\csc\left(x + \dfrac{\pi}{4}\right)$

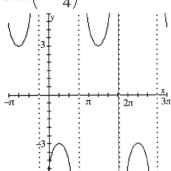

33. $y = \dfrac{1}{2}\cot\left(x + \dfrac{\pi}{4}\right)$

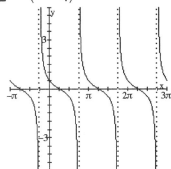

35. (a) $L = \dfrac{3}{\cos\theta} + \dfrac{4}{\sin\theta} = 3\sec\theta + 4\csc\theta$

     (b) Graph:

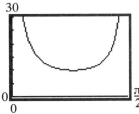

     (c) Use MINIMUM to find the least value: L is least when $\theta = 0.83$.

     (d) $L = \dfrac{3}{\cos 0.83} + \dfrac{4}{\sin 0.83} \approx 9.86$ feet

6.6 Sinusoidal Graphs; Sinusoidal Curve Fitting

1. $y = 2\sin x$

    This is in the form $y = A\sin(\omega x)$ where $A = 2$ and $\omega = 1$.

    Thus, the amplitude is $|A| = |2| = 2$ and the period is $T = \dfrac{2\pi}{\omega} = \dfrac{2\pi}{1} = 2\pi$.

3. $y = -4\cos(2x)$

    This is in the form $y = A\cos(\omega x)$ where $A = -4$ and $\omega = 2$.

    Thus, the amplitude is $|A| = |-4| = 4$ and the period is $T = \dfrac{2\pi}{\omega} = \dfrac{2\pi}{2} = \pi$.

5. $y = 6\sin(\pi x)$

    This is in the form $y = A\sin(\omega x)$ where $A = 6$ and $\omega = \pi$.

    Thus, the amplitude is $|A| = |6| = 6$ and the period is $T = \dfrac{2\pi}{\omega} = \dfrac{2\pi}{\pi} = 2$.

7. $y = -\frac{1}{2}\cos\left(\frac{3}{2}x\right)$

     This is in the form $y = A\cos(\omega\, x)$ where $A = -\frac{1}{2}$ and $\omega = \frac{3}{2}$.

     Thus, the amplitude is $|A| = \left|-\frac{1}{2}\right| = \frac{1}{2}$ and the period is $T = \dfrac{2\pi}{\omega} = \dfrac{2\pi}{\frac{3}{2}} = \dfrac{4\pi}{3}$.

9. $y = \frac{5}{3}\sin\left(-\frac{2\pi}{3}x\right) = -\frac{5}{3}\sin\left(\frac{2\pi}{3}x\right)$

     This is in the form $y = A\sin(\omega\, x)$ where $A = -\frac{5}{3}$ and $\omega = \frac{2\pi}{3}$.

     Thus, the amplitude is $|A| = \left|-\frac{5}{3}\right| = \frac{5}{3}$ and the period is $T = \dfrac{2\pi}{\omega} = \dfrac{2\pi}{\frac{2\pi}{3}} = 3$.

11.   F   13.   A   15.   H

17.   C   19.   J

21.   A   23.   D   25.   B

27. $y = 5\sin(4x)$ $A = 5;\;\; T = \dfrac{\pi}{2}$ 29. $y = 5\cos(\pi x)$ $A = 5;\;\; T = 2$

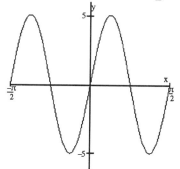 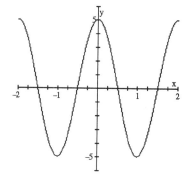

31. $y = -2\cos(2\pi x)$ $A = -2;\;\; T = 1$ 33. $y = -4\sin\left(\frac{1}{2}x\right)$ $A = -4;\;\; T = 4\pi$

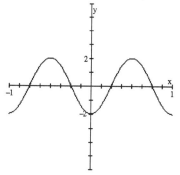 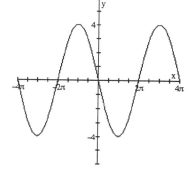

35. $y = \frac{3}{2}\sin\left(-\frac{2}{3}x\right) = -\frac{3}{2}\sin\left(\frac{2}{3}x\right)$ $A = -\frac{3}{2};$ $T = 3\pi$

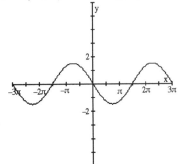

37.   The graph is a cosine graph with an amplitude of 5 and a period of 8. Find ω:

$$8 = \frac{2\pi}{\omega} \;\rightarrow\; 8\omega = 2\pi \;\rightarrow\; \omega = \frac{2\pi}{8} = \frac{\pi}{4}$$

The equation is: $y = 5\cos\left(\frac{\pi}{4}x\right)$.

39.   The graph is a reflected cosine graph with an amplitude of 3 and a period of 4π.
Find ω:

$$4\pi = \frac{2\pi}{\omega} \;\rightarrow\; 4\pi\omega = 2\pi \;\rightarrow\; \omega = \frac{2\pi}{4\pi} = \frac{1}{2}$$

The equation is: $y = -3\cos\left(\frac{1}{2}x\right)$.

41.   The graph is a sine graph with an amplitude of $\frac{3}{4}$ and a period of 1. Find ω:

$$1 = \frac{2\pi}{\omega} \;\rightarrow\; \omega = 2\pi$$

The equation is: $y = \frac{3}{4}\sin(2\pi x)$.

43.   The graph is a reflected sine graph with an amplitude of 1 and a period of $\frac{4\pi}{3}$.
Find ω:

$$\frac{4\pi}{3} = \frac{2\pi}{\omega} \;\rightarrow\; 4\pi\omega = 6\pi \;\rightarrow\; \omega = \frac{6\pi}{4\pi} = \frac{3}{2}$$

The equation is: $y = -\sin\left(\frac{3}{2}x\right)$.

45.   The graph is a reflected cosine graph with an amplitude of 2 and a period of $\frac{4}{3}$.
Find ω:

$$\frac{4}{3} = \frac{2\pi}{\omega} \;\rightarrow\; 4\omega = 6\pi \;\rightarrow\; \omega = \frac{6\pi}{4} = \frac{3\pi}{2}$$

The equation is: $y = -2\cos\left(\frac{3\pi}{2}x\right)$.

47. The graph is a sine graph with an amplitude of 3 and a period of 4. Find ω:
$$4 = \frac{2\pi}{\omega} \;\rightarrow\; 4\omega = 2\pi \;\rightarrow\; \omega = \frac{2\pi}{4} = \frac{\pi}{2}$$
The equation is: $y = 3\sin\left(\frac{\pi}{2}x\right)$.

49. The graph is a reflected cosine graph with an amplitude of 4 and a period of $\frac{2\pi}{3}$. Find ω:
$$\frac{2\pi}{3} = \frac{2\pi}{\omega} \;\rightarrow\; 2\pi\omega = 6\pi \;\rightarrow\; \omega = \frac{6\pi}{2\pi} = 3$$
The equation is: $y = -4\cos(3x)$.

51. $y = 4\sin(2x - \pi)$
Amplitude: $|A| = |4| = 4$
Period: $T = \frac{2\pi}{\omega} = \frac{2\pi}{2} = \pi$
Phase Shift: $\frac{\phi}{\omega} = \frac{\pi}{2}$

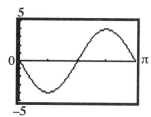

53. $y = 2\cos\left(3x + \frac{\pi}{2}\right)$
Amplitude: $|A| = |2| = 2$
Period: $T = \frac{2\pi}{\omega} = \frac{2\pi}{3}$
Phase Shift: $\frac{\phi}{\omega} = \frac{-\frac{\pi}{2}}{3} = -\frac{\pi}{6}$

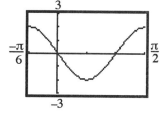

55. $y = -3\sin\left(2x + \frac{\pi}{2}\right)$
Amplitude: $|A| = |-3| = 3$
Period: $T = \frac{2\pi}{\omega} = \frac{2\pi}{2} = \pi$
Phase Shift: $\frac{\phi}{\omega} = \frac{-\frac{\pi}{2}}{2} = -\frac{\pi}{4}$

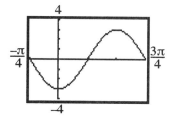

57. $y = 4\sin(\pi x + 2)$
Amplitude: $|A| = |4| = 4$
Period: $T = \frac{2\pi}{\omega} = \frac{2\pi}{\pi} = 2$
Phase Shift: $\frac{\phi}{\omega} = \frac{-2}{\pi}$

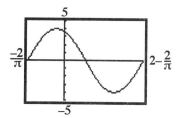

59. $y = 3\cos(\pi x - 2) + 1$

   Amplitude: $|A| = |3| = 3$

   Period: $T = \dfrac{2\pi}{\omega} = \dfrac{2\pi}{\pi} = 2$

   Phase Shift: $\dfrac{\phi}{\omega} = \dfrac{2}{\pi}$

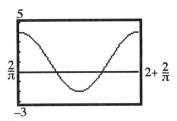

61. $y = 3\sin\left(-2x + \dfrac{\pi}{2}\right) - 2 = -3\sin\left(2x - \dfrac{\pi}{2}\right) - 2$

   Amplitude: $|A| = |-3| = 3$

   Period: $T = \dfrac{2\pi}{\omega} = \dfrac{2\pi}{2} = \pi$

   Phase Shift: $\dfrac{\phi}{\omega} = \dfrac{\frac{\pi}{2}}{2} = \dfrac{\pi}{4}$

63. $|A| = 3$; $T = \pi$; $\omega = \dfrac{2\pi}{T} = \dfrac{2\pi}{\pi} = 2$; $\qquad y = 3\sin(2x)$

65. $|A| = 3$; $T = 2$; $\omega = \dfrac{2\pi}{T} = \dfrac{2\pi}{2} = \pi$; $\qquad y = 3\sin(\pi x)$

67. $|A| = 2$; $T = \pi$; $\dfrac{\phi}{\omega} = \dfrac{1}{2}$; $\omega = \dfrac{2\pi}{T} = \dfrac{2\pi}{\pi} = 2$; $\dfrac{\phi}{\omega} = \dfrac{\phi}{2} = \dfrac{1}{2} \rightarrow \phi = 1$

   $y = 2\sin(2x - 1) = 2\sin\left[2\left(x - \dfrac{1}{2}\right)\right]$

69. $|A| = 3$; $T = 3\pi$; $\dfrac{\phi}{\omega} = -\dfrac{1}{3}$; $\omega = \dfrac{2\pi}{T} = \dfrac{2\pi}{3\pi} = \dfrac{2}{3}$;

   $\dfrac{\phi}{\omega} = \dfrac{\phi}{\frac{2}{3}} = -\dfrac{1}{3} \rightarrow \phi = -\dfrac{1}{3} \cdot \dfrac{2}{3} = -\dfrac{2}{9}$ $\qquad y = 3\sin\left(\dfrac{2}{3}x + \dfrac{2}{9}\right) = 3\sin\left[\dfrac{2}{3}\left(x + \dfrac{1}{3}\right)\right]$

71. $I = 220\sin(60\pi t)$, $t \geq 0$

   Period: $T = \dfrac{2\pi}{\omega} = \dfrac{2\pi}{60\pi} = \dfrac{1}{30}$

   Amplitude: $|A| = |220| = 220$

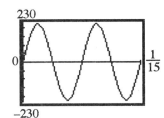

73. $I = 120\sin\left(30\pi t - \dfrac{\pi}{3}\right),\quad t \ge 0$

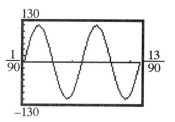

Period: $T = \dfrac{2\pi}{\omega} = \dfrac{2\pi}{30\pi} = \dfrac{1}{15}$

Amplitude: $|A| = |120| = 120$

Phase Shift: $\dfrac{\phi}{\omega} = \dfrac{\frac{\pi}{3}}{30\pi} = \dfrac{1}{90}$

75. $V = 220\sin(120\pi t)$

   (a)   Amplitude: $|A| = |220| = 220$

       Period: $T = \dfrac{2\pi}{\omega} = \dfrac{2\pi}{120\pi} = \dfrac{1}{60}$

   (b)

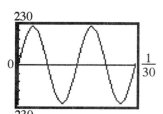

   (c) $V = IR$

       $220\sin(120\pi t) = 10I$

       $22\sin(120\pi t) = I$

   (d)   Amplitude: $|A| = |22| = 22$

       Period: $T = \dfrac{2\pi}{\omega} = \dfrac{2\pi}{120\pi} = \dfrac{1}{60}$

   (e)

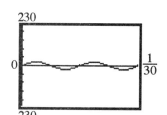

77. (a) $P = \dfrac{V^2}{R} = \dfrac{\left(V_0 \sin(2\pi f t)\right)^2}{R} = \dfrac{V_0^2 \sin^2(2\pi f t)}{R}$

   (b)   The graph is the reflected cosine graph translated up a distance equivalent to the

      amplitude. The period is $\dfrac{1}{2f}$, so $\omega = 4\pi f$. The amplitude is $\dfrac{1}{2} \cdot \dfrac{V_0^2}{R} = \dfrac{V_0^2}{2R}$.

      The equation is: $P = -\dfrac{V_0^2}{2R}\cos(4\pi f t) + \dfrac{V_0^2}{2R} = \dfrac{V_0^2}{R} \cdot \dfrac{1}{2}\left(1 - \cos(4\pi f t)\right)$

   (c)   Comparing the formulas:

      $\sin^2(2\pi f t) = \dfrac{1}{2}\left(1 - \cos(4\pi f t)\right)$

79. (a)   Draw a scatter diagram:

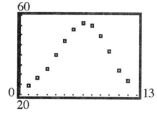

(b) Amplitude : $A = \dfrac{56.0 - 24.2}{2} = \dfrac{31.8}{2} = 15.9$

Vertical Shift : $\dfrac{56.0 + 24.2}{2} = \dfrac{80.2}{2} = 40.1$

$\omega = \dfrac{2\pi}{12} = \dfrac{\pi}{6}$

Phase shift (use $y = 24.2$, $x = 1$):

$$24.2 = 15.9\sin\left(\dfrac{\pi}{6}\cdot 1 - \phi\right) + 40.1$$

$$-15.9 = 15.9\sin\left(\dfrac{\pi}{6} - \phi\right)$$

$$-1 = \sin\left(\dfrac{\pi}{6} - \phi\right)$$

$$\dfrac{-\pi}{2} = \dfrac{\pi}{6} - \phi$$

$$\phi = \dfrac{2\pi}{3}$$

Thus, $y = 15.9\sin\left(\dfrac{\pi}{6}x - \dfrac{2\pi}{3}\right) + 40.1$

(c)

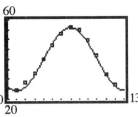

(e)

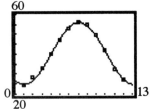

(d) $y = 15.62\sin(0.517x - 2.096) + 40.377$

81. (a) Draw a scatter diagram:

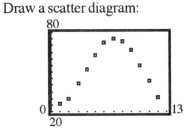

(b) Amplitude : $A = \dfrac{75.4 - 25.5}{2} = \dfrac{49.9}{2} = 24.95$

Vertical Shift : $\dfrac{75.4 + 25.5}{2} = \dfrac{100.9}{2} = 50.45$

$\omega = \dfrac{2\pi}{12} = \dfrac{\pi}{6}$

Phase shift (use $y = 25.5$, $x = 1$):

$$25.5 = 24.95\sin\left(\frac{\pi}{6}\cdot 1 - \phi\right) + 50.45$$

$$-24.95 = 24.95\sin\left(\frac{\pi}{6} - \phi\right)$$

$$-1 = \sin\left(\frac{\pi}{6} - \phi\right)$$

$$\frac{-\pi}{2} = \frac{\pi}{6} - \phi$$

$$\phi = \frac{2\pi}{3}$$

Thus, $y = 24.95\sin\left(\frac{\pi}{6}x - \frac{2\pi}{3}\right) + 50.45$

(c)

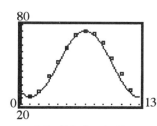

(e)

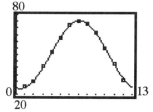

(d) $y = 25.693\sin(0.476x - 1.814) + 49.854$

83.   (a) $3.6333 + 12.5 = 16.1333$ hours which is at 4:08 p.m.

(b)   Amplitude : $A = \dfrac{8.2 - (-0.6)}{2} = \dfrac{8.8}{2} = 4.4$

Vertical Shift : $\dfrac{8.2 + (-0.6)}{2} = \dfrac{7.6}{2} = 3.8$

$\omega = \dfrac{2\pi}{12.5} = \dfrac{\pi}{6.25}$

Phase shift (use $y = -0.6$, $x = 10.1333$):

$$-0.6 = 4.4\sin\left(\frac{\pi}{6.25}\cdot 10.1333 - \phi\right) + 3.8$$

$$-4.4 = 4.4\sin\left(\frac{\pi}{6.25}\cdot 10.1333 - \phi\right)$$

$$-1 = \sin\left(\frac{10.1333\pi}{6.25} - \phi\right)$$

$$\frac{-\pi}{2} = \frac{10.1333\pi}{6.25} - \phi$$

$$\phi = 6.6643$$

Thus, $y = 4.4\sin\left(\dfrac{\pi}{6.25}x - 6.6643\right) + 3.8$

(c)

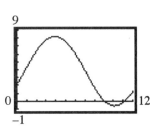

(d) $y = 4.4\sin\left(\frac{\pi}{6.25}(16.1333) - 6.6643\right) + 3.8 = 8.2$ feet

85. (a)   Amplitude : $A = \dfrac{12.75 - 10.583}{2} = \dfrac{2.167}{2} = 1.0835$

Vertical Shift : $\dfrac{12.75 + 10.583}{2} = \dfrac{23.333}{2} = 11.6665$

$\omega = \dfrac{2\pi}{365}$

Phase shift (use $y = 10.583, x = 356$):

$$10.583 = 1.0835\sin\left(\frac{2\pi}{365} \cdot 356 - \phi\right) + 11.6665$$

$$-1.0835 = 1.0835\sin\left(\frac{2\pi}{365} \cdot 356 - \phi\right)$$

$$-1 = \sin\left(\frac{712\pi}{365} - \phi\right)$$

$$\frac{-\pi}{2} = \frac{712\pi}{365} - \phi$$

$$\phi = 7.6991$$

Thus, $y = 1.0835\sin\left(\dfrac{2\pi}{365}x - 7.6991\right) + 11.6665$

(b)

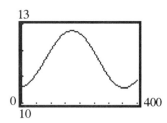

(c) $y = 1.0835\sin\left(\dfrac{2\pi}{365}(91) - 7.691\right) + 11.6665 = 11.83$ hours

87. (a)   Amplitude : $A = \dfrac{16.233 - 5.45}{2} = \dfrac{10.783}{2} = 5.3915$

Vertical Shift : $\dfrac{16.233 + 5.45}{2} = \dfrac{21.683}{2} = 10.8415$

$\omega = \dfrac{2\pi}{365}$

Phase shift (use $y = 5.45$, $x = 356$):

$$5.45 = 5.3915\sin\left(\dfrac{2\pi}{365} \cdot 356 - \phi\right) + 10.8415$$

$$-5.3915 = 5.3915\sin\left(\dfrac{2\pi}{365} \cdot 356 - \phi\right)$$

$$-1 = \sin\left(\dfrac{712\pi}{365} - \phi\right)$$

$$\dfrac{-\pi}{2} = \dfrac{712\pi}{365} - \phi$$

$$\phi = 7.6991$$

Thus, $y = 5.3915\sin\left(\dfrac{2\pi}{365}x - 7.6991\right) + 10.8415$

(b)

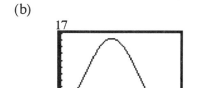

(d) $y = 5.3915\sin\left(\dfrac{2\pi}{365}(91) - 7.6991\right) + 10.8415 = 11.65$ hours

6 Chapter Review

1. $135° = 135 \cdot \dfrac{\pi}{180}$ radian $= \dfrac{3\pi}{4}$ radians

3. $18° = 18 \cdot \dfrac{\pi}{180}$ radian $= \dfrac{\pi}{10}$ radians

5. $\dfrac{3\pi}{4} = \dfrac{3\pi}{4} \cdot \dfrac{180}{\pi}$ degrees $= 135°$

7. $-\dfrac{5\pi}{2} = -\dfrac{5\pi}{2} \cdot \dfrac{180}{\pi}$ degrees $= -450°$

9. $\tan \dfrac{\pi}{4} - \sin \dfrac{\pi}{6} = 1 - \dfrac{1}{2} = \dfrac{1}{2}$

11. $3\sin 45° - 4\tan \dfrac{\pi}{6} = 3 \cdot \dfrac{\sqrt{2}}{2} - 4 \cdot \dfrac{\sqrt{3}}{3} = \dfrac{3\sqrt{2}}{2} - \dfrac{4\sqrt{3}}{3}$

13. $6\cos \dfrac{3\pi}{4} + 2\tan\left(-\dfrac{\pi}{3}\right) = 6\left(-\dfrac{\sqrt{2}}{2}\right) + 2\left(-\sqrt{3}\right) = -3\sqrt{2} - 2\sqrt{3}$

15. $\sec\left(-\dfrac{\pi}{3}\right) - \cot\left(-\dfrac{5\pi}{4}\right) = \sec \dfrac{\pi}{3} + \cot \dfrac{5\pi}{4} = 2 + 1 = 3$

17. $\tan \pi + \sin \pi = 0 + 0 = 0$

19. $\cos 180° - \tan(-45°) = -1 - (-1) = -1 + 1 = 0$

21. $\sin^2 20° + \dfrac{1}{\sec^2 20°} = \sin^2 20° + \cos^2 20° = 1$

23. $\sec 50°\cos 50° = \dfrac{1}{\cos 50°} \cdot \cos 50° = 1$

25. $\dfrac{\cos 400°}{\cos(-40°)} = \dfrac{\cos(40° + 360°)}{\cos 40°} = \dfrac{\cos 40°}{\cos 40°} = 1$

27. $\sin \theta = -\dfrac{4}{5}, \quad \cos \theta > 0, \quad \Rightarrow \quad \theta$ in quadrant IV
     Solve for $\cos \theta$:
     $$\sin^2 \theta + \cos^2 \theta = 1$$
     $$\cos^2 \theta = 1 - \sin^2 \theta$$
     $$\cos \theta = \pm\sqrt{1 - \sin^2 \theta}$$
     Since θ is in quadrant IV, $\cos \theta > 0$.
     $\cos \theta = \sqrt{1 - \sin^2 \theta} = \sqrt{1 - \left(-\dfrac{4}{5}\right)^2} = \sqrt{1 - \dfrac{16}{25}} = \sqrt{\dfrac{9}{25}} = \dfrac{3}{5}$
     $\tan \theta = \dfrac{\sin \theta}{\cos \theta} = \dfrac{-\frac{4}{5}}{\frac{3}{5}} = -\dfrac{4}{5} \cdot \dfrac{5}{3} = -\dfrac{4}{3}$ $\qquad$ $\sec \theta = \dfrac{1}{\cos \theta} = \dfrac{1}{\frac{3}{5}} = \dfrac{5}{3}$
     $\csc \theta = \dfrac{1}{\sin \theta} = \dfrac{1}{-\frac{4}{5}} = -\dfrac{5}{4}$ $\qquad\qquad$ $\cot \theta = \dfrac{1}{\tan \theta} = \dfrac{1}{-\frac{4}{3}} = -\dfrac{3}{4}$

29. $\tan \theta = \dfrac{12}{5}, \quad \sin \theta < 0, \quad \Rightarrow \quad \theta$ in quadrant III
     Solve for $\sec \theta$:
     $$\sec^2 \theta = 1 + \tan^2 \theta$$
     $$\sec \theta = \pm\sqrt{1 + \tan^2 \theta}$$

Since θ is in quadrant III, $\sec\theta < 0$.

$\sec\theta = -\sqrt{1 + \tan^2\theta} = -\sqrt{1 + \left(\frac{12}{5}\right)^2} = -\sqrt{1 + \frac{144}{25}} = -\sqrt{\frac{169}{25}} = -\frac{13}{5}$

$\cos\theta = -\frac{5}{13}$

$\sin\theta = -\sqrt{1 - \cos^2\theta} = -\sqrt{1 - \left(-\frac{5}{13}\right)^2} = -\sqrt{1 - \frac{25}{169}} = -\sqrt{\frac{144}{169}} = -\frac{12}{13}$

$\csc\theta = \dfrac{1}{\sin\theta} = \dfrac{1}{-\frac{12}{13}} = -\dfrac{13}{12}$ $\qquad\qquad$ $\cot\theta = \dfrac{1}{\tan\theta} = \dfrac{1}{\frac{12}{5}} = \dfrac{5}{12}$

31. $\sec\theta = -\frac{5}{4}$, $\quad \tan\theta < 0$, $\quad\Rightarrow\quad$ θ in quadrant II

Solve for $\cos\theta$:

$$\cos\theta = \frac{1}{\sec\theta} = \frac{1}{-\frac{5}{4}} = -\frac{4}{5}$$

Solve for $\sin\theta$:

$$\sin^2\theta + \cos^2\theta = 1$$
$$\sin^2\theta = 1 - \cos^2\theta$$
$$\sin\theta = \pm\sqrt{1 - \cos^2\theta}$$

Since θ is in quadrant II, $\sin\theta > 0$.

$\sin\theta = \sqrt{1 - \cos^2\theta} = \sqrt{1 - \left(-\frac{4}{5}\right)^2} = \sqrt{1 - \frac{16}{25}} = \sqrt{\frac{9}{25}} = \frac{3}{5}$

$\tan\theta = \dfrac{\sin\theta}{\cos\theta} = \dfrac{\frac{3}{5}}{-\frac{4}{5}} = \dfrac{3}{5}\cdot -\dfrac{5}{4} = -\dfrac{3}{4}$ $\qquad$ $\cot\theta = \dfrac{1}{\tan\theta} = \dfrac{1}{-\frac{3}{4}} = -\dfrac{4}{3}$

$\csc\theta = \dfrac{1}{\sin\theta} = \dfrac{1}{\frac{3}{5}} = \dfrac{5}{3}$

33. $\sin\theta = \frac{12}{13}$, $\quad\theta$ in quadrant II

Solve for $\cos\theta$:

$$\sin^2\theta + \cos^2\theta = 1$$
$$\cos^2\theta = 1 - \sin^2\theta$$
$$\cos\theta = \pm\sqrt{1 - \sin^2\theta}$$

Since θ is in quadrant II, $\cos\theta < 0$.

$\cos\theta = -\sqrt{1 - \sin^2\theta} = -\sqrt{1 - \left(\frac{12}{13}\right)^2} = -\sqrt{1 - \frac{144}{169}} = -\sqrt{\frac{25}{169}} = -\frac{5}{13}$

$\tan\theta = \dfrac{\sin\theta}{\cos\theta} = \dfrac{\frac{12}{13}}{-\frac{5}{13}} = \dfrac{12}{13}\cdot -\dfrac{13}{5} = -\dfrac{12}{5}$ $\qquad$ $\sec\theta = \dfrac{1}{\cos\theta} = \dfrac{1}{-\frac{5}{13}} = -\dfrac{13}{5}$

$\csc\theta = \dfrac{1}{\sin\theta} = \dfrac{1}{\frac{12}{13}} = \dfrac{13}{12}$ $\qquad\qquad$ $\cot\theta = \dfrac{1}{\tan\theta} = \dfrac{1}{-\frac{12}{5}} = -\dfrac{5}{12}$

35. $\sin\theta = -\frac{5}{13}$, $\frac{3\pi}{2} < \theta < 2\pi$, θ in quadrant IV

Solve for $\cos\theta$:

$$\sin^2\theta + \cos^2\theta = 1$$
$$\cos^2\theta = 1 - \sin^2\theta$$
$$\cos\theta = \pm\sqrt{1-\sin^2\theta}$$

Since θ is in quadrant IV, $\cos\theta > 0$.

$$\cos\theta = \sqrt{1-\sin^2\theta} = \sqrt{1-\left(-\frac{5}{13}\right)^2} = \sqrt{1-\frac{25}{169}} = \sqrt{\frac{144}{169}} = \frac{12}{13}$$

$$\tan\theta = \frac{\sin\theta}{\cos\theta} = \frac{-\frac{5}{13}}{\frac{12}{13}} = -\frac{5}{13}\cdot\frac{13}{12} = -\frac{5}{12} \qquad \sec\theta = \frac{1}{\cos\theta} = \frac{1}{\frac{12}{13}} = \frac{13}{12}$$

$$\csc\theta = \frac{1}{\sin\theta} = \frac{1}{-\frac{5}{13}} = -\frac{13}{5} \qquad \cot\theta = \frac{1}{\tan\theta} = \frac{1}{-\frac{5}{12}} = -\frac{12}{5}$$

37. $\tan\theta = \frac{1}{3}$, $180° < \theta < 270°$, $\Rightarrow$ θ in quadrant III

Solve for $\sec\theta$:

$$\sec^2\theta = 1 + \tan^2\theta$$
$$\sec\theta = \pm\sqrt{1+\tan^2\theta}$$

Since θ is in quadrant III, $\sec\theta < 0$.

$$\sec\theta = -\sqrt{1+\tan^2\theta} = -\sqrt{1+\left(\frac{1}{3}\right)^2} = -\sqrt{1+\frac{1}{9}} = -\sqrt{\frac{10}{9}} = -\frac{\sqrt{10}}{3}$$

$$\cos\theta = \frac{1}{\sec\theta} = \frac{1}{-\frac{\sqrt{10}}{3}} = -\frac{3}{\sqrt{10}}\cdot\frac{\sqrt{10}}{\sqrt{10}} = -\frac{3\sqrt{10}}{10}$$

$$\sin\theta = -\sqrt{1-\cos^2\theta} = -\sqrt{1-\left(-\frac{3\sqrt{10}}{10}\right)^2} = -\sqrt{1-\frac{90}{100}} = -\sqrt{\frac{10}{100}} = -\frac{\sqrt{10}}{10}$$

$$\csc\theta = \frac{1}{\sin\theta} = \frac{1}{-\frac{\sqrt{10}}{10}} = -\frac{10}{\sqrt{10}} = -\sqrt{10} \qquad \cot\theta = \frac{1}{\tan\theta} = \frac{1}{\frac{1}{3}} = 3$$

39. $\sec\theta = 3$, $\frac{3\pi}{2} < \theta < 2\pi$, $\Rightarrow$ θ in quadrant IV

Solve for $\cos\theta$:

$$\cos\theta = \frac{1}{\sec\theta} = \frac{1}{3}$$

Solve for $\sin\theta$:

$$\sin^2\theta + \cos^2\theta = 1 \;\rightarrow\; \sin^2\theta = 1 - \cos^2\theta \;\rightarrow\; \sin\theta = \pm\sqrt{1-\cos^2\theta}$$

Since θ is in quadrant IV, $\sin\theta < 0$.

$$\sin\theta = -\sqrt{1-\cos^2\theta} = -\sqrt{1-\left(\frac{1}{3}\right)^2} = -\sqrt{1-\frac{1}{9}} = -\sqrt{\frac{8}{9}} = -\frac{2\sqrt{2}}{3}$$

$$\tan\theta = \frac{\sin\theta}{\cos\theta} = \frac{-\frac{2\sqrt{2}}{3}}{\frac{1}{3}} = -\frac{2\sqrt{2}}{3}\cdot\frac{3}{1} = -2\sqrt{2} \qquad \cot\theta = \frac{1}{\tan\theta} = \frac{1}{-2\sqrt{2}} = -\frac{\sqrt{2}}{4}$$

$$\csc\theta = \frac{1}{\sin\theta} = \frac{1}{-\frac{2\sqrt{2}}{3}} = -\frac{3}{2\sqrt{2}} = -\frac{3\sqrt{2}}{4}$$

41. $\cot \theta = -2$, $\frac{\pi}{2} < \theta < \pi$, $\Rightarrow$ θ in quadrant II

$\tan \theta = \dfrac{1}{\cot \theta} = \dfrac{1}{-2} = -\dfrac{1}{2}$

Solve for $\sec \theta$:

$$\sec^2 \theta = 1 + \tan^2 \theta$$

$$\sec \theta = \pm \sqrt{1 + \tan^2 \theta}$$

Since θ is in quadrant II, $\sec \theta < 0$.

$\sec \theta = -\sqrt{1 + \tan^2 \theta} = -\sqrt{1 + \left(-\frac{1}{2}\right)^2} = -\sqrt{1 + \frac{1}{4}} = -\sqrt{\frac{5}{4}} = -\frac{\sqrt{5}}{2}$

$\cos \theta = \dfrac{1}{\sec \theta} = \dfrac{1}{-\frac{\sqrt{5}}{2}} = -\dfrac{2}{\sqrt{5}} \cdot \dfrac{\sqrt{5}}{\sqrt{5}} = -\dfrac{2\sqrt{5}}{5}$

$\sin \theta = \sqrt{1 - \cos^2 \theta} = \sqrt{1 - \left(-\frac{2\sqrt{5}}{5}\right)^2} = \sqrt{1 - \frac{20}{25}} = \sqrt{\frac{5}{25}} = \dfrac{\sqrt{5}}{5}$

$\csc \theta = \dfrac{1}{\sin \theta} = \dfrac{1}{\frac{\sqrt{5}}{5}} = \dfrac{5}{\sqrt{5}} = \sqrt{5}$

43. $y = 2\sin(4x)$

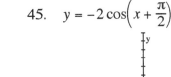

45. $y = -2\cos\left(x + \dfrac{\pi}{2}\right)$

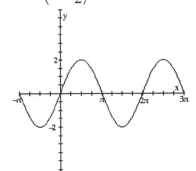

47. $y = \tan(x + \pi)$

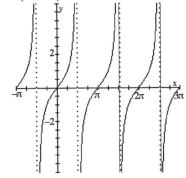

49. $y = -2\tan(3x)$

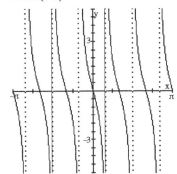

51. $y = \cot\left(x + \dfrac{\pi}{8}\right)$

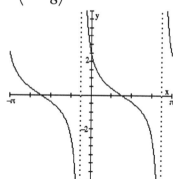

53. $y = \sec\left(x - \dfrac{\pi}{4}\right)$

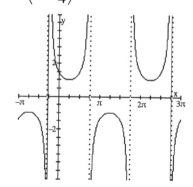

55. $y = 4\cos x$

Amplitude $= 4$
Period $= 2\pi$

57. $y = -8\sin\left(\dfrac{\pi}{2}x\right)$

Amplitude $= 8$
Period $= 4$

59. $y = 4\sin(3x)$

Amplitude: $|A| = |4| = 4$

Period: $T = \dfrac{2\pi}{\omega} = \dfrac{2\pi}{3}$

Phase Shift: $\dfrac{\phi}{\omega} = \dfrac{0}{3} = 0$

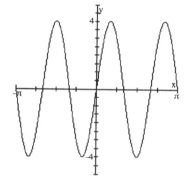

61. $y = -2\sin\left(\dfrac{\pi}{2}x + \dfrac{1}{2}\right)$

Amplitude: $|A| = |-2| = 2$

Period: $T = \dfrac{2\pi}{\omega} = \dfrac{2\pi}{\dfrac{\pi}{2}} = 4$

Phase Shift: $\dfrac{\phi}{\omega} = \dfrac{-\dfrac{1}{2}}{\dfrac{\pi}{2}} = -\dfrac{1}{\pi}$

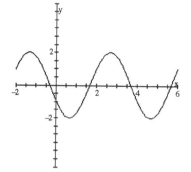

63. $y = \frac{1}{2}\sin\left(\frac{3}{2}x - \pi\right)$

Amplitude: $|A| = \left|\frac{1}{2}\right| = \frac{1}{2}$

Period: $T = \dfrac{2\pi}{\omega} = \dfrac{2\pi}{\frac{3}{2}} = \dfrac{4\pi}{3}$

Phase Shift: $\dfrac{\phi}{\omega} = \dfrac{\pi}{\frac{3}{2}} = \dfrac{2\pi}{3}$

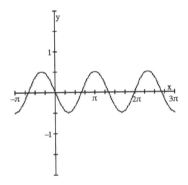

65. $y = -\frac{2}{3}\cos(\pi x - 6) + \frac{2}{3}$

Amplitude: $|A| = \left|-\frac{2}{3}\right| = \frac{2}{3}$

Period: $T = \dfrac{2\pi}{\omega} = \dfrac{2\pi}{\pi} = 2$

Phase Shift: $\dfrac{\phi}{\omega} = \dfrac{6}{\pi}$

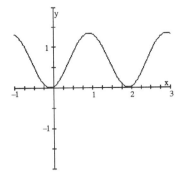

67. The graph is a cosine graph with an amplitude of 5 and a period of 8π. Find ω:

$$8\pi = \dfrac{2\pi}{\omega} \;\rightarrow\; 8\pi\omega = 2\pi \;\rightarrow\; \omega = \dfrac{2\pi}{8\pi} = \dfrac{1}{4}$$

The equation is: $y = 5\cos\left(\dfrac{1}{4}x\right)$.

69. The graph is a reflected cosine graph with an amplitude of 6 and a period of 8. Find ω:

$$8 = \dfrac{2\pi}{\omega} \;\rightarrow\; 8\omega = 2\pi \;\rightarrow\; \omega = \dfrac{2\pi}{8} = \dfrac{\pi}{4}$$

The equation is: $y = -6\cos\left(\dfrac{\pi}{4}x\right)$.

71. $r = 2$ feet, $\theta = 30°$ or $\theta = \dfrac{\pi}{6}$

$$s = r\theta = 2 \cdot \dfrac{\pi}{6} = \dfrac{\pi}{3} \text{ feet}$$

73. $v = 180$ mi / hr, $d = \dfrac{1}{2}$ mile, $r = \dfrac{1}{4}$ mile

$$\omega = \dfrac{v}{r} = \dfrac{180 \text{ mi / hr}}{\frac{1}{4} \text{ mi}} = 720 \text{ rad / hr} = \dfrac{720 \text{ rad}}{\text{hr}} \cdot \dfrac{1 \text{ rev}}{2\pi \text{ rad}} = \dfrac{360 \text{ rev}}{\pi \text{ hr}} \approx 114.6 \text{ rev / hr}$$

75. Since there are two lights on opposite sides and the light is seen every 5 seconds, the beacon makes 1 revolution every 10 seconds.

$$\omega = \frac{1 \text{ rev}}{10 \text{ sec}} \cdot \frac{2\pi}{1 \text{ rev}} = \frac{\pi}{5} \text{ radians / second}$$

77. $E(t) = 120\sin(120\pi t), \quad t \geq 0$

    (a) The maximum value of E is the amplitude which is 120.

    (b) Period $= \dfrac{2\pi}{120\pi} = \dfrac{1}{60}$

    (c) Graphing:

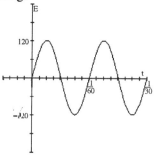

79. (a) Draw a scatter diagram:

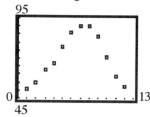

    (b) Amplitude: $A = \dfrac{90 - 51}{2} = \dfrac{39}{2} = 19.5$

    Vertical Shift: $\dfrac{90 + 51}{2} = \dfrac{141}{2} = 70.5$

    $\omega = \dfrac{2\pi}{12} = \dfrac{\pi}{6}$

    Phase shift (use $y = 51$, $x = 1$):

    $$51 = 19.5\sin\left(\frac{\pi}{6} \cdot 1 - \phi\right) + 70.5$$

    $$-19.5 = 19.5\sin\left(\frac{\pi}{6} - \phi\right)$$

    $$-1 = \sin\left(\frac{\pi}{6} - \phi\right)$$

    $$\frac{-\pi}{2} = \frac{\pi}{6} - \phi$$

    $$\phi = \frac{2\pi}{3}$$

    Thus, $y = 19.5\sin\left(\dfrac{\pi}{6}x - \dfrac{2\pi}{3}\right) + 70.5$

(c)

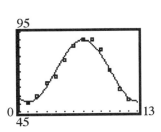

(e)

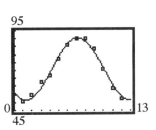

(d) $y = 19.518\sin(0.541x - 2.283) + 71.01$

81. (a)   Amplitude: $A = \dfrac{13.367 - 9.667}{2} = \dfrac{3.7}{2} = 1.85$

Vertical Shift: $\dfrac{13.367 + 9.667}{2} = \dfrac{23.034}{2} = 11.517$

$\omega = \dfrac{2\pi}{365}$

Phase shift (use $y = 9.667, x = 355$):

$$9.667 = 1.85\sin\left(\dfrac{2\pi}{365}\cdot 355 - \phi\right) + 11.517$$

$$-1.85 = 1.85\sin\left(\dfrac{2\pi}{365}\cdot 355 - \phi\right)$$

$$-1 = \sin\left(\dfrac{710\pi}{365} - \phi\right)$$

$$\dfrac{-\pi}{2} = \dfrac{710\pi}{365} - \phi$$

$$\phi = 7.6818$$

Thus, $y = 1.85\sin\left(\dfrac{2\pi}{365}x - 7.6818\right) + 11.517$

(b)

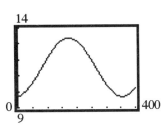

(c) $y = 1.85\sin\left(\dfrac{2\pi}{365}(91) - 7.6818\right) + 11.517 = 11.83$ hours

Chapter 7

Analytic Trigonometry

7.1 Trigonometric Identities

1. $\csc\theta\cdot\cos\theta = \dfrac{1}{\sin\theta}\cdot\cos\theta = \dfrac{\cos\theta}{\sin\theta} = \cot\theta$

3. $1+\tan^{2}(-\theta) = 1+(-\tan\theta)^{2} = 1+\tan^{2}\theta = \sec^{2}\theta$

5. $\cos\theta(\tan\theta+\cot\theta) = \cos\theta\left(\dfrac{\sin\theta}{\cos\theta}+\dfrac{\cos\theta}{\sin\theta}\right) = \cos\theta\left(\dfrac{\sin^{2}\theta+\cos^{2}\theta}{\cos\theta\sin\theta}\right) = \dfrac{1}{\sin\theta} = \csc\theta$

7. $\tan\theta\cot\theta - \cos^{2}\theta = \tan\theta\cdot\dfrac{1}{\tan\theta} - \cos^{2}\theta = 1-\cos^{2}\theta = \sin^{2}\theta$

9. $(\sec\theta-1)(\sec\theta+1) = \sec^{2}\theta-1 = \tan^{2}\theta$

11. $(\sec\theta+\tan\theta)(\sec\theta-\tan\theta) = \sec^{2}\theta-\tan^{2}\theta = 1$

13. $\cos^{2}\theta(1+\tan^{2}\theta) = \cos^{2}\theta\cdot\sec^{2}\theta = \cos^{2}\theta\cdot\dfrac{1}{\cos^{2}\theta} = 1$

15. $(\sin\theta+\cos\theta)^{2}+(\sin\theta-\cos\theta)^{2}$
$$= \sin^{2}\theta+2\sin\theta\cos\theta+\cos^{2}\theta+\sin^{2}\theta-2\sin\theta\cos\theta+\cos^{2}\theta$$
$$= 2\sin^{2}\theta+2\cos^{2}\theta = 2(\sin^{2}\theta+\cos^{2}\theta) = 2\cdot1 = 2$$

17. $\sec^{4}\theta-\sec^{2}\theta = \sec^{2}\theta(\sec^{2}\theta-1) = (\tan^{2}\theta+1)\tan^{2}\theta = \tan^{4}\theta+\tan^{2}\theta$

19. $\sec\theta-\tan\theta = \dfrac{1}{\cos\theta}-\dfrac{\sin\theta}{\cos\theta} = \dfrac{1-\sin\theta}{\cos\theta}\cdot\dfrac{1+\sin\theta}{1+\sin\theta} = \dfrac{1-\sin^{2}\theta}{\cos\theta(1+\sin\theta)}$
$$= \dfrac{\cos^{2}\theta}{\cos\theta(1+\sin\theta)} = \dfrac{\cos\theta}{1+\sin\theta}$$

21. $3\sin^{2}\theta+4\cos^{2}\theta = 3\sin^{2}\theta+3\cos^{2}\theta+\cos^{2}\theta = 3(\sin^{2}\theta+\cos^{2}\theta)+\cos^{2}\theta$
$$= 3\cdot1+\cos^{2}\theta = 3+\cos^{2}\theta$$

23. $1 - \dfrac{\cos^2 \theta}{1 + \sin \theta} = 1 - \dfrac{1 - \sin^2 \theta}{1 + \sin \theta} = 1 - \dfrac{(1 - \sin \theta)(1 + \sin \theta)}{1 + \sin \theta} = 1 - 1 + \sin \theta = \sin \theta$

25. $\dfrac{1 + \tan \theta}{1 - \tan \theta} = \dfrac{1 + \dfrac{1}{\cot \theta}}{1 - \dfrac{1}{\cot \theta}} = \dfrac{\dfrac{\cot \theta + 1}{\cot \theta}}{\dfrac{\cot \theta - 1}{\cot \theta}} = \dfrac{\cot \theta + 1}{\cot \theta} \cdot \dfrac{\cot \theta}{\cot \theta - 1} = \dfrac{\cot \theta + 1}{\cot \theta - 1}$

27. $\dfrac{\sec \theta}{\csc \theta} + \dfrac{\sin \theta}{\cos \theta} = \dfrac{\dfrac{1}{\cos \theta}}{\dfrac{1}{\sin \theta}} + \dfrac{\sin \theta}{\cos \theta} = \dfrac{\sin \theta}{\cos \theta} + \dfrac{\sin \theta}{\cos \theta} = \tan \theta + \tan \theta = 2 \tan \theta$

29. $\dfrac{1 + \sin \theta}{1 - \sin \theta} = \dfrac{1 + \dfrac{1}{\csc \theta}}{1 - \dfrac{1}{\csc \theta}} = \dfrac{\dfrac{\csc \theta + 1}{\csc \theta}}{\dfrac{\csc \theta - 1}{\csc \theta}} = \dfrac{\csc \theta + 1}{\csc \theta} \cdot \dfrac{\csc \theta}{\csc \theta - 1} = \dfrac{\csc \theta + 1}{\csc \theta - 1}$

31. $\dfrac{1 - \sin \theta}{\cos \theta} + \dfrac{\cos \theta}{1 - \sin \theta} = \dfrac{(1 - \sin \theta)^2 + \cos^2 \theta}{\cos \theta (1 - \sin \theta)} = \dfrac{1 - 2 \sin \theta + \sin^2 \theta + \cos^2 \theta}{\cos \theta (1 - \sin \theta)}$

$\qquad = \dfrac{1 - 2 \sin \theta + 1}{\cos \theta (1 - \sin \theta)} = \dfrac{2 - 2 \sin \theta}{\cos \theta (1 - \sin \theta)} = \dfrac{2(1 - \sin \theta)}{\cos \theta (1 - \sin \theta)} = \dfrac{2}{\cos \theta} = 2 \sec \theta$

33. $\dfrac{\sin \theta}{\sin \theta - \cos \theta} = \dfrac{\sin \theta}{\sin \theta - \cos \theta} \cdot \dfrac{\dfrac{1}{\sin \theta}}{\dfrac{1}{\sin \theta}} = \dfrac{1}{1 - \dfrac{\cos \theta}{\sin \theta}} = \dfrac{1}{1 - \cot \theta}$

35. $(\sec \theta - \tan \theta)^2 = \sec^2 \theta - 2 \sec \theta \tan \theta + \tan^2 \theta = \dfrac{1}{\cos^2 \theta} - 2 \cdot \dfrac{1}{\cos \theta} \cdot \dfrac{\sin \theta}{\cos \theta} + \dfrac{\sin^2 \theta}{\cos^2 \theta}$

$\qquad = \dfrac{1 - 2 \sin \theta + \sin^2 \theta}{\cos^2 \theta} = \dfrac{(1 - \sin \theta)(1 - \sin \theta)}{1 - \sin^2 \theta} = \dfrac{(1 - \sin \theta)(1 - \sin \theta)}{(1 - \sin \theta)(1 + \sin \theta)} = \dfrac{1 - \sin \theta}{1 + \sin \theta}$

37. $\dfrac{\cos \theta}{1 - \tan \theta} + \dfrac{\sin \theta}{1 - \cot \theta} = \dfrac{\cos \theta}{1 - \dfrac{\sin \theta}{\cos \theta}} + \dfrac{\sin \theta}{1 - \dfrac{\cos \theta}{\sin \theta}} = \dfrac{\cos \theta}{\dfrac{\cos \theta - \sin \theta}{\cos \theta}} + \dfrac{\sin \theta}{\dfrac{\sin \theta - \cos \theta}{\sin \theta}}$

$\qquad = \dfrac{\cos^2 \theta}{\cos \theta - \sin \theta} + \dfrac{\sin^2 \theta}{\sin \theta - \cos \theta} = \dfrac{\cos^2 \theta - \sin^2 \theta}{\cos \theta - \sin \theta}$

$\qquad = \dfrac{(\cos \theta - \sin \theta)(\cos \theta + \sin \theta)}{\cos \theta - \sin \theta} = \cos \theta + \sin \theta = \sin \theta + \cos \theta$

39. $\tan \theta + \dfrac{\cos \theta}{1 + \sin \theta} = \dfrac{\sin \theta}{\cos \theta} + \dfrac{\cos \theta}{1 + \sin \theta} = \dfrac{\sin \theta (1 + \sin \theta) + \cos^2 \theta}{\cos \theta (1 + \sin \theta)}$

$\qquad = \dfrac{\sin \theta + \sin^2 \theta + \cos^2 \theta}{\cos \theta (1 + \sin \theta)} = \dfrac{\sin \theta + 1}{\cos \theta (1 + \sin \theta)} = \dfrac{1}{\cos \theta} = \sec \theta$

41. $\dfrac{\tan\theta + \sec\theta - 1}{\tan\theta - \sec\theta + 1} = \dfrac{\tan\theta + (\scc\theta - 1)}{\tan\theta - (\sec\theta - 1)} \cdot \dfrac{\tan\theta + (\sec\theta - 1)}{\tan\theta + (\sec\theta - 1)}$

$$= \dfrac{\tan^2\theta + 2\tan\theta(\sec\theta - 1) + \sec^2\theta - 2\sec\theta + 1}{\tan^2\theta - (\sec^2\theta - 2\sec\theta + 1)}$$

$$= \dfrac{\sec^2\theta - 1 + 2\tan\theta(\sec\theta - 1) + \sec^2\theta - 2\sec\theta + 1}{\sec^2\theta - 1 - \sec^2\theta + 2\sec\theta - 1}$$

$$= \dfrac{2\sec^2\theta - 2\sec\theta + 2\tan\theta(\sec\theta - 1)}{2\sec\theta - 2}$$

$$= \dfrac{2\sec\theta(\sec\theta - 1) + 2\tan\theta(\sec\theta - 1)}{2\sec\theta - 2}$$

$$= \dfrac{2(\sec\theta - 1)(\sec\theta + \tan\theta)}{2(\sec\theta - 1)} = \sec\theta + \tan\theta = \tan\theta + \sec\theta$$

43. $\dfrac{\tan\theta - \cot\theta}{\tan\theta + \cot\theta} = \dfrac{\dfrac{\sin\theta}{\cos\theta} - \dfrac{\cos\theta}{\sin\theta}}{\dfrac{\sin\theta}{\cos\theta} + \dfrac{\cos\theta}{\sin\theta}} = \dfrac{\dfrac{\sin^2\theta - \cos^2\theta}{\cos\theta\sin\theta}}{\dfrac{\sin^2\theta + \cos^2\theta}{\cos\theta\sin\theta}} = \dfrac{\sin^2\theta - \cos^2\theta}{1} = \sin^2\theta - \cos^2\theta$

45. $\dfrac{\tan\theta - \cot\theta}{\tan\theta + \cot\theta} + 1 = \dfrac{\dfrac{\sin\theta}{\cos\theta} - \dfrac{\cos\theta}{\sin\theta}}{\dfrac{\sin\theta}{\cos\theta} + \dfrac{\cos\theta}{\sin\theta}} + 1 = \dfrac{\dfrac{\sin^2\theta - \cos^2\theta}{\cos\theta\sin\theta}}{\dfrac{\sin^2\theta + \cos^2\theta}{\cos\theta\sin\theta}} + 1 = \dfrac{\sin^2\theta - \cos^2\theta}{1} + 1$

$$= \sin^2\theta - \cos^2\theta + 1 = \sin^2\theta + (1 - \cos^2\theta) = \sin^2\theta + \sin^2\theta = 2\sin^2\theta$$

47. $\dfrac{\sec\theta + \tan\theta}{\cot\theta + \cos\theta} = \dfrac{\dfrac{1}{\cos\theta} + \dfrac{\sin\theta}{\cos\theta}}{\dfrac{\cos\theta}{\sin\theta} + \cos\theta} = \dfrac{\dfrac{1 + \sin\theta}{\cos\theta}}{\dfrac{\cos\theta + \cos\theta\sin\theta}{\sin\theta}} = \dfrac{1 + \sin\theta}{\cos\theta} \cdot \dfrac{\sin\theta}{\cos\theta(1 + \sin\theta)}$

$$= \dfrac{\sin\theta}{\cos\theta} \cdot \dfrac{1}{\cos\theta} = \tan\theta\sec\theta$$

49. $\dfrac{1 - \tan^2\theta}{1 + \tan^2\theta} + 1 = \dfrac{1 - \tan^2\theta + 1 + \tan^2\theta}{1 + \tan^2\theta} = \dfrac{2}{\sec^2\theta} = 2 \cdot \dfrac{1}{\sec^2\theta} = 2\cos^2\theta$

51. $\dfrac{\sec\theta - \csc\theta}{\sec\theta\csc\theta} = \dfrac{\dfrac{1}{\cos\theta} - \dfrac{1}{\sin\theta}}{\dfrac{1}{\cos\theta} \cdot \dfrac{1}{\sin\theta}} = \dfrac{\dfrac{\sin\theta - \cos\theta}{\cos\theta\sin\theta}}{\dfrac{1}{\cos\theta\sin\theta}} = \sin\theta - \cos\theta$

53. $\sec\theta - \cos\theta - \sin\theta\tan\theta = \dfrac{1}{\cos\theta} - \cos\theta - \sin\theta \cdot \dfrac{\sin\theta}{\cos\theta} = \dfrac{1 - \cos^2\theta - \sin^2\theta}{\cos\theta}$

$$= \dfrac{\sin^2\theta - \sin^2\theta}{\cos\theta} = 0$$

55. $\dfrac{1}{1 - \sin\theta} + \dfrac{1}{1 + \sin\theta} = \dfrac{1 + \sin\theta + 1 - \sin\theta}{(1 - \sin\theta)(1 + \sin\theta)} = \dfrac{2}{1 - \sin^2\theta} = \dfrac{2}{\cos^2\theta} = 2\sec^2\theta$

57. $\dfrac{\sec\theta}{1-\sin\theta} = \dfrac{\sec\theta}{1-\sin\theta}\cdot\dfrac{1+\sin\theta}{1+\sin\theta} = \dfrac{\sec\theta(1+\sin\theta)}{1-\sin^2\theta} = \dfrac{\sec\theta(1+\sin\theta)}{\cos^2\theta}$

$\qquad = \dfrac{1}{\cos\theta}\cdot\dfrac{1+\sin\theta}{\cos^2\theta} = \dfrac{1+\sin\theta}{\cos^3\theta}$

59. $\dfrac{(\sec\theta-\tan\theta)^2+1}{\csc\theta(\sec\theta-\tan\theta)} = \dfrac{\sec^2\theta-2\sec\theta\tan\theta+\tan^2\theta+1}{\csc\theta(\sec\theta-\tan\theta)} = \dfrac{2\sec^2\theta-2\sec\theta\tan\theta}{\csc\theta(\sec\theta-\tan\theta)}$

$\qquad = \dfrac{2\sec\theta(\sec\theta-\tan\theta)}{\csc\theta(\sec\theta-\tan\theta)} = \dfrac{2\sec\theta}{\csc\theta} = \dfrac{2\cdot\dfrac{1}{\cos\theta}}{\dfrac{1}{\sin\theta}} = 2\cdot\dfrac{1}{\cos\theta}\cdot\dfrac{\sin\theta}{1} = 2\tan\theta$

61. $\dfrac{\sin\theta+\cos\theta}{\cos\theta} - \dfrac{\sin\theta-\cos\theta}{\sin\theta} = \dfrac{\sin\theta}{\cos\theta}+\dfrac{\cos\theta}{\cos\theta}-\dfrac{\sin\theta}{\sin\theta}+\dfrac{\cos\theta}{\sin\theta} = \dfrac{\sin\theta}{\cos\theta}+1-1+\dfrac{\cos\theta}{\sin\theta}$

$\qquad = \dfrac{\sin^2\theta+\cos^2\theta}{\cos\theta\sin\theta} = \dfrac{1}{\cos\theta\sin\theta} = \sec\theta\csc\theta$

63. $\dfrac{\sin^3\theta+\cos^3\theta}{\sin\theta+\cos\theta} = \dfrac{(\sin\theta+\cos\theta)(\sin^2\theta-\sin\theta\cos\theta+\cos^2\theta)}{\sin\theta+\cos\theta} = 1-\sin\theta\cos\theta$

65. $\dfrac{\cos^2\theta-\sin^2\theta}{1-\tan^2\theta} = \dfrac{\cos^2\theta-\sin^2\theta}{1-\dfrac{\sin^2\theta}{\cos^2\theta}} = \dfrac{\cos^2\theta-\sin^2\theta}{\dfrac{\cos^2\theta-\sin^2\theta}{\cos^2\theta}} = \cos^2\theta$

67. $\dfrac{(2\cos^2\theta-1)^2}{\cos^4\theta-\sin^4\theta} = \dfrac{\left[2\cos^2\theta-(\sin^2\theta+\cos^2\theta)\right]^2}{(\cos^2\theta-\sin^2\theta)(\cos^2\theta+\sin^2\theta)}$

$\qquad = \dfrac{(\cos^2\theta-\sin^2\theta)^2}{(\cos^2\theta-\sin^2\theta)(\cos^2\theta+\sin^2\theta)} = \dfrac{\cos^2\theta-\sin^2\theta}{\cos^2\theta+\sin^2\theta}$

$\qquad = \cos^2\theta-\sin^2\theta = 1-\sin^2\theta-\sin^2\theta = 1-2\sin^2\theta$

69. $\dfrac{1+\sin\theta+\cos\theta}{1+\sin\theta-\cos\theta} = \dfrac{(1+\sin\theta)+\cos\theta}{(1+\sin\theta)-\cos\theta}\cdot\dfrac{(1+\sin\theta)+\cos\theta}{(1+\sin\theta)+\cos\theta}$

$\qquad = \dfrac{1+2\sin\theta+\sin^2\theta+2\cos\theta(1+\sin\theta)+\cos^2\theta}{1+2\sin\theta+\sin^2\theta-\cos^2\theta}$

$\qquad = \dfrac{1+2\sin\theta+\sin^2\theta+2\cos\theta(1+\sin\theta)+(1-\sin^2\theta)}{1+2\sin\theta+\sin^2\theta-(1-\sin^2\theta)}$

$\qquad = \dfrac{2+2\sin\theta+2\cos\theta(1+\sin\theta)}{2\sin\theta+2\sin^2\theta} = \dfrac{2(1+\sin\theta)+2\cos\theta(1+\sin\theta)}{2\sin\theta(1+\sin\theta)}$

$\qquad = \dfrac{2(1+\sin\theta)(1+\cos\theta)}{2\sin\theta(1+\sin\theta)} = \dfrac{1+\cos\theta}{\sin\theta}$

71. $(a\sin\theta+b\cos\theta)^2+(a\cos\theta-b\sin\theta)^2$

$\qquad = a^2\sin^2\theta+2ab\sin\theta\cos\theta+b^2\cos^2\theta+a^2\cos^2\theta-2ab\sin\theta\cos\theta+b^2\sin^2\theta$

$\qquad = a^2(\sin^2\theta+\cos^2\theta)+b^2(\sin^2\theta+\cos^2\theta) = a^2+b^2$

73. $\dfrac{\tan \alpha + \tan \beta}{\cot \alpha + \cot \beta} = \dfrac{\tan \alpha + \tan \beta}{\dfrac{1}{\tan \alpha} + \dfrac{1}{\tan \beta}} = \dfrac{\tan \alpha + \tan \beta}{\dfrac{\tan \beta + \tan \alpha}{\tan \alpha \tan \beta}}$

$$= (\tan \alpha + \tan \beta) \cdot \dfrac{\tan \alpha \tan \beta}{\tan \alpha + \tan \beta} = \tan \alpha \tan \beta$$

75. $(\sin \alpha + \cos \beta)^2 + (\cos \beta + \sin \alpha)(\cos \beta - \sin \alpha)$

$$= \sin^2 \alpha + 2\sin \alpha \cos \beta + \cos^2 \beta + \cos^2 \beta - \sin^2 \alpha$$
$$= 2\sin \alpha \cos \beta + 2\cos^2 \beta = 2\cos \beta(\sin \alpha + \cos \beta)$$

77. $\ln|\sec \theta| = \ln\left|\dfrac{1}{\cos \theta}\right| = \ln|\cos \theta|^{-1} = -\ln|\cos \theta|$

79. $\ln|1 + \cos \theta| + \ln|1 - \cos \theta| = \ln\big(|1 + \cos \theta| \cdot |1 - \cos\theta|\big) = \ln\big|1 - \cos^2 \theta\big|$

$$= \ln\big|\sin^2 \theta\big| = 2\ln|\sin \theta|$$

7.2 Sum and Difference Formulas

1. $\sin\left(\dfrac{5\pi}{12}\right) = \sin\left(\dfrac{3\pi}{12} + \dfrac{2\pi}{12}\right) = \sin\dfrac{\pi}{4}\cos\dfrac{\pi}{6} + \cos\dfrac{\pi}{4}\sin\dfrac{\pi}{6} = \dfrac{\sqrt{2}}{2}\cdot\dfrac{\sqrt{3}}{2} + \dfrac{\sqrt{2}}{2}\cdot\dfrac{1}{2}$

$$= \dfrac{1}{4}\left(\sqrt{6} + \sqrt{2}\right)$$

3. $\cos\dfrac{7\pi}{12} = \cos\left(\dfrac{4\pi}{12} + \dfrac{3\pi}{12}\right) = \cos\dfrac{\pi}{3}\cos\dfrac{\pi}{4} - \sin\dfrac{\pi}{3}\sin\dfrac{\pi}{4} = \dfrac{1}{2}\cdot\dfrac{\sqrt{2}}{2} - \dfrac{\sqrt{3}}{2}\cdot\dfrac{\sqrt{2}}{2}$

$$= \dfrac{1}{4}\left(\sqrt{2} - \sqrt{6}\right)$$

5. $\cos 165° = \cos(120° + 45°) = \cos 120°\cos 45° - \sin 120°\sin 45°$

$$= -\dfrac{1}{2}\cdot\dfrac{\sqrt{2}}{2} - \dfrac{\sqrt{3}}{2}\cdot\dfrac{\sqrt{2}}{2} = -\dfrac{1}{4}\left(\sqrt{2} + \sqrt{6}\right)$$

7. $\tan 15° = \tan(45° - 30°) = \dfrac{\tan 45° - \tan 30°}{1 + \tan 45°\tan 30°} = \dfrac{1 - \dfrac{\sqrt{3}}{3}}{1 + 1\cdot\dfrac{\sqrt{3}}{3}} = \dfrac{\dfrac{3 - \sqrt{3}}{3}}{\dfrac{3 + \sqrt{3}}{3}}$

$$= \dfrac{3 - \sqrt{3}}{3 + \sqrt{3}}\cdot\dfrac{3 - \sqrt{3}}{3 - \sqrt{3}} = \dfrac{9 - 6\sqrt{3} + 3}{9 - 3} = \dfrac{12 - 6\sqrt{3}}{6} = \dfrac{6\left(2 - \sqrt{3}\right)}{6} = 2 - \sqrt{3}$$

9. $\sin\dfrac{17\pi}{12} = \sin\left(\dfrac{15\pi}{12} + \dfrac{2\pi}{12}\right) = \sin\dfrac{5\pi}{4}\cos\dfrac{\pi}{6} + \cos\dfrac{5\pi}{4}\sin\dfrac{\pi}{6} = -\dfrac{\sqrt{2}}{2}\cdot\dfrac{\sqrt{3}}{2} + -\dfrac{\sqrt{2}}{2}\cdot\dfrac{1}{2}$

$= -\dfrac{1}{4}\left(\sqrt{6} + \sqrt{2}\right)$

11. $\sec\left(-\dfrac{\pi}{12}\right) = \dfrac{1}{\cos\left(-\dfrac{\pi}{12}\right)} = \dfrac{1}{\cos\left(\dfrac{3\pi}{12} - \dfrac{4\pi}{12}\right)} = \dfrac{1}{\cos\dfrac{\pi}{4}\cos\dfrac{\pi}{3} + \sin\dfrac{\pi}{4}\sin\dfrac{\pi}{3}}$

$= \dfrac{1}{\dfrac{\sqrt{2}}{2}\cdot\dfrac{1}{2} + \dfrac{\sqrt{2}}{2}\cdot\dfrac{\sqrt{3}}{2}} = \dfrac{1}{\dfrac{\sqrt{2}+\sqrt{6}}{4}} = \dfrac{4}{\sqrt{2}+\sqrt{6}}\cdot\dfrac{\sqrt{2}-\sqrt{6}}{\sqrt{2}-\sqrt{6}}$

$= \dfrac{4\left(\sqrt{2}-\sqrt{6}\right)}{2-6} = \sqrt{6} - \sqrt{2}$

13. $\sin 20°\cos 10° + \cos 20°\sin 10° = \sin(20° + 10°) = \sin 30° = \dfrac{1}{2}$

15. $\cos 70°\cos 20° - \sin 70°\sin 20° = \cos(70° + 20°) = \cos 90° = 0$

17. $\dfrac{\tan 20° + \tan 25°}{1 - \tan 20°\tan 25°} = \tan(20° + 25°) = \tan 45° = 1$

19. $\sin\dfrac{\pi}{12}\cos\dfrac{7\pi}{12} - \cos\dfrac{\pi}{12}\sin\dfrac{7\pi}{12} = \sin\left(\dfrac{\pi}{12} - \dfrac{7\pi}{12}\right) = \sin\left(-\dfrac{\pi}{2}\right) = -1$

21. $\cos\dfrac{\pi}{12}\cos\dfrac{5\pi}{12} + \sin\dfrac{\pi}{12}\sin\dfrac{5\pi}{12} = \cos\left(\dfrac{\pi}{12} - \dfrac{5\pi}{12}\right) = \cos\left(-\dfrac{\pi}{3}\right) = \cos\dfrac{\pi}{3} = \dfrac{1}{2}$

23. $\sin\alpha = \dfrac{3}{5},\ 0 < \alpha < \dfrac{\pi}{2}; \qquad \cos\beta = \dfrac{2\sqrt{5}}{5},\ -\dfrac{\pi}{2} < \beta < 0$

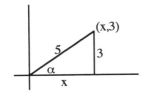
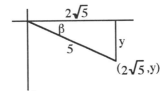

$x^2 + 3^2 = 5^2,\ x > 0$

$\qquad x^2 = 25 - 9 = 16,\ x > 0$

$\qquad x = 4$

$\qquad \cos\alpha = \dfrac{4}{5},\ \tan\alpha = \dfrac{3}{4}$

$\left(2\sqrt{5}\right)^2 + y^2 = 5^2,\ y < 0$

$\qquad y^2 = 25 - 20 = 5.\ y < 0$

$\qquad y = -\sqrt{5}$

$\qquad \sin\beta = -\dfrac{\sqrt{5}}{5},\ \tan\beta = \dfrac{-\sqrt{5}}{2\sqrt{5}} = -\dfrac{1}{2}$

(a) $\sin(\alpha + \beta) = \sin\alpha\cos\beta + \cos\alpha\sin\beta = \dfrac{3}{5}\cdot\dfrac{2\sqrt{5}}{5} + \dfrac{4}{5}\cdot -\dfrac{\sqrt{5}}{5} = \dfrac{6\sqrt{5} - 4\sqrt{5}}{25} = \dfrac{2\sqrt{5}}{25}$

(b) $\cos(\alpha + \beta) = \cos\alpha\cos\beta - \sin\alpha\sin\beta = \dfrac{4}{5}\cdot\dfrac{2\sqrt{5}}{5} - \dfrac{3}{5}\cdot-\dfrac{\sqrt{5}}{5} = \dfrac{8\sqrt{5}+3\sqrt{5}}{25} = \dfrac{11\sqrt{5}}{25}$

(c) $\sin(\alpha - \beta) = \sin\alpha\cos\beta - \cos\alpha\sin\beta = \dfrac{3}{5}\cdot\dfrac{2\sqrt{5}}{5} - \dfrac{4}{5}\cdot-\dfrac{\sqrt{5}}{5} = \dfrac{6\sqrt{5}+4\sqrt{5}}{25}$

$$= \dfrac{10\sqrt{5}}{25} = \dfrac{2\sqrt{5}}{5}$$

(d) $\tan(\alpha - \beta) = \dfrac{\tan\alpha - \tan\beta}{1 + \tan\alpha\tan\beta} = \dfrac{\dfrac{3}{4} - \dfrac{-1}{2}}{1 + \dfrac{3}{4}\cdot\dfrac{-1}{2}} = \dfrac{\dfrac{5}{4}}{\dfrac{5}{8}} = 2$

25. $\tan\alpha = -\dfrac{4}{3}, \dfrac{\pi}{2} < \alpha < \pi;$ $\cos\beta = \dfrac{1}{2}, 0 < \beta < \dfrac{\pi}{2}$

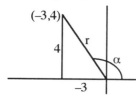

$r^2 = (-3)^2 + 4^2 = 25$

$r = 5$

$\sin\alpha = \dfrac{4}{5}, \cos\alpha = \dfrac{-3}{5}$

$1^2 + y^2 = 2^2, \ y > 0$

$y^2 = 4 - 1 = 3. \ y > 0$

$y = \sqrt{3}$

$\sin\beta = \dfrac{\sqrt{3}}{2}, \quad \tan\beta = \dfrac{\sqrt{3}}{1} = \sqrt{3}$

(a) $\sin(\alpha + \beta) = \sin\alpha\cos\beta + \cos\alpha\sin\beta = \dfrac{4}{5}\cdot\dfrac{1}{2} + \dfrac{-3}{5}\cdot\dfrac{\sqrt{3}}{2} = \dfrac{4 - 3\sqrt{3}}{10}$

(b) $\cos(\alpha + \beta) = \cos\alpha\cos\beta - \sin\alpha\sin\beta = \dfrac{-3}{5}\cdot\dfrac{1}{2} - \dfrac{4}{5}\cdot\dfrac{\sqrt{3}}{2} = \dfrac{-3 - 4\sqrt{3}}{10}$

(c) $\sin(\alpha - \beta) = \sin\alpha\cos\beta - \cos\alpha\sin\beta = \dfrac{4}{5}\cdot\dfrac{1}{2} - \dfrac{-3}{5}\cdot\dfrac{\sqrt{3}}{2} = \dfrac{4 + 3\sqrt{3}}{10}$

(d) $\tan(\alpha - \beta) = \dfrac{\tan\alpha - \tan\beta}{1 + \tan\alpha\tan\beta} = \dfrac{\dfrac{-4}{3} - \sqrt{3}}{1 + \dfrac{-4}{3}\cdot\sqrt{3}} = \dfrac{\dfrac{-4 - 3\sqrt{3}}{3}}{\dfrac{3 - 4\sqrt{3}}{3}} = \dfrac{-4 - 3\sqrt{3}}{3 - 4\sqrt{3}}\cdot\dfrac{3 + 4\sqrt{3}}{3 + 4\sqrt{3}}$

$$= \dfrac{-48 - 25\sqrt{3}}{-39} = \dfrac{48 + 25\sqrt{3}}{39}$$

27. $\sin\alpha = \dfrac{5}{13},\ -\dfrac{3\pi}{2} < \alpha < -\pi;\qquad \tan\beta = -\sqrt{3},\ \dfrac{\pi}{2} < \beta < \pi$

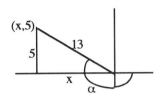

$x^2 + 5^2 = 13^2,\ x < 0$

$\qquad x^2 = 169 - 25 = 144,\ x < 0$

$\qquad x = -12$

$\cos\alpha = \dfrac{-12}{13},\ \tan\alpha = -\dfrac{5}{12}$

$r^2 = (-1)^2 + \sqrt{3}^{\,2} = 4$

$r = 2$

$\sin\beta = \dfrac{\sqrt{3}}{2},\quad \cos\beta = \dfrac{-1}{2}$

(a) $\sin(\alpha + \beta) = \sin\alpha\cos\beta + \cos\alpha\sin\beta = \dfrac{5}{13}\cdot\dfrac{-1}{2} + \dfrac{-12}{13}\cdot\dfrac{\sqrt{3}}{2} = \dfrac{-5 - 12\sqrt{3}}{26}$

(b) $\cos(\alpha + \beta) = \cos\alpha\cos\beta - \sin\alpha\sin\beta = \dfrac{-12}{13}\cdot\dfrac{-1}{2} - \dfrac{5}{13}\cdot\dfrac{\sqrt{3}}{2} = \dfrac{12 - 5\sqrt{3}}{26}$

(c) $\sin(\alpha - \beta) = \sin\alpha\cos\beta - \cos\alpha\sin\beta = \dfrac{5}{13}\cdot\dfrac{-1}{2} - \dfrac{-12}{13}\cdot\dfrac{\sqrt{3}}{2} = \dfrac{-5 + 12\sqrt{3}}{26}$

(d) $\tan(\alpha - \beta) = \dfrac{\tan\alpha - \tan\beta}{1 + \tan\alpha\tan\beta} = \dfrac{\dfrac{-5}{12} - \left(-\sqrt{3}\right)}{1 + \dfrac{-5}{12}\cdot\left(-\sqrt{3}\right)} = \dfrac{\dfrac{-5 + 12\sqrt{3}}{12}}{\dfrac{12 + 5\sqrt{3}}{12}}$

$\qquad\qquad = \dfrac{-5 + 12\sqrt{3}}{12 + 5\sqrt{3}}\cdot\dfrac{12 - 5\sqrt{3}}{12 - 5\sqrt{3}} = \dfrac{-240 + 169\sqrt{3}}{69}$

29. $\sin\theta = \dfrac{1}{3},\quad \theta$ in quadrant II

(a) $\cos\theta = -\sqrt{1 - \sin^2\theta} = -\sqrt{1 - \left(\dfrac{1}{3}\right)^2} = -\sqrt{1 - \dfrac{1}{9}} = -\sqrt{\dfrac{8}{9}} = -\dfrac{2\sqrt{2}}{3}$

(b) $\sin\left(\theta + \dfrac{\pi}{6}\right) = \sin\theta\cos\dfrac{\pi}{6} + \cos\theta\sin\dfrac{\pi}{6} = \dfrac{1}{3}\cdot\dfrac{\sqrt{3}}{2} + \dfrac{-2\sqrt{2}}{3}\cdot\dfrac{1}{2} = \dfrac{\sqrt{3} - 2\sqrt{2}}{6}$

(c) $\cos\left(\theta - \dfrac{\pi}{3}\right) = \cos\theta\cos\dfrac{\pi}{3} + \sin\theta\sin\dfrac{\pi}{3} = \dfrac{-2\sqrt{2}}{3}\cdot\dfrac{1}{2} + \dfrac{1}{3}\cdot\dfrac{\sqrt{3}}{2} = \dfrac{-2\sqrt{2} + \sqrt{3}}{6}$

(d) $\tan\left(\theta + \dfrac{\pi}{4}\right) = \dfrac{\tan\theta + \tan\dfrac{\pi}{4}}{1 - \tan\theta\tan\dfrac{\pi}{4}} = \dfrac{\dfrac{-1}{2\sqrt{2}} + 1}{1 - \dfrac{-1}{2\sqrt{2}}\cdot 1} = \dfrac{\dfrac{-1 + 2\sqrt{2}}{2\sqrt{2}}}{\dfrac{2\sqrt{2} + 1}{2\sqrt{2}}}$

$\qquad\qquad = \dfrac{2\sqrt{2} - 1}{2\sqrt{2} + 1}\cdot\dfrac{2\sqrt{2} - 1}{2\sqrt{2} - 1} = \dfrac{9 - 4\sqrt{2}}{7}$

31. $\sin\left(\dfrac{\pi}{2} + \theta\right) = \sin\dfrac{\pi}{2}\cos\theta + \cos\dfrac{\pi}{2}\sin\theta = 1\cdot\cos\theta + 0\cdot\sin\theta = \cos\theta$

33. $\sin(\pi - \theta) = \sin\pi\cos\theta - \cos\pi\sin\theta = 0\cdot\cos\theta - (-1)\sin\theta = \sin\theta$

35. $\sin(\pi + \theta) = \sin\pi\cos\theta + \cos\pi\sin\theta = 0\cdot\cos\theta + (-1)\sin\theta = -\sin\theta$

37. $\tan(\pi - \theta) = \dfrac{\tan\pi - \tan\theta}{1 + \tan\pi\tan\theta} = \dfrac{0 - \tan\theta}{1 + 0\cdot\tan\theta} = \dfrac{-\tan\theta}{1} = -\tan\theta$

39. $\sin\left(\dfrac{3\pi}{2} + \theta\right) = \sin\dfrac{3\pi}{2}\cos\theta + \cos\dfrac{3\pi}{2}\sin\theta = -1\cdot\cos\theta + 0\cdot\sin\theta = -\cos\theta$

41. $\sin(\alpha + \beta) + \sin(\alpha - \beta) = \sin\alpha\cos\beta + \cos\alpha\sin\beta + \sin\alpha\cos\beta - \cos\alpha\sin\beta$
$$= 2\sin\alpha\cos\beta$$

43. $\dfrac{\sin(\alpha + \beta)}{\sin\alpha\cos\beta} = \dfrac{\sin\alpha\cos\beta + \cos\alpha\sin\beta}{\sin\alpha\cos\beta} = \dfrac{\sin\alpha\cos\beta}{\sin\alpha\cos\beta} + \dfrac{\cos\alpha\sin\beta}{\sin\alpha\cos\beta} = 1 + \cot\alpha\tan\beta$

45. $\dfrac{\cos(\alpha + \beta)}{\cos\alpha\cos\beta} = \dfrac{\cos\alpha\cos\beta - \sin\alpha\sin\beta}{\cos\alpha\cos\beta} = \dfrac{\cos\alpha\cos\beta}{\cos\alpha\cos\beta} - \dfrac{\sin\alpha\sin\beta}{\cos\alpha\cos\beta} = 1 - \tan\alpha\tan\beta$

47. $\dfrac{\sin(\alpha + \beta)}{\sin(\alpha - \beta)} = \dfrac{\sin\alpha\cos\beta + \cos\alpha\sin\beta}{\sin\alpha\cos\beta - \cos\alpha\sin\beta} = \dfrac{\dfrac{\sin\alpha\cos\beta}{\cos\alpha\cos\beta} + \dfrac{\cos\alpha\sin\beta}{\cos\alpha\cos\beta}}{\dfrac{\sin\alpha\cos\beta}{\cos\alpha\cos\beta} - \dfrac{\cos\alpha\sin\beta}{\cos\alpha\cos\beta}} = \dfrac{\tan\alpha + \tan\beta}{\tan\alpha - \tan\beta}$

49. $\cot(\alpha + \beta) = \dfrac{\cos(\alpha + \beta)}{\sin(\alpha + \beta)} = \dfrac{\cos\alpha\cos\beta - \sin\alpha\sin\beta}{\sin\alpha\cos\beta + \cos\alpha\sin\beta}$
$$= \dfrac{\dfrac{\cos\alpha\cos\beta}{\sin\alpha\sin\beta} - \dfrac{\sin\alpha\sin\beta}{\sin\alpha\sin\beta}}{\dfrac{\sin\alpha\cos\beta}{\sin\alpha\sin\beta} + \dfrac{\cos\alpha\sin\beta}{\sin\alpha\sin\beta}} = \dfrac{\cot\alpha\cot\beta - 1}{\cot\beta + \cot\alpha}$$

51. $\sec(\alpha + \beta) = \dfrac{1}{\cos(\alpha + \beta)} = \dfrac{1}{\cos\alpha\cos\beta - \sin\alpha\sin\beta}$
$$= \dfrac{\dfrac{1}{\sin\alpha\sin\beta}}{\dfrac{\cos\alpha\cos\beta}{\sin\alpha\sin\beta} - \dfrac{\sin\alpha\sin\beta}{\sin\alpha\sin\beta}} = \dfrac{\csc\alpha\csc\beta}{\cot\alpha\cot\beta - 1}$$

53. $\sin(\alpha - \beta)\sin(\alpha + \beta) = \left(\sin\alpha\cos\beta - \cos\alpha\sin\beta\right)\left(\sin\alpha\cos\beta + \cos\alpha\sin\beta\right)$
$$= \sin^2\alpha\cos^2\beta - \cos^2\alpha\sin^2\beta = \sin^2\alpha(1 - \sin^2\beta) - (1 - \sin^2\alpha)\sin^2\beta$$
$$\sin^2\alpha - \sin^2\alpha\sin^2\beta - \sin^2\beta + \sin^2\alpha\sin^2\beta = \sin^2\alpha - \sin^2\beta$$

55. $\sin(\theta + k\pi) = \sin\theta\cos k\pi + \cos\theta\sin k\pi = \sin\theta(-1)^k + \cos\theta\cdot 0$
$$= (-1)^k\sin\theta,\ k \text{ any integer}$$

57. $\dfrac{\sin(x+h)-\sin x}{h} = \dfrac{\sin x \cos h + \cos x \sin h - \sin x}{h} = \dfrac{\cos x \sin h - \sin x + \sin x \cos h}{h}$

$$= \cos x \cdot \frac{\sin h}{h} - \sin x \cdot \frac{1-\cos h}{h}$$

59. $\tan\left(\dfrac{\pi}{2}-\theta\right) = \dfrac{\tan\dfrac{\pi}{2}-\tan\theta}{1+\tan\dfrac{\pi}{2}\tan\theta}$ This is impossible because $\tan\dfrac{\pi}{2}$ is undefined.

$$\tan\left(\frac{\pi}{2}-\theta\right) = \frac{\sin\left(\dfrac{\pi}{2}-\theta\right)}{\cos\left(\dfrac{\pi}{2}-\theta\right)} = \frac{\cos\theta}{\sin\theta} = \cot\theta$$

61. $\tan\theta = \tan(\theta_2 - \theta_1) = \dfrac{\tan\theta_2 - \tan\theta_1}{1+\tan\theta_2 \tan\theta_1} = \dfrac{m_2 - m_1}{1+m_2 m_1}$

7.3 Double-Angle and Half-Angle Formulas

1. $\sin\theta = \dfrac{3}{5}, \quad 0 < \theta < \dfrac{\pi}{2}; \qquad$ thus, $\ 0 < \dfrac{\theta}{2} < \dfrac{\pi}{4} \ $ or $\ \dfrac{\theta}{2} \ $ is in quadrant I.

$y = 3, \ r = 5$

$x^2 + 3^2 = 5^2, \ x > 0 \ \rightarrow \ x^2 = 25 - 9 = 16, \ x > 0 \ \rightarrow \ x = 4$

$\cos\theta = \dfrac{4}{5}$

(a) $\sin(2\theta) = 2\sin\theta\cos\theta = 2\cdot\dfrac{3}{5}\cdot\dfrac{4}{5} = \dfrac{24}{25}$

(b) $\cos(2\theta) = \cos^2\theta - \sin^2\theta = \left(\dfrac{4}{5}\right)^2 - \left(\dfrac{3}{5}\right)^2 = \dfrac{16}{25} - \dfrac{9}{25} = \dfrac{7}{25}$

(c) $\sin\dfrac{\theta}{2} = \sqrt{\dfrac{1-\cos\theta}{2}} = \sqrt{\dfrac{1-\frac{4}{5}}{2}} = \sqrt{\dfrac{\frac{1}{5}}{2}} = \sqrt{\dfrac{1}{10}} = \dfrac{1}{\sqrt{10}} = \dfrac{\sqrt{10}}{10}$

(d) $\cos\dfrac{\theta}{2} = \sqrt{\dfrac{1+\cos\theta}{2}} = \sqrt{\dfrac{1+\frac{4}{5}}{2}} = \sqrt{\dfrac{\frac{9}{5}}{2}} = \sqrt{\dfrac{9}{10}} = \dfrac{3}{\sqrt{10}} = \dfrac{3\sqrt{10}}{10}$

3. $\tan\theta = \dfrac{4}{3}, \quad \pi < \theta < \dfrac{3\pi}{2}; \qquad$ thus, $\ \dfrac{\pi}{2} < \dfrac{\theta}{2} < \dfrac{3\pi}{4} \ $ or $\ \dfrac{\theta}{2} \ $ is in quadrant II.

$x = -3, \ y = -4$

$r^2 = (-3)^2 + (-4)^2 = 9 + 16 = 25 \ \rightarrow \ r = 5$

$\sin\theta = -\dfrac{4}{5}, \quad \cos\theta = -\dfrac{3}{5}$

(a) $\sin(2\theta) = 2\sin\theta\cos\theta = 2\cdot\dfrac{-4}{5}\cdot\dfrac{-3}{5} = \dfrac{24}{25}$

(b) $\cos(2\theta) = \cos^2\theta - \sin^2\theta = \left(\dfrac{-3}{5}\right)^2 - \left(\dfrac{-4}{5}\right)^2 = \dfrac{9}{25} - \dfrac{16}{25} = -\dfrac{7}{25}$

(c) $\sin\dfrac{\theta}{2} = \sqrt{\dfrac{1-\cos\theta}{2}} = \sqrt{\dfrac{1-\frac{-3}{5}}{2}} = \sqrt{\dfrac{\frac{8}{5}}{2}} = \sqrt{\dfrac{4}{5}} = \dfrac{2}{\sqrt{5}} = \dfrac{2\sqrt{5}}{5}$

(d) $\cos\dfrac{\theta}{2} = -\sqrt{\dfrac{1+\cos\theta}{2}} = -\sqrt{\dfrac{1+\frac{-3}{5}}{2}} = -\sqrt{\dfrac{\frac{2}{5}}{2}} = -\sqrt{\dfrac{1}{5}} = -\dfrac{1}{\sqrt{5}} = -\dfrac{\sqrt{5}}{5}$

5. $\cos\theta = \dfrac{-\sqrt{6}}{3}$, $\dfrac{\pi}{2} < \theta < \pi$; thus, $\dfrac{\pi}{4} < \dfrac{\theta}{2} < \dfrac{\pi}{2}$ or $\dfrac{\theta}{2}$ is in quadrant I.

$x = -\sqrt{6}$, $r = 3$

$(-\sqrt{6})^2 + y^2 = 3^2 \;\rightarrow\; y^2 = 9 - 6 = 3 \;\rightarrow\; y = \sqrt{3}$

$\sin\theta = \dfrac{\sqrt{3}}{3}$

(a) $\sin(2\theta) = 2\sin\theta\cos\theta = 2\cdot\dfrac{\sqrt{3}}{3}\cdot\dfrac{-\sqrt{6}}{3} = \dfrac{-2\sqrt{18}}{9} = \dfrac{-6\sqrt{2}}{9} = \dfrac{-2\sqrt{2}}{3}$

(b) $\cos(2\theta) = \cos^2\theta - \sin^2\theta = \left(\dfrac{-\sqrt{6}}{3}\right)^2 - \left(\dfrac{\sqrt{3}}{3}\right)^2 = \dfrac{6}{9} - \dfrac{3}{9} = \dfrac{3}{9} = \dfrac{1}{3}$

(c) $\sin\dfrac{\theta}{2} = \sqrt{\dfrac{1-\cos\theta}{2}} = \sqrt{\dfrac{1-\frac{-\sqrt{6}}{3}}{2}} = \sqrt{\dfrac{\frac{3+\sqrt{6}}{3}}{2}} = \sqrt{\dfrac{3+\sqrt{6}}{6}}$

(d) $\cos\dfrac{\theta}{2} = \sqrt{\dfrac{1+\cos\theta}{2}} = \sqrt{\dfrac{1+\frac{-\sqrt{6}}{3}}{2}} = \sqrt{\dfrac{\frac{3-\sqrt{6}}{3}}{2}} = \sqrt{\dfrac{3-\sqrt{6}}{6}}$

7. $\sec\theta = 3$, $\sin\theta > 0$; $0 < \theta < \dfrac{\pi}{2}$; thus, $0 < \dfrac{\theta}{2} < \dfrac{\pi}{4}$ or $\dfrac{\theta}{2}$ is in quadrant I.

$x = 1$, $r = 3$

$1^2 + y^2 = 3^2 \;\rightarrow\; y^2 = 9 - 1 = 8 \;\rightarrow\; y = 2\sqrt{2}$

$\sin\theta = \dfrac{2\sqrt{2}}{3}$, $\cos\theta = \dfrac{1}{3}$

(a) $\sin(2\theta) = 2\sin\theta\cos\theta = 2\cdot\dfrac{2\sqrt{2}}{3}\cdot\dfrac{1}{3} = \dfrac{4\sqrt{2}}{9}$

(b) $\cos(2\theta) = \cos^2\theta - \sin^2\theta = \left(\dfrac{1}{3}\right)^2 - \left(\dfrac{2\sqrt{2}}{3}\right)^2 = \dfrac{1}{9} - \dfrac{8}{9} = \dfrac{-7}{9}$

(c) $\sin\dfrac{\theta}{2} = \sqrt{\dfrac{1-\cos\theta}{2}} = \sqrt{\dfrac{1-\frac{1}{3}}{2}} = \sqrt{\dfrac{\frac{2}{3}}{2}} = \sqrt{\dfrac{1}{3}} = \dfrac{1}{\sqrt{3}} = \dfrac{\sqrt{3}}{3}$

(d) $\cos\dfrac{\theta}{2} = \sqrt{\dfrac{1+\cos\theta}{2}} = \sqrt{\dfrac{1+\frac{1}{3}}{2}} = \sqrt{\dfrac{\frac{4}{3}}{2}} = \sqrt{\dfrac{2}{3}} = \dfrac{\sqrt{6}}{3}$

9. $\cot\theta = -2$, $\sec\theta < 0$; $\dfrac{\pi}{2} < \theta < \pi$;   thus, $\dfrac{\pi}{4} < \dfrac{\theta}{2} < \dfrac{\pi}{2}$ or $\dfrac{\theta}{2}$ is in quadrant I.

$x = -2$, $y = 1$

$r^2 = (-2)^2 + 1^2 = 4 + 1 = 5$ $\rightarrow$ $r = \sqrt{5}$

$\sin\theta = \dfrac{1}{\sqrt{5}} = \dfrac{\sqrt{5}}{5}$, $\cos\theta = -\dfrac{2}{\sqrt{5}} = \dfrac{-2\sqrt{5}}{5}$

(a) $\sin(2\theta) = 2\sin\theta\cos\theta = 2\cdot\dfrac{\sqrt{5}}{5}\cdot\dfrac{-2\sqrt{5}}{5} = \dfrac{-20}{25} = \dfrac{-4}{5}$

(b) $\cos(2\theta) = \cos^2\theta - \sin^2\theta = \left(\dfrac{-2\sqrt{5}}{5}\right)^2 - \left(\dfrac{\sqrt{5}}{5}\right)^2 = \dfrac{20}{25} - \dfrac{5}{25} = \dfrac{15}{25} = \dfrac{3}{5}$

(c) $\sin\dfrac{\theta}{2} = \sqrt{\dfrac{1-\cos\theta}{2}} = \sqrt{\dfrac{1-\dfrac{-2\sqrt{5}}{5}}{2}} = \sqrt{\dfrac{\dfrac{5+2\sqrt{5}}{5}}{2}} = \sqrt{\dfrac{5+2\sqrt{5}}{10}}$

(d) $\cos\dfrac{\theta}{2} = \sqrt{\dfrac{1+\cos\theta}{2}} = \sqrt{\dfrac{1+\dfrac{-2\sqrt{5}}{5}}{2}} = \sqrt{\dfrac{\dfrac{5-2\sqrt{5}}{5}}{2}} = \sqrt{\dfrac{5-2\sqrt{5}}{10}}$

11. $\tan\theta = -3$, $\sin\theta < 0$; $\dfrac{3\pi}{2} < \theta < 2\pi$;   thus, $\dfrac{3\pi}{4} < \dfrac{\theta}{2} < \pi$ or $\dfrac{\theta}{2}$ is in quadrant II.

$x = 1$, $y = -3$

$r^2 = 1^2 + (-3)^2 = 1 + 9 = 10$ $\rightarrow$ $r = \sqrt{10}$

$\sin\theta = \dfrac{-3}{\sqrt{10}} = \dfrac{-3\sqrt{10}}{10}$, $\cos\theta = \dfrac{1}{\sqrt{10}} = \dfrac{\sqrt{10}}{10}$

(a) $\sin(2\theta) = 2\sin\theta\cos\theta = 2\cdot\dfrac{-3\sqrt{10}}{10}\cdot\dfrac{\sqrt{10}}{10} = \dfrac{-60}{100} = \dfrac{-3}{5}$

(b) $\cos(2\theta) = \cos^2\theta - \sin^2\theta = \left(\dfrac{\sqrt{10}}{10}\right)^2 - \left(\dfrac{-3\sqrt{10}}{10}\right)^2 = \dfrac{10}{100} - \dfrac{90}{100} = \dfrac{-80}{100} = \dfrac{-4}{5}$

(c) $\sin\dfrac{\theta}{2} = \sqrt{\dfrac{1-\cos\theta}{2}} = \sqrt{\dfrac{1-\dfrac{\sqrt{10}}{10}}{2}} = \sqrt{\dfrac{\dfrac{10-\sqrt{10}}{10}}{2}} = \sqrt{\dfrac{10-\sqrt{10}}{20}} = \dfrac{1}{2}\sqrt{\dfrac{10-\sqrt{10}}{5}}$

(d) $\cos\dfrac{\theta}{2} = -\sqrt{\dfrac{1+\cos\theta}{2}} = -\sqrt{\dfrac{1+\dfrac{\sqrt{10}}{10}}{2}} = -\sqrt{\dfrac{\dfrac{10+\sqrt{10}}{10}}{2}} = -\sqrt{\dfrac{10+\sqrt{10}}{20}}$

$= -\dfrac{1}{2}\sqrt{\dfrac{10+\sqrt{10}}{5}}$

13. $\sin 22.5° = \sin\dfrac{45°}{2} = \sqrt{\dfrac{1-\cos 45°}{2}} = \sqrt{\dfrac{1-\dfrac{\sqrt{2}}{2}}{2}} = \sqrt{\dfrac{2-\sqrt{2}}{4}} = \dfrac{\sqrt{2-\sqrt{2}}}{2}$

15. $\tan\dfrac{7\pi}{8} = \tan\dfrac{\frac{7\pi}{4}}{2} = -\sqrt{\dfrac{1 - \cos\frac{7\pi}{4}}{1 + \cos\frac{7\pi}{4}}} = -\sqrt{\dfrac{1 - \frac{\sqrt{2}}{2}}{1 + \frac{\sqrt{2}}{2}}} = -\sqrt{\dfrac{2 - \sqrt{2}}{2 + \sqrt{2}} \cdot \dfrac{2 - \sqrt{2}}{2 - \sqrt{2}}}$

$$= -\sqrt{\dfrac{6 - 4\sqrt{2}}{2}} = -\sqrt{3 - 2\sqrt{2}}$$

17. $\cos 165° = \cos\dfrac{330°}{2} = -\sqrt{\dfrac{1 + \cos 330°}{2}} = -\sqrt{\dfrac{1 + \frac{\sqrt{3}}{2}}{2}} = -\sqrt{\dfrac{2 + \sqrt{3}}{4}} = -\dfrac{\sqrt{2 + \sqrt{3}}}{2}$

19. $\sec\dfrac{15\pi}{8} = \dfrac{1}{\cos\frac{15\pi}{8}} = \dfrac{1}{\cos\frac{\frac{15\pi}{4}}{2}} = \dfrac{1}{\sqrt{\dfrac{1 + \cos\frac{15\pi}{4}}{2}}} = \dfrac{1}{\sqrt{\dfrac{1 + \frac{\sqrt{2}}{2}}{2}}} = \dfrac{1}{\sqrt{\dfrac{2 + \sqrt{2}}{4}}}$

$$= \dfrac{2}{\sqrt{2 + \sqrt{2}}} \cdot \dfrac{\sqrt{2 + \sqrt{2}}}{\sqrt{2 + \sqrt{2}}} = \dfrac{2\sqrt{2 + \sqrt{2}}}{2 + \sqrt{2}} \cdot \dfrac{2 - \sqrt{2}}{2 - \sqrt{2}}$$

$$= \dfrac{2\left(2 - \sqrt{2}\right)\sqrt{2 + \sqrt{2}}}{2} = \left(2 - \sqrt{2}\right)\sqrt{2 + \sqrt{2}}$$

21. $\sin\left(\dfrac{-\pi}{8}\right) = \sin\left(\dfrac{\frac{-\pi}{4}}{2}\right) = -\sqrt{\dfrac{1 - \cos\left(\frac{-\pi}{4}\right)}{2}} = -\sqrt{\dfrac{1 - \frac{\sqrt{2}}{2}}{2}} = -\sqrt{\dfrac{2 - \sqrt{2}}{4}} = -\dfrac{\sqrt{2 - \sqrt{2}}}{2}$

23. $\sin^4\theta = \left(\sin^2\theta\right)^2 = \left(\dfrac{1 - \cos 2\theta}{2}\right)^2 = \dfrac{1}{4}\left(1 - 2\cos 2\theta + \cos^2 2\theta\right)$

$$= \dfrac{1}{4} - \dfrac{1}{2}\cos 2\theta + \dfrac{1}{4}\cos^2 2\theta = \dfrac{1}{4} - \dfrac{1}{2}\cos 2\theta + \dfrac{1}{4}\left(\dfrac{1 + \cos 4\theta}{2}\right)$$

$$= \dfrac{1}{4} - \dfrac{1}{2}\cos 2\theta + \dfrac{1}{8} + \dfrac{1}{8}\cos 4\theta = \dfrac{3}{8} - \dfrac{1}{2}\cos 2\theta + \dfrac{1}{8}\cos 4\theta$$

25. $\sin 4\theta = \sin 2(2\theta) = 2\sin 2\theta\cos 2\theta = 2(2\sin\theta\cos\theta)\left(1 - 2\sin^2\theta\right)$

$$= \cos\theta\left(4\sin\theta - 8\sin^3\theta\right)$$

27. Use the result of problem 25 to help solve the problem:

$$\sin 5\theta = \sin(4\theta + \theta) = \sin 4\theta \cos\theta + \cos 4\theta \sin\theta$$

$$= \cos\theta\left(4\sin\theta - 8\sin^3\theta\right)\cos\theta + \cos 2(2\theta)\sin\theta$$

$$= \cos^2\theta\left(4\sin\theta - 8\sin^3\theta\right) + \left(1 - 2\sin^2 2\theta\right)\sin\theta$$

$$= \left(1 - \sin^2\theta\right)\left(4\sin\theta - 8\sin^3\theta\right) + \sin\theta\left(1 - 2(2\sin\theta\cos\theta)^2\right)$$

$$= 4\sin\theta - 12\sin^3\theta + 8\sin^5\theta + \sin\theta\left(1 - 8\sin^2\theta\cos^2\theta\right)$$

$$= 4\sin\theta - 12\sin^3\theta + 8\sin^5\theta + \sin\theta - 8\sin^3\theta\left(1 - \sin^2\theta\right)$$

$$= 5\sin\theta - 12\sin^3\theta + 8\sin^5\theta - 8\sin^3\theta + 8\sin^5\theta$$

$$= 5\sin\theta - 20\sin^3\theta + 16\sin^5\theta$$

29. $\cos^4\theta - \sin^4\theta = \left(\cos^2\theta + \sin^2\theta\right)\left(\cos^2\theta - \sin^2\theta\right) = 1 \cdot \cos 2\theta = \cos 2\theta$

31. $\cot(2\theta) = \dfrac{1}{\tan(2\theta)} = \dfrac{1}{\dfrac{2\tan\theta}{1-\tan^2\theta}} = \dfrac{1-\tan^2\theta}{2\tan\theta} = \dfrac{1 - \dfrac{1}{\cot^2\theta}}{\dfrac{2}{\cot\theta}} = \dfrac{\dfrac{\cot^2\theta - 1}{\cot^2\theta}}{\dfrac{2}{\cot\theta}} = \dfrac{\cot^2\theta - 1}{2\cot\theta}$

33. $\sec(2\theta) = \dfrac{1}{\cos(2\theta)} = \dfrac{1}{2\cos^2\theta - 1} = \dfrac{1}{\dfrac{2}{\sec^2\theta} - 1} = \dfrac{1}{\dfrac{2 - \sec^2\theta}{\sec^2\theta}} = \dfrac{\sec^2\theta}{2 - \sec^2\theta}$

35. $\cos^2(2\theta) - \sin^2(2\theta) = \cos(2(2\theta)) = \cos(4\theta)$

37. $\dfrac{\cos(2\theta)}{1+\sin(2\theta)} = \dfrac{\cos^2\theta - \sin^2\theta}{1 + 2\sin\theta\cos\theta} = \dfrac{(\cos\theta - \sin\theta)(\cos\theta + \sin\theta)}{\cos^2\theta + \sin^2\theta + 2\sin\theta\cos\theta}$

$$= \dfrac{(\cos\theta - \sin\theta)(\cos\theta + \sin\theta)}{(\cos\theta + \sin\theta)(\cos\theta + \sin\theta)} = \dfrac{\cos\theta - \sin\theta}{\cos\theta + \sin\theta} = \dfrac{\dfrac{\cos\theta}{\sin\theta} - \dfrac{\sin\theta}{\sin\theta}}{\dfrac{\cos\theta}{\sin\theta} + \dfrac{\sin\theta}{\sin\theta}} = \dfrac{\cot\theta - 1}{\cot\theta + 1}$$

39. $\sec^2\dfrac{\theta}{2} = \dfrac{1}{\cos^2\dfrac{\theta}{2}} = \dfrac{1}{\dfrac{1+\cos\theta}{2}} = \dfrac{2}{1+\cos\theta}$

41. $\cot^2\dfrac{\theta}{2} = \dfrac{1}{\tan^2\dfrac{\theta}{2}} = \dfrac{1}{\dfrac{1-\cos\theta}{1+\cos\theta}} = \dfrac{1+\cos\theta}{1-\cos\theta} = \dfrac{1+\dfrac{1}{\sec\theta}}{1-\dfrac{1}{\sec\theta}} = \dfrac{\dfrac{\sec\theta+1}{\sec\theta}}{\dfrac{\sec\theta-1}{\sec\theta}} = \dfrac{\sec\theta+1}{\sec\theta-1}$

43. $\dfrac{1-\tan^2\dfrac{\theta}{2}}{1+\tan^2\dfrac{\theta}{2}} = \dfrac{1-\dfrac{1-\cos\theta}{1+\cos\theta}}{1+\dfrac{1-\cos\theta}{1+\cos\theta}} = \dfrac{\dfrac{1+\cos\theta - (1-\cos\theta)}{1+\cos\theta}}{\dfrac{1+\cos\theta + 1 - \cos\theta}{1+\cos\theta}} = \dfrac{2\cos\theta}{2} = \cos\theta$

45. $\dfrac{\sin(3\theta)}{\sin\theta} - \dfrac{\cos(3\theta)}{\cos\theta} = \dfrac{\sin 3\theta\cos\theta - \cos 3\theta\sin\theta}{\sin\theta\cos\theta} = \dfrac{\sin(3\theta - \theta)}{\sin\theta\cos\theta}$

$\qquad = \dfrac{\sin 2\theta}{\sin\theta\cos\theta} = \dfrac{2\sin\theta\cos\theta}{\sin\theta\cos\theta} = 2$

47. $\tan 3\theta = \tan(2\theta + \theta) = \dfrac{\tan 2\theta + \tan\theta}{1 - \tan 2\theta\tan\theta} = \dfrac{\dfrac{2\tan\theta}{1 - \tan^2\theta} + \tan\theta}{1 - \dfrac{2\tan\theta}{1 - \tan^2\theta}\cdot\tan\theta}$

$\qquad = \dfrac{\dfrac{2\tan\theta + \tan\theta - \tan^3\theta}{1 - \tan^2\theta}}{\dfrac{1 - \tan^2\theta - 2\tan^2\theta}{1 - \tan^2\theta}} = \dfrac{3\tan\theta - \tan^3\theta}{1 - 3\tan^2\theta}$

49. If $x = 2\tan\theta$, then:

$\sin 2\theta = 2\sin\theta\cos\theta = \dfrac{2\sin\theta}{\cos\theta}\cdot\dfrac{\cos^2\theta}{1} = \dfrac{2\cdot\dfrac{\sin\theta}{\cos\theta}}{\dfrac{1}{\cos^2\theta}} = \dfrac{2\tan\theta}{\sec^2\theta} = \dfrac{2\tan\theta}{1 + \tan^2\theta}\cdot\dfrac{4}{4}$

$\qquad = \dfrac{4(2\tan\theta)}{4 + (2\tan\theta)^2} = \dfrac{4x}{4 + x^2}$

51. Solve for C:

$\dfrac{1}{2}\sin^2\theta + C = -\dfrac{1}{4}\cos 2\theta$

$C = -\dfrac{1}{4}\cos 2\theta - \dfrac{1}{2}\sin^2\theta = -\dfrac{1}{4}\left(\cos 2\theta + 2\sin^2\theta\right)$

$\qquad = -\dfrac{1}{4}\left(1 - 2\sin^2\theta + 2\sin^2\theta\right) = -\dfrac{1}{4}(1) = -\dfrac{1}{4}$

53. $f(x) = \sin^2 x = \dfrac{1 - \cos 2x}{2}$

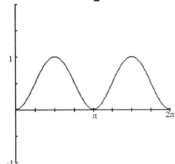

55. $\sin\dfrac{\pi}{24} = \sin\dfrac{\dfrac{\pi}{12}}{2} = \sqrt{\dfrac{1-\cos\dfrac{\pi}{12}}{2}} = \sqrt{\dfrac{1-\left(\dfrac{1}{4}\left(\sqrt{6}+\sqrt{2}\right)\right)}{2}} = \sqrt{\dfrac{1}{2}-\dfrac{1}{8}\left(\sqrt{6}+\sqrt{2}\right)}$

$\qquad = \sqrt{\dfrac{8-2\left(\sqrt{6}+\sqrt{2}\right)}{16}} = \dfrac{\sqrt{8-2\left(\sqrt{6}+\sqrt{2}\right)}}{4}$

$\cos\dfrac{\pi}{24} = \cos\dfrac{\dfrac{\pi}{12}}{2} = \sqrt{\dfrac{1+\cos\dfrac{\pi}{12}}{2}} = \sqrt{\dfrac{1+\left(\dfrac{1}{4}\left(\sqrt{6}+\sqrt{2}\right)\right)}{2}} = \sqrt{\dfrac{1}{2}+\dfrac{1}{8}\left(\sqrt{6}+\sqrt{2}\right)}$

$\qquad = \sqrt{\dfrac{8+2\left(\sqrt{6}+\sqrt{2}\right)}{16}} = \dfrac{\sqrt{8+2\left(\sqrt{6}+\sqrt{2}\right)}}{4}$

57. $\sin^3\theta + \sin^3(\theta+120°) + \sin^3(\theta+240°)$

$\qquad = \sin^3\theta + \left(\sin\theta\cos120°+\cos\theta\sin120°\right)^3 + \left(\sin\theta\cos240°+\cos\theta\sin240°\right)^3$

$\qquad = \sin^3\theta + \left(\dfrac{-1}{2}\sin\theta+\dfrac{\sqrt{3}}{2}\cos\theta\right)^3 + \left(\dfrac{-1}{2}\sin\theta-\dfrac{\sqrt{3}}{2}\cos\theta\right)^3$

$\qquad = \sin^3\theta + \dfrac{1}{8}\left(-\sin^3\theta+3\sqrt{3}\sin^2\theta\cos\theta-9\sin\theta\cos^2\theta+3\sqrt{3}\cos^3\theta\right)$

$\qquad\qquad -\dfrac{1}{8}\left(\sin^3\theta+3\sqrt{3}\sin^2\theta\cos\theta+9\sin\theta\cos^2\theta+3\sqrt{3}\cos^3\theta\right)$

$\qquad = \sin^3\theta - \dfrac{1}{8}\sin^3\theta + \dfrac{3\sqrt{3}}{8}\sin^2\theta\cos\theta - \dfrac{9}{8}\sin\theta\cos^2\theta + \dfrac{3\sqrt{3}}{8}\cos^3\theta$

$\qquad\qquad -\dfrac{1}{8}\sin^3\theta - \dfrac{3\sqrt{3}}{8}\sin^2\theta\cos\theta - \dfrac{9}{8}\sin\theta\cos^2\theta - \dfrac{3\sqrt{3}}{8}\cos^3\theta$

$\qquad = \dfrac{3}{4}\sin^3\theta - \dfrac{9}{4}\sin\theta\cos^2\theta = \dfrac{3}{4}\left(\sin^3\theta-3\sin\theta(1-\sin^2\theta)\right)$

$\qquad = \dfrac{3}{4}\left(\sin^3\theta-3\sin\theta+3\sin^3\theta\right) = \dfrac{3}{4}\left(4\sin^3\theta-3\sin\theta\right) = -\dfrac{3}{4}\sin3\theta$

$\qquad$ (See the formula for $\sin3\theta$ on page 463 of the text.)

59. $\dfrac{1}{2}\left(\ln|1-\cos2\theta|-\ln2\right) = \dfrac{1}{2}\ln\left|\dfrac{1-\cos2\theta}{2}\right| = \ln\left|\dfrac{1-\cos2\theta}{2}\right|^{\frac{1}{2}} = \ln\left|\sin^2\theta\right|^{\frac{1}{2}} = \ln|\sin\theta|$

61. (a) $R(\theta) = \dfrac{v_0^2\sqrt{2}}{16}\cos\theta(\sin\theta-\cos\theta) = \dfrac{v_0^2\sqrt{2}}{16}\left(\cos\theta\sin\theta-\cos^2\theta\right)$

$\qquad\qquad = \dfrac{v_0^2\sqrt{2}}{16}\cdot\dfrac{1}{2}\left(2\cos\theta\sin\theta-2\cos^2\theta\right) = \dfrac{v_0^2\sqrt{2}}{32}\left(\sin2\theta-2\left(\dfrac{1+\cos2\theta}{2}\right)\right)$

$\qquad\qquad = \dfrac{v_0^2\sqrt{2}}{32}\left(\sin2\theta-1-\cos2\theta\right) = \dfrac{v_0^2\sqrt{2}}{32}\left(\sin2\theta-\cos2\theta-1\right)$

(b)

(c) Using the MAXIMUM feature on the calculator:
R has the largest value when $\theta = 67.5°$.

63.
$$z = \tan\frac{\alpha}{2} = \frac{1 - \cos\alpha}{\sin\alpha}$$
$$z\sin\alpha = 1 - \cos\alpha$$
$$z\sin\alpha = 1 - \sqrt{1 - \sin^2\alpha}$$
$$z\sin\alpha - 1 = -\sqrt{1 - \sin^2\alpha}$$
$$z^2\sin^2\alpha - 2z\sin\alpha + 1 = 1 - \sin^2\alpha$$
$$z^2\sin^2\alpha + \sin^2\alpha = 2z\sin\alpha$$
$$\sin^2\alpha(z^2 + 1) = 2z\sin\alpha$$
$$\sin\alpha(z^2 + 1) = 2z$$
$$\sin\alpha = \frac{2z}{z^2 + 1}$$

7.4 Product-to-Sum and Sum-to-Product Formulas

For Problems 1-10, use the formulas:
$$\sin\alpha\sin\beta = \tfrac{1}{2}\big[\cos(\alpha - \beta) - \cos(\alpha + \beta)\big]$$
$$\cos\alpha\cos\beta = \tfrac{1}{2}\big[\cos(\alpha - \beta) + \cos(\alpha + \beta)\big]$$
$$\sin\alpha\cos\beta = \tfrac{1}{2}\big[\sin(\alpha + \beta) + \sin(\alpha - \beta)\big]$$

1. $\sin(4\theta)\sin(2\theta) = \tfrac{1}{2}\big[\cos(4\theta - 2\theta) - \cos(4\theta + 2\theta)\big] = \tfrac{1}{2}\big[\cos2\theta - \cos6\theta\big]$

3. $\sin(4\theta)\cos(2\theta) = \tfrac{1}{2}\big[\sin(4\theta + 2\theta) + \sin(4\theta - 2\theta)\big] = \tfrac{1}{2}\big[\sin6\theta + \sin2\theta\big]$

5. $\cos(3\theta)\cos(5\theta) = \tfrac{1}{2}\big[\cos(3\theta - 5\theta) + \cos(3\theta + 5\theta)\big] = \tfrac{1}{2}\big[\cos(-2\theta) + \cos8\theta\big]$
$$= \tfrac{1}{2}\big[\cos2\theta + \cos8\theta\big]$$

7. $\sin\theta\sin(2\theta) = \tfrac{1}{2}\big[\cos(\theta - 2\theta) - \cos(\theta + 2\theta)\big] = \tfrac{1}{2}\big[\cos(-\theta) - \cos3\theta\big]$
$$= \tfrac{1}{2}\big[\cos\theta - \cos3\theta\big]$$

9. $\sin\dfrac{3\theta}{2}\cos\dfrac{\theta}{2} = \dfrac{1}{2}\left[\sin\left(\dfrac{3\theta}{2}+\dfrac{\theta}{2}\right) + \sin\left(\dfrac{3\theta}{2}-\dfrac{\theta}{2}\right)\right] = \dfrac{1}{2}\left[\sin2\theta + \sin\theta\right]$

For Problems 11-18, use the formulas:

$$\sin\alpha + \sin\beta = 2\sin\dfrac{\alpha+\beta}{2}\cos\dfrac{\alpha-\beta}{2}$$

$$\sin\alpha - \sin\beta = 2\sin\dfrac{\alpha-\beta}{2}\cos\dfrac{\alpha+\beta}{2}$$

$$\cos\alpha + \cos\beta = 2\cos\dfrac{\alpha+\beta}{2}\cos\dfrac{\alpha-\beta}{2}$$

$$\cos\alpha - \cos\beta = -2\sin\dfrac{\alpha+\beta}{2}\sin\dfrac{\alpha-\beta}{2}$$

11. $\sin(4\theta) - \sin(2\theta) = 2\sin\dfrac{4\theta-2\theta}{2}\cos\dfrac{4\theta+2\theta}{2} = 2\sin\theta\cos3\theta$

13. $\cos(2\theta) + \cos(4\theta) = 2\cos\dfrac{2\theta+4\theta}{2}\cos\dfrac{2\theta-4\theta}{2} = 2\cos3\theta\cos(-\theta) = 2\cos3\theta\cos\theta$

15. $\sin\theta + \sin(3\theta) = 2\sin\dfrac{\theta+3\theta}{2}\cos\dfrac{\theta-3\theta}{2} = 2\sin2\theta\cos(-\theta) = 2\sin2\theta\cos\theta$

17. $\cos\dfrac{\theta}{2} - \cos\dfrac{3\theta}{2} = -2\sin\dfrac{\dfrac{\theta}{2}+\dfrac{3\theta}{2}}{2}\sin\dfrac{\dfrac{\theta}{2}-\dfrac{3\theta}{2}}{2} = -2\sin\theta\sin\left(\dfrac{-\theta}{2}\right)$

$$= -2\sin\theta\left(-\sin\dfrac{\theta}{2}\right) = 2\sin\theta\sin\dfrac{\theta}{2}$$

19. $\dfrac{\sin\theta + \sin(3\theta)}{2\sin(2\theta)} = \dfrac{2\sin(2\theta)\cos(-\theta)}{2\sin(2\theta)} = \cos(-\theta) = \cos\theta$

21. $\dfrac{\sin(4\theta) + \sin(2\theta)}{\cos(4\theta) + \cos(2\theta)} = \dfrac{2\sin(3\theta)\cos\theta}{2\cos(3\theta)\cos\theta} = \dfrac{\sin(3\theta)}{\cos(3\theta)} = \tan(3\theta)$

23. $\dfrac{\cos\theta - \cos(3\theta)}{\sin\theta + \sin(3\theta)} = \dfrac{-2\sin(2\theta)\sin(-\theta)}{2\sin(2\theta)\cos(-\theta)} = \dfrac{-(-\sin\theta)}{\cos\theta} = \tan\theta$

25. $\sin\theta\left[\sin\theta + \sin(3\theta)\right] = \sin\theta\left[2\sin(2\theta)\cos(-\theta)\right] = \cos\theta\left[2\sin(2\theta)\sin\theta\right]$

$$= \cos\theta\left[2\cdot\dfrac{1}{2}(\cos\theta - \cos(3\theta))\right] = \cos\theta(\cos\theta - \cos(3\theta))$$

27. $\dfrac{\sin(4\theta) + \sin(8\theta)}{\cos(4\theta) + \cos(8\theta)} = \dfrac{2\sin(6\theta)\cos(-2\theta)}{2\cos(6\theta)\cos(-2\theta)} = \dfrac{\sin(6\theta)}{\cos(6\theta)} = \tan(6\theta)$

29. $\dfrac{\sin(4\theta) + \sin(8\theta)}{\sin(4\theta) - \sin(8\theta)} = \dfrac{2\sin(6\theta)\cos(-2\theta)}{2\sin(-2\theta)\cos(6\theta)} = \dfrac{\sin(6\theta)\cos(2\theta)}{-\sin(2\theta)\cos(6\theta)}$

$\qquad = -\tan(6\theta)\cot(2\theta) = -\dfrac{\tan(6\theta)}{\tan(2\theta)}$

31. $\dfrac{\sin\alpha + \sin\beta}{\sin\alpha - \sin\beta} = \dfrac{2\sin\dfrac{\alpha+\beta}{2}\cos\dfrac{\alpha-\beta}{2}}{2\sin\dfrac{\alpha-\beta}{2}\cos\dfrac{\alpha+\beta}{2}} = \tan\dfrac{\alpha+\beta}{2}\cot\dfrac{\alpha-\beta}{2}$

33. $\dfrac{\sin\alpha + \sin\beta}{\cos\alpha + \cos\beta} = \dfrac{2\sin\dfrac{\alpha+\beta}{2}\cos\dfrac{\alpha-\beta}{2}}{2\cos\dfrac{\alpha+\beta}{2}\cos\dfrac{\alpha-\beta}{2}} = \tan\dfrac{\alpha+\beta}{2}$

35. $1 + \cos(2\theta) + \cos(4\theta) + \cos(6\theta) = \cos 0 + \cos(6\theta) + \cos(2\theta) + \cos(4\theta)$

$\qquad = 2\cos(3\theta)\cos(-3\theta) + 2\cos(3\theta)\cos(-\theta) = 2\cos^2(3\theta) + 2\cos(3\theta)\cos\theta$

$\qquad = 2\cos(3\theta)\big(\cos(3\theta) + \cos\theta\big) = 2\cos(3\theta)2\cos(2\theta)\cos\theta$

$\qquad = 4\cos\theta\cos(2\theta)\cos(3\theta)$

37. (a) $\quad y = \sin\big[2\pi(852)t\big] + \sin\big[2\pi(1209)t\big]$

$\qquad = 2\sin\dfrac{2\pi(852)t + 2\pi(1209)t}{2}\cos\dfrac{2\pi(852)t - 2\pi(1209)t}{2}$

$\qquad = 2\sin(2061\pi t)\cos(357\pi t)$

(b) The maximum value of y is 2.

(c)

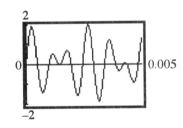

39. $\sin 2\alpha + \sin 2\beta + \sin 2\gamma$

$\qquad = 2\sin\dfrac{2\alpha + 2\beta}{2}\cos\dfrac{2\alpha - 2\beta}{2} + \sin 2\gamma$

$\qquad = 2\sin(\alpha + \beta)\cos(\alpha - \beta) + 2\sin\gamma\cos\gamma$

$\qquad = 2\sin(\pi - \gamma)\cos(\alpha - \beta) + 2\sin\gamma\cos\gamma$

$\qquad = 2\sin\gamma\cos(\alpha - \beta) + 2\sin\gamma\cos\gamma = 2\sin\gamma\big[\cos(\alpha - \beta) + \cos\gamma\big]$

$\qquad = 2\sin\gamma\left(2\cos\dfrac{\alpha - \beta + \gamma}{2}\cos\dfrac{\alpha - \beta - \gamma}{2}\right)$

$\qquad = 4\sin\gamma\cos\left(\dfrac{\pi}{2} - \beta\right)\cos\left(\alpha - \dfrac{\pi}{2}\right) = 4\sin\gamma\sin\beta\sin\alpha$

$\qquad = 4\sin\alpha\sin\beta\sin\gamma$

41.   Add the two sum formulas for $\sin(\alpha + \beta)$ and $\sin(\alpha - \beta)$ and solve:

$$\sin(\alpha + \beta) = \sin\alpha\cos\beta + \cos\alpha\sin\beta$$
$$\sin(\alpha - \beta) = \sin\alpha\cos\beta - \cos\alpha\sin\beta$$
$$\sin(\alpha + \beta) + \sin(\alpha - \beta) = 2\sin\alpha\cos\beta$$
$$\sin\alpha\cos\beta = \tfrac{1}{2}\left[\sin(\alpha + \beta) + \sin(\alpha - \beta)\right]$$

43. $2\cos\dfrac{\alpha + \beta}{2}\cos\dfrac{\alpha - \beta}{2} = 2\cdot\dfrac{1}{2}\left[\cos\left(\dfrac{\alpha + \beta}{2} - \dfrac{\alpha - \beta}{2}\right) + \cos\left(\dfrac{\alpha + \beta}{2} + \dfrac{\alpha - \beta}{2}\right)\right]$

$$= \cos\dfrac{2\beta}{2} + \cos\dfrac{2\alpha}{2} = \cos\beta + \cos\alpha$$

Therefore, $\cos\alpha + \cos\beta = 2\cos\dfrac{\alpha + \beta}{2}\cos\dfrac{\alpha - \beta}{2}$

7.5   The Inverse Trigonometric Functions (I)

1. $\sin^{-1}0$

We are finding the angle θ, $\dfrac{-\pi}{2} \le \theta \le \dfrac{\pi}{2}$, whose sine equals 0.

$$\sin\theta = 0 \quad -\dfrac{\pi}{2} \le \theta \le \dfrac{\pi}{2}$$
$$\theta = 0$$
$$\sin^{-1}0 = 0$$

3. $\sin^{-1}(-1)$

We are finding the angle θ, $\dfrac{-\pi}{2} \le \theta \le \dfrac{\pi}{2}$, whose sine equals -1.

$$\sin\theta = -1 \quad -\dfrac{\pi}{2} \le \theta \le \dfrac{\pi}{2}$$
$$\theta = -\dfrac{\pi}{2}$$
$$\sin^{-1}(-1) = -\dfrac{\pi}{2}$$

5. $\tan^{-1}0$

We are finding the angle θ, $\dfrac{-\pi}{2} < \theta < \dfrac{\pi}{2}$, whose tangent equals 0.

$$\tan\theta = 0 \quad -\dfrac{\pi}{2} < \theta < \dfrac{\pi}{2}$$
$$\theta = 0$$
$$\tan^{-1}0 = 0$$

7. $\sin^{-1}\dfrac{\sqrt{2}}{2}$

We are finding the angle θ, $\dfrac{-\pi}{2} \le \theta \le \dfrac{\pi}{2}$, whose sine equals $\dfrac{\sqrt{2}}{2}$.

$$\sin\theta = \dfrac{\sqrt{2}}{2} \qquad -\dfrac{\pi}{2} \le \theta \le \dfrac{\pi}{2}$$

$$\theta = \dfrac{\pi}{4}$$

$$\sin^{-1}\dfrac{\sqrt{2}}{2} = \dfrac{\pi}{4}$$

9. $\tan^{-1}\sqrt{3}$

We are finding the angle θ, $\dfrac{-\pi}{2} < \theta < \dfrac{\pi}{2}$, whose tangent equals $\sqrt{3}$.

$$\tan\theta = \sqrt{3} \qquad -\dfrac{\pi}{2} < \theta < \dfrac{\pi}{2}$$

$$\theta = \dfrac{\pi}{3}$$

$$\tan^{-1}\sqrt{3} = \dfrac{\pi}{3}$$

11. $\cos^{-1}\left(-\dfrac{\sqrt{3}}{2}\right)$

We are finding the angle θ, $0 \le \theta \le \pi$, whose cosine equals $-\dfrac{\sqrt{3}}{2}$.

$$\cos\theta = -\dfrac{\sqrt{3}}{2} \qquad 0 \le \theta \le \pi$$

$$\theta = \dfrac{5\pi}{6}$$

$$\cos^{-1}\left(-\dfrac{\sqrt{3}}{2}\right) = \dfrac{5\pi}{6}$$

13. $\cot^{-1}\sqrt{3}$

We are finding the angle θ, $0 < \theta < \pi$, whose cotangent equals $\sqrt{3}$.

$$\cot\theta = \sqrt{3} \qquad 0 < \theta < \pi$$

$$\theta = \dfrac{\pi}{6}$$

$$\cot^{-1}\sqrt{3} = \dfrac{\pi}{6}$$

15. $\csc^{-1}(-1)$

We are finding the angle θ, $\dfrac{-\pi}{2} \le \theta \le \dfrac{\pi}{2}$, $\theta \ne 0$, whose cosecant equals -1.

$$\csc \theta = -1 \qquad -\frac{\pi}{2} \le \theta \le \frac{\pi}{2}, \ \theta \ne 0$$

$$\theta = -\frac{\pi}{2}$$

$$\csc^{-1}(-1) = -\frac{\pi}{2}$$

17. $\sec^{-1}\left(\dfrac{2\sqrt{3}}{3}\right)$

We are finding the angle θ, $0 \le \theta \le \pi$, $\theta \ne \dfrac{\pi}{2}$, whose secant equals $\dfrac{2\sqrt{3}}{3}$.

$$\sec \theta = \frac{2\sqrt{3}}{3} \qquad 0 \le \theta \le \pi, \ \theta \ne \frac{\pi}{2}$$

$$\theta = \frac{\pi}{6}$$

$$\sec^{-1}\left(\frac{2\sqrt{3}}{3}\right) = \frac{\pi}{6}$$

19. $\cot^{-1}\left(-\dfrac{\sqrt{3}}{3}\right)$

We are finding the angle θ, $0 < \theta < \pi$, whose cotangent equals $-\dfrac{\sqrt{3}}{3}$.

$$\cot \theta = -\frac{\sqrt{3}}{3} \qquad 0 < \theta < \pi$$

$$\theta = \frac{2\pi}{3}$$

$$\cot^{-1}\left(-\frac{\sqrt{3}}{3}\right) = \frac{2\pi}{3}$$

For Problems 21-44, set the calculator's mode to radians.

21. $\sin^{-1} 0.1 = 0.10$

23. $\tan^{-1} 5 = 1.37$

25. $\cos^{-1} \frac{7}{8} = 0.51$

27. $\tan^{-1}(-0.4) = -0.38$

29. $\sin^{-1}(-0.12) = -0.12$

31. $\cos^{-1} \dfrac{\sqrt{2}}{3} = 1.08$

33. $\sec^{-1} 4 = \cos^{-1} \frac{1}{4} = 1.32$

35. $\cot^{-1} 2 = \tan^{-1} \frac{1}{2} = 0.46$

37. $\csc^{-1}(-3) = \sin^{-1}\left(-\frac{1}{3}\right) = -0.34$

39. $\cot^{-1}\left(-\sqrt{5}\right) = \tan^{-1}\left(-\frac{1}{\sqrt{5}}\right) = -0.42$

41. $\csc^{-1}\left(-\frac{3}{2}\right) = \sin^{-1}\left(-\frac{2}{3}\right) = -0.73$

43. $\cot^{-1}\left(-\frac{3}{2}\right) = \tan^{-1}\left(-\frac{2}{3}\right) = -0.59$

45. $\sin\left[\sin^{-1}(0.54)\right] = 0.54$

47. $\cos^{-1}\left[\cos\left(\frac{4\pi}{5}\right)\right] = \frac{4\pi}{5}$

49. $\tan\left[\tan^{-1}(-3.5)\right] = -3.5$

51. $\sin^{-1}\left[\sin\left(-\frac{3\pi}{7}\right)\right] = -\frac{3\pi}{7}$

53. $1530 \text{ ft} \cdot \dfrac{1 \text{ mile}}{5280 \text{ feet}} = 0.29 \text{ mile}$

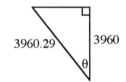

$$\cos\theta = \frac{3960}{3960.29}$$
$$\theta = 0.0121 \text{ radians}$$
$$s = r\theta = 3960(0.0121) = 47.92 \text{ miles}$$
$$\frac{2\pi(2710)}{24} = \frac{47.92}{t}$$
$$t = 0.0675 \text{ hour} = 4.05 \text{ minutes}$$

55. $\cos(\cos^{-1} x) = x$ when $-1 \le x \le 1$.

57. $\cos^{-1}(\cos x) = x$ when $0 \le x \le \pi$.

59. $y = \sec^{-1} x$

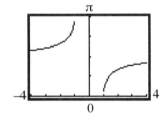

7.6 The Inverse Trigonometric Functions (II)

1. $\cos\left(\sin^{-1}\dfrac{\sqrt{2}}{2}\right)$

Find the angle θ, $\dfrac{-\pi}{2} \le \theta \le \dfrac{\pi}{2}$, whose sine equals $\dfrac{\sqrt{2}}{2}$.

$$\sin\theta = \dfrac{\sqrt{2}}{2} \qquad -\dfrac{\pi}{2} \le \theta \le \dfrac{\pi}{2}$$

$$\theta = \dfrac{\pi}{4}$$

$$\cos\left(\sin^{-1}\dfrac{\sqrt{2}}{2}\right) = \cos\dfrac{\pi}{4} = \dfrac{\sqrt{2}}{2}$$

3. $\tan\left(\cos^{-1}\left(-\dfrac{\sqrt{3}}{2}\right)\right)$

Find the angle θ, $0 \le \theta \le \pi$, whose cosine equals $-\dfrac{\sqrt{3}}{2}$.

$$\cos\theta = -\dfrac{\sqrt{3}}{2} \qquad 0 \le \theta \le \pi$$

$$\theta = \dfrac{5\pi}{6}$$

$$\tan\left(\cos^{-1}\left(-\dfrac{\sqrt{3}}{2}\right)\right) = \tan\dfrac{5\pi}{6} = -\dfrac{\sqrt{3}}{3}$$

5. $\sec\left(\cos^{-1}\dfrac{1}{2}\right)$

Find the angle θ, $0 \le \theta \le \pi$, whose cosine equals $\dfrac{1}{2}$.

$$\cos\theta = \dfrac{1}{2} \qquad 0 \le \theta \le \pi$$

$$\theta = \dfrac{\pi}{3}$$

$$\sec\left(\cos^{-1}\dfrac{1}{2}\right) = \sec\dfrac{\pi}{3} = 2$$

7. $\csc\left(\tan^{-1}1\right)$

Find the angle θ, $\dfrac{-\pi}{2} < \theta < \dfrac{\pi}{2}$, whose tangent equals 1.

$$\tan\theta = 1 \qquad -\dfrac{\pi}{2} < \theta < \dfrac{\pi}{2}$$

$$\theta = \dfrac{\pi}{4}$$

$$\csc\left(\tan^{-1}1\right) = \csc\dfrac{\pi}{4} = \sqrt{2}$$

9. $\sin\left(\tan^{-1}(-1)\right)$

Find the angle θ, $\dfrac{-\pi}{2} < \theta < \dfrac{\pi}{2}$, whose tangent equals -1.

$$\tan\theta = -1 \qquad -\dfrac{\pi}{2} < \theta < \dfrac{\pi}{2}$$

$$\theta = -\dfrac{\pi}{4}$$

$$\sin\left(\tan^{-1}(-1)\right) = \sin\left(-\dfrac{\pi}{4}\right) = -\dfrac{\sqrt{2}}{2}$$

11. $\sec\left(\sin^{-1}\left(-\dfrac{1}{2}\right)\right)$

Find the angle θ, $\dfrac{-\pi}{2} \le \theta \le \dfrac{\pi}{2}$, whose sine equals $-\dfrac{1}{2}$.

$$\sin\theta = -\dfrac{1}{2} \qquad -\dfrac{\pi}{2} \le \theta \le \dfrac{\pi}{2}$$

$$\theta = -\dfrac{\pi}{6}$$

$$\sec\left(\sin^{-1}\left(-\dfrac{1}{2}\right)\right) = \sec\left(-\dfrac{\pi}{6}\right) = \dfrac{2\sqrt{3}}{3}$$

13. $\cos^{-1}\left(\cos\dfrac{5\pi}{4}\right) = \cos^{-1}\left(-\dfrac{\sqrt{2}}{2}\right)$

Find the angle θ, $0 \le \theta \le \pi$, whose cosine equals $-\dfrac{\sqrt{2}}{2}$.

$$\cos\theta = -\dfrac{\sqrt{2}}{2} \qquad 0 \le \theta \le \pi$$

$$\theta = \dfrac{3\pi}{4}$$

15. $\sin^{-1}\left(\sin\left(-\dfrac{7\pi}{6}\right)\right) = \sin^{-1}\dfrac{1}{2}$

Find the angle θ, $\dfrac{-\pi}{2} \le \theta \le \dfrac{\pi}{2}$, whose sine equals $\dfrac{1}{2}$.

$$\sin\theta = \dfrac{1}{2} \qquad -\dfrac{\pi}{2} \le \theta \le \dfrac{\pi}{2}$$

$$\theta = \dfrac{\pi}{6}$$

17. $\tan\left(\sin^{-1}\frac{1}{3}\right)$

Since $\sin\theta = \frac{1}{3}$, $-\frac{\pi}{2} \le \theta \le \frac{\pi}{2}$, let $y = 1$ and $r = 3$. Solve for x:

$x^2 + 1 = 9$

$x^2 = 8$

$x = \pm\sqrt{8} = \pm 2\sqrt{2}$

Since θ is in quadrant I, $x = 2\sqrt{2}$.

$\tan\left(\sin^{-1}\frac{1}{3}\right) = \tan\theta = \frac{y}{x} = \frac{1}{2\sqrt{2}} \cdot \frac{\sqrt{2}}{\sqrt{2}} = \frac{\sqrt{2}}{4}$

19. $\sec\left(\tan^{-1}\frac{1}{2}\right)$

Since $\tan\theta = \frac{1}{2}$, $-\frac{\pi}{2} < \theta < \frac{\pi}{2}$, let $x = 2$ and $y = 1$. Solve for r:

$2^2 + 1 = r^2$

$r^2 = 5$

$r = \sqrt{5}$

θ is in quadrant I.

$\sec\left(\tan^{-1}\frac{1}{2}\right) = \sec\theta = \frac{r}{x} = \frac{\sqrt{5}}{2}$

21. $\cot\left(\sin^{-1}\left(-\frac{\sqrt{2}}{3}\right)\right)$

Since $\sin\theta = -\frac{\sqrt{2}}{3}$, $-\frac{\pi}{2} \le \theta \le \frac{\pi}{2}$, let $y = -\sqrt{2}$ and $r = 3$. Solve for x:

$x^2 + 2 = 9$

$x^2 = 7$

$x = \pm\sqrt{7}$

Since θ is in quadrant IV, $x = \sqrt{7}$.

$\cot\left(\sin^{-1}\left(-\frac{\sqrt{2}}{3}\right)\right) = \cot\theta = \frac{x}{y} = \frac{\sqrt{7}}{-\sqrt{2}} \cdot \frac{\sqrt{2}}{\sqrt{2}} = -\frac{\sqrt{14}}{2}$

23. $\sin\left(\tan^{-1}(-3)\right)$

Since $\tan\theta = -3$, $-\frac{\pi}{2} < \theta < \frac{\pi}{2}$, let $x = 1$ and $y = -3$. Solve for r:

$1 + 9 = r^2 \;\rightarrow\; r^2 = 10 \;\rightarrow\; r = \sqrt{10}$

θ is in quadrant IV.

$\sin\left(\tan^{-1}(-3)\right) = \sin\theta = \frac{y}{r} = \frac{-3}{\sqrt{10}} \cdot \frac{\sqrt{10}}{\sqrt{10}} = -\frac{3\sqrt{10}}{10}$

25. $\sec\left(\sin^{-1}\left(\dfrac{2\sqrt{5}}{5}\right)\right)$

Since $\sin\theta = \dfrac{2\sqrt{5}}{5}$, $-\dfrac{\pi}{2} \le \theta \le \dfrac{\pi}{2}$, let $y = 2\sqrt{5}$ and $r = 5$. Solve for x:

$$x^2 + 20 = 25$$
$$x^2 = 5$$
$$x = \pm\sqrt{5}$$

Since θ is in quadrant I, $x = \sqrt{5}$.

$$\sec\left(\sin^{-1}\left(\dfrac{2\sqrt{5}}{5}\right)\right) = \sec\theta = \dfrac{r}{x} = \dfrac{5}{\sqrt{5}} \cdot \dfrac{\sqrt{5}}{\sqrt{5}} = \sqrt{5}$$

27. $\sin^{-1}\left(\cos\dfrac{3\pi}{4}\right) = \sin^{-1}\left(-\dfrac{\sqrt{2}}{2}\right) = -\dfrac{\pi}{4}$

29. $\sin\left(\sin^{-1}\dfrac{1}{2} + \cos^{-1}0\right) = \sin\left(\dfrac{\pi}{6} + \dfrac{\pi}{2}\right) = \sin\dfrac{\pi}{6}\cos\dfrac{\pi}{2} + \cos\dfrac{\pi}{6}\sin\dfrac{\pi}{2} = \dfrac{1}{2}\cdot 0 + \dfrac{\sqrt{3}}{2}\cdot 1 = \dfrac{\sqrt{3}}{2}$

31. $\sin\left[\sin^{-1}\dfrac{3}{5} - \cos^{-1}\left(-\dfrac{4}{5}\right)\right]$

Let $\alpha = \sin^{-1}\dfrac{3}{5}$ and $\beta = \cos^{-1}\left(-\dfrac{4}{5}\right)$. α is in quadrant I; β is in quadrant II.

Then $\sin\alpha = \dfrac{3}{5}$, $-\dfrac{\pi}{2} \le \alpha \le \dfrac{\pi}{2}$, and $\cos\beta = -\dfrac{4}{5}$, $0 \le \beta \le \pi$.

$$\cos\alpha = \sqrt{1 - \sin^2\alpha} = \sqrt{1 - \left(\dfrac{3}{5}\right)^2} = \sqrt{1 - \dfrac{9}{25}} = \sqrt{\dfrac{16}{25}} = \dfrac{4}{5}$$

$$\sin\beta = \sqrt{1 - \cos^2\beta} = \sqrt{1 - \left(-\dfrac{4}{5}\right)^2} = \sqrt{1 - \dfrac{16}{25}} = \sqrt{\dfrac{9}{25}} = \dfrac{3}{5}$$

$$\sin\left[\sin^{-1}\dfrac{3}{5} - \cos^{-1}\left(-\dfrac{4}{5}\right)\right] = \sin(\alpha - \beta) = \sin\alpha\cos\beta - \cos\alpha\sin\beta$$

$$= \dfrac{3}{5}\cdot -\dfrac{4}{5} - \dfrac{4}{5}\cdot\dfrac{3}{5} = -\dfrac{12}{25} - \dfrac{12}{25} = -\dfrac{24}{25}$$

33. $\cos\left(\tan^{-1}\dfrac{4}{3} + \cos^{-1}\dfrac{5}{13}\right)$

Let $\alpha = \tan^{-1}\dfrac{4}{3}$ and $\beta = \cos^{-1}\dfrac{5}{13}$. α is in quadrant I; β is in quadrant I.

Then $\tan\alpha = \dfrac{4}{3}$, $-\dfrac{\pi}{2} < \alpha < \dfrac{\pi}{2}$, and $\cos\beta = \dfrac{5}{13}$, $0 \le \beta \le \pi$.

$$\sec\alpha = \sqrt{1 + \tan^2\alpha} = \sqrt{1 + \left(\dfrac{4}{3}\right)^2} = \sqrt{1 + \dfrac{16}{9}} = \sqrt{\dfrac{25}{9}} = \dfrac{5}{3}; \quad \cos\alpha = \dfrac{3}{5}$$

$$\sin\alpha = \sqrt{1 - \cos^2\alpha} = \sqrt{1 - \left(\dfrac{3}{5}\right)^2} = \sqrt{1 - \dfrac{9}{25}} = \sqrt{\dfrac{16}{25}} = \dfrac{4}{5}$$

$$\sin\beta = \sqrt{1 - \cos^2\beta} = \sqrt{1 - \left(\dfrac{5}{13}\right)^2} = \sqrt{1 - \dfrac{25}{169}} = \sqrt{\dfrac{144}{169}} = \dfrac{12}{13}$$

$$\cos\left(\tan^{-1}\dfrac{4}{3} + \cos^{-1}\dfrac{5}{13}\right) = \cos(\alpha + \beta) = \cos\alpha\cos\beta - \sin\alpha\sin\beta$$

$$= \dfrac{3}{5}\cdot\dfrac{5}{13} - \dfrac{4}{5}\cdot\dfrac{12}{13} = \dfrac{15}{65} - \dfrac{48}{65} = -\dfrac{33}{65}$$

35. $\sec\left(\sin^{-1}\frac{5}{13} - \tan^{-1}\frac{3}{4}\right)$

Let $\alpha = \sin^{-1}\frac{5}{13}$ and $\beta = \tan^{-1}\frac{3}{4}$. α is in quadrant I; β is in quadrant I.

Then $\sin\alpha = \frac{5}{13}$, $-\frac{\pi}{2} \le \alpha \le \frac{\pi}{2}$, and $tan\beta = \frac{3}{4}$, $-\frac{\pi}{2} < \beta < \frac{\pi}{2}$.

$$\cos\alpha = \sqrt{1 - \sin^2\alpha} = \sqrt{1 - \left(\frac{5}{13}\right)^2} = \sqrt{1 - \frac{25}{169}} = \sqrt{\frac{144}{169}} = \frac{12}{13}$$

$$\sec\beta = \sqrt{1 + \tan^2\beta} = \sqrt{1 + \left(\frac{3}{4}\right)^2} = \sqrt{1 + \frac{9}{16}} = \sqrt{\frac{25}{16}} = \frac{5}{4}; \quad \cos\beta = \frac{4}{5}$$

$$\sin\beta = \sqrt{1 - \cos^2\beta} = \sqrt{1 - \left(\frac{4}{5}\right)^2} = \sqrt{1 - \frac{16}{25}} = \sqrt{\frac{9}{25}} = \frac{3}{5}$$

$$\sec\left[\sin^{-1}\frac{5}{13} - \tan^{-1}\frac{3}{4}\right] = \frac{1}{\cos\left[\sin^{-1}\frac{5}{13} - \tan^{-1}\frac{3}{4}\right]} = \frac{1}{\cos(\alpha - \beta)}$$

$$= \frac{1}{\cos\alpha\cos\beta + \sin\alpha\sin\beta} = \frac{1}{\frac{12}{13}\cdot\frac{4}{5} + \frac{5}{13}\cdot\frac{3}{5}}$$

$$= \frac{1}{\frac{48}{65} + \frac{15}{65}} = \frac{1}{\frac{63}{65}} = \frac{65}{63}$$

37. $\cot\left(\sec^{-1}\frac{5}{3} + \frac{\pi}{6}\right)$

Let $\alpha = \sec^{-1}\frac{5}{3}$. α is in quadrant I.

Then $\sec\alpha = \frac{5}{3}$, $0 \le \alpha \le \pi$.

$$\tan\alpha = \sqrt{\sec^2\alpha - 1} = \sqrt{\left(\frac{5}{3}\right)^2 - 1} = \sqrt{\frac{25}{9} - 1} = \sqrt{\frac{16}{9}} = \frac{4}{3}$$

$$\cot\left[\sec^{-1}\frac{5}{3} + \frac{\pi}{6}\right] = \frac{1}{\tan\left[\sec^{-1}\frac{5}{3} + \frac{\pi}{6}\right]} = \frac{1}{\tan\left(\alpha + \frac{\pi}{6}\right)} = \frac{1}{\frac{\tan\alpha + \tan\frac{\pi}{6}}{1 - \tan\alpha\tan\frac{\pi}{6}}} = \frac{1 - \tan\alpha\tan\frac{\pi}{6}}{\tan\alpha + \tan\frac{\pi}{6}}$$

$$= \frac{1 - \frac{4}{3}\cdot\frac{\sqrt{3}}{3}}{\frac{4}{3} + \frac{\sqrt{3}}{3}} = \frac{1 - \frac{4\sqrt{3}}{9}}{\frac{4 + \sqrt{3}}{3}} = \frac{9 - 4\sqrt{3}}{9}\cdot\frac{3}{4 + \sqrt{3}} = \frac{9 - 4\sqrt{3}}{3(4 + \sqrt{3})}\cdot\frac{4 - \sqrt{3}}{4 - \sqrt{3}}$$

$$= \frac{48 - 25\sqrt{3}}{3(13)} = \frac{48 - 25\sqrt{3}}{39}$$

39. $\sin\left(2\sin^{-1}\frac{1}{2}\right) = \sin\left(2\left(\frac{\pi}{6}\right)\right) = \sin\frac{\pi}{3} = \frac{\sqrt{3}}{2}$

41. $\cos\left(2\sin^{-1}\frac{3}{5}\right) = 1 - 2\sin^2\left(\sin^{-1}\frac{3}{5}\right) = 1 - 2\left(\frac{3}{5}\right)^2 = 1 - 2\left(\frac{9}{25}\right) = 1 - \frac{18}{25} = \frac{7}{25}$

43. $\tan\left[2\cos^{-1}\left(-\frac{3}{5}\right)\right]$

Let $\alpha = \cos^{-1}\left(-\frac{3}{5}\right)$. α is in quadrant II.

Then $\cos\alpha = -\frac{3}{5}$, $0 \le \alpha \le \pi$.

$$\sec\alpha = -\frac{5}{3}; \quad \tan\alpha = -\sqrt{\sec^2\alpha - 1} = -\sqrt{\left(-\frac{5}{3}\right)^2 - 1} = -\sqrt{\frac{25}{9} - 1} = -\sqrt{\frac{16}{9}} = -\frac{4}{3}$$

$$\tan\left[2\cos^{-1}\left(-\frac{3}{5}\right)\right] = \tan 2\alpha = \frac{2\tan\alpha}{1-\tan^2\alpha} = \frac{2\left(-\frac{4}{3}\right)}{1-\left(-\frac{4}{3}\right)^2} = \frac{-\frac{8}{3}}{1-\frac{16}{9}} = \frac{-\frac{8}{3}}{-\frac{7}{9}} = -\frac{8}{3}\cdot-\frac{9}{7} = \frac{24}{7}$$

45. $\sin\left(2\cos^{-1}\frac{4}{5}\right)$

Let $\alpha = \cos^{-1}\frac{4}{5}$. α is in quadrant I.

Then $\cos\alpha = \frac{4}{5}$, $0 \le \alpha \le \pi$.

$$\sin\alpha = \sqrt{1-\cos^2\alpha} = \sqrt{1-\left(\frac{4}{5}\right)^2} = \sqrt{1-\frac{16}{25}} = \sqrt{\frac{9}{25}} = \frac{3}{5}$$

$$\sin\left[2\cos^{-1}\frac{4}{5}\right] = \sin 2\alpha = 2\sin\alpha\cos\alpha = 2\cdot\frac{3}{5}\cdot\frac{4}{5} = \frac{24}{25}$$

47. $\sin^2\left[\frac{1}{2}\cos^{-1}\frac{3}{5}\right] = \frac{1-\cos\left(\cos^{-1}\frac{3}{5}\right)}{2} = \frac{1-\frac{3}{5}}{2} = \frac{\frac{2}{5}}{2} = \frac{1}{5}$

49. $\sec\left(2\tan^{-1}\frac{3}{4}\right)$

Let $\alpha = \tan^{-1}\frac{3}{4}$. α is in quadrant I.

Then $\tan\alpha = \frac{3}{4}$, $-\frac{\pi}{2} < \alpha < \frac{\pi}{2}$.

$$\sec\alpha = \sqrt{\tan^2\alpha + 1} = \sqrt{\left(\frac{3}{4}\right)^2 + 1} = \sqrt{\frac{9}{16} + 1} = \sqrt{\frac{25}{16}} = \frac{5}{4}; \quad \cos\alpha = \frac{4}{5}$$

$$\sec\left[2\tan^{-1}\frac{3}{4}\right] = \sec 2\alpha = \frac{1}{\cos 2\alpha} = \frac{1}{2\cos^2\alpha - 1} = \frac{1}{2\left(\frac{4}{5}\right)^2 - 1} = \frac{1}{2\cdot\frac{16}{25} - 1} = \frac{1}{\frac{7}{25}} = \frac{25}{7}$$

51. $\cot^2\left(\frac{1}{2}\tan^{-1}\frac{4}{3}\right)$

Let $\alpha = \tan^{-1}\frac{4}{3}$. α is in quadrant I.

Then $\tan\alpha = \frac{4}{3}$, $-\frac{\pi}{2} < \alpha < \frac{\pi}{2}$.

$$\sec\alpha = \sqrt{\tan^2\alpha + 1} = \sqrt{\left(\frac{4}{3}\right)^2 + 1} = \sqrt{\frac{16}{9} + 1} = \sqrt{\frac{25}{9}} = \frac{5}{3}; \quad \cos\alpha = \frac{3}{5}$$

$$\cot^2\left[\frac{1}{2}\tan^{-1}\frac{4}{3}\right] = \cot^2\left(\frac{1}{2}\alpha\right) = \frac{1}{\tan^2\left(\frac{1}{2}\alpha\right)} = \frac{1}{\frac{1-\cos\alpha}{1+\cos\alpha}} = \frac{1+\cos\alpha}{1-\cos\alpha} = \frac{1+\frac{3}{5}}{1-\frac{3}{5}} = \frac{\frac{8}{5}}{\frac{2}{5}} = 4$$

53. $\cos\left(\cos^{-1}u + \sin^{-1}v\right)$

   Let $\alpha = \cos^{-1}u$ and $\beta = \sin^{-1}v$.

   Then $\cos\alpha = u,\ 0 \le \alpha \le \pi,$ and $\sin\beta = v,\ -\dfrac{\pi}{2} \le \beta \le \dfrac{\pi}{2}$

   $$\sin\alpha = \sqrt{1-\cos^2\alpha} = \sqrt{1-u^2}$$
   $$\cos\beta = \sqrt{1-\sin^2\beta} = \sqrt{1-v^2}$$

   $\cos\left(\cos^{-1}u + \sin^{-1}v\right) = \cos(\alpha+\beta) = \cos\alpha\cos\beta - \sin\alpha\sin\beta = u\sqrt{1-v^2} - v\sqrt{1-u^2}$

55. $\sin\left(\tan^{-1}u - \sin^{-1}v\right)$

   Let $\alpha = \tan^{-1}u$ and $\beta = \sin^{-1}v$.

   Then $\tan\alpha = u,\ -\dfrac{\pi}{2} < \alpha < \dfrac{\pi}{2},$ and $\sin\beta = v,\ -\dfrac{\pi}{2} \le \beta \le \dfrac{\pi}{2}$

   $$\sec\alpha = \sqrt{\tan^2\alpha + 1} = \sqrt{u^2 + 1};\qquad \cos\alpha = \frac{1}{\sqrt{u^2+1}}$$

   $$\sin\alpha = \sqrt{1-\cos^2\alpha} = \sqrt{1 - \frac{1}{u^2+1}} = \sqrt{\frac{u^2+1-1}{u^2+1}} = \sqrt{\frac{u^2}{u^2+1}} = \frac{u}{\sqrt{u^2+1}}$$

   $$\cos\beta = \sqrt{1-\sin^2\beta} = \sqrt{1-v^2}$$

   $\sin\left(\tan^{-1}u - \sin^{-1}v\right) = \sin(\alpha-\beta) = \sin\alpha\cos\beta - \cos\alpha\sin\beta$

   $$= \frac{u}{\sqrt{u^2+1}}\cdot\sqrt{1-v^2} - \frac{1}{\sqrt{u^2+1}}\cdot v = \frac{u\sqrt{1-v^2} - v}{\sqrt{u^2+1}}$$

57. $\tan\left(\sin^{-1}u - \cos^{-1}v\right)$

   Let $\alpha = \sin^{-1}u$ and $\beta = \cos^{-1}v$.

   Then $\sin\alpha = u,\ -\dfrac{\pi}{2} \le \alpha \le \dfrac{\pi}{2},$ and $\cos\beta = v,\ 0 \le \beta \le \pi$

   $$\cos\alpha = \sqrt{1-\sin^2\alpha} = \sqrt{1-u^2};\qquad \tan\alpha = \frac{\sin\alpha}{\cos\alpha} = \frac{u}{\sqrt{1-u^2}}$$

   $$\sin\beta = \sqrt{1-\cos^2\beta} = \sqrt{1-v^2};\qquad \tan\beta = \frac{\sin\beta}{\cos\beta} = \frac{\sqrt{1-v^2}}{v}$$

   $$\tan\left(\sin^{-1}u - \cos^{-1}v\right) = \tan(\alpha-\beta) = \frac{\tan\alpha - \tan\beta}{1+\tan\alpha\tan\beta} = \frac{\dfrac{u}{\sqrt{1-u^2}} - \dfrac{\sqrt{1-v^2}}{v}}{1 + \dfrac{u}{\sqrt{1-u^2}}\cdot\dfrac{\sqrt{1-v^2}}{v}}$$

   $$= \frac{\dfrac{uv - \sqrt{1-u^2}\sqrt{1-v^2}}{v\sqrt{1-u^2}}}{\dfrac{v\sqrt{1-u^2} + u\sqrt{1-v^2}}{v\sqrt{1-u^2}}} = \frac{uv - \sqrt{1-u^2}\sqrt{1-v^2}}{v\sqrt{1-u^2} + u\sqrt{1-v^2}}$$

59. Show that $\sec\left(\tan^{-1}v\right) = \sqrt{1+v^2}$.

    Let $\alpha = \tan^{-1}v$. Then $\tan\alpha = v$, $-\dfrac{\pi}{2} < \alpha < \dfrac{\pi}{2}$.

    $$\sec\left(\tan^{-1}v\right) = \sec\alpha = \sqrt{1+\tan^2\alpha} = \sqrt{1+v^2}$$

61. Show that $\tan\left(\cos^{-1}v\right) = \dfrac{\sqrt{1-v^2}}{v}$.

    Let $\alpha = \cos^{-1}v$. Then $\cos\alpha = v$, $0 \le \alpha \le \pi$.

    $$\tan\left(\cos^{-1}v\right) = \tan\alpha = \frac{\sin\alpha}{\cos\alpha} = \frac{\sqrt{1-\cos^2\alpha}}{\cos\alpha} = \frac{\sqrt{1-v^2}}{v}$$

63. Show that $\cos\left(\sin^{-1}v\right) = \sqrt{1-v^2}$.

    Let $\alpha = \sin^{-1}v$. Then $\sin\alpha = v$, $-\dfrac{\pi}{2} \le \alpha \le \dfrac{\pi}{2}$.

    $$\cos\left(\sin^{-1}v\right) = \cos\alpha = \sqrt{1-\sin^2\alpha} = \sqrt{1-v^2}$$

65. Show that $\sin^{-1}v + \cos^{-1}v = \dfrac{\pi}{2}$.

    Let $\alpha = \sin^{-1}v$ and $\beta = \cos^{-1}v$.

    Then $\sin\alpha = v = \cos\beta$, and since $\sin\alpha = \cos\left(\dfrac{\pi}{2}-\alpha\right)$, $\cos\left(\dfrac{\pi}{2}-\alpha\right) = \cos\beta$. If

    $v \ge 0$, then $0 \le \alpha \le \dfrac{\pi}{2}$, so that $\left(\dfrac{\pi}{2}-\alpha\right)$ and β both lie in the interval $\left[0,\dfrac{\pi}{2}\right]$. If

    $v < 0$, then $-\dfrac{\pi}{2} \le \alpha < 0$, so that $\left(\dfrac{\pi}{2}-\alpha\right)$ and β both lie in the interval $\left[\dfrac{\pi}{2},\pi\right]$.

    Either way, $\cos\left(\dfrac{\pi}{2}-\alpha\right) = \cos\beta$ implies $\dfrac{\pi}{2}-\alpha = \beta$, or $\alpha + \beta = \dfrac{\pi}{2}$. Thus,

    $\sin^{-1}v + \cos^{-1}v = \dfrac{\pi}{2}$.

67. Show that $\tan^{-1}\left(\dfrac{1}{v}\right) = \dfrac{\pi}{2} - \tan^{-1}v$, if $v > 0$.

    Let $\alpha = \tan^{-1}\left(\dfrac{1}{v}\right)$ and $\beta = \tan^{-1}v$. Because $\dfrac{1}{v}$ must be defined, $v \ne 0$ and so

    $\alpha, \beta \ne 0$. Then $\tan\alpha = \dfrac{1}{v} = \dfrac{1}{\tan\beta} = \cot\beta$, and since $\tan\alpha = \cot\left(\dfrac{\pi}{2}-\alpha\right)$,

    $\cot\left(\dfrac{\pi}{2}-\alpha\right) = \cot\beta$. Because $v > 0, 0 < \alpha < \dfrac{\pi}{2}$ and so $\dfrac{\pi}{2}-\alpha$ and β both lie in the

    interval $\left(0,\dfrac{\pi}{2}\right)$. Then, $\cot\left(\dfrac{\pi}{2}-\alpha\right) = \cot\beta$ implies $\dfrac{\pi}{2}-\alpha = \beta$ or $\alpha = \dfrac{\pi}{2}-\beta$.

    Thus, $\tan^{-1}\left(\dfrac{1}{v}\right) = \dfrac{\pi}{2} - \tan^{-1}v$, if $v > 0$.

69. $\sin\left(\sin^{-1}v + \cos^{-1}v\right) = \sin\left(\sin^{-1}v\right)\cos\left(\cos^{-1}v\right) + \cos\left(\sin^{-1}v\right)\sin\left(\cos^{-1}v\right)$

$$= v \cdot v + \sqrt{1-v^2}\sqrt{1-v^2} = v^2 + 1 - v^2 = 1$$

71. (a) $D = 24\left[1 - \dfrac{\cos^{-1}\left(\tan\left(23.5 \cdot \dfrac{\pi}{180}\right)\tan\left(29.75 \cdot \dfrac{\pi}{180}\right)\right)}{\pi}\right] = 13.92$ hours

(b) $D = 24\left[1 - \dfrac{\cos^{-1}\left(\tan\left(0 \cdot \dfrac{\pi}{180}\right)\tan\left(29.75 \cdot \dfrac{\pi}{180}\right)\right)}{\pi}\right] = 12$ hours

(c) $D = 24\left[1 - \dfrac{\cos^{-1}\left(\tan\left(22.8 \cdot \dfrac{\pi}{180}\right)\tan\left(29.75 \cdot \dfrac{\pi}{180}\right)\right)}{\pi}\right] = 13.85$ hours

73. (a) $D = 24\left(1 - \dfrac{\cos^{-1}\left(\tan\left(23.5 \cdot \dfrac{\pi}{180}\right)\tan\left(21.3 \cdot \dfrac{\pi}{180}\right)\right)}{\pi}\right) = 13.30$ hours

(b) $D = 24\left(1 - \dfrac{\cos^{-1}\left(\tan\left(0 \cdot \dfrac{\pi}{180}\right)\tan\left(21.3 \cdot \dfrac{\pi}{180}\right)\right)}{\pi}\right) = 12$ hours

(c) $D = 24\left(1 - \dfrac{\cos^{-1}\left(\tan\left(22.8 \cdot \dfrac{\pi}{180}\right)\tan\left(21.3 \cdot \dfrac{\pi}{180}\right)\right)}{\pi}\right) = 13.26$ hours

75. a) $D = 24\left(1 - \dfrac{\cos^{-1}\left(\tan\left(23.5 \cdot \dfrac{\pi}{180}\right)\tan\left(0 \cdot \dfrac{\pi}{180}\right)\right)}{\pi}\right) = 12$ hours

(b) $D = 24\left(1 - \dfrac{\cos^{-1}\left(\tan\left(0 \cdot \dfrac{\pi}{180}\right)\tan\left(0 \cdot \dfrac{\pi}{180}\right)\right)}{\pi}\right) = 12$ hours

(c) $D = 24\left(1 - \dfrac{\cos^{-1}\left(\tan\left(22.8 \cdot \dfrac{\pi}{180}\right)\tan\left(0 \cdot \dfrac{\pi}{180}\right)\right)}{\pi}\right) = 12$ hours

7.7 Trigonometric Equations (I)

1. $\sin\theta = \dfrac{1}{2}$

    $\theta = \dfrac{\pi}{6} + 2k\pi$ or $\theta = \dfrac{5\pi}{6} + 2k\pi$, where k is any integer

    The solutions on the interval $[0, 2\pi)$ are $\theta = \dfrac{\pi}{6}, \dfrac{5\pi}{6}$

3. $\tan\theta = -\dfrac{\sqrt{3}}{3}$

    $\theta = \dfrac{5\pi}{6} + k\pi$, where k is any integer

    The solutions on the interval $[0, 2\pi)$ are $\theta = \dfrac{5\pi}{6}, \dfrac{11\pi}{6}$

5. $\cos\theta = 0$

    $\theta = \dfrac{\pi}{2} + 2k\pi$ or $\theta = \dfrac{3\pi}{2} + 2k\pi$, where k is any integer

    The solutions on the interval $[0, 2\pi)$ are $\theta = \dfrac{\pi}{2}, \dfrac{3\pi}{2}$

7. $\sin 3\theta = -1$

    $3\theta = \dfrac{3\pi}{2} + 2k\pi \;\;\rightarrow\;\; \theta = \dfrac{\pi}{2} + \dfrac{2k\pi}{3}$, where k is any integer

    The solutions on the interval $[0, 2\pi)$ are $\theta = \dfrac{\pi}{2}, \dfrac{7\pi}{6}, \dfrac{11\pi}{6}$

9. $\cos 2\theta = -\dfrac{1}{2}$

    $2\theta = \dfrac{2\pi}{3} + 2k\pi \;\;\rightarrow\;\; \theta = \dfrac{\pi}{3} + k\pi$, where k is any integer

    $2\theta = \dfrac{4\pi}{3} + 2k\pi \;\;\rightarrow\;\; \theta = \dfrac{2\pi}{3} + k\pi$, where k is any integer

    The solutions on the interval $[0, 2\pi)$ are $\theta = \dfrac{\pi}{3}, \dfrac{2\pi}{3}, \dfrac{4\pi}{3}, \dfrac{5\pi}{3}$

11. $\sec\dfrac{3\theta}{2} = -2$

    $\dfrac{3\theta}{2} = \dfrac{2\pi}{3} + 2k\pi \;\;\rightarrow\;\; \theta = \dfrac{4\pi}{9} + \dfrac{4k\pi}{3}$, where k is any integer

    $\dfrac{3\theta}{2} = \dfrac{4\pi}{3} + 2k\pi \;\;\rightarrow\;\; \theta = \dfrac{8\pi}{9} + \dfrac{4k\pi}{3}$, where k is any integer

    The solutions on the interval $[0, 2\pi)$ are $\theta = \dfrac{4\pi}{9}, \dfrac{8\pi}{9}, \dfrac{16\pi}{9}$

13. $\cos\left(2\theta - \dfrac{\pi}{2}\right) = -1$

$2\theta - \dfrac{\pi}{2} = \pi + 2k\pi \quad \rightarrow \quad 2\theta = \dfrac{3\pi}{2} + 2k\pi \quad \rightarrow \quad \theta = \dfrac{3\pi}{4} + k\pi, \ k \text{ is any integer}$

The solutions on the interval $[0, 2\pi)$ are $\theta = \dfrac{3\pi}{4}, \dfrac{7\pi}{4}$.

15. $\tan\left(\dfrac{\theta}{2} + \dfrac{\pi}{3}\right) = 1$

$\dfrac{\theta}{2} + \dfrac{\pi}{3} = \dfrac{\pi}{4} + k\pi \quad \rightarrow \quad \dfrac{\theta}{2} = -\dfrac{\pi}{12} + k\pi \quad \rightarrow \quad \theta = -\dfrac{\pi}{6} + 2k\pi, \ k \text{ is any integer}$

The solutions on the interval $[0, 2\pi)$ are $\theta = \dfrac{11\pi}{6}$.

17. $2\sin\theta + 1 = 0 \quad \rightarrow \quad 2\sin\theta = -1 \quad \rightarrow \quad \sin\theta = -\dfrac{1}{2}$

$\theta = \dfrac{7\pi}{6} + 2k\pi \quad \text{or} \quad \theta = \dfrac{11\pi}{6} + 2k\pi, \ k \text{ is any integer}$

The solutions on the interval $[0, 2\pi)$ are $\theta = \dfrac{7\pi}{6}, \dfrac{11\pi}{6}$.

19. $\tan\theta + 1 = 0 \quad \rightarrow \quad \tan\theta = -1$

$\theta = \dfrac{3\pi}{4} + k\pi, \ k \text{ is any integer}$

The solutions on the interval $[0, 2\pi)$ are $\theta = \dfrac{3\pi}{4}, \dfrac{7\pi}{4}$.

21. $4\sec\theta + 6 = -2 \quad \rightarrow \quad 4\sec\theta = -8 \quad \rightarrow \quad \sec\theta = -2$

$\theta = \dfrac{2\pi}{3} + 2k\pi \quad \text{or} \quad \theta = \dfrac{4\pi}{3} + 2k\pi, \ k \text{ is any integer}$

The solutions on the interval $[0, 2\pi)$ are $\theta = \dfrac{2\pi}{3}, \dfrac{4\pi}{3}$.

23. $3\sqrt{2}\cos\theta + 2 = -1 \quad \rightarrow \quad 3\sqrt{2}\cos\theta = -3 \quad \rightarrow \quad \cos\theta = -\dfrac{1}{\sqrt{2}} = -\dfrac{\sqrt{2}}{2}$

$\theta = \dfrac{3\pi}{4} + 2k\pi \quad \text{or} \quad \theta = \dfrac{5\pi}{4} + 2k\pi, \ k \text{ is any integer}$

The solutions on the interval $[0, 2\pi)$ are $\theta = \dfrac{3\pi}{4}, \dfrac{5\pi}{4}$.

25. $\sin\theta = 0.4$

$\theta = 0.4115168 \ \text{ or } \ \theta = \pi - 0.4115168 = 2.7300758$

$\theta \approx 0.41, 2.73$

27. $\tan\theta = 5$

$\theta = 1.3734008 \ \text{ or } \ \theta = \pi + 1.3734008 = 4.5149934$

$\theta \approx 1.37, 4.51$

29. $\cos\theta = -0.9$

$\theta = 2.6905658$ or $\theta = 2\pi - 2.6905658 = 3.5926195$

$\theta \approx 2.69, 3.59$

31. $\sec\theta = -4 \quad\rightarrow\quad \cos\theta = -\dfrac{1}{4}$

$\theta = 1.8234766$ or $\theta = 2\pi - 1.8234766 = 4.4597087$

$\theta \approx 1.82, 4.46$

33. Use Snell's Law to solve:

$$\frac{\sin 40°}{\sin\theta_2} = 1.33$$

$$\sin 40° = 1.33\sin\theta_2$$

$$\sin\theta_2 = \frac{\sin 40°}{1.33} = 0.4833$$

$$\theta_2 = \sin^{-1} 0.4833 = 28.9°$$

35. Calculate the index of refraction for each:

$\theta_1 = 10°, \quad \theta_2 = 7°45' = 7.75° \qquad \dfrac{\sin\theta_1}{\sin\theta_2} = \dfrac{\sin 10°}{\sin 7.75°} \approx 1.2877$

$\theta_1 = 20°, \quad \theta_2 = 15°30' = 15.5° \qquad \dfrac{\sin\theta_1}{\sin\theta_2} = \dfrac{\sin 20°}{\sin 15.5°} \approx 1.2798$

$\theta_1 = 30°, \quad \theta_2 = 22°30' = 22.5° \qquad \dfrac{\sin\theta_1}{\sin\theta_2} = \dfrac{\sin 30°}{\sin 22.5°} \approx 1.3066$

$\theta_1 = 40°, \quad \theta_2 = 29°0' = 29° \qquad \dfrac{\sin\theta_1}{\sin\theta_2} = \dfrac{\sin 40°}{\sin 29°} \approx 1.3259$

$\theta_1 = 50°, \quad \theta_2 = 35°0' = 35° \qquad \dfrac{\sin\theta_1}{\sin\theta_2} = \dfrac{\sin 50°}{\sin 35°} \approx 1.3356$

$\theta_1 = 60°, \quad \theta_2 = 40°30' = 40.5° \qquad \dfrac{\sin\theta_1}{\sin\theta_2} = \dfrac{\sin 60°}{\sin 40.5°} \approx 1.3335$

$\theta_1 = 70°, \quad \theta_2 = 45°30' = 45.5° \qquad \dfrac{\sin\theta_1}{\sin\theta_2} = \dfrac{\sin 70°}{\sin 45.5°} \approx 1.3175$

$\theta_1 = 80°, \quad \theta_2 = 50°0' = 50° \qquad \dfrac{\sin\theta_1}{\sin\theta_2} = \dfrac{\sin 80°}{\sin 50°} \approx 1.2856$

The results range from 1.28 to 1.34 and are surprisingly close to Snell's Law.

37. Calculate the index of refraction:

$\theta_1 = 40°, \quad \theta_2 = 26° \qquad \dfrac{\sin\theta_1}{\sin\theta_2} = \dfrac{\sin 40°}{\sin 26°} \approx 1.47$

39. If θ is the original angle of incidence and ϕ is the angle of refraction, then $\dfrac{\sin\theta}{\sin\phi} = n_2$. The angle of incidence of the emerging beam is also ϕ, and the index of refraction is $\dfrac{1}{n_2}$. Thus, θ is the angle of refraction of the emerging beam. The two beams are parallel since the original angle of incidence and the angle of refraction of the emerging beam are equal.

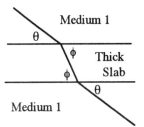

7.8 Trigonometric Equations (II)

1. $2\cos^2\theta + \cos\theta = 0$
$\cos\theta(2\cos\theta + 1) = 0$
$\cos\theta = 0$ or $2\cos\theta + 1 = 0$
$2\cos\theta = -1$
$\cos\theta = -\dfrac{1}{2}$
$\theta = \dfrac{\pi}{2}, \dfrac{3\pi}{2}$ or $\theta = \dfrac{2\pi}{3}, \dfrac{4\pi}{3}$

3. $2\sin^2\theta - \sin\theta - 1 = 0$
$(2\sin\theta + 1)(\sin\theta - 1) = 0$
$2\sin\theta + 1 = 0$ or $\sin\theta - 1 = 0$
$2\sin\theta = -1$ $\sin\theta = 1$
$\sin\theta = -\dfrac{1}{2}$
$\theta = \dfrac{7\pi}{6}, \dfrac{11\pi}{6}$ or $\theta = \dfrac{\pi}{2}$

5. $(\tan\theta - 1)(\sec\theta - 1) = 0$
$\tan\theta - 1 = 0$ or $\sec\theta - 1 = 0$
$\tan\theta = 1$ $\sec\theta = 1$
$\theta = \dfrac{\pi}{4}, \dfrac{5\pi}{4}$ or $\theta = 0$

7. $\cos\theta = \sin\theta$
$\dfrac{\sin\theta}{\cos\theta} = 1$
$\tan\theta = 1$
$\theta = \dfrac{\pi}{4}, \dfrac{5\pi}{4}$

9.
$$\tan \theta = 2\sin \theta$$
$$\frac{\sin \theta}{\cos \theta} = 2\sin \theta$$
$$\sin \theta = 2\sin \theta \cos \theta$$
$$0 = 2\sin \theta \cos \theta - \sin \theta$$
$$0 = \sin \theta (2\cos \theta - 1)$$
$$2\cos \theta - 1 = 0 \quad \text{or} \quad \sin \theta = 0$$
$$\cos \theta = \frac{1}{2}$$
$$\theta = \frac{\pi}{3}, \frac{5\pi}{3} \quad \text{or} \quad \theta = 0, \pi$$

11.
$$\sin \theta = \csc \theta$$
$$\sin \theta = \frac{1}{\sin \theta}$$
$$\sin^2 \theta = 1$$
$$\sin \theta = \pm 1$$
$$\theta = \frac{\pi}{2}, \frac{3\pi}{2}$$

13.
$$\cos(2\theta) = \cos \theta$$
$$2\cos^2 \theta - 1 = \cos \theta$$
$$2\cos^2 \theta - \cos \theta - 1 = 0$$
$$(2\cos \theta + 1)(\cos \theta - 1) = 0$$
$$2\cos \theta + 1 = 0 \quad \text{or} \quad \cos \theta - 1 = 0$$
$$\cos \theta = -\frac{1}{2} \qquad \cos \theta = 1$$
$$\theta = \frac{2\pi}{3}, \frac{4\pi}{3} \quad \text{or} \quad \theta = 0$$

15.
$$\sin(2\theta) + \sin(4\theta) = 0$$
$$\sin(2\theta) + 2\sin(2\theta)\cos(2\theta) = 0$$
$$\sin(2\theta)\big(1 + 2\cos(2\theta)\big) = 0$$
$$1 + 2\cos(2\theta) = 0 \quad \text{or} \quad \sin(2\theta) = 0$$
$$\cos(2\theta) = -\frac{1}{2}$$
$$2\theta = \frac{2\pi}{3} + 2k\pi \quad \rightarrow \quad \theta = \frac{\pi}{3} + k\pi$$
$$2\theta = \frac{4\pi}{3} + 2k\pi \quad \rightarrow \quad \theta = \frac{2\pi}{3} + k\pi$$
$$2\theta = 0 + 2k\pi \quad \rightarrow \quad \theta = k\pi$$
$$2\theta = \pi + 2k\pi \quad \rightarrow \quad \theta = \frac{\pi}{2} + k\pi$$
$$\theta = 0, \frac{\pi}{3}, \frac{\pi}{2}, \frac{2\pi}{3}, \pi, \frac{4\pi}{3}, \frac{3\pi}{2}, \frac{5\pi}{3}$$

17. $\cos(4\theta) - \cos(6\theta) = 0$

$\cos(5\theta - \theta) - \cos(5\theta + \theta) = 0$

$-2\sin(5\theta)\sin(-\theta) = 0$

$2\sin(5\theta)\sin\theta = 0$

$\sin(5\theta) = 0$ or $\sin\theta = 0$

$5\theta = 0 + 2k\pi \quad \rightarrow \quad \theta = \dfrac{2k\pi}{5}$

$5\theta = \pi + 2k\pi \quad \rightarrow \quad \theta = \dfrac{\pi}{5} + \dfrac{2k\pi}{5}$

$\theta = 0 + 2k\pi$

$\theta = \pi + 2k\pi$

$\theta = 0, \dfrac{\pi}{5}, \dfrac{2\pi}{5}, \dfrac{3\pi}{5}, \dfrac{4\pi}{5}, \pi, \dfrac{6\pi}{5}, \dfrac{7\pi}{5}, \dfrac{8\pi}{5}, \dfrac{9\pi}{5}$

19. $1 + \sin\theta = 2\cos^2\theta$

$1 + \sin\theta = 2(1 - \sin^2\theta)$

$1 + \sin\theta = 2 - 2\sin^2\theta$

$2\sin^2\theta + \sin\theta - 1 = 0$

$(2\sin\theta - 1)(\sin\theta + 1) = 0$

$2\sin\theta - 1 = 0$ or $\sin\theta + 1 = 0$

$\sin\theta = \dfrac{1}{2} \qquad \sin\theta = -1$

$\theta = \dfrac{\pi}{6}, \dfrac{5\pi}{6} \quad$ or $\theta = \dfrac{3\pi}{2}$

21. $\tan^2\theta = \tfrac{3}{2}\sec\theta$

$\sec^2\theta - 1 = \dfrac{3}{2}\sec\theta$

$2\sec^2\theta - 2 = 3\sec\theta$

$2\sec^2\theta - 3\sec\theta - 2 = 0$

$(2\sec\theta + 1)(\sec\theta - 2) = 0$

$2\sec\theta + 1 = 0$ or $\sec\theta - 2 = 0$

$\sec\theta = -\dfrac{1}{2} \qquad$ has no solution

$\sec\theta = 2 \quad \rightarrow \quad \theta = \dfrac{\pi}{3}, \dfrac{5\pi}{3}$

23. $3 - \sin\theta = \cos(2\theta)$

$3 - \sin\theta = 1 - 2\sin^2\theta$

$2\sin^2\theta - \sin\theta + 2 = 0$

This is a quadratic equation in $\sin\theta$. The discriminant is
$b^2 - 4ac = 1 - 16 = -15 < 0$. The equation has no real solutions.

25. $\sec^2 \theta + \tan \theta = 0$
$\tan^2 \theta + 1 + \tan \theta = 0$
This is a quadratic equation in $\tan \theta$. The discriminant is $b^2 - 4ac = 1 - 4 = -3 < 0$.
The equation has no real solutions.

27. $\sin \theta - \sqrt{3} \cos \theta = 1$
Divide each side by 2:
$$\frac{1}{2} \sin \theta - \frac{\sqrt{3}}{2} \cos \theta = \frac{1}{2}$$
Rewrite in the difference of two angles form where
$$\cos \phi = \frac{1}{2} \text{ and } \sin \phi = \frac{\sqrt{3}}{2} \text{ and } \phi = \frac{\pi}{3}:$$
$$\sin \theta \cos \phi - \cos \theta \sin \phi = \frac{1}{2}$$
$$\sin(\theta - \phi) = \frac{1}{2}$$
$$\theta - \phi = \frac{\pi}{6} \quad \text{or} \quad \theta - \phi = \frac{5\pi}{6}$$
$$\theta - \frac{\pi}{3} = \frac{\pi}{6} \quad \text{or} \quad \theta - \frac{\pi}{3} = \frac{5\pi}{6}$$
$$\theta = \frac{\pi}{2} \quad \text{or} \quad \theta = \frac{7\pi}{6}$$

29. $\tan(2\theta) + 2\sin \theta = 0$
$$\frac{\sin(2\theta)}{\cos(2\theta)} + 2\sin \theta = 0$$
$$\frac{\sin 2\theta + 2\sin \theta \cos 2\theta}{\cos 2\theta} = 0$$
$$2\sin \theta \cos \theta + 2\sin \theta (2\cos^2 \theta - 1) = 0$$
$$2\sin \theta \left(\cos \theta + 2\cos^2 \theta - 1 \right) = 0$$
$$2\sin \theta \left(2\cos^2 \theta + \cos \theta - 1 \right) = 0$$
$$2\sin \theta (2\cos \theta - 1)(\cos \theta + 1) = 0$$
$$2\cos \theta - 1 = 0 \quad \text{or} \quad 2\sin \theta = 0 \quad \text{or} \quad \cos \theta + 1 = 0$$
$$\cos \theta = \frac{1}{2} \qquad \sin \theta = 0 \qquad \cos \theta = -1$$
$$\theta = \frac{\pi}{3}, \frac{5\pi}{3} \quad \text{or} \quad \theta = 0, \pi \quad \text{or} \quad \theta = \pi$$

$$\theta = 0, \frac{\pi}{3}, \pi, \frac{5\pi}{3}$$

31. $\sin\theta + \cos\theta = \sqrt{2}$

Divide each side by $\sqrt{2}$:

$$\frac{1}{\sqrt{2}}\sin\theta + \frac{1}{\sqrt{2}}\cos\theta = 1$$

Rewrite in the sum of two angles form where $\cos\phi = \frac{1}{\sqrt{2}}$ and $\sin\phi = \frac{1}{\sqrt{2}}$ and $\phi = \frac{\pi}{4}$:

$$\sin\theta\cos\phi + \cos\theta\sin\phi = 1$$
$$\sin(\theta + \phi) = 1$$
$$\theta + \phi = \frac{\pi}{2}$$
$$\theta + \frac{\pi}{4} = \frac{\pi}{2}$$
$$\theta = \frac{\pi}{4}$$

33. Use INTERSECT to solve:

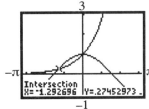 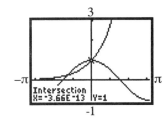

$x = -1.29,\ 0$

35. Use INTERSECT to solve:

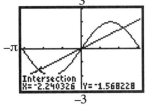 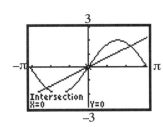 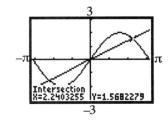

$x = -2.24,\ 0,\ 2.24$

37. Use INTERSECT to solve:

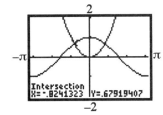 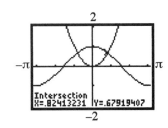

$x = -0.82,\ 0.82$

39. $x + 5\cos x = 0$
Find the intersection of
$y_1 = x + 5\cos x$ and $y_2 = 0$:

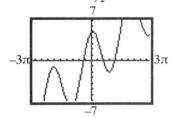

$x = -1.31, 1.98, 3.84$

41. $22x - 17\sin x = 3$
Find the intersection of
$y_1 = 22x - 17\sin x$ and $y_2 = 3$:

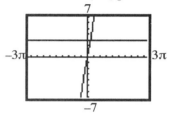

$x = 0.52$

43. $\sin x + \cos x = x$
Find the intersection of
$y_1 = \sin x + \cos x$ and $y_2 = x$:

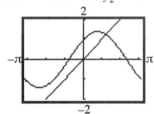

$x = 1.26$

45. $x^2 - 2\cos x = 0$
Find the intersection of
$y_1 = x^2 - 2\cos x$ and $y_2 = 0$:

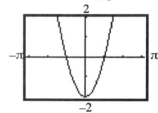

$x = -1.02, 1.02$

47. $x^2 - 2\sin 2x = 3x$
Find the intersection of
$y_1 = x^2 - 2\sin 2x$ and $y_2 = 3x$:

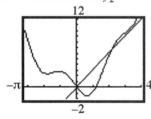

$x = 0, 2.15$

49. $6\sin x - e^x = 2, \ x > 0$
Find the intersection of
$y_1 = 6\sin x - e^x$ and $y_2 = 2$:

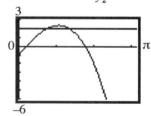

$x = 0.76, 1.35$

51. (a) Solve: $\quad \cos(2\theta) + \cos\theta = 0, \ \ 0° < \theta < 90°$

$$2\cos^2\theta - 1 + \cos\theta = 0$$
$$2\cos^2\theta + \cos\theta - 1 = 0$$
$$(2\cos\theta - 1)(\cos\theta + 1) = 0$$
$$2\cos\theta - 1 = 0 \ \ \text{or} \ \ \cos\theta + 1 = 0$$
$$\cos\theta = \frac{1}{2} \qquad \cos\theta = -1$$
$$\theta = 60°, 300° \ \ \text{or} \ \ \theta = 180°$$

The solution is $60°$.

(b)   Solve: $\cos(2\theta) + \cos\theta = 0, \quad 0° < \theta < 90°$

$$2\cos\frac{3\theta}{2}\cos\frac{\theta}{2} = 0$$

$$\cos\frac{3\theta}{2} = 0 \quad \text{or} \quad \cos\frac{\theta}{2} = 0$$

$$\frac{3\theta}{2} = 90° \quad \rightarrow \quad \theta = 60°$$

$$\frac{3\theta}{2} = 270° \quad \rightarrow \quad \theta = 180°$$

$$\frac{\theta}{2} = 90° \quad \rightarrow \quad \theta = 180°$$

$$\frac{\theta}{2} = 270° \quad \rightarrow \quad \theta = 540°$$

The solution is 60°.

(c) $A(60°) = 16\sin 60°(\cos 60° + 1) = 16 \cdot \dfrac{\sqrt{3}}{2}\left(\dfrac{1}{2}+1\right) = 12\sqrt{3} \text{ in}^2 \approx 20.78 \text{ in}^2$

(d)   Graph and use the MAXIMUM feature:

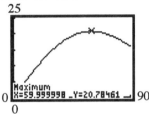

The maximum area is approximately 20.78 in² when the angle is 60°.

53.   Graph:

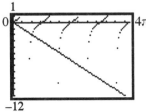

The first two positive solutions are 2.03 and 4.91.

55.   (a) $107 = \dfrac{(34.8)^2 \sin 2\theta}{9.8}$

$$\sin 2\theta = \dfrac{107(9.8)}{(34.8)^2} = 0.8659$$

$$2\theta = \sin^{-1} 0.8659 = 59.98° \text{ or } 120.02°$$

$$\theta = 29.99° \text{ or } 60.01°$$

(b)   Graph and use the MAXIMUM feature:
        The maximum distance is 123.58 meters when the angle is 45°.

(c) Graph:

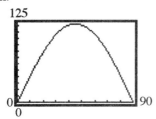

7 Chapter Review

1. $\tan\theta\cot\theta - \sin^2\theta = \tan\theta \cdot \dfrac{1}{\tan\theta} - \sin^2\theta = 1 - \sin^2\theta = \cos^2\theta$

3. $\cos^2\theta(1 + \tan^2\theta) = \cos^2\theta\cdot\sec^2\theta = \cos^2\theta\cdot\dfrac{1}{\cos^2\theta} = 1$

5. $4\cos^2\theta + 3\sin^2\theta = \cos^2\theta + 3\cos^2\theta + 3\sin^2\theta = \cos^2\theta + 3(\cos^2\theta + \sin^2\theta)$
    $= \cos^2\theta + 3\cdot 1 = \cos^2\theta + 3 = 3 + \cos^2\theta$

7. $\dfrac{1 - \cos\theta}{\sin\theta} + \dfrac{\sin\theta}{1 - \cos\theta} = \dfrac{(1 - \cos\theta)^2 + \sin^2\theta}{\sin\theta(1 - \cos\theta)} = \dfrac{1 - 2\cos\theta + \cos^2\theta + \sin^2\theta}{\sin\theta(1 - \cos\theta)}$
    $= \dfrac{1 - 2\cos\theta + 1}{\sin\theta(1 - \cos\theta)} = \dfrac{2 - 2\cos\theta}{\sin\theta(1 - \cos\theta)} = \dfrac{2(1 - \cos\theta)}{\sin\theta(1 - \cos\theta)} = \dfrac{2}{\sin\theta} = 2\csc\theta$

9. $\dfrac{\cos\theta}{\cos\theta - \sin\theta} = \dfrac{\cos\theta}{\cos\theta - \sin\theta}\cdot\dfrac{\dfrac{1}{\cos\theta}}{\dfrac{1}{\cos\theta}} = \dfrac{1}{1 - \dfrac{\sin\theta}{\cos\theta}} = \dfrac{1}{1 - \tan\theta}$

11. $\dfrac{\csc\theta}{1 + \csc\theta} = \dfrac{\dfrac{1}{\sin\theta}}{1 + \dfrac{1}{\sin\theta}} = \dfrac{\dfrac{1}{\sin\theta}}{\dfrac{\sin\theta + 1}{\sin\theta}} = \dfrac{1}{1 + \sin\theta}\cdot\dfrac{1 - \sin\theta}{1 - \sin\theta} = \dfrac{1 - \sin\theta}{1 - \sin^2\theta} = \dfrac{1 - \sin\theta}{\cos^2\theta}$

13. $\csc\theta - \sin\theta = \dfrac{1}{\sin\theta} - \sin\theta = \dfrac{1 - \sin^2\theta}{\sin\theta} = \dfrac{\cos^2\theta}{\sin\theta} = \cos\theta\cdot\dfrac{\cos\theta}{\sin\theta} = \cos\theta\cot\theta$

15. $\dfrac{1 - \sin\theta}{\sec\theta} = \cos\theta(1 - \sin\theta) = \cos\theta(1 - \sin\theta)\cdot\dfrac{1 + \sin\theta}{1 + \sin\theta} = \dfrac{\cos\theta(1 - \sin^2\theta)}{1 + \sin\theta}$
    $= \dfrac{\cos\theta(\cos^2\theta)}{1 + \sin\theta} = \dfrac{\cos^3\theta}{1 + \sin\theta}$

17. $\cot\theta - \tan\theta = \dfrac{\cos\theta}{\sin\theta} - \dfrac{\sin\theta}{\cos\theta} = \dfrac{\cos^2\theta - \sin^2\theta}{\sin\theta\cos\theta} = \dfrac{1 - \sin^2\theta - \sin^2\theta}{\sin\theta\cos\theta} = \dfrac{1 - 2\sin^2\theta}{\sin\theta\cos\theta}$

19. $\dfrac{\cos(\alpha+\beta)}{\cos\alpha\sin\beta} = \dfrac{\cos\alpha\cos\beta-\sin\alpha\sin\beta}{\cos\alpha\sin\beta} = \dfrac{\cos\alpha\cos\beta}{\cos\alpha\sin\beta} - \dfrac{\sin\alpha\sin\beta}{\cos\alpha\sin\beta} = \cot\beta - \tan\alpha$

21. $\dfrac{\cos(\alpha-\beta)}{\cos\alpha\cos\beta} = \dfrac{\cos\alpha\cos\beta+\sin\alpha\sin\beta}{\cos\alpha\cos\beta} = \dfrac{\cos\alpha\cos\beta}{\cos\alpha\cos\beta} + \dfrac{\sin\alpha\sin\beta}{\cos\alpha\cos\beta} = 1 + \tan\alpha\tan\beta$

23. $(1+\cos\theta)\left(\tan\dfrac{\theta}{2}\right) = (1+\cos\theta)\cdot\dfrac{\sin\theta}{1+\cos\theta} = \sin\theta$

25. $2\cot\theta\cot(2\theta) = 2\cdot\dfrac{\cos\theta}{\sin\theta}\cdot\dfrac{\cos 2\theta}{\sin 2\theta} = \dfrac{2\cos\theta(\cos^2\theta-\sin^2\theta)}{\sin\theta 2\sin\theta\cos\theta} = \dfrac{\cos^2\theta-\sin^2\theta}{\sin^2\theta}$

$= \cot^2\theta - 1$

27. $1 - 8\sin^2\theta\cos^2\theta = 1 - 2(2\sin\theta\cos\theta)^2 = 1 - 2\sin^2(2\theta) = \cos(4\theta)$

29. $\dfrac{\sin(2\theta)+\sin(4\theta)}{\cos(2\theta)+\cos(4\theta)} = \dfrac{2\sin(3\theta)\cos(-\theta)}{2\cos(3\theta)\cos(-\theta)} = \dfrac{\sin(3\theta)}{\cos(3\theta)} = \tan(3\theta)$

31. $\dfrac{\cos(2\theta)-\cos(4\theta)}{\cos(2\theta)+\cos(4\theta)} - \tan\theta\tan(3\theta) = \dfrac{-2\sin(3\theta)\sin(-\theta)}{2\cos(3\theta)\cos(-\theta)} - \tan\theta\tan(3\theta)$

$= \dfrac{2\sin(3\theta)\sin\theta}{2\cos(3\theta)\cos\theta} - \tan\theta\tan(3\theta) = \tan(3\theta)\tan\theta - \tan\theta\tan(3\theta) = 0$

33. $\sin 165^\circ = \sin(120^\circ + 45^\circ) = \sin 120^\circ\cos 45^\circ + \cos 120^\circ\sin 45^\circ$

$= \dfrac{\sqrt{3}}{2}\cdot\dfrac{\sqrt{2}}{2} + \dfrac{1}{2}\cdot\dfrac{\sqrt{2}}{2} = \dfrac{1}{4}\left(\sqrt{6}+\sqrt{2}\right)$

35. $\cos\left(\dfrac{5\pi}{12}\right) = \cos\left(\dfrac{3\pi}{12}+\dfrac{2\pi}{12}\right) = \cos\dfrac{\pi}{4}\cos\dfrac{\pi}{6} - \sin\dfrac{\pi}{4}\sin\dfrac{\pi}{6} = \dfrac{\sqrt{2}}{2}\cdot\dfrac{\sqrt{3}}{2} - \dfrac{\sqrt{2}}{2}\cdot\dfrac{1}{2}$

$= \dfrac{1}{4}\left(\sqrt{6}-\sqrt{2}\right)$

37. $\cos 80^\circ\cos 20^\circ + \sin 80^\circ\sin 20^\circ = \cos(80^\circ - 20^\circ) = \cos 60^\circ = \dfrac{1}{2}$

39. $\tan\dfrac{\pi}{8} = \tan\dfrac{\frac{\pi}{4}}{2} = \sqrt{\dfrac{1-\cos\frac{\pi}{4}}{1+\cos\frac{\pi}{4}}} = \sqrt{\dfrac{1-\frac{\sqrt{2}}{2}}{1+\frac{\sqrt{2}}{2}}} = \sqrt{\dfrac{2-\sqrt{2}}{2+\sqrt{2}}\cdot\dfrac{2-\sqrt{2}}{2-\sqrt{2}}}$

$= \sqrt{\dfrac{6-4\sqrt{2}}{2}} = \sqrt{3-2\sqrt{2}}$

41. $\sin\alpha = \dfrac{4}{5},\ 0 < \alpha < \dfrac{\pi}{2};\qquad \sin\beta = \dfrac{5}{13},\ \dfrac{\pi}{2} < \beta < \pi$

$\cos\alpha = \dfrac{3}{5},\ \tan\alpha = \dfrac{4}{3},\ \cos\beta = -\dfrac{12}{13},\ \tan\beta = -\dfrac{5}{12},\quad 0 < \dfrac{\alpha}{2} < \dfrac{\pi}{4},\ \dfrac{\pi}{4} < \dfrac{\beta}{2} < \dfrac{\pi}{2}$

(a) $\sin(\alpha+\beta) = \sin\alpha\cos\beta + \cos\alpha\sin\beta = \dfrac{4}{5}\cdot\dfrac{-12}{13} + \dfrac{3}{5}\cdot\dfrac{5}{13} = \dfrac{-48+15}{65} = -\dfrac{33}{65}$

(b) $\cos(\alpha+\beta) = \cos\alpha\cos\beta - \sin\alpha\sin\beta = \dfrac{3}{5}\cdot\dfrac{-12}{13} - \dfrac{4}{5}\cdot\dfrac{5}{13} = \dfrac{-36-20}{65} = -\dfrac{56}{65}$

(c) $\sin(\alpha-\beta) = \sin\alpha\cos\beta - \cos\alpha\sin\beta = \dfrac{4}{5}\cdot\dfrac{-12}{13} - \dfrac{3}{5}\cdot\dfrac{5}{13} = \dfrac{-48-15}{65} = -\dfrac{63}{65}$

(d) $\tan(\alpha+\beta) = \dfrac{\tan\alpha + \tan\beta}{1 - \tan\alpha\tan\beta} = \dfrac{\dfrac{4}{3} + \left(-\dfrac{5}{12}\right)}{1 - \dfrac{4}{3}\cdot\left(-\dfrac{5}{12}\right)} = \dfrac{\dfrac{11}{12}}{\dfrac{14}{9}} = \dfrac{11}{12}\cdot\dfrac{9}{14} = \dfrac{33}{56}$

(e) $\sin(2\alpha) = 2\sin\alpha\cos\alpha = 2\cdot\dfrac{4}{5}\cdot\dfrac{3}{5} = \dfrac{24}{25}$

(f) $\cos(2\beta) = \cos^2\beta - \sin^2\beta = \left(-\dfrac{12}{13}\right)^2 - \left(\dfrac{5}{13}\right)^2 = \dfrac{144}{169} - \dfrac{25}{169} = \dfrac{119}{169}$

(g) $\sin\dfrac{\beta}{2} = \sqrt{\dfrac{1-\cos\beta}{2}} = \sqrt{\dfrac{1-\left(-\frac{12}{13}\right)}{2}} = \sqrt{\dfrac{\frac{25}{13}}{2}} = \sqrt{\dfrac{25}{26}} = \dfrac{5}{\sqrt{26}} = \dfrac{5\sqrt{26}}{26}$

(h) $\cos\dfrac{\alpha}{2} = \sqrt{\dfrac{1+\cos\alpha}{2}} = \sqrt{\dfrac{1+\frac{3}{5}}{2}} = \sqrt{\dfrac{\frac{8}{5}}{2}} = \sqrt{\dfrac{4}{5}} = \dfrac{2}{\sqrt{5}} = \dfrac{2\sqrt{5}}{5}$

43. $\sin\alpha = -\dfrac{3}{5},\ \pi < \alpha < \dfrac{3\pi}{2};\qquad \cos\beta = \dfrac{12}{13},\ \dfrac{3\pi}{2} < \beta < 2\pi$

$\cos\alpha = -\dfrac{4}{5},\ \tan\alpha = \dfrac{3}{4},\ \sin\beta = -\dfrac{5}{13},\ \tan\beta = -\dfrac{5}{12},\quad \dfrac{\pi}{2} < \dfrac{\alpha}{2} < \dfrac{3\pi}{4},\ \dfrac{3\pi}{4} < \dfrac{\beta}{2} < \pi$

(a) $\sin(\alpha+\beta) = \sin\alpha\cos\beta + \cos\alpha\sin\beta = \dfrac{-3}{5}\cdot\dfrac{12}{13} + \dfrac{-4}{5}\cdot\dfrac{-5}{13} = \dfrac{-36+20}{65} = -\dfrac{16}{65}$

(b) $\cos(\alpha+\beta) = \cos\alpha\cos\beta - \sin\alpha\sin\beta = \dfrac{-4}{5}\cdot\dfrac{12}{13} - \dfrac{-3}{5}\cdot\dfrac{-5}{13} = \dfrac{-48-15}{65} = -\dfrac{63}{65}$

(c) $\sin(\alpha-\beta) = \sin\alpha\cos\beta - \cos\alpha\sin\beta = \dfrac{-3}{5}\cdot\dfrac{12}{13} - \dfrac{-4}{5}\cdot\dfrac{-5}{13} = \dfrac{-36-20}{65} = -\dfrac{56}{65}$

(d) $\tan(\alpha+\beta) = \dfrac{\tan\alpha + \tan\beta}{1 - \tan\alpha\tan\beta} = \dfrac{\dfrac{3}{4} + \left(-\dfrac{5}{12}\right)}{1 - \dfrac{3}{4}\cdot\left(-\dfrac{5}{12}\right)} = \dfrac{\dfrac{1}{3}}{\dfrac{21}{16}} = \dfrac{1}{3}\cdot\dfrac{16}{21} = \dfrac{16}{63}$

(e) $\sin(2\alpha) = 2\sin\alpha\cos\alpha = 2\cdot\dfrac{-3}{5}\cdot\dfrac{-4}{5} = \dfrac{24}{25}$

(f) $\cos(2\beta) = \cos^2\beta - \sin^2\beta = \left(\dfrac{12}{13}\right)^2 - \left(\dfrac{-5}{13}\right)^2 = \dfrac{144}{169} - \dfrac{25}{169} = \dfrac{119}{169}$

(g) $\sin\dfrac{\beta}{2} = \sqrt{\dfrac{1-\cos\beta}{2}} = \sqrt{\dfrac{1-\left(\frac{12}{13}\right)}{2}} = \sqrt{\dfrac{\frac{1}{13}}{2}} = \sqrt{\dfrac{1}{26}} = \dfrac{1}{\sqrt{26}} = \dfrac{\sqrt{26}}{26}$

(h) $\cos\dfrac{\alpha}{2} = -\sqrt{\dfrac{1+\cos\alpha}{2}} = -\sqrt{\dfrac{1+\frac{-4}{5}}{2}} = -\sqrt{\dfrac{\frac{1}{5}}{2}} = -\sqrt{\dfrac{1}{10}} = -\dfrac{1}{\sqrt{10}} = -\dfrac{\sqrt{10}}{10}$

45. $\tan\alpha = \dfrac{3}{4}, \ \pi < \alpha < \dfrac{3\pi}{2}; \qquad \tan\beta = \dfrac{12}{5}, \ 0 < \beta < \dfrac{\pi}{2}$

$\sin\alpha = -\dfrac{3}{5}, \cos\alpha = -\dfrac{4}{5}, \sin\beta = \dfrac{12}{13}, \cos\beta = \dfrac{5}{13}, \ \dfrac{\pi}{2} < \dfrac{\alpha}{2} < \dfrac{3\pi}{4}, \ 0 < \dfrac{\beta}{2} < \dfrac{\pi}{4}$

(a) $\sin(\alpha+\beta) = \sin\alpha\cos\beta + \cos\alpha\sin\beta = \dfrac{-3}{5}\cdot\dfrac{5}{13} + \dfrac{-4}{5}\cdot\dfrac{12}{13} = \dfrac{-15-48}{65} = -\dfrac{63}{65}$

(b) $\cos(\alpha+\beta) = \cos\alpha\cos\beta - \sin\alpha\sin\beta = \dfrac{-4}{5}\cdot\dfrac{5}{13} - \dfrac{-3}{5}\cdot\dfrac{12}{13} = \dfrac{-20+36}{65} = \dfrac{16}{65}$

(c) $\sin(\alpha-\beta) = \sin\alpha\cos\beta - \cos\alpha\sin\beta = \dfrac{-3}{5}\cdot\dfrac{5}{13} - \dfrac{-4}{5}\cdot\dfrac{12}{13} = \dfrac{-15+48}{65} = \dfrac{33}{65}$

(d) $\tan(\alpha+\beta) = \dfrac{\tan\alpha + \tan\beta}{1 - \tan\alpha\tan\beta} = \dfrac{\dfrac{3}{4} + \left(\dfrac{12}{5}\right)}{1 - \dfrac{3}{4}\cdot\left(\dfrac{12}{5}\right)} = \dfrac{\dfrac{15+48}{20}}{-\dfrac{4}{5}} = \dfrac{63}{20}\cdot\dfrac{-5}{4} = -\dfrac{63}{16}$

(e) $\sin(2\alpha) = 2\sin\alpha\cos\alpha = 2\cdot\dfrac{-3}{5}\cdot\dfrac{-4}{5} = \dfrac{24}{25}$

(f) $\cos(2\beta) = \cos^2\beta - \sin^2\beta = \left(\dfrac{5}{13}\right)^2 - \left(\dfrac{12}{13}\right)^2 = \dfrac{25}{169} - \dfrac{144}{169} = -\dfrac{119}{169}$

(g) $\sin\dfrac{\beta}{2} = \sqrt{\dfrac{1-\cos\beta}{2}} = \sqrt{\dfrac{1-\left(\frac{5}{13}\right)}{2}} = \sqrt{\dfrac{\frac{8}{13}}{2}} = \sqrt{\dfrac{4}{13}} = \dfrac{2}{\sqrt{13}} = \dfrac{2\sqrt{13}}{13}$

(h) $\cos\dfrac{\alpha}{2} = -\sqrt{\dfrac{1+\cos\alpha}{2}} = -\sqrt{\dfrac{1+\frac{-4}{5}}{2}} = -\sqrt{\dfrac{\frac{1}{5}}{2}} = -\sqrt{\dfrac{1}{10}} = -\dfrac{1}{\sqrt{10}} = -\dfrac{\sqrt{10}}{10}$

47. $\sec\alpha = 2, \ -\dfrac{\pi}{2} < \alpha < 0; \qquad \sec\beta = 3, \ \dfrac{3\pi}{2} < \beta < 2\pi$

$\sin\alpha = -\dfrac{\sqrt{3}}{2}, \cos\alpha = \dfrac{1}{2}, \tan\alpha = -\sqrt{3}, \sin\beta = -\dfrac{2\sqrt{2}}{3}, \cos\beta = \dfrac{1}{3}, \tan\beta = -2\sqrt{2},$

$-\dfrac{\pi}{4} < \dfrac{\alpha}{2} < 0, \ \dfrac{3\pi}{4} < \dfrac{\beta}{2} < \pi$

(a) $\sin(\alpha+\beta) = \sin\alpha\cos\beta + \cos\alpha\sin\beta = \dfrac{-\sqrt{3}}{2}\cdot\dfrac{1}{3} + \dfrac{1}{2}\cdot\dfrac{-2\sqrt{2}}{3} = \dfrac{-\sqrt{3} - 2\sqrt{2}}{6}$

(b) $\cos(\alpha+\beta) = \cos\alpha\cos\beta - \sin\alpha\sin\beta = \dfrac{1}{2}\cdot\dfrac{1}{3} - \dfrac{-\sqrt{3}}{2}\cdot\dfrac{-2\sqrt{2}}{3} = \dfrac{1-2\sqrt{6}}{6}$

(c) $\sin(\alpha-\beta) = \sin\alpha\cos\beta - \cos\alpha\sin\beta = \dfrac{-\sqrt{3}}{2}\cdot\dfrac{1}{3} - \dfrac{1}{2}\cdot\dfrac{-2\sqrt{2}}{3} = \dfrac{-\sqrt{3} + 2\sqrt{2}}{6}$

(d) $\tan(\alpha+\beta) = \dfrac{\tan\alpha + \tan\beta}{1 - \tan\alpha\tan\beta} = \dfrac{-\sqrt{3} + \left(-2\sqrt{2}\right)}{1 - \left(-\sqrt{3}\right)\left(-2\sqrt{2}\right)} = \dfrac{-\sqrt{3} - 2\sqrt{2}}{1 - 2\sqrt{6}}\cdot\dfrac{1 + 2\sqrt{6}}{1 + 2\sqrt{6}}$

$= \dfrac{-9\sqrt{3} - 8\sqrt{2}}{-23} = \dfrac{9\sqrt{3} + 8\sqrt{2}}{23}$

(e) $\sin(2\alpha) = 2\sin\alpha\cos\alpha = 2\cdot\dfrac{-\sqrt{3}}{2}\cdot\dfrac{1}{2} = \dfrac{-\sqrt{3}}{2}$

(f) $\cos(2\beta) = \cos^2\beta - \sin^2\beta = \left(\dfrac{1}{3}\right)^2 - \left(\dfrac{-2\sqrt{2}}{3}\right)^2 = \dfrac{1}{9} - \dfrac{8}{9} = -\dfrac{7}{9}$

(g) $\sin\dfrac{\beta}{2} = \sqrt{\dfrac{1-\cos\beta}{2}} = \sqrt{\dfrac{1-\left(\frac{1}{3}\right)}{2}} = \sqrt{\dfrac{\frac{2}{3}}{2}} = \sqrt{\dfrac{1}{3}} = \dfrac{1}{\sqrt{3}} = \dfrac{\sqrt{3}}{3}$

(h) $\cos\dfrac{\alpha}{2} = \sqrt{\dfrac{1+\cos\alpha}{2}} = \sqrt{\dfrac{1+\frac{1}{2}}{2}} = \sqrt{\dfrac{\frac{3}{2}}{2}} = \sqrt{\dfrac{3}{4}} = \dfrac{\sqrt{3}}{2}$

49. $\sin\alpha = -\dfrac{2}{3}, \ \pi < \alpha < \dfrac{3\pi}{2}; \quad \cos\beta = -\dfrac{2}{3}, \ \pi < \beta < \dfrac{3\pi}{2}$

$\cos\alpha = -\dfrac{\sqrt{5}}{3}, \ \tan\alpha = \dfrac{2\sqrt{5}}{5}, \ \sin\beta = -\dfrac{\sqrt{5}}{3}, \ \tan\beta = \dfrac{\sqrt{5}}{2}, \dfrac{\pi}{2} < \dfrac{\alpha}{2} < \dfrac{3\pi}{4}, \dfrac{\pi}{2} < \dfrac{\beta}{2} < \dfrac{3\pi}{4}$

(a) $\sin(\alpha+\beta) = \sin\alpha\cos\beta + \cos\alpha\sin\beta = \dfrac{-2}{3}\cdot\dfrac{-2}{3} + \dfrac{-\sqrt{5}}{3}\cdot\dfrac{-\sqrt{5}}{3} = \dfrac{4+5}{9} = 1$

(b) $\cos(\alpha+\beta) = \cos\alpha\cos\beta - \sin\alpha\sin\beta = \dfrac{-\sqrt{5}}{3}\cdot\dfrac{-2}{3} - \dfrac{-2}{3}\cdot\dfrac{-\sqrt{5}}{3} = \dfrac{2\sqrt{5}-2\sqrt{5}}{9} = 0$

(c) $\sin(\alpha-\beta) = \sin\alpha\cos\beta - \cos\alpha\sin\beta = \dfrac{-2}{3}\cdot\dfrac{-2}{3} - \dfrac{-\sqrt{5}}{3}\cdot\dfrac{-\sqrt{5}}{3} = \dfrac{4-5}{9} = -\dfrac{1}{9}$

(d) $\tan(\alpha+\beta) = \dfrac{\tan\alpha+\tan\beta}{1-\tan\alpha\tan\beta} = \dfrac{\dfrac{2\sqrt{5}}{5}+\dfrac{\sqrt{5}}{2}}{1-\dfrac{2\sqrt{5}}{5}\cdot\dfrac{\sqrt{5}}{2}} = \dfrac{\dfrac{4\sqrt{5}+5\sqrt{5}}{10}}{\dfrac{10-10}{10}} = \dfrac{\dfrac{9\sqrt{5}}{10}}{0}$ Undefined

(e) $\sin(2\alpha) = 2\sin\alpha\cos\alpha = 2\cdot\dfrac{-2}{3}\cdot\dfrac{-\sqrt{5}}{3} = \dfrac{4\sqrt{5}}{9}$

(f) $\cos(2\beta) = \cos^2\beta - \sin^2\beta = \left(-\dfrac{2}{3}\right)^2 - \left(\dfrac{-\sqrt{5}}{3}\right)^2 = \dfrac{4}{9} - \dfrac{5}{9} = -\dfrac{1}{9}$

(g) $\sin\dfrac{\beta}{2} = \sqrt{\dfrac{1-\cos\beta}{2}} = \sqrt{\dfrac{1-\left(\frac{-2}{3}\right)}{2}} = \sqrt{\dfrac{\frac{5}{3}}{2}} = \sqrt{\dfrac{5}{6}} = \dfrac{\sqrt{30}}{6}$

(h) $\cos\dfrac{\alpha}{2} = -\sqrt{\dfrac{1+\cos\alpha}{2}} = -\sqrt{\dfrac{1+\frac{-\sqrt{5}}{3}}{2}} = -\sqrt{\dfrac{\frac{3-\sqrt{5}}{3}}{2}} = -\sqrt{\dfrac{3-\sqrt{5}}{6}} = -\dfrac{\sqrt{18-6\sqrt{5}}}{6}$

51. $\sin^{-1}1$

We are finding the angle θ, $\dfrac{-\pi}{2} \leq \theta \leq \dfrac{\pi}{2}$, whose sine equals 1.

$\sin\theta = 1 \qquad -\dfrac{\pi}{2} \leq \theta \leq \dfrac{\pi}{2}$

$\theta = \dfrac{\pi}{2}$

$\sin^{-1}1 = \dfrac{\pi}{2}$

53. $\tan^{-1}1$

We are finding the angle θ, $\dfrac{-\pi}{2} < \theta < \dfrac{\pi}{2}$, whose tangent equals 1.

$$\tan\theta = 1 \qquad -\frac{\pi}{2} < \theta < \frac{\pi}{2}$$

$$\theta = \frac{\pi}{4}$$

$$\tan^{-1}1 = \frac{\pi}{4}$$

55. $\cos^{-1}\left(-\dfrac{\sqrt{3}}{2}\right)$

We are finding the angle θ, $0 \le \theta \le \pi$, whose cosine equals $-\dfrac{\sqrt{3}}{2}$.

$$\cos\theta = -\frac{\sqrt{3}}{2} \qquad 0 \le \theta \le \pi$$

$$\theta = \frac{5\pi}{6}$$

$$\cos^{-1}\left(-\frac{\sqrt{3}}{2}\right) = \frac{5\pi}{6}$$

57. $\sin\left(\cos^{-1}\dfrac{\sqrt{2}}{2}\right)$

Find the angle θ, $0 \le \theta \le \pi$, whose cosine equals $\dfrac{\sqrt{2}}{2}$.

$$\cos\theta = \frac{\sqrt{2}}{2} \qquad 0 \le \theta \le \pi$$

$$\theta = \frac{\pi}{4}$$

$$\sin\left(\cos^{-1}\frac{\sqrt{2}}{2}\right) = \sin\frac{\pi}{4} = \frac{\sqrt{2}}{2}$$

59. $\tan\left(\sin^{-1}\left(-\dfrac{\sqrt{3}}{2}\right)\right)$

Find the angle θ, $-\dfrac{\pi}{2} \le \theta \le \dfrac{\pi}{2}$, whose sine equals $-\dfrac{\sqrt{3}}{2}$.

$$\sin\theta = -\frac{\sqrt{3}}{2} \qquad -\frac{\pi}{2} \le \theta \le \frac{\pi}{2}$$

$$\theta = -\frac{\pi}{3}$$

$$\tan\left(\sin^{-1}\left(-\frac{\sqrt{3}}{2}\right)\right) = \tan\left(-\frac{\pi}{3}\right) = -\sqrt{3}$$

61. $\sec\left(\tan^{-1}\dfrac{\sqrt{3}}{3}\right)$

Find the angle θ, $-\dfrac{\pi}{2} < \theta < \dfrac{\pi}{2}$, whose tangent is $\dfrac{\sqrt{3}}{3}$

$$\tan\theta = \dfrac{\sqrt{3}}{3}, \qquad -\dfrac{\pi}{2} < \theta < \dfrac{\pi}{2}$$

$$\theta = \dfrac{\pi}{6}$$

$$\sec\left(\tan^{-1}\dfrac{\sqrt{3}}{3}\right) = \sec\dfrac{\pi}{6} = \dfrac{2\sqrt{3}}{3}$$

63. $\sin\left(\tan^{-1}\dfrac{3}{4}\right)$

Since $\tan\theta = \dfrac{3}{4}$, $-\dfrac{\pi}{2} < \theta < \dfrac{\pi}{2}$, let $x = 4$ and $y = 3$. Solve for r:

$$16 + 9 = r^2$$
$$r^2 = 25$$
$$r = 5$$

θ is in quadrant I.

$$\sin\left(\tan^{-1}\dfrac{3}{4}\right) = \sin\theta = \dfrac{y}{r} = \dfrac{3}{5}$$

65. $\tan\left(\sin^{-1}\left(-\dfrac{4}{5}\right)\right)$

Since $\sin\theta = -\dfrac{4}{5}$, $-\dfrac{\pi}{2} \le \theta \le \dfrac{\pi}{2}$, let $y = -4$ and $r = 5$. Solve for x:

$$x^2 + 16 = 25$$
$$x^2 = 9$$
$$x = \pm 3$$

Since θ is in quadrant IV, $x = 3$.

$$\tan\left(\sin^{-1}\left(-\dfrac{4}{5}\right)\right) = \tan\theta = \dfrac{y}{x} = \dfrac{-4}{3}$$

67. $\sin^{-1}\left(\cos\dfrac{2\pi}{3}\right) = \sin^{-1}\left(-\dfrac{1}{2}\right) = -\dfrac{\pi}{6}$

69. $\tan^{-1}\left(\tan\dfrac{7\pi}{4}\right) = \tan^{-1}(-1) = -\dfrac{\pi}{4}$

71. $\cos\left(\sin^{-1}\frac{3}{5} - \cos^{-1}\frac{1}{2}\right)$

Let $\alpha = \sin^{-1}\frac{3}{5}$ and $\beta = \cos^{-1}\frac{1}{2}$. α is in quadrant I; β is in quadrant I.

Then $\sin\alpha = \frac{3}{5},\ -\frac{\pi}{2} \le \alpha \le \frac{\pi}{2}$, and $\cos\beta = \frac{1}{2}, 0 \le \beta \le \pi$.

$$\cos\alpha = \sqrt{1 - \sin^2\alpha} = \sqrt{1 - \left(\tfrac{3}{5}\right)^2} = \sqrt{1 - \tfrac{9}{25}} = \sqrt{\tfrac{16}{25}} = \tfrac{4}{5}$$

$$\sin\beta = \sqrt{1 - \cos^2\beta} = \sqrt{1 - \left(\tfrac{1}{2}\right)^2} = \sqrt{1 - \tfrac{1}{4}} = \sqrt{\tfrac{3}{4}} = \tfrac{\sqrt{3}}{2}$$

$$\cos\left[\sin^{-1}\tfrac{3}{5} - \cos^{-1}\tfrac{1}{2}\right] = \cos(\alpha - \beta) = \cos\alpha\cos\beta + \sin\alpha\sin\beta$$

$$= \frac{4}{5}\cdot\frac{1}{2} + \frac{3}{5}\cdot\frac{\sqrt{3}}{2} = \frac{4 + 3\sqrt{3}}{10}$$

73. $\tan\left[\sin^{-1}\left(-\tfrac{1}{2}\right) - \tan^{-1}\tfrac{3}{4}\right]$

Let $\alpha = \sin^{-1}\left(-\tfrac{1}{2}\right)$ and $\beta = \tan^{-1}\tfrac{3}{4}$. α is in quadrant IV; β is in quadrant I.

Then $\sin\alpha = -\frac{1}{2},\ -\frac{\pi}{2} \le \alpha \le \frac{\pi}{2}$, and $\tan\beta = \frac{3}{4}, -\frac{\pi}{2} < \beta < \frac{\pi}{2}$.

$$\cos\alpha = \sqrt{1 - \sin^2\alpha} = \sqrt{1 - \left(-\tfrac{1}{2}\right)^2} = \sqrt{1 - \tfrac{1}{4}} = \sqrt{\tfrac{3}{4}} = \tfrac{\sqrt{3}}{2};\quad \tan\alpha = -\frac{1}{\sqrt{3}} = -\frac{\sqrt{3}}{3}$$

$$\tan\left[\sin^{-1}\left(-\tfrac{1}{2}\right) - \tan^{-1}\tfrac{3}{4}\right] = \tan(\alpha - \beta) = \frac{\tan\alpha - \tan\beta}{1 + \tan\alpha\tan\beta} = \frac{-\dfrac{\sqrt{3}}{3} - \dfrac{3}{4}}{1 + \left(-\dfrac{\sqrt{3}}{3}\right)\cdot\dfrac{3}{4}}$$

$$= \frac{\dfrac{-4\sqrt{3} - 9}{12}}{1 - \dfrac{3\sqrt{3}}{12}} = \frac{-9 - 4\sqrt{3}}{12 - 3\sqrt{3}}\cdot\frac{12 + 3\sqrt{3}}{12 + 3\sqrt{3}}$$

$$= \frac{-144 - 75\sqrt{3}}{117} = \frac{-48 - 25\sqrt{3}}{39}$$

75. $\sin\left[2\cos^{-1}\left(-\tfrac{3}{5}\right)\right]$

Let $\alpha = \cos^{-1}\left(-\tfrac{3}{5}\right)$. α is in quadrant II.

Then $\cos\alpha = -\frac{3}{5},\ 0 \le \alpha \le \pi$.

$$\sin\alpha = \sqrt{1 - \cos^2\alpha} = \sqrt{1 - \left(-\tfrac{3}{5}\right)^2} = \sqrt{1 - \tfrac{9}{25}} = \sqrt{\tfrac{16}{25}} = \tfrac{4}{5}$$

$$\sin\left[2\cos^{-1}\tfrac{4}{5}\right] = \sin 2\alpha = 2\sin\alpha\cos\alpha = 2\cdot\frac{4}{5}\cdot -\frac{3}{5} = -\frac{24}{25}$$

77. $\cos\theta = \dfrac{1}{2}$

$$\theta = \frac{\pi}{3} + 2k\pi \ \text{ or } \ \theta = \frac{5\pi}{3} + 2k\pi, \text{ where } k \text{ is any integer}$$

The solutions on the interval $[0, 2\pi)$ are $\theta = \dfrac{\pi}{3}, \dfrac{5\pi}{3}$

79. $2\cos\theta + \sqrt{2} = 0 \quad \rightarrow \quad 2\cos\theta = -\sqrt{2} \quad \rightarrow \quad \cos\theta = -\dfrac{\sqrt{2}}{2}$

$\theta = \dfrac{3\pi}{4} + 2k\pi \quad$ or $\quad \theta = \dfrac{5\pi}{4} + 2k\pi, \;\; k$ is any integer

The solutions on the interval $[0, 2\pi)$ are $\theta = \dfrac{3\pi}{4}, \dfrac{5\pi}{4}$.

81. $\sin 2\theta + 1 = 0 \quad \rightarrow \quad \sin 2\theta = -1$

$2\theta = \dfrac{3\pi}{2} + 2k\pi \quad \rightarrow \quad \theta = \dfrac{3\pi}{4} + k\pi$, where k is any integer

The solutions on the interval $[0, 2\pi)$ are $\theta = \dfrac{3\pi}{4}, \dfrac{7\pi}{4}$

83. $\tan 2\theta = 0$

$2\theta = 0 + k\pi \quad \rightarrow \quad \theta = \dfrac{k\pi}{2}$, where k is any integer

The solutions on the interval $[0, 2\pi)$ are $\theta = 0, \dfrac{\pi}{2}, \pi, \dfrac{3\pi}{2}$.

85. $\sin\theta = 0.9$

$\theta = 1.1197695 \;$ or $\; \theta = \pi - 1.1197695 = 2.0218231$

$\theta \approx 1.12, 2.02$

87.
$$\sin\theta = \tan\theta$$
$$\sin\theta = \dfrac{\sin\theta}{\cos\theta}$$
$$\sin\theta\cos\theta = \sin\theta$$
$$\sin\theta\cos\theta - \sin\theta = 0$$
$$\sin\theta(\cos\theta - 1) = 0$$
$$\cos\theta - 1 = 0 \;\; \text{or} \;\; \sin\theta = 0$$
$$\cos\theta = 1$$
$$\theta = 0, \pi$$

89.
$$\sin\theta + \sin(2\theta) = 0$$
$$\sin\theta + 2\sin\theta\cos\theta = 0$$
$$\sin\theta(1 + 2\cos\theta) = 0$$
$$1 + 2\cos\theta = 0 \qquad \text{or} \;\; \sin\theta = 0$$
$$\cos\theta = -\dfrac{1}{2}$$
$$\theta = \dfrac{2\pi}{3}, \dfrac{4\pi}{3} \;\; \text{or} \;\; \theta = 0, \pi$$

91. $\sin(2\theta) - \cos\theta - 2\sin\theta + 1 = 0$
$2\sin\theta\cos\theta - \cos\theta - 2\sin\theta + 1 = 0$
$\cos\theta(2\sin\theta - 1) - 1(2\sin\theta - 1) = 0$
$(2\sin\theta - 1)(\cos\theta - 1) = 0$
$$\sin\theta = \frac{1}{2} \text{ or } \cos\theta = 1$$
$$\theta = \frac{\pi}{6}, \frac{5\pi}{6} \text{ or } \theta = 0$$

93. $2\sin^2\theta - 3\sin\theta + 1 = 0$
$(2\sin\theta - 1)(\sin\theta - 1) = 0$
$2\sin\theta - 1 = 0 \text{ or } \sin\theta - 1 = 0$
$$\sin\theta = \frac{1}{2} \qquad \sin\theta = 1$$
$$\theta = \frac{\pi}{6}, \frac{5\pi}{6} \text{ or } \theta = \frac{\pi}{2}$$

95. $\sin\theta - \cos\theta = 1$
Divide each side by $\sqrt{2}$:
$$\frac{1}{\sqrt{2}}\sin\theta - \frac{1}{\sqrt{2}}\cos\theta = \frac{1}{\sqrt{2}}$$
Rewrite in the difference of two angles form where
$$\cos\phi = \frac{1}{\sqrt{2}} \text{ and } \sin\phi = \frac{1}{\sqrt{2}} \text{ and } \phi = \frac{\pi}{4}:$$
$$\sin\theta\cos\phi - \cos\theta\sin\phi = \frac{1}{\sqrt{2}}$$
$$\sin(\theta - \phi) = \frac{\sqrt{2}}{2}$$
$$\theta - \phi = \frac{\pi}{4} \text{ or } \theta - \phi = \frac{3\pi}{4}$$
$$\theta - \frac{\pi}{4} = \frac{\pi}{4} \text{ or } \theta - \frac{\pi}{4} = \frac{3\pi}{4}$$
$$\theta = \frac{\pi}{2} \text{ or } \theta = \pi$$

97. $2x = 5\cos x$
Find the intersection of
$y_1 = 2x$ and $y_2 = 5\cos x$:

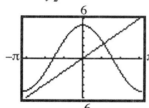

$x = 1.11$

99. $2\sin x + 3\cos x = 4x$
Find the intersection of
$y_1 = 2\sin x + 3\cos x$ and $y_2 = 4x$:

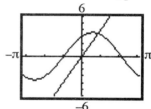

$x = 0.87$

101. $\sin x = \ln x$

Find the intersection of $y_1 = \sin x$ and $y_2 = \ln x$:

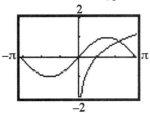

$x = 2.22$

Chapter 8

Applications of Trigonometric Functions

8.1 Right Triangle Trigonometry

1. opposite = 5; adjacent = 12
   Find the hypotenuse:
   $$5^2 + 12^2 = (\text{hypotenuse})^2$$
   $$(\text{hypotenuse})^2 = 25 + 144 = 169$$
   $$\text{hypotenuse} = 13$$

   $\sin\theta = \dfrac{\text{opp}}{\text{hyp}} = \dfrac{5}{13}$ $\cos\theta = \dfrac{\text{adj}}{\text{hyp}} = \dfrac{12}{13}$ $\tan\theta = \dfrac{\text{opp}}{\text{adj}} = \dfrac{5}{12}$

   $\csc\theta = \dfrac{\text{hyp}}{\text{opp}} = \dfrac{13}{5}$ $\sec\theta = \dfrac{\text{hyp}}{\text{adj}} = \dfrac{13}{12}$ $\cot\theta = \dfrac{\text{adj}}{\text{opp}} = \dfrac{12}{5}$

3. opposite = 2; adjacent = 3
   Find the hypotenuse:
   $$2^2 + 3^2 = (\text{hypotenuse})^2$$
   $$(\text{hypotenuse})^2 = 4 + 9 = 13$$
   $$\text{hypotenuse} = \sqrt{13}$$

   $\sin\theta = \dfrac{\text{opp}}{\text{hyp}} = \dfrac{2}{\sqrt{13}} = \dfrac{2\sqrt{13}}{13}$ $\cos\theta = \dfrac{\text{adj}}{\text{hyp}} = \dfrac{3}{\sqrt{13}} = \dfrac{3\sqrt{13}}{13}$ $\tan\theta = \dfrac{\text{opp}}{\text{adj}} = \dfrac{2}{3}$

   $\csc\theta = \dfrac{\text{hyp}}{\text{opp}} = \dfrac{\sqrt{13}}{2}$ $\sec\theta = \dfrac{\text{hyp}}{\text{adj}} = \dfrac{\sqrt{13}}{3}$ $\cot\theta = \dfrac{\text{adj}}{\text{opp}} = \dfrac{3}{2}$

5. adjacent = 2; hypotenuse = 4
   Find the opposite side:
   $$(\text{opposite})^2 + 2^2 = 4^2$$
   $$(\text{opposite})^2 = 16 - 4 = 12$$
   $$\text{opposite} = \sqrt{12} = 2\sqrt{3}$$

   $\sin\theta = \dfrac{\text{opp}}{\text{hyp}} = \dfrac{2\sqrt{3}}{4} = \dfrac{\sqrt{3}}{2}$ $\cos\theta = \dfrac{\text{adj}}{\text{hyp}} = \dfrac{2}{4} = \dfrac{1}{2}$ $\tan\theta = \dfrac{\text{opp}}{\text{adj}} = \dfrac{2\sqrt{3}}{2} = \sqrt{3}$

   $\csc\theta = \dfrac{\text{hyp}}{\text{opp}} = \dfrac{4}{2\sqrt{3}} = \dfrac{2\sqrt{3}}{3}$ $\sec\theta = \dfrac{\text{hyp}}{\text{adj}} = \dfrac{4}{2} = 2$ $\cot\theta = \dfrac{\text{adj}}{\text{opp}} = \dfrac{2}{2\sqrt{3}} = \dfrac{\sqrt{3}}{3}$

7. opposite = $\sqrt{2}$; adjacent = 1
   Find the hypotenuse:
   $$\sqrt{2}^2 + 1^2 = (\text{hypotenuse})^2$$
   $$(\text{hypotenuse})^2 = 2 + 1 = 3$$
   $$\text{hypotenuse} = \sqrt{3}$$

   $\sin\theta = \dfrac{\text{opp}}{\text{hyp}} = \dfrac{\sqrt{2}}{\sqrt{3}} = \dfrac{\sqrt{6}}{3}$ $\cos\theta = \dfrac{\text{adj}}{\text{hyp}} = \dfrac{1}{\sqrt{3}} = \dfrac{\sqrt{3}}{3}$ $\tan\theta = \dfrac{\text{opp}}{\text{adj}} = \dfrac{\sqrt{2}}{1} = \sqrt{2}$

   $\csc\theta = \dfrac{\text{hyp}}{\text{opp}} = \dfrac{\sqrt{3}}{\sqrt{2}} = \dfrac{\sqrt{6}}{2}$ $\sec\theta = \dfrac{\text{hyp}}{\text{adj}} = \dfrac{\sqrt{3}}{1} = \sqrt{3}$ $\cot\theta = \dfrac{\text{adj}}{\text{opp}} = \dfrac{1}{\sqrt{2}} = \dfrac{\sqrt{2}}{2}$

9. opposite = 1; hypotenuse = $\sqrt{5}$
   Find the adjacent side:
   $$1^2 + (\text{adjacent})^2 = \sqrt{5}^2$$
   $$(\text{adjacent})^2 = 5 - 1 = 4$$
   $$\text{adjacent} = 2$$

   $\sin\theta = \dfrac{\text{opp}}{\text{hyp}} = \dfrac{1}{\sqrt{5}} = \dfrac{\sqrt{5}}{5}$ $\cos\theta = \dfrac{\text{adj}}{\text{hyp}} = \dfrac{2}{\sqrt{5}} = \dfrac{2\sqrt{5}}{5}$ $\tan\theta = \dfrac{\text{opp}}{\text{adj}} = \dfrac{1}{2}$

   $\csc\theta = \dfrac{\text{hyp}}{\text{opp}} = \dfrac{\sqrt{5}}{1} = \sqrt{5}$ $\sec\theta = \dfrac{\text{hyp}}{\text{adj}} = \dfrac{\sqrt{5}}{2}$ $\cot\theta = \dfrac{\text{adj}}{\text{opp}} = \dfrac{2}{1} = 2$

11. $\sin 38° - \cos 52° = \sin 38° - \sin(90° - 52°) = \sin 38° - \sin 38° = 0$

13. $\dfrac{\cos 10°}{\sin 80°} = \dfrac{\sin(90° - 10°)}{\sin 80°} = \dfrac{\sin 80°}{\sin 80°} = 1$

15. $1 - \cos^2 20° - \cos^2 70° = \sin^2 20° - \sin^2(90° - 70°) = \sin^2 20° - \sin^2 20° = 0$

17. $\tan 20° - \dfrac{\cos 70°}{\cos 20°} = \tan 20° - \dfrac{\sin(90° - 70°)}{\cos 20°} = \tan 20° - \dfrac{\sin 20°}{\cos 20°} = \tan 20° - \tan 20° = 0$

19. $\cos 35° \sin 55° + \sin 35° \cos 55° = \sin(55° + 35°) = \sin 90° = 1$

21. Given: $\sin\theta = \frac{1}{3}$

    (a) $\cos(90° - \theta) = \sin\theta = \frac{1}{3}$

    (b) $\cos^2\theta = 1 - \sin^2\theta = 1 - \left(\frac{1}{3}\right)^2 = 1 - \frac{1}{9} = \frac{8}{9}$

    (c) $\csc\theta = \dfrac{1}{\sin\theta} = \dfrac{1}{\frac{1}{3}} = 3$

    (d) $\sec\left(\dfrac{\pi}{2} - \theta\right) = \csc\theta = \dfrac{1}{\sin\theta} = \dfrac{1}{\frac{1}{3}} = 3$

23. Given: $\tan \theta = 4$

    (a) $\sec^2 \theta = 1 + \tan^2 \theta = 1 + 4^2 = 1 + 16 = 17$

    (b) $\cot \theta = \dfrac{1}{\tan \theta} = \dfrac{1}{4}$

    (c) $\cot\left(\dfrac{\pi}{2} - \theta\right) = \tan \theta = 4$

    (d) $\csc^2 \theta = 1 + \cot^2 \theta = 1 + \dfrac{1}{\tan^2 \theta} = 1 + \dfrac{1}{4^2} = 1 + \dfrac{1}{16} = \dfrac{17}{16}$

25. Given: $\csc \theta = 4$

    (a) $\sin \theta = \dfrac{1}{\csc \theta} = \dfrac{1}{4}$

    (b) $\cot^2 \theta = \csc^2 \theta - 1 = 4^2 - 1 = 16 - 1 = 15$

    (c) $\sec(90° - \theta) = \csc \theta = 4$

    (d) $\sec^2 \theta = 1 + \tan^2 \theta = 1 + \dfrac{1}{\cot^2 \theta} = 1 + \dfrac{1}{\csc^2 \theta - 1} = 1 + \dfrac{1}{4^2 - 1} = 1 + \dfrac{1}{15} = \dfrac{16}{15}$

27. Given: $\sin \theta = 0.3$

    $\sin \theta + \cos\left(\dfrac{\pi}{2} - \theta\right) = \sin \theta + \sin \theta = 0.3 + 0.3 = 0.6$

29. $b = 5, \ \beta = 20°$

    $\sin \beta = \dfrac{b}{c} \ \rightarrow \ \sin 20° = \dfrac{5}{c} \ \rightarrow \ c = \dfrac{5}{\sin 20°} = \dfrac{5}{0.3420} \approx 14.62$

    $\tan \beta = \dfrac{b}{a} \ \rightarrow \ \tan 20° = \dfrac{5}{a} \ \rightarrow \ a = \dfrac{5}{\tan 20°} = \dfrac{5}{0.3640} \approx 13.74$

    $\alpha = 90° - \beta = 90° - 20° = 70°$

31. $a = 6, \ \beta = 40°$

    $\cos \beta = \dfrac{a}{c} \ \rightarrow \ \cos 40° = \dfrac{6}{c} \ \rightarrow \ c = \dfrac{6}{\cos 40°} = \dfrac{6}{0.7660} \approx 7.83$

    $\tan \beta = \dfrac{b}{a} \ \rightarrow \ \tan 40° = \dfrac{b}{6} \ \rightarrow \ b = 6 \tan 40° = 6(0.8391) \approx 5.03$

    $\alpha = 90° - \beta = 90° - 40° = 50°$

33. $b = 4, \ \alpha = 10°$

    $\tan \alpha = \dfrac{a}{b} \ \rightarrow \ \tan 10° = \dfrac{a}{4} \ \rightarrow \ a = 4 \tan 10° = 4(0.1763) \approx 0.71$

    $\cos \alpha = \dfrac{b}{c} \ \rightarrow \ \cos 10° = \dfrac{4}{c} \ \rightarrow \ c = \dfrac{4}{\cos 10°} = \dfrac{4}{0.9848} \approx 4.06$

    $\beta = 90° - \alpha = 90° - 10° = 80°$

35. $a = 5, \ b = 3$

    $c^2 = a^2 + b^2 = 5^2 + 3^2 = 25 + 9 = 34 \ \rightarrow \ c = \sqrt{34} \approx 5.83$

    $\tan \alpha = \dfrac{a}{b} = \dfrac{5}{3} = 1.6667 \ \rightarrow \ \alpha \approx 59.0°$

    $\beta = 90° - \alpha = 90° - 59.0° = 31.0°$

37. $a = 2, \ c = 5$

$$c^2 = a^2 + b^2 \quad \rightarrow \quad b^2 = c^2 - a^2 = 5^2 - 2^2 = 25 - 4 = 21 \quad \rightarrow \quad b = \sqrt{21} \approx 4.58$$

$$\sin\alpha = \frac{a}{c} = \frac{2}{5} = 0.4000 \quad \rightarrow \quad \alpha \approx 23.6°$$

$$\beta = 90° - \alpha = 90° - 23.6° = 66.4°$$

39. $c = 8, \ \alpha = 35°$

$$\sin 35° = \frac{a}{8} \quad \rightarrow \quad a = 8\sin 35° = 8(0.5736) \approx 4.59 \text{ in.}$$

$$\cos 35° = \frac{b}{8} \quad \rightarrow \quad b = 8\cos 35° = 8(0.8192) \approx 6.55 \text{ in.}$$

41. $\alpha = 25°, \ a = 5$

$$\sin 25° = \frac{5}{c} \quad \rightarrow \quad c = \frac{5}{\sin 25°} = \frac{5}{0.4226} \approx 11.83 \text{ in.}$$

$\alpha = 25°, \ b = 5$

$$\cos 25° = \frac{5}{c} \quad \rightarrow \quad c = \frac{5}{\cos 25°} = \frac{5}{0.9063} \approx 5.52 \text{ in.}$$

43. $c = 5, \ a = 2$

$$\sin\alpha = \frac{2}{5} = 0.4000 \quad \rightarrow \quad \alpha \approx 23.6 \quad \rightarrow \quad \beta = 90° - \alpha = 90° - 23.6° \approx 66.4°$$

45. $\tan 35° = \dfrac{a}{100} \quad \rightarrow \quad a = 100\tan 35° = 100(0.7002) \approx 70 \text{ feet}$

47. $\tan 85.361° = \dfrac{a}{80} \quad \rightarrow \quad a = 80\tan 85.361° = 80(12.3239) \approx 985.9 \text{ feet}$

49.

$$\tan 20° = \frac{50}{x} \quad \rightarrow \quad x = \frac{50}{\tan 20°} = \frac{50}{0.3640} \approx 137 \text{ meters}$$

51.

$$\sin 70° = \frac{x}{22} \quad \rightarrow \quad x = 22\sin 70° = 22(0.9397) \approx 20.67 \text{ feet}$$

53. opposite side $= 10$ feet, adjacent side $= 35$ feet

$\tan \theta = \dfrac{10}{35} = 0.2857 \quad \rightarrow \quad \theta = \tan^{-1}\left(\tfrac{10}{35}\right) \approx 15.9°$

55. Let h represent the height of Lincoln's face.

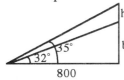

$\tan 32° = \dfrac{b}{800} \quad \rightarrow \quad b = 800 \tan 32° = 800(0.6249) \approx 499.9$

$\tan 35° = \dfrac{b+h}{800} \quad \rightarrow \quad b + h = 800 \tan 35° = 800(0.7002) \approx 560.2$

$h = (b+h) - b = 560.2 - 499.9 \approx 60.3$ feet

57.

$\sin 21° = \dfrac{190}{x} \quad \rightarrow \quad x = \dfrac{190}{\sin 21°} = \dfrac{190}{0.3584} \approx 530$ ft.

59.

$\tan 35.1° = \dfrac{x}{789} \quad \rightarrow \quad x = 789 \tan 35.1° = 789(0.7028) \approx 555$ ft

61. (a) $\tan 15° = \dfrac{30}{x} \quad \rightarrow \quad x = \dfrac{30}{\tan 15°} = \dfrac{30}{0.2679} \approx 111.96$ feet

The truck is traveling at 111.96 ft/sec.

$\dfrac{111.96 \text{ ft}}{\text{sec}} \cdot \dfrac{1 \text{ mile}}{5280 \text{ ft}} \cdot \dfrac{3600 \text{ sec}}{\text{hr}} \approx 76.34$ mi / hr

(b) $\tan 20° = \dfrac{30}{x} \quad \rightarrow \quad x = \dfrac{30}{\tan 20°} = \dfrac{30}{0.3640} \approx 82.42$ feet

The truck is traveling at 82.42 ft/sec.

$\dfrac{82.42 \text{ ft}}{\text{sec}} \cdot \dfrac{1 \text{ mile}}{5280 \text{ ft}} \cdot \dfrac{3600 \text{ sec}}{\text{hr}} \approx 56.20$ mi / hr

(c)   A ticket is issued for traveling at a speed of 60 mi/hr or more.

$\dfrac{60 \text{ mi}}{\text{hr}} \cdot \dfrac{5280 \text{ ft}}{\text{mi}} \cdot \dfrac{1\text{hr}}{3600 \text{ sec}} = 88$ ft / sec.

If $\tan \theta < \dfrac{30}{88}$, the trooper should issue a ticket.

A ticket is issued if $\theta < 18.8°$.

63. Find angle θ: (see the figure)

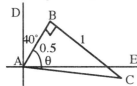

$\tan \theta = \dfrac{1}{0.5} = 2 \quad \rightarrow \quad \theta = 63.4°$

$\angle DAC = 40° + 63.4° = 103.4°$

$\angle EAC = 103.4° - 90° = 13.4°$

The bearing the control tower should use is S76.6°E.

65. $\tan \alpha = \dfrac{10-6}{15} = \dfrac{4}{15} \quad \rightarrow \quad \alpha = 14.9°$

67. Makc thc following obscrvations related to the figure:

$$\tan \theta = \frac{1}{z} \quad \rightarrow \quad z = \frac{1}{\tan \theta}$$

$$\sin \theta = \frac{1}{x} \quad \rightarrow \quad x = \frac{1}{\sin \theta}$$

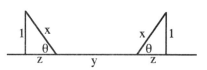

$$2z + y = 8 \quad \rightarrow \quad y = 8 - 2z = 8 - 2\left(\frac{1}{\tan \theta}\right)$$

(a)   The time equation is:

$$T = \frac{x}{3} + \frac{y}{8} + \frac{x}{3} = \frac{2x}{3} + \frac{y}{8} = \frac{2 \cdot \dfrac{1}{\sin \theta}}{3} + \frac{8 - \dfrac{2}{\tan \theta}}{8} = \frac{2}{3\sin \theta} + 1 - \frac{1}{4 \tan \theta}$$

(b)   Graph:

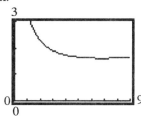

The least time occurs when the angle is approximately 68°. The least time is approximately 1.62 hours. The distance and time on the paved road is:

$$8 - \frac{2}{\tan \theta} = 8 - \frac{2}{\tan 68°} \approx 7.2 \text{ miles}$$

$$t = \frac{7.2}{8} = 0.9 \text{ hours}$$

69. (a) $T = \dfrac{1500}{300} + \dfrac{500}{100} = 5 + 5 = 10$ minutes

(b) $T = \dfrac{500}{100} + \dfrac{1500}{100} = 5 + 15 = 20$ minutes

(c) $\tan \theta = \dfrac{500}{x} \quad \rightarrow \quad x = \dfrac{500}{\tan \theta}$

$\sin \theta = \dfrac{500}{\text{distance in sand}} \quad \rightarrow \quad \text{distance in sand} = \dfrac{500}{\sin \theta}$

$$T = \frac{1500 - x}{300} + \frac{\text{distance in sand}}{100} = \frac{1500 - \dfrac{500}{\tan \theta}}{300} + \frac{\dfrac{500}{\sin \theta}}{100} = 5 - \frac{5}{3 \tan \theta} + \frac{5}{\sin \theta}$$

(d)   1000 feet along the paved path leaves an additional 500 feet in the direction of the path, so the angle of the path across the sand is 45°.

$$T = 5 - \frac{5}{3 \tan 45°} + \frac{5}{\sin 45°} = 5 - \frac{5}{3 \cdot 1} + \frac{5}{\dfrac{\sqrt{2}}{2}} = 5 - \frac{5}{3} + \frac{10}{\sqrt{2}} \approx 10.4 \text{ minutes}$$

(e)   Graph:

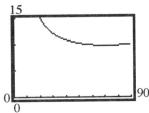

The time is least when the angle is approximately 70.53°. The least time is approximately 9.71 minutes. The value of x is:

$$x = \frac{500}{\tan 70.53°} \approx 176.8 \text{ feet}$$

71.   The length of the highway $= x + y + z$

$$\sin 40° = \frac{1}{x} \quad \rightarrow \quad x = \frac{1}{\sin 40°} \approx 1.56 \text{ mi}$$

$$\sin 50° = \frac{1}{z} \quad \rightarrow \quad z = \frac{1}{\sin 50°} \approx 1.31 \text{ mi}$$

$$\tan 40° = \frac{1}{a} \quad \rightarrow \quad a = \frac{1}{\tan 40°} \approx 1.19 \text{ mi}$$

$$\tan 50° = \frac{1}{b} \quad \rightarrow \quad b = \frac{1}{\tan 50°} \approx 0.84 \text{ mi}$$

$$a + y + b = 3 \quad \rightarrow \quad y = 3 - a - b = 3 - 1.19 - 0.84 = 0.97 \text{ mi}$$

The length of the highway is: $1.56 + 0.97 + 1.31 = 3.84$ miles.

73.   In order to see George's head and feet the camera must be x feet from George. Solve:

$$\tan 20° = \frac{4}{x} \quad \rightarrow \quad x = \frac{4}{\tan 20°} = 10.99 \text{ feet}$$

The camera will need to be moved back 1 foot to see George's feet.

75.   Find θ: (see figure)

$$\cos \theta = \frac{3960}{3960 + \dfrac{362}{5280}} \approx 0.99998269$$

$$\theta = 0.00588439 \text{ radians}$$

Find the arc length from the base of the lighthouse
    to the horizon:

$s = r\theta = 3960(0.00588439) \approx 23.3$ miles

The distance from the ship to the horizon point is $40 - 23.3 = 16.7$ miles.

$$\theta = \frac{s}{r} = \frac{16.7}{3960} = 0.00421717$$

If h is the height of the ship, $\cos 0.00421717 = \dfrac{3960}{3960 + h}$.

Solve for h:

$$(3960 + h)\cos 0.00421717 = 3960$$

$$h = \frac{3960}{\cos 0.00421717} - 3960 \approx 0.0352 \text{ miles or } \approx 186 \text{ feet}$$

The ship would have to be 186 feet tall to see the lighthouse from 40 miles away.

The distance from the plane to the horizon point is $120 - 23.3 = 96.7$ miles.

$$\theta = \frac{s}{r} = \frac{96.7}{3960} = 0.02441919$$

If h is the height of the plane, $\cos 0.02441919 = \dfrac{3960}{3960 + h}$.

Solve for h:

$$(3960 + h)\cos 0.02441919 = 3960$$

$$h = \frac{3960}{\cos 0.02441919} - 3960 \approx 1.18 \text{ miles or } \approx 6230 \text{ feet}$$

A plane at an altitude of 6230 feet could see the lighthouse from 120 miles away. The brochure understates the distance from which the lighthouse can be seen.

77. (a) $|OA| = |OC| = 1;\quad \angle OAC = \angle OCA;$

$$\angle OAC + \angle OAC + 180° - \theta = 180°$$

$$2(\angle OAC) = \theta$$

$$\angle OAC = \frac{\theta}{2}$$

(b) $\sin\theta = \dfrac{|CD|}{|OC|} = |CD|$ $\qquad \cos\theta = \dfrac{|OD|}{|OC|} = |OD|$

(c) $\tan\dfrac{\theta}{2} = \dfrac{|CD|}{|AD|} = \dfrac{|CD|}{1 + |OD|} = \dfrac{\sin\theta}{1 + \cos\theta}$

79. $h = x \cdot \dfrac{h}{x} = x\tan\theta;\quad h = (1 - x) \cdot \dfrac{h}{1 - x} = (1 - x)\tan n\theta$

$$x\tan\theta = (1 - x)\tan n\theta$$

$$x\tan\theta = \tan n\theta - x\tan n\theta$$

$$x(\tan\theta + \tan n\theta) = \tan n\theta$$

$$x = \frac{\tan n\theta}{\tan\theta + \tan n\theta}$$

81. $\sin\theta = \dfrac{y}{1} = y;\qquad \cos\theta = \dfrac{x}{1} = x$

(a) $A = 2xy = 2\cos\theta\sin\theta$

(b) $2\cos\theta\sin\theta = 2\sin\theta\cos\theta = \sin(2\theta)$

(c)   The largest value of the sine function is 1. Solve:

$$\sin(2\theta) = 1$$

$$2\theta = \frac{\pi}{2}$$

$$\theta = \frac{\pi}{4}$$

(d) $x = \cos\dfrac{\pi}{4} = \dfrac{\sqrt{2}}{2}$ $y = \sin\dfrac{\pi}{4} = \dfrac{\sqrt{2}}{2}$

The dimensions are $\sqrt{2}$ by $\dfrac{\sqrt{2}}{2}$.

8.2   The Law of Sines

1. $c = 5,\ \beta = 45°,\ \gamma = 95°$
$\alpha = 180° - \beta - \gamma = 180° - 45° - 95° = 40°$

$$\frac{\sin\alpha}{a} = \frac{\sin\gamma}{c} \rightarrow \frac{\sin 40°}{a} = \frac{\sin 95°}{5} \rightarrow a = \frac{5\sin 40°}{\sin 95°} \approx 3.23$$

$$\frac{\sin\beta}{b} = \frac{\sin\gamma}{c} \rightarrow \frac{\sin 45°}{b} = \frac{\sin 95°}{5} \rightarrow b = \frac{5\sin 45°}{\sin 95°} \approx 3.55$$

3. $b = 3,\ \alpha = 50°,\ \gamma = 85°$
$\beta = 180° - \alpha - \gamma = 180° - 50° - 85° = 45°$

$$\frac{\sin\alpha}{a} = \frac{\sin\beta}{b} \rightarrow \frac{\sin 50°}{a} = \frac{\sin 45°}{3} \rightarrow a = \frac{3\sin 50°}{\sin 45°} \approx 3.25$$

$$\frac{\sin\gamma}{c} = \frac{\sin\beta}{b} \rightarrow \frac{\sin 85°}{c} = \frac{\sin 45°}{3} \rightarrow c = \frac{3\sin 85°}{\sin 45°} \approx 4.23$$

5. $b = 7,\ \alpha = 40°,\ \beta = 45°$
$\gamma = 180° - \alpha - \beta = 180° - 40° - 45° = 95°$

$$\frac{\sin\alpha}{a} = \frac{\sin\beta}{b} \rightarrow \frac{\sin 40°}{a} = \frac{\sin 45°}{7} \rightarrow a = \frac{7\sin 40°}{\sin 45°} \approx 6.36$$

$$\frac{\sin\gamma}{c} = \frac{\sin\beta}{b} \rightarrow \frac{\sin 95°}{c} = \frac{\sin 45°}{7} \rightarrow c = \frac{7\sin 95°}{\sin 45°} \approx 9.86$$

7. $b = 2,\ \beta = 40°,\ \gamma = 100°$
$\alpha = 180° - \beta - \gamma = 180° - 40° - 100° = 40°$

$$\frac{\sin\alpha}{a} = \frac{\sin\beta}{b} \rightarrow \frac{\sin 40°}{a} = \frac{\sin 40°}{2} \rightarrow a = \frac{2\sin 40°}{\sin 40°} = 2$$

$$\frac{\sin\gamma}{c} = \frac{\sin\beta}{b} \rightarrow \frac{\sin 100°}{c} = \frac{\sin 40°}{2} \rightarrow c = \frac{2\sin 100°}{\sin 40°} \approx 3.06$$

9. $\alpha = 40°,\ \beta = 20°,\ a = 2$
$\gamma = 180° - \alpha - \beta = 180° - 40° - 20° = 120°$

$$\frac{\sin\alpha}{a} = \frac{\sin\beta}{b} \quad\rightarrow\quad \frac{\sin 40°}{2} = \frac{\sin 20°}{b} \quad\rightarrow\quad b = \frac{2\sin 20°}{\sin 40°} \approx 1.06$$

$$\frac{\sin\gamma}{c} = \frac{\sin\alpha}{a} \quad\rightarrow\quad \frac{\sin 120°}{c} = \frac{\sin 40°}{2} \quad\rightarrow\quad c = \frac{2\sin 120°}{\sin 40°} \approx 2.69$$

11. $\beta = 70°,\ \gamma = 10°,\ b = 5$
$\alpha = 180° - \beta - \gamma = 180° - 70° - 10° = 100°$

$$\frac{\sin\alpha}{a} = \frac{\sin\beta}{b} \quad\rightarrow\quad \frac{\sin 100°}{a} = \frac{\sin 70°}{5} \quad\rightarrow\quad a = \frac{5\sin 100°}{\sin 70°} \approx 5.24$$

$$\frac{\sin\gamma}{c} = \frac{\sin\beta}{b} \quad\rightarrow\quad \frac{\sin 10°}{c} = \frac{\sin 70°}{5} \quad\rightarrow\quad c = \frac{5\sin 10°}{\sin 70°} \approx 0.92$$

13. $\alpha = 110°,\ \gamma = 30°,\ c = 3$
$\beta = 180° - \alpha - \gamma = 180° - 110° - 30° = 40°$

$$\frac{\sin\alpha}{a} = \frac{\sin\gamma}{c} \quad\rightarrow\quad \frac{\sin 110°}{a} = \frac{\sin 30°}{3} \quad\rightarrow\quad a = \frac{3\sin 110°}{\sin 30°} \approx 5.64$$

$$\frac{\sin\gamma}{c} = \frac{\sin\beta}{b} \quad\rightarrow\quad \frac{\sin 30°}{3} = \frac{\sin 40°}{b} \quad\rightarrow\quad b = \frac{3\sin 40°}{\sin 30°} \approx 3.86$$

15. $\alpha = 40°,\ \beta = 40°,\ c = 2$
$\gamma = 180° - \alpha - \beta = 180° - 40° - 40° = 100°$

$$\frac{\sin\alpha}{a} = \frac{\sin\gamma}{c} \quad\rightarrow\quad \frac{\sin 40°}{a} = \frac{\sin 100°}{2} \quad\rightarrow\quad a = \frac{2\sin 40°}{\sin 100°} \approx 1.31$$

$$\frac{\sin\beta}{b} = \frac{\sin\gamma}{c} \quad\rightarrow\quad \frac{\sin 40°}{b} = \frac{\sin 100°}{2} \quad\rightarrow\quad b = \frac{2\sin 40°}{\sin 100°} \approx 1.31$$

17. $a = 3,\ b = 2,\ \alpha = 50°$

$$\frac{\sin\beta}{b} = \frac{\sin\alpha}{a} \quad\rightarrow\quad \frac{\sin\beta}{2} = \frac{\sin 50°}{3} \quad\rightarrow\quad \sin\beta = \frac{2\sin 50°}{3} = 0.5107$$

$$\rightarrow\quad \beta = 30.7° \ \text{ or } \ \beta = 149.3°$$

The second value is discarded because $\alpha + \beta > 180°$.
$\gamma = 180° - \alpha - \beta = 180° - 50° - 30.7° = 99.3°$

$$\frac{\sin\gamma}{c} = \frac{\sin\alpha}{a} \quad\rightarrow\quad \frac{\sin 99.3°}{c} = \frac{\sin 50°}{3} \quad\rightarrow\quad c = \frac{3\sin 99.3°}{\sin 50°} \approx 3.86$$

One triangle: $\beta \approx 30.7°,\ \gamma \approx 99.3°,\ c \approx 3.86$

19. $b = 5, \; c = 3, \; \beta = 100°$

$$\frac{\sin\beta}{b} = \frac{\sin\gamma}{c} \;\rightarrow\; \frac{\sin 100°}{5} = \frac{\sin\gamma}{3} \;\rightarrow\; \sin\gamma = \frac{3\sin 100°}{5} = 0.5909$$

$$\rightarrow\; \gamma = 36.2° \;\text{ or }\; \gamma = 143.8°$$

The second value is discarded because $\beta + \gamma > 180°$.

$\alpha = 180° - \beta - \gamma = 180° - 100° - 36.2° = 43.8°$

$$\frac{\sin\beta}{b} = \frac{\sin\alpha}{a} \;\rightarrow\; \frac{\sin 100°}{5} = \frac{\sin 43.8°}{a} \;\rightarrow\; a = \frac{5\sin 43.8°}{\sin 100°} \approx 3.51$$

One triangle: $\alpha \approx 43.8°, \; \gamma \approx 36.2°, \; a \approx 3.51$

21. $a = 4, \; b = 5, \; \alpha = 60°$

$$\frac{\sin\beta}{b} = \frac{\sin\alpha}{a} \;\rightarrow\; \frac{\sin\beta}{5} = \frac{\sin 60°}{4} \;\rightarrow\; \sin\beta = \frac{5\sin 60°}{4} = 1.0825$$

There is no angle β for which $\sin\beta > 1$. Therefore, there is no triangle with the given measurements.

23. $b = 4, \; c = 6, \; \beta = 20°$

$$\frac{\sin\beta}{b} = \frac{\sin\gamma}{c} \;\rightarrow\; \frac{\sin 20°}{4} = \frac{\sin\gamma}{6} \;\rightarrow\; \sin\gamma = \frac{6\sin 20°}{4} = 0.5130$$

$$\rightarrow\; \gamma_1 = 30.9° \;\text{ or }\; \gamma_2 = 149.1°$$

For both values, $\beta + \gamma < 180°$. Therefore, there are two triangles.

$\alpha_1 = 180° - \beta - \gamma_1 = 180° - 20° - 30.9° = 129.1°$

$$\frac{\sin\beta}{b} = \frac{\sin\alpha_1}{a_1} \;\rightarrow\; \frac{\sin 20°}{4} = \frac{\sin 129.1°}{a_1} \;\rightarrow\; a_1 = \frac{4\sin 129.1°}{\sin 20°} \approx 9.08$$

$\alpha_2 = 180° - \beta - \gamma_2 = 180° - 20° - 149.1° = 10.9°$

$$\frac{\sin\beta}{b} = \frac{\sin\alpha_2}{a_2} \;\rightarrow\; \frac{\sin 20°}{4} = \frac{\sin 10.9°}{a_2} \;\rightarrow\; a_2 = \frac{4\sin 10.9°}{\sin 20°} \approx 2.21$$

Two triangles: $\alpha_1 \approx 129.1°, \; \gamma_1 \approx 30.9°, \; a_1 \approx 9.08$

$\text{or}\;\; \alpha_2 \approx 10.9°, \; \gamma_2 \approx 149.1°, \; a_2 \approx 2.21$

25. $a = 2, \; c = 1, \; \gamma = 100°$

$$\frac{\sin\gamma}{c} = \frac{\sin\alpha}{a} \;\rightarrow\; \frac{\sin 100°}{1} = \frac{\sin\alpha}{2} \;\rightarrow\; \sin\alpha = \frac{2\sin 100°}{1} = 1.9696$$

There is no angle α for which $\sin\alpha > 1$. Therefore, there is no triangle with the given measurements.

27. $a = 2, \ c = 1, \ \gamma = 25°$

$$\frac{\sin \alpha}{a} = \frac{\sin \gamma}{c} \quad \rightarrow \quad \frac{\sin \alpha}{2} = \frac{\sin 25°}{1} \quad \rightarrow \quad \sin \alpha = \frac{2 \sin 25°}{1} = 0.8452$$

$$\rightarrow \quad \alpha_1 = 57.7° \ \text{ or } \ \alpha_2 = 122.3°$$

For both values, $\alpha + \gamma < 180°$. Therefore, there are two triangles.

$\beta_1 = 180° - \alpha_1 - \gamma = 180° - 57.7° - 25° = 97.3°$

$$\frac{\sin \beta_1}{b_1} = \frac{\sin \gamma}{c} \quad \rightarrow \quad \frac{\sin 97.3°}{b_1} = \frac{\sin 25°}{1} \quad \rightarrow \quad b_1 = \frac{1 \sin 97.3°}{\sin 25°} \approx 2.35$$

$\beta_2 = 180° - \alpha_2 - \gamma = 180° - 122.3° - 25° = 32.7°$

$$\frac{\sin \beta_2}{b_2} = \frac{\sin \gamma}{c} \quad \rightarrow \quad \frac{\sin 32.7°}{b_2} = \frac{\sin 25°}{1} \quad \rightarrow \quad b_2 = \frac{1 \sin 32.7°}{\sin 25°} \approx 1.28$$

Two triangles: $\alpha_1 \approx 57.7°, \ \beta_1 \approx 97.3°, \ b_1 \approx 2.35$
    or $\alpha_2 \approx 122.3°, \ \beta_2 \approx 32.7°, \ b_2 \approx 1.28$

29. (a) Find γ ; then use the Law of Sines:

$\gamma = 180° - 60° - 55° = 65°$

$$\frac{\sin 55°}{a} = \frac{\sin 65°}{150} \quad \rightarrow \quad a = \frac{150 \sin 55°}{\sin 65°} \approx 135.6 \ \text{miles}$$

$$\frac{\sin 60°}{b} = \frac{\sin 65°}{150} \quad \rightarrow \quad b = \frac{150 \sin 60°}{\sin 65°} \approx 143.3 \ \text{miles}$$

(b) $t = \dfrac{a}{r} = \dfrac{135.6}{200} \approx 0.68$ hours or ≈ 41 minutes

31. $\angle CAB = 180° - 25° = 155° \qquad \angle ABC = 180° - 155° - 15° = 10°$

Let c represent the distance from A to B.

$$\frac{\sin 15°}{c} = \frac{\sin 10°}{1000} \quad \rightarrow \quad c = \frac{1000 \sin 15°}{\sin 10°} \approx 1490.5 \ \text{feet}$$

The length of the proposed ski lift is approximately 1490 feet.

33. Find the distance from B to the plane:

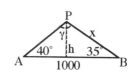

$$\gamma = 180° - 40° - 35° = 105° \qquad (\gamma = \angle APB)$$

$$\frac{\sin 40°}{x} = \frac{\sin 105°}{1000} \quad \rightarrow \quad x = \frac{1000 \sin 40°}{\sin 105°} \approx 665.5 \ \text{feet}$$

Find the height:

$$\sin 35° = \frac{h}{x} = \frac{h}{665.5} \quad \rightarrow \quad h = 665.5(\sin 35°) = 381.7 \ \text{feet}$$

The plane is 381.7 feet high.

35. (a) $\angle ABC = 180° - 40° = 140°$

Find the angle at city C:

$$\frac{\sin C}{150} = \frac{\sin 140°}{300} \quad \rightarrow \quad \sin C = \frac{150 \sin 140°}{300} = 0.3214 \quad \rightarrow \quad C \approx 18.7°$$

Find the angle at city A :

$$A = 180° - 140° - 18.7° = 21.3°$$

$$\frac{\sin 21.3°}{y} = \frac{\sin 140°}{300} \quad \rightarrow \quad y = \frac{300 \sin 21.3°}{\sin 140°} \approx 170 \text{ miles}$$

The distance from city B to city C is approximately 170 miles.

(b) To find the angle to turn, subtract angle C from 180°:

$$180° - 18.7° = 161.3°$$

The pilot needs to turn through an angle of 161.3° to return to city A.

37. Find angle β ($\angle ACB$):

$$\frac{\sin \beta}{123} = \frac{\sin 60°}{184.5} \quad \rightarrow \quad \sin \beta = \frac{123 \sin 60°}{184.5} = 0.5774$$

$$\beta \approx 35.3°$$

$$\angle CAB = 180° - 60° - 35.3° \approx 84.7°$$

Find the perpendicular distance:

$$\sin 84.7° = \frac{h}{184.5}$$

$$h = 184.5 \sin 84.7° = 183.7 \text{ feet}$$

39. $\alpha = 180° - 140° = 40°$ $\beta = 180° - 135° = 45°$

$$\gamma = 180° - 40° - 45° = 95°$$

$$\frac{\sin 40°}{a} = \frac{\sin 95°}{2} \quad \rightarrow \quad a = \frac{2 \sin 40°}{\sin 95°} \approx 1.290 \text{ mi}$$

$$\frac{\sin 45°}{b} = \frac{\sin 95°}{2} \quad \rightarrow \quad b = \frac{2 \sin 45°}{\sin 95°} \approx 1.420 \text{ mi}$$

$$\overline{BE} = 1.290 - 0.125 = 1.165 \text{ mi}$$

$$\overline{AD} = 1.420 - 0.125 = 1.295 \text{ mi}$$

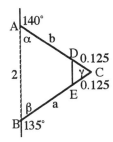

For the isosceles triangle,

$$\angle CDE = \angle CED = \frac{180° - 95°}{2} = 42.5°$$

$$\frac{\sin 95°}{DE} = \frac{\sin 42.5°}{0.125} \quad \rightarrow \quad DE = \frac{0.125 \sin 95°}{\sin 42.5°} \approx 0.184 \text{ miles}$$

The length of the highway is $1.165 + 1.295 + 0.184 = 2.644$ miles.

41. $\angle ABD = 180° - 30° = 150°$ $\quad \gamma = 180° - 150° - 20° = 10°$

$\dfrac{\sin 150°}{y} = \dfrac{\sin 10°}{1} \quad \rightarrow \quad y = \dfrac{1\sin 150°}{\sin 10°} \approx 2.88$ mi

$\dfrac{\sin \beta}{2.88} = \dfrac{\sin 20°}{1} \quad \rightarrow \quad \sin \beta = \dfrac{2.88\sin 20°}{1} \approx 0.9850$

$\beta \approx 80°$

$\alpha = 180° - 80° - 30° = 70°$

$\dfrac{\sin 70°}{x} = \dfrac{\sin 30°}{1} \quad \rightarrow \quad x = \dfrac{\sin 70°}{\sin 30°} \approx 1.88$ mi

The ship is about 1.88 miles from the harbor.

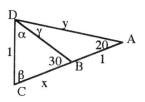

43. Using the Law of Sines:

$\dfrac{\sin 46.27°}{x} = \dfrac{\sin(90° - 46.27°)}{y + 100}$

$\qquad (y + 100)\sin 46.27° = x\sin 43.73°$

$\qquad y\sin 46.27° + 100\sin 46.27° = x\sin 43.73°$

$\qquad y = \dfrac{x\sin 43.73° - 100\sin 46.27°}{\sin 46.27°}$

$\dfrac{\sin 40.3°}{x} = \dfrac{\sin(90° - 40.3°)}{y + 200}$

$\qquad (y + 200)\sin 40.3° = x\sin 49.7°$

$\qquad y\sin 40.3° + 200\sin 40.3° = x\sin 49.7°$

$\qquad y = \dfrac{x\sin 49.7° - 200\sin 40.3°}{\sin 40.3°}$

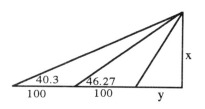

Set the two equations equal to each other and solve:

$\dfrac{x\sin 43.73° - 100\sin 46.27°}{\sin 46.27°} = \dfrac{x\sin 49.7° - 200\sin 40.3°}{\sin 40.3°}$

$x\sin 43.73°\cdot \sin 40.3° - 100\sin 46.27°\cdot \sin 40.3°$

$\qquad = x\sin 49.7°\cdot \sin 46.27° - 200\sin 40.3°\cdot \sin 46.27°$

$x\sin 43.73°\cdot \sin 40.3° - x\sin 49.7°\cdot \sin 46.27°$

$\qquad = 100\sin 46.27°\cdot \sin 40.3° - 200\sin 40.3°\cdot \sin 46.27°$

$x = \dfrac{100\sin 46.27°\cdot \sin 40.3° - 200\sin 40.3°\cdot \sin 46.27°}{\sin 43.73°\cdot \sin 40.3° - \sin 49.7°\cdot \sin 46.27°} \approx 449.36$ feet

45. Using the Law of Sines:

$\dfrac{\sin 30°}{h} = \dfrac{\sin 60°}{x} \quad \rightarrow \quad x = \dfrac{h\sin 60°}{\sin 30°}$

$\dfrac{\sin 20°}{h} = \dfrac{\sin 70°}{x + 40} \quad \rightarrow \quad x = \dfrac{h\sin 70°}{\sin 20°} - 40$

$\dfrac{h\sin 60°}{\sin 30°} = \dfrac{h\sin 70°}{\sin 20°} - 40$

$h\left(\dfrac{\sin 60°}{\sin 30°} - \dfrac{\sin 70°}{\sin 20°}\right) = -40$

$h = \dfrac{-40}{\dfrac{\sin 60°}{\sin 30°} - \dfrac{\sin 70°}{\sin 20°}} \approx 39.4$ feet

47. Find the distance from B to the helicopter:

$$\gamma = 180° - 40° - 25° = 115° \qquad (\gamma = \angle APB)$$

$$\frac{\sin 40°}{x} = \frac{\sin 115°}{100} \quad \rightarrow \quad x = \frac{100\sin 40°}{\sin 115°} \approx 70.9 \text{ feet}$$

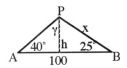

Find the height:

$$\sin 25° = \frac{h}{x} = \frac{h}{70.9} \quad \rightarrow \quad h = 70.9(\sin 25°) \approx 30 \text{ feet}$$

The helicopter is about 30 feet high.

49. $$\frac{a-b}{c} = \frac{a}{c} - \frac{b}{c} = \frac{\sin\alpha}{\sin\gamma} - \frac{\sin\beta}{\sin\gamma} = \frac{\sin\alpha - \sin\beta}{\sin\gamma} = \frac{2\sin\dfrac{\alpha-\beta}{2}\cos\dfrac{\alpha+\beta}{2}}{\sin\left(2\cdot\dfrac{\gamma}{2}\right)}$$

$$= \frac{2\sin\dfrac{\alpha-\beta}{2}\cos\dfrac{\alpha+\beta}{2}}{2\sin\dfrac{\gamma}{2}\cos\dfrac{\gamma}{2}} = \frac{\sin\dfrac{\alpha-\beta}{2}\cos\left(\dfrac{\pi}{2}-\dfrac{\gamma}{2}\right)}{\sin\dfrac{\gamma}{2}\cos\dfrac{\gamma}{2}} = \frac{\sin\dfrac{\alpha-\beta}{2}\sin\dfrac{\gamma}{2}}{\sin\dfrac{\gamma}{2}\cos\dfrac{\gamma}{2}}$$

$$= \frac{\sin\dfrac{\alpha-\beta}{2}}{\cos\dfrac{\gamma}{2}} = \frac{\sin\frac{1}{2}(\alpha-\beta)}{\cos\frac{1}{2}\gamma}$$

51. Derive the Law of Tangents:

$$\frac{a-b}{a+b} = \frac{\dfrac{a-b}{c}}{\dfrac{a+b}{c}} = \frac{\dfrac{\sin\frac{1}{2}(\alpha-\beta)}{\cos\frac{1}{2}\gamma}}{\dfrac{\cos\frac{1}{2}(\alpha-\beta)}{\sin\frac{1}{2}\gamma}} = \frac{\sin\frac{1}{2}(\alpha-\beta)}{\cos\frac{1}{2}\gamma}\cdot\frac{\sin\frac{1}{2}\gamma}{\cos\frac{1}{2}(\alpha-\beta)} = \tan\frac{1}{2}(\alpha-\beta)\tan\frac{1}{2}\gamma$$

$$= \tan\tfrac{1}{2}(\alpha-\beta)\tan\tfrac{1}{2}\big(\pi-(\alpha+\beta)\big) = \tan\tfrac{1}{2}(\alpha-\beta)\tan\left(\frac{\pi}{2}-\left(\frac{\alpha+\beta}{2}\right)\right)$$

$$= \tan\tfrac{1}{2}(\alpha-\beta)\cot\left(\frac{\alpha+\beta}{2}\right) = \frac{\tan\frac{1}{2}(\alpha-\beta)}{\tan\frac{1}{2}(\alpha+\beta)}$$

8.3   The Law of Cosines

1. $a = 2, \ c = 4, \ \beta = 45°$ $\qquad b^2 = a^2 + c^2 - 2ac\cos\beta$

$$b^2 = 2^2 + 4^2 - 2\cdot 2\cdot 4\cos 45° = 20 - 16\cdot\frac{\sqrt{2}}{2} = 20 - 8\sqrt{2} \approx 8.6863$$

$$b \approx 2.95$$

$$a^2 = b^2 + c^2 - 2bc\cos\alpha \quad \rightarrow \quad 2bc\cos\alpha = b^2 + c^2 - a^2 \quad \rightarrow \quad \cos\alpha = \frac{b^2 + c^2 - a^2}{2bc}$$

$$\cos\alpha = \frac{2.95^2 + 4^2 - 2^2}{2(2.95)(4)} = \frac{20.6863}{23.6} \approx 0.8765$$

$$\alpha \approx 28.8°$$

$$c^2 = a^2 + b^2 - 2ab\cos\gamma \quad \rightarrow \quad \cos\gamma = \frac{a^2 + b^2 - c^2}{2ab}$$

$$\cos\gamma = \frac{2^2 + 2.95^2 - 4^2}{2(2)(2.95)} = \frac{-3.2975}{11.8} \approx -0.2794$$

$$\gamma = 106.2°$$

3. $a = 2, \; b = 3, \; \gamma = 95°$ $c^2 = a^2 + b^2 - 2ab\cos\gamma$

$$c^2 = 2^2 + 3^2 - 2\cdot2\cdot3\cos95° = 13 - 12\cdot(-0.0872) \approx 14.0459$$

$$c \approx 3.75$$

$$a^2 = b^2 + c^2 - 2bc\cos\alpha \quad \rightarrow \quad \cos\alpha = \frac{b^2 + c^2 - a^2}{2bc}$$

$$\cos\alpha = \frac{3^2 + 3.75^2 - 2^2}{2(3)(3.75)} \approx 0.8472$$

$$\alpha \approx 32.1°$$

$$b^2 = a^2 + c^2 - 2ac\cos\beta \quad \rightarrow \quad \cos\beta = \frac{a^2 + c^2 - b^2}{2ac}$$

$$\cos\beta = \frac{2^2 + 3.75^2 - 3^2}{2(2)(3.75)} \approx 0.6042$$

$$\beta = 52.8°$$

5. $a = 6, \; b = 5, \; c = 8$

$$a^2 = b^2 + c^2 - 2bc\cos\alpha \quad \rightarrow \quad \cos\alpha = \frac{b^2 + c^2 - a^2}{2bc}$$

$$\cos\alpha = \frac{5^2 + 8^2 - 6^2}{2(5)(8)} \approx 0.6625$$

$$\alpha \approx 48.5°$$

$$b^2 = a^2 + c^2 - 2ac\cos\beta \quad \rightarrow \quad \cos\beta = \frac{a^2 + c^2 - b^2}{2ac}$$

$$\cos\beta = \frac{6^2 + 8^2 - 5^2}{2(6)(8)} \approx 0.7813$$

$$\beta = 38.6°$$

$$c^2 = a^2 + b^2 - 2ab\cos\gamma \quad \rightarrow \quad \cos\gamma = \frac{a^2 + b^2 - c^2}{2ab}$$

$$\cos\gamma = \frac{6^2 + 5^2 - 8^2}{2(6)(5)} \approx -0.0500$$

$$\gamma = 92.9°$$

7. $a = 9, \ b = 6, \ c = 4$

$a^2 = b^2 + c^2 - 2bc\cos\alpha \quad \rightarrow \quad \cos\alpha = \dfrac{b^2 + c^2 - a^2}{2bc}$

$\cos\alpha = \dfrac{6^2 + 4^2 - 9^2}{2(6)(4)} \approx -0.6042$

$\alpha \approx 127.2°$

$b^2 = a^2 + c^2 - 2ac\cos\beta \quad \rightarrow \quad \cos\beta = \dfrac{a^2 + c^2 - b^2}{2ac}$

$\cos\beta = \dfrac{9^2 + 4^2 - 6^2}{2(9)(4)} \approx 0.8472$

$\beta = 32.1°$

$c^2 = a^2 + b^2 - 2ab\cos\gamma \quad \rightarrow \quad \cos\gamma = \dfrac{a^2 + b^2 - c^2}{2ab}$

$\cos\gamma = \dfrac{9^2 + 6^2 - 4^2}{2(9)(6)} \approx 0.9352$

$\gamma = 20.7°$

9. $a = 3, \ b = 4, \ \gamma = 40°$

$c^2 = a^2 + b^2 - 2ab\cos\gamma$

$c^2 = 3^2 + 4^2 - 2\cdot 3\cdot 4\cos 40° \approx 6.6149$

$c \approx 2.57$

$a^2 = b^2 + c^2 - 2bc\cos\alpha \quad \rightarrow \quad \cos\alpha = \dfrac{b^2 + c^2 - a^2}{2bc}$

$\cos\alpha = \dfrac{4^2 + 2.57^2 - 3^2}{2(4)(2.57)} \approx 0.6617$

$\alpha \approx 48.6°$

$b^2 = a^2 + c^2 - 2ac\cos\beta \quad \rightarrow \quad \cos\beta = \dfrac{a^2 + c^2 - b^2}{2ac}$

$\cos\beta = \dfrac{3^2 + 2.57^2 - 4^2}{2(3)(2.57)} \approx -0.0256$

$\beta = 91.5°$

11. $b = 1, \ c = 3, \ \alpha = 80°$

$a^2 = b^2 + c^2 - 2bc\cos\alpha$

$a^2 = 1^2 + 3^2 - 2\cdot 1\cdot 3\cos 80° \approx 8.9581$

$a \approx 2.99$

$c^2 = a^2 + b^2 - 2ab\cos\gamma \quad \rightarrow \quad \cos\gamma = \dfrac{a^2 + b^2 - c^2}{2ab}$

$\cos\gamma = \dfrac{2.99^2 + 1^2 - 3^2}{2(2.99)(1)} \approx 0.1572$

$\gamma \approx 81.0°$

$$b^2 = a^2 + c^2 - 2ac\cos\beta \quad \rightarrow \quad \cos\beta = \frac{a^2 + c^2 - b^2}{2ac}$$

$$\cos\beta = \frac{2.99^2 + 3^2 - 1^2}{2(2.99)(3)} \approx 0.9443$$

$$\beta = 19.2°$$

13. $a = 3, \; c = 2, \; \beta = 110°$

$$b^2 = a^2 + c^2 - 2ac\cos\beta$$

$$b^2 = 3^2 + 2^2 - 2\cdot 3\cdot 2\cos110° \approx 17.1042$$

$$b \approx 4.14$$

$$a^2 = b^2 + c^2 - 2bc\cos\alpha \quad \rightarrow \quad \cos\alpha = \frac{b^2 + c^2 - a^2}{2bc}$$

$$\cos\alpha = \frac{4.14^2 + 2^2 - 3^2}{2(4.14)(2)} \approx 0.7331$$

$$\alpha \approx 42.9°$$

$$c^2 = a^2 + b^2 - 2ab\cos\gamma \quad \rightarrow \quad \cos\gamma = \frac{a^2 + b^2 - c^2}{2ab}$$

$$\cos\gamma = \frac{3^2 + 4.14^2 - 2^2}{2(3)(4.14)} \approx 0.8913$$

$$\gamma = 27.0°$$

15. $a = 2, \; b = 2, \; \gamma = 50°$

$$c^2 = a^2 + b^2 - 2ab\cos\gamma$$

$$c^2 = 2^2 + 2^2 - 2\cdot 2\cdot 2\cos50° \approx 2.8577$$

$$c \approx 1.69$$

$$a^2 = b^2 + c^2 - 2bc\cos\alpha \quad \rightarrow \quad \cos\alpha = \frac{b^2 + c^2 - a^2}{2bc}$$

$$\cos\alpha = \frac{2^2 + 1.69^2 - 2^2}{2(2)(1.69)} \approx 0.4225$$

$$\alpha \approx 65.0°$$

$$b^2 = a^2 + c^2 - 2ac\cos\beta \quad \rightarrow \quad \cos\beta = \frac{a^2 + c^2 - b^2}{2ac}$$

$$\cos\beta = \frac{2^2 + 1.69^2 - 2^2}{2(2)(1.69)} \approx 0.4225$$

$$\beta = 65.0°$$

17. $a = 12, \; b = 13, \; c = 5$

$$a^2 = b^2 + c^2 - 2bc\cos\alpha \quad \rightarrow \quad \cos\alpha = \frac{b^2 + c^2 - a^2}{2bc}$$

$$\cos\alpha = \frac{13^2 + 5^2 - 12^2}{2(13)(5)} \approx 0.3846$$

$$\alpha \approx 67.4°$$

$$b^2 = a^2 + c^2 - 2ac\cos\beta \quad \rightarrow \quad \cos\beta = \frac{a^2 + c^2 - b^2}{2ac}$$

$$\cos\beta = \frac{12^2 + 5^2 - 13^2}{2(12)(5)} = 0$$

$$\beta = 90°$$

$$c^2 = a^2 + b^2 - 2ab\cos\gamma \quad \rightarrow \quad \cos\gamma = \frac{a^2 + b^2 - c^2}{2ab}$$

$$\cos\gamma = \frac{12^2 + 13^2 - 5^2}{2(12)(13)} \approx 0.9231$$

$$\gamma = 22.6°$$

19. $a = 2, \ b = 2, \ c = 2$

$$a^2 = b^2 + c^2 - 2bc\cos\alpha \quad \rightarrow \quad \cos\alpha = \frac{b^2 + c^2 - a^2}{2bc}$$

$$\cos\alpha = \frac{2^2 + 2^2 - 2^2}{2(2)(2)} = 0.5$$

$$\alpha = 60°$$

$$b^2 = a^2 + c^2 - 2ac\cos\beta \quad \rightarrow \quad \cos\beta = \frac{a^2 + c^2 - b^2}{2ac}$$

$$\cos\beta = \frac{2^2 + 2^2 - 2^2}{2(2)(2)} = 0.5$$

$$\beta = 60°$$

$$c^2 = a^2 + b^2 - 2ab\cos\gamma \quad \rightarrow \quad \cos\gamma = \frac{a^2 + b^2 - c^2}{2ab}$$

$$\cos\gamma = \frac{2^2 + 2^2 - 2^2}{2(2)(2)} = 0.5$$

$$\gamma = 60°$$

21. $a = 5, \ b = 8, \ c = 9$

$$a^2 = b^2 + c^2 - 2bc\cos\alpha \quad \rightarrow \quad \cos\alpha = \frac{b^2 + c^2 - a^2}{2bc}$$

$$\cos\alpha = \frac{8^2 + 9^2 - 5^2}{2(8)(9)} \approx 0.8333$$

$$\alpha \approx 33.6°$$

$$b^2 = a^2 + c^2 - 2ac\cos\beta \quad \rightarrow \quad \cos\beta = \frac{a^2 + c^2 - b^2}{2ac}$$

$$\cos\beta = \frac{5^2 + 9^2 - 8^2}{2(5)(9)} \approx 0.4667$$

$$\beta \approx 62.2°$$

$$c^2 = a^2 + b^2 - 2ab\cos\gamma \quad \rightarrow \quad \cos\gamma = \frac{a^2 + b^2 - c^2}{2ab}$$

$$\cos\gamma = \frac{5^2 + 8^2 - 9^2}{2(5)(8)} \approx 0.1000$$

$$\gamma = 84.3°$$

23. $a = 10,\ b = 8,\ c = 5$

$$a^2 = b^2 + c^2 - 2bc\cos\alpha \quad \rightarrow \quad \cos\alpha = \frac{b^2 + c^2 - a^2}{2bc}$$

$$\cos\alpha = \frac{8^2 + 5^2 - 10^2}{2(8)(5)} \approx -0.1375$$

$$\alpha \approx 97.9°$$

$$b^2 = a^2 + c^2 - 2ac\cos\beta \quad \rightarrow \quad \cos\beta = \frac{a^2 + c^2 - b^2}{2ac}$$

$$\cos\beta = \frac{10^2 + 5^2 - 8^2}{2(10)(5)} \approx 0.6100$$

$$\beta \approx 52.4°$$

$$c^2 = a^2 + b^2 - 2ab\cos\gamma \quad \rightarrow \quad \cos\gamma = \frac{a^2 + b^2 - c^2}{2ab}$$

$$\cos\gamma = \frac{10^2 + 8^2 - 5^2}{2(10)(8)} \approx 0.8688$$

$$\gamma = 29.7°$$

25. Find the third side of the triangle using the Law of Cosines:

$$a = 50,\ b = 70,\ \gamma = 70°$$
$$c^2 = a^2 + b^2 - 2ab\cos\gamma$$
$$c^2 = 50^2 + 70^2 - 2 \cdot 50 \cdot 70\cos 70° \approx 5005.86$$
$$c \approx 70.75$$

The houses are approximately 70.75 feet apart.

27. (a) After 15 minutes, the plane would have flown $220(0.25) = 55$ miles.
    Find the third side of the triangle:

$$a = 55,\ b = 330,\ \gamma = 10°$$
$$c^2 = a^2 + b^2 - 2ab\cos\gamma$$
$$c^2 = 55^2 + 330^2 - 2 \cdot 55 \cdot 330\cos 10° \approx 76176.48$$
$$c \approx 276$$

Find the measure of the angle opposite the 330 side:

$$\cos\beta = \frac{a^2 + c^2 - b^2}{2ac}$$
$$\cos\beta = \frac{55^2 + 276^2 - 330^2}{2(55)(276)} \approx -0.9782$$
$$\beta \approx 168°$$

The pilot should turn through an angle of $180° - 168° = 12°$.

   (b) If the total trip is to be done in 90 minutes, and 15 minutes were used already, then there are 75 minutes or 1.25 hours to complete the trip. The plane must travel 276 miles in 1.25 hours.

$$r = \frac{276}{1.25} = 220.8 \text{ miles / hour}$$

The pilot must maintain a speed of 220.8 mi/hr to complete the whole trip in 90 minutes.

29. (a) Find x in the figure:

$$x^2 = 60.5^2 + 90^2 - 2(60.5)90\cos 45° \approx 4059.86$$

$$x \approx 63.7 \text{ feet}$$

It is about 63.7 feet from the pitching rubber
to first base.

(b) Use the Pythagorean Theorem to find y in the figure:

$$90^2 + 90^2 = (60.5 + y)^2$$

$$8100 + 8100 = (60.5 + y)^2$$

$$16200 = (60.5 + y)^2$$

$$60.5 + y = 127.3$$

$$y = 66.8 \text{ feet}$$

It is about 66.8 feet from the pitching rubber to second base.

(c) Find β in the figure by using the Law of Cosines:

$$\cos \beta = \frac{60.5^2 + 63.7^2 - 90^2}{2(60.5)(63.7)} \approx -0.0496$$

$$\beta \approx 92.8°$$

The pitcher needs to turn through an angle of 92.8° to face first base.

31. (a) Find x by using the Law of Cosines:

$$x^2 = 500^2 + 100^2 - 2(500)100\cos 80° \approx 242,635$$

$$x \approx 492.6 \text{ feet}$$

The guy wire needs to be about 492.6 feet long.

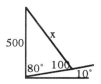

(b) Use the Pythagorean Theorem to find the value of y:

$$y^2 = 100^2 + 250^2 = 72500$$

$$y = 269.3 \text{ feet}$$

The guy wire needs to be about 269.3 feet long.

33. Find x by using the Law of Cosines:

$$x^2 = 400^2 + 90^2 - 2(400)90\cos 45° \approx 117,188.3$$

$$x \approx 342.3 \text{ feet}$$

It is approximately 342.3 feet from dead center
to third base.

35. Use the Law of Cosines:

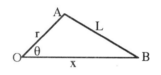

$$L^2 = x^2 + r^2 - 2xr\cos\theta$$
$$x^2 - 2xr\cos\theta + r^2 - L^2 = 0$$
$$x = \frac{2r\cos\theta + \sqrt{(2r\cos\theta)^2 - 4(1)(r^2 - L^2)}}{2(1)}$$
$$x = \frac{2r\cos\theta + \sqrt{4r^2\cos^2\theta - 4(r^2 - L^2)}}{2}$$
$$x = r\cos\theta + \sqrt{r^2\cos^2\theta + L^2 - r^2}$$

37. $$\cos\frac{\gamma}{2} = \sqrt{\frac{1 + \cos\gamma}{2}} = \sqrt{\frac{1 + \dfrac{a^2 + b^2 - c^2}{2ab}}{2}} = \sqrt{\frac{2ab + a^2 + b^2 - c^2}{4ab}} = \sqrt{\frac{(a + b)^2 - c^2}{4ab}}$$
$$= \sqrt{\frac{(a + b + c)(a + b - c)}{4ab}} = \sqrt{\frac{2s(2s - c - c)}{4ab}} = \sqrt{\frac{4s(s - c)}{4ab}} = \sqrt{\frac{s(s - c)}{ab}}$$

39. $$\frac{\cos\alpha}{a} + \frac{\cos\beta}{b} + \frac{\cos\gamma}{c} = \frac{b^2 + c^2 - a^2}{2bca} + \frac{a^2 + c^2 - b^2}{2acb} + \frac{a^2 + b^2 - c^2}{2abc}$$
$$= \frac{b^2 + c^2 - a^2 + a^2 + c^2 - b^2 + a^2 + b^2 - c^2}{2abc} = \frac{a^2 + b^2 + c^2}{2abc}$$

8.4 The Area of a Triangle

1. $a = 2$, $c = 4$, $\beta = 45°$
$$A = \frac{1}{2}ac\sin\beta = \frac{1}{2}(2)(4)\sin 45° \approx 2.83$$

3. $a = 2$, $b = 3$, $\gamma = 95°$
$$A = \frac{1}{2}ab\sin\gamma = \frac{1}{2}(2)(3)\sin 95° \approx 2.99$$

5. $a = 6$, $b = 5$, $c = 8$
$$s = \frac{1}{2}(a + b + c) = \frac{1}{2}(6 + 5 + 8) = \frac{19}{2}$$
$$A = \sqrt{s(s - a)(s - b)(s - c)} = \sqrt{\frac{19}{2}\left(\frac{7}{2}\right)\left(\frac{9}{2}\right)\left(\frac{3}{2}\right)} = \sqrt{\frac{3591}{16}} \approx 14.98$$

7. $a = 9$, $b = 6$, $c = 4$
$$s = \frac{1}{2}(a + b + c) = \frac{1}{2}(9 + 6 + 4) = \frac{19}{2}$$
$$A = \sqrt{s(s - a)(s - b)(s - c)} = \sqrt{\frac{19}{2}\left(\frac{1}{2}\right)\left(\frac{7}{2}\right)\left(\frac{11}{2}\right)} = \sqrt{\frac{1463}{16}} \approx 9.56$$

9. $a = 3, \ b = 4, \ \gamma = 40°$

$A = \dfrac{1}{2}ab\sin\gamma = \dfrac{1}{2}(3)(4)\sin 40° \approx 3.86$

11. $b = 1, \ c = 3, \ \alpha = 80°$

$A = \dfrac{1}{2}bc\sin\alpha = \dfrac{1}{2}(1)(3)\sin 80° \approx 1.48$

13. $a = 3, \ c = 2, \ \beta = 110°$

$A = \dfrac{1}{2}ac\sin\beta = \dfrac{1}{2}(3)(2)\sin 110° \approx 2.82$

15. $a = 2, \ b = 2, \ \gamma = 50°$

$A = \dfrac{1}{2}ab\sin\gamma = \dfrac{1}{2}(2)(2)\sin 50° \approx 1.53$

17. $a = 12, \ b = 13, \ c = 5$

$s = \dfrac{1}{2}(a + b + c) = \dfrac{1}{2}(12 + 13 + 5) = 15$

$A = \sqrt{s(s - a)(s - b)(s - c)} = \sqrt{15(3)(2)(10)} = \sqrt{900} \approx 30$

19. $a = 2, \ b = 2, \ c = 2$

$s = \dfrac{1}{2}(a + b + c) = \dfrac{1}{2}(2 + 2 + 2) = 3$

$A = \sqrt{s(s - a)(s - b)(s - c)} = \sqrt{3(1)(1)(1)} = \sqrt{3} \approx 1.73$

21. $a = 5, \ b = 8, \ c = 9$

$s = \dfrac{1}{2}(a + b + c) = \dfrac{1}{2}(5 + 8 + 9) = 11$

$A = \sqrt{s(s - a)(s - b)(s - c)} = \sqrt{11(6)(3)(2)} = \sqrt{396} \approx 19.90$

23. $a = 10, \ b = 8, \ c = 5$

$s = \dfrac{1}{2}(a + b + c) = \dfrac{1}{2}(10 + 8 + 5) = \dfrac{23}{2}$

$A = \sqrt{s(s - a)(s - b)(s - c)} = \sqrt{\dfrac{23}{2}\left(\dfrac{3}{2}\right)\left(\dfrac{7}{2}\right)\left(\dfrac{13}{2}\right)} = \sqrt{\dfrac{6279}{16}} \approx 19.81$

25.   Area of a sector $= \dfrac{1}{2}r^2\theta$ where θ is in radians.

$\theta = 70° \cdot \dfrac{\pi}{180} = \dfrac{7\pi}{18}$

Area of the sector $= \dfrac{1}{2} \cdot 8^2 \cdot \dfrac{7\pi}{18} = \dfrac{112\pi}{9} \approx 39.10$ square feet

Area of the triangle $= \dfrac{1}{2} \cdot 8 \cdot 8\sin 70° = 32\sin 70° \approx 30.07$ square feet

Area of the segment $= 39.10 - 30.07 = 9.03$ square feet

27. Find the area of the lot using Heron's Formula:

$a = 100, \; b = 50, \; c = 75$

$s = \dfrac{1}{2}(a + b + c) = \dfrac{1}{2}(100 + 50 + 75) = \dfrac{225}{2}$

$A = \sqrt{s(s-a)(s-b)(s-c)} = \sqrt{\dfrac{225}{2}\left(\dfrac{25}{2}\right)\left(\dfrac{125}{2}\right)\left(\dfrac{75}{2}\right)} = \sqrt{\dfrac{52{,}734{,}375}{16}} \approx 1815.46$

The cost is \$3 times the area:

Cost $= \$3(1815.46) = \$5446.38$

29. The area of the shaded region = the area of the semicircle – the area of the triangle.

Area of the semicircle $= \dfrac{1}{2}\pi r^2 = \dfrac{1}{2}\pi(4)^2 = 8\pi$ square centimeters

The triangle is a right triangle. Find the other leg:

$6^2 + b^2 = 8^2 \;\rightarrow\; b^2 = 64 - 36 = 28 \;\rightarrow\; b = \sqrt{28} = 2\sqrt{7}$

Area of the triangle $= \dfrac{1}{2}\cdot 6 \cdot 2\sqrt{7} = 6\sqrt{7}$ square centimeters

Area of the shaded region $= 8\pi - 6\sqrt{7} \approx 9.26$ square centimeters

31. Use the Law of Sines in the area of the triangle formula:

$A = \dfrac{1}{2}ab\sin\gamma = \dfrac{1}{2}a\sin\gamma\left(\dfrac{a\sin\beta}{\sin\alpha}\right) = \dfrac{a^2\sin\beta\sin\gamma}{2\sin\alpha}$

33. $\alpha = 40^\circ, \; \beta = 20^\circ, \; a = 2 \qquad \gamma = 180^\circ - \alpha - \beta = 180^\circ - 40^\circ - 20^\circ = 120^\circ$

$A = \dfrac{a^2\sin\beta\sin\gamma}{2\sin\alpha} = \dfrac{2^2\sin 20^\circ\sin 120^\circ}{2\sin 40^\circ} = \dfrac{4(0.3420)(0.8660)}{2(0.6428)} \approx 0.92$

35. $\beta = 70^\circ, \; \gamma = 10^\circ, \; b = 5 \qquad \alpha = 180^\circ - \beta - \gamma = 180^\circ - 70^\circ - 10^\circ = 100^\circ$

$A = \dfrac{b^2\sin\alpha\sin\gamma}{2\sin\beta} = \dfrac{5^2\sin 100^\circ\sin 10^\circ}{2\sin 70^\circ} = \dfrac{25(0.9848)(0.1736)}{2(0.9397)} \approx 2.27$

37. $\alpha = 110^\circ, \; \gamma = 30^\circ, \; c = 3 \qquad \beta = 180^\circ - \alpha - \gamma = 180^\circ - 110^\circ - 30^\circ = 40^\circ$

$A = \dfrac{c^2\sin\alpha\sin\beta}{2\sin\gamma} = \dfrac{3^2\sin 110^\circ\sin 40^\circ}{2\sin 30^\circ} = \dfrac{9(0.9397)(0.6428)}{2(0.5000)} \approx 5.44$

39. The area is the sum of the area of a triangle and a sector.

Area of the triangle $= \dfrac{1}{2}r\cdot r\sin(\pi - \theta) = \dfrac{1}{2}r^2\sin(\pi - \theta)$

Area of the sector $= \dfrac{1}{2}r^2\theta$

$A = \dfrac{1}{2}r^2\sin(\pi - \theta) + \dfrac{1}{2}r^2\theta = \dfrac{1}{2}r^2\big(\sin(\pi - \theta) + \theta\big)$

$= \dfrac{1}{2}r^2\big(\sin\pi\cos\theta - \cos\pi\sin\theta + \theta\big) = \dfrac{1}{2}r^2\big(0 + \sin\theta + \theta\big) = \dfrac{1}{2}r^2\big(\theta + \sin\theta\big)$

41. (a)   Area $\Delta OAC = \frac{1}{2}|OC|\cdot|AC| = \frac{1}{2}\cdot\frac{|OC|}{1}\cdot\frac{|AC|}{1} = \frac{1}{2}\cos\alpha\sin\alpha = \frac{1}{2}\sin\alpha\cos\alpha$

(b)   Area $\Delta OCB = \frac{1}{2}|OC|\cdot|BC| = \frac{1}{2}\cdot|OB|^2\cdot\frac{|OC|}{|OB|}\cdot\frac{|BC|}{|OB|} = \frac{1}{2}|OB|^2\cos\beta\sin\beta$

$$= \frac{1}{2}|OB|^2\sin\beta\cos\beta$$

(c)   Area $\Delta OAB = \frac{1}{2}|BD|\cdot|OA| = \frac{1}{2}|BD|\cdot1 = \frac{1}{2}\cdot|OB|\cdot\frac{|BD|}{|OB|} = \frac{1}{2}|OB|\sin(\alpha+\beta)$

(d) $\dfrac{\cos\alpha}{\cos\beta} = \dfrac{\dfrac{|OC|}{|OA|}}{\dfrac{|OC|}{|OB|}} = \dfrac{|OC|}{1}\cdot\dfrac{|OB|}{|OC|} = |OB|$

(e)   Area ΔOAB = Area ΔOAC + Area ΔOCB

$$\frac{1}{2}|OB|\sin(\alpha+\beta) = \frac{1}{2}\sin\alpha\cos\alpha + \frac{1}{2}|OB|^2\sin\beta\cos\beta$$

$$\frac{\cos\alpha}{\cos\beta}\sin(\alpha+\beta) = \sin\alpha\cos\alpha + \frac{\cos^2\alpha}{\cos^2\beta}\sin\beta\cos\beta$$

$$\sin(\alpha+\beta) = \frac{\cos\beta}{\cos\alpha}\sin\alpha\cos\alpha + \frac{\cos\alpha}{\cos\beta}\sin\beta\cos\beta$$

$$\sin(\alpha+\beta) = \sin\alpha\cos\beta + \cos\alpha\sin\beta$$

43.   The grazing area must be considered in sections. A_1 represents $\frac{3}{4}$ of a circle:

$$A_1 = \frac{3}{4}\pi(100)^2 = 7500\pi \approx 23,562 \text{ square feet}$$

Angles are needed to find A_2 and A_3: (see the figure)

In ΔABC, $\angle CBA = 45°$, $AB = 10$, $AC = 90$

Find $\angle BCA$:

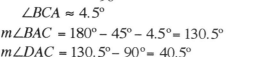 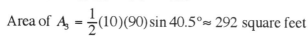

$$\frac{\sin\angle CBA}{90} = \frac{\sin\angle BCA}{10} \quad\rightarrow\quad \frac{\sin 45°}{90} = \frac{\sin\angle BCA}{10}$$

$$\sin\angle BCA = \frac{10\sin 45°}{90} \approx 0.0786$$

$$\angle BCA \approx 4.5°$$

$m\angle BAC = 180° - 45° - 4.5° = 130.5°$

$m\angle DAC = 130.5° - 90° = 40.5°$

Area of $A_3 = \frac{1}{2}(10)(90)\sin 40.5° \approx 292$ square feet

Area of sector $A_2 = \frac{1}{2}(90)^2\left(49.5°\cdot\frac{\pi}{180}\right) \approx 3499$ square feet

Since the cow can go in either direction around the barn , A_2 and A_3 must be doubled. Total grazing area is : $23,562 + 2(3499) + 2(292) = 31,144$ square feet

45. $h_1 = \dfrac{2K}{a}$, $h_2 = \dfrac{2K}{b}$, $h_3 = \dfrac{2K}{c}$ where K is the area of the triangle .

$$\dfrac{1}{h_1} + \dfrac{1}{h_2} + \dfrac{1}{h_3} = \dfrac{a}{2K} + \dfrac{b}{2K} + \dfrac{c}{2K} = \dfrac{a+b+c}{2K} = \dfrac{2s}{2K} = \dfrac{s}{K}$$

47. $h = \dfrac{a \sin\beta \sin\gamma}{\sin\alpha}$ where h is the altitude to side a.

In $\triangle OAB$, c is opposite angle AOB. The two adjacent angles are $\dfrac{\alpha}{2}$ and $\dfrac{\beta}{2}$.

Then $r = \dfrac{c \cdot \sin\dfrac{\alpha}{2}\sin\dfrac{\beta}{2}}{\sin(\angle AOB)}$

$\angle AOB = \pi - \left(\dfrac{\alpha}{2} + \dfrac{\beta}{2}\right)$

$\sin(\angle AOB) = \sin\left(\pi - \left(\dfrac{\alpha}{2} + \dfrac{\beta}{2}\right)\right) = \sin\left(\dfrac{\alpha}{2} + \dfrac{\beta}{2}\right) = \sin\left(\dfrac{\alpha + \beta}{2}\right) = \cos\left(\dfrac{\pi}{2} - \left(\dfrac{\alpha + \beta}{2}\right)\right)$

$\qquad = \cos\left(\dfrac{\pi - (\alpha + \beta)}{2}\right) = \cos\dfrac{\gamma}{2}$

Thus, $r = \dfrac{c \cdot \sin\dfrac{\alpha}{2}\sin\dfrac{\beta}{2}}{\cos\dfrac{\gamma}{2}}$

49. Use the result of Problem 48:

$$\cot\dfrac{\alpha}{2} + \cot\dfrac{\beta}{2} + \cot\dfrac{\gamma}{2} = \dfrac{s-a}{r} + \dfrac{s-b}{r} + \dfrac{s-c}{r} = \dfrac{s-a+s-b+s-c}{r}$$

$$= \dfrac{3s - (a+b+c)}{r} = \dfrac{3s - 2s}{r} = \dfrac{s}{r}$$

8.5 Simple Harmonic Motion; Damped Motion

1. $d = -5\cos(\pi t)$

3. $d = -6\cos(2t)$

5. $d = -5\sin(\pi t)$

7. $d = -6\sin(2t)$

9. $d = 5\sin(3t)$
   (a) Simple harmonic
   (b) 5 meters
   (c) $\dfrac{2\pi}{3}$ seconds
   (d) $\dfrac{3}{2\pi}$ oscillation/second

11. $d = 6\cos(\pi t)$
   (a) Simple harmonic
   (b) 6 meters
   (c) 2 seconds
   (d) $\dfrac{1}{2}$ oscillation/second

13. $d = -3\sin\left(\frac{1}{2}t\right)$

   (a) Simple harmonic
   (b) 3 meters
   (c) 4π seconds
   (d) $\dfrac{1}{4\pi}$ oscillation/second

15. $d = 6 + 2\cos(2\pi t)$

   (a) Simple harmonic
   (b) 2 meters
   (c) 1 second
   (d) 1 oscillation/second

17. (a)

$$d = -10e^{-0.7t/2(25)}\cos\left(\sqrt{\left(\frac{2\pi}{5}\right)^2 - \frac{(0.7)^2}{4(25)^2}}\, t\right)$$

$$d = -10e^{-0.7t/50}\cos\left(\sqrt{\frac{4\pi^2}{25} - \frac{0.49}{2500}}\, t\right)$$

(b)

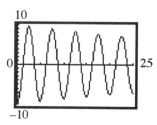

19. (a)

$$d = -18e^{-0.6\, t/2(30)}\cos\left(\sqrt{\left(\frac{\pi}{2}\right)^2 - \frac{(0.6)^2}{4(30)^2}}\, t\right)$$

$$d = -18e^{-0.6\, t/60}\cos\left(\sqrt{\frac{\pi^2}{4} - \frac{0.36}{3600}}\, t\right)$$

(b)

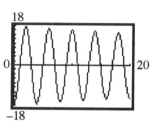

21. (a)

$$d = -5e^{-0.8\, t/2(10)}\cos\left(\sqrt{\left(\frac{2\pi}{3}\right)^2 - \frac{(0.8)^2}{4(10)^2}}\, t\right)$$

$$d = -5e^{-0.8\, t/20}\cos\left(\sqrt{\frac{4\pi^2}{9} - \frac{0.64}{400}}\, t\right)$$

(b)

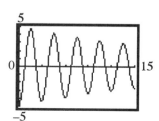

23. (a) Damped motion with a bob of mass 20 kg and a damping factor of 0.7.
   (b) 20 meters downward
   (c) Graph:

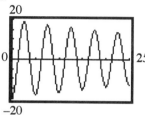

   (d) The maximum displacement after one oscillation is 18.32 meters.
   (e) It approaches zero, since $e^{-0.7\, t/40} \to 0$ as $t \to \infty$.

25. (a) Damped motion with a bob of mass 40 kg and a damping factor of 0.6.
    (b) 30 meters downward
    (c) Graph:

    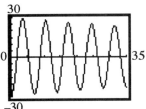

    (d) The maximum displacement after one oscillation is 28.46 meters.
    (e) It approaches zero, since $e^{-0.6\,t/80} \to 0$ as $t \to \infty$.

27. (a) Damped motion with a bob of mass 15 kg and a damping factor of 0.9.
    (b) 15 meters downward
    (c) Graph:

    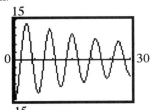

    (d) The maximum displacement after one oscillation is 12.53 meters.
    (e) It approaches zero, since $e^{-0.9\,t/30} \to 0$ as $t \to \infty$.

29. (a) Graph:

    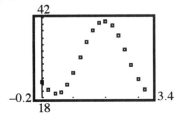

    (b) Amplitude $= \dfrac{40.93 - 19.06}{2} = 10.935$

    Vertical Shift $= \dfrac{40.93 + 19.06}{2} = 29.995$

    Period $= 3$ seconds (estimate from data) $\omega = \dfrac{2\pi}{3}$

Find the horizontal shift (use y = 19.06, x = 0.4)

$$19.06 = 10.935\sin\left(\frac{2\pi}{3}(0.4) - \phi\right) + 29.995$$

$$-10.935 = 10.935\sin\left(\frac{2\pi}{3}(0.4) - \phi\right)$$

$$-1 = \sin\left(\frac{2\pi}{3}(0.4) - \phi\right)$$

$$\frac{-\pi}{2} = \frac{2\pi}{3}(0.4) - \phi$$

$$\phi = 2.4086$$

$$y = 10.935\sin\left(\frac{2\pi}{3}x - 2.4086\right) + 29.995$$

(c)

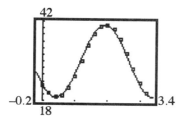

(d) $y = 11.04305\sin(2.07608x - 2.47156) + 29.87711$

(e)

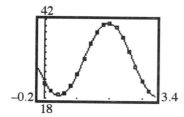

(f) $\frac{2\pi}{T} = 2.07608 \quad \rightarrow \quad T = \frac{2\pi}{2.07608} \approx 3.03$ seconds

31. (a)   Graph:

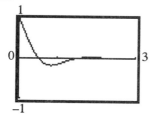

    (b)   The graph of V touches the graph of $y = e^{-1.9t}$ when $t = 0, 2$.

        The graph of V touches the graph of $y = -e^{-1.9t}$ when $t = 1, 3$.

    (c) $-0.1 < V < 0.1$ for $0.43 < t < 0.60$ or $t > 1.15$.

33. The graph will lie between the bounding curves $y = \pm x$, $y = \pm x^2$, $y = \pm x^3$, respectively, touching them at odd multiples of $\frac{\pi}{2}$. The x-intercepts of each graph are the multiples of π.

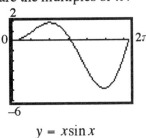
$y = x \sin x$

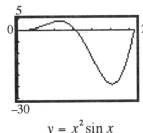
$y = x^2 \sin x$

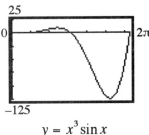
$y = x^3 \sin x$

8 Chapter Review

1. $c = 10$, $\beta = 20°$

$\sin\beta = \dfrac{b}{c}$ → $\sin 20° = \dfrac{b}{10}$ → $b = 10\sin 20° = 10(0.3420) \approx 3.42$

$\cos\beta = \dfrac{a}{c}$ → $\cos 20° = \dfrac{a}{10}$ → $a = 10\cos 20° = 10(0.9397) \approx 9.40$

$\alpha = 90° - \beta = 90° - 20° = 70°$

3. $b = 2$, $c = 5$

$c^2 = a^2 + b^2$ → $a^2 = c^2 - b^2 = 5^2 - 2^2 = 25 - 4 = 21$ → $a = \sqrt{21} \approx 4.58$

$\sin\beta = \dfrac{b}{c} = \dfrac{2}{5} = 0.4000$ → $\beta \approx 23.6°$

$\alpha = 90° - \beta = 90° - 23.6° = 66.4°$

5. $\alpha = 50°$, $\beta = 30°$, $a = 1$

$\gamma = 180° - \alpha - \beta = 180° - 50° - 30° = 100°$

$\dfrac{\sin\alpha}{a} = \dfrac{\sin\beta}{b}$ → $\dfrac{\sin 50°}{1} = \dfrac{\sin 30°}{b}$ → $b = \dfrac{1\sin 30°}{\sin 50°} \approx 0.65$

$\dfrac{\sin\gamma}{c} = \dfrac{\sin\alpha}{a}$ → $\dfrac{\sin 100°}{c} = \dfrac{\sin 50°}{1}$ → $c = \dfrac{1\sin 100°}{\sin 50°} \approx 1.29$

7. $a = 5$, $c = 2$, $\alpha = 100°$

$\dfrac{\sin\gamma}{c} = \dfrac{\sin\alpha}{a}$ → $\dfrac{\sin\gamma}{2} = \dfrac{\sin 100°}{5}$ → $\sin\gamma = \dfrac{2\sin 100°}{5} \approx 0.3939$

→ $\gamma = 23.2°$ or $\gamma = 156.8°$

The second value is discarded because $\alpha + \gamma > 180°$.

$\beta = 180° - \alpha - \gamma = 180° - 100° - 23.2° = 56.8°$

$\dfrac{\sin\beta}{b} = \dfrac{\sin\alpha}{a}$ → $\dfrac{\sin 56.8°}{b} = \dfrac{\sin 100°}{5}$ → $b = \dfrac{5\sin 56.8°}{\sin 100°} \approx 4.25$

9. $a = 3,\ c = 1,\ \gamma = 110°$

$\dfrac{\sin \gamma}{c} = \dfrac{\sin \alpha}{a} \quad \rightarrow \quad \dfrac{\sin 110°}{1} = \dfrac{\sin \alpha}{3} \quad \rightarrow \quad \sin \alpha = \dfrac{3 \sin 110°}{1} \approx 2.8191$

There is no angle α for which $\sin \alpha > 1$. Therefore, there is no triangle with the given measurements.

11. $a = 3,\ c = 1,\ \beta = 100°$

$b^2 = a^2 + c^2 - 2ac\cos\beta$

$b^2 = 3^2 + 1^2 - 2 \cdot 3 \cdot 1 \cos 100° \approx 11.0419$

$b \approx 3.32$

$a^2 = b^2 + c^2 - 2bc\cos\alpha \quad \rightarrow \quad \cos\alpha = \dfrac{b^2 + c^2 - a^2}{2bc}$

$\cos\alpha = \dfrac{3.32^2 + 1^2 - 3^2}{2(3.32)(1)} \approx 0.4552$

$\alpha \approx 62.9°$

$c^2 = a^2 + b^2 - 2ab\cos\gamma \quad \rightarrow \quad \cos\gamma = \dfrac{a^2 + b^2 - c^2}{2ab}$

$\cos\gamma = \dfrac{3^2 + 3.32^2 - 1^2}{2(3)(3.32)} \approx 0.9549$

$\gamma = 17.3°$

13. $a = 2,\ b = 3,\ c = 1$

$a^2 = b^2 + c^2 - 2bc\cos\alpha \quad \rightarrow \quad \cos\alpha = \dfrac{b^2 + c^2 - a^2}{2bc}$

$\cos\alpha = \dfrac{3^2 + 1^2 - 2^2}{2(3)(1)} = 1.000$

$\alpha \approx 0°$

No triangle exists with an angle of 0°.

15. $a = 1,\ b = 3,\ \gamma = 40°$

$c^2 = a^2 + b^2 - 2ab\cos\gamma$

$c^2 = 1^2 + 3^2 - 2 \cdot 1 \cdot 3 \cos 40° \approx 5.4037$

$c \approx 2.32$

$a^2 = b^2 + c^2 - 2bc\cos\alpha \quad \rightarrow \quad \cos\alpha = \dfrac{b^2 + c^2 - a^2}{2bc}$

$\cos\alpha = \dfrac{3^2 + 2.32^2 - 1^2}{2(3)(2.32)} \approx 0.9614$

$\alpha \approx 16.0°$

$b^2 = a^2 + c^2 - 2ac\cos\beta \quad \rightarrow \quad \cos\beta = \dfrac{a^2 + c^2 - b^2}{2ac}$

$\cos\beta = \dfrac{1^2 + 2.32^2 - 3^2}{2(1)(2.32)} \approx -0.5641$

$\beta = 124.3°$

17. $a = 5, \ b = 3, \ \alpha = 80°$

$$\frac{\sin\beta}{b} = \frac{\sin\alpha}{a} \quad \rightarrow \quad \frac{\sin\beta}{3} = \frac{\sin 80°}{5} \quad \rightarrow \quad \sin\beta = \frac{3\sin 80°}{5} = 0.5909$$

$$\rightarrow \quad \beta = 36.2° \quad \text{or} \quad \beta = 143.8°$$

The second value is discarded because $\alpha + \beta > 180°$.

$\gamma = 180° - \alpha - \beta = 180° - 80° - 36.2° = 63.8°$

$$\frac{\sin\gamma}{c} = \frac{\sin\alpha}{a} \quad \rightarrow \quad \frac{\sin 63.8°}{c} = \frac{\sin 80°}{5} \quad \rightarrow \quad c = \frac{5\sin 63.8°}{\sin 80°} \approx 4.56$$

19. $a = 1, \ b = \frac{1}{2}, \ c = \frac{4}{3}$

$$a^2 = b^2 + c^2 - 2bc\cos\alpha \quad \rightarrow \quad \cos\alpha = \frac{b^2 + c^2 - a^2}{2bc}$$

$$\cos\alpha = \frac{\left(\frac{1}{2}\right)^2 + \left(\frac{4}{3}\right)^2 - 1^2}{2\left(\frac{1}{2}\right)\left(\frac{4}{3}\right)} \approx 0.7708$$

$\alpha \approx 39.6°$

$$b^2 = a^2 + c^2 - 2ac\cos\beta \quad \rightarrow \quad \cos\beta = \frac{a^2 + c^2 - b^2}{2ac}$$

$$\cos\beta = \frac{1^2 + \left(\frac{4}{3}\right)^2 - \left(\frac{1}{2}\right)^2}{2(1)\left(\frac{4}{3}\right)} \approx 0.9479$$

$\beta \approx 18.6°$

$\gamma = 180° - \alpha - \beta = 180° - 39.6° - 18.6° = 121.8°$

21. $a = 3, \ b = 4, \ \alpha = 10°$

$$\frac{\sin\beta}{b} = \frac{\sin\alpha}{a} \quad \rightarrow \quad \frac{\sin\beta}{4} = \frac{\sin 10°}{3} \quad \rightarrow \quad \sin\beta = \frac{4\sin 10°}{3} \approx 0.2315$$

$$\rightarrow \quad \beta_1 \approx 13.4° \quad \text{or} \quad \beta_2 \approx 166.6°$$

For both values, $\alpha + \beta < 180°$. Therefore, there are two triangles.

$\gamma_1 = 180° - \alpha - \beta_1 = 180° - 10° - 13.4° \approx 156.6°$

$$\frac{\sin\alpha}{a} = \frac{\sin\gamma_1}{c_1} \quad \rightarrow \quad \frac{\sin 10°}{3} = \frac{\sin 156.6°}{c_1} \quad \rightarrow \quad c_1 = \frac{3\sin 156.6°}{\sin 10°} \approx 6.86$$

$\gamma_2 = 180° - \alpha - \beta_2 = 180° - 10° - 166.6° \approx 3.4°$

$$\frac{\sin\alpha}{a} = \frac{\sin\gamma_2}{c_2} \quad \rightarrow \quad \frac{\sin 10°}{3} = \frac{\sin 3.4°}{c_2} \quad \rightarrow \quad c_2 = \frac{3\sin 3.4°}{\sin 10°} \approx 1.02$$

Two triangles: $\beta_1 \approx 13.4°, \ \gamma_1 \approx 156.6°, \ c_1 \approx 6.86$

$\qquad\qquad$ or $\beta_2 \approx 166.6°, \ \gamma_2 \approx 3.4°, \ c_2 \approx 1.02$

23. $b = 4, \ c = 5, \ \alpha = 70°$

$a^2 = b^2 + c^2 - 2bc\cos\alpha$

$a^2 = 4^2 + 5^2 - 2 \cdot 4 \cdot 5\cos70° \approx 27.3192$

$a \approx 5.23$

$c^2 = a^2 + b^2 - 2ab\cos\gamma \quad \rightarrow \quad \cos\gamma = \dfrac{a^2 + b^2 - c^2}{2ab}$

$\cos\gamma = \dfrac{5.23^2 + 4^2 - 5^2}{2(5.23)(4)} \approx 0.4386$

$\gamma \approx 64.0°$

$\beta = 180° - \alpha - \gamma = 180° - 70° - 64° \approx 46.0°$

25. $a = 2, \ b = 3, \ \gamma = 40°$

$A = \dfrac{1}{2}ab\sin\gamma = \dfrac{1}{2}(2)(3)\sin40° \approx 1.93$

27. $b = 4, \ c = 10, \ \alpha = 70°$

$A = \dfrac{1}{2}bc\sin\alpha = \dfrac{1}{2}(4)(10)\sin70° \approx 18.79$

29. $a = 4, \ b = 3, \ c = 5$

$s = \dfrac{1}{2}(a + b + c) = \dfrac{1}{2}(4 + 3 + 5) = 6$

$A = \sqrt{s(s - a)(s - b)(s - c)} = \sqrt{6(2)(3)(1)} = \sqrt{36} = 6$

31. $a = 4, \ b = 2, \ c = 5$

$s = \dfrac{1}{2}(a + b + c) = \dfrac{1}{2}(4 + 2 + 5) = \dfrac{11}{2}$

$A = \sqrt{s(s - a)(s - b)(s - c)} = \sqrt{\dfrac{11}{2}\left(\dfrac{3}{2}\right)\left(\dfrac{7}{2}\right)\left(\dfrac{1}{2}\right)} = \sqrt{\dfrac{231}{16}} \approx 3.80$

33. $\alpha = 50°, \ \beta = 30°, \ a = 1 \qquad \gamma = 180° - \alpha - \beta = 180° - 50° - 30° = 100°$

$A = \dfrac{a^2 \sin\beta \sin\gamma}{2\sin\alpha} = \dfrac{1^2 \sin30° \sin100°}{2\sin50°} = \dfrac{1(0.5000)(0.9848)}{2(0.7660)} \approx 0.32$

35. Use right triangle methods:

$\tan65° = \dfrac{500}{b} \quad \rightarrow \quad b = \dfrac{500}{\tan65°} \approx 233.15$

$\tan25° = \dfrac{500}{a + b} \quad \rightarrow \quad a + b = \dfrac{500}{\tan25°} \approx 1072.25$

$a = 1072.25 - 233.15 = 839.1 \text{ feet}$

The lake is approximately 839 feet long.

37. $\tan25° = \dfrac{b}{50} \quad \rightarrow \quad b = 50\tan25° = 50(0.4663) \approx 23.3 \text{ feet}$

39. 1454 ft ≈ 0.2754 miles

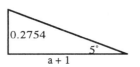

$\tan 5° = \dfrac{0.2754}{a+1}$

$a+1 = \dfrac{0.2754}{\tan 5°} \approx 3.15 \quad \rightarrow \quad a \approx 2.15$ miles

The boat is about 2.15 miles offshore.

41. $\angle ABC = 180° - 20° = 160°$

Find the angle at city C:

$\dfrac{\sin C}{100} = \dfrac{\sin 160°}{300} \quad \rightarrow \quad \sin C = \dfrac{100 \sin 160°}{300} \approx 0.1140 \quad \rightarrow \quad C \approx 6.55°$

Find the angle at city A :

$A = 180° - 160° - 6.55° = 13.45°$

$\dfrac{\sin 13.45°}{y} = \dfrac{\sin 160°}{300} \quad \rightarrow \quad y = \dfrac{300 \sin 13.45°}{\sin 160°} \approx 204$ miles

The distance from city B to city C is approximately 204 miles.

43. Draw a line perpendicular to the shore:

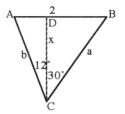

   (a) $\angle ACB = 12° + 30° = 42°$

$\angle ABC = 90° - 30° = 60°$

$\angle CAB = 90° - 12° = 78°$

$\dfrac{\sin 60°}{b} = \dfrac{\sin 42°}{2} \quad \rightarrow \quad b = \dfrac{2 \sin 60°}{\sin 42°} \approx 2.59$ miles

   (b) $\dfrac{\sin 78°}{a} = \dfrac{\sin 42°}{2} \quad \rightarrow \quad a = \dfrac{2 \sin 78°}{\sin 42°} \approx 2.92$ miles

   (c) $\cos 12° = \dfrac{x}{2.59} \quad \rightarrow \quad x = 2.59 \cos 12° \approx 2.53$ miles

45. (a) After 4 hours, the yacht would have sailed $18(4) = 72$ miles.

Find the third side of the triangle determines the distance from the island:

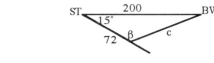

$a = 72, \quad b = 200, \quad \gamma = 15°$

$c^2 = a^2 + b^2 - 2ab \cos \gamma$

$c^2 = 72^2 + 200^2 - 2 \cdot 72 \cdot 200 \cos 15° \approx 17365.34$

$c \approx 131.8$ miles

The yacht is about 131.8 miles from the island.

   (b) Find the measure of the angle opposite the 200 side:

$\cos \beta = \dfrac{a^2 + c^2 - b^2}{2ac}$

$\cos \beta = \dfrac{72^2 + 131.8^2 - 200^2}{2(72)(131.8)} \approx -0.9192$

$\beta \approx 156.8°$

The yacht should turn through an angle of $180° - 156.8° = 23.2°$ to correct its course.

(c)   The original trip would have taken: $t = \dfrac{200}{18} \approx 11.1$ hours.

The actual trip takes: $t = 4 + \dfrac{131.8}{18} \approx 4 + 7.3 \approx 11.3$ hours.

The trip take about 0.2 hour or 12 minutes longer.

47.   Find the lengths of the two unknown sides of the middle triangle:

$x^2 = 100^2 + 125^2 - 2(100)(125)\cos 50° \approx 9555.31$

$x \approx 97.75$ feet

$y^2 = 70^2 + 50^2 - 2(70)(50)\cos 100° \approx 8615.54$

$y \approx 92.82$ feet

Find the areas of the three triangles:

$A_1 = \dfrac{1}{2}(100)(125)\sin 50° \approx 4787.78 \text{ ft}^2$

$A_2 = \dfrac{1}{2}(50)(70)\sin 100° \approx 1723.41 \text{ ft}^2$

$s = \dfrac{1}{2}(50 + 97.75 + 92.82) = 120.285$

$A_3 = \sqrt{120.285(70.285)(22.535)(27.465)} \approx 2287.47$

The approximate area of the lake is $4787.78 + 1723.41 + 2287.47 = 8798.66$ sq.ft.

49.   Area of the segment = area of the sector - area of the triangle.

Area of sector $= \dfrac{1}{2}r^2\theta = \dfrac{1}{2}\cdot 6^2\left(50\cdot\dfrac{\pi}{180}\right) \approx 15.708 \text{ in}^2$

Area of triangle $= \dfrac{1}{2}ab\sin\theta = \dfrac{1}{2}\cdot 6\cdot 6\sin 50° \approx 13.789 \text{ in}^2$

Area of segment $= 15.708 - 13.789 = 1.919 \text{ in}^2$

51.   Extend the tangent line until it meets a line extended through the centers of the pulleys. Label these extensions x and y. The distance between the points of tangency is z. Two similar triangles are formed. Therefore:

$\dfrac{24 + y}{y} = \dfrac{6.5}{2.5}$

where $24 + y$ is the hypotenuse of the larger triangle and y is the hypotenuse of the smaller triangle .

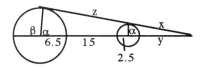

Solve for y:

$6.5y = 2.5(24 + y)$

$6.5y = 60 + 2.5y$

$4y = 60$

$y = 15$

Use the Pythagorean Theorem to find x:

$x^2 + 2.5^2 = 15^2$

$x^2 = 225 - 6.25 = 218.75$

$x \approx 14.79$

Use the Pythagorean Theorem to find z:

$$(z + 14.79)^2 + 6.5^2 = (24 + 15)^2$$
$$(z + 14.79)^2 = 1521 - 42.25 = 1478.75$$
$$z + 14.79 = 38.45$$
$$z = 23.66$$

Find α:

$$\cos\alpha = \frac{2.5}{15} \approx 0.1667 \quad\rightarrow\quad \alpha \approx 1.4033 \text{ radians}$$

$\beta = \pi - 1.4033 \approx 1.7383$ radians
The arc length on the larger pulley is: $6.5(1.7383) = 11.30$ inches.
The arc length on the smaller pulley is $2.5(1.4033) = 3.51$ inches.
The distance between the points of tangency is 23.66 inches.
The length of the belt is: $2(11.30 + 3.51 + 23.66) = 76.94$ inches.

53. $d = 6\sin(2t)$
   (a) Simple harmonic
   (b) 6 feet
   (c) π seconds
   (d) $\dfrac{1}{\pi}$ oscillation/second

55. $d = -2\cos(\pi t)$
   (a) Simple harmonic
   (b) 2 feet
   (c) 2 seconds
   (d) $\dfrac{1}{2}$ oscillation/second

57. (a)

$$d = -15e^{-0.75t/2(40)}\cos\left(\sqrt{\left(\frac{2\pi}{5}\right)^2 - \frac{(0.75)^2}{4(40)^2}}\,t\right)$$

$$d = -15e^{-0.75t/80}\cos\left(\sqrt{\frac{4\pi^2}{25} - \frac{0.5625}{6400}}\,t\right)$$

(b)

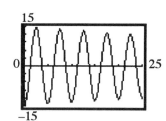

59. (a) Damped motion with a bob of mass 20 kg and a damping factor of 0.6.
   (b) 15 meters downward
   (c) Graph:

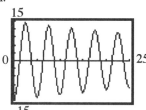

   (d) The maximum displacement after one oscillation is 13.92 meters.
   (e) It approaches zero, since $e^{-0.6\,t/40} \to 0$ as $t \to \infty$.

Chapter 9

Polar Coordinates; Vectors

9.1 Polar Coordinates

1. A   3. C   5. B   7. A

9. $(3, 90°)$

11. $(-2, 0)$

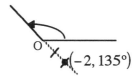

13. $\left(6, \frac{\pi}{6}\right)$

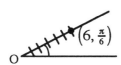

15. $(-2, 135°)$

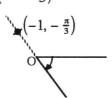

17. $\left(-1, -\frac{\pi}{3}\right)$

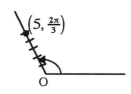

19. $(-2, -\pi)$

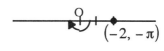

21. $\left(5, \frac{2\pi}{3}\right)$

(a) $r > 0$, $-2\pi \le \theta < 0$ $\left(5, -\frac{4\pi}{3}\right)$

(b) $r < 0$, $0 \le \theta < 2\pi$ $\left(-5, \frac{5\pi}{3}\right)$

(c) $r > 0$, $2\pi \le \theta < 4\pi$ $\left(5, \frac{8\pi}{3}\right)$

23. $(-2, 3\pi)$

$(-2, 3\pi)$

(a) $r > 0$, $-2\pi \le \theta < 0$ $\quad(2, -2\pi)$
(b) $r < 0$, $0 \le \theta < 2\pi$ $\quad(-2, \pi)$
(c) $r > 0$, $2\pi \le \theta < 4\pi$ $\quad(2, 2\pi)$

25. $\left(1, \dfrac{\pi}{2}\right)$

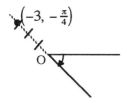
$\left(1, \dfrac{\pi}{2}\right)$

(a) $r > 0$, $-2\pi \le \theta < 0$ $\quad\left(1, -\dfrac{3\pi}{2}\right)$

(b) $r < 0$, $0 \le \theta < 2\pi$ $\quad\left(-1, \dfrac{3\pi}{2}\right)$

(c) $r > 0$, $2\pi \le \theta < 4\pi$ $\quad\left(1, \dfrac{5\pi}{2}\right)$

27. $\left(-3, -\dfrac{\pi}{4}\right)$

$\left(-3, -\dfrac{\pi}{4}\right)$

(a) $r > 0$, $-2\pi \le \theta < 0$ $\quad\left(3, -\dfrac{5\pi}{4}\right)$

(b) $r < 0$, $0 \le \theta < 2\pi$ $\quad\left(-3, \dfrac{7\pi}{4}\right)$

(c) $r > 0$, $2\pi \le \theta < 4\pi$ $\quad\left(3, \dfrac{11\pi}{4}\right)$

29. $x = r\cos\theta = 3\cos\dfrac{\pi}{2} = 3\cdot 0 = 0$

$y = r\sin\theta = 3\sin\dfrac{\pi}{2} = 3\cdot 1 = 3$

The rectangular coordinates of the point $\left(3, \dfrac{\pi}{2}\right)$ are $(0, 3)$.

31. $x = r\cos\theta = -2\cos 0 = -2\cdot 1 = -2$
$y = r\sin\theta = -2\sin 0 = -2\cdot 0 = 0$
The rectangular coordinates of the point $(-2, 0)$ are $(-2, 0)$.

33. $x = r\cos\theta = 6\cos 150° = 6\cdot\dfrac{-\sqrt{3}}{2} = -3\sqrt{3}$

$y = r\sin\theta = 6\sin 150° = 6\cdot\dfrac{1}{2} = 3$

The rectangular coordinates of the point $(6, 150°)$ are $\left(-3\sqrt{3}, 3\right)$.

35. $x = r\cos\theta = -2\cos\dfrac{3\pi}{4} = -2\cdot\dfrac{-\sqrt{2}}{2} = \sqrt{2}$

   $y = r\sin\theta = -2\sin\dfrac{3\pi}{4} = -2\cdot\dfrac{\sqrt{2}}{2} = -\sqrt{2}$

   The rectangular coordinates of the point $\left(-2, \dfrac{3\pi}{4}\right)$ are $\left(\sqrt{2}, -\sqrt{2}\right)$.

37. $x = r\cos\theta = -1\cos\dfrac{-\pi}{3} = -1\cdot\dfrac{1}{2} = -\dfrac{1}{2}$

   $y = r\sin\theta = -1\sin\dfrac{-\pi}{3} = -1\cdot\dfrac{-\sqrt{3}}{2} = \dfrac{\sqrt{3}}{2}$

   The rectangular coordinates of the point $\left(-1, \dfrac{-\pi}{3}\right)$ are $\left(-\dfrac{1}{2}, \dfrac{\sqrt{3}}{2}\right)$.

39. $x = r\cos\theta = -2\cos(-180°) = -2\cdot-1 = 2$
   $y = r\sin\theta = -2\sin(-180°) = -2\cdot 0 = 0$

   The rectangular coordinates of the point $(-2, -180°)$ are $(2, 0)$.

41. $x = r\cos\theta = 7.5\cos110° = 7.5(-0.3420) = -2.57$
   $y = r\sin\theta = 7.5\sin110° = 7.5(0.9397) = 7.05$

   The rectangular coordinates of the point $(7.5, 110°)$ are $(-2.57, 7.05)$.

43. $x = r\cos\theta = 6.3\cos3.8 = 6.3(-0.7910) = -4.98$
   $y = r\sin\theta = 6.3\sin3.8 = 6.3(-0.6119) = -3.85$

   The rectangular coordinates of the point $(6.3, 3.8)$ are $(-4.98, -3.85)$.

45. $r = \sqrt{x^2 + y^2} = \sqrt{3^2 + 0^2} = \sqrt{9} = 3$ $\theta = \tan^{-1}\dfrac{y}{x} = \tan^{-1}\dfrac{0}{3} = \tan^{-1}0 = 0$
   Polar coordinates of the point $(3, 0)$ are $(3, 0)$.

47. $r = \sqrt{x^2 + y^2} = \sqrt{(-1)^2 + 0^2} = \sqrt{1} = 1$ $\theta = \tan^{-1}\dfrac{y}{x} = \tan^{-1}\dfrac{0}{-1} = \tan^{-1}0 = 0$
   The point lies on the negative x-axis thus $\theta = \pi$.
   Polar coordinates of the point $(-1, 0)$ are $(1, \pi)$.

49. The point $(1, -1)$ lies in quadrant IV.
   $r = \sqrt{x^2 + y^2} = \sqrt{1^2 + (-1)^2} = \sqrt{2}$ $\theta = \tan^{-1}\dfrac{y}{x} = \tan^{-1}\dfrac{-1}{1} = \tan^{-1}(-1) = -\dfrac{\pi}{4}$
   Polar coordinates of the point $(1, -1)$ are $\left(\sqrt{2}, -\dfrac{\pi}{4}\right)$.

51. The point $\left(\sqrt{3}, 1\right)$ lies in quadrant I.

$$r = \sqrt{x^2 + y^2} = \sqrt{\left(\sqrt{3}\right)^2 + 1^2} = \sqrt{4} = 2 \qquad \theta = \tan^{-1}\frac{y}{x} = \tan^{-1}\frac{1}{\sqrt{3}} = \frac{\pi}{6}$$

Polar coordinates of the point $\left(\sqrt{3}, 1\right)$ are $\left(2, \frac{\pi}{6}\right)$.

53. The point $(1.3, -2.1)$ lies in quadrant IV.

$$r = \sqrt{x^2 + y^2} = \sqrt{1.3^2 + (-2.1)^2} = \sqrt{6.1} \approx 2.47$$

$$\theta = \tan^{-1}\frac{y}{x} = \tan^{-1}\frac{-2.1}{1.3} = \tan^{-1}(-1.6154) = -1.02$$

Polar coordinates of the point $(1.3, -2.1)$ are $(2.47, -1.02)$.

55. The point $(8.3, 4.2)$ lies in quadrant I.

$$r = \sqrt{x^2 + y^2} = \sqrt{8.3^2 + 4.2^2} = \sqrt{86.53} \approx 9.30$$

$$\theta = \tan^{-1}\frac{y}{x} = \tan^{-1}\frac{4.2}{8.3} = \tan^{-1}0.5060 \approx 0.47$$

Polar coordinates of the point $(8.3, 4.2)$ are $(9.30, 0.47)$.

57.
$$2x^2 + 2y^2 = 3$$
$$2\left(x^2 + y^2\right) = 3$$
$$2r^2 = 3$$
$$r^2 = \frac{3}{2}$$

59.
$$x^2 = 4y$$
$$(r\cos\theta)^2 = 4r\sin\theta$$
$$r^2\cos^2\theta - 4r\sin\theta = 0$$

61.
$$2xy = 1$$
$$2(r\cos\theta)(r\sin\theta) = 1$$
$$2r^2\sin\theta\cos\theta = 1$$
$$r^2\sin 2\theta = 1$$

63.
$$x = 4$$
$$r\cos\theta = 4$$

65.
$$r = \cos\theta$$
$$r^2 = r\cos\theta$$
$$x^2 + y^2 = x$$
$$x^2 - x + y^2 = 0$$

67.
$$r^2 = \cos\theta$$
$$r^3 = r\cos\theta$$
$$\left(x^2 + y^2\right)^{\frac{3}{2}} = x$$
$$\left(x^2 + y^2\right)^{\frac{3}{2}} - x = 0$$

69.
$$r = 2$$
$$\sqrt{x^2 + y^2} = 2$$
$$x^2 + y^2 = 4$$

71.
$$r = \frac{4}{1 - \cos\theta}$$
$$r(1 - \cos\theta) = 4$$
$$r - r\cos\theta = 4$$
$$\sqrt{x^2 + y^2} - x = 4$$
$$\sqrt{x^2 + y^2} = x + 4$$
$$x^2 + y^2 = x^2 + 8x + 16$$
$$y^2 = 8(x + 2)$$

73. Rewrite the polar coordinates in rectangular form:
$$P_1 = (r_1, \theta_1) \rightarrow P_1 = (r_1\cos\theta_1, r_1\sin\theta_1)$$
$$P_2 = (r_2, \theta_2) \rightarrow P_2 = (r_2\cos\theta_2, r_2\sin\theta_2)$$
$$d = \sqrt{(r_2\cos\theta_2 - r_1\cos\theta_1)^2 + (r_2\sin\theta_2 - r_1\sin\theta_1)^2}$$
$$= \sqrt{r_2^2\cos^2\theta_2 - 2r_1r_2\cos\theta_2\cos\theta_1 + r_1^2\cos^2\theta_1 + r_2^2\sin^2\theta_2 - 2r_1r_2\sin\theta_2\sin\theta_1 + r_1^2\sin^2\theta_1}$$
$$= \sqrt{r_2^2(\cos^2\theta_2 + \sin^2\theta_2) + r_1^2(\cos^2\theta_1 + \sin^2\theta_1) - 2r_1r_2(\cos\theta_2\cos\theta_1 + \sin\theta_2\sin\theta_1)}$$
$$= \sqrt{r_2^2 + r_1^2 - 2r_1r_2\cos(\theta_2 - \theta_1)}$$

9.2   Polar Equations and Graphs

1. $r = 4$

The equation is of the form $r = a$, $a > 0$. It is a circle, center at the pole and radius 4. Transform to rectangular form:
$$r = 4$$
$$r^2 = 16$$
$$x^2 + y^2 = 16$$

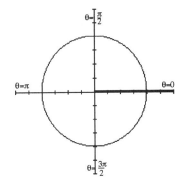

3. $\theta = \dfrac{\pi}{3}$

   The equation is of the form $\theta = \alpha$. It is a
   line, passing through the pole at an angle of
   $\dfrac{\pi}{3}$.

   Transform to rectangular form:

   $$\theta = \frac{\pi}{3}$$
   $$\tan \theta = \tan \frac{\pi}{3}$$
   $$\frac{y}{x} = \sqrt{3}$$
   $$y = \sqrt{3}x$$

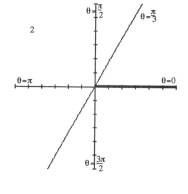

5. $r \sin \theta = 4$

   The equation is of the form $r \sin \theta = b$. It is
   a horizontal line, 4 units above the pole.
   Transform to rectangular form:
   $$r \sin \theta = 4$$
   $$y = 4$$

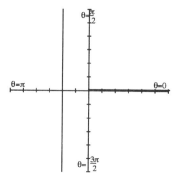

7. $r \cos \theta = -2$

   The equation is of the form $r \cos \theta = a$. It is
   a vertical line, 2 units to the left of the pole.
   Transform to rectangular form:
   $$r \cos \theta = -2$$
   $$x = -2$$

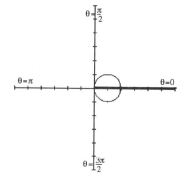

9. $r = 2 \cos \theta$

   The equation is of the form
   $r = 2a \cos \theta$, $a > 0$. It is a circle, passing
   through the pole, and center on the polar axis.
   Transform to rectangular form:
   $$r = 2 \cos \theta$$
   $$r^2 = 2r \cos \theta$$
   $$x^2 + y^2 = 2x$$
   $$x^2 - 2x + y^2 = 0$$
   $$(x - 1)^2 + y^2 = 1$$

11. $r = -4\sin\theta$
The equation is of the form
$r = 2a\sin\theta,\ a > 0$. It is a circle, passing
through the pole, and center on the line
$\theta = \dfrac{\pi}{2}$.
Transform to rectangular form:
$$r = -4\sin\theta$$
$$r^2 = -4r\sin\theta$$
$$x^2 + y^2 = -4y$$
$$x^2 + y^2 + 4y = 0$$
$$x^2 + (y+2)^2 = 4$$

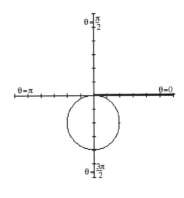

13. $r\sec\theta = 4$
Transform to rectangular form:
$$r\sec\theta = 4$$
$$r \cdot \frac{1}{\cos\theta} = 4$$
$$r = 4\cos\theta$$
$$r^2 = 4r\cos\theta$$
$$x^2 + y^2 = 4x$$
$$x^2 - 4x + y^2 = 0$$
$$(x-2)^2 + y^2 = 4$$
The equation is a circle, passing through the
pole, center on the polar axis and radius 2.

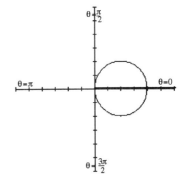

15. $r\csc\theta = -2$
Transform to rectangular form:
$$r\csc\theta = -2$$
$$r \cdot \frac{1}{\sin\theta} = -2$$
$$r = -2\sin\theta$$
$$r^2 = -2r\sin\theta$$
$$x^2 + y^2 = -2y$$
$$x^2 + y^2 + 2y = 0$$
$$x^2 + (y+1)^2 = 1$$
The equation is a circle, passing through the
pole, center on the line $\theta = \dfrac{\pi}{2}$ and radius 1.

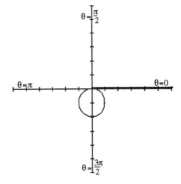

17.   E   19.   F   21.   H   23.   D

25.   D   27.   F   29.   A

31. $r = 2 + 2\cos\theta$ The graph will be a cardioid. Check for symmetry:

Polar axis: Replace θ by $-\theta$. The result is $r = 2 + 2\cos(-\theta) = 2 + 2\cos\theta$.
The graph is symmetric with respect to the polar axis.

The line $\theta = \dfrac{\pi}{2}$: Replace θ by $\pi - \theta$.

$$r = 2 + 2\cos(\pi - \theta) = 2 + 2(\cos\pi\cos\theta + \sin\pi\sin\theta)$$
$$= 2 + 2(-\cos\theta + 0) = 2 - 2\cos\theta$$

The test fails.

The pole: Replace r by $-r$. $-r = 2 + 2\cos\theta$. The test fails.

Due to symmetry to the polar axis, assign values to θ from 0 to π.

θ	0	$\dfrac{\pi}{6}$	$\dfrac{\pi}{3}$	$\dfrac{\pi}{2}$	$\dfrac{2\pi}{3}$	$\dfrac{5\pi}{6}$	π
$r = 2 + 2\cos\theta$	4	$2 + \sqrt{3} \approx 3.7$	3	2	1	$2 - \sqrt{3} \approx 0.3$	0

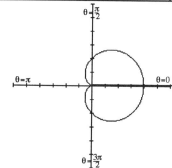

33. $r = 3 - 3\sin\theta$ The graph will be a cardioid. Check for symmetry:

Polar axis: Replace θ by $-\theta$. The result is $r = 3 - 3\sin(-\theta) = 3 + 3\sin\theta$.
The test fails.

The line $\theta = \dfrac{\pi}{2}$: Replace θ by $\pi - \theta$.

$$r = 3 - 3\sin(\pi - \theta) = 3 - 3(\sin\pi\cos\theta - \cos\pi\sin\theta)$$
$$= 3 - 3(0 + \sin\theta) = 3 - 3\sin\theta$$

The graph is symmetric with respect to the line $\theta = \dfrac{\pi}{2}$.

The pole: Replace r by $-r$. $-r = 3 - 3\sin\theta$. The test fails.

Due to symmetry to the line $\theta = \dfrac{\pi}{2}$, assign values to θ from $-\dfrac{\pi}{2}$ to $\dfrac{\pi}{2}$.

θ	$-\dfrac{\pi}{2}$	$-\dfrac{\pi}{3}$	$-\dfrac{\pi}{6}$	0	$\dfrac{\pi}{6}$	$\dfrac{\pi}{3}$	$\dfrac{\pi}{2}$
$r = 3 - 3\sin\theta$	6	$3 + \dfrac{3\sqrt{3}}{2} \approx 5.6$	$\dfrac{9}{2}$	3	$\dfrac{3}{2}$	$3 - \dfrac{3\sqrt{3}}{2} \approx 0.4$	0

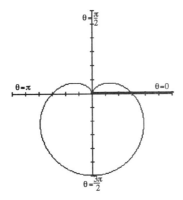

35. $r = 2 + \sin\theta$ The graph will be a limacon without an inner loop.

Check for symmetry:

Polar axis: Replace θ by $-\theta$. The result is $r = 2 + \sin(-\theta) = 2 - \sin\theta$.
The test fails.

The line $\theta = \dfrac{\pi}{2}$: Replace θ by $\pi - \theta$.

$$r = 2 + \sin(\pi - \theta) = 2 + (\sin\pi\cos\theta - \cos\pi\sin\theta)$$
$$= 2 + (0 + \sin\theta) = 2 + \sin\theta$$

The graph is symmetric with respect to the line $\theta = \dfrac{\pi}{2}$.

The pole: Replace r by $-r$. $-r = 2 + \sin\theta$. The test fails.

Due to symmetry to the line $\theta = \dfrac{\pi}{2}$, assign values to θ from $-\dfrac{\pi}{2}$ to $\dfrac{\pi}{2}$.

θ	$-\dfrac{\pi}{2}$	$-\dfrac{\pi}{3}$	$-\dfrac{\pi}{6}$	0	$\dfrac{\pi}{6}$	$\dfrac{\pi}{3}$	$\dfrac{\pi}{2}$
$r = 2 + \sin\theta$	1	$2 - \dfrac{\sqrt{3}}{2} \approx 1.1$	$\dfrac{3}{2}$	2	$\dfrac{5}{2}$	$2 + \dfrac{\sqrt{3}}{2} \approx 2.9$	3

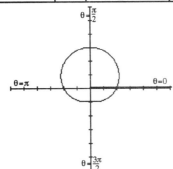

37. $r = 4 - 2\cos\theta$ The graph will be a limacon without an inner loop.

Check for symmetry:

Polar axis: Replace θ by $-\theta$. The result is $r = 4 - 2\cos(-\theta) = 4 - 2\cos\theta$.
The graph is symmetric with respect to the polar axis.

The line $\theta = \dfrac{\pi}{2}$: Replace θ by $\pi - \theta$.

$$r = 4 - 2\cos(\pi - \theta) = 4 - 2(\cos\pi\cos\theta + \sin\pi\sin\theta)$$
$$= 4 - 2(-\cos\theta + 0) = 4 + 2\cos\theta$$

The test fails.

The pole: Replace r by $-r$. $-r = 4 - 2\cos\theta$. The test fails.

Due to symmetry to the polar axis, assign values to θ from 0 to π.

θ	0	$\dfrac{\pi}{6}$	$\dfrac{\pi}{3}$	$\dfrac{\pi}{2}$	$\dfrac{2\pi}{3}$	$\dfrac{5\pi}{6}$	π
$r = 4 - 2\cos\theta$	2	$4 - \sqrt{3} \approx 2.3$	3	4	5	$4 + \sqrt{3} \approx 5.7$	6

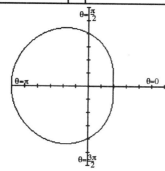

39. $r = 1 + 2\sin\theta$ The graph will be a limacon with an inner loop.
Check for symmetry:

Polar axis: Replace θ by $-\theta$. The result is $r = 1 + 2\sin(-\theta) = 1 - 2\sin\theta$.
The test fails.

The line $\theta = \dfrac{\pi}{2}$: Replace θ by $\pi - \theta$.

$$r = 1 + 2\sin(\pi - \theta) = 1 + 2(\sin\pi\cos\theta - \cos\pi\sin\theta)$$
$$= 1 + 2(0 + \sin\theta) = 1 + 2\sin\theta$$

The graph is symmetric with respect to the line $\theta = \dfrac{\pi}{2}$.

The pole: Replace r by $-r$. $-r = 1 + 2\sin\theta$. The test fails.

Due to symmetry to the line $\theta = \dfrac{\pi}{2}$, assign values to θ from $-\dfrac{\pi}{2}$ to $\dfrac{\pi}{2}$.

θ	$-\dfrac{\pi}{2}$	$-\dfrac{\pi}{3}$	$-\dfrac{\pi}{6}$	0	$\dfrac{\pi}{6}$	$\dfrac{\pi}{3}$	$\dfrac{\pi}{2}$
$r = 1 + 2\sin\theta$	-1	$1 - \sqrt{3} \approx -0.7$	0	1	2	$1 + \sqrt{3} \approx 2.7$	3

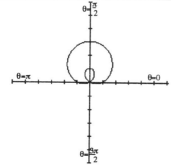

41. $r = 2 - 3\cos\theta$ The graph will be a limacon with an inner loop.
Check for symmetry:
Polar axis: Replace θ by $-\theta$. The result is $r = 2 - 3\cos(-\theta) = 2 - 3\cos\theta$.
The graph is symmetric with respect to the polar axis.

The line $\theta = \dfrac{\pi}{2}$: Replace θ by $\pi - \theta$.
$$r = 2 - 3\cos(\pi - \theta) = 2 - 3(\cos\pi\cos\theta + \sin\pi\sin\theta)$$
$$= 2 - 3(-\cos\theta + 0) = 2 + 3\cos\theta$$
The test fails.

The pole: Replace r by $-r$. $-r = 2 - 3\cos\theta$. The test fails.

Due to symmetry to the polar axis, assign values to θ from 0 to π.

θ	0	$\dfrac{\pi}{6}$	$\dfrac{\pi}{3}$	$\dfrac{\pi}{2}$	$\dfrac{2\pi}{3}$	$\dfrac{5\pi}{6}$	π
$r = 2 - 3\cos\theta$	-1	$2 - \dfrac{3\sqrt{3}}{2} \approx -0.6$	$\dfrac{1}{2}$	2	$\dfrac{7}{2}$	$2 + \dfrac{3\sqrt{3}}{2} \approx 4.6$	5

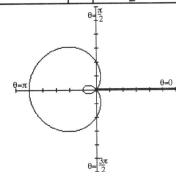

43. $r = 3\cos(2\theta)$ The graph will be a rose with four petals. Check for symmetry:
Polar axis: Replace θ by $-\theta$. $r = 3\cos(2(-\theta)) = 3\cos(-2\theta) = 3\cos(2\theta)$.
The graph is symmetric with respect to the polar axis.

The line $\theta = \dfrac{\pi}{2}$: Replace θ by $\pi - \theta$.
$$r = 3\cos(2(\pi - \theta)) = 3\cos(2\pi - 2\theta)$$
$$= 3(\cos 2\pi\cos 2\theta + \sin 2\pi\sin 2\theta) = 3(\cos 2\theta + 0) = 3\cos 2\theta$$

The graph is symmetric with respect to the line $\theta = \dfrac{\pi}{2}$.

The pole: Since the graph is symmetric to both the polar axis and the line

$\theta = \dfrac{\pi}{2}$, it is also symmetric to the pole.

Due to symmetry, assign values to θ from 0 to $\dfrac{\pi}{2}$.

θ	0	$\dfrac{\pi}{6}$	$\dfrac{\pi}{4}$	$\dfrac{\pi}{3}$	$\dfrac{\pi}{2}$
$r = 3\cos 2\theta$	3	$\dfrac{3}{2}$	0	$-\dfrac{3}{2}$	-3

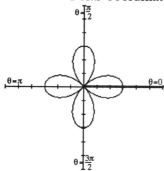

45. $r = 4\sin(3\theta)$ The graph will be a rose with three petals. Check for symmetry:

Polar axis: Replace θ by $-\theta$. $r = 4\sin(3(-\theta)) = 4\sin(-3\theta) = -4\sin 3\theta$.
The test fails.

The line $\theta = \dfrac{\pi}{2}$: Replace θ by $\pi - \theta$.

$$r = 4\sin(3(\pi - \theta)) = 4\sin(3\pi - 3\theta)$$
$$= 4(\sin 3\pi \cos 3\theta - \cos 3\pi \sin 3\theta) = 4(0 + \sin 3\theta) = 4\sin 3\theta$$

The graph is symmetric with respect to the line $\theta = \dfrac{\pi}{2}$.

The pole: Replace r by $-r$. $-r = 4\sin 3\theta$. The test fails.

Due to symmetry to the line $\theta = \dfrac{\pi}{2}$, assign values to θ from $-\dfrac{\pi}{2}$ to $\dfrac{\pi}{2}$.

θ	$-\dfrac{\pi}{2}$	$-\dfrac{\pi}{3}$	$-\dfrac{\pi}{4}$	$-\dfrac{\pi}{6}$	0	$\dfrac{\pi}{6}$	$\dfrac{\pi}{4}$	$\dfrac{\pi}{3}$	$\dfrac{\pi}{2}$
$r = 4\sin 3\theta$	4	0	$-2\sqrt{2} \approx -2.8$	-4	0	4	$2\sqrt{2} \approx 2.8$	0	-4

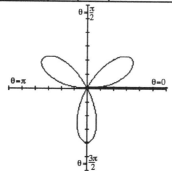

47. $r^2 = 9\cos(2\theta)$ The graph will be a lemniscate. Check for symmetry:

Polar axis: Replace θ by $-\theta$. $r^2 = 9\cos(2(-\theta)) = 9\cos(-2\theta) = 9\cos(2\theta)$.
The graph is symmetric with respect to the polar axis.

The line $\theta = \dfrac{\pi}{2}$: Replace θ by $\pi - \theta$.

$$r^2 = 9\cos(2(\pi - \theta)) = 9\cos(2\pi - 2\theta)$$
$$= 9(\cos 2\pi \cos 2\theta + \sin 2\pi \sin 2\theta) = 9(\cos 2\theta + 0) = 9\cos 2\theta$$

The graph is symmetric with respect to the line $\theta = \dfrac{\pi}{2}$.

The pole: Since the graph is symmetric to both the polar axis and the line $\theta = \dfrac{\pi}{2}$, it is also symmetric to the pole.

Due to symmetry, assign values to θ from 0 to $\dfrac{\pi}{2}$.

θ	0	$\dfrac{\pi}{6}$	$\dfrac{\pi}{4}$	$\dfrac{\pi}{3}$	$\dfrac{\pi}{2}$
$r = \pm\sqrt{9\cos 2\theta}$	± 3	$\pm\dfrac{3\sqrt{2}}{2} \approx \pm 2.1$	0	not defined	not defined

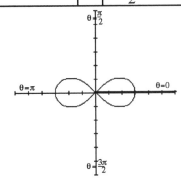

49. $r = 2^{\theta}$ The graph will be a spiral. Check for symmetry:

Polar axis: Replace θ by $-\theta$. $r = 2^{-\theta}$. The test fails.

The line $\theta = \dfrac{\pi}{2}$: Replace θ by $\pi - \theta$. $r = 2^{\pi-\theta}$. The test fails.

The pole: Replace r by $-r$. $-r = 2^{\theta}$. The test fails.

θ	$-\pi$	$-\dfrac{\pi}{2}$	$-\dfrac{\pi}{4}$	0	$\dfrac{\pi}{4}$	$\dfrac{\pi}{2}$	π	$\dfrac{3\pi}{2}$	2π
$r = 2^{\theta}$	0.1	0.3	0.6	1	1.7	3.0	8.8	26.2	77.9

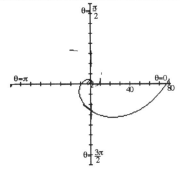

51. $r = 1 - \cos\theta$ The graph will be a cardioid. Check for symmetry:

Polar axis: Replace θ by $-\theta$. The result is $r = 1 - \cos(-\theta) = 1 - \cos\theta$.
The graph is symmetric with respect to the polar axis.

The line $\theta = \dfrac{\pi}{2}$: Replace θ by $\pi - \theta$.

$$r = 1 - \cos(\pi - \theta) = 1 - (\cos\pi\cos\theta + \sin\pi\sin\theta)$$
$$= 1 - (-\cos\theta + 0) = 1 + \cos\theta \qquad \text{The test fails}$$

The pole: Replace r by $-r$. $-r = 1 - \cos\theta$. The test fails.

Due to symmetry to the polar axis, assign values to θ from 0 to π.

θ	0	$\dfrac{\pi}{6}$	$\dfrac{\pi}{3}$	$\dfrac{\pi}{2}$	$\dfrac{2\pi}{3}$	$\dfrac{5\pi}{6}$	π
$r = 1 - \cos\theta$	0	$1 - \dfrac{\sqrt{3}}{2} \approx 0.1$	$\dfrac{1}{2}$	1	$\dfrac{3}{2}$	$1 + \dfrac{\sqrt{3}}{2} \approx 1.9$	2

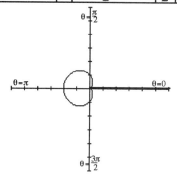

53. $r = 1 - 3\cos\theta$ The graph will be a limacon with an inner loop.

Check for symmetry:

Polar axis: Replace θ by $-\theta$. The result is $r = 1 - 3\cos(-\theta) = 1 - 3\cos\theta$.

                    The graph is symmetric with respect to the polar axis.

The line $\theta = \dfrac{\pi}{2}$: Replace θ by $\pi - \theta$.

$$r = 1 - 3\cos(\pi - \theta) = 1 - 3(\cos\pi\cos\theta + \sin\pi\sin\theta)$$
$$= 1 - 3(-\cos\theta + 0) = 1 + 3\cos\theta$$

            The test fails.

The pole: Replace r by $-r$. $-r = 1 - 3\cos\theta$. The test fails.

Due to symmetry to the polar axis, assign values to θ from 0 to π.

θ	0	$\dfrac{\pi}{6}$	$\dfrac{\pi}{3}$	$\dfrac{\pi}{2}$	$\dfrac{2\pi}{3}$	$\dfrac{5\pi}{6}$	π
$r = 1 - 3\cos\theta$	-2	$1 - \dfrac{3\sqrt{3}}{2} \approx -1.6$	$-\dfrac{1}{2}$	1	$\dfrac{5}{2}$	$1 + \dfrac{3\sqrt{3}}{2} \approx 3.6$	4

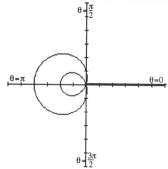

55. $r = \dfrac{2}{1 - \cos\theta}$ Check for symmetry:

Polar axis: Replace θ by $-\theta$. The result is $r = \dfrac{2}{1 - \cos(-\theta)} = \dfrac{2}{1 - \cos\theta}$.
The graph is symmetric with respect to the polar axis.

The line $\theta = \dfrac{\pi}{2}$: Replace θ by $\pi - \theta$.

$$r = \dfrac{2}{1 - \cos(\pi - \theta)} = \dfrac{2}{1 - (\cos\pi\cos\theta + \sin\pi\sin\theta)}$$

$$= \dfrac{2}{1 - (-\cos\theta + 0)} = \dfrac{2}{1 + \cos\theta}$$

The test fails.

The pole: Replace r by $-r$. $-r = \dfrac{2}{1 - \cos\theta}$. The test fails.

Due to symmetry to the polar axis, assign values to θ from 0 to π.

θ	0	$\dfrac{\pi}{6}$	$\dfrac{\pi}{3}$	$\dfrac{\pi}{2}$	$\dfrac{2\pi}{3}$	$\dfrac{5\pi}{6}$	π
$r = \dfrac{2}{1 - \cos\theta}$	undefined	$\dfrac{2}{1 - \dfrac{\sqrt{3}}{2}} \approx 14.9$	4	2	$\dfrac{4}{3}$	$\dfrac{2}{1 + \dfrac{\sqrt{3}}{2}} \approx 1.1$	1

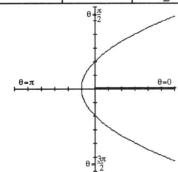

57. $r = \dfrac{1}{3 - 2\cos\theta}$ Check for symmetry:

Polar axis: Replace θ by $-\theta$. The result is $r = \dfrac{1}{3 - 2\cos(-\theta)} = \dfrac{1}{3 - 2\cos\theta}$.
The graph is symmetric with respect to the polar axis.

The line $\theta = \dfrac{\pi}{2}$: Replace θ by $\pi - \theta$.

$$r = \dfrac{1}{3 - 2\cos(\pi - \theta)} = \dfrac{1}{3 - 2(\cos\pi\cos\theta + \sin\pi\sin\theta)}$$

$$= \dfrac{1}{3 - 2(-\cos\theta + 0)} = \dfrac{1}{3 + 2\cos\theta}$$

The test fails.

The pole: Replace r by $-r$. $-r = \dfrac{1}{3 - 2\cos\theta}$. The test fails.

Due to symmetry to the polar axis, assign values to θ from 0 to π.

θ	0	$\dfrac{\pi}{6}$	$\dfrac{\pi}{3}$	$\dfrac{\pi}{2}$	$\dfrac{2\pi}{3}$	$\dfrac{5\pi}{6}$	π
$r = \dfrac{1}{3-2\cos\theta}$	1	$\dfrac{1}{3-\sqrt{3}} \approx 0.8$	$\dfrac{1}{2}$	$\dfrac{1}{3}$	$\dfrac{1}{4}$	$\dfrac{1}{3+\sqrt{3}} \approx 0.2$	$\dfrac{1}{5}$

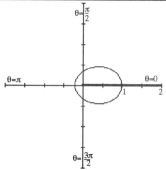

59. $r = \theta$, $\theta \geq 0$ Check for symmetry:

Polar axis: Replace θ by $-\theta$. $r = -\theta$. The test fails.

The line $\theta = \dfrac{\pi}{2}$: Replace θ by $\pi - \theta$. $r = \pi - \theta$. The test fails.

The pole: Replace r by $-r$. $-r = \theta$. The test fails.

θ	0	$\dfrac{\pi}{6}$	$\dfrac{\pi}{3}$	$\dfrac{\pi}{2}$	π	$\dfrac{3\pi}{2}$	2π
$r = \theta$	0	$\dfrac{\pi}{6} \approx 0.5$	$\dfrac{\pi}{3} \approx 1.0$	$\dfrac{\pi}{2} \approx 1.6$	$\pi \approx 3.1$	$\dfrac{3\pi}{2} \approx 4.7$	$2\pi \approx 6.3$

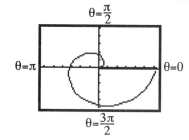

61. $r = \csc\theta - 2 = \dfrac{1}{\sin\theta} - 2$, $0 < \theta < \pi$ Check for symmetry:

Polar axis: Replace θ by $-\theta$. $r = \csc(-\theta) - 2 = -\csc\theta - 2$.
                   The test fails.

The line $\theta = \dfrac{\pi}{2}$: Replace θ by $\pi - \theta$.

$$r = \csc(\pi - \theta) - 2 = \dfrac{1}{\sin(\pi - \theta)} - 2$$

$$= \dfrac{1}{\sin\pi\cos\theta - \cos\pi\sin\theta} - 2 = \dfrac{1}{\sin\theta} - 2 = \csc\theta - 2$$

The graph is symmetric with respect to the line $\theta = \dfrac{\pi}{2}$.

The pole: Replace r by $-r$. $-r = \csc\theta - 2$. The test fails.

Due to symmetry, assign values to θ from 0 to $\dfrac{\pi}{2}$.

θ	0	$\dfrac{\pi}{6}$	$\dfrac{\pi}{4}$	$\dfrac{\pi}{3}$	$\dfrac{\pi}{2}$
$r = \csc\theta - 2$	not defined	0	$\sqrt{2} - 2 \approx -0.6$	$\dfrac{2\sqrt{3}}{3} - 2 \approx -0.8$	-1

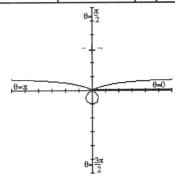

63. $r = \tan\theta$, $-\dfrac{\pi}{2} < \theta < \dfrac{\pi}{2}$ Check for symmetry:

Polar axis: Replace θ by $-\theta$. $r = \tan(-\theta) = -\tan\theta$.   The test fails.

The line $\theta = \dfrac{\pi}{2}$: Replace θ by $\pi - \theta$.

$$r = \tan(\pi - \theta) = \frac{\tan\pi - \tan\theta}{1 + \tan\pi\tan\theta} = \frac{-\tan\theta}{1} = -\tan\theta$$

The test fails.

The pole: Replace r by $-r$. $-r = \tan\theta$.   The test fails.

θ	$-\dfrac{\pi}{3}$	$-\dfrac{\pi}{4}$	$-\dfrac{\pi}{6}$	0	$\dfrac{\pi}{6}$	$\dfrac{\pi}{4}$	$\dfrac{\pi}{3}$
$r = \tan\theta$	$-\sqrt{3} \approx -1.7$	-1	$-\dfrac{\sqrt{3}}{3} \approx -0.6$	0	$\dfrac{\sqrt{3}}{3} \approx 0.6$	1	$\sqrt{3} \approx 1.7$

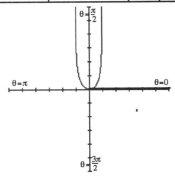

65.   Convert the equation to rectangular form:

$$r\sin\theta = a$$
$$y = a$$

The graph of $r\sin\theta = a$ is a horizontal line a units above the pole if $a > 0$, and $|a|$ units below the pole, if $a < 0$.

67.   Convert the equation to rectangular form:

$$r = 2a\sin\theta, \, a > 0$$
$$r^2 = 2a\,r\sin\theta$$
$$x^2 + y^2 = 2ay$$
$$x^2 + y^2 - 2ay = 0$$
$$x^2 + (y - a)^2 = a^2$$

Circle: radius a, center at rectangular coordinates $(0, a)$.

69.   Convert the equation to rectangular form:

$$r = 2a\cos\theta, \, a > 0$$
$$r^2 = 2ar\cos\theta$$
$$x^2 + y^2 = 2ax$$
$$x^2 - 2ax + y^2 = 0$$
$$(x - a)^2 + y^2 = a^2$$

Circle: radius a, center at rectangular coordinates $(a, 0)$.

71.   (a) $r^2 = \cos\theta$: $r^2 = \cos(\pi - \theta)$ $\rightarrow$ $r^2 = -\cos\theta$ Test fails.

                      $(-r)^2 = \cos(-\theta)$ $\rightarrow$ $r^2 = \cos\theta$ New test works.

     (b) $r^2 = \sin\theta$: $r^2 = \sin(\pi - \theta)$ $\rightarrow$ $r^2 = \sin\theta$ Test works.

                      $(-r)^2 = \sin(-\theta)$ $\rightarrow$ $r^2 = -\sin\theta$ New test fails.

9.3   The Complex Plane; DeMoivre's Theorem

1. $r = \sqrt{x^2 + y^2} = \sqrt{1^2 + 1^2} = \sqrt{2}$

     $\tan\theta = \dfrac{y}{x} = 1$ $\rightarrow$ $\theta = 45°$

     The polar form of $z = 1 + i$ is

     $z = r(\cos\theta + i\sin\theta) = \sqrt{2}(\cos 45° + i\sin 45°)$

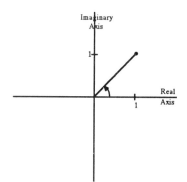

3. $r = \sqrt{x^2 + y^2} = \sqrt{\left(\sqrt{3}\right)^2 + (-1)^2} = \sqrt{4} = 2$

$\tan\theta = \dfrac{y}{x} = \dfrac{-1}{\sqrt{3}} = -\dfrac{\sqrt{3}}{3} \quad \rightarrow \quad \theta = 330°$

The polar form of $z = \sqrt{3} - i$ is
$z = r(\cos\theta + i\sin\theta)$
$\quad = 2(\cos 330° + i\sin 330°)$

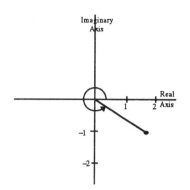

5. $r = \sqrt{x^2 + y^2} = \sqrt{0^2 + (-3)^2} = \sqrt{9} = 3$

$\tan\theta = \dfrac{y}{x} = \dfrac{-3}{0} = \text{undefined} \quad \rightarrow \quad \theta = 270°$

The polar form of $z = -3i$ is
$z = r(\cos\theta + i\sin\theta)$
$\quad = 3(\cos 270° + i\sin 270°)$

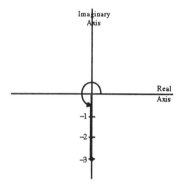

7. $r = \sqrt{x^2 + y^2} = \sqrt{4^2 + (-4)^2} = \sqrt{32} = 4\sqrt{2}$

$\tan\theta = \dfrac{y}{x} = \dfrac{-4}{4} = -1 \quad \rightarrow \quad \theta = 315°$

The polar form of $z = 4 - 4i$ is
$z = r(\cos\theta + i\sin\theta)$
$\quad = 4\sqrt{2}(\cos 315° + i\sin 315°)$

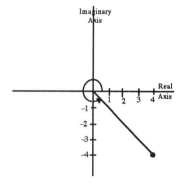

9. $r = \sqrt{x^2 + y^2} = \sqrt{3^2 + (-4)^2} = \sqrt{25} = 5$

$\tan\theta = \dfrac{y}{x} = \dfrac{-4}{3} \quad \rightarrow \quad \theta = 306.9°$

The polar form of $z = 3 - 4i$ is
$z = r(\cos\theta + i\sin\theta)$
$\quad = 5(\cos 306.9° + i\sin 306.9°)$

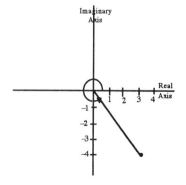

11. $r = \sqrt{x^2 + y^2} = \sqrt{(-2)^2 + 3^2} = \sqrt{13}$

$\tan\theta = \dfrac{y}{x} = \dfrac{3}{-2} = -\dfrac{3}{2} \quad \rightarrow \quad \theta = 123.7°$

The polar form of $z = -2 + 3i$ is

$z = r(\cos\theta + i\sin\theta)$

$\quad = \sqrt{13}(\cos 123.7° + i\sin 123.7°)$

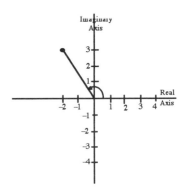

13. $2(\cos 120° + i\sin 120°) = 2\left(-\dfrac{1}{2} + \dfrac{\sqrt{3}}{2}i\right) = -1 + \sqrt{3}\ i$

15. $4\left(\cos\dfrac{7\pi}{4} + i\sin\dfrac{7\pi}{4}\right) = 4\left(\dfrac{\sqrt{2}}{2} - \dfrac{\sqrt{2}}{2}i\right) = 2\sqrt{2} - 2\sqrt{2}\ i$

17. $3\left(\cos\dfrac{3\pi}{2} + i\sin\dfrac{3\pi}{2}\right) = 3(0 - 1i) = -3i$

19. $0.2(\cos 100° + i\sin 100°) = 0.2(-0.1736 + 0.9848i) = -0.0347 + 0.1970\ i$

21. $2\left(\cos\dfrac{\pi}{18} + i\sin\dfrac{\pi}{18}\right) = 2(0.9848 + 0.1736i) = 1.9696 + 0.3472i$

23. $zw = 2(\cos 40° + i\sin 40°) \cdot 4(\cos 20° + i\sin 20°)$

$\quad = 2\cdot 4(\cos(40° + 20°) + i\sin(40° + 20°)) = 8(\cos 60° + i\sin 60°)$

$\dfrac{z}{w} = \dfrac{2(\cos 40° + i\sin 40°)}{4(\cos 20° + i\sin 20°)} = \dfrac{2}{4}(\cos(40° - 20°) + i\sin(40° - 20°))$

$\quad = \dfrac{1}{2}(\cos 20° + i\sin 20°)$

25. $zw = 3(\cos 130° + i\sin 130°) \cdot 4(\cos 270° + i\sin 270°)$

$\quad = 3\cdot 4(\cos(130° + 270°) + i\sin(130° + 270°)) = 12(\cos 400° + i\sin 400°)$

$\quad = 12(\cos(400° - 360°) + i\sin(400° - 360°)) = 12(\cos 40° + i\sin 40°)$

$\dfrac{z}{w} = \dfrac{3(\cos 130° + i\sin 130°)}{4(\cos 270° + i\sin 270°)} = \dfrac{3}{4}(\cos(130° - 270°) + i\sin(130° - 270°))$

$\quad = \dfrac{3}{4}(\cos(-140°) + i\sin(-140°)) = \dfrac{3}{4}(\cos 220° + i\sin 220°)$

27. $zw = 2\left(\cos\frac{\pi}{8} + i\sin\frac{\pi}{8}\right) \cdot 2\left(\cos\frac{\pi}{10} + i\sin\frac{\pi}{10}\right) = 2\cdot 2\left(\cos\left(\frac{\pi}{8} + \frac{\pi}{10}\right) + i\sin\left(\frac{\pi}{8} + \frac{\pi}{10}\right)\right)$

$= 4\left(\cos\frac{9\pi}{40} + i\sin\frac{9\pi}{40}\right)$

$\dfrac{z}{w} = \dfrac{2\left(\cos\frac{\pi}{8} + i\sin\frac{\pi}{8}\right)}{2\left(\cos\frac{\pi}{10} + i\sin\frac{\pi}{10}\right)} = \dfrac{2}{2}\left(\cos\left(\frac{\pi}{8} - \frac{\pi}{10}\right) + i\sin\left(\frac{\pi}{8} - \frac{\pi}{10}\right)\right) = \cos\frac{\pi}{40} + i\sin\frac{\pi}{40}$

29. $z = 2 + 2i \quad r = \sqrt{2^2 + 2^2} = \sqrt{8} = 2\sqrt{2} \quad \tan\theta = \frac{2}{2} = 1 \quad \theta = 45°$

$z = 2\sqrt{2}(\cos 45° + i\sin 45°)$

$w = \sqrt{3} - i \quad r = \sqrt{\left(\sqrt{3}\right)^2 + (-1)^2} = \sqrt{4} = 2 \quad \tan\theta = \frac{-1}{\sqrt{3}} = -\frac{\sqrt{3}}{3} \quad \theta = 330°$

$w = 2(\cos 330° + i\sin 330°)$

$zw = 2\sqrt{2}(\cos 45° + i\sin 45°) \cdot 2(\cos 330° + i\sin 330°)$

$= 2\sqrt{2}\cdot 2(\cos(45° + 330°) + i\sin(45° + 330°)) = 4\sqrt{2}(\cos 375° + i\sin 375°)$

$= 4\sqrt{2}(\cos(375° - 360°) + i\sin(375° - 360°)) = 4\sqrt{2}(\cos 15° + i\sin 15°)$

$\dfrac{z}{w} = \dfrac{2\sqrt{2}(\cos 45° + i\sin 45°)}{2(\cos 330° + i\sin 330°)} = \dfrac{2\sqrt{2}}{2}(\cos(45° - 330°) + i\sin(45° - 330°))$

$= \sqrt{2}(\cos(-285°) + i\sin(-285°)) = \sqrt{2}(\cos 75° + i\sin 75°)$

31. $\left[4(\cos 40° + i\sin 40°)\right]^3 = 4^3(\cos(3\cdot 40°) + i\sin(3\cdot 40°)) = 64(\cos 120° + i\sin 120°)$

$= 64\left(-\frac{1}{2} + \frac{\sqrt{3}}{2}i\right) = -32 + 32\sqrt{3}\,i$

33. $\left[2\left(\cos\frac{\pi}{10} + i\sin\frac{\pi}{10}\right)\right]^5 = 2^5\left(\cos\left(5\cdot\frac{\pi}{10}\right) + i\sin\left(5\cdot\frac{\pi}{10}\right)\right) = 32\left(\cos\frac{\pi}{2} + i\sin\frac{\pi}{2}\right)$

$= 32(0 + 1\,i) = 0 + 32\,i$

35. $\left[\sqrt{3}(\cos 10° + i\sin 10°)\right]^6 = \sqrt{3}^6(\cos(6\cdot 10°) + i\sin(6\cdot 10°)) = 27(\cos 60° + i\sin 60°)$

$= 27\left(\frac{1}{2} + \frac{\sqrt{3}}{2}i\right) = \frac{27}{2} + \frac{27\sqrt{3}}{2}i$

37. $\left[\sqrt{5}\left(\cos\frac{3\pi}{16} + i\sin\frac{3\pi}{16}\right)\right]^4 = \sqrt{5}^4\left(\cos\left(4\cdot\frac{3\pi}{16}\right) + i\sin\left(4\cdot\frac{3\pi}{16}\right)\right)$

$= 25\left(\cos\frac{3\pi}{4} + i\sin\frac{3\pi}{4}\right) = 25\left(-\frac{\sqrt{2}}{2} + \frac{\sqrt{2}}{2}i\right) = -\frac{25\sqrt{2}}{2} + \frac{25\sqrt{2}}{2}i$

39. $1 - i$ $r = \sqrt{1^2 + (-1)^2} = \sqrt{2}$ $\tan \theta = \dfrac{-1}{1} = -1$ $\theta = \dfrac{7\pi}{4}$

$$1 - i = \sqrt{2}\left(\cos \dfrac{7\pi}{4} + i\sin \dfrac{7\pi}{4}\right)$$

$$(1 - i)^5 = \left[\sqrt{2}\left(\cos \dfrac{7\pi}{4} + i\sin \dfrac{7\pi}{4}\right)\right]^5 = \sqrt{2}^5\left(\cos\left(5\cdot\dfrac{7\pi}{4}\right) + i\sin\left(5\cdot\dfrac{7\pi}{4}\right)\right)$$

$$= 4\sqrt{2}\left(\cos\dfrac{35\pi}{4} + i\sin\dfrac{35\pi}{4}\right) = 4\sqrt{2}\left(-\dfrac{\sqrt{2}}{2} + \dfrac{\sqrt{2}}{2}\,i\right) = -4 + 4\,i$$

41. $\sqrt{2} - i$ $r = \sqrt{\left(\sqrt{2}\right)^2 + (-1)^2} = \sqrt{3}$ $\tan \theta = \dfrac{-1}{\sqrt{2}} = -\dfrac{\sqrt{2}}{2}$ $\theta = 324.7°$

$$\sqrt{2} - i = \sqrt{3}(\cos 324.7° + i\sin 324.7°)$$

$$\left(\sqrt{2} - i\right)^6 = \left[\sqrt{3}(\cos 324.7° + i\sin 324.7°)\right]^6 = \sqrt{3}^6\left(\cos(6\cdot 324.7°) + i\sin(6\cdot 324.7°)\right)$$

$$= 27(\cos 1948.2 + i\sin 1948.2) = 27(-0.8499 + 0.5270\,i)$$

$$= -22.95 + 14.23\,i$$

43. $1 + i$ $r = \sqrt{1^2 + 1^2} = \sqrt{2}$ $\tan \theta = \dfrac{1}{1} = 1$ $\theta = 45°$

$$1 + i = \sqrt{2}(\cos 45° + i\sin 45°)$$

The three complex cube roots of $1 + i = \sqrt{2}(\cos 45° + i\sin 45°)$ are:

$$z_k = \sqrt[3]{\sqrt{2}}\left[\cos\left(\dfrac{45°}{3} + \dfrac{360° k}{3}\right) + i\sin\left(\dfrac{45°}{3} + \dfrac{360° k}{3}\right)\right]$$

$$= \sqrt[6]{2}\left[\cos(15° + 120° k) + i\sin(15° + 120° k)\right]$$

$$z_0 = \sqrt[6]{2}\left[\cos(15° + 120°\cdot 0) + i\sin(15° + 120°\cdot 0)\right] = \sqrt[6]{2}(\cos 15° + i\sin 15°)$$

$$z_1 = \sqrt[6]{2}\left[\cos(15° + 120°\cdot 1) + i\sin(15° + 120°\cdot 1)\right] = \sqrt[6]{2}(\cos 135° + i\sin 135°)$$

$$z_2 = \sqrt[6]{2}\left[\cos(15° + 120°\cdot 2) + i\sin(15° + 120°\cdot 2)\right] = \sqrt[6]{2}(\cos 255° + i\sin 255°)$$

45. $4 - 4\sqrt{3}\,i$ $r = \sqrt{4^2 + \left(-4\sqrt{3}\right)^2} = \sqrt{64} = 8$ $\tan \theta = \dfrac{-4\sqrt{3}}{4} = -\sqrt{3}$ $\theta = 300°$

$$4 - 4\sqrt{3}\,i = 8(\cos 300° + i\sin 300°)$$

The four complex fourth roots of $4 - 4\sqrt{3}\,i = 8(\cos 300° + i\sin 300°)$ are:

$$z_k = \sqrt[4]{8}\left[\cos\left(\dfrac{300°}{4} + \dfrac{360° k}{4}\right) + i\sin\left(\dfrac{300°}{4} + \dfrac{360° k}{4}\right)\right]$$

$$= \sqrt[4]{8}\left[\cos(75° + 90° k) + i\sin(75° + 90° k)\right]$$

$$z_0 = \sqrt[4]{8}\left[\cos(75° + 90°\cdot 0) + i\sin(75° + 90°\cdot 0)\right] = \sqrt[4]{8}(\cos 75° + i\sin 75°)$$

$$z_1 = \sqrt[4]{8}\left[\cos(75° + 90°\cdot 1) + i\sin(75° + 90°\cdot 1)\right] = \sqrt[4]{8}(\cos 165° + i\sin 165°)$$

$$z_2 = \sqrt[4]{8}\left[\cos(75° + 90°\cdot 2) + i\sin(75° + 90°\cdot 2)\right] = \sqrt[4]{8}(\cos 255° + i\sin 255°)$$

$$z_3 = \sqrt[4]{8}\left[\cos(75° + 90°\cdot 3) + i\sin(75° + 90°\cdot 3)\right] = \sqrt[4]{8}(\cos 345° + i\sin 345°)$$

47. $-16i$ $r = \sqrt{0^2 + (-16)^2} = \sqrt{256} = 16$ $\tan\theta = \dfrac{-16}{0} =$ undefined $\theta = 270°$

$-16i = 16(\cos 270° + i\sin 270°)$

The four complex fourth roots of $-16i = 16(\cos 270° + i\sin 270°)$ are:

$$z_k = \sqrt[4]{16}\left[\cos\left(\frac{270°}{4} + \frac{360°\,k}{4}\right) + i\sin\left(\frac{270°}{4} + \frac{360°\,k}{4}\right)\right]$$

$$= 2\left[\cos(67.5° + 90°k) + i\sin(67.5° + 90°\,k)\right]$$

$$z_0 = 2\left[\cos(67.5° + 90°\cdot 0) + i\sin(67.5° + 90°\cdot 0)\right] = 2(\cos 67.5° + i\sin 67.5°)$$

$$z_1 = 2\left[\cos(67.5° + 90°\cdot 1) + i\sin(67.5° + 90°\cdot 1)\right] = 2(\cos 157.5° + i\sin 157.5°)$$

$$z_2 = 2\left[\cos(67.5° + 90°\cdot 2) + i\sin(67.5° + 90°\cdot 2)\right] = 2(\cos 247.5° + i\sin 247.5°)$$

$$z_3 = 2\left[\cos(67.5° + 90°\cdot 3) + i\sin(67.5° + 90°\cdot 3)\right] = 2(\cos 337.5° + i\sin 337.5°)$$

49. i $r = \sqrt{0^2 + 1^2} = \sqrt{1} = 1$ $\tan\theta = \dfrac{1}{0} =$ undefined $\theta = 90°$

$i = 1(\cos 90° + i\sin 90°)$

The five complex fifth roots of $i = 1(\cos 90° + i\sin 90°)$ are:

$$z_k = \sqrt[5]{1}\left[\cos\left(\frac{90°}{5} + \frac{360°\,k}{5}\right) + i\sin\left(\frac{90°}{5} + \frac{360°\,k}{5}\right)\right]$$

$$= 1\left[\cos(18° + 72°k) + i\sin(18° + 72°\,k)\right]$$

$$z_0 = 1\left[\cos(18° + 72°\cdot 0) + i\sin(18° + 72°\cdot 0)\right] = \cos 18° + i\sin 18°$$

$$z_1 = 1\left[\cos(18° + 72°\cdot 1) + i\sin(18° + 72°\cdot 1)\right] = \cos 90° + i\sin 90°$$

$$z_2 = 1\left[\cos(18° + 72°\cdot 2) + i\sin(18° + 72°\cdot 2)\right] = \cos 162° + i\sin 162°$$

$$z_3 = 1\left[\cos(18° + 72°\cdot 3) + i\sin(18° + 72°\cdot 3)\right] = \cos 234° + i\sin 234°$$

$$z_4 = 1\left[\cos(18° + 72°\cdot 4) + i\sin(18° + 72°\cdot 4)\right] = \cos 306° + i\sin 306°$$

51. $1 = 1 + 0i$ $r = \sqrt{1^2 + 0^2} = \sqrt{1} = 1$ $\tan\theta = \dfrac{0}{1} = 0$ $\theta = 0°$

$1 + 0i = 1(\cos 0° + i\sin 0°)$

The four complex fourth roots of $1 + 0i = 1(\cos 0° + i\sin 0°)$ are:

$$z_k = \sqrt[4]{1}\left[\cos\left(\frac{0°}{4} + \frac{360°\,k}{4}\right) + i\sin\left(\frac{0°}{4} + \frac{360°\,k}{4}\right)\right]$$

$$= 1\left[\cos(90°k) + i\sin(90°\,k)\right]$$

$$z_0 = \cos(90°\cdot 0) + i\sin(90°\cdot 0) = \cos 0° + i\sin 0° = 1 + 0i = 1$$

$$z_1 = \cos(90°\cdot 1) + i\sin(90°\cdot 1) = \cos 90° + i\sin 90° = 0 + 1i = i$$

$$z_2 = \cos(90°\cdot 2) + i\sin(90°\cdot 2) = \cos 180° + i\sin 180° = -1 + 0i = -1$$

$$z_3 = \cos(90°\cdot 3) + i\sin(90°\cdot 3) = \cos 270° + i\sin 270° = 0 - 1i = -i$$

The complex roots are: $1, i, -1, -i$.

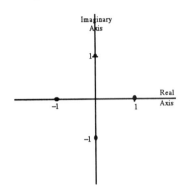

53. Let $w = r(\cos\theta + i\sin\theta)$ be a complex number. If $w \neq 0$, there are n distinct n th roots of w, given by the formula:

$$z_k = \sqrt[n]{r}\left(\cos\left(\frac{\theta}{n} + \frac{2k\pi}{n}\right) + i\sin\left(\frac{\theta}{n} + \frac{2k\pi}{n}\right)\right), \text{ where } k = 0, 1, 2, \ldots, n-1$$

$$|z_k| = \sqrt[n]{r} \text{ for all } k$$

55. Examining the formula for the distinct complex n th roots of the complex number $w = r(\cos\theta + i\sin\theta)$,

$$z_k = \sqrt[n]{r}\left(\cos\left(\frac{\theta}{n} + \frac{2k\pi}{n}\right) + i\sin\left(\frac{\theta}{n} + \frac{2k\pi}{n}\right)\right), \text{ where } k = 0, 1, 2, \ldots, n-1$$

we see that the z_k are spaced apart by an angle of $\frac{2\pi}{n}$.

9.4 Vectors

1. **v + w**

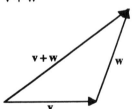

3. **3v**

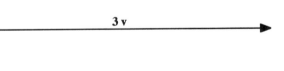

5. **v** – **w**

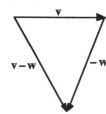

7. **3v** + **u** – **2w**

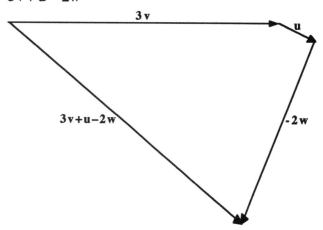

9. True

11. False **C = –F + E – D**

13. False **D – E = H + G**

15. True

17. If $\|\mathbf{v}\| = 4$, then $\|3\mathbf{v}\| = |3|\|\mathbf{v}\| = 3(4) = 12$.

19. $P = (0, 0), Q = (3, 4)$ $\quad \mathbf{v} = (3 - 0)\mathbf{i} + (4 - 0)\mathbf{j} = 3\mathbf{i} + 4\mathbf{j}$

21. $P = (3, 2), Q = (5, 6)$ $\quad \mathbf{v} = (5 - 3)\mathbf{i} + (6 - 2)\mathbf{j} = 2\mathbf{i} + 4\mathbf{j}$

23. $P = (-2, -1), Q = (6, -2)$ $\quad \mathbf{v} = (6 - (-2))\mathbf{i} + (-2 - (-1))\mathbf{j} = 8\mathbf{i} - \mathbf{j}$

25. $P = (1, 0), Q = (0, 1)$ $\quad \mathbf{v} = (0 - 1)\mathbf{i} + (1 - 0)\mathbf{j} = -\mathbf{i} + \mathbf{j}$

27. For $\mathbf{v} = 3\mathbf{i} - 4\mathbf{j}$, $\|\mathbf{v}\| = \sqrt{3^2 + (-4)^2} = \sqrt{25} = 5$

29. For $\mathbf{v} = \mathbf{i} - \mathbf{j}$, $\|\mathbf{v}\| = \sqrt{1^2 + (-1)^2} = \sqrt{2}$

31. For $\mathbf{v} = -2\mathbf{i} + 3\mathbf{j}$, $\|\mathbf{v}\| = \sqrt{(-2)^2 + 3^2} = \sqrt{13}$

33. $\mathbf{v} = 3\mathbf{i} - 5\mathbf{j}$, $\mathbf{w} = -2\mathbf{i} + 3\mathbf{j}$
$$2\mathbf{v} + 3\mathbf{w} = 2(3\mathbf{i} - 5\mathbf{j}) + 3(-2\mathbf{i} + 3\mathbf{j}) = 6\mathbf{i} - 10\mathbf{j} - 6\mathbf{i} + 9\mathbf{j} = -\mathbf{j}$$

35. $\mathbf{v} = 3\mathbf{i} - 5\mathbf{j}$, $\mathbf{w} = -2\mathbf{i} + 3\mathbf{j}$
$$\|\mathbf{v} - \mathbf{w}\| = \|(3\mathbf{i} - 5\mathbf{j}) - (-2\mathbf{i} + 3\mathbf{j})\| = \|5\mathbf{i} - 8\mathbf{j}\| = \sqrt{5^2 + (-8)^2} = \sqrt{89}$$

37. $\mathbf{v} = 3\mathbf{i} - 5\mathbf{j}$, $\mathbf{w} = -2\mathbf{i} + 3\mathbf{j}$
$$\|\mathbf{v}\| - \|\mathbf{w}\| = |3\mathbf{i} - 5\mathbf{j}| - |-2\mathbf{i} + 3\mathbf{j}| = \sqrt{3^2 + (-5)^2} - \sqrt{(-2)^2 + 3^2} = \sqrt{34} - \sqrt{13}$$

39. $\mathbf{u} = \dfrac{\mathbf{v}}{\|\mathbf{v}\|} = \dfrac{5\mathbf{i}}{\|5\mathbf{i}\|} = \dfrac{5\mathbf{i}}{\sqrt{25+0}} = \dfrac{5\mathbf{i}}{5} = \mathbf{i}$

41. $\mathbf{u} = \dfrac{\mathbf{v}}{\|\mathbf{v}\|} = \dfrac{3\mathbf{i}-4\mathbf{j}}{\|3\mathbf{i}-4\mathbf{j}\|} = \dfrac{3\mathbf{i}-4\mathbf{j}}{\sqrt{3^2+(-4)^2}} = \dfrac{3\mathbf{i}-4\mathbf{j}}{\sqrt{25}} = \dfrac{3\mathbf{i}-4\mathbf{j}}{5} = \dfrac{3}{5}\mathbf{i} - \dfrac{4}{5}\mathbf{j}$

43. $\mathbf{u} = \dfrac{\mathbf{v}}{\|\mathbf{v}\|} = \dfrac{\mathbf{i}-\mathbf{j}}{\|\mathbf{i}-\mathbf{j}\|} = \dfrac{\mathbf{i}-\mathbf{j}}{\sqrt{1^2+(-1)^2}} = \dfrac{\mathbf{i}-\mathbf{j}}{\sqrt{2}} = \dfrac{1}{\sqrt{2}}\mathbf{i} - \dfrac{1}{\sqrt{2}}\mathbf{j} = \dfrac{\sqrt{2}}{2}\mathbf{i} - \dfrac{\sqrt{2}}{2}\mathbf{j}$

45. Let $\mathbf{v} = a\mathbf{i} + b\mathbf{j}$. We want $\|\mathbf{v}\| = 4$ and $a = 2b$.

$\|\mathbf{v}\| = \sqrt{a^2+b^2} = \sqrt{(2b)^2+b^2} = \sqrt{5b^2}$

$\sqrt{5b^2} = 4 \;\rightarrow\; 5b^2 = 16 \;\rightarrow\; b^2 = \dfrac{16}{5} \;\rightarrow\; b = \pm\sqrt{\dfrac{16}{5}} = \pm\dfrac{4}{\sqrt{5}} = \pm\dfrac{4\sqrt{5}}{5}$

$a = 2b = \pm\dfrac{8\sqrt{5}}{5}$

$\mathbf{v} = \dfrac{8\sqrt{5}}{5}\mathbf{i} + \dfrac{4\sqrt{5}}{5}\mathbf{j}$ or $\mathbf{v} = -\dfrac{8\sqrt{5}}{5}\mathbf{i} - \dfrac{4\sqrt{5}}{5}\mathbf{j}$

47. $\mathbf{v} = 2\mathbf{i}-\mathbf{j}, \;\; \mathbf{w} = x\mathbf{i}+3\mathbf{j} \quad \|\mathbf{v}+\mathbf{w}\| = 5$

$\|\mathbf{v}+\mathbf{w}\| = \|2\mathbf{i}-\mathbf{j}+x\mathbf{i}+3\mathbf{j}\| = \|(2+x)\mathbf{i}+2\mathbf{j}\| = \sqrt{(2+x)^2+2^2}$

$= \sqrt{x^2+4x+4+4} = \sqrt{x^2+4x+8}$

Solve for x:

$\sqrt{x^2+4x+8} = 5$

$x^2+4x+8 = 25$

$x^2+4x-17 = 0$

$x = \dfrac{-4\pm\sqrt{16-4(1)(-17)}}{2(1)} = \dfrac{-4\pm\sqrt{84}}{2} = \dfrac{-4\pm2\sqrt{21}}{2} = -2\pm\sqrt{21}$

$x = -2+\sqrt{21} \approx 2.58 \;\; \text{or} \;\; x = -2-\sqrt{21} \approx -6.58$

49. $\|\mathbf{v}\| = 5, \;\; \alpha = 60°$

$\mathbf{v} = \|\mathbf{v}\|(\cos\alpha\,\mathbf{i} + \sin\alpha\,\mathbf{j}) = 5(\cos 60°\,\mathbf{i} + \sin 60°\,\mathbf{j}) = 5\left(\dfrac{1}{2}\mathbf{i} + \dfrac{\sqrt{3}}{2}\mathbf{j}\right) = \dfrac{5}{2}\mathbf{i} + \dfrac{5\sqrt{3}}{2}\mathbf{j}$

51. $\|\mathbf{v}\| = 14, \;\; \alpha = 120°$

$\mathbf{v} = \|\mathbf{v}\|(\cos\alpha\,\mathbf{i} + \sin\alpha\,\mathbf{j}) = 14(\cos 120°\,\mathbf{i} + \sin 120°\,\mathbf{j}) = 14\left(-\dfrac{1}{2}\mathbf{i} + \dfrac{\sqrt{3}}{2}\mathbf{j}\right) = -7\mathbf{i} + 7\sqrt{3}\mathbf{j}$

53. $\|\mathbf{v}\| = 25, \quad \alpha = 330°$

$$\mathbf{v} = \|\mathbf{v}\|(\cos\alpha\mathbf{i} + \sin\alpha\mathbf{j}) = 25(\cos 330°\,\mathbf{i} + \sin 330°\,\mathbf{j}) = 25\left(\frac{\sqrt{3}}{2}\mathbf{i} - \frac{1}{2}\mathbf{j}\right) = \frac{25\sqrt{3}}{2}\mathbf{i} - \frac{25}{2}\mathbf{j}$$

55. $\mathbf{F} = 40\left(\cos 30°\,\mathbf{i} + \sin 30°\,\mathbf{j}\right) = 40\left(\frac{\sqrt{3}}{2}\mathbf{i} + \frac{1}{2}\mathbf{j}\right) = 20\sqrt{3}\,\mathbf{i} + 20\mathbf{j}$

57. $\mathbf{F}_1 = 40\left(\cos 30°\,\mathbf{i} + \sin 30°\,\mathbf{j}\right) = 40\left(\frac{\sqrt{3}}{2}\mathbf{i} + \frac{1}{2}\mathbf{j}\right) = 20\sqrt{3}\,\mathbf{i} + 20\mathbf{j}$

$$\mathbf{F}_2 = 60\left(\cos(-45°)\mathbf{i} + \sin(-45°)\mathbf{j}\right) = 60\left(\frac{\sqrt{2}}{2}\mathbf{i} - \frac{\sqrt{2}}{2}\mathbf{j}\right) = 30\sqrt{2}\,\mathbf{i} - 30\sqrt{2}\mathbf{j}$$

$$\mathbf{F}_1 + \mathbf{F}_2 = 20\sqrt{3}\,\mathbf{i} + 20\mathbf{j} + 30\sqrt{2}\,\mathbf{i} - 30\sqrt{2}\mathbf{j} = \left(20\sqrt{3} + 30\sqrt{2}\right)\mathbf{i} + \left(20 - 30\sqrt{2}\right)\mathbf{j}$$

59.   Let $\mathbf{F}_1$ be the tension on the left cable and $\mathbf{F}_2$ be the tension on the right cable.
Let $\mathbf{F}_3$ represent the force of the weight of the box.
$\mathbf{F}_1 = \|\mathbf{F}_1\|(\cos 155°\mathbf{i} + \sin 155°\mathbf{j}) = \|\mathbf{F}_1\|(-0.9063\mathbf{i} + 0.4226\mathbf{j})$
$\mathbf{F}_2 = \|\mathbf{F}_2\|(\cos 40°\mathbf{i} + \sin 40°\mathbf{j}) = \|\mathbf{F}_2\|(0.7660\mathbf{i} + 0.6428\mathbf{j})$
$\mathbf{F}_3 = -1000\,\mathbf{j}$
For equilibrium, the sum of the force vectors must be zero.
$\mathbf{F}_1 + \mathbf{F}_2 + \mathbf{F}_3 = -0.9063\|\mathbf{F}_1\|\mathbf{i} + 0.4226\|\mathbf{F}_1\|\mathbf{j} + 0.7660\|\mathbf{F}_2\|\mathbf{i} + 0.6428\|\mathbf{F}_2\|\mathbf{j} - 1000\mathbf{j}$

$$= \left(-0.9063\|\mathbf{F}_1\| + 0.7660\|\mathbf{F}_2\|\right)\mathbf{i} + \left(0.4226\|\mathbf{F}_1\| + 0.6428\|\mathbf{F}_2\| - 1000\right)\mathbf{j}$$

$$= 0$$

Set the $\mathbf{i}$ and $\mathbf{j}$ components equal to zero and solve:

$$\begin{cases} -0.9063\|\mathbf{F}_1\| + 0.7660\|\mathbf{F}_2\| = 0 \quad \rightarrow \quad \|\mathbf{F}_2\| = \dfrac{0.9063}{0.7660}\|\mathbf{F}_1\| = 1.1832\|\mathbf{F}_1\| \\ 0.4226\|\mathbf{F}_1\| + 0.6428\|\mathbf{F}_2\| - 1000 = 0 \end{cases}$$

$0.4226\|\mathbf{F}_1\| + 0.6428\left(1.1832\|\mathbf{F}_1\|\right) - 1000 = 0$

$1.1832\|\mathbf{F}_1\| = 1000$

$\|\mathbf{F}_1\| = 845.2$ pounds

$\|\mathbf{F}_2\| = 1.1832(845.2) = 1000$ pounds

The tension in the left cable is about 845.2 pounds and the tension in the right cable is about 1000 pounds.

61. Let $\mathbf{F}_1$ be the tension on the left end of the rope and $\mathbf{F}_2$ be the tension on the right end of the rope. Let $\mathbf{F}_3$ represent the force of the weight of the tightrope walker.

$\mathbf{F}_1 = |\mathbf{F}_1|(\cos 175.8° \mathbf{i} + \sin 175.8° \mathbf{j}) = \|\mathbf{F}_1\|(-0.9973\mathbf{i} + 0.0732\mathbf{j})$

$\mathbf{F}_2 = |\mathbf{F}_2|(\cos 3.7° \mathbf{i} + \sin 3.7° \mathbf{j}) = |\mathbf{F}_2|(0.9979\mathbf{i} + 0.0645\mathbf{j})$

$\mathbf{F}_3 = -150\mathbf{j}$

For equilibrium, the sum of the force vectors must be zero.

$\mathbf{F}_1 + \mathbf{F}_2 + \mathbf{F}_3 = -0.9973|\mathbf{F}_1\|\mathbf{i} + 0.0732|\mathbf{F}_1\|\mathbf{j} + 0.9979|\mathbf{F}_2|\mathbf{i} + 0.0645|\mathbf{F}_2|\mathbf{j} - 150\mathbf{j}$

$= (-0.9973|\mathbf{F}_1\| + 0.9979\|\mathbf{F}_2\|)\mathbf{i} + (0.0732\|\mathbf{F}_1\| + 0.0645\|\mathbf{F}_2\| - 150)\mathbf{j}$

$= 0$

Set the $\mathbf{i}$ and $\mathbf{j}$ components equal to zero and solve:

$\begin{cases} -0.9973\|\mathbf{F}_1\| + 0.9979|\mathbf{F}_2| = 0 \quad \rightarrow \quad |\mathbf{F}_2| = \dfrac{0.9973}{0.9979}\|\mathbf{F}_1\| = 0.9994\|\mathbf{F}_1\| \\ 0.0732|\mathbf{F}_1\| + 0.0645\|\mathbf{F}_2\| - 150 = 0 \end{cases}$

$0.0732|\mathbf{F}_1\| + 0.0645(0.9994|\mathbf{F}_1\|) - 150 = 0$

$0.1377\|\mathbf{F}_1\| = 150$

$|\mathbf{F}_1\| = 1089.3$ pounds

$|\mathbf{F}_2| = 0.9994(1089.3) = 1088.6$ pounds

The tension in the left end of the rope is about 1089.3 pounds and the tension in the right end of the rope is about 1088.6 pounds.

63. The given forces are:

$\mathbf{F}_1 = -3\mathbf{i}$

$\mathbf{F}_2 = -\mathbf{i} + 4\mathbf{j}$

$\mathbf{F}_3 = 4\mathbf{i} - 2\mathbf{j}$

$\mathbf{F}_4 = -4\mathbf{j}$

A vector $\mathbf{x} = a\mathbf{i} + b\mathbf{j}$ needs to be added for equilibrium. Find vector $\mathbf{x} = a\mathbf{i} + b\mathbf{j}$:

$\mathbf{F}_1 + \mathbf{F}_2 + \mathbf{F}_3 + \mathbf{F}_4 + \mathbf{x} = \mathbf{0}$

$-3\mathbf{i} + (-\mathbf{i} + 4\mathbf{j}) + (4\mathbf{i} - 2\mathbf{j}) + (-4\mathbf{j}) + (a\mathbf{i} + b\mathbf{j}) = \mathbf{0}$

$0\mathbf{i} - 2\mathbf{j} + (a\mathbf{i} + b\mathbf{j}) = \mathbf{0}$

$a\mathbf{i} + (-2 + b)\mathbf{j} = \mathbf{0}$

$a = 0$

$-2 + b = 0 \quad \rightarrow \quad b = 2$

Therefore, $\mathbf{x} = 2\mathbf{j}$.

9.5 The Dot Product

1. $\mathbf{v} = \mathbf{i} - \mathbf{j}, \quad \mathbf{w} = \mathbf{i} + \mathbf{j}$

   (a) $\mathbf{v} \bullet \mathbf{w} = 1(1) + (-1)(1) = 1 - 1 = 0$

   (b) $\cos\theta = \dfrac{\mathbf{v} \bullet \mathbf{w}}{\|\mathbf{v}\|\|\mathbf{w}\|} = \dfrac{0}{\sqrt{1^2 + (-1)^2}\sqrt{1^2 + 1^2}} = \dfrac{0}{\sqrt{2}\sqrt{2}} = \dfrac{0}{2} = 0 \quad \rightarrow \quad \theta = 90°$

   (c) The vectors are orthogonal.

3. $\mathbf{v} = 2\mathbf{i} + \mathbf{j}, \quad \mathbf{w} = \mathbf{i} + 2\mathbf{j}$

   (a) $\mathbf{v} \bullet \mathbf{w} = 2(1) + 1(2) = 2 + 2 = 4$

   (b) $\cos\theta = \dfrac{\mathbf{v} \bullet \mathbf{w}}{\|\mathbf{v}\|\|\mathbf{w}\|} = \dfrac{4}{\sqrt{2^2 + 1^2}\sqrt{1^2 + 2^2}} = \dfrac{4}{\sqrt{5}\sqrt{5}} = \dfrac{4}{5} = 0.8 \quad \rightarrow \quad \theta = 36.87°$

   (c) The vectors are neither parallel nor orthogonal.

5. $\mathbf{v} = \sqrt{3}\,\mathbf{i} - \mathbf{j}, \quad \mathbf{w} = \mathbf{i} + \mathbf{j}$

   (a) $\mathbf{v} \bullet \mathbf{w} = \sqrt{3}(1) + (-1)(1) = \sqrt{3} - 1$

   (b) $\cos\theta = \dfrac{\mathbf{v} \bullet \mathbf{w}}{\|\mathbf{v}\|\|\mathbf{w}\|} = \dfrac{\sqrt{3} - 1}{\sqrt{\left(\sqrt{3}\right)^2 + (-1)^2}\sqrt{1^2 + 1^2}} = \dfrac{\sqrt{3} - 1}{\sqrt{4}\sqrt{2}} = \dfrac{\sqrt{3} - 1}{2\sqrt{2}} = \dfrac{\sqrt{6} - \sqrt{2}}{4}$

   $\theta = 75°$

   (c) The vectors are neither parallel nor orthogonal.

7. $\mathbf{v} = 3\mathbf{i} + 4\mathbf{j}, \quad \mathbf{w} = 4\mathbf{i} + 3\mathbf{j}$

   (a) $\mathbf{v} \bullet \mathbf{w} = 3(4) + 4(3) = 12 + 12 = 24$

   (b) $\cos\theta = \dfrac{\mathbf{v} \bullet \mathbf{w}}{\|\mathbf{v}\|\|\mathbf{w}\|} = \dfrac{24}{\sqrt{3^2 + 4^2}\sqrt{4^2 + 3^2}} = \dfrac{24}{\sqrt{25}\sqrt{25}} = \dfrac{24}{25} = 0.96 \quad \rightarrow \quad \theta = 16.26°$

   (c) The vectors are neither parallel nor orthogonal.

9. $\mathbf{v} = 4\mathbf{i}, \quad \mathbf{w} = \mathbf{j}$

   (a) $\mathbf{v} \bullet \mathbf{w} = 4(0) + 0(1) = 0 + 0 = 0$

   (b) $\cos\theta = \dfrac{\mathbf{v} \bullet \mathbf{w}}{\|\mathbf{v}\|\|\mathbf{w}\|} = \dfrac{0}{\sqrt{4^2 + 0^2}\sqrt{0^2 + 1^2}} = \dfrac{0}{4 \cdot 1} = \dfrac{0}{4} = 0 \quad \rightarrow \quad \theta = 90°$

   (c) The vectors are orthogonal.

11. $\mathbf{v} = \mathbf{i} - a\mathbf{j}, \quad \mathbf{w} = 2\mathbf{i} + 3\mathbf{j}$

    Two vectors are orthogonal is the dot product is zero. Solve for a:

    $\mathbf{v} \bullet \mathbf{w} = 1(2) + (-a)(3) = 2 - 3a$

    $2 - 3a = 0$

    $3a = 2$

    $a = \dfrac{2}{3}$

13. $\mathbf{v} = 2\mathbf{i} - 3\mathbf{j}, \quad \mathbf{w} = \mathbf{i} - \mathbf{j}$

$$\mathbf{v_1} = \text{proj}_w \mathbf{v} = \frac{\mathbf{v} \cdot \mathbf{w}}{\|\mathbf{w}\|^2}\mathbf{w} = \frac{2(1) + (-3)(-1)}{\sqrt{1^2 + (-1)^2}^2}(\mathbf{i} - \mathbf{j}) = \frac{5}{2}(\mathbf{i} - \mathbf{j}) = \frac{5}{2}\mathbf{i} - \frac{5}{2}\mathbf{j}$$

$$\mathbf{v_2} = \mathbf{v} - \mathbf{v_1} = (2\mathbf{i} - 3\mathbf{j}) - \left(\frac{5}{2}\mathbf{i} - \frac{5}{2}\mathbf{j}\right) = -\frac{1}{2}\mathbf{i} - \frac{1}{2}\mathbf{j}$$

15. $\mathbf{v} = \mathbf{i} - \mathbf{j}, \quad \mathbf{w} = \mathbf{i} + 2\mathbf{j}$

$$\mathbf{v_1} = \text{proj}_w \mathbf{v} = \frac{\mathbf{v} \cdot \mathbf{w}}{\|\mathbf{w}\|^2}\mathbf{w} = \frac{1(1) + (-1)(2)}{\sqrt{1^2 + 2^2}^2}(\mathbf{i} + 2\mathbf{j}) = \frac{-1}{5}(\mathbf{i} + 2\mathbf{j}) = -\frac{1}{5}\mathbf{i} - \frac{2}{5}\mathbf{j}$$

$$\mathbf{v_2} = \mathbf{v} - \mathbf{v_1} = (\mathbf{i} - \mathbf{j}) - \left(-\frac{1}{5}\mathbf{i} - \frac{2}{5}\mathbf{j}\right) = \frac{6}{5}\mathbf{i} - \frac{3}{5}\mathbf{j}$$

17. $\mathbf{v} = 3\mathbf{i} + \mathbf{j}, \quad \mathbf{w} = -2\mathbf{i} - \mathbf{j}$

$$\mathbf{v_1} = \text{proj}_w \mathbf{v} = \frac{\mathbf{v} \cdot \mathbf{w}}{\|\mathbf{w}\|^2}\mathbf{w} = \frac{3(-2) + 1(-1)}{\sqrt{(-2)^2 + (-1)^2}^2}(-2\mathbf{i} - \mathbf{j}) = \frac{-7}{5}(-2\mathbf{i} - \mathbf{j}) = \frac{14}{5}\mathbf{i} + \frac{7}{5}\mathbf{j}$$

$$\mathbf{v_2} = \mathbf{v} - \mathbf{v_1} = (3\mathbf{i} + \mathbf{j}) - \left(\frac{14}{5}\mathbf{i} + \frac{7}{5}\mathbf{j}\right) = \frac{1}{5}\mathbf{i} - \frac{2}{5}\mathbf{j}$$

19.   Let $\mathbf{v_a}$ = the velocity of the plane in still air.

$\mathbf{v_w}$ = the velocity of the wind.

$\mathbf{v_g}$ = the velocity of the plane relative to the ground.

$\mathbf{v_g} = \mathbf{v_a} + \mathbf{v_w}$

$$\mathbf{v_a} = 550(\cos 225°\mathbf{i} + \sin 225° \,\mathbf{j}) = 550\left(-\frac{\sqrt{2}}{2}\mathbf{i} - \frac{\sqrt{2}}{2}\,\mathbf{j}\right) = -275\sqrt{2}\mathbf{i} - 275\sqrt{2}\,\mathbf{j}$$

$\mathbf{v_w} = 80\mathbf{i}$

$$\mathbf{v_g} = \mathbf{v_a} + \mathbf{v_w} = -275\sqrt{2}\,\mathbf{i} - 275\sqrt{2}\,\mathbf{j} + 80\mathbf{i} = \left(80 - 275\sqrt{2}\right)\mathbf{i} - 275\sqrt{2}\,\mathbf{j}$$

The speed of the plane relative to the ground is:

$$\|\mathbf{v_g}\| = \sqrt{\left(80 - 275\sqrt{2}\right)^2 + \left(-275\sqrt{2}\right)^2} = \sqrt{6400 - 44000\sqrt{2} + 151250 + 151250}$$

$$= \sqrt{246674.6} \approx 496.7 \text{ miles per hour}$$

To find the direction, find the angle between $\mathbf{v_g}$ and a convenient vector such as due south, $-\mathbf{j}$.

$$\cos\theta = \frac{\mathbf{v_g} \cdot -\mathbf{j}}{\|\mathbf{v_g}\|\,|-\mathbf{j}|} = \frac{\left(80 - 275\sqrt{2}\right) \cdot 0 + \left(-275\sqrt{2}\right)(-1)}{496.7\sqrt{0^2 + (-1)^2}} = \frac{275\sqrt{2}}{496.7} \approx 0.7829$$

$\theta \approx 38.5°$

The plane is traveling with a ground speed of about 496.7 miles per hour in a direction of 38.5° west of south.

21. Let the positive x-axis point downstream, so that the velocity of the current is $\mathbf{v}_c = 3\mathbf{i}$.
    Let $\mathbf{v}_w$ = the velocity of the boat in the water.
    Let $\mathbf{v}_g$ = the velocity of the boat relative to the land.
    Then $\mathbf{v}_g = \mathbf{v}_w + \mathbf{v}_c$
    The speed of the boat is $\|\mathbf{v}_w\| = 20$; we need to find the direction.

    Let $\mathbf{v}_w = a\mathbf{i} + b\mathbf{j}$ so $\|\mathbf{v}_w\| = \sqrt{a^2 + b^2} = 20 \;\rightarrow\; a^2 + b^2 = 400.$
    Let $\mathbf{v}_g = k\mathbf{j}$.
    Since $\mathbf{v}_g = \mathbf{v}_w + \mathbf{v}_c,\quad k\mathbf{j} = a\mathbf{i} + b\mathbf{j} + 3\mathbf{i} \;\rightarrow\; k\mathbf{j} = (a+3)\mathbf{i} + b\mathbf{j}$
    $a + 3 = 0$ and $k = b \;\rightarrow\; a = -3$
    $a^2 + b^2 = 400 \;\rightarrow\; 9 + b^2 = 400 \;\rightarrow\; b^2 = 391 \;\rightarrow\; k = b \approx 19.77$
    $\mathbf{v}_w = -3\mathbf{i} + 19.77\mathbf{j}$ and $\mathbf{v}_g = 19.77\mathbf{j}$

    Find the angle between $\mathbf{v}_w$ and $\mathbf{j}$:
    $$\cos\theta = \frac{\mathbf{v}_w \bullet \mathbf{j}}{\|\mathbf{v}_w\|\|\mathbf{j}\|} = \frac{-3\cdot 0 + 19.77(1)}{20\sqrt{0^2 + 1^2}} = \frac{19.77}{20} \approx 0.9885$$
    $$\theta \approx 8.7°$$

    The heading of the boat needs to be 8.7° upstream.
    The velocity of the boat directly across the river is 19.77 kilometers per hour. The time to cross the river is: $t = \dfrac{0.5}{19.77} \approx 0.025$ hours or $t \approx 1.5$ minutes.

23. Split the force into the components going down the hill and perpendicular to the hill.

    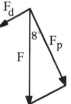

    $$\mathbf{F}_d = \mathbf{F}\sin 8° = 5300\sin 8°$$
    $$= 5300(0.1392) \approx 738 \text{ pounds}$$
    $$\mathbf{F}_p = \mathbf{F}\cos 8° = 5300\cos 8°$$
    $$= 5300(0.9903) \approx 5249 \text{ pounds}$$

    The force required to keep the car from rolling down the hill is about 738 pounds.
    The force perpendicular to the hill is approximately 5249 pounds.

25. Let $\mathbf{v}_a$ = the velocity of the plane in still air.
    $\mathbf{v}_w$ = the velocity of the wind.
    $\mathbf{v}_g$ = the velocity of the plane relative to the ground.
    $\mathbf{v}_g = \mathbf{v}_a + \mathbf{v}_w$

    $$\mathbf{v}_a = 500(\cos 45°\mathbf{i} + \sin 45°\mathbf{j}) = 500\left(\frac{\sqrt{2}}{2}\mathbf{i} + \frac{\sqrt{2}}{2}\mathbf{j}\right) = 250\sqrt{2}\,\mathbf{i} + 250\sqrt{2}\,\mathbf{j}$$

    $$\mathbf{v}_w = 60(\cos 120°\mathbf{i} + \sin 120°\mathbf{j}) = 60\left(-\frac{1}{2}\mathbf{i} + \frac{\sqrt{3}}{2}\mathbf{j}\right) = -30\mathbf{i} + 30\sqrt{3}\,\mathbf{j}$$

    $$\mathbf{v}_g = \mathbf{v}_a + \mathbf{v}_w = 250\sqrt{2}\,\mathbf{i} + 250\sqrt{2}\,\mathbf{j} - 30\mathbf{i} + 30\sqrt{3}\,\mathbf{j}$$
    $$= \left(-30 + 250\sqrt{2}\right)\mathbf{i} + \left(250\sqrt{2} + 30\sqrt{3}\right)\mathbf{j}$$

The speed of the plane relative to the ground is:

$$\left\| \mathbf{v_g} \right\| = \sqrt{\left(-30 + 250\sqrt{2}\right)^2 + \left(250\sqrt{2} + 30\sqrt{3}\right)^2}$$

$$= \sqrt{269129.1} \approx 518.8 \text{ kilometers per hour}$$

To find the direction, find the angle between $\mathbf{v_g}$ and a convenient vector such as due north, $\mathbf{j}$.

$$\cos\theta = \frac{\mathbf{v_g} \cdot \mathbf{j}}{\left\| \mathbf{v_g} \right\| \left\| \mathbf{j} \right\|} = \frac{\left(-30 + 250\sqrt{2}\right) \cdot 0 + \left(250\sqrt{2} + 30\sqrt{3}\right)(1)}{518.8\sqrt{0^2 + 1^2}} = \frac{250\sqrt{2} + 30\sqrt{3}}{518.8}$$

$$= \frac{405.5}{518.8} \approx 0.7816$$

$$\theta \approx 38.6°$$

The plane is traveling with a ground speed of about 518.8 kilometers per hour in a direction of 38.6° east of north.

27. Let the positive x-axis point downstream, so that the velocity of the current is $\mathbf{v_c} = 3\mathbf{i}$.

Let $\mathbf{v_w}$ = the velocity of the boat in the water.

Let $\mathbf{v_g}$ = the velocity of the boat relative to the land.

Then $\mathbf{v_g} = \mathbf{v_w} + \mathbf{v_c}$

The speed of the boat is $\left\| \mathbf{v_w} \right\| = 20$; its direction is directly across the river, so

Let $\mathbf{v_w} = 20\mathbf{j}$.

$$\mathbf{v_g} = \mathbf{v_w} + \mathbf{v_c} = 20\mathbf{j} + 3\mathbf{i} = 3\mathbf{i} + 20\mathbf{j}$$

Let $\left\| \mathbf{v_g} \right\| = \sqrt{3^2 + 20^2} = \sqrt{409} \approx 20.2 \text{ miles per hour}$

Find the angle between $\mathbf{v_g}$ and $\mathbf{j}$:

$$\cos\theta = \frac{\mathbf{v_g} \cdot \mathbf{j}}{\left\| \mathbf{v_g} \right\| \left\| \mathbf{j} \right\|} = \frac{3 \cdot 0 + 20(1)}{20.2\sqrt{0^2 + 1^2}} = \frac{20}{20.2} \approx 0.9901$$

$$\theta \approx 8.1°$$

The heading of the boat will be 8.1° downstream.

29. $\mathbf{F} = 3\left(\cos 60°\mathbf{i} + \sin 60°\mathbf{j}\right) = 3\left(\frac{1}{2}\mathbf{i} + \frac{\sqrt{3}}{2}\mathbf{j}\right) = \frac{3}{2}\mathbf{i} + \frac{3\sqrt{3}}{2}\mathbf{j}$

$W = \mathbf{F} \cdot AB = \left(\frac{3}{2}\mathbf{i} + \frac{3\sqrt{3}}{2}\mathbf{j}\right) \cdot 2\mathbf{i} = \frac{3}{2}(2) + \frac{3\sqrt{3}}{2} \cdot 0 = 3 \text{ foot - pounds}$

31. $\mathbf{F} = 20\left(\cos 30°\mathbf{i} + \sin 30°\mathbf{j}\right) = 20\left(\frac{\sqrt{3}}{2}\mathbf{i} + \frac{1}{2}\mathbf{j}\right) = 10\sqrt{3}\mathbf{i} + 10\mathbf{j}$

$W = \mathbf{F} \cdot AB = \left(10\sqrt{3}\mathbf{i} + 10\mathbf{j}\right) \cdot 100\mathbf{i} = 10\sqrt{3}(100) + 10 \cdot 0 = 1732 \text{ foot - pounds}$

33.   Let $\mathbf{u} = a_1\mathbf{i} + b_1\mathbf{j}, \quad \mathbf{v} = a_2\mathbf{i} + b_2\mathbf{j}, \quad \mathbf{w} = a_3\mathbf{i} + b_3\mathbf{j}$

$$\mathbf{u} \bullet (\mathbf{v} + \mathbf{w}) = (a_1\mathbf{i} + b_1\mathbf{j}) \bullet (a_2\mathbf{i} + b_2\mathbf{j} + a_3\mathbf{i} + b_3\mathbf{j}) = (a_1\mathbf{i} + b_1\mathbf{j}) \bullet (a_2\mathbf{i} + a_3\mathbf{i} + b_2\mathbf{j} + b_3\mathbf{j})$$
$$= (a_1\mathbf{i} + b_1\mathbf{j}) \bullet ((a_2 + a_3)\mathbf{i} + (b_2 + b_3)\mathbf{j}) = a_1(a_2 + a_3) + b_1(b_2 + b_3)$$
$$= a_1a_2 + a_1a_3 + b_1b_2 + b_1b_3 = a_1a_2 + b_1b_2 + a_1a_3 + b_1b_3$$
$$= (a_1\mathbf{i} + b_1\mathbf{j}) \bullet (a_2\mathbf{i} + b_2\mathbf{j}) + (a_1\mathbf{i} + b_1\mathbf{j}) \bullet (a_3\mathbf{i} + b_3\mathbf{j}) = \mathbf{u} \bullet \mathbf{v} + \mathbf{u} \bullet \mathbf{w}$$

35.   Let $\mathbf{v} = a\mathbf{i} + b\mathbf{j}$.

Since $\mathbf{v}$ is a unit vector, $\|\mathbf{v}\| = \sqrt{a^2 + b^2} = 1$ or $a^2 + b^2 = 1$

If α is the angle between $\mathbf{v}$ and $\mathbf{i}$, then $\cos\alpha = \dfrac{\mathbf{v} \bullet \mathbf{i}}{\|\mathbf{v}\|\|\mathbf{i}\|}$ or $\cos\alpha = \dfrac{(a\mathbf{i} + b\mathbf{j}) \bullet \mathbf{i}}{1 \cdot 1} = a$.

$$a^2 + b^2 = 1$$
$$\cos^2\alpha + b^2 = 1$$
$$b^2 = 1 - \cos^2\alpha$$
$$b^2 = \sin^2\alpha$$
$$b = \sin\alpha$$

Thus, $\mathbf{v} = \cos\alpha\,\mathbf{i} + \sin\alpha\,\mathbf{j}$

37.   Let $\mathbf{v} = a\mathbf{i} + b\mathbf{j}$.

$$\text{proj}_{\mathbf{i}}\mathbf{v} = \dfrac{\mathbf{v} \bullet \mathbf{i}}{\|\mathbf{i}\|^2}\mathbf{i} = \dfrac{(a\mathbf{i} + b\mathbf{j}) \bullet \mathbf{i}}{\sqrt{1^2 + 0^2}}\mathbf{i} = \dfrac{a(1) + b(0)}{1}\mathbf{i} = a\mathbf{i}$$
$$\mathbf{v} \bullet \mathbf{i} = a, \quad \mathbf{v} \bullet \mathbf{j} = b,$$
$$\mathbf{v} = (\mathbf{v} \bullet \mathbf{i})\mathbf{i} + (\mathbf{v} \bullet \mathbf{j})\mathbf{j}$$

39. $(\mathbf{v} - \alpha\mathbf{w}) \bullet \mathbf{w} = \mathbf{v} \bullet \mathbf{w} - \alpha\mathbf{w} \bullet \mathbf{w} = \mathbf{v} \bullet \mathbf{w} - \alpha\|\mathbf{w}\|^2 = \mathbf{v} \bullet \mathbf{w} - \dfrac{\mathbf{v} \bullet \mathbf{w}}{\|\mathbf{w}\|^2}\|\mathbf{w}\|^2 = 0$

Therefore the vectors are orthogonal.

41.   If $\mathbf{F}$ is orthogonal to $\mathbf{AB}$, then $W = \mathbf{F} \bullet \mathbf{AB} = (\mathbf{i} + \mathbf{j})(\mathbf{i} - \mathbf{j}) = 1 \cdot 1 + 1(-1) = 0$

9.6   Vectors in Space

1. $y = 0$ is the set of all points in the xz-plane.

3. $z = 2$ is the set of all points of the form $(x, y, 2)$; the plane two units above the xy-plane.

5. $x = -4$ is the set of all points of the form $(-4, y, z)$; the plane four units to the left of yz-plane.

7. $x = 1$ and $y = 2$ is the set of all points of the form $(1, 2, z)$; a line parallel to the z-axis.

9. $d = \sqrt{(4-0)^2 + (1-0)^2 + (2-0)^2} = \sqrt{16 + 1 + 4} = \sqrt{21}$

11. $d = \sqrt{(0-(-1))^2 + (-2-2)^2 + (1-(-3))^2} = \sqrt{1 + 16 + 16} = \sqrt{33}$

13. $d = \sqrt{(3-4)^2 + (2-(-2))^2 + (1-(-2))^2} = \sqrt{1 + 16 + 9} = \sqrt{26}$

15.   The bottom of the box is formed by the vertices $(0, 0, 0)$, $(2, 0, 0)$, $(0, 1, 0)$, and $(2, 1, 0)$. The top of the box is formed by the vertices $(0, 0, 3)$, $(2, 0, 3)$, $(0, 1, 3)$, and $(2, 1, 3)$.

17.   The bottom of the box is formed by the vertices $(1, 2, 3)$, $(3, 2, 3)$, $(3, 4, 3)$, and $(1, 4, 3)$. The top of the box is formed by the vertices $(3, 4, 5)$, $(1, 2, 5)$, $(3, 2, 5)$, and $(1, 4, 5)$.

19.   The bottom of the box is formed by the vertices $(-1, 0, 2)$, $(4, 0, 2)$, $(-1, 2, 2)$, and $(4, 2, 2)$. The top of the box is formed by the vertices $(4, 2, 5)$, $(-1, 0, 5)$, $(4, 0, 5)$, and $(-1, 2, 5)$.

21. $\mathbf{v} = (3-0)\mathbf{i} + (4-0)\mathbf{j} + (-1-0)\mathbf{k} = 3\mathbf{i} + 4\mathbf{j} - 1\mathbf{k}$

23. $\mathbf{v} = (5-3)\mathbf{i} + (6-2)\mathbf{j} + (0-(-1))\mathbf{k} = 2\mathbf{i} + 4\mathbf{j} + 1\mathbf{k}$

25. $\mathbf{v} = (6-(-2))\mathbf{i} + (-2-(-1))\mathbf{j} + (4-4)\mathbf{k} = 8\mathbf{i} - 1\mathbf{j}$

27. $\|\mathbf{v}\| = \sqrt{3^2 + (-6)^2 + (-2)^2} = \sqrt{9 + 36 + 4} = \sqrt{49} = 7$

29. $\|\mathbf{v}\| = \sqrt{1^2 + (-1)^2 + 1^2} = \sqrt{1 + 1 + 1} = \sqrt{3}$

31. $\|\mathbf{v}\| = \sqrt{(-2)^2 + 3^2 + (-3)^2} = \sqrt{4 + 9 + 9} = \sqrt{22}$

33. $2\mathbf{v} + 3\mathbf{w} = 2(3\mathbf{i} - 5\mathbf{j} + 2\mathbf{k}) + 3(-2\mathbf{i} + 3\mathbf{j} - 2\mathbf{k}) = 6\mathbf{i} - 10\mathbf{j} + 4\mathbf{k} - 6\mathbf{i} + 9\mathbf{j} - 6\mathbf{k}$
      $= 0\mathbf{i} - 1\mathbf{j} - 2\mathbf{k}$

35. $\|\mathbf{v} - \mathbf{w}\| = \|(3\mathbf{i} - 5\mathbf{j} + 2\mathbf{k}) - (-2\mathbf{i} + 3\mathbf{j} - 2\mathbf{k})\| = |3\mathbf{i} - 5\mathbf{j} + 2\mathbf{k} + 2\mathbf{i} - 3\mathbf{j} + 2\mathbf{k}|$
      $= \|5\mathbf{i} - 8\mathbf{j} + 4\mathbf{k}\| = \sqrt{5^2 + (-8)^2 + 4^2} = \sqrt{25 + 64 + 16} = \sqrt{105}$

37. $\|\mathbf{v}\| - \|\mathbf{w}\| = |3\mathbf{i} - 5\mathbf{j} + 2\mathbf{k}| - \|-2\mathbf{i} + 3\mathbf{j} - 2\mathbf{k}\|$
      $= \sqrt{3^2 + (-5)^2 + 2^2} - \sqrt{(-2)^2 + 3^2 + (-2)^2} = \sqrt{38} - \sqrt{17}$

39. $\mathbf{u} = \dfrac{\mathbf{v}}{\|\mathbf{v}\|} = \dfrac{5\mathbf{i}}{\sqrt{5^2 + 0^2 + 0^2}} = \dfrac{5\mathbf{i}}{5} = \mathbf{i}$

41. $\mathbf{u} = \dfrac{\mathbf{v}}{\|\mathbf{v}\|} = \dfrac{3\mathbf{i} - 6\mathbf{j} - 2\mathbf{k}}{\sqrt{3^2 + (-6)^2 + (-2)^2}} = \dfrac{3\mathbf{i} - 6\mathbf{j} - 2\mathbf{k}}{7} = \dfrac{3}{7}\mathbf{i} - \dfrac{6}{7}\mathbf{j} - \dfrac{2}{7}\mathbf{k}$

43. $\mathbf{u} = \dfrac{\mathbf{v}}{\|\mathbf{v}\|} = \dfrac{\mathbf{i} + \mathbf{j} + \mathbf{k}}{\sqrt{1^2 + 1^2 + 1^2}} = \dfrac{\mathbf{i} + \mathbf{j} + \mathbf{k}}{\sqrt{3}} = \dfrac{1}{\sqrt{3}}\mathbf{i} + \dfrac{1}{\sqrt{3}}\mathbf{j} + \dfrac{1}{\sqrt{3}}\mathbf{k} = \dfrac{\sqrt{3}}{3}\mathbf{i} + \dfrac{\sqrt{3}}{3}\mathbf{j} + \dfrac{\sqrt{3}}{3}\mathbf{k}$

45. $\mathbf{v} \cdot \mathbf{w} = (\mathbf{i} - \mathbf{j}) \cdot (\mathbf{i} + \mathbf{j} + \mathbf{k}) = 1 \cdot 1 + (-1)(1) + 0 \cdot 1 = 1 - 1 + 0 = 0$

     $\cos\theta = \dfrac{\mathbf{v} \cdot \mathbf{w}}{\|\mathbf{v}\|\|\mathbf{w}\|} = \dfrac{0}{\sqrt{1^2 + (-1)^2 + 0^2}\,\sqrt{1^2 + 1^2 + 1^2}} = \dfrac{0}{\sqrt{2}\sqrt{3}} = \dfrac{0}{\sqrt{6}} = 0$

     $\theta = \dfrac{\pi}{2} = 90°$

47. $\mathbf{v} \cdot \mathbf{w} = (2\mathbf{i} + \mathbf{j} - 3\mathbf{k}) \cdot (\mathbf{i} + 2\mathbf{j} + 2\mathbf{k}) = 2 \cdot 1 + 1(2) + (-3)(2) = 2 + 2 - 6 = -2$

     $\cos\theta = \dfrac{\mathbf{v} \cdot \mathbf{w}}{\|\mathbf{v}\|\|\mathbf{w}\|} = \dfrac{-2}{\sqrt{2^2 + 1^2 + (-3)^2}\,\sqrt{1^2 + 2^2 + 2^2}} = \dfrac{-2}{\sqrt{14}\sqrt{9}} = \dfrac{-2}{3\sqrt{14}} \approx -0.1782$

     $\theta = 1.75 = 100.3°$

49. $\mathbf{v} \cdot \mathbf{w} = (3\mathbf{i} - \mathbf{j} + 2\mathbf{k}) \cdot (\mathbf{i} + \mathbf{j} - \mathbf{k}) = 3 \cdot 1 + (-1)(1) + 2(-1) = 3 - 1 - 2 = 0$

     $\cos\theta = \dfrac{\mathbf{v} \cdot \mathbf{w}}{\|\mathbf{v}\|\|\mathbf{w}\|} = \dfrac{0}{\sqrt{3^2 + (-1)^2 + 2^2}\,\sqrt{1^2 + 1^2 + (-1)^2}} = \dfrac{0}{\sqrt{14}\sqrt{3}} = 0$

     $\theta = \dfrac{\pi}{2} = 90°$

51. $\mathbf{v} \cdot \mathbf{w} = (3\mathbf{i} + 4\mathbf{j} + \mathbf{k}) \cdot (6\mathbf{i} + 8\mathbf{j} + 2\mathbf{k}) = 3 \cdot 6 + 4 \cdot 8 + 1 \cdot 2 = 18 + 32 + 2 = 52$

     $\cos\theta = \dfrac{\mathbf{v} \cdot \mathbf{w}}{\|\mathbf{v}\|\|\mathbf{w}\|} = \dfrac{52}{\sqrt{3^2 + 4^2 + 1^2}\,\sqrt{6^2 + 8^2 + 2^2}} = \dfrac{52}{\sqrt{26}\sqrt{104}} = \dfrac{52}{52} = 1$

     $\theta = 0 = 0°$

53. $\cos\alpha = \dfrac{a}{\|\mathbf{v}\|} = \dfrac{3}{\sqrt{3^2 + (-6)^2 + (-2)^2}} = \dfrac{3}{\sqrt{49}} = \dfrac{3}{7} \;\rightarrow\; \alpha \approx 64.6°$

     $\cos\beta = \dfrac{b}{\|\mathbf{v}\|} = \dfrac{-6}{\sqrt{3^2 + (-6)^2 + (-2)^2}} = \dfrac{-6}{\sqrt{49}} = \dfrac{-6}{7} \;\rightarrow\; \beta \approx 149.0°$

     $\cos\gamma = \dfrac{c}{\|\mathbf{v}\|} = \dfrac{-2}{\sqrt{3^2 + (-6)^2 + (-2)^2}} = \dfrac{-2}{\sqrt{49}} = \dfrac{-2}{7} \;\rightarrow\; \gamma \approx 106.6°$

     $\mathbf{v} = 7(\cos 64.6°\, \mathbf{i} + \cos 149.0°\, \mathbf{j} + \cos 106.6°\, \mathbf{k})$

55. $\cos\alpha = \dfrac{a}{\|\mathbf{v}\|} = \dfrac{1}{\sqrt{1^2 + 1^2 + 1^2}} = \dfrac{1}{\sqrt{3}} = \dfrac{\sqrt{3}}{3} \quad \rightarrow \quad \alpha \approx 54.7°$

$\cos\beta = \dfrac{b}{\|\mathbf{v}\|} = \dfrac{1}{\sqrt{1^2 + 1^2 + 1^2}} = \dfrac{1}{\sqrt{3}} = \dfrac{\sqrt{3}}{3} \quad \rightarrow \quad \beta \approx 54.7°$

$\cos\gamma = \dfrac{c}{\|\mathbf{v}\|} = \dfrac{1}{\sqrt{1^2 + 1^2 + 1^2}} = \dfrac{1}{\sqrt{3}} = \dfrac{\sqrt{3}}{3} \quad \rightarrow \quad \gamma \approx 54.7°$

$\mathbf{v} = \sqrt{3}\left(\cos 54.7°\mathbf{i} + \cos 54.7°\,\mathbf{j} + \cos 54.7°\mathbf{k}\right)$

57. $\cos\alpha = \dfrac{a}{\|\mathbf{v}\|} = \dfrac{1}{\sqrt{1^2 + 1^2 + 0^2}} = \dfrac{1}{\sqrt{2}} = \dfrac{\sqrt{2}}{2} \quad \rightarrow \quad \alpha = 45°$

$\cos\beta = \dfrac{b}{\|\mathbf{v}\|} = \dfrac{1}{\sqrt{1^2 + 1^2 + 0^2}} = \dfrac{1}{\sqrt{2}} = \dfrac{\sqrt{2}}{2} \quad \rightarrow \quad \beta = 45°$

$\cos\gamma = \dfrac{c}{\|\mathbf{v}\|} = \dfrac{0}{\sqrt{1^2 + 1^2 + 0^2}} = \dfrac{0}{\sqrt{2}} = 0 \quad \rightarrow \quad \gamma = 90°$

$\mathbf{v} = \sqrt{2}\left(\cos 45°\mathbf{i} + \cos 45°\,\mathbf{j} + \cos 90°\mathbf{k}\right)$

59. $\cos\alpha = \dfrac{a}{\|\mathbf{v}\|} = \dfrac{3}{\sqrt{3^2 + (-5)^2 + 2^2}} = \dfrac{3}{\sqrt{38}} \quad \rightarrow \quad \alpha \approx 60.9°$

$\cos\beta = \dfrac{b}{\|\mathbf{v}\|} = \dfrac{-5}{\sqrt{3^2 + (-5)^2 + 2^2}} = \dfrac{-5}{\sqrt{38}} \quad \rightarrow \quad \beta \approx 144.2°$

$\cos\gamma = \dfrac{c}{\|\mathbf{v}\|} = \dfrac{2}{\sqrt{3^2 + (-5)^2 + 2^2}} = \dfrac{2}{\sqrt{38}} \quad \rightarrow \quad \gamma \approx 71.1°$

$\mathbf{v} = \sqrt{38}\left(\cos 60.9°\mathbf{i} + \cos 144.2°\,\mathbf{j} + \cos 71.1°\mathbf{k}\right)$

61. $d(P_0, P) = \sqrt{(x - x_0)^2 + (y - y_0)^2 + (z - z_0)^2} = r$

$(x - x_0)^2 + (y - y_0)^2 + (z - z_0)^2 = r^2$

63. $(x - 1)^2 + (y - 2)^2 + (z - 2)^2 = 4$

65. $x^2 + y^2 + z^2 + 2x - 2y = 2$

$x^2 + 2x + y^2 - 2y + z^2 = 2$

$x^2 + 2x + 1 + y^2 - 2y + 1 + z^2 = 2 + 1 + 1$

$(x + 1)^2 + (y - 1)^2 + (z - 0)^2 = 4$

Center: $(-1, 1, 0)$; Radius: 2

67. $$x^2 + y^2 + z^2 - 4x + 4y + 2z = 0$$
$$x^2 - 4x + y^2 + 4y + z^2 + 2z = 0$$
$$x^2 - 4x + 4 + y^2 + 4y + 4 + z^2 + 2z + 1 = 4 + 4 + 1$$
$$(x - 2)^2 + (y + 2)^2 + (z + 1)^2 = 9$$
Center: $(2, -2, -1)$; Radius: 3

69. $$2x^2 + 2y^2 + 2z^2 - 8x + 4z = -1$$
$$x^2 - 4x + y^2 + z^2 + 2z = \frac{-1}{2}$$
$$x^2 - 4x + 4 + y^2 + z^2 + 2z + 1 = \frac{-1}{2} + 4 + 1$$
$$(x - 2)^2 + (y - 0)^2 + (z + 1)^2 = \frac{9}{2}$$
Center: $(2, 0, -1)$; Radius: $\frac{3\sqrt{2}}{2}$

71. Write the force as a vector:
$$\cos\alpha = \frac{2}{\sqrt{2^2 + 1^2 + 2^2}} = \frac{2}{\sqrt{9}} = \frac{2}{3}; \quad \cos\beta = \frac{1}{3}; \quad \cos\gamma = \frac{2}{3}$$
$$\mathbf{F} = 3\left(\frac{2}{3}\mathbf{i} + \frac{1}{3}\mathbf{j} + \frac{2}{3}\mathbf{k}\right)$$
$$W = 3\left(\frac{2}{3}\mathbf{i} + \frac{1}{3}\mathbf{j} + \frac{2}{3}\mathbf{k}\right) \bullet 2\mathbf{j} = 3\left(\frac{1}{3} \cdot 2\right) = 2 \text{ joules}$$

73. $W = \mathbf{F} \bullet \mathbf{AB} = (2\mathbf{i} - \mathbf{j} - \mathbf{k}) \bullet (3\mathbf{i} + 2\mathbf{j} - 5\mathbf{k}) = 2 \cdot 3 + (-1)(2) + (-1)(-5) = 9$

9.7   The Cross Product

1. $\begin{vmatrix} 3 & 4 \\ 1 & 2 \end{vmatrix} = 3 \cdot 2 - 1 \cdot 4 = 6 - 4 = 2$

3. $\begin{vmatrix} 6 & 5 \\ -2 & -1 \end{vmatrix} = 6(-1) - (-2)(5) = -6 + 10 = 4$

5. $\begin{vmatrix} A & B & C \\ 2 & 1 & 4 \\ 1 & 3 & 1 \end{vmatrix} = \begin{vmatrix} 1 & 4 \\ 3 & 1 \end{vmatrix}A - \begin{vmatrix} 2 & 4 \\ 1 & 1 \end{vmatrix}B + \begin{vmatrix} 2 & 1 \\ 1 & 3 \end{vmatrix}C = (1 - 12)A - (2 - 4)B + (6 - 1)C$
$$= -11A + 2B + 5C$$

7. $\begin{vmatrix} A & B & C \\ -1 & 3 & 5 \\ 5 & 0 & -2 \end{vmatrix} = \begin{vmatrix} 3 & 5 \\ 0 & -2 \end{vmatrix}A - \begin{vmatrix} -1 & 5 \\ 5 & -2 \end{vmatrix}B + \begin{vmatrix} -1 & 3 \\ 5 & 0 \end{vmatrix}C = (-6 - 0)A - (2 - 25)B + (0 - 15)C$
$$= -6A + 23B - 15C$$

9. (a) $\mathbf{v} \times \mathbf{w} = \begin{vmatrix} \mathbf{i} & \mathbf{j} & \mathbf{k} \\ 2 & -3 & 1 \\ 3 & -2 & -1 \end{vmatrix} = \begin{vmatrix} -3 & 1 \\ -2 & -1 \end{vmatrix} \mathbf{i} - \begin{vmatrix} 2 & 1 \\ 3 & -1 \end{vmatrix} \mathbf{j} + \begin{vmatrix} 2 & -3 \\ 3 & -2 \end{vmatrix} \mathbf{k} = 5\mathbf{i} + 5\mathbf{j} + 5\mathbf{k}$

    (b) $\mathbf{w} \times \mathbf{v} = \begin{vmatrix} \mathbf{i} & \mathbf{j} & \mathbf{k} \\ 3 & -2 & -1 \\ 2 & -3 & 1 \end{vmatrix} = \begin{vmatrix} -2 & -1 \\ -3 & 1 \end{vmatrix} \mathbf{i} - \begin{vmatrix} 3 & -1 \\ 2 & 1 \end{vmatrix} \mathbf{j} + \begin{vmatrix} 3 & -2 \\ 2 & -3 \end{vmatrix} \mathbf{k} = -5\mathbf{i} - 5\mathbf{j} - 5\mathbf{k}$

    (c) $\mathbf{w} \times \mathbf{w} = \begin{vmatrix} \mathbf{i} & \mathbf{j} & \mathbf{k} \\ 3 & -2 & -1 \\ 3 & -2 & -1 \end{vmatrix} = \begin{vmatrix} -2 & -1 \\ -2 & -1 \end{vmatrix} \mathbf{i} - \begin{vmatrix} 3 & -1 \\ 3 & -1 \end{vmatrix} \mathbf{j} + \begin{vmatrix} 3 & -2 \\ 3 & -2 \end{vmatrix} \mathbf{k} = 0\mathbf{i} + 0\mathbf{j} + 0\mathbf{k} = \mathbf{0}$

    (d) $\mathbf{v} \times \mathbf{v} = \begin{vmatrix} \mathbf{i} & \mathbf{j} & \mathbf{k} \\ 2 & -3 & 1 \\ 2 & -3 & 1 \end{vmatrix} = \begin{vmatrix} -3 & 1 \\ -3 & 1 \end{vmatrix} \mathbf{i} - \begin{vmatrix} 2 & 1 \\ 2 & 1 \end{vmatrix} \mathbf{j} + \begin{vmatrix} 2 & -3 \\ 2 & -3 \end{vmatrix} \mathbf{k} = 0\mathbf{i} + 0\mathbf{j} + 0\mathbf{k} = \mathbf{0}$

11. (a) $\mathbf{v} \times \mathbf{w} = \begin{vmatrix} \mathbf{i} & \mathbf{j} & \mathbf{k} \\ 1 & 1 & 0 \\ 2 & 1 & 1 \end{vmatrix} = \begin{vmatrix} 1 & 0 \\ 1 & 1 \end{vmatrix} \mathbf{i} - \begin{vmatrix} 1 & 0 \\ 2 & 1 \end{vmatrix} \mathbf{j} + \begin{vmatrix} 1 & 1 \\ 2 & 1 \end{vmatrix} \mathbf{k} = 1\mathbf{i} - 1\mathbf{j} - 1\mathbf{k}$

    (b) $\mathbf{w} \times \mathbf{v} = \begin{vmatrix} \mathbf{i} & \mathbf{j} & \mathbf{k} \\ 2 & 1 & 1 \\ 1 & 1 & 0 \end{vmatrix} = \begin{vmatrix} 1 & 1 \\ 1 & 0 \end{vmatrix} \mathbf{i} - \begin{vmatrix} 2 & 1 \\ 1 & 0 \end{vmatrix} \mathbf{j} + \begin{vmatrix} 2 & 1 \\ 1 & 1 \end{vmatrix} \mathbf{k} = -1\mathbf{i} + 1\mathbf{j} + 1\mathbf{k}$

    (c) $\mathbf{w} \times \mathbf{w} = \begin{vmatrix} \mathbf{i} & \mathbf{j} & \mathbf{k} \\ 2 & 1 & 1 \\ 2 & 1 & 1 \end{vmatrix} = \begin{vmatrix} 1 & 1 \\ 1 & 1 \end{vmatrix} \mathbf{i} - \begin{vmatrix} 2 & 1 \\ 2 & 1 \end{vmatrix} \mathbf{j} + \begin{vmatrix} 2 & 1 \\ 2 & 1 \end{vmatrix} \mathbf{k} = 0\mathbf{i} + 0\mathbf{j} + 0\mathbf{k} = \mathbf{0}$

    (d) $\mathbf{v} \times \mathbf{v} = \begin{vmatrix} \mathbf{i} & \mathbf{j} & \mathbf{k} \\ 1 & 1 & 0 \\ 1 & 1 & 0 \end{vmatrix} = \begin{vmatrix} 1 & 0 \\ 1 & 0 \end{vmatrix} \mathbf{i} - \begin{vmatrix} 1 & 0 \\ 1 & 0 \end{vmatrix} \mathbf{j} + \begin{vmatrix} 1 & 1 \\ 1 & 1 \end{vmatrix} \mathbf{k} = 0\mathbf{i} + 0\mathbf{j} + 0\mathbf{k} = \mathbf{0}$

13. (a) $\mathbf{v} \times \mathbf{w} = \begin{vmatrix} \mathbf{i} & \mathbf{j} & \mathbf{k} \\ 2 & -1 & 2 \\ 0 & 1 & -1 \end{vmatrix} = \begin{vmatrix} -1 & 2 \\ 1 & -1 \end{vmatrix} \mathbf{i} - \begin{vmatrix} 2 & 2 \\ 0 & -1 \end{vmatrix} \mathbf{j} + \begin{vmatrix} 2 & -1 \\ 0 & 1 \end{vmatrix} \mathbf{k} = -1\mathbf{i} + 2\mathbf{j} + 2\mathbf{k}$

    (b) $\mathbf{w} \times \mathbf{v} = \begin{vmatrix} \mathbf{i} & \mathbf{j} & \mathbf{k} \\ 0 & 1 & -1 \\ 2 & -1 & 2 \end{vmatrix} = \begin{vmatrix} 1 & -1 \\ -1 & 2 \end{vmatrix} \mathbf{i} - \begin{vmatrix} 0 & -1 \\ 2 & 2 \end{vmatrix} \mathbf{j} + \begin{vmatrix} 0 & 1 \\ 2 & -1 \end{vmatrix} \mathbf{k} = 1\mathbf{i} - 2\mathbf{j} - 2\mathbf{k}$

    (c) $\mathbf{w} \times \mathbf{w} = \begin{vmatrix} \mathbf{i} & \mathbf{j} & \mathbf{k} \\ 0 & 1 & -1 \\ 0 & 1 & -1 \end{vmatrix} = \begin{vmatrix} 1 & -1 \\ 1 & -1 \end{vmatrix} \mathbf{i} - \begin{vmatrix} 0 & -1 \\ 0 & -1 \end{vmatrix} \mathbf{j} + \begin{vmatrix} 0 & 1 \\ 0 & 1 \end{vmatrix} \mathbf{k} = 0\mathbf{i} + 0\mathbf{j} + 0\mathbf{k} = \mathbf{0}$

    (d) $\mathbf{v} \times \mathbf{v} = \begin{vmatrix} \mathbf{i} & \mathbf{j} & \mathbf{k} \\ 2 & -1 & 2 \\ 2 & -1 & 2 \end{vmatrix} = \begin{vmatrix} -1 & 2 \\ -1 & 2 \end{vmatrix} \mathbf{i} - \begin{vmatrix} 2 & 2 \\ 2 & 2 \end{vmatrix} \mathbf{j} + \begin{vmatrix} 2 & -1 \\ 2 & -1 \end{vmatrix} \mathbf{k} = 0\mathbf{i} + 0\mathbf{j} + 0\mathbf{k} = \mathbf{0}$

15. (a) $\mathbf{v} \times \mathbf{w} = \begin{vmatrix} \mathbf{i} & \mathbf{j} & \mathbf{k} \\ 1 & -1 & -1 \\ 4 & 0 & -3 \end{vmatrix} = \begin{vmatrix} -1 & -1 \\ 0 & -3 \end{vmatrix} \mathbf{i} - \begin{vmatrix} 1 & -1 \\ 4 & -3 \end{vmatrix} \mathbf{j} + \begin{vmatrix} 1 & -1 \\ 4 & 0 \end{vmatrix} \mathbf{k} = 3\mathbf{i} - 1\mathbf{j} + 4\mathbf{k}$

    (b) $\mathbf{w} \times \mathbf{v} = \begin{vmatrix} \mathbf{i} & \mathbf{j} & \mathbf{k} \\ 4 & 0 & -3 \\ 1 & -1 & -1 \end{vmatrix} = \begin{vmatrix} 0 & -3 \\ -1 & -1 \end{vmatrix} \mathbf{i} - \begin{vmatrix} 4 & -3 \\ 1 & -1 \end{vmatrix} \mathbf{j} + \begin{vmatrix} 4 & 0 \\ 1 & -1 \end{vmatrix} \mathbf{k} = -3\mathbf{i} + 1\mathbf{j} - 4\mathbf{k}$

    (c) $\mathbf{w} \times \mathbf{w} = \begin{vmatrix} \mathbf{i} & \mathbf{j} & \mathbf{k} \\ 4 & 0 & -3 \\ 4 & 0 & -3 \end{vmatrix} = \begin{vmatrix} 0 & -3 \\ 0 & -3 \end{vmatrix} \mathbf{i} - \begin{vmatrix} 4 & -3 \\ 4 & -3 \end{vmatrix} \mathbf{j} + \begin{vmatrix} 4 & 0 \\ 4 & 0 \end{vmatrix} \mathbf{k} = 0\mathbf{i} + 0\mathbf{j} + 0\mathbf{k} = \mathbf{0}$

    (d) $\mathbf{v} \times \mathbf{v} = \begin{vmatrix} \mathbf{i} & \mathbf{j} & \mathbf{k} \\ 1 & -1 & -1 \\ 1 & -1 & -1 \end{vmatrix} = \begin{vmatrix} -1 & -1 \\ -1 & -1 \end{vmatrix} \mathbf{i} - \begin{vmatrix} 1 & -1 \\ 1 & -1 \end{vmatrix} \mathbf{j} + \begin{vmatrix} 1 & -1 \\ 1 & -1 \end{vmatrix} \mathbf{k} = 0\mathbf{i} + 0\mathbf{j} + 0\mathbf{k} = \mathbf{0}$

17. $\mathbf{u} \times \mathbf{v} = \begin{vmatrix} \mathbf{i} & \mathbf{j} & \mathbf{k} \\ 2 & -3 & 1 \\ -3 & 3 & 2 \end{vmatrix} = \begin{vmatrix} -3 & 1 \\ 3 & 2 \end{vmatrix} \mathbf{i} - \begin{vmatrix} 2 & 1 \\ -3 & 2 \end{vmatrix} \mathbf{j} + \begin{vmatrix} 2 & -3 \\ -3 & 3 \end{vmatrix} \mathbf{k} = -9\mathbf{i} - 7\mathbf{j} - 3\mathbf{k}$

19. $\mathbf{v} \times \mathbf{u} = \begin{vmatrix} \mathbf{i} & \mathbf{j} & \mathbf{k} \\ -3 & 3 & 2 \\ 2 & -3 & 1 \end{vmatrix} = \begin{vmatrix} 3 & 2 \\ -3 & 1 \end{vmatrix} \mathbf{i} - \begin{vmatrix} -3 & 2 \\ 2 & 1 \end{vmatrix} \mathbf{j} + \begin{vmatrix} -3 & 3 \\ 2 & -3 \end{vmatrix} \mathbf{k} = 9\mathbf{i} + 7\mathbf{j} + 3\mathbf{k}$

21. $\mathbf{v} \times \mathbf{v} = \begin{vmatrix} \mathbf{i} & \mathbf{j} & \mathbf{k} \\ -3 & 3 & 2 \\ -3 & 3 & 2 \end{vmatrix} = \begin{vmatrix} 3 & 2 \\ 3 & 2 \end{vmatrix} \mathbf{i} - \begin{vmatrix} -3 & 2 \\ -3 & 2 \end{vmatrix} \mathbf{j} + \begin{vmatrix} -3 & 3 \\ -3 & 3 \end{vmatrix} \mathbf{k} = 0\mathbf{i} + 0\mathbf{j} + 0\mathbf{k} = \mathbf{0}$

23. $(3\mathbf{u}) \times \mathbf{v} = \begin{vmatrix} \mathbf{i} & \mathbf{j} & \mathbf{k} \\ 6 & -9 & 3 \\ -3 & 3 & 2 \end{vmatrix} = \begin{vmatrix} -9 & 3 \\ 3 & 2 \end{vmatrix} \mathbf{i} - \begin{vmatrix} 6 & 3 \\ -3 & 2 \end{vmatrix} \mathbf{j} + \begin{vmatrix} 6 & -9 \\ -3 & 3 \end{vmatrix} \mathbf{k} = -27\mathbf{i} - 21\mathbf{j} - 9\mathbf{k}$

25. $\mathbf{u} \times (2\mathbf{v}) = \begin{vmatrix} \mathbf{i} & \mathbf{j} & \mathbf{k} \\ 2 & -3 & 1 \\ -6 & 6 & 4 \end{vmatrix} = \begin{vmatrix} -3 & 1 \\ 6 & 4 \end{vmatrix} \mathbf{i} - \begin{vmatrix} 2 & 1 \\ -6 & 4 \end{vmatrix} \mathbf{j} + \begin{vmatrix} 2 & -3 \\ -6 & 6 \end{vmatrix} \mathbf{k} = -18\mathbf{i} - 14\mathbf{j} - 6\mathbf{k}$

27. $\mathbf{u} \bullet (\mathbf{u} \times \mathbf{v}) = \mathbf{u} \bullet \begin{vmatrix} \mathbf{i} & \mathbf{j} & \mathbf{k} \\ 2 & -3 & 1 \\ -3 & 3 & 2 \end{vmatrix} = \mathbf{u} \cdot \left(\begin{vmatrix} -3 & 1 \\ 3 & 2 \end{vmatrix} \mathbf{i} - \begin{vmatrix} 2 & 1 \\ -3 & 2 \end{vmatrix} \mathbf{j} + \begin{vmatrix} 2 & -3 \\ -3 & 3 \end{vmatrix} \mathbf{k} \right)$

    $= (2\mathbf{i} - 3\mathbf{j} + \mathbf{k}) \bullet (-9\mathbf{i} - 7\mathbf{j} - 3\mathbf{k}) = 2(-9) + (-3)(-7) + 1(-3) = -18 + 21 - 3 = 0$

29. $\mathbf{u} \bullet (\mathbf{v} \times \mathbf{w}) = \mathbf{u} \bullet \begin{vmatrix} \mathbf{i} & \mathbf{j} & \mathbf{k} \\ -3 & 3 & 2 \\ 1 & 1 & 3 \end{vmatrix} = \mathbf{u} \bullet \left(\begin{vmatrix} 3 & 2 \\ 1 & 3 \end{vmatrix} \mathbf{i} - \begin{vmatrix} -3 & 2 \\ 1 & 3 \end{vmatrix} \mathbf{j} + \begin{vmatrix} -3 & 3 \\ 1 & 1 \end{vmatrix} \mathbf{k} \right)$

    $= (2\mathbf{i} - 3\mathbf{j} + \mathbf{k}) \bullet (7\mathbf{i} + 11\mathbf{j} - 6\mathbf{k}) = 2 \cdot 7 + (-3)(11) + 1(-6) = 14 - 33 - 6 = -25$

31. $\mathbf{v} \cdot (\mathbf{u} \times \mathbf{w}) = \mathbf{v} \cdot \begin{vmatrix} \mathbf{i} & \mathbf{j} & \mathbf{k} \\ 2 & -3 & 1 \\ 1 & 1 & 3 \end{vmatrix} = \mathbf{v} \cdot \left(\begin{vmatrix} -3 & 1 \\ 1 & 3 \end{vmatrix} \mathbf{i} - \begin{vmatrix} 2 & 1 \\ 1 & 3 \end{vmatrix} \mathbf{j} + \begin{vmatrix} 2 & -3 \\ 1 & 1 \end{vmatrix} \mathbf{k} \right)$

$= (-3\mathbf{i} + 3\mathbf{j} + 2\mathbf{k}) \cdot (-10\mathbf{i} - 5\mathbf{j} + 5\mathbf{k}) = -3(-10) + 3(-5) + 2 \cdot 5 = 30 - 15 + 10 = 25$

33. $\mathbf{u} \times (\mathbf{v} \times \mathbf{v}) = \mathbf{u} \times \begin{vmatrix} \mathbf{i} & \mathbf{j} & \mathbf{k} \\ -3 & 3 & 2 \\ -3 & 3 & 2 \end{vmatrix} = \mathbf{u} \times \left(\begin{vmatrix} 3 & 2 \\ 3 & 2 \end{vmatrix} \mathbf{i} - \begin{vmatrix} -3 & 2 \\ -3 & 2 \end{vmatrix} \mathbf{j} + \begin{vmatrix} -3 & 3 \\ -3 & 3 \end{vmatrix} \mathbf{k} \right)$

$= (2\mathbf{i} - 3\mathbf{j} + \mathbf{k}) \times (0\mathbf{i} + 0\mathbf{j} + 0\mathbf{k}) = \begin{vmatrix} \mathbf{i} & \mathbf{j} & \mathbf{k} \\ 2 & -3 & 1 \\ 0 & 0 & 0 \end{vmatrix}$

$= \begin{vmatrix} -3 & 1 \\ 0 & 0 \end{vmatrix} \mathbf{i} - \begin{vmatrix} 2 & 1 \\ 0 & 0 \end{vmatrix} \mathbf{j} + \begin{vmatrix} 2 & -3 \\ 0 & 0 \end{vmatrix} \mathbf{k} = 0\mathbf{i} + 0\mathbf{j} + 0\mathbf{k} = \mathbf{0}$

35. $\mathbf{u} \times \mathbf{v} = \begin{vmatrix} \mathbf{i} & \mathbf{j} & \mathbf{k} \\ 2 & -3 & 1 \\ -3 & 3 & 2 \end{vmatrix} = \begin{vmatrix} -3 & 1 \\ 3 & 2 \end{vmatrix} \mathbf{i} - \begin{vmatrix} 2 & 1 \\ -3 & 2 \end{vmatrix} \mathbf{j} + \begin{vmatrix} 2 & -3 \\ -3 & 3 \end{vmatrix} \mathbf{k} = -9\mathbf{i} - 7\mathbf{j} - 3\mathbf{k}$ is orthogonal

to both $\mathbf{u}$ and $\mathbf{v}$.

37. A vector that is orthogonal to both $\mathbf{u}$ and $\mathbf{i} + \mathbf{j}$ is $\mathbf{u} \times (\mathbf{i} + \mathbf{j})$.

$\mathbf{u} \times (\mathbf{i} + \mathbf{j}) = \begin{vmatrix} \mathbf{i} & \mathbf{j} & \mathbf{k} \\ 2 & -3 & 1 \\ 1 & 1 & 0 \end{vmatrix} = \begin{vmatrix} -3 & 1 \\ 1 & 0 \end{vmatrix} \mathbf{i} - \begin{vmatrix} 2 & 1 \\ 1 & 0 \end{vmatrix} \mathbf{j} + \begin{vmatrix} 2 & -3 \\ 1 & 1 \end{vmatrix} \mathbf{k} = -1\mathbf{i} + 1\mathbf{j} + 5\mathbf{k}$

39. $\mathbf{u} = P_1 P_2 = 1\mathbf{i} + 2\mathbf{j} + 3\mathbf{k}$ $\mathbf{v} = P_1 P_3 = -2\mathbf{i} + 3\mathbf{j} + 0\mathbf{k}$

$\mathbf{u} \times \mathbf{v} = \begin{vmatrix} \mathbf{i} & \mathbf{j} & \mathbf{k} \\ 1 & 2 & 3 \\ -2 & 3 & 0 \end{vmatrix} = \begin{vmatrix} 2 & 3 \\ 3 & 0 \end{vmatrix} \mathbf{i} - \begin{vmatrix} 1 & 3 \\ -2 & 0 \end{vmatrix} \mathbf{j} + \begin{vmatrix} 1 & 2 \\ -2 & 3 \end{vmatrix} \mathbf{k} = -9\mathbf{i} - 6\mathbf{j} + 7\mathbf{k}$

Area $= \|\mathbf{u} \times \mathbf{v}\| = \sqrt{(-9)^2 + (-6)^2 + 7^2} = \sqrt{166} \approx 12.9$

41. $\mathbf{u} = P_1 P_2 = -3\mathbf{i} + 1\mathbf{j} + 4\mathbf{k}$ $\mathbf{v} = P_1 P_3 = -1\mathbf{i} - 4\mathbf{j} + 3\mathbf{k}$

$\mathbf{u} \times \mathbf{v} = \begin{vmatrix} \mathbf{i} & \mathbf{j} & \mathbf{k} \\ -3 & 1 & 4 \\ -1 & -4 & 3 \end{vmatrix} = \begin{vmatrix} 1 & 4 \\ -4 & 3 \end{vmatrix} \mathbf{i} - \begin{vmatrix} -3 & 4 \\ -1 & 3 \end{vmatrix} \mathbf{j} + \begin{vmatrix} -3 & 1 \\ -1 & -4 \end{vmatrix} \mathbf{k} = 19\mathbf{i} + 5\mathbf{j} + 13\mathbf{k}$

Area $= \|\mathbf{u} \times \mathbf{v}\| = \sqrt{19^2 + 5^2 + 13^2} = \sqrt{555} \approx 23.6$

43. $\mathbf{u} = P_1 P_2 = 0\mathbf{i} + 1\mathbf{j} + 1\mathbf{k}$ $\mathbf{v} = P_1 P_3 = -3\mathbf{i} + 2\mathbf{j} - 2\mathbf{k}$

$\mathbf{u} \times \mathbf{v} = \begin{vmatrix} \mathbf{i} & \mathbf{j} & \mathbf{k} \\ 0 & 1 & 1 \\ -3 & 2 & -2 \end{vmatrix} = \begin{vmatrix} 1 & 1 \\ 2 & -2 \end{vmatrix} \mathbf{i} - \begin{vmatrix} 0 & 1 \\ -3 & -2 \end{vmatrix} \mathbf{j} + \begin{vmatrix} 0 & 1 \\ -3 & 2 \end{vmatrix} \mathbf{k} = -4\mathbf{i} - 3\mathbf{j} + 3\mathbf{k}$

Area $= \|\mathbf{u} \times \mathbf{v}\| = \sqrt{(-4)^2 + (-3)^2 + 3^2} = \sqrt{34} \approx 5.8$

45. $\mathbf{u} = P_1P_2 = 3\mathbf{i} + 0\mathbf{j} - 2\mathbf{k}$ $\mathbf{v} = P_1P_3 = 5\mathbf{i} - 7\mathbf{j} + 3\mathbf{k}$

$$\mathbf{u} \times \mathbf{v} = \begin{vmatrix} \mathbf{i} & \mathbf{j} & \mathbf{k} \\ 3 & 0 & -2 \\ 5 & -7 & 3 \end{vmatrix} = \begin{vmatrix} 0 & -2 \\ -7 & 3 \end{vmatrix}\mathbf{i} - \begin{vmatrix} 3 & -2 \\ 5 & 3 \end{vmatrix}\mathbf{j} + \begin{vmatrix} 3 & 0 \\ 5 & -7 \end{vmatrix}\mathbf{k} = -14\mathbf{i} - 19\mathbf{j} - 21\mathbf{k}$$

Area $= \|\mathbf{u} \times \mathbf{v}\| = \sqrt{(-14)^2 + (-19)^2 + (-21)^2} = \sqrt{998} \approx 31.6$

47. $\mathbf{v} \times \mathbf{w} = \begin{vmatrix} \mathbf{i} & \mathbf{j} & \mathbf{k} \\ 1 & 3 & -2 \\ -2 & 1 & 3 \end{vmatrix} = \begin{vmatrix} 3 & -2 \\ 1 & 3 \end{vmatrix}\mathbf{i} - \begin{vmatrix} 1 & -2 \\ -2 & 3 \end{vmatrix}\mathbf{j} + \begin{vmatrix} 1 & 3 \\ -2 & 1 \end{vmatrix}\mathbf{k} = 11\mathbf{i} + 1\mathbf{j} + 7\mathbf{k}$

$|\mathbf{v} \times \mathbf{w}| = \sqrt{11^2 + 1^2 + 7^2} = \sqrt{171}$

$\mathbf{u} = \dfrac{\mathbf{v} \times \mathbf{w}}{\|\mathbf{v} \times \mathbf{w}\|} = \dfrac{11\mathbf{i} + 1\mathbf{j} + 7\mathbf{k}}{\sqrt{171}} = \dfrac{11}{\sqrt{171}}\mathbf{i} + \dfrac{1}{\sqrt{171}}\mathbf{j} + \dfrac{7}{\sqrt{171}}\mathbf{k}$

49. Prove: $\mathbf{u} \times \mathbf{v} = -(\mathbf{v} \times \mathbf{u})$

Let $\mathbf{u} = a_1\mathbf{i} + b_1\mathbf{j} + c_1\mathbf{k}$ and $\mathbf{v} = a_2\mathbf{i} + b_2\mathbf{j} + c_2\mathbf{k}$

$$\mathbf{u} \times \mathbf{v} = \begin{vmatrix} \mathbf{i} & \mathbf{j} & \mathbf{k} \\ a_1 & b_1 & c_1 \\ a_2 & b_2 & c_2 \end{vmatrix} = \begin{vmatrix} b_1 & c_1 \\ b_2 & c_2 \end{vmatrix}\mathbf{i} - \begin{vmatrix} a_1 & c_1 \\ a_2 & c_2 \end{vmatrix}\mathbf{j} + \begin{vmatrix} a_1 & b_1 \\ a_2 & b_2 \end{vmatrix}\mathbf{k}$$

$= (b_1c_2 - b_2c_1)\mathbf{i} - (a_1c_2 - a_2c_1)\mathbf{j} + (a_1b_2 - a_2b_1)\mathbf{k}$

$= -(b_2c_1 - b_1c_2)\mathbf{i} + (a_2c_1 - a_1c_2)\mathbf{j} - (a_2b_1 - a_1b_2)\mathbf{k}$

$= -\big((b_2c_1 - b_1c_2)\mathbf{i} - (a_2c_1 - a_1c_2)\mathbf{j} + (a_2b_1 - a_1b_2)\mathbf{k}\big)$

$= -\left(\begin{vmatrix} b_2 & c_2 \\ b_1 & c_1 \end{vmatrix}\mathbf{i} - \begin{vmatrix} a_2 & c_2 \\ a_1 & c_1 \end{vmatrix}\mathbf{j} + \begin{vmatrix} a_2 & b_2 \\ a_1 & b_1 \end{vmatrix}\mathbf{k}\right) = -\begin{vmatrix} \mathbf{i} & \mathbf{j} & \mathbf{k} \\ a_2 & b_2 & c_2 \\ a_1 & b_1 & c_1 \end{vmatrix} = -(\mathbf{v} \times \mathbf{u})$

51. Prove: $\|\mathbf{u} \times \mathbf{v}\|^2 = \|\mathbf{u}\|^2 \|\mathbf{v}\|^2 - (\mathbf{u} \cdot \mathbf{v})^2$

Let $\mathbf{u} = a_1\mathbf{i} + b_1\mathbf{j} + c_1\mathbf{k}$ and $\mathbf{v} = a_2\mathbf{i} + b_2\mathbf{j} + c_2\mathbf{k}$

$$\mathbf{u} \times \mathbf{v} = \begin{vmatrix} \mathbf{i} & \mathbf{j} & \mathbf{k} \\ a_1 & b_1 & c_1 \\ a_2 & b_2 & c_2 \end{vmatrix} = \begin{vmatrix} b_1 & c_1 \\ b_2 & c_2 \end{vmatrix}\mathbf{i} - \begin{vmatrix} a_1 & c_1 \\ a_2 & c_2 \end{vmatrix}\mathbf{j} + \begin{vmatrix} a_1 & b_1 \\ a_2 & b_2 \end{vmatrix}\mathbf{k}$$

$= (b_1c_2 - b_2c_1)\mathbf{i} - (a_1c_2 - a_2c_1)\mathbf{j} + (a_1b_2 - a_2b_1)\mathbf{k}$

$\|\mathbf{u} \times \mathbf{v}\|^2 = (b_1c_2 - b_2c_1)^2 + (a_1c_2 - a_2c_1)^2 + (a_1b_2 - a_2b_1)^2$

$= b_1^2c_2^2 - 2b_1b_2c_1c_2 + b_2^2c_1^2 + a_1^2c_2^2 - 2a_1a_2c_1c_2 + a_2^2c_1^2$
$\qquad\qquad + a_1^2b_2^2 - 2a_1a_2b_1b_2 + a_2^2b_1^2$

$= a_1^2b_2^2 + a_1^2c_2^2 + a_2^2b_1^2 + a_2^2c_1^2 + b_1^2c_2^2 + b_2^2c_1^2 - 2a_1a_2b_1b_2$
$\qquad\qquad - 2a_1a_2c_1c_2 - 2b_1b_2c_1c_2$

$$\left| \mathbf{u} \right|^2 = a_1^2 + b_1^2 + c_1^2$$

$$\left\| \mathbf{v} \right\|^2 = a_2^2 + b_2^2 + c_2^2$$

$$(\mathbf{u} \cdot \mathbf{v})^2 = \left(a_1 a_2 + b_1 b_2 + c_1 c_2 \right)^2$$

$$\left| \mathbf{u} \right|^2 \left\| \mathbf{v} \right\|^2 - (\mathbf{u} \cdot \mathbf{v})^2 = \left(a_1^2 + b_1^2 + c_1^2 \right)\left(a_2^2 + b_2^2 + c_2^2 \right) - \left(a_1 a_2 + b_1 b_2 + c_1 c_2 \right)^2$$

$$= a_1^2 a_2^2 + a_1^2 b_2^2 + a_1^2 c_2^2 + b_1^2 a_2^2 + b_1^2 b_2^2 + b_1^2 c_2^2 + c_1^2 a_2^2 + c_1^2 b_2^2 + c_1^2 c_2^2$$

$$- \left(\begin{array}{l} a_1^2 a_2^2 + a_1 a_2 b_1 b_2 + a_1 a_2 c_1 c_2 + a_1 a_2 b_1 b_2 + b_1^2 b_2^2 + b_1 b_2 c_1 c_2 \\ \qquad\qquad + a_1 a_2 c_1 c_2 + b_1 b_2 c_1 c_2 + c_1^2 c_2^2 \end{array} \right)$$

$$= a_1^2 b_2^2 + a_1^2 c_2^2 + a_2^2 b_1^2 + a_2^2 c_1^2 + b_1^2 c_2^2 + b_2^2 c_1^2 - 2 a_1 a_2 b_1 b_2$$

$$- 2 a_1 a_2 c_1 c_2 - 2 b_1 b_2 c_1 c_2$$

53. If $\mathbf{u}$ and $\mathbf{v}$ are orthogonal, then $\mathbf{u} \cdot \mathbf{v} = 0$. From problem 51, then:

$$\left| \mathbf{u} \times \mathbf{v} \right|^2 = \left\| \mathbf{u} \right\|^2 \left| \mathbf{v} \right|^2 - (\mathbf{u} \cdot \mathbf{v})^2 = \left\| \mathbf{u} \right\|^2 \left| \mathbf{v} \right|^2 - (0)^2 = \left| \mathbf{u} \right|^2 \left\| \mathbf{v} \right\|^2$$

$$\left| \mathbf{u} \times \mathbf{v} \right| = \left\| \mathbf{u} \right\| \left\| \mathbf{v} \right\|$$

9   Chapter Review

1. $\left(3, \dfrac{\pi}{6} \right)$

$$x = 3\cos\frac{\pi}{6} = \frac{3\sqrt{3}}{2}$$

$$y = 3\sin\frac{\pi}{6} = \frac{3}{2}$$

$$\left(\frac{3\sqrt{3}}{2}, \frac{3}{2} \right)$$

3. $\left(-2, \dfrac{4\pi}{3} \right)$

$$x = -2\cos\frac{4\pi}{3} = 1$$

$$y = -2\sin\frac{4\pi}{3} = \sqrt{3}$$

$$\left(1, \sqrt{3} \right)$$

5. $\left(-3, -\dfrac{\pi}{2} \right)$

$$x = -3\cos\left(-\frac{\pi}{2} \right) = 0$$

$$y = -3\sin\left(-\frac{\pi}{2} \right) = 3$$

$$(0, 3)$$

7. The point $(-3, 3)$ lies in quadrant II.

$$r = \sqrt{x^2 + y^2} = \sqrt{(-3)^2 + 3^2} = 3\sqrt{2} \qquad \theta = \tan^{-1}\frac{y}{x} = \tan^{-1}\frac{3}{-3} = \tan^{-1}(-1) = -\frac{\pi}{4}$$

Polar coordinates of the point $(-3, 3)$ are $\left(-3\sqrt{2}, -\dfrac{\pi}{4} \right)$ or $\left(3\sqrt{2}, \dfrac{3\pi}{4} \right)$.

9.   The point $(0, -2)$ lies on the negative y-axis.

$$r = \sqrt{x^2 + y^2} = \sqrt{0^2 + (-2)^2} = 2 \qquad \theta = \tan^{-1}\frac{y}{x} = \tan^{-1}\frac{-2}{0} = -\frac{\pi}{2}$$

Polar coordinates of the point $(0, -2)$ are $\left(2, -\frac{\pi}{2}\right)$ or $\left(-2, \frac{\pi}{2}\right)$.

11. The point $(3, 4)$ lies in quadrant I.

$$r = \sqrt{x^2 + y^2} = \sqrt{3^2 + 4^2} = 5 \qquad \theta = \tan^{-1}\frac{y}{x} = \tan^{-1}\frac{4}{3} \approx 0.93$$

Polar coordinates of the point $(3, 4)$ are $(5, 0.93)$ or $(-5, 4.07)$.

13.
$$3x^2 + 3y^2 = 6y$$
$$x^2 + y^2 = 2y$$
$$r^2 = 2r\sin\theta$$
$$r^2 - 2r\sin\theta = 0$$

15.
$$2x^2 - y^2 = \frac{y}{x}$$
$$2x^2 + 2y^2 - 3y^2 = \frac{y}{x}$$
$$2\left(x^2 + y^2\right) - 3y^2 = \frac{y}{x}$$
$$2r^2 - 3(r\sin\theta)^2 = \tan\theta$$
$$2r^2 - 3r^2\sin^2\theta - \tan\theta = 0$$
$$r^2(2 - 3\sin\theta) - \tan\theta = 0$$

17.
$$x\left(x^2 + y^2\right) = 4$$
$$r\cos\theta\left(r^2\right) = 4$$
$$r^3\cos\theta = 4$$

19.
$$r = 2\sin\theta$$
$$r^2 = 2r\sin\theta$$
$$x^2 + y^2 = 2y$$
$$x^2 + y^2 - 2y = 0$$

21.
$$r = 5$$
$$r^2 = 25$$
$$x^2 + y^2 = 25$$

23.
$$r\cos\theta + 3r\sin\theta = 6$$
$$x + 3y = 6$$

25. $r = 4\cos\theta$ The graph will be a circle. Check for symmetry:

Polar axis: Replace θ by $-\theta$. The result is $r = 4\cos(-\theta) = 4\cos\theta$.
The graph is symmetric with respect to the polar axis.

The line $\theta = \frac{\pi}{2}$: Replace θ by $\pi - \theta$.

$$r = 4\cos(\pi - \theta) = 4(\cos\pi\cos\theta + \sin\pi\sin\theta)$$
$$= 4(-\cos\theta + 0) = -4\cos\theta$$

The test fails.

The pole: Replace r by $-r$. $-r = 4\cos\theta$. The test fails.

Due to symmetry to the polar axis, assign values to θ from 0 to π.

θ	0	$\frac{\pi}{6}$	$\frac{\pi}{3}$	$\frac{\pi}{2}$	$\frac{2\pi}{3}$	$\frac{5\pi}{6}$	π
$r = 4\cos\theta$	4	$2\sqrt{3} \approx 3.5$	2	0	-2	$-2\sqrt{3} \approx -3.5$	-4

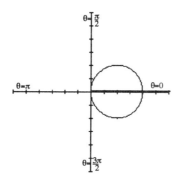

27. $r = 3 - 3\sin\theta$ The graph will be a cardioid. Check for symmetry:

Polar axis: Replace θ by $-\theta$. The result is $r = 3 - 3\sin(-\theta) = 3 + 3\sin\theta$.
The test fails.

The line $\theta = \dfrac{\pi}{2}$: Replace θ by $\pi - \theta$.

$$r = 3 - 3\sin(\pi - \theta) = 3 - 3(\sin\pi\cos\theta - \cos\pi\sin\theta)$$
$$= 3 - 3(0 + \sin\theta) = 3 - 3\sin\theta$$

The graph is symmetric with respect to the line $\theta = \dfrac{\pi}{2}$.

The pole: Replace r by $-r$. $-r = 3 - 3\sin\theta$. The test fails.

Due to symmetry to the line $\theta = \dfrac{\pi}{2}$, assign values to θ from $-\dfrac{\pi}{2}$ to $\dfrac{\pi}{2}$.

θ	$-\dfrac{\pi}{2}$	$-\dfrac{\pi}{3}$	$-\dfrac{\pi}{6}$	0	$\dfrac{\pi}{6}$	$\dfrac{\pi}{3}$	$\dfrac{\pi}{2}$
$r = 3 - 3\sin\theta$	6	$3 + \dfrac{3\sqrt{3}}{2} \approx 5.6$	$\dfrac{9}{2}$	3	$\dfrac{3}{2}$	$3 - \dfrac{3\sqrt{3}}{2} \approx 0.4$	0

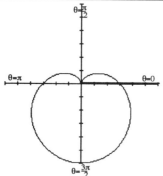

29. $r = 4 - \cos\theta$ The graph will be a limacon without an inner loop.
Check for symmetry:

Polar axis: Replace θ by $-\theta$. The result is $r = 4 - \cos(-\theta) = 4 - \cos\theta$.
The graph is symmetric with respect to the polar axis.

The line $\theta = \dfrac{\pi}{2}$: Replace θ by $\pi - \theta$.

$$r = 4 - \cos(\pi - \theta) = 4 - (\cos\pi\cos\theta + \sin\pi\sin\theta)$$
$$= 4 - (-\cos\theta + 0) = 4 + \cos\theta$$

The test fails.

The pole: Replace r by $-r$. $-r = 4 - \cos\theta$. The test fails.
Due to symmetry to the polar axis, assign values to θ from 0 to π.

θ	0	$\dfrac{\pi}{6}$	$\dfrac{\pi}{3}$	$\dfrac{\pi}{2}$	$\dfrac{2\pi}{3}$	$\dfrac{5\pi}{6}$	π
$r = 4 - \cos\theta$	3	$4 - \dfrac{\sqrt{3}}{2} \approx 3.1$	$\dfrac{7}{2}$	4	$\dfrac{9}{2}$	$4 + \dfrac{\sqrt{3}}{2} \approx 4.9$	5

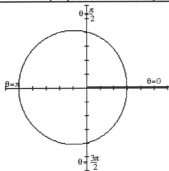

31. $r = \sqrt{x^2 + y^2} = \sqrt{(-1)^2 + (-1)^2} = \sqrt{2}$

   $\tan\theta = \dfrac{y}{x} = \dfrac{-1}{-1} = 1 \quad \rightarrow \quad \theta = 225°$

   The polar form of $z = -1 - i$ is
   $z = r(\cos\theta + i\sin\theta) = \sqrt{2}(\cos 225° + i\sin 225°)$

33. $r = \sqrt{x^2 + y^2} = \sqrt{4^2 + (-3)^2} = \sqrt{25} = 5$

   $\tan\theta = \dfrac{y}{x} = \dfrac{-3}{4} \quad \rightarrow \quad \theta = 323.1°$

   The polar form of $z = 4 - 3i$ is
   $z = r(\cos\theta + i\sin\theta) = 5(\cos 323.1° + i\sin 323.1°)$

35. $2(\cos 150° + i\sin 150°) = 2\left(-\dfrac{\sqrt{3}}{2} + \dfrac{1}{2}i\right) = -\sqrt{3} + i$

37. $3\left(\cos\dfrac{2\pi}{3} + i\sin\dfrac{2\pi}{3}\right) = 3\left(-\dfrac{1}{2} + \dfrac{\sqrt{3}}{2}i\right) = -\dfrac{3}{2} + \dfrac{3\sqrt{3}}{2}i$

39. $0.1(\cos 350° + i\sin 350°) = 0.1(0.9848 - 0.1736i) = 0.0985 - 0.0174i$

41. $zw = (\cos 80° + i\sin 80°) \cdot (\cos 50° + i\sin 50°)$

   $= 1 \cdot 1(\cos(80° + 50°) + i\sin(80° + 50°)) = \cos 130° + i\sin 130°$

   $\dfrac{z}{w} = \dfrac{(\cos 80° + i\sin 80°)}{(\cos 50° + i\sin 50°)} = \dfrac{1}{1}(\cos(80° - 50°) + i\sin(80° - 50°))$

   $= \cos 30° + i\sin 30°$

43. $zw = 3\left(\cos\dfrac{9\pi}{5} + i\sin\dfrac{9\pi}{5}\right) \cdot 2\left(\cos\dfrac{\pi}{5} + i\sin\dfrac{\pi}{5}\right) = 3 \cdot 2\left(\cos\left(\dfrac{9\pi}{5} + \dfrac{\pi}{5}\right) + i\sin\left(\dfrac{9\pi}{5} + \dfrac{\pi}{5}\right)\right)$

$= 6(\cos 2\pi + i\sin 2\pi) = 6(\cos 0 + i\sin 0)$

$\dfrac{z}{w} = \dfrac{3\left(\cos\dfrac{9\pi}{5} + i\sin\dfrac{9\pi}{5}\right)}{2\left(\cos\dfrac{\pi}{5} + i\sin\dfrac{\pi}{5}\right)} = \dfrac{3}{2}\left(\cos\left(\dfrac{9\pi}{5} - \dfrac{\pi}{5}\right) + i\sin\left(\dfrac{9\pi}{5} - \dfrac{\pi}{5}\right)\right) = \dfrac{3}{2}\left(\cos\dfrac{8\pi}{5} + i\sin\dfrac{8\pi}{5}\right)$

45. $zw = 5(\cos 10° + i\sin 10°) \cdot (\cos 355° + i\sin 355°)$

$= 5 \cdot 1(\cos(10° + 355°) + i\sin(10° + 355°)) = 5(\cos 365° + i\sin 365°)$

$= 5(\cos 5° + i\sin 5°)$

$\dfrac{z}{w} = \dfrac{5(\cos 10° + i\sin 10°)}{(\cos 355° + i\sin 355°)} = \dfrac{5}{1}(\cos(10° - 355°) + i\sin(10° - 355°))$

$= 5(\cos(-345°) + i\sin(-345°)) = 5(\cos 15° + i\sin 15°)$

47. $\left[3(\cos 20° + i\sin 20°)\right]^3 = 3^3(\cos(3 \cdot 20°) + i\sin(3 \cdot 20°)) = 27(\cos 60° + i\sin 60°)$

$= 27\left(\dfrac{1}{2} + \dfrac{\sqrt{3}}{2}i\right) = \dfrac{27}{2} + \dfrac{27\sqrt{3}}{2}i$

49. $\left[\sqrt{2}\left(\cos\dfrac{5\pi}{8} + i\sin\dfrac{5\pi}{8}\right)\right]^4 = \sqrt{2}^4\left(\cos\left(4 \cdot \dfrac{5\pi}{8}\right) + i\sin\left(4 \cdot \dfrac{5\pi}{8}\right)\right)$

$= 4\left(\cos\dfrac{5\pi}{2} + i\sin\dfrac{5\pi}{2}\right) = 4(0 + 1i) = 4i$

51. $1 - \sqrt{3}\,i$ $r = \sqrt{1^2 + \left(-\sqrt{3}\right)^2} = 2$ $\tan\theta = \dfrac{-\sqrt{3}}{1} = -\sqrt{3}$ $\theta = 300°$

$1 - \sqrt{3}\,i = 2(\cos 300° + i\sin 300°)$

$\left(1 - \sqrt{3}\,i\right)^6 = \left[2(\cos 300° + i\sin 300°)\right]^6 = 2^6(\cos(6 \cdot 300°) + i\sin(6 \cdot 300°))$

$= 64(\cos 1800° + i\sin 1800°) = 64(\cos 0° + i\sin 0°)$

$= 64 + 0i = 64$

53. $3 + 4i$ $r = \sqrt{3^2 + 4^2} = 5$ $\tan\theta = \dfrac{4}{3}$ $\theta = 53.1°$

$3 + 4i = 5(\cos 53.1° + i\sin 53.1°)$

$(3 + 4i)^4 = \left[5^4(\cos(4 \cdot 53.1°) + i\sin(4 \cdot 53.1°))\right]^4 = 625(\cos(212.4°) + i\sin(212.4°))$

$= 625(-0.8443 + i(-0.5358)) = -527.7 - 334.9i$

55. $27 + 0i$ $r = \sqrt{27^2 + 0^2} = 27$ $\tan \theta = \dfrac{0}{27} = 0$ $\theta = 0°$

$27 + 0i = 27(\cos 0° + i \sin 0°)$

The three complex cube roots of $27 = 27(\cos 0° + i \sin 0°)$ are:

$$z_k = \sqrt[3]{27}\left[\cos\left(\frac{0°}{3} + \frac{360°k}{3}\right) + i\sin\left(\frac{0°}{3} + \frac{360°k}{3}\right)\right]$$

$$= 3\left[\cos(120°k) + i\sin(120°k)\right]$$

$$z_0 = 3\left[\cos(120° \cdot 0) + i\sin(120° \cdot 0)\right] = 3(\cos 0° + i\sin 0°) = 3$$

$$z_1 = 3\left[\cos(120° \cdot 1) + i\sin(120° \cdot 1)\right] = 3(\cos 120° + i\sin 120°) = -\frac{3}{2} + \frac{3\sqrt{3}}{2}i$$

$$z_2 = 3\left[\cos(120° \cdot 2) + i\sin(120° \cdot 2)\right] = 3(\cos 240° + i\sin 240°) = -\frac{3}{2} - \frac{3\sqrt{3}}{2}i$$

57. $P = (1, -2), Q = (3, -6)$ $\mathbf{v} = (3 - 1)\mathbf{i} + (-6 - (-2))\mathbf{j} = 2\mathbf{i} - 4\mathbf{j}$

$\|\mathbf{v}\| = \sqrt{2^2 + (-4)^2} = \sqrt{20} = 2\sqrt{5}$

59. $P = (0, -2), Q = (-1, 1)$ $\mathbf{v} = (-1 - 0)\mathbf{i} + (1 - (-2))\mathbf{j} = -1\mathbf{i} + 3\mathbf{j}$

$\|\mathbf{v}\| = \sqrt{(-1)^2 + 3^2} = \sqrt{10}$

61. $\mathbf{v} = (3 - 6)\mathbf{i} + (0 - 2)\mathbf{j} + (2 - 1)\mathbf{k} = -3\mathbf{i} - 2\mathbf{j} + \mathbf{k}$

$\|\mathbf{v}\| = \sqrt{(-3)^2 + (-2)^2 + 1^2} = \sqrt{14}$

63. $\mathbf{v} = (2 - (-1))\mathbf{i} + (0 - 0)\mathbf{j} + (0 - 1)\mathbf{k} = 3\mathbf{i} - \mathbf{k}$

$\|\mathbf{v}\| = \sqrt{3^2 + 0^2 + (-1)^2} = \sqrt{10}$

65. $\mathbf{v} = -2\mathbf{i} + \mathbf{j}, \quad \mathbf{w} = 4\mathbf{i} - 3\mathbf{j}$

$4\mathbf{v} - 3\mathbf{w} = 4(-2\mathbf{i} + \mathbf{j}) - 3(4\mathbf{i} - 3\mathbf{j}) = -8\mathbf{i} + 4\mathbf{j} - 12\mathbf{i} + 9\mathbf{j} = -20\mathbf{i} + 13\mathbf{j}$

67. $\mathbf{v} = -2\mathbf{i} + \mathbf{j}$

$\|\mathbf{v}\| = |-2\mathbf{i} + \mathbf{j}| = \sqrt{(-2)^2 + 1^2} = \sqrt{5}$

69. $\mathbf{v} = -2\mathbf{i} + \mathbf{j}, \quad \mathbf{w} = 4\mathbf{i} - 3\mathbf{j}$

$\|\mathbf{v}\| + \|\mathbf{w}\| = \|-2\mathbf{i} + \mathbf{j}\| + \|4\mathbf{i} - 3\mathbf{j}\| = \sqrt{(-2)^2 + 1^2} + \sqrt{4^2 + (-3)^2} = \sqrt{5} + 5$

71. $\mathbf{u} = \dfrac{\mathbf{v}}{\|\mathbf{v}\|} = \dfrac{-2\mathbf{i} + \mathbf{j}}{\|-2\mathbf{i} + \mathbf{j}\|} = \dfrac{-2\mathbf{i} + \mathbf{j}}{\sqrt{(-2)^2 + 1^2}} = \dfrac{-2\mathbf{i} + \mathbf{j}}{\sqrt{5}} = -\dfrac{2\sqrt{5}}{5}\mathbf{i} + \dfrac{\sqrt{5}}{5}\mathbf{j}$

73. $4\mathbf{v} - 3\mathbf{w} = 4(3\mathbf{i} + \mathbf{j} - 2\mathbf{k}) - 3(-3\mathbf{i} + 2\mathbf{j} - \mathbf{k}) = 12\mathbf{i} + 4\mathbf{j} - 8\mathbf{k} + 9\mathbf{i} - 6\mathbf{j} + 3\mathbf{k}$

$= 21\mathbf{i} - 2\mathbf{j} - 5\mathbf{k}$

75. $\left|\,\mathbf{v} - \mathbf{w}\,\right\| = \left\|(3\mathbf{i} + \mathbf{j} - 2\mathbf{k}) - (-3\mathbf{i} + 2\mathbf{j} - \mathbf{k})\right| = \left\|3\mathbf{i} + \mathbf{j} - 2\mathbf{k} + 3\mathbf{i} - 2\mathbf{j} + \mathbf{k}\right\|$

$= \left\|6\mathbf{i} - \mathbf{j} - \mathbf{k}\right\| = \sqrt{6^2 + (-1)^2 + (-1)^2} = \sqrt{36 + 1 + 1} = \sqrt{38}$

77. $\left|\,\mathbf{v}\,\right\| - \left|\,\mathbf{w}\,\right\| = \left|3\mathbf{i} + \mathbf{j} - 2\mathbf{k}\right\| - \left\|-3\mathbf{i} + 2\mathbf{j} - \mathbf{k}\right\|$

$= \sqrt{3^2 + 1^2 + (-2)^2} - \sqrt{(-3)^2 + 2^2 + (-1)^2} = \sqrt{14} - \sqrt{14} = 0$

79. $\mathbf{v} \times \mathbf{w} = \begin{vmatrix} \mathbf{i} & \mathbf{j} & \mathbf{k} \\ 3 & 1 & -2 \\ -3 & 2 & -1 \end{vmatrix} = \begin{vmatrix} 1 & -2 \\ 2 & -1 \end{vmatrix}\mathbf{i} - \begin{vmatrix} 3 & -2 \\ -3 & -1 \end{vmatrix}\mathbf{j} + \begin{vmatrix} 3 & 1 \\ -3 & 2 \end{vmatrix}\mathbf{k} = 3\mathbf{i} + 9\mathbf{j} + 9\mathbf{k}$

81. Same direction:

$$\frac{\mathbf{v}}{\left|\,\mathbf{v}\,\right\|} = \frac{3\mathbf{i} + \mathbf{j} - 2\mathbf{k}}{\sqrt{3^2 + 1^2 + (-2)^2}} = \frac{3\mathbf{i} + \mathbf{j} - 2\mathbf{k}}{\sqrt{14}} = \frac{3\sqrt{14}}{14}\mathbf{i} + \frac{\sqrt{14}}{14}\mathbf{j} - \frac{\sqrt{14}}{7}\mathbf{k}$$

Opposite direction:

$$\frac{-\mathbf{v}}{\left|\,\mathbf{v}\,\right\|} = -\frac{3\sqrt{14}}{14}\mathbf{i} - \frac{\sqrt{14}}{14}\mathbf{j} + \frac{\sqrt{14}}{7}\mathbf{k}$$

83. $\mathbf{v} = -2\mathbf{i} + \mathbf{j}, \quad \mathbf{w} = 4\mathbf{i} - 3\mathbf{j}$

$\mathbf{v} \cdot \mathbf{w} = -2(4) + 1(-3) = -8 - 3 = -11$

$\cos\theta = \dfrac{\mathbf{v} \cdot \mathbf{w}}{\left\|\mathbf{v}\right\|\left\|\mathbf{w}\right\|} = \dfrac{-11}{\sqrt{(-2)^2 + 1^2}\sqrt{4^2 + (-3)^2}} = \dfrac{-11}{\sqrt{5} \cdot 5} = \dfrac{-11}{5\sqrt{5}} = -0.9839$

$\theta = 169.7°$

85. $\mathbf{v} = \mathbf{i} - 3\mathbf{j}, \quad \mathbf{w} = -\mathbf{i} + \mathbf{j}$

$\mathbf{v} \cdot \mathbf{w} = 1(-1) + (-3)(1) = -1 - 3 = -4$

$\cos\theta = \dfrac{\mathbf{v} \cdot \mathbf{w}}{\left\|\mathbf{v}\right\|\left\|\mathbf{w}\right\|} = \dfrac{-4}{\sqrt{1^2 + (-3)^2}\sqrt{(-1)^2 + 1^2}} = \dfrac{-4}{\sqrt{10}\sqrt{2}} = \dfrac{-2}{\sqrt{5}} = -0.8944$

$\theta = 153.4°$

87. $\mathbf{v} \cdot \mathbf{w} = (\mathbf{i} + \mathbf{j} + \mathbf{k}) \cdot (\mathbf{i} - \mathbf{j} + \mathbf{k}) = 1 \cdot 1 + 1(-1) + 1 \cdot 1 = 1 - 1 + 1 = 1$

$\cos\theta = \dfrac{\mathbf{v} \cdot \mathbf{w}}{\left\|\mathbf{v}\right\|\left\|\mathbf{w}\right\|} = \dfrac{1}{\sqrt{1^2 + 1^2 + 1^2}\sqrt{1^2 + (-1)^2 + 1^2}} = \dfrac{1}{\sqrt{3}\sqrt{3}} = \dfrac{1}{3}$

$\theta \approx 70.5°$

89. $\mathbf{v} \cdot \mathbf{w} = (4\mathbf{i} - \mathbf{j} + 2\mathbf{k}) \cdot (\mathbf{i} - 2\mathbf{j} - 3\mathbf{k}) = 4 \cdot 1 + (-1)(-2) + 2(-3) = 4 + 2 - 6 = 0$

$\cos\theta = \dfrac{\mathbf{v} \cdot \mathbf{w}}{\left\|\mathbf{v}\right\|\left\|\mathbf{w}\right\|} = \dfrac{0}{\sqrt{4^2 + (-1)^2 + 2^2}\sqrt{1^2 + (-2)^2 + (-3)^2}} = 0$

$\theta = 90°$

91. $\text{proj}_{\mathbf{w}}\mathbf{v} = \dfrac{\mathbf{v} \cdot \mathbf{w}}{\left\|\mathbf{w}\right\|^2}\mathbf{w} = \dfrac{(2\mathbf{i} + 3\mathbf{j}) \cdot (3\mathbf{i} + \mathbf{j})}{\sqrt{3^2 + 1^2}^2}(3\mathbf{i} + \mathbf{j}) = \dfrac{2 \cdot 3 + 3 \cdot 1}{10}(3\mathbf{i} + \mathbf{j}) = \dfrac{27}{10}\mathbf{i} + \dfrac{9}{10}\mathbf{j}$

93. $\cos\alpha = \dfrac{a}{\|\mathbf{v}\|} = \dfrac{3}{\sqrt{3^2 + (-4)^2 + 2^2}} = \dfrac{3}{\sqrt{29}} \quad\rightarrow\quad \alpha \approx 56.1°$

$\cos\beta = \dfrac{b}{\|\mathbf{v}\|} = \dfrac{-4}{\sqrt{3^2 + (-4)^2 + 2^2}} = \dfrac{-4}{\sqrt{29}} \quad\rightarrow\quad \beta \approx 138.0°$

$\cos\gamma = \dfrac{c}{\|\mathbf{v}\|} = \dfrac{2}{\sqrt{3^2 + (-4)^2 + 2^2}} = \dfrac{2}{\sqrt{29}} \quad\rightarrow\quad \gamma \approx 68.2°$

95. $\mathbf{u} = P_1P_2 = 1\mathbf{i} + 2\mathbf{j} + 3\mathbf{k} \qquad \mathbf{v} = P_1P_3 = 5\mathbf{i} + 4\mathbf{j} + 1\mathbf{k}$

$\mathbf{u} \times \mathbf{v} = \begin{vmatrix} \mathbf{i} & \mathbf{j} & \mathbf{k} \\ 1 & 2 & 3 \\ 5 & 4 & 1 \end{vmatrix} = \begin{vmatrix} 2 & 3 \\ 4 & 1 \end{vmatrix}\mathbf{i} - \begin{vmatrix} 1 & 3 \\ 5 & 1 \end{vmatrix}\mathbf{j} + \begin{vmatrix} 1 & 2 \\ 5 & 4 \end{vmatrix}\mathbf{k} = -10\mathbf{i} + 14\mathbf{j} - 6\mathbf{k}$

Area $= \|\mathbf{u} \times \mathbf{v}\| = \sqrt{(-10)^2 + 14^2 + (-6)^2} = \sqrt{332} \approx 18.2$

97. Let the positive x-axis point downstream, so that the velocity of the current is $\mathbf{v}_c = 2\mathbf{i}$.

Let $\mathbf{v}_w$ = the velocity of the swimmer in the water.

Let $\mathbf{v}_g$ = the velocity of the swimmer relative to the land.

Then $\mathbf{v}_g = \mathbf{v}_w + \mathbf{v}_c$

The speed of the swimmer is $\|\mathbf{v}_w\| = 5$; its direction is directly across the river, so

Let $\mathbf{v}_w = 5\mathbf{j}$.

$\mathbf{v}_g = \mathbf{v}_w + \mathbf{v}_c = 5\mathbf{j} + 2\mathbf{i} = 2\mathbf{i} + 5\mathbf{j}$

Let $\|\mathbf{v}_g\| = \sqrt{2^2 + 5^2} = \sqrt{29} \approx 5.4$ miles per hour .

Since the river is 1 mile wide, it takes the swimmer 0.2 hours to cross the river. The swimmer will end up $(0.2)(2) = 0.4$ miles downstream.

99. Let $\mathbf{F}_1$ be the tension on the left cable and $\mathbf{F}_2$ be the tension on the right cable.

Let $\mathbf{F}_3$ represent the force of the weight of the box.

$\mathbf{F}_1 = \|\mathbf{F}_1\|(\cos 140°\mathbf{i} + \sin 140°\mathbf{j}) = \|\mathbf{F}_1\|(-0.7660\mathbf{i} + 0.6428\mathbf{j})$

$\mathbf{F}_2 = \|\mathbf{F}_2\|(\cos 30°\mathbf{i} + \sin 30°\mathbf{j}) = \|\mathbf{F}_2\|(0.8660\mathbf{i} + 0.5000\mathbf{j})$

$\mathbf{F}_3 = -2000\mathbf{j}$

For equilibrium, the sum of the force vectors must be zero.

$\mathbf{F}_1 + \mathbf{F}_2 + \mathbf{F}_3 = -0.7660\|\mathbf{F}_1\|\mathbf{i} + 0.6428\|\mathbf{F}_1\|\mathbf{j} + 0.8660\|\mathbf{F}_2\|\mathbf{i} + 0.5000\|\mathbf{F}_2\|\mathbf{j} - 2000\mathbf{j}$

$= \left(-0.7660\|\mathbf{F}_1\| + 0.8660\|\mathbf{F}_2\|\right)\mathbf{i} + \left(0.6428\|\mathbf{F}_1\| + 0.5000\|\mathbf{F}_2\| - 2000\right)\mathbf{j}$

$= 0$

Set the **i** and **j** components equal to zero and solve:

$$\begin{cases} -0.7660\|\mathbf{F}_1\| + 0.8660\|\mathbf{F}_2\| = 0 \quad \rightarrow \quad \|\mathbf{F}_2\| = \dfrac{0.7660}{0.8660}\|\mathbf{F}_1\| = 0.8845\|\mathbf{F}_1\| \\ 0.6428\|\mathbf{F}_1\| + 0.5000\|\mathbf{F}_2\| - 2000 = 0 \end{cases}$$

$$0.6428\|\mathbf{F}_1\| + 0.5000\big(0.8845\|\mathbf{F}_1\|\big) - 2000 = 0$$

$$1.0851\|\mathbf{F}_1\| = 2000$$

$$\|\mathbf{F}_1\| \approx 1843 \text{ pounds}$$

$$\|\mathbf{F}_2\| = 0.8845(1843) = 1630 \text{ pounds}$$

The tension in the left cable is about 1843 pounds and the tension in the right cable is about 1630 pounds.

Analytic Geometry

10.2 The Parabola

1. B   3. E   5. H   7. C

9. F   11. G   13. D   15. B

17. The focus is (4, 0) and the vertex is (0, 0). Both lie on the horizontal line $y = 0$. $a = 4$ and since (4, 0) is to the right of (0, 0), the parabola opens to the right. The equation of the parabola is:

$$y^2 = 4ax$$
$$y^2 = 4 \cdot 4 \cdot x$$
$$y^2 = 16x$$

Letting $x = 4$, we find $y^2 = 64$ or $y = \pm 8$.
The points (4, 8) and (4, –8) define the latus rectum.

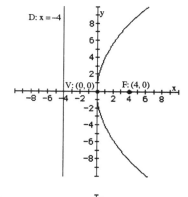

19. The focus is (0, –3) and the vertex is (0, 0). Both lie on the vertical line $x = 0$. $a = 3$ and since (0, –3) is below (0, 0), the parabola opens down. The equation of the parabola is:

$$x^2 = -4ay$$
$$x^2 = -4 \cdot 3 \cdot y$$
$$x^2 = -12y$$

Letting $y = -3$, we find $x^2 = 36$ or $x = \pm 6$.
The points (6, 3) and (6, –3) define the latus rectum.

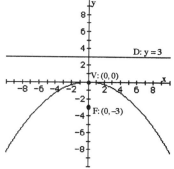

21. The focus is (–2, 0) and the directrix is $x = 2$. The vertex is (0, 0). $a = 2$ and since (–2, 0) is to the left of (0, 0), the parabola opens to the left. The equation of the parabola is:

$$y^2 = -4ax$$
$$y^2 = -4 \cdot 2 \cdot x$$
$$y^2 = -8x$$

Letting $x = -2$, we find $y^2 = 16$ or $y = \pm 4$. The points (–2, 4) and (–2, –4) define the latus rectum.

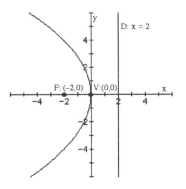

23. The directrix is $y = -\frac{1}{2}$ and the vertex is (0, 0). The focus is $\left(0, \frac{1}{2}\right)$. $a = \frac{1}{2}$ and since $\left(0, \frac{1}{2}\right)$ is above (0, 0), the parabola opens up. The equation of the parabola is:

$$x^2 = 4ay$$
$$x^2 = 4 \cdot \frac{1}{2} \cdot y$$
$$x^2 = 2y$$

Letting $y = \frac{1}{2}$, we find $x^2 = 1$ or $x = \pm 1$.

The points $\left(1, \frac{1}{2}\right)$ and $\left(-1, \frac{1}{2}\right)$ define the latus rectum.

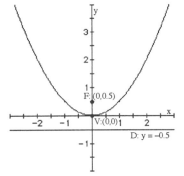

25. The focus is (2, –5) and the vertex is (2, –3). Both lie on the vertical line $x = 2$. $a = 2$ and since (2, –5) is below (2, –3), the parabola opens down. The equation of the parabola is:

$$(x - h)^2 = -4a(y - k)$$
$$(x - 2)^2 = -4 \cdot 2 \cdot (y - (-3))$$
$$(x - 2)^2 = -8(y + 3)$$

Letting $y = -5$, we find $(x - 2)^2 = 16$ or $x - 2 = \pm 4$. So, $x = 6$ or $x = -2$. The points (6, –5) and (–2, –5) define the latus rectum.

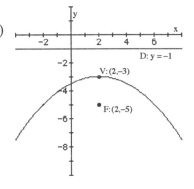

27. Vertex: (0,0). Since the axis of symmetry is vertical, the parabola opens up or down. Since (2, 3) is above (0, 0), the parabola opens up. The equation has the form $x^2 = 4ay$. Substitute the coordinates of (2, 3) into the equation to find a:

$$2^2 = 4a \cdot 3$$
$$4 = 12a$$
$$a = \frac{1}{3}$$

The equation of the parabola is: $x^2 = \frac{4}{3}y$. The focus is $\left(0, \frac{1}{3}\right)$. Letting $y = \frac{1}{3}$, we find $x^2 = \frac{4}{9}$ or $x = \pm\frac{2}{3}$.

The points $\left(\frac{2}{3}, \frac{1}{3}\right)$ and $\left(-\frac{2}{3}, \frac{1}{3}\right)$ define the latus rectum.

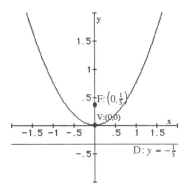

29. The directrix is $y = 2$ and the focus is $(-3, 4)$. This is a vertical case, so the vertex is $(-3, 3)$. $a = 1$ and since $(-3, 4)$ is above $y = 2$, the parabola opens up. The equation of the parabola is:

$$(x - h)^2 = 4a(y - k)$$
$$(x - (-3))^2 = 4 \cdot 1 \cdot (y - 3)$$
$$(x + 3)^2 = 4(y - 3)$$

Letting $y = 4$, we find $(x + 3)^2 = 4$ or $x + 3 = \pm 2$. So, $x = -1$ or $x = -5$. The points $(-1, 4)$ and $(-5, 4)$ define the latus rectum.

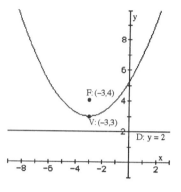

31. The directrix is $x = 1$ and the focus is $(-3, -2)$. This is a horizontal case, so the vertex is $(-1, -2)$. $a = 2$ and since $(-3, -2)$ is to the left of $x = 1$, the parabola opens to the left. The equation of the parabola is:

$$(y - k)^2 = -4a(x - h)$$
$$(y - (-2))^2 = -4 \cdot 2 \cdot (x - (-1))$$
$$(y + 2)^2 = -8(x + 1)$$

Letting $x = -3$, we find $(y + 2)^2 = 16$ or $y + 2 = \pm 4$. So, $y = 2$ or $y = -6$. The points $(-3, 2)$ and $(-3, -6)$ define the latus rectum.

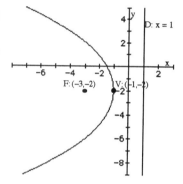

33. The equation $x^2 = 4y$ is in the form $x^2 = 4ay$ where $4a = 4$ or $a = 1$. Thus, we have:

   Vertex: $(0, 0)$
   Focus: $(0, 1)$
   Directrix: $y = -1$

To graph, enter: $y_1 = x^2 / 4$

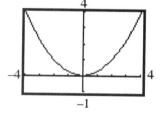

35. The equation $y^2 = -16x$ is in the form $y^2 = -4ax$ where $-4a = -16$ or $a = 4$. Thus, we have:

   Vertex: $(0, 0)$
   Focus: $(-4, 0)$
   Directrix: $x = 4$

To graph, enter: $y_1 = \sqrt{-16x}$; $y_2 = -\sqrt{-16x}$

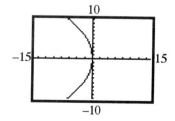

37.   The equation $(y-2)^2 = 8(x+1)$ is in the form
$(y-k)^2 = 4a(x-h)$ where $4a = 8$ or $a = 2$,
$h = -1$, and $k = 2$. Thus, we have:

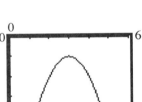

Vertex: $(-1, 2)$
Focus: $(1, 2)$
Directrix: $x = -3$
To graph, enter:
$$y_1 = 2 + \sqrt{8(x+1)}; \quad y_2 = 2 - \sqrt{8(x+1)}$$

39.   The equation $(x-3)^2 = -(y+1)$ is in the form
$(x-h)^2 = -4a(y-k)$ where $-4a = -1$ or $a = \frac{1}{4}$,
$h = 3$, and $k = -1$. Thus, we have:

Vertex: $(3, -1)$
Focus: $\left(3, -\frac{5}{4}\right)$
Directrix: $y = -\frac{3}{4}$
To graph, enter: $y_1 = -1 - (x-3)^2$

41.   The equation $(y+3)^2 = 8(x-2)$ is in the form
$(y-k)^2 = 4a(x-h)$ where $4a = 8$ or $a = 2$,
$h = 2$, and $k = -3$. Thus, we have:

Vertex: $(2, -3)$
Focus: $(4, -3)$
Directrix: $x = 0$
To graph, enter:
$$y_1 = -3 + \sqrt{8(x-2)}; \quad y_2 = -3 - \sqrt{8(x-2)}$$

43.   Complete the square to put in standard form:
$$y^2 - 4y + 4x + 4 = 0$$
$$y^2 - 4y + 4 = -4x$$
$$(y-2)^2 = -4x$$

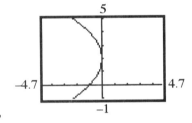

The equation is in the form $(y-k)^2 = -4a(x-h)$
where $-4a = -4$ or $a = 1$, $h = 0$, and $k = 2$. Thus,
we have:

Vertex: $(0, 2)$
Focus: $(-1, 2)$
Directrix: $x = 1$
To graph, enter: $y_1 = 2 + \sqrt{-4x}; \quad y_2 = 2 - \sqrt{-4x}$

45. Complete the square to put in standard form:

$$x^2 + 8x = 4y - 8$$

$$x^2 + 8x + 16 = 4y - 8 + 16$$

$$(x + 4)^2 = 4(y + 2)$$

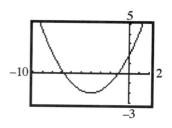

The equation is in the form $(x - h)^2 = 4a(y - k)$ where $4a = 4$ or $a = 1$, $h = -4$, and $k = -2$. Thus, we have:

Vertex: $(-4, -2)$
Focus: $(-4, -1)$
Directrix: $y = -3$

To graph, enter: $y_1 = -2 + (x + 4)^2 / 4$

47. Complete the square to put in standard form:

$$y^2 + 2y - x = 0$$

$$y^2 + 2y + 1 = x + 1$$

$$(y + 1)^2 = x + 1$$

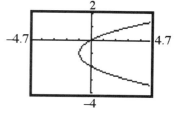

The equation is in the form $(y - k)^2 = 4a(x - h)$ where $4a = 1$ or $a = \frac{1}{4}$, $h = -1$, and $k = -1$. Thus, we have:

Vertex: $(-1, -1)$
Focus: $\left(-\frac{3}{4}, -1\right)$
Directrix: $x = -\frac{5}{4}$

To graph, enter: $y_1 = -1 + \sqrt{x + 1}$; $y_2 = -1 - \sqrt{x + 1}$

49. Complete the square to put in standard form:

$$x^2 - 4x = y + 4$$

$$x^2 - 4x + 4 = y + 4 + 4$$

$$(x - 2)^2 = y + 8$$

The equation is in the form $(x - h)^2 = 4a(y - k)$ where $4a = 1$ or $a = \frac{1}{4}$, $h = 2$, and $k = -8$. Thus, we have:

Vertex: $(2, -8)$
Focus: $\left(2, -\frac{31}{4}\right)$
Directrix: $y = -\frac{33}{4}$

To graph, enter: $y_1 = -8 + (x - 2)^2$

51. $(y - 1)^2 = c(x - 0)$
    $(y - 1)^2 = cx$
    $(2 - 1)^2 = c(1)$
    $1 = c$
    $(y - 1)^2 = x$

53. $(y - 1)^2 = c(x - 2)$
    $(0 - 1)^2 = c(1 - 2)$
    $1 = -c$
    $c = -1$
    $(y - 1)^2 = -(x - 2)$

55. $(x - 0)^2 = c(y - 1)$
    $(2 - 0)^2 = c(2 - 1)$
    $4 = c$
    $x^2 = 4(y - 1)$

57. $(y - 0)^2 = c(x - (-2))$
    $y^2 = c(x + 2)$
    $1^2 = c(0 + 2)$
    $1 = 2c$
    $c = \frac{1}{2}$
    $y^2 = \frac{1}{2}(x + 2)$

59. Set up the problem so that the vertex of the parabola is at $(0, 0)$ and it opens up. Then the equation of the parabola has the form: $x^2 = 4ay$. Since the parabola is 10 feet across and 4 feet deep, the points $(5, 4)$ and $(-5, 4)$ are on the parabola. Substitute and solve for a:
    $5^2 = 4a(4)$
    $25 = 16a$
    $a = \frac{25}{16}$
    a is the distance from the vertex to the focus. Thus, the receiver (located at the focus) is $\frac{25}{16} = 1.5625$ feet, or 18.75 inches from the base of the dish, along the axis of the parabola.

61. Set up the problem so that the vertex of the parabola is at $(0, 0)$ and it opens up. Then the equation of the parabola has the form: $x^2 = 4ay$. Since the parabola is 4 inches across and 1 inch deep, the points $(2, 1)$ and $(-2, 1)$ are on the parabola. Substitute and solve for a:
    $2^2 = 4a(1)$
    $4 = 4a$
    $a = 1$
    a is the distance from the vertex to the focus. Thus, the bulb (located at the focus) should be 1 inch, from the vertex.

63. Set up the problem so that the vertex of the parabola is at $(0, 0)$ and it opens up. Then the equation of the parabola has the form: $x^2 = cy$.
    The point $(300, 80)$ is a point on the parabola. Solve for c and find the equation:
    $300^2 = c(80)$
    $c = 1125$
    $x^2 = 1125y$

    Since the height of the cable, 150 feet from the center, is to be found, the point $(150, h)$ is a point on the parabola. Solve for h:
    $150^2 = 1125h$
    $22500 = 1125h$
    $h = 20$
    The height of the cable, 150 feet from the center, is 20 feet.

65. Set up the problem so that the vertex of the parabola is at (0, 0) and it opens up. Then the equation of the parabola has the form: $x^2 = 4ay$. a is the distance from the vertex to the focus (where the source is located), so $a = 2$. Since the opening is 5 feet across, there is a point (2.5, y) on the parabola. Solve for y:

$$x^2 = 8y$$
$$2.5^2 = 8y$$
$$6.25 = 8y$$
$$y = 0.78125 \text{ feet}$$

The depth of the searchlight should be 0.78125 feet.

67. Set up the problem so that the vertex of the parabola is at (0, 0) and it opens up. Then the equation of the parabola has the form: $x^2 = 4ay$. Since the parabola is 20 feet across and 6 feet deep, the points (10, 6) and (−10, 6) are on the parabola. Substitute and solve for a:

$$10^2 = 4a(6)$$
$$100 = 24a$$
$$a \approx 4.17 \text{ feet}$$

The heat source will be concentrated 4.17 feet from the base, along the axis of symmetry.

69. Set up the problem so that the vertex of the parabola is at (0, 0) and it opens down. Then the equation of the parabola has the form: $x^2 = cy$.
The point (60, −25) is a point on the parabola. Solve for c and find the equation:

$$60^2 = c(-25)$$
$$c = -144$$
$$x^2 = -144y$$

To find the height of the bridge, 10 feet from the center, the point (10, y) is a point on the parabola. Solve for y:

$$10^2 = -144y$$
$$100 = -144y$$
$$y = -0.69$$

The height of the bridge, 10 feet from the center, is $25 - 0.69 = 24.31$ feet.

To find the height of the bridge, 30 feet from the center, the point (30, y) is a point on the parabola. Solve for y:

$$30^2 = -144y$$
$$900 = -144y$$
$$y = -6.25$$

The height of the bridge, 30 feet from the center, is $25 - 6.25 = 18.75$ feet.

To find the height of the bridge, 50 feet from the center, the point $(50, y)$ is a point on the parabola. Solve for y:

$$50^2 = -144y$$
$$2500 = -144y$$
$$y = -17.36$$

The height of the bridge, 50 feet from the center, is $25 - 17.36 = 7.64$ feet.

71. $Ax^2 + Ey = 0 \quad A \neq 0, \ E \neq 0$

$$Ax^2 = -Ey$$
$$x^2 = \frac{-E}{A}y$$

This is the equation of a parabola with vertex at $(0, 0)$ and axis of symmetry being the y-axis. The focus is $\left(0, \frac{-E}{4A}\right)$. The directrix is $y = \frac{E}{4A}$.

73. $Ax^2 + Dx + Ey + F = 0 \quad A \neq 0$

     (a)   If $E \neq 0$, then:

$$Ax^2 + Dx = -Ey - F$$
$$A\left(x^2 + \frac{D}{A}x + \frac{D^2}{4A^2}\right) = -Ey - F + \frac{D^2}{4A}$$
$$\left(x + \frac{D}{2A}\right)^2 = \frac{1}{A}\left(-Ey - F + \frac{D^2}{4A}\right)$$
$$\left(x + \frac{D}{2A}\right)^2 = \frac{-E}{A}\left(y + \frac{F}{E} - \frac{D^2}{4AE}\right)$$
$$\left(x + \frac{D}{2A}\right)^2 = \frac{-E}{A}\left(y - \frac{D^2 - 4AF}{4AE}\right)$$

This is the equation of a parabola whose vertex is $\left(\frac{-D}{2A}, \frac{D^2 - 4AF}{4AE}\right)$.

     (b)   If $E = 0$, then

$$Ax^2 + Dx + F = 0$$
$$x = \frac{-D \pm \sqrt{D^2 - 4AF}}{2A}$$

If $D^2 - 4AF = 0$, then $x = \frac{-D}{2A}$ is a vertical line .

     (c)   If $E = 0$, then

$$Ax^2 + Dx + F = 0$$
$$x = \frac{-D \pm \sqrt{D^2 - 4AF}}{2A}$$

If $D^2 - 4AF > 0$,

then $x = \frac{-D + \sqrt{D^2 - 4AF}}{2A}$ or $x = \frac{-D - \sqrt{D^2 - 4AF}}{2A}$ are two vertical lines .

(d)   If $E = 0$, then
$$Ax^2 + Dx + F = 0$$
$$x = \frac{-D \pm \sqrt{D^2 - 4AF}}{2A}$$
If $D^2 - 4AF < 0$, there is no real solution. The graph contains no points.

10.3 The Ellipse

1.   C   3.   B   5.   C   7.   D

9. $\dfrac{x^2}{25} + \dfrac{y^2}{4} = 1$
   The center of the ellipse is at the origin.
   $a = 5$, $b = 2$. The vertices are $(5, 0)$ and $(-5, 0)$.
   Find the value of c:
   $$c^2 = a^2 - b^2 = 25 - 4 = 21$$
   $$c = \sqrt{21}$$
   The foci are $\left(\sqrt{21}, 0\right)$ and $\left(-\sqrt{21}, 0\right)$.
   To graph, enter:
   $$y_1 = 2\sqrt{(1 - x^2 / 25)};\ \ y_1 = -2\sqrt{(1 - x^2 / 25)}$$

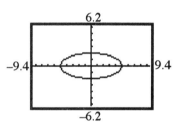

11. $\dfrac{x^2}{9} + \dfrac{y^2}{25} = 1$
   The center of the ellipse is at the origin.
   $a = 5$, $b = 3$. The vertices are $(0, 5)$ and $(0, -5)$.
   Find the value of c:
   $$c^2 = a^2 - b^2 = 25 - 9 = 16$$
   $$c = 4$$
   The foci are $(0, 4)$ and $(0, -4)$.
   To graph, enter:
   $$y_1 = 5\sqrt{(1 - x^2 / 9)};\ \ y_1 = -5\sqrt{(1 - x^2 / 9)}$$

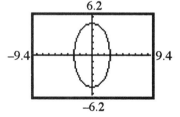

13. $4x^2 + y^2 = 16$

Divide by 16 to put in standard form:

$$\frac{4x^2}{16} + \frac{y^2}{16} = \frac{16}{16} \quad \rightarrow \quad \frac{x^2}{4} + \frac{y^2}{16} = 1$$

The center of the ellipse is at the origin.

$a = 4$, $b = 2$. The vertices are $(0, 4)$ and $(0, -4)$.

Find the value of c:

$$c^2 = a^2 - b^2 = 16 - 4 = 12$$

$$c = \sqrt{12} = 2\sqrt{3}$$

The foci are $\left(0, 2\sqrt{3}\right)$ and $\left(0, -2\sqrt{3}\right)$.

To graph, enter:

$$y_1 = \sqrt{(16 - 4x^2)}; \quad y_1 = -\sqrt{(16 - 4x^2)}$$

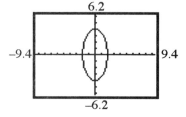

15. $4y^2 + x^2 = 8$

Divide by 8 to put in standard form:

$$\frac{4y^2}{8} + \frac{x^2}{8} = \frac{8}{8} \quad \rightarrow \quad \frac{x^2}{8} + \frac{y^2}{2} = 1$$

The center of the ellipse is at the origin.

$a = \sqrt{8} = 2\sqrt{2}$, $b = \sqrt{2}$. The vertices are $\left(2\sqrt{2}, 0\right)$ and $\left(-2\sqrt{2}, 0\right)$. Find the value of c:

$$c^2 = a^2 - b^2 = 8 - 2 = 6$$

$$c = \sqrt{6}$$

The foci are $\left(\sqrt{6}, 0\right)$ and $\left(-\sqrt{6}, 0\right)$.

To graph, enter:

$$y_1 = \sqrt{(2 - x^2 / 4)}; \quad y_1 = -\sqrt{(2 - x^2 / 4)}$$

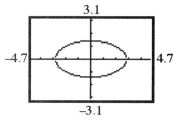

17. $x^2 + y^2 = 16$

This is the equation of a circle whose center is at $(0, 0)$ and radius $= 4$.

To graph, enter: $y_1 = \sqrt{(16 - x^2)}; \quad y_1 = -\sqrt{(16 - x^2)}$

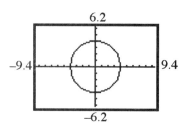

19. Center: $(0, 0)$; Focus: $(3, 0)$; Vertex: $(5, 0)$;
Major axis is the x-axis; $a = 5$; $c = 3$. Find b:
$$b^2 = a^2 - c^2 = 25 - 9 = 16$$
$$b = 4$$
Write the equation: $\dfrac{x^2}{25} + \dfrac{y^2}{16} = 1$

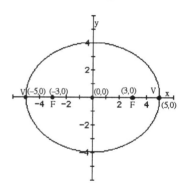

21. Center: $(0, 0)$; Focus: $(0, -4)$; Vertex: $(0, 5)$;
Major axis is the y-axis; $a = 5$; $c = 4$. Find b:
$$b^2 = a^2 - c^2 = 25 - 16 = 9$$
$$b = 3$$
Write the equation: $\dfrac{x^2}{9} + \dfrac{y^2}{25} = 1$

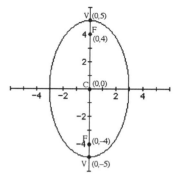

23. Foci: $(\pm 2, 0)$; Length of major axis is 6.
Center: $(0, 0)$; Major axis is the x-axis;
$a = 3$; $c = 2$. Find b:
$$b^2 = a^2 - c^2 = 9 - 4 = 5$$
$$b = \sqrt{5}$$
Write the equation: $\dfrac{x^2}{9} + \dfrac{y^2}{5} = 1$

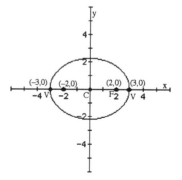

25. Foci: $(0, \pm 3)$; x-intercepts are ± 2. Center: $(0, 0)$;
Major axis is the y-axis; $c = 3$; $b = 2$. Find a:
$$a^2 = b^2 + c^2 = 4 + 9 = 13$$
$$a = \sqrt{13}$$
Write the equation: $\dfrac{x^2}{4} + \dfrac{y^2}{13} = 1$

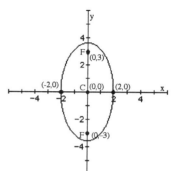

27. Center: $(0, 0)$; Vertex: $(0, 4)$; $b = 1$; Major axis is the y-axis; $a = 4$; $b = 1$.

    Write the equation: $\dfrac{x^2}{1} + \dfrac{y^2}{16} = 1$

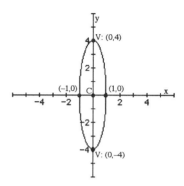

29. $\dfrac{(x+1)^2}{4} + \dfrac{(y-1)^2}{1} = 1$

31. $\dfrac{(x-1)^2}{1} + \dfrac{y^2}{4} = 1$

33. The equation $\dfrac{(x-3)^2}{4} + \dfrac{(y+1)^2}{9} = 1$ is in the form $\dfrac{(x-h)^2}{b^2} + \dfrac{(y-k)^2}{a^2} = 1$ (major axis parallel to the y-axis) where $a = 3$, $b = 2$, $h = 3$, and $k = -1$. Solving for c:

$$c^2 = a^2 - b^2 = 9 - 4 = 5$$
$$c = \sqrt{5}$$

Thus, we have:

    Center: $(3, -1)$

    Foci: $\left(3, -1 + \sqrt{5}\right)$, $\left(3, -1 - \sqrt{5}\right)$

    Vertices: $(3, 2)$, $(3, -4)$

To graph, enter: $y_1 = -1 + 3\sqrt{1 - (x-3)^2 / 4}$;

$$y_2 = -1 - 3\sqrt{1 - (x-3)^2 / 4}$$

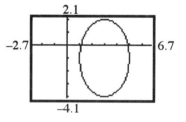

35. Divide by 16 to put the equation in standard form:

$$(x+5)^2 + 4(y-4)^2 = 16$$
$$\frac{(x+5)^2}{16} + \frac{4(y-4)^2}{16} = \frac{16}{16}$$
$$\frac{(x+5)^2}{16} + \frac{(y-4)^2}{4} = 1$$

The equation is in the form $\dfrac{(x-h)^2}{a^2} + \dfrac{(y-k)^2}{b^2} = 1$ (major axis parallel to the x-axis) where $a = 4$, $b = 2$, $h = -5$, and $k = 4$. Solving for c:

$$c^2 = a^2 - b^2 = 16 - 4 = 12$$
$$c = \sqrt{12} = 2\sqrt{3}$$

Thus, we have:
    Center: $(-5, 4)$
    Foci: $\left(-5 - 2\sqrt{3}, 4\right)$, $\left(-5 + 2\sqrt{3}, 4\right)$
    Vertices: $(-9, 4)$, $(-1, 4)$
To graph, enter: $y_1 = 4 + 2\sqrt{1 - (x + 5)^2 / 16}$;
    $y_2 = 4 - 2\sqrt{1 - (x + 5)^2 / 16}$

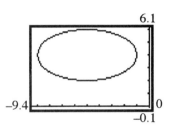

37. Complete the square to put the equation in standard form:
$$x^2 + 4x + 4y^2 - 8y + 4 = 0$$
$$(x^2 + 4x + 4) + 4(y^2 - 2y + 1) = -4 + 4 + 4$$
$$(x + 2)^2 + 4(y - 1)^2 = 4$$
$$\frac{(x + 2)^2}{4} + \frac{4(y - 1)^2}{4} = \frac{4}{4}$$
$$\frac{(x + 2)^2}{4} + \frac{(y - 1)^2}{1} = 1$$

The equation is in the form $\dfrac{(x - h)^2}{a^2} + \dfrac{(y - k)^2}{b^2} = 1$ (major axis parallel to the x-axis)
where $a = 2$, $b = 1$, $h = -2$, and $k = 1$. Solving for c:
$$c^2 = a^2 - b^2 = 4 - 1 = 3$$
$$c = \sqrt{3}$$

Thus, we have:
    Center: $(-2, 1)$
    Foci: $\left(-2 - \sqrt{3}, 1\right)$, $\left(-2 + \sqrt{3}, 1\right)$
    Vertices: $(-4, 1)$, $(0, 1)$
To graph, enter: $y_1 = 1 + \sqrt{1 - (x + 2)^2 / 4}$;
    $y_2 = 1 - \sqrt{1 - (x + 2)^2 / 4}$

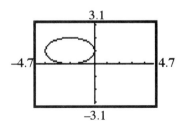

39. Complete the square to put the equation in standard form:
$$2x^2 + 3y^2 - 8x + 6y + 5 = 0$$
$$2(x^2 - 4x) + 3(y^2 + 2y) = -5$$
$$2(x^2 - 4x + 4) + 3(y^2 + 2y + 1) = -5 + 8 + 3$$
$$2(x - 2)^2 + 3(y + 1)^2 = 6$$
$$\frac{2(x - 2)^2}{6} + \frac{3(y + 1)^2}{6} = \frac{6}{6}$$
$$\frac{(x - 2)^2}{3} + \frac{(y + 1)^2}{2} = 1$$

The equation is in the form $\dfrac{(x - h)^2}{a^2} + \dfrac{(y - k)^2}{b^2} = 1$ (major axis parallel to the x-axis)
where $a = \sqrt{3}$, $b = \sqrt{2}$, $h = 2$, and $k = -1$. Solving for c:
$$c^2 = a^2 - b^2 = 3 - 2 = 1 \quad \rightarrow \quad c = 1$$

Thus, we have:
    Center: $(2, -1)$
    Foci: $(1, -1), (3, -1)$
    Vertices: $\left(2 - \sqrt{3}, -1\right), \left(2 + \sqrt{3}, -1\right)$

To graph, enter: $y_1 = -1 + \sqrt{2 - 2(x-2)^2 / 3}$;

    $y_2 = -1 - \sqrt{2 - 2(x-2)^2 / 3}$

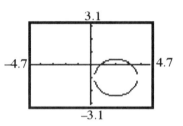

41. Complete the square to put the equation in standard form:

$$9x^2 + 4y^2 - 18x + 16y - 11 = 0$$
$$9(x^2 - 2x) + 4(y^2 + 4y) = 11$$
$$9(x^2 - 2x + 1) + 4(y^2 + 4y + 4) = 11 + 9 + 16$$
$$9(x - 1)^2 + 4(y + 2)^2 = 36$$
$$\frac{9(x-1)^2}{36} + \frac{4(y+2)^2}{36} = \frac{36}{36}$$
$$\frac{(x-1)^2}{4} + \frac{(y+2)^2}{9} = 1$$

The equation is in the form $\dfrac{(x-h)^2}{b^2} + \dfrac{(y-k)^2}{a^2} = 1$ (major axis parallel to the y-axis)

where $a = 3$, $b = 2$, $h = 1$, and $k = -2$. Solving for c:

    $c^2 = a^2 - b^2 = 9 - 4 = 5$

    $c = \sqrt{5}$

Thus, we have:
    Center: $(1, -2)$
    Foci: $\left(1, -2 + \sqrt{5}\right), \left(1, -2 - \sqrt{5}\right)$
    Vertices: $(1, 1), (1, -5)$

To graph, enter: $y_1 = -2 + 3\sqrt{1 - (x-1)^2 / 4}$;

    $y_2 = -2 - 3\sqrt{1 - (x-1)^2 / 4}$

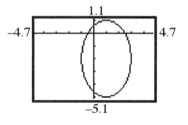

43. Complete the square to put the equation in standard form:

$$4x^2 + y^2 + 4y = 0$$
$$4x^2 + y^2 + 4y + 4 = 4$$
$$4x^2 + (y + 2)^2 = 4$$
$$\frac{4x^2}{4} + \frac{(y+2)^2}{4} = \frac{4}{4}$$
$$\frac{x^2}{1} + \frac{(y+2)^2}{4} = 1$$

The equation is in the form $\dfrac{(x-h)^2}{b^2} + \dfrac{(y-k)^2}{a^2} = 1$ (major axis parallel to the y-axis)

where $a = 2$, $b = 1$, $h = 0$, and $k = -2$. Solving for c:

    $c^2 = a^2 - b^2 = 4 - 1 = 3 \quad \rightarrow \quad c = \sqrt{3}$

Thus, we have:
   Center: $(0, -2)$
   Foci: $\left(0, -2 + \sqrt{3}\right)$, $\left(0, -2 - \sqrt{3}\right)$
   Vertices: $(0, 0)$, $(0, -4)$
To graph, enter: $y_1 = -2 + 2\sqrt{1 - x^2}$;

   $y_2 = -2 + 2\sqrt{1 - x^2}$

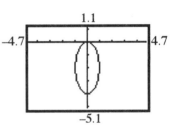

45. Center: $(2, -2)$; Vertex: $(7, -2)$; Focus: $(4, -2)$;
    Major axis parallel to the x-axis; $a = 5$; $c = 2$.
    Find b:
       $b^2 = a^2 - c^2 = 25 - 4 = 21$
       $b = \sqrt{21}$
    Write the equation: $\dfrac{(x - 2)^2}{25} + \dfrac{(y + 2)^2}{21} = 1$

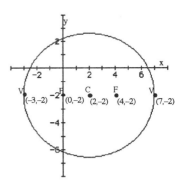

47. Vertices: $(4, 3)$, $(4, 9)$; Focus: $(4, 8)$;
    Center: $(4, 6)$; Major axis parallel to the y-axis;
    $a = 3$; $c = 2$. Find b:
       $b^2 = a^2 - c^2 = 9 - 4 = 5$
       $b = \sqrt{5}$
    Write the equation: $\dfrac{(x - 4)^2}{5} + \dfrac{(y - 6)^2}{9} = 1$

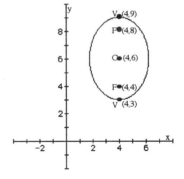

49. Foci: $(5, 1)$, $(-1, 1)$; length of the major axis $= 8$;
    Center: $(2, 1)$; Major axis parallel to the x-axis;
    $a = 4$; $c = 3$. Find b:
       $b^2 = a^2 - c^2 = 16 - 9 = 7$
       $b = \sqrt{7}$
    Write the equation: $\dfrac{(x - 2)^2}{16} + \dfrac{(y - 1)^2}{7} = 1$

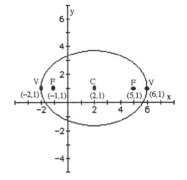

51. Center: $(1, 2)$; Focus: $(4, 2)$; contains the point
     $(1, 3)$; Major axis parallel to the x-axis; $c = 3$.

     The equation has the form: $\dfrac{(x-1)^2}{a^2} + \dfrac{(y-2)^2}{b^2} = 1$

     Since the point $(1, 3)$ is on the curve:

     $$\frac{0}{a^2} + \frac{1}{b^2} = 1 \;\rightarrow\; \frac{1}{b^2} = 1 \;\rightarrow\; b^2 = 1 \;\rightarrow\; b = 1$$

     Find a:

     $$a^2 = b^2 + c^2 = 1 + 9 = 10 \;\rightarrow\; a = \sqrt{10}$$

     Write the equation: $\dfrac{(x-1)^2}{10} + \dfrac{(y-2)^2}{1} = 1$

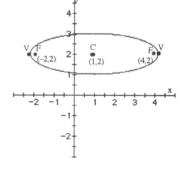

53. Center: $(1, 2)$; Vertex: $(4, 2)$; contains the point
     $(1, 3)$; Major axis parallel to the x-axis; $a = 3$.

     The equation has the form: $\dfrac{(x-1)^2}{a^2} + \dfrac{(y-2)^2}{b^2} = 1$

     Since the point $(1, 3)$ is on the curve:

     $$\frac{0}{9} + \frac{1}{b^2} = 1 \;\rightarrow\; \frac{1}{b^2} = 1 \;\rightarrow\; b^2 = 1 \;\rightarrow\; b = 1$$

     Write the equation: $\dfrac{(x-1)^2}{9} + \dfrac{(y-2)^2}{1} = 1$

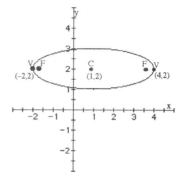

55. Rewrite the equation:

$$
\begin{aligned}
y &= \sqrt{16 - 4x^2} \\
y^2 &= 16 - 4x^2, && y \ge 0 \\
4x^2 + y^2 &= 16, && y \ge 0 \\
\frac{x^2}{4} + \frac{y^2}{16} &= 1, && y \ge 0
\end{aligned}
$$

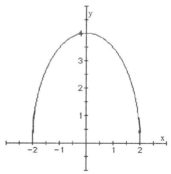

57. Rewrite the equation:

$$
\begin{aligned}
y &= -\sqrt{64 - 16x^2} \\
y^2 &= 64 - 16x^2, && y \le 0 \\
16x^2 + y^2 &= 64, && y \le 0 \\
\frac{x^2}{4} + \frac{y^2}{64} &= 1, && y \le 0
\end{aligned}
$$

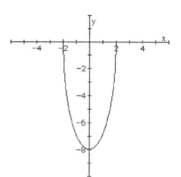

59. The center of the ellipse is $(0, 0)$. The length of the major axis is 20, so $a = 10$. The length of half the minor axis is 6, so $b = 6$. The ellipse is situated with its major axis on the x-axis. The equation is: $\dfrac{x^2}{100} + \dfrac{y^2}{36} = 1$.

61. Assume that the half ellipse formed by the gallery is centered at $(0, 0)$. Since the hall is 100 feet long, $2a = 100$ or $a = 50$. The distance from the center to the foci is 25 feet, so $c = 25$. Find the height of the gallery which is b:
$$b^2 = a^2 - c^2 = 2500 - 625 = 1875 \rightarrow b = \sqrt{1875} \approx 43.3$$
The ceiling will be 43.3 feet high in the center.

63. Place the semielliptical arch so that the x-axis coincides with the water and the y-axis passes through the center of the arch. Since the bridge has a span of 120 feet, the length of the major axis is 120, or $2a = 120$ or $a = 60$. The maximum height of the bridge is 25 feet, so $b = 25$. The equation is: $\dfrac{x^2}{3600} + \dfrac{y^2}{625} = 1$.
The height 10 feet from the center:
$$\dfrac{10^2}{3600} + \dfrac{y^2}{625} = 1 \rightarrow \dfrac{y^2}{625} = 1 - \dfrac{100}{3600} \rightarrow y^2 = 625 \cdot \dfrac{3500}{3600} \rightarrow y \approx 24.65 \text{ feet}$$
The height 30 feet from the center:
$$\dfrac{30^2}{3600} + \dfrac{y^2}{625} = 1 \rightarrow \dfrac{y^2}{625} = 1 - \dfrac{900}{3600} \rightarrow y^2 = 625 \cdot \dfrac{2700}{3600} \rightarrow y \approx 21.65 \text{ feet}$$
The height 50 feet from the center:
$$\dfrac{50^2}{3600} + \dfrac{y^2}{625} = 1 \rightarrow \dfrac{y^2}{625} = 1 - \dfrac{2500}{3600} \rightarrow y^2 = 625 \cdot \dfrac{1100}{3600} \rightarrow y \approx 13.82 \text{ feet}$$

65. Place the semielliptical arch so that the x-axis coincides with the major axis and the y-axis passes through the center of the arch. Since the ellipse is 40 feet wide, the length of the major axis is 40, or $2a = 40$ or $a = 20$. The height is 15 feet at the center, so $b = 15$. The equation is: $\dfrac{x^2}{400} + \dfrac{y^2}{225} = 1$.
The height 10 feet either side of the center:
$$\dfrac{10^2}{400} + \dfrac{y^2}{225} = 1 \rightarrow \dfrac{y^2}{225} = 1 - \dfrac{100}{400} \rightarrow y^2 = 225 \cdot \dfrac{3}{4} \rightarrow y \approx 12.99 \text{ feet}$$
The height 20 feet either side of the center:
$$\dfrac{20^2}{400} + \dfrac{y^2}{225} = 1 \rightarrow \dfrac{y^2}{225} = 1 - \dfrac{400}{400} \rightarrow y^2 = 225 \cdot 0 \rightarrow y \approx 0 \text{ feet}$$

67. Since the mean distance is 93 million miles, $a = 93$ million. The length of the major axis is 186 million. The perihelion is 186 million – 94.5 million = 91.5 million miles. The distance from the center of the ellipse to the sun (focus) is 93 million – 91.5 million = 1.5 million miles; therefore, $c = 1.5$ million. Find b:
$$b^2 = a^2 - c^2 = \left(93 \times 10^6\right)^2 - \left(1.5 \times 10^6\right)^2 = 8.64675 \times 10^{15} \rightarrow b = 92.99 \times 10^6$$
The equation of the orbit is: $\dfrac{x^2}{\left(93 \times 10^6\right)^2} + \dfrac{y^2}{\left(92.99 \times 10^6\right)^2} = 1$.

69. The mean distance is 507 million – 23.2 million = 483.8 million miles.
     The perihelion is 483.8 million – 23.2 million = 460.6 million miles.
     Since $a = 483.8 \times 10^6$ and $c = 23.2 \times 10^6$, we can find b:
     $$b^2 = a^2 - c^2 = (483.8 \times 10^6)^2 - (23.2 \times 10^6)^2 = 2.335242 \times 10^{17} \;\rightarrow\; b = 483.2 \times 10^6$$
     The equation of the orbit of Jupiter is: $\dfrac{x^2}{(483.8 \times 10^6)^2} + \dfrac{y^2}{(483.2 \times 10^6)^2} = 1$.

71. If the x-axis is placed along the 100 foot length and the y-axis is placed along the 50
     foot length, the equation for the ellipse is: $\dfrac{x^2}{50^2} + \dfrac{y^2}{25^2} = 1$.
     Find y when x = 40:
     $$\frac{40^2}{50^2} + \frac{y^2}{25^2} = 1 \rightarrow \frac{y^2}{625} = 1 - \frac{1600}{2500} \rightarrow y^2 = 625 \cdot \frac{9}{25} \rightarrow y \approx 15\,\text{feet}$$
     The width 10 feet from the side is 30 feet.

73. (a) Put the equation in standard ellipse form:
     $$Ax^2 + Cy^2 + F = 0 \qquad A \neq 0,\, C \neq 0,\, F \neq 0$$
     $$Ax^2 + Cy^2 = -F$$
     $$\frac{Ax^2}{-F} + \frac{Cy^2}{-F} = 1$$
     $$\frac{x^2}{(-F/A)} + \frac{y^2}{(-F/C)} = 1 \qquad \text{where } -F/A \text{ and } -F/C \text{ are positive}$$
     This is the equation of an ellipse with center at (0, 0).

     (b) If $A = C$, the equation becomes:
     $$Ax^2 + Ay^2 = -F \;\rightarrow\; x^2 + y^2 = \frac{-F}{A}$$
     This is the equation of a circle with center at (0, 0) and radius of $\sqrt{\dfrac{-F}{A}}$.

10.4 The Hyperbola

1. B   3. A   5. B   7. C

9. Center: (0, 0); Focus: (3, 0); Vertex: (1, 0);
    Transverse axis is the x-axis; $a = 1$; $c = 3$. Find b:
    $$b^2 = c^2 - a^2 = 9 - 1 = 8$$
    $$b = \sqrt{8} = 2\sqrt{2}$$
    Write the equation: $\dfrac{x^2}{1} - \dfrac{y^2}{8} = 1$
    To graph, enter: $y_1 = \sqrt{8(x^2 - 1)}$; $y_2 = -\sqrt{8(x^2 - 1)}$

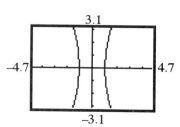

11. Center: $(0, 0)$; Focus: $(0, -6)$; Vertex: $(0, 4)$

Transverse axis is the y-axis; $a = 4$; $c = 6$. Find b:

$$b^2 = c^2 - a^2 = 36 - 16 = 20$$

$$b = \sqrt{20} = 2\sqrt{5}$$

Write the equation: $\dfrac{y^2}{16} - \dfrac{x^2}{20} = 1$

To graph, enter:

$$y_1 = 4\sqrt{1 + x^2 / 20} \; ; \; y_2 = -4\sqrt{1 + x^2 / 20}$$

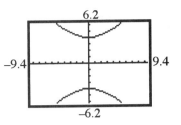

13. Foci: $(-5, 0)$, $(5, 0)$; Vertex: $(3, 0)$ Center: $(0, 0)$;

Transverse axis is the x-axis; $a = 3$; $c = 5$. Find b:

$$b^2 = c^2 - a^2 = 25 - 9 = 16 \quad \rightarrow \quad b = 4$$

Write the equation: $\dfrac{x^2}{9} - \dfrac{y^2}{16} = 1$

To graph, enter:

$$y_1 = 4\sqrt{x^2 / 9 - 1} \; ; \; y_2 = -4\sqrt{x^2 / 9 - 1}$$

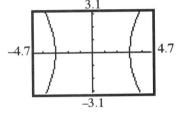

15. Vertices: $(0, -6)$, $(0, 6)$; Asymptote: $y = 2x$;

Center: $(0, 0)$; Transverse axis is the y-axis; $a = 6$.
Find b using the slope of the asymptote:

$$\frac{a}{b} = \frac{6}{b} = 2 \quad \rightarrow \quad 2b = 6 \quad \rightarrow \quad b = 3$$

Write the equation: $\dfrac{y^2}{36} - \dfrac{x^2}{9} = 1$

To graph, enter:

$$y_1 = 6\sqrt{1 + x^2 / 9} \; ; \; y_2 = -6\sqrt{1 + x^2 / 9}$$

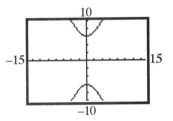

17. Foci: $(-4, 0)$, $(4, 0)$; Asymptote: $y = -x$;

Center: $(0, 0)$; Transverse axis is the x-axis; $c = 4$.
Using the slope of the asymptote:

$$-\frac{b}{a} = -1 \quad \rightarrow \quad -b = -a \quad \rightarrow \quad b = a$$

Find b:

$$b^2 = c^2 - a^2$$

$$a^2 + b^2 = c^2 \qquad (c = 4)$$

$$b^2 + b^2 = 16$$

$$2b^2 = 16$$

$$b^2 = 8 \quad \rightarrow \quad b = \sqrt{8} = 2\sqrt{2}$$

$$a = \sqrt{8} = 2\sqrt{2} \qquad (a = b)$$

Write the equation: $\dfrac{x^2}{8} - \dfrac{y^2}{8} = 1$

To graph, enter: $y_1 = \sqrt{x^2 - 8} \; ; \; y_2 = -\sqrt{x^2 - 8}$

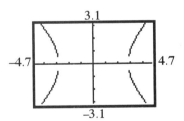

19. $\dfrac{x^2}{25} - \dfrac{y^2}{9} = 1$

The center of the hyperbola is at (0, 0).

$a = 5,\ b = 3$. The vertices are (5, 0) and (–5, 0).

Find the value of c:

$$c^2 = a^2 + b^2 = 25 + 9 = 34$$

$$c = \sqrt{34}$$

The foci are $\left(\sqrt{34},0\right)$ and $\left(-\sqrt{34},0\right)$.

The transverse axis is $y = 0$.

The asymptotes are $y = \frac{3}{5}x$ and $y = -\frac{3}{5}x$.

To graph, enter:

$$y_1 = 3\sqrt{(x^2/25 - 1)};\ \ y_2 = -3\sqrt{(x^2/25 - 1)}$$

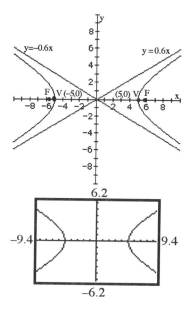

21. $4x^2 - y^2 = 16$

Divide both sides by 16 to put in standard form:

$$\frac{4x^2}{16} - \frac{y^2}{16} = \frac{16}{16}$$

$$\frac{x^2}{4} - \frac{y^2}{16} = 1$$

The center of the hyperbola is at (0, 0).

$a = 2,\ b = 4$. The vertices are (2, 0) and (–2, 0).

Find the value of c:

$$c^2 = a^2 + b^2 = 4 + 16 = 20$$

$$c = \sqrt{20} = 2\sqrt{5}$$

The foci are $\left(2\sqrt{5},0\right)$ and $\left(-2\sqrt{5},0\right)$.

The transverse axis is $y = 0$.

The asymptotes are $y = 2x$ and $y = -2x$.

To graph, enter:

$$y_1 = 4\sqrt{(x^2/4 - 1)};\ \ y_2 = -4\sqrt{(x^2/4 - 1)}$$

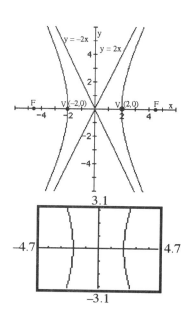

23. $y^2 - 9x^2 = 9$
Divide both sides by 9 to put in standard form:

$$\frac{y^2}{9} - \frac{9x^2}{9} = \frac{9}{9}$$

$$\frac{y^2}{9} - \frac{x^2}{1} = 1$$

The center of the hyperbola is at $(0, 0)$.
$a = 3$, $b = 1$. The vertices are $(0, 3)$ and $(0, -3)$.
Find the value of c:

$$c^2 = a^2 + b^2 = 9 + 1 = 10$$

$$c = \sqrt{10}$$

The foci are $\left(0, \sqrt{10}\right)$ and $\left(0, -\sqrt{10}\right)$.

The transverse axis is $x = 0$.
The asymptotes are $y = 3x$ and $y = -3x$.
To graph, enter:

$$y_1 = \sqrt{(9x^2 + 9)}; \ y_2 = -\sqrt{(9x^2 + 9)}$$

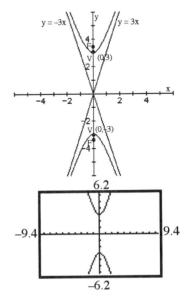

25. $y^2 - x^2 = 25$
Divide both sides by 25 to put in standard form:

$$\frac{y^2}{25} - \frac{x^2}{25} = 1$$

The center of the hyperbola is at $(0, 0)$.
$a = 5$, $b = 5$. The vertices are $(0, 5)$ and $(0, -5)$.
Find the value of c:

$$c^2 = a^2 + b^2 = 25 + 25 = 50$$

$$c = \sqrt{50} = 5\sqrt{2}$$

The foci are $\left(0, 5\sqrt{2}\right)$ and $\left(0, -5\sqrt{2}\right)$.

The transverse axis is $x = 0$.
The asymptotes are $y = x$ and $y = -x$.
To graph, enter:

$$y_1 = \sqrt{(x^2 + 25)}; \ y_2 = -\sqrt{(x^2 + 25)}$$

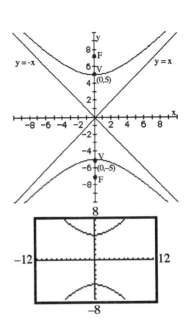

27. $x^2 - y^2 = 1$

29. $\dfrac{y^2}{36} - \dfrac{x^2}{9} = 1$

31. Center: $(4, -1)$; Focus: $(7, -1)$; Vertex: $(6, -1)$;
    Transverse axis is parallel to the x-axis;
    $a = 2$; $c = 3$. Find b:
    $$b^2 = c^2 - a^2 = 9 - 4 = 5$$
    $$b = \sqrt{5}$$
    Write the equation: $\dfrac{(x-4)^2}{4} - \dfrac{(y+1)^2}{5} = 1$

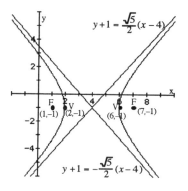

33. Center: $(-3, -4)$; Focus: $(-3, -8)$; Vertex: $(-3, -2)$;
    Transverse axis is parallel to the y-axis;
    $a = 2$; $c = 4$. Find b:
    $$b^2 = c^2 - a^2 = 16 - 4 = 12$$
    $$b = \sqrt{12} = 2\sqrt{3}$$
    Write the equation: $\dfrac{(y+4)^2}{4} - \dfrac{(x+3)^2}{12} = 1$

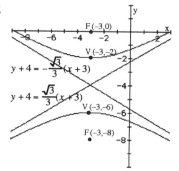

35. Foci: $(3, 7), (7, 7)$; Vertex: $(6, 7)$; Center: $(5, 7)$;
    Transverse axis is parallel to the x-axis;
    $a = 1$; $c = 2$. Find b:
    $$b^2 = c^2 - a^2 = 4 - 1 = 3$$
    $$b = \sqrt{3}$$
    Write the equation: $\dfrac{(x-5)^2}{1} - \dfrac{(y-7)^2}{3} = 1$

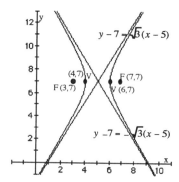

37. Vertices: $(-1, -1)$, $(3, -1)$; Center: $(1, -1)$;
    Transverse axis is parallel to the x-axis; $a = 2$.

    Asymptote: $\dfrac{x-1}{2} = \dfrac{y+1}{3}$ Using the slope of the asymptote:

    $$\frac{b}{a} = \frac{b}{2} = \frac{3}{2} \rightarrow b = 3$$

    Write the equation: $\dfrac{(x-1)^2}{4} - \dfrac{(y+1)^2}{9} = 1$

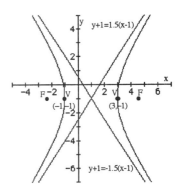

39. $\dfrac{(x-2)^2}{4} - \dfrac{(y+3)^2}{9} = 1$

    The center of the hyperbola is at $(2, -3)$.
    $a = 2$, $b = 3$. The vertices are $(0, -3)$ and $(4, -3)$.
    Find the value of c:

    $$c^2 = a^2 + b^2 = 4 + 9 = 13$$
    $$c = \sqrt{13}$$

    Foci: $\left(2 - \sqrt{13}, -3\right)$ and $\left(2 + \sqrt{13}, -3\right)$.
    Transverse axis: $y = -3$.
    Asymptotes: $y + 3 = \frac{3}{2}(x - 2)$, $y + 3 = -\frac{3}{2}(x - 2)$.

    To graph, enter: $y_1 = -3 + 3\sqrt{((x-2)^2/4 - 1)}$;
    $$y_2 = -3 - 3\sqrt{((x-2)^2/4 - 1)}$$

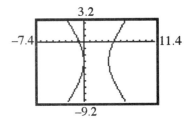

41. $(y - 2)^2 - 4(x + 2)^2 = 4$
    Divide both sides by 4 to put in standard form:

    $$\frac{(y-2)^2}{4} - \frac{(x+2)^2}{1} = 1$$

    The center of the hyperbola is at $(-2, 2)$.
    $a = 2$, $b = 1$. The vertices are $(-2, 4)$ and $(-2, 0)$.
    Find the value of c:

    $$c^2 = a^2 + b^2 = 4 + 1 = 5$$
    $$c = \sqrt{5}$$

    Foci: $\left(-2, 2 - \sqrt{5}\right)$ and $\left(-2, 2 + \sqrt{5}\right)$.
    Transverse axis: $x = -2$.
    Asymptotes: $y - 2 = 2(x + 2)$, $y - 2 = -2(x + 2)$.

    To graph, enter: $y_1 = 2 + 2\sqrt{((x+2)^2 + 1)}$;
    $$y_2 = 2 - 2\sqrt{((x+2)^2 + 1)}$$

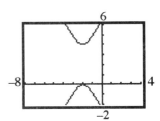

43. $(x+1)^2 - (y+2)^2 = 4$

Divide both sides by 4 to put in standard form:

$$\frac{(x+1)^2}{4} - \frac{(y+2)^2}{4} = 1$$

The center of the hyperbola is at $(-1, -2)$.

$a = 2, \ b = 2$. The vertices are $(-3, -2)$ and $(1, -2)$.

Find the value of c:

$$c^2 = a^2 + b^2 = 4 + 4 = 8$$

$$c = \sqrt{8} = 2\sqrt{2}$$

Foci: $\left(-1 - 2\sqrt{2}, -2\right)$ and $\left(-1 + 2\sqrt{2}, -2\right)$.

Transverse axis: $y = -2$.

Asymptotes: $y + 2 = x + 1, \ y + 2 = -(x+1)$.

To graph, enter: $y_1 = -2 + 2\sqrt{((x+1)^2 / 4 - 1)}$;

$$y_2 = -2 - 2\sqrt{((x+1)^2 / 4 - 1)}$$

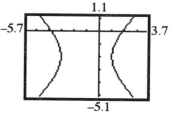

45. Complete the square to put in standard form:

$$x^2 - y^2 - 2x - 2y - 1 = 0$$

$$(x^2 - 2x + 1) - (y^2 + 2y + 1) = 1 + 1 - 1$$

$$(x - 1)^2 - (y + 1)^2 = 1$$

The center of the hyperbola is at $(1, -1)$.

$a = 1, \ b = 1$. The vertices are $(0, -1)$ and $(2, -1)$.

Find the value of c:

$$c^2 = a^2 + b^2 = 1 + 1 = 2$$

$$c = \sqrt{2}$$

Foci: $\left(1 - \sqrt{2}, -1\right)$ and $\left(1 + \sqrt{2}, -1\right)$.

Transverse axis: $y = -1$.

Asymptotes: $y + 1 = x - 1, \ y + 1 = -(x - 1)$.

To graph, enter: $y_1 = -1 + \sqrt{((x - 1)^2 - 1)}$;

$$y_2 = -1 - \sqrt{((x - 1)^2 - 1)}$$

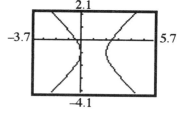

47. Complete the square to put in standard form:
$$y^2 - 4x^2 - 4y - 8x - 4 = 0$$
$$(y^2 - 4y + 4) - 4(x^2 + 2x + 1) = 4 + 4 - 4$$
$$(y - 2)^2 - 4(x + 1)^2 = 4$$
$$\frac{(y - 2)^2}{4} - \frac{(x + 1)^2}{1} = 1$$

The center of the hyperbola is at $(-1, 2)$.
$a = 2$, $b = 1$. The vertices are $(-1, 4)$ and $(-1, 0)$.
Find the value of c:
$$c^2 = a^2 + b^2 = 4 + 1 = 5$$
$$c = \sqrt{5}$$
Foci: $\left(-1, 2 - \sqrt{5}\right)$ and $\left(-1, 2 + \sqrt{5}\right)$.

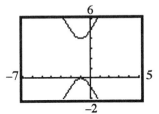

Transverse axis: $x = -1$.
Asymptotes: $y - 2 = 2(x + 1)$, $y - 2 = -2(x + 1)$.

To graph, enter: $y_1 = 2 + 2\sqrt{((x + 1)^2 + 1)}$;
$$y_2 = 2 - 2\sqrt{((x + 1)^2 + 1)}$$

49. Complete the square to put in standard form:
$$4x^2 - y^2 - 24x - 4y + 16 = 0$$
$$4(x^2 - 6x + 9) - (y^2 + 4y + 4) = -16 + 36 - 4$$
$$4(x - 3)^2 - (y + 2)^2 = 16$$
$$\frac{(x - 3)^2}{4} - \frac{(y + 2)^2}{16} = 1$$

The center of the hyperbola is at $(3, -2)$.
$a = 2$, $b = 4$. The vertices are $(1, -2)$ and $(5, -2)$.
Find the value of c:
$$c^2 = a^2 + b^2 = 4 + 16 = 20$$
$$c = \sqrt{20} = 2\sqrt{5}$$
Foci: $\left(3 - 2\sqrt{5}, -2\right)$ and $\left(3 + 2\sqrt{5}, -2\right)$.

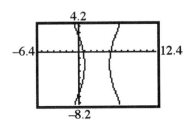

Transverse axis: $y = -2$.
Asymptotes: $y + 2 = 2(x - 3)$, $y + 2 = -2(x - 3)$.

To graph, enter: $y_1 = -2 + 4\sqrt{((x - 3)^2 / 4 - 1)}$;
$$y_2 = -2 - 4\sqrt{((x - 3)^2 / 4 - 1)}$$

51.   Complete the square to put in standard form:

$$y^2 - 4x^2 - 16x - 2y - 19 = 0$$

$$(y^2 - 2y + 1) - 4(x^2 + 4x + 4) = 19 + 1 - 16$$

$$(y - 1)^2 - 4(x + 2)^2 = 4$$

$$\frac{(y-1)^2}{4} - \frac{(x+2)^2}{1} = 1$$

The center of the hyperbola is at $(-2, 1)$.

$a = 2,\ b = 1$. The vertices are $(-2, 3)$ and $(-2, -1)$.

Find the value of c:

$$c^2 = a^2 + b^2 = 4 + 1 = 5$$

$$c = \sqrt{5}$$

Foci: $\left(-2, 1 - \sqrt{5}\right)$ and $\left(-2, 1 + \sqrt{5}\right)$.

Transverse axis: $x = -2$.

Asymptotes: $y - 1 = 2(x + 2),\ y - 1 = -2(x + 2)$.

To graph, enter: $y_1 = 1 + 2\sqrt{((x + 2)^2 + 1)}$;

$$y_2 = 1 - 2\sqrt{((x + 2)^2 + 1)}$$

53.   Rewrite the equation:

$$y = \sqrt{16 + 4x^2}$$

$$y^2 = 16 + 4x^2, \quad y \geq 0$$

$$y^2 - 4x^2 = 16, \quad\quad y \geq 0$$

$$\frac{y^2}{16} - \frac{x^2}{4} = 1, \quad\quad y \geq 0$$

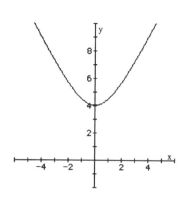

55.   Rewrite the equation:

$$y = -\sqrt{-25 + x^2}$$

$$y^2 = -25 + x^2, \quad y \leq 0$$

$$x^2 - y^2 = 25, \quad\quad y \leq 0$$

$$\frac{x^2}{25} - \frac{y^2}{25} = 1, \quad\quad y \leq 0$$

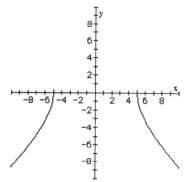

57. (a) Set up a coordinate system so that the two stations lie on the x-axis and the origin is midway between them. The ship lies on a hyperbola whose foci are the locations of the two stations. Since the time difference is 0.00038 seconds and the speed of the signal is 186,000 miles per second, the difference in the distances of the ships from each station is:

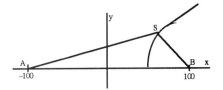

$$\text{distance} = (186{,}000)(0.00038) = 70.68 \text{ miles}$$

The difference of the distances from the ship to each station, 70.68, equals $2a$, so $a = 35.34$ and the vertex of the corresponding hyperbola is at (35.34, 0). Since the focus is at (100, 0), following this hyperbola, the ship would reach shore 64.66 miles from the master station.

(b) The ship should follow a hyperbola with a vertex at (80, 0). For this hyperbola, $a = 80$, so the constant difference of the distances from the ship to each station is 160. The time difference the ship should look for is:

$$\text{time} = \frac{160}{186{,}000} = 0.00086 \text{ seconds}$$

(c) Find the equation of the hyperbola with vertex at (80, 0) and a focus at (100, 0). The form of the equation of the hyperbola is:

$$\frac{x^2}{a^2} - \frac{y^2}{b^2} = 1 \quad \text{where } a = 80.$$

Since $c = 100$ and $b^2 = c^2 - a^2 \rightarrow b^2 = 100^2 - 80^2 = 3600.$

The equation of the hyperbola is: $\dfrac{x^2}{6400} - \dfrac{y^2}{3600} = 1.$

Since the ship is 50 miles off shore, we have $y = 50$. Solve the equation for x:

$$\frac{x^2}{6400} - \frac{50^2}{3600} = 1 \rightarrow \frac{x^2}{6400} = 1 + \frac{2500}{3600} = \frac{61}{36} \rightarrow x^2 = 6400 \cdot \frac{61}{36}$$
$$x \approx 104 \text{ miles}$$

The ship's location is (104, 50).

59. (a) Set up a rectangular coordinate system so that the two devices lie on the x-axis and the origin is midway between them. The devices serve as foci to the hyperbola so $c = \dfrac{2000}{2} = 1000$. Since the explosion occurs 200 feet from point B, the vertex of the hyperbola is (800, 0); therefore, $a = 800$. Finding b:

$$b^2 = c^2 - a^2 \rightarrow b^2 = 1000^2 - 800^2 = 360000 \rightarrow b = 600$$

The equation of the hyperbola is:

$$\frac{x^2}{800^2} - \frac{y^2}{600^2} = 1$$

If $x = 1000$, find y:

$$\frac{1000^2}{800^2} - \frac{y^2}{600^2} = 1 \rightarrow \frac{y^2}{600^2} = \frac{1000^2}{800^2} - 1 = \frac{600^2}{800^2} \rightarrow y^2 = 600^2 \cdot \frac{600^2}{800^2}$$
$$y = 450 \text{ feet}$$

The second detonation should take place 450 feet north of point B.

61. If the eccentricity is close to 1, then $c \approx a$ and $b \approx 0$. When b is close to 0, the hyperbola is very narrow, because the slopes of the asymptotes are close to 0. If the eccentricity is very large, then c is much larger than a and b is very large. The result is a hyperbola that is very wide.

63. $\frac{x^2}{4} - y^2 = 1$ $(a = 2, \ b = 1)$

    is a hyperbola with horizontal transverse axis,

    centered at (0, 0) and has asymptotes: $y = \pm \frac{1}{2}x$

    $y^2 - \frac{x^2}{4} = 1$ $(a = 1, \ b = 2)$

    is a hyperbola with vertical transverse axis, centered

    at (0, 0) and has asymptotes: $y = \pm \frac{1}{2}x$

    Since the two hyperbolas have the same asymptotes, they are conjugate.

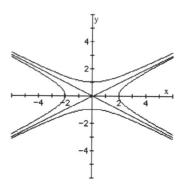

65. Put the equation in standard hyperbola form:

    $$Ax^2 + Cy^2 + F = 0 \qquad A \neq 0, \ C \neq 0, \ F \neq 0$$
    $$Ax^2 + Cy^2 = -F$$
    $$\frac{Ax^2}{-F} + \frac{Cy^2}{-F} = 1$$
    $$\frac{x^2}{-F/A} + \frac{y^2}{-F/C} = 1$$

    Since $-F/A$ and $-F/C$ have opposite signs, this is a hyperbola with center at (0, 0).

10.5 Rotation of Axes; General Form of a Conic

1. $x^2 + 4x + y + 3 = 0$
   $A = 1$ and $C = 0$; $AC = (1)(0) = 0$. Since $AC = 0$, the equation defines a parabola.

3. $6x^2 + 3y^2 - 12x + 6y = 0$
   $A = 6$ and $C = 3$; $AC = (6)(3) = 18$. Since $AC > 0$ and $A \neq C$, the equation defines an ellipse.

5. $3x^2 - 2y^2 + 6x + 4 = 0$
   $A = 3$ and $C = -2$; $AC = (3)(-2) = -6$. Since $AC < 0$, the equation defines a hyperbola.

7. $2y^2 - x^2 - y + x = 0$

$A = -1$ and $C = 2$; $AC = (-1)(2) = -2$. Since $AC < 0$, the equation defines a hyperbola.

9. $x^2 + y^2 - 8x + 4y = 0$

$A = 1$ and $C = 1$; $AC = (1)(1) = 1$. Since $AC > 0$ and $A = C$, the equation defines a circle.

11. $x^2 + 4xy + y^2 - 3 = 0$

$A = 1, B = 4$, and $C = 1$; $\cot 2\theta = \dfrac{A - C}{B} = \dfrac{1-1}{4} = \dfrac{0}{4} = 0 \;\rightarrow\; 2\theta = \dfrac{\pi}{2} \;\rightarrow\; \theta = \dfrac{\pi}{4}$

$x = x' \cos\dfrac{\pi}{4} - y' \sin\dfrac{\pi}{4} = \dfrac{\sqrt{2}}{2}x' - \dfrac{\sqrt{2}}{2}y' = \dfrac{\sqrt{2}}{2}(x' - y')$

$y = x' \sin\dfrac{\pi}{4} + y' \cos\dfrac{\pi}{4} = \dfrac{\sqrt{2}}{2}x' + \dfrac{\sqrt{2}}{2}y' = \dfrac{\sqrt{2}}{2}(x' + y')$

13. $5x^2 + 6xy + 5y^2 - 8 = 0$

$A = 5, B = 6$, and $C = 5$; $\cot 2\theta = \dfrac{A - C}{B} = \dfrac{5-5}{6} = \dfrac{0}{6} = 0 \;\rightarrow\; 2\theta = \dfrac{\pi}{2} \;\rightarrow\; \theta = \dfrac{\pi}{4}$

$x = x' \cos\dfrac{\pi}{4} - y' \sin\dfrac{\pi}{4} = \dfrac{\sqrt{2}}{2}x' - \dfrac{\sqrt{2}}{2}y' = \dfrac{\sqrt{2}}{2}(x' - y')$

$y = x' \sin\dfrac{\pi}{4} + y' \cos\dfrac{\pi}{4} = \dfrac{\sqrt{2}}{2}x' + \dfrac{\sqrt{2}}{2}y' = \dfrac{\sqrt{2}}{2}(x' + y')$

15. $13x^2 - 6\sqrt{3}xy + 7y^2 - 16 = 0$

$A = 13, B = -6\sqrt{3}$, and $C = 7$; $\cot 2\theta = \dfrac{A - C}{B} = \dfrac{13-7}{-6\sqrt{3}} = \dfrac{6}{-6\sqrt{3}} = -\dfrac{\sqrt{3}}{3}$

$2\theta = \dfrac{2\pi}{3} \;\rightarrow\; \theta = \dfrac{\pi}{3}$

$x = x' \cos\dfrac{\pi}{3} - y' \sin\dfrac{\pi}{3} = \dfrac{1}{2}x' - \dfrac{\sqrt{3}}{2}y' = \dfrac{1}{2}\left(x' - \sqrt{3}y'\right)$

$y = x' \sin\dfrac{\pi}{3} + y' \cos\dfrac{\pi}{3} = \dfrac{\sqrt{3}}{2}x' + \dfrac{1}{2}y' = \dfrac{1}{2}\left(\sqrt{3}x' + y'\right)$

17. $4x^2 - 4xy + y^2 - 8\sqrt{5}x - 16\sqrt{5}y = 0$

$A = 4, B = -4$, and $C = 1$; $\cot 2\theta = \dfrac{A - C}{B} = \dfrac{4-1}{-4} = -\dfrac{3}{4}$; $\cos 2\theta = -\dfrac{3}{5}$

$\sin\theta = \sqrt{\dfrac{1 - \left(-\dfrac{3}{5}\right)}{2}} = \sqrt{\dfrac{4}{5}} = \dfrac{2}{\sqrt{5}} = \dfrac{2\sqrt{5}}{5}$; $\cos\theta = \sqrt{\dfrac{1 + \left(-\dfrac{3}{5}\right)}{2}} = \sqrt{\dfrac{1}{5}} = \dfrac{1}{\sqrt{5}} = \dfrac{\sqrt{5}}{5}$

$x = x' \cos\theta - y' \sin\theta = \dfrac{\sqrt{5}}{5}x' - \dfrac{2\sqrt{5}}{5}y' = \dfrac{\sqrt{5}}{5}(x' - 2y')$

$y = x' \sin\theta + y' \cos\theta = \dfrac{2\sqrt{5}}{5}x' + \dfrac{\sqrt{5}}{5}y' = \dfrac{\sqrt{5}}{5}(2x' + y')$

19. $25x^2 - 36xy + 40y^2 - 12\sqrt{13}\,x - 8\sqrt{13}\,y = 0$

$A = 25, B = -36$, and $C = 40$; $\cot 2\theta = \dfrac{A-C}{B} = \dfrac{25-40}{-36} = \dfrac{5}{12}$; $\cos 2\theta = \dfrac{5}{13}$

$\sin\theta = \sqrt{\dfrac{1-\dfrac{5}{13}}{2}} = \sqrt{\dfrac{4}{13}} = \dfrac{2}{\sqrt{13}} = \dfrac{2\sqrt{13}}{13}$; $\cos\theta = \sqrt{\dfrac{1+\dfrac{5}{13}}{2}} = \sqrt{\dfrac{9}{13}} = \dfrac{3}{\sqrt{13}} = \dfrac{3\sqrt{13}}{13}$

$x = x'\cos\theta - y'\sin\theta = \dfrac{3\sqrt{13}}{13}x' - \dfrac{2\sqrt{13}}{13}y' = \dfrac{\sqrt{13}}{13}(3x' - 2y')$

$y = x'\sin\theta + y'\cos\theta = \dfrac{2\sqrt{13}}{13}x' + \dfrac{3\sqrt{13}}{13}y' = \dfrac{\sqrt{13}}{13}(2x' + 3y')$

21.

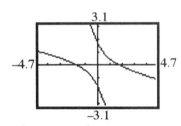

$x^2 + 4xy + y^2 - 3 = 0$; $\theta = 45°$ (see Problem 11)

$\left(\dfrac{\sqrt{2}}{2}(x'-y')\right)^2 + 4\left(\dfrac{\sqrt{2}}{2}(x'-y')\right)\left(\dfrac{\sqrt{2}}{2}(x'+y')\right) + \left(\dfrac{\sqrt{2}}{2}(x'+y')\right)^2 - 3 = 0$

$\dfrac{1}{2}\left(x'^2 - 2x'y' + y'^2\right) + 2\left(x'^2 - y'^2\right) + \dfrac{1}{2}\left(x'^2 + 2x'y' + y'^2\right) - 3 = 0$

$\dfrac{1}{2}x'^2 - x'y' + \dfrac{1}{2}y'^2 + 2x'^2 - 2y'^2 + \dfrac{1}{2}x'^2 + x'y' + \dfrac{1}{2}y'^2 = 3$

$3x'^2 - y'^2 = 3$

$\dfrac{x'^2}{1} - \dfrac{y'^2}{3} = 1$

Hyperbola; center at the origin, transverse axis is the x'-axis, vertices at $(\pm 1, 0)$.

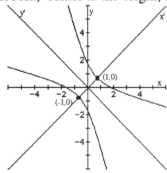

23.

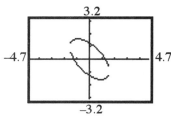

$5x^2 + 6xy + 5y^2 - 8 = 0;\quad \theta = 45°$ (see Problem 13)

$$5\left(\frac{\sqrt{2}}{2}(x'-y')\right)^2 + 6\left(\frac{\sqrt{2}}{2}(x'-y')\right)\left(\frac{\sqrt{2}}{2}(x'+y')\right) + 5\left(\frac{\sqrt{2}}{2}(x'+y')\right)^2 - 8 = 0$$

$$\frac{5}{2}\left(x'^2 - 2x'y' + y'^2\right) + 3\left(x'^2 - y'^2\right) + \frac{5}{2}\left(x'^2 + 2x'y' + y'^2\right) - 8 = 0$$

$$\frac{5}{2}x'^2 - 5x'y' + \frac{5}{2}y'^2 + 3x'^2 - 3y'^2 + \frac{5}{2}x'^2 + 5x'y' + \frac{5}{2}y'^2 = 8$$

$$8x'^2 + 2y'^2 = 8$$

$$\frac{x'^2}{1} + \frac{y'^2}{4} = 1$$

Ellipse; center at the origin, major axis is the y'-axis, vertices at $(0, \pm 2)$.

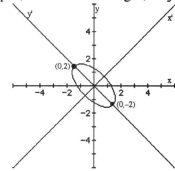

25.

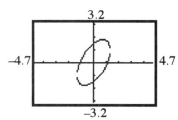

$13x^2 - 6\sqrt{3}\,xy + 7y^2 - 16 = 0;\quad \theta = 60°$ (see Problem 15)

$$13\left(\tfrac{1}{2}\left(x'-\sqrt{3}y'\right)\right)^2 - 6\sqrt{3}\left(\tfrac{1}{2}\left(x'-\sqrt{3}y'\right)\right)\left(\tfrac{1}{2}\left(\sqrt{3}\,x'+y'\right)\right) + 7\left(\tfrac{1}{2}\left(\sqrt{3}\,x'+y'\right)\right)^2 - 16 = 0$$

$$\tfrac{13}{4}\left(x'^2 - 2\sqrt{3}\,x'\,y' + 3y'^2\right) - \tfrac{3\sqrt{3}}{2}\left(\sqrt{3}\,x'^2 - 2x'\,y' - \sqrt{3}y'^2\right) + \tfrac{7}{4}\left(3x'^2 + 2\sqrt{3}\,x'\,y' + y'^2\right) = 16$$

$$\tfrac{13}{4}x'^2 - \tfrac{13\sqrt{3}}{2}x'\,y' + \tfrac{39}{4}y'^2 - \tfrac{9}{2}x'^2 + 3\sqrt{3}x'\,y' + \tfrac{9}{2}y'^2 + \tfrac{21}{4}x'^2 + \tfrac{7\sqrt{3}}{2}x'\,y' + \tfrac{7}{4}y'^2 = 16$$

$$4x'^2 + 16y'^2 = 16$$

$$\frac{x'^2}{4} + \frac{y'^2}{1} = 1$$

Ellipse; center at the origin, major axis is the x'-axis, vertices at (±2, 0).

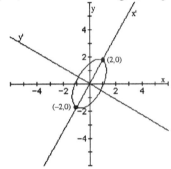

27.

$$4x^2 - 4xy + y^2 - 8\sqrt{5}\,x - 16\sqrt{5}\,y = 0; \quad \theta = 63.4° \text{ (see Problem 17)}$$

$$4\left(\frac{\sqrt{5}}{5}(x'-2y')\right)^2 - 4\left(\frac{\sqrt{5}}{5}(x'-2y')\right)\left(\frac{\sqrt{5}}{5}(2x'+y')\right) + \left(\frac{\sqrt{5}}{5}(2x'+y')\right)^2$$

$$- 8\sqrt{5}\left(\frac{\sqrt{5}}{5}(x'-2y')\right) - 16\sqrt{5}\left(\frac{\sqrt{5}}{5}(2x'+y')\right) = 0$$

$$\tfrac{4}{5}\left(x'^2 - 4x'\,y' + 4y'^2\right) - \tfrac{4}{5}\left(2x'^2 - 3x'\,y' - 2y'^2\right) + \tfrac{1}{5}\left(4x'^2 + 4x'\,y' + y'^2\right)$$

$$- 8x' + 16y' - 32x' - 16y' = 0$$

$$\tfrac{4}{5}x'^2 - \tfrac{16}{5}x'\,y' + \tfrac{16}{5}y'^2 - \tfrac{8}{5}x'^2 + \tfrac{12}{5}x'\,y' + \tfrac{8}{5}y'^2 + \tfrac{4}{5}x'^2 + \tfrac{4}{5}x'\,y' + \tfrac{1}{5}y'^2 - 40x' = 0$$

$$5y'^2 - 40x' = 0$$

$$y'^2 = 8x'$$

Parabola; vertex at the origin, focus at (2, 0).

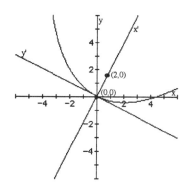

29.

$$25x^2 - 36xy + 40y^2 - 12\sqrt{13}x - 8\sqrt{13}y = 0; \quad \theta \approx 33.7° \text{ (see Problem 19)}$$

$$25\left(\frac{\sqrt{13}}{13}(3x' - 2y')\right)^2 - 36\left(\frac{\sqrt{13}}{13}(3x' - 2y')\right)\left(\frac{\sqrt{13}}{13}(2x' + 3y')\right) + 40\left(\frac{\sqrt{13}}{13}(2x' + 3y')\right)^2$$

$$- 12\sqrt{13}\left(\frac{\sqrt{13}}{13}(3x' - 2y')\right) - 8\sqrt{13}\left(\frac{\sqrt{13}}{13}(2x' + 3y')\right) = 0$$

$$\frac{25}{13}\left(9x'^2 - 12x'y' + 4y'^2\right) - \frac{36}{13}\left(6x'^2 + 5x'y' - 6y'^2\right) + \frac{40}{13}\left(4x'^2 + 12x'y' + 9y'^2\right)$$

$$- 36x' + 24y' - 16x' - 24y' = 0$$

$$\frac{225}{13}x'^2 - \frac{300}{13}x'y' + \frac{100}{13}y'^2 - \frac{216}{13}x'^2 - \frac{180}{13}x'y' + \frac{216}{13}y'^2$$

$$+ \frac{160}{13}x'^2 + \frac{480}{13}x'y' + \frac{360}{13}y'^2 - 52x' = 0$$

$$13x'^2 + 52y'^2 - 52x' = 0$$

$$x'^2 - 4x' + 4y'^2 = 0$$

$$(x' - 2)^2 + 4y'^2 = 4$$

$$\frac{(x' - 2)^2}{4} + \frac{y'^2}{1} = 1$$

Ellipse; center at (2, 0), major axis is the x'-axis, vertices at (4, 0) and (0, 0).

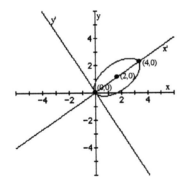

31.

$$16x^2 + 24xy + 9y^2 - 130x + 90y = 0$$

$$A = 16, B = 24, \text{ and } C = 9; \quad \cot 2\theta = \frac{A - C}{B} = \frac{16 - 9}{24} = \frac{7}{24} \quad \rightarrow \quad \cos 2\theta = \frac{7}{25}$$

$$\sin\theta = \sqrt{\frac{1 - \frac{7}{25}}{2}} = \sqrt{\frac{9}{25}} = \frac{3}{5}; \quad \cos\theta = \sqrt{\frac{1 + \frac{7}{25}}{2}} = \sqrt{\frac{16}{25}} = \frac{4}{5} \quad \rightarrow \quad \theta \approx 36.9$$

$$x = x'\cos\theta - y'\sin\theta = \frac{4}{5}x' - \frac{3}{5}y' = \frac{1}{5}(4x' - 3y')$$

$$y = x'\sin\theta + y'\cos\theta = \frac{3}{5}x' + \frac{4}{5}y' = \frac{1}{5}(3x' + 4y')$$

$$16\left(\frac{1}{5}(4x' - 3y')\right)^2 + 24\left(\frac{1}{5}(4x' - 3y')\right)\left(\frac{1}{5}(3x' + 4y')\right) + 9\left(\frac{1}{5}(3x' + 4y')\right)^2$$

$$- 130\left(\frac{1}{5}(4x' - 3y')\right) + 90\left(\frac{1}{5}(3x' + 4y')\right) = 0$$

$$\frac{16}{25}\left(16x'^2 - 24x'y' + 9y'^2\right) + \frac{24}{25}\left(12x'^2 + 7x'y' - 12y'^2\right) + \frac{9}{25}\left(9x'^2 + 24x'y' + 16y'^2\right)$$

$$- 104x' + 78y' + 54x' + 72y' = 0$$

$$\frac{256}{25}x'^2 - \frac{384}{25}x'y' + \frac{144}{25}y'^2 + \frac{288}{25}x'^2 + \frac{168}{25}x'y' - \frac{288}{25}y'^2$$

$$+ \frac{81}{25}x'^2 + \frac{216}{25}x'y' + \frac{144}{25}y'^2 - 50x' + 150y' = 0$$

$$25x'^2 - 50x' + 150y' = 0$$

$$x'^2 - 2x' = -6y'$$

$$(x' - 1)^2 = -6y' + 1$$

$$(x' - 1)^2 = -6\left(y' - \frac{1}{6}\right)$$

Parabola; vertex at $\left(1, \dfrac{1}{6}\right)$, focus at $\left(1, -\dfrac{4}{3}\right)$.

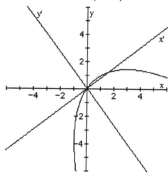

33. $A = 1.$ $B = 3,$ $C = -2$ $B^2 - 4AC = 3^2 - 4(1)(-2) = 17 > 0;$ hyperbola

35. $A = 1.$ $B = -7,$ $C = 3$ $B^2 - 4AC = (-7)^2 - 4(1)(3) = 37 > 0;$ hyperbola

37. $A = 9.$ $B = 12,$ $C = 4$ $B^2 - 4AC = 12^2 - 4(9)(4) = 0;$ parabola

39. $A = 10.$ $B = -12,$ $C = 4$ $B^2 - 4AC = (-12)^2 - 4(10)(4) = -16 < 0;$ ellipse

41. $A = 3.$ $B = -2,$ $C = 1$ $B^2 - 4AC = (-2)^2 - 4(3)(1) = -8 < 0;$ ellipse

43.   See equation 6 on page 678.

$$A' = A\cos^2\theta + B\sin\theta\cos\theta + C\sin^2\theta$$
$$B' = B(\cos^2\theta - \sin^2\theta) + 2(C - A)(\sin\theta\cos\theta)$$
$$C' = A\sin^2\theta - B\sin\theta\cos\theta + C\cos^2\theta$$
$$D' = D\cos\theta + E\sin\theta$$
$$E' = -D\sin\theta + E\cos\theta$$
$$F' = F$$

45. $B'^2 - 4A'C'$

$$= \left[B(\cos^2\theta - \sin^2\theta) + 2(C - A)\sin\theta\cos\theta\right]^2$$
$$- 4\left(A\cos^2\theta + B\sin\theta\cos\theta + C\sin^2\theta\right)\left(A\sin^2\theta - B\sin\theta\cos\theta + C\cos^2\theta\right)$$
$$= B^2\left(\cos^4\theta - 2\cos^2\theta\sin^2\theta + \sin^4\theta\right) + 4B(C - A)\sin\theta\cos\theta(\cos^2\theta - \sin^2\theta)$$
$$+ 4(C - A)^2\sin^2\theta\cos^2\theta - 4\left[A^2\sin^2\theta\cos^2\theta - AB\sin\theta\cos^3\theta + AC\cos^4\theta\right.$$
$$+ AB\sin^3\theta\cos\theta - B^2\sin^2\theta\cos^2\theta + BC\sin\theta\cos^3\theta + AC\sin^4\theta$$
$$\left. - BC\sin^3\theta\cos\theta + C^2\sin^2\theta\cos^2\theta\right]$$

$$= B^2\left(\cos^4\theta - 2\cos^2\theta\sin^2\theta + \sin^4\theta + 4\sin^2\theta\cos^2\theta\right)$$
$$+ BC\left(4\sin\theta\cos\theta(\cos^2\theta - \sin^2\theta) - 4\sin\theta\cos^3\theta + 4\sin^3\theta\cos\theta\right)$$
$$- AB\left(4\sin\theta\cos\theta(\cos^2\theta - \sin^2\theta) - 4\sin\theta\cos^3\theta + 4\sin^3\theta\cos\theta\right)$$
$$+ 4C^2\left(\sin^2\theta\cos^2\theta - \sin^2\theta\cos^2\theta\right) - 4AC\left(2\sin^2\theta\cos^2\theta + \cos^4\theta + \sin^4\theta\right)$$
$$+ 4A^2\left(\sin^2\theta\cos^2\theta - \sin^2\theta\cos^2\theta\right)$$
$$= B^2\left(\cos^4\theta + 2\sin^2\theta\cos^2\theta + \sin^4\theta\right) - 4AC\left(\cos^4\theta + 2\sin^2\theta\cos^2\theta + \sin^4\theta\right)$$
$$= B^2\left(\cos^2\theta + \sin^2\theta\right)^2 - 4AC\left(\cos^2\theta + \sin^2\theta\right)^2$$
$$= B^2 - 4AC$$

47.
$$d^2 = (y_2 - y_1)^2 + (x_2 - x_1)^2$$
$$= \left(x_2'\sin\theta + y_2'\cos\theta - x_1'\sin\theta - y_1'\cos\theta\right)^2$$
$$\qquad\qquad + \left(x_2'\cos\theta - y_2'\sin\theta - x_1'\cos\theta + y_1'\sin\theta\right)^2$$
$$= \left((x_2' - x_1')\sin\theta + (y_2' - y_1')\cos\theta\right)^2 + \left((x_2' - x_1')\cos\theta - (y_2' - y_1')\sin\theta\right)^2$$
$$= (x_2' - x_1')^2\sin^2\theta + 2(x_2' - x_1')(y_2' - y_1')\sin\theta\cos\theta + (y_2' - y_1')^2\cos^2\theta$$
$$\qquad + (x_2' - x_1')^2\cos^2\theta - 2(x_2' - x_1')(y_2' - y_1')\sin\theta\cos\theta + (y_2' - y_1')^2\sin^2\theta$$
$$= (x_2' - x_1')^2\sin^2\theta + (x_2' - x_1')^2\cos^2\theta + (y_2' - y_1')^2\cos^2\theta + (y_2' - y_1')^2\sin^2\theta$$
$$= (x_2' - x_1')^2 + (y_2' - y_1')^2$$

10.6 Polar Equations of Conics

1. $e = 1$; $p = 1$; parabola; directrix is perpendicular to the polar axis and 1 unit to the right of the pole.

3. $r = \dfrac{4}{2 - 3\sin\theta} = \dfrac{4}{2\left(1 - \dfrac{3}{2}\sin\theta\right)} = \dfrac{2}{1 - \dfrac{3}{2}\sin\theta}$; $ep = 2$, $e = \dfrac{3}{2}$; $p = \dfrac{4}{3}$

Hyperbola; directrix is parallel to the polar axis and $\dfrac{4}{3}$ units below the pole.

5. $r = \dfrac{3}{4 - 2\cos\theta} = \dfrac{3}{4\left(1 - \dfrac{1}{2}\cos\theta\right)} = \dfrac{\dfrac{3}{4}}{1 - \dfrac{1}{2}\cos\theta}$; $ep = \dfrac{3}{4}$, $e = \dfrac{1}{2}$; $p = \dfrac{3}{2}$

Ellipse; directrix is perpendicular to the polar axis and $\dfrac{3}{2}$ units to the left of the pole.

7. $r = \dfrac{1}{1 + \cos \theta}$

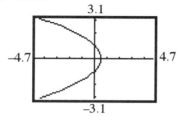

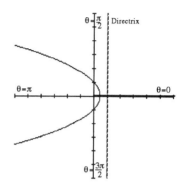

$ep = 1, \ e = 1, \ p = 1$

Parabola; directrix is perpendicular to the polar axis 1 unit to the right of the pole; vertex is $\left(\dfrac{1}{2}, 0\right)$.

9. $r = \dfrac{8}{4 + 3\sin \theta}$

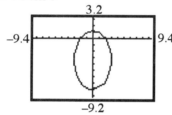

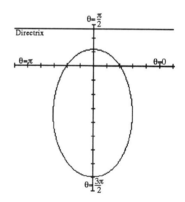

$$r = \dfrac{8}{4\left(1 + \dfrac{3}{4}\sin \theta\right)} = \dfrac{2}{1 + \dfrac{3}{4}\sin \theta}$$

$ep = 2, \ e = \dfrac{3}{4}, \ p = \dfrac{8}{3}$

Ellipse; directrix is parallel to the polar axis $\dfrac{8}{3}$ units above the pole; vertices are $\left(\dfrac{8}{7}, \dfrac{\pi}{2}\right)$ and $\left(8, \dfrac{3\pi}{2}\right)$.

11. $r = \dfrac{9}{3 - 6\cos\theta}$

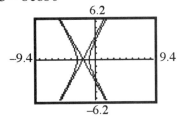

$r = \dfrac{9}{3(1 - 2\cos\theta)} = \dfrac{3}{1 - 2\cos\theta}$

$ep = 3,\ e = 2,\ p = \dfrac{3}{2}$

Hyperbola; directrix is perpendicular to the polar axis $\dfrac{3}{2}$ units to the left of the pole; vertices are $(-3, 0)$ and $(1, \pi)$.

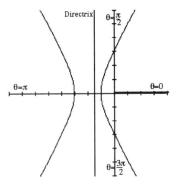

13. $r = \dfrac{8}{2 - \sin\theta}$

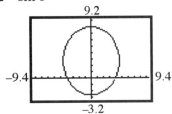

$r = \dfrac{8}{2\left(1 - \dfrac{1}{2}\sin\theta\right)} = \dfrac{4}{1 - \dfrac{1}{2}\sin\theta}$

$ep = 4,\ e = \dfrac{1}{2},\ p = 8$

Ellipse; directrix is parallel to the polar axis 8 units below the pole; vertices are

$\left(8, \dfrac{\pi}{2}\right)$ and $\left(\dfrac{8}{3}, \dfrac{3\pi}{2}\right)$.

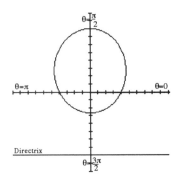

15. $r(3 - 2\sin\theta) = 6 \quad \rightarrow \quad r = \dfrac{6}{3 - 2\sin\theta}$

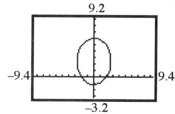

$r = \dfrac{6}{3\left(1 - \dfrac{2}{3}\sin\theta\right)} = \dfrac{2}{1 - \dfrac{2}{3}\sin\theta}$

$ep = 2, \quad e = \dfrac{2}{3}, \quad p = 3$

Ellipse; directrix is parallel to the polar axis 3 units below the pole; vertices are

$\left(6, \dfrac{\pi}{2}\right)$ and $\left(\dfrac{6}{5}, \dfrac{3\pi}{2}\right)$.

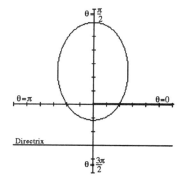

17. $r = \dfrac{6\sec\theta}{2\sec\theta - 1} = \dfrac{6}{2 - \cos\theta}$

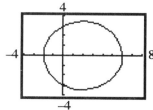

$r = \dfrac{6}{2\left(1 - \dfrac{1}{2}\cos\theta\right)} = \dfrac{3}{1 - \dfrac{1}{2}\cos\theta}$

$ep = 3, \quad e = \dfrac{1}{2}, \quad p = 6$

Ellipse; directrix is perpendicular to the polar axis 6 units to the left of the pole; vertices are $(6, 0)$ and $(2, \pi)$.

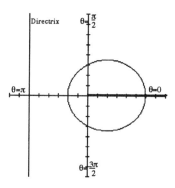

19. $r = \dfrac{1}{1 + \cos\theta}$

$r + r\cos\theta = 1$

$\quad\quad r = 1 - r\cos\theta$

$\quad\quad r^2 = (1 - r\cos\theta)^2$

$\quad x^2 + y^2 = (1 - x)^2$

$\quad x^2 + y^2 = 1 - 2x + x^2$

$y^2 + 2x - 1 = 0$

21. $r = \dfrac{8}{4 + 3\sin\theta}$

$4r + 3r\sin\theta = 8$

$\quad\quad 4r = 8 - 3r\sin\theta$

$\quad\quad 16r^2 = (8 - 3r\sin\theta)^2$

$\quad 16(x^2 + y^2) = (8 - 3y)^2$

$\quad 16x^2 + 16y^2 = 64 - 48y + 9y^2$

$16x^2 + 7y^2 + 48y - 64 = 0$

23. $$r = \frac{9}{3 - 6\cos\theta}$$
$$3r - 6r\cos\theta = 9$$
$$3r = 9 + 6r\cos\theta$$
$$r = 3 + 2r\cos\theta$$
$$r^2 = (3 + 2r\cos\theta)^2$$
$$x^2 + y^2 = (3 + 2x)^2$$
$$x^2 + y^2 = 9 + 12x + 4x^2$$
$$3x^2 - y^2 + 12x + 9 = 0$$

25. $$r = \frac{8}{2 - \sin\theta}$$
$$2r - r\sin\theta = 8$$
$$2r = 8 + r\sin\theta$$
$$4r^2 = (8 + r\sin\theta)^2$$
$$4(x^2 + y^2) = (8 + y)^2$$
$$4x^2 + 4y^2 = 64 + 16y + y^2$$
$$4x^2 + 3y^2 - 16y - 64 = 0$$

27. $$r(3 - 2\sin\theta) = 6$$
$$3r - 2r\sin\theta = 6$$
$$3r = 6 + 2r\sin\theta$$
$$9r^2 = (6 + 2r\sin\theta)^2$$
$$9(x^2 + y^2) = (6 + 2y)^2$$
$$9x^2 + 9y^2 = 36 + 24y + 4y^2$$
$$9x^2 + 5y^2 - 24y - 36 = 0$$

29. $$r = \frac{6\sec\theta}{2\sec\theta - 1}$$
$$r = \frac{6}{2 - \cos\theta}$$
$$2r - r\cos\theta = 6$$
$$2r = 6 + r\cos\theta$$
$$4r^2 = (6 + r\cos\theta)^2$$
$$4(x^2 + y^2) = (6 + x)^2$$
$$4x^2 + 4y^2 = 36 + 12x + x^2$$
$$3x^2 + 4y^2 - 12x - 36 = 0$$

31. $$r = \frac{ep}{1 + e\sin\theta}$$
$$e = 1; \quad p = 1$$
$$r = \frac{1}{1 + \sin\theta}$$

33. $$r = \frac{ep}{1 - e\cos\theta}$$
$$e = \frac{4}{5}; \quad p = 3$$
$$r = \frac{\frac{12}{5}}{1 - \frac{4}{5}\cos\theta} = \frac{12}{5 - 4\cos\theta}$$

35. $$r = \frac{ep}{1 - e\sin\theta}$$
$$e = 6; \quad p = 2$$
$$r = \frac{12}{1 - 6\sin\theta}$$

37. $$d(F, P) = e \cdot d(D, P) \qquad d(D, P) = p - r\cos\theta$$
$$r = e(p - r\cos\theta)$$
$$r = ep - er\cos\theta$$
$$r + er\cos\theta = ep$$
$$r(1 + e\cos\theta) = ep$$
$$r = \frac{ep}{1 + e\cos\theta}$$

39. $d(F, P) = e \cdot d(D, P)$ $d(D, P) = p + r \sin \theta$

$$r = e(p + r \sin \theta)$$
$$r = ep + er \sin \theta$$
$$r - er \sin \theta = ep$$
$$r(1 - e \sin \theta) = ep$$
$$r = \frac{ep}{1 - e \sin \theta}$$

10.7 Plane Curves and Parametric Equations

1. $x = 3t + 2, \ y = t + 1, \ 0 \le t \le 4$

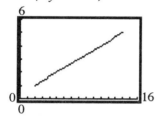

$$x = 3(y - 1) + 2$$
$$x = 3y - 3 + 2$$
$$x = 3y - 1$$
$$x - 3y + 1 = 0$$

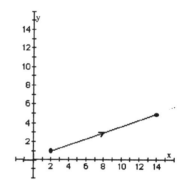

3. $x = t + 2, \ y = \sqrt{t}, \ t \ge 0$

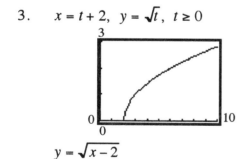

$$y = \sqrt{x - 2}$$

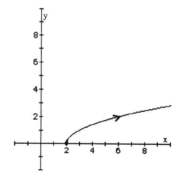

5. $x = t^2 + 4$, $y = t^2 - 4$, $-\infty < t < \infty$

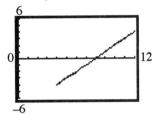

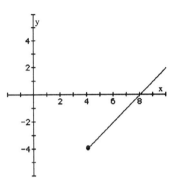

$y = (x - 4) - 4$

$y = x - 8$

For $-\infty < t < 0$ the movement is to the left. For $0 < t < \infty$ the movement is to the right.

7. $x = 3t^2$, $y = t + 1$, $-\infty < t < \infty$

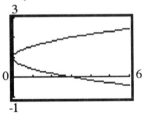

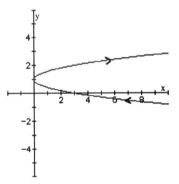

$x = 3(y - 1)^2$

9. $x = 2e^t$, $y = 1 + e^t$, $t \geq 0$

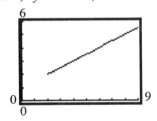

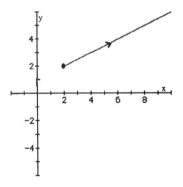

$y = 1 + \dfrac{x}{2}$

$2y = 2 + x$

11. $x = \sqrt{t}$, $y = t^{\frac{3}{2}}$, $t \geq 0$

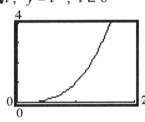

$y = \left(x^2\right)^{\frac{3}{2}}$

$y = x^3$

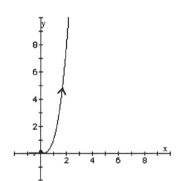

13. $x = 2\cos t$, $y = 3\sin t$, $0 \leq t \leq 2\pi$

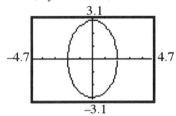

$\dfrac{x}{2} = \cos t \qquad \dfrac{y}{3} = \sin t$

$\left(\dfrac{x}{2}\right)^2 + \left(\dfrac{y}{3}\right)^2 = \cos^2 t + \sin^2 t = 1$

$\dfrac{x^2}{4} + \dfrac{y^2}{9} = 1$

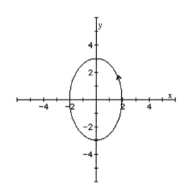

15. $x = 2\cos t$, $y = 3\sin t$, $-\pi \leq t \leq 0$

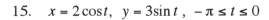

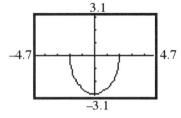

$\dfrac{x}{2} = \cos t \qquad \dfrac{y}{3} = \sin t$

$\left(\dfrac{x}{2}\right)^2 + \left(\dfrac{y}{3}\right)^2 = \cos^2 t + \sin^2 t = 1$

$\dfrac{x^2}{4} + \dfrac{y^2}{9} = 1$

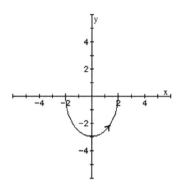

17. $x = \sec t, \quad y = \tan t, \quad 0 \le t \le \dfrac{\pi}{4}$

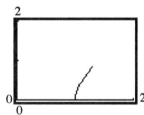

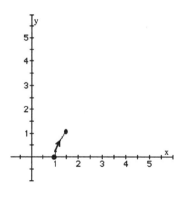

$\sec^2 t = 1 + \tan^2 t$

$x^2 = 1 + y^2$

$x^2 - y^2 = 1$

19. $x = \sin^2 t, \quad y = \cos^2 t, \quad 0 \le t \le 2\pi$

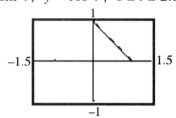

$\sin^2 t + \cos^2 t = 1$

$x + y = 1$

21.   (a)   Use equation (2):

$x = 50 \cos 90° = 0$

$y = -\dfrac{1}{2}(32)t^2 + (50 \sin 90°)t + 6 = -16t^2 + 50t + 6$

(b)   The ball is in the air until $y = 0$. Solve:

$-16t^2 + 50t + 6 = 0$

$t = \dfrac{-50 \pm \sqrt{50^2 - 4(-16)(6)}}{2(-16)} = \dfrac{-50 \pm \sqrt{2884}}{-32} \approx -0.12 \text{ or } 3.24$

The ball is in the air for about 3.24 seconds. (The negative solution is extraneous.)

(c)   The maximum height occurs at the vertex of the quadratic function.

$t = \dfrac{-b}{2a} = \dfrac{-50}{2(-16)} = 1.5625 \text{ seconds}$

Evaluate the function to find the maximum height:

$-16(1.5625)^2 + 50(1.5625) + 6 = 45.0625$

The maximum height is 45.0625 feet.

(d) (We use $x = 3$ so that the line is not on top of the y-axis.)

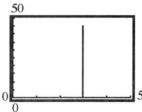

23. (a) Train: Use equation (2) with $g = 2$, $v_0 = 0$, $h = 0$

$$x_1 = \frac{1}{2}(2)t^2 + 0 \cdot t + 0 = t^2$$

$$y_1 = 1$$

Bill:

$$x_2 = 5(t - 5)$$

$$y_2 = 3$$

(b) Bill will catch the train if $x_1 = x_2$.

$$t^2 = 5(t - 5)$$

$$t^2 = 5t - 25$$

$$t^2 - 5t + 25 = 0$$

Since $b^2 - 4ac = (-5)^2 - 4(1)(25) = 25 - 100 = -75 < 0$, the equation has no real solution. Thus, Bill will not catch the train.

(c)

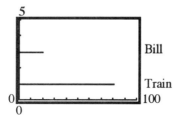

25. (a) Use equation (2):

$$x = (145\cos 20°)t$$

$$y = -\frac{1}{2}(32)t^2 + (145\sin 20°)t + 5$$

(b) The ball is in the air until $y = 0$. Solve:

$$-16t^2 + (145\sin 20°)t + 5 = 0$$

$$t = \frac{-145\sin 20° \pm \sqrt{(145\sin 20°)^2 - 4(-16)(5)}}{2(-16)}$$

$$= \frac{-49.59 \pm \sqrt{2779.46}}{-32} \approx -0.10 \text{ or } 3.20$$

The ball is in the air for about 3.20 seconds. (The negative solution is extraneous.)

(c) The maximum height occurs at the vertex of the quadratic function.

$$t = \frac{-b}{2a} = \frac{-145\sin 20°}{2(-16)} \approx 1.55 \text{ seconds}$$

Evaluate the function to find the maximum height:

$$-16(1.55)^2 + 145\sin 20°(1.55) + 5 = 43.43$$

The maximum height is about 43.43 feet.

(d) Find the horizontal displacement:

$$x = (145\cos 20°)(3.20) \approx 436 \text{ feet}$$

(e)

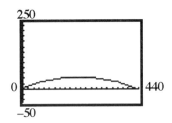

27. (a) Use equation (2):

$$x = (40\cos 45°)t$$

$$y = -\frac{1}{2}(9.8)t^2 + (40\sin 45°)t + 300$$

(b) The ball is in the air until $y = 0$. Solve:

$$-4.9t^2 + (40\sin 45°)t + 300 = 0$$

$$t = \frac{-20\sqrt{2} \pm \sqrt{\left(20\sqrt{2}\right)^2 - 4(-4.9)(300)}}{2(-4.9)}$$

$$= \frac{-20\sqrt{2} \pm \sqrt{6680}}{-9.8} \approx -5.45 \text{ or } 11.23$$

The ball is in the air for about 11.23 seconds. (The negative solution is extraneous.)

(c) The maximum height occurs at the vertex of the quadratic function.

$$t = \frac{-b}{2a} = \frac{-20\sqrt{2}}{2(-4.9)} \approx 2.89 \text{ seconds}$$

Evaluate the function to find the maximum height:

$$-4.9(2.89)^2 + 20\sqrt{2}(2.89) + 300 = 340.8 \text{ meters}$$

(d) Find the horizontal displacement:

$$x = (40\cos 45°)(11.23) \approx 317.6 \text{ meters}$$

(e)

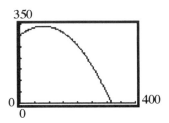

29. (a)   At $t = 0$, the Paseo is 5 miles from the intersection (at $(0, 0)$) traveling east (along the x-axis) at 40 mph. Thus, $x = 40t - 5$ describes the position of the Paseo as a function of time. The Bonneville, at $t = 0$, is 4 miles from the intersection traveling north (along the y-axis) at 30 mph. Thus, $y = 30t - 4$ describes the position of the Bonneville as a function of time.

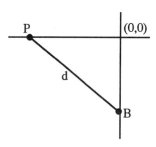

   Let d represent the distance between the cars. Use the Pythagorean Theorem to find the distance:
   $$d = \sqrt{(40t - 5)^2 + (30t - 4)^2}$$

   (b)   (Note this is a function graph not a parametric graph.)

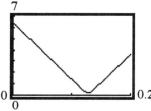

   (c)   The minimum distance between the cars is 0.2 miles and occurs at 0.128 seconds.
   (d)   From part (a):

   Paseo: $x = 40t - 5$ Bonneville: $x = 0$
   $y = 0$ $y = 30t - 4$

   (e)

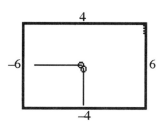

31. $x = t, \ y = 4t - 1$ $x = t + 1, \ y = 4t + 3$

33. $x = t, \ y = t^2 + 1$ $x = t - 1, \ y = t^2 - 2t + 2$

35. $x = t, \ y = t^3$ $x = \sqrt[3]{t}, \ y = t$

37. $x = t^{\frac{3}{2}}, \ y = t$ $x = t, \ y = t^{\frac{2}{3}}$

39. $x = t + 2, \ y = t; \ \ 0 \le t \le 5$

41. $x = 3\cos t, \ y = 2\sin t; \ \ 0 \le t \le 2\pi$

43. $x = 2\cos\omega t$, $y = -3\sin\omega t$

$\dfrac{2\pi}{\omega} = 2$ $\rightarrow$ $\omega = \pi$

$x = 2\cos\pi t$, $y = -3\sin\pi t$, $0 \le t \le 2$

45. $x = -2\sin\omega t$, $y = 3\cos\omega t$

$\dfrac{2\pi}{\omega} = 1$ $\rightarrow$ $\omega = 2\pi$

$x = -2\sin 2\pi t$, $y = 3\cos 2\pi t$, $0 \le t \le 1$

47. C_1 C_2

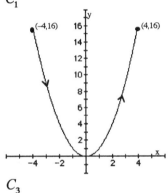   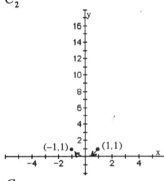

C_3 C_4

       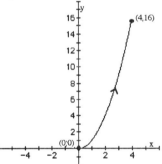

49. $x = \left(x_2 - x_1\right)t + x_1$, $y = \left(y_2 - y_1\right)t + y_1$, $-\infty < t < \infty$

$\dfrac{x - x_1}{x_2 - x_1} = t$

$y = \left(y_2 - y_1\right)\left(\dfrac{x - x_1}{x_2 - x_1}\right) + y_1$

$y - y_1 = \left(\dfrac{y_2 - y_1}{x_2 - x_1}\right)\left(x - x_1\right)$

This is the two point form for the equation of a line. Its orientation is from
$\left(x_1, y_1\right)$ to $\left(x_2, y_2\right)$.

51. $x = t\sin t, \quad y = t\cos t$

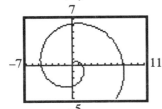

53. $x = 4\sin t - 2\sin(2t), \quad y = 4\cos t - 2\cos(2t)$

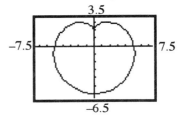

55. (a) $x(t) = \cos^3 t, \quad y(t) = \sin^3 t, \quad 0 \le t \le 2\pi$

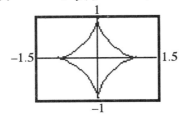

(b) $\cos t = x^{\frac{1}{3}}, \quad \sin t = y^{\frac{1}{3}}$

$$\cos^2 t + \sin^2 t = \left(x^{\frac{1}{3}}\right)^2 + \left(y^{\frac{1}{3}}\right)^2$$

$$x^{\frac{2}{3}} + y^{\frac{2}{3}} = 1$$

10 Chapter Review

1. $y^2 = -16x$
This is a parabola.
$a = 4$

    Vertex: (0, 0)
    Focus: (–4, 0)
    Directrix: $x = 4$

3. $\dfrac{x^2}{25} - y^2 = 1$
This is a hyperbola.
$a = 5, \; b = 1$. Find the value of c:

$$c^2 = a^2 + b^2 = 25 + 1 = 26 \quad \rightarrow \quad c = \sqrt{26}$$

    Center: (0, 0)
    Vertices: (5, 0), (–5, 0)
    Foci: $\left(\sqrt{26}, 0\right), \left(-\sqrt{26}, 0\right)$
    Asymptotes: $y = \tfrac{1}{5}x; \quad y = -\tfrac{1}{5}x$

5. $\dfrac{y^2}{25} + \dfrac{x^2}{16} = 1$
This is an ellipse.
$a = 5,\ b = 4$. Find the value of c:

$$c^2 = a^2 - b^2 = 25 - 16 = 9 \quad \rightarrow \quad c = 3$$

Center:	(0, 0)
Vertices:	(0, 5), (0, −5)
Foci:	(0, 3), (0, −3)

7. $x^2 + 4y = 4$
This is a parabola.
Write in standard form:

$$x^2 = -4y + 4$$
$$x^2 = -4(y - 1)$$

$a = 1$

Vertex:	(0, 1)
Focus:	(0, 0)
Directrix:	$y = 2$

9. $4x^2 - y^2 = 8$
This is a hyperbola.
Write in standard form:

$$\dfrac{x^2}{2} - \dfrac{y^2}{8} = 1$$

$a = \sqrt{2},\ b = \sqrt{8} = 2\sqrt{2}$. Find the value of c:

$$c^2 = a^2 + b^2 = 2 + 8 = 10 \quad \rightarrow \quad c = \sqrt{10}$$

Center:	(0, 0)
Vertices:	$\left(-\sqrt{2}, 0\right), \left(\sqrt{2}, 0\right)$
Foci:	$\left(-\sqrt{10}, 0\right), \left(\sqrt{10}, 0\right)$
Asymptotes:	$y = 2x;\ \ y = -2x$

11. $x^2 - 4x = 2y$
This is a parabola.
Write in standard form:

$$x^2 - 4x + 4 = 2y + 4$$
$$(x - 2)^2 = 2(y + 2)$$

$a = \frac{1}{2}$

Vertex:	(2, −2)
Focus:	$\left(2, -\frac{3}{2}\right)$
Directrix:	$y = -\frac{5}{2}$

13. $y^2 - 4y - 4x^2 + 8x = 4$
     This is a hyperbola.
     Write in standard form:
     $$(y^2 - 4y + 4) - 4(x^2 - 2x + 1) = 4 + 4 - 4$$
     $$(y - 2)^2 - 4(x - 1)^2 = 4$$
     $$\frac{(y - 2)^2}{4} - \frac{(x - 1)^2}{1} = 1$$
     $a = 2$, $b = 1$. Find the value of c:
     $$c^2 = a^2 + b^2 = 4 + 1 = 5 \quad \rightarrow \quad c = \sqrt{5}$$

Center:	$(1, 2)$
Vertices:	$(1, 0)$, $(1, 4)$
Foci:	$\left(1, 2 - \sqrt{5}\right)$, $\left(1, 2 + \sqrt{5}\right)$
Asymptotes:	$y - 2 = 2(x - 1); \quad y - 2 = -2(x - 1)$

15. $4x^2 + 9y^2 - 16x - 18y = 11$
     This is an ellipse.
     Write in standard form:
     $$4x^2 + 9y^2 - 16x - 18y = 11$$
     $$4(x^2 - 4x + 4) + 9(y^2 - 2y + 1) = 11 + 16 + 9$$
     $$4(x - 2)^2 + 9(y - 1)^2 = 36$$
     $$\frac{(x - 2)^2}{9} + \frac{(y - 1)^2}{4} = 1$$
     $a = 3$, $b = 2$. Find the value of c:
     $$c^2 = a^2 - b^2 = 9 - 4 = 5 \quad \rightarrow \quad c = \sqrt{5}$$

Center:	$(2, 1)$
Vertices:	$(-1, 1)$, $(5, 1)$
Foci:	$\left(2 - \sqrt{5}, 1\right)$, $\left(2 + \sqrt{5}, 1\right)$

17. $4x^2 - 16x + 16y + 32 = 0$
     This is a parabola.
     Write in standard form:
     $$4(x^2 - 4x + 4) = -16y - 32 + 16$$
     $$4(x - 2)^2 = -16(y + 1)$$
     $$(x - 2)^2 = -4(y + 1)$$
     $a = 1$

Vertex:	$(2, -1)$
Focus:	$(2, -2)$
Directrix:	$y = 0$

19. $9x^2 + 4y^2 - 18x + 8y = 23$
This is an ellipse.
Write in standard form:
$$9(x^2 - 2x + 1) + 4(y^2 + 2y + 1) = 23 + 9 + 4$$
$$9(x - 1)^2 + 4(y + 1)^2 = 36$$
$$\frac{(x - 1)^2}{4} + \frac{(y + 1)^2}{9} = 1$$
$a = 3,\ b = 2$. Find the value of c:
$$c^2 = a^2 - b^2 = 9 - 4 = 5 \quad \rightarrow \quad c = \sqrt{5}$$
Center: $(1, -1)$
Vertices: $(1, -4), (1, 2)$
Foci: $\left(1, -1 - \sqrt{5}\right), \left(1, -1 + \sqrt{5}\right)$

21.   Parabola: The focus is $(-2, 0)$ and the directrix is
$x = 2$. The vertex is $(0, 0)$. $a = 2$ and since $(-2, 0)$
is to the left of $(0, 0)$, the parabola opens to the left.
The equation of the parabola is:
$$y^2 = -4ax$$
$$y^2 = -4 \cdot 2 \cdot x$$
$$y^2 = -8x$$

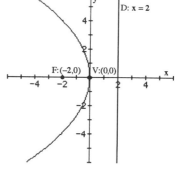

23.   Hyperbola: Center: $(0, 0)$; Focus: $(0, 4)$;
Vertex: $(0, -2)$; Transverse axis is the y-axis;
$a = 2;\ c = 4$. Find b:
$$b^2 = c^2 - a^2 = 16 - 4 = 12$$
$$b = \sqrt{12} = 2\sqrt{3}$$
Write the equation: $\dfrac{y^2}{4} - \dfrac{x^2}{12} = 1$

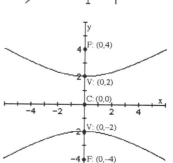

25.   Ellipse: Foci: $(-3, 0), (3, 0)$; Vertex: $(4, 0)$;
Center: $(0, 0)$; Major axis is the x-axis;
$a = 4;\ c = 3$. Find b:
$$b^2 = a^2 - c^2 = 16 - 9 = 7$$
$$b = \sqrt{7}$$
Write the equation: $\dfrac{x^2}{16} + \dfrac{y^2}{7} = 1$

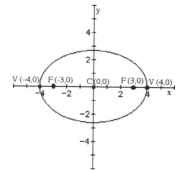

27. Parabola: The focus is $(2, -4)$ and the vertex is $(2, -3)$. Both lie on the vertical line $x = 2$. $a = 1$ and since $(2, -4)$ is below $(2, -3)$, the parabola opens down. The equation of the parabola is:

$$(x - h)^2 = -4a(y - k)$$
$$(x - 2)^2 = -4 \cdot 1 \cdot (y - (-3))$$
$$(x - 2)^2 = -4(y + 3)$$

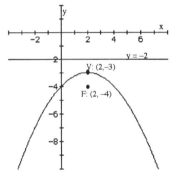

29. Hyperbola: Center: $(-2, -3)$; Focus: $(-4, -3)$; Vertex: $(-3, -3)$; Transverse axis is parallel to the x-axis; $a = 1$; $c = 2$. Find b:

$$b^2 = c^2 - a^2 = 4 - 1 = 3$$
$$b = \sqrt{3}$$

Write the equation: $\dfrac{(x + 2)^2}{1} - \dfrac{(y + 3)^2}{3} = 1$

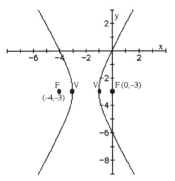

31. Ellipse: Foci: $(-4, 2), (-4, 8)$; Vertex: $(-4, 10)$; Center: $(-4, 5)$; Major axis is parallel to the y-axis; $a = 5$; $c = 3$. Find b:

$$b^2 = a^2 - c^2 = 25 - 9 = 16$$
$$b = 4$$

Write the equation: $\dfrac{(x + 4)^2}{16} + \dfrac{(y - 5)^2}{25} = 1$

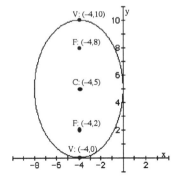

33. Hyperbola: Center: $(-1, 2)$; $a = 3$; $c = 4$; Transverse axis parallel to the x-axis; Find b:

$$b^2 = c^2 - a^2 = 16 - 9 = 7$$
$$b = \sqrt{7}$$

Write the equation: $\dfrac{(x + 1)^2}{9} - \dfrac{(y - 2)^2}{7} = 1$

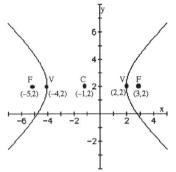

35. Hyperbola: Vertices: $(0, 1), (6, 1)$;
Asymptote: $3y + 2x - 9 = 0$; Center: $(3, 1)$;
Transverse axis is parallel to the x-axis; $a = 3$; The
slope of the asymptote is $-\frac{2}{3}$; Find b:

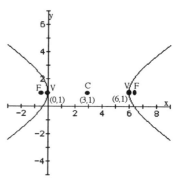

$$\frac{-b}{a} = \frac{-b}{3} = \frac{-2}{3} \rightarrow -3b = -6 \rightarrow b = 2$$

Write the equation: $\dfrac{(x-3)^2}{9} - \dfrac{(y-1)^2}{4} = 1$

37. $y^2 + 4x + 3y - 8 = 0$
$A = 0$ and $C = 1$; $AC = (0)(1) = 0$. Since $AC = 0$, the equation defines a parabola.

39. $x^2 + 2y^2 + 4x - 8y + 2 = 0$
$A = 1$ and $C = 2$; $AC = (1)(2) = 2$. Since $AC > 0$ and $A \neq C$, the equation defines
an ellipse.

41. $9x^2 - 12xy + 4y^2 + 8x + 12y = 0$
$A = 9$. $B = -12$, $C = 4$ $B^2 - 4AC = (-12)^2 - 4(9)(4) = 0$; parabola

43. $4x^2 + 10xy + 4y^2 - 9 = 0$
$A = 4$, $B = 10$, $C = 4$ $B^2 - 4AC = 10^2 - 4(4)(4) = 36 > 0$; hyperbola

45. $x^2 - 2xy + 3y^2 + 2x + 4y - 1 = 0$
$A = 1$. $B = -2$, $C = 3$ $B^2 - 4AC = (-2)^2 - 4(1)(3) = -8 < 0$; ellipse

47. $2x^2 + 5xy + 2y^2 - \frac{9}{2} = 0$

$A = 2, B = 5$, and $C = 2$; $\cot 2\theta = \dfrac{A - C}{B} = \dfrac{2 - 2}{5} = 0 \rightarrow 2\theta = \dfrac{\pi}{2} \rightarrow \theta = \dfrac{\pi}{4}$

$x = x'\cos\theta - y'\sin\theta = \dfrac{\sqrt{2}}{2}x' - \dfrac{\sqrt{2}}{2}y' = \dfrac{\sqrt{2}}{2}(x' - y')$

$y = x'\sin\theta + y'\cos\theta = \dfrac{\sqrt{2}}{2}x' + \dfrac{\sqrt{2}}{2}y' = \dfrac{\sqrt{2}}{2}(x' + y')$

$2\left(\dfrac{\sqrt{2}}{2}(x' - y')\right)^2 + 5\left(\dfrac{\sqrt{2}}{2}(x' - y')\right)\left(\dfrac{\sqrt{2}}{2}(x' + y')\right) + 2\left(\dfrac{\sqrt{2}}{2}(x' + y')\right)^2 - \dfrac{9}{2} = 0$

$\left(x'^2 - 2x'y' + y'^2\right) + \dfrac{5}{2}\left(x'^{2} - y'^2\right) + \left(x'^2 + 2x'y' + y'^2\right) - \dfrac{9}{2} = 0$

$\dfrac{9}{2}x'^2 - \dfrac{1}{2}y'^2 = \dfrac{9}{2}$

$9x'^2 - y'^2 = 9$

$\dfrac{x'^2}{1} - \dfrac{y'^2}{9} = 1$

Hyperbola; center at $(0, 0)$, transverse axis is the x'-axis, vertices at $(\pm 1, 0)$.

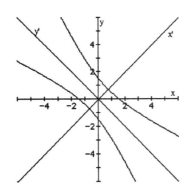

49. $6x^2 + 4xy + 9y^2 - 20 = 0$

$A = 6, B = 4,$ and $C = 9;$ $\cot 2\theta = \dfrac{A-C}{B} = \dfrac{6-9}{4} = \dfrac{-3}{4}$ $\rightarrow$ $\cos 2\theta = \dfrac{-3}{5}$

$\sin\theta = \sqrt{\dfrac{1-\dfrac{-3}{5}}{2}} = \sqrt{\dfrac{4}{5}} = \dfrac{2\sqrt{5}}{5};$ $\cos\theta = \sqrt{\dfrac{1+\dfrac{-3}{5}}{2}} = \sqrt{\dfrac{1}{5}} = \dfrac{\sqrt{5}}{5}$ $\rightarrow$ $\theta \approx 63.4°$

$x = x'\cos\theta - y'\sin\theta = \dfrac{\sqrt{5}}{5}x' - \dfrac{2\sqrt{5}}{5}y' = \dfrac{\sqrt{5}}{5}(x' - 2y')$

$y = x'\sin\theta + y'\cos\theta = \dfrac{2\sqrt{5}}{5}x' + \dfrac{\sqrt{5}}{5}y' = \dfrac{\sqrt{5}}{5}(2x' + y')$

$6\left(\dfrac{\sqrt{5}}{5}(x' - 2y')\right)^2 + 4\left(\dfrac{\sqrt{5}}{5}(x' - 2y')\right)\left(\dfrac{\sqrt{5}}{5}(2x' + y')\right) + 9\left(\dfrac{\sqrt{5}}{5}(2x' + y')\right)^2 - 20 = 0$

$\dfrac{6}{5}\left(x'^2 - 4x'y' + 4y'^2\right) + \dfrac{4}{5}\left(2x'^2 - 3x'y' - 2y'^2\right) + \dfrac{9}{5}\left(4x'^2 + 4x'y' + y'^2\right) - 20 = 0$

$\dfrac{6}{5}x'^2 - \dfrac{24}{5}x'y' + \dfrac{24}{5}y'^2 + \dfrac{8}{5}x'^2 - \dfrac{12}{5}x'y' - \dfrac{8}{5}y'^2 + \dfrac{36}{5}x'^2 + \dfrac{36}{5}x'y' + \dfrac{9}{5}y'^2 = 20$

$10x'^2 + 5y'^2 = 20$

$\dfrac{x'^2}{2} + \dfrac{y'^2}{4} = 1$

Ellipse; center at the origin, major axis is the y'-axis, vertices at $(0, \pm 2)$.

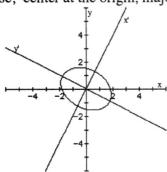

51. $4x^2 - 12xy + 9y^2 + 12x + 8y = 0$

$A = 4$, $B = -12$, and $C = 9$; $\cot 2\theta = \dfrac{A-C}{B} = \dfrac{4-9}{-12} = \dfrac{5}{12}$ $\rightarrow$ $\cos 2\theta = \dfrac{5}{13}$

$\sin\theta = \sqrt{\dfrac{1-\dfrac{5}{13}}{2}} = \sqrt{\dfrac{4}{13}} = \dfrac{2\sqrt{13}}{13}$; $\cos\theta = \sqrt{\dfrac{1+\dfrac{5}{13}}{2}} = \sqrt{\dfrac{9}{13}} = \dfrac{3\sqrt{13}}{13}$ $\rightarrow$ $\theta \approx 33.7°$

$x = x'\cos\theta - y'\sin\theta = \dfrac{3\sqrt{13}}{13}x' - \dfrac{2\sqrt{13}}{13}y' = \dfrac{\sqrt{13}}{13}(3x' - 2y')$

$y = x'\sin\theta + y'\cos\theta = \dfrac{2\sqrt{13}}{13}x' + \dfrac{3\sqrt{13}}{13}y' = \dfrac{\sqrt{13}}{13}(2x' + 3y')$

$4\left(\dfrac{\sqrt{13}}{13}(3x' - 2y')\right)^2 - 12\left(\dfrac{\sqrt{13}}{13}(3x' - 2y')\right)\left(\dfrac{\sqrt{13}}{13}(2x' + 3y')\right) + 9\left(\dfrac{\sqrt{13}}{13}(2x' + 3y')\right)^2$

$+ 12\left(\dfrac{\sqrt{13}}{13}(3x' - 2y')\right) + 8\left(\dfrac{\sqrt{13}}{13}(2x' + 3y')\right) = 0$

$\dfrac{4}{13}\left(9x'^2 - 12x'y' + 4y'^2\right) - \dfrac{12}{13}\left(6x'^2 + 5x'y' - 6y'^2\right) + \dfrac{9}{13}\left(4x'^2 + 12x'y' + 9y'^2\right)$

$+ \dfrac{36\sqrt{13}}{13}x' - \dfrac{24\sqrt{13}}{13}y' + \dfrac{16\sqrt{13}}{13}x' + \dfrac{24\sqrt{13}}{13}y' = 0$

$\dfrac{36}{13}x'^2 - \dfrac{48}{13}x'y' + \dfrac{16}{13}y'^2 - \dfrac{72}{13}x'^2 - \dfrac{60}{13}x'y' + \dfrac{72}{13}y'^2$

$+ \dfrac{36}{13}x'^2 + \dfrac{108}{13}x'y' + \dfrac{81}{13}y'^2 + 4\sqrt{13}x' = 0$

$13y'^2 + 4\sqrt{13}x' = 0$

$y'^2 = -\dfrac{4\sqrt{13}}{13}x'$

Parabola; vertex at the origin,

focus at $\left(-\dfrac{\sqrt{13}}{13}, 0\right)$.

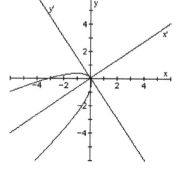

53. $r = \dfrac{4}{1 - \cos\theta}$

$ep = 4$, $e = 1$, $p = 4$

Parabola; directrix is perpendicular to the polar axis 4 units to the left of the pole; vertex is $(2, \pi)$.

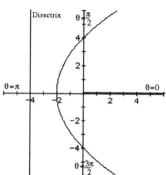

55. $r = \dfrac{6}{2 - \sin\theta} = \dfrac{3}{1 - \frac{1}{2}\sin\theta}$

$ep = 3,\ e = \dfrac{1}{2},\ p = 6$

Ellipse; directrix is parallel to the polar axis 6 units below the pole; vertices are

$\left(6, \dfrac{\pi}{2}\right)$ and $\left(2, \dfrac{3\pi}{2}\right)$.

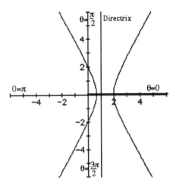

57. $r = \dfrac{8}{4 + 8\cos\theta} = \dfrac{2}{1 + 2\cos\theta}$

$ep = 2,\ e = 2,\ p = 1$

Hyperbola; directrix is perpendicular to the polar axis 1 unit to the right of the pole;

vertices are $\left(\dfrac{2}{3}, 0\right)$ and $(-2, \pi)$.

59.
$$r = \frac{4}{1 - \cos\theta}$$
$$r - r\cos\theta = 4$$
$$r = 4 + r\cos\theta$$
$$r^2 = (4 + r\cos\theta)^2$$
$$x^2 + y^2 = (4 + x)^2$$
$$x^2 + y^2 = 16 + 8x + x^2$$
$$y^2 - 8x - 16 = 0$$

61.
$$r = \frac{8}{4 + 8\cos\theta}$$
$$4r + 8r\cos\theta = 8$$
$$4r = 8 - 8r\cos\theta$$
$$r = 2 - 2r\cos\theta$$
$$r^2 = (2 - 2r\cos\theta)^2$$
$$x^2 + y^2 = (2 - 2x)^2$$
$$x^2 + y^2 = 4 - 8x + 4x^2$$
$$3x^2 - y^2 - 8x + 4 = 0$$

63. $x = 4t - 2,\ y = 1 - t,\ -\infty < t < \infty$

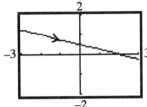

$x = 4(1 - y) - 2$
$x = 4 - 4y - 2$
$x + 4y = 2$

65. $x = 3\sin t, \ y = 4\cos t + 2, \ 0 \le t \le 2\pi$

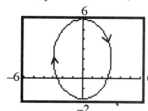

$\dfrac{x}{3} = \sin t, \quad \dfrac{y-2}{4} = \cos t$

$\sin^2 t + \cos^2 t = 1$

$\left(\dfrac{x}{3}\right)^2 + \left(\dfrac{y-2}{4}\right)^2 = 1$

$\dfrac{x^2}{9} + \dfrac{(y-2)^2}{16} = 1$

67. $x = \sec^2 t, \ y = \tan^2 t, \ 0 \le t \le \dfrac{\pi}{4}$

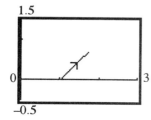

$\tan^2 t + 1 = \sec^2 t$

$y + 1 = x$

69. Write the equation in standard form: $4x^2 + 9y^2 = 36 \ \rightarrow \ \dfrac{x^2}{9} + \dfrac{y^2}{4} = 1$

The center of the ellipse is $(0, 0)$. The major axis is the x-axis.

$a = 3; \ b = 2; \ c^2 = a^2 - b^2 = 9 - 4 = 5 \ \rightarrow \ c = \sqrt{5}$.

For the ellipse:

Vertices:	$(-3, 0), (3, 0)$	Foci:	$\left(-\sqrt{5}, 0\right), \left(\sqrt{5}, 0\right)$

For the hyperbola:

Foci:	$(-3, 0), (3, 0)$	Vertices:	$\left(-\sqrt{5}, 0\right), \left(\sqrt{5}, 0\right)$
Center:	$(0, 0)$		

$a = \sqrt{5}; \ c = 3; \ b^2 = c^2 - a^2 = 9 - 5 = 4 \ \rightarrow \ b = 2$

The equation of the hyperbola is: $\dfrac{x^2}{5} - \dfrac{y^2}{4} = 1$

71. Let (x, y) be any point in the collection of points.

The distance from (x, y) to $(3, 0) = \sqrt{(x-3)^2 + y^2}$.

The distance from (x, y) to the line $x = \dfrac{16}{3}$ is $\left| x - \dfrac{16}{3} \right|$.

Relating the distances, we have:

$$\sqrt{(x-3)^2 + y^2} = \frac{3}{4}\left| x - \frac{16}{3} \right|$$

$$(x-3)^2 + y^2 = \frac{9}{16}\left(x - \frac{16}{3} \right)^2$$

$$x^2 - 6x + 9 + y^2 = \frac{9}{16}\left(x^2 - \frac{32}{3}x + \frac{256}{9} \right)$$

$$16x^2 - 96x + 144 + 16y^2 = 9x^2 - 96x + 256$$
$$7x^2 + 16y^2 = 112$$
$$\frac{7x^2}{112} + \frac{16y^2}{112} = 1$$
$$\frac{x^2}{16} + \frac{y^2}{7} = 1$$

The set of points is an ellipse.

73. Locate the parabola so that the vertex is at $(0, 0)$ and opens up. It then has the equation: $x^2 = 4ay$. Since the light source is located at the focus and is 1 foot from the base, $a = 1$. The diameter is 2, so the point $(1, y)$ is located on the parabola. Solve for y:
$$1^2 = 4(1)y$$
$$1 = 4y$$
$$y = 0.25 \text{ feet}$$
The mirror should be 0.25 feet deep or 3 inches deep.

75. Place the semielliptical arch so that the x-axis coincides with the water and the y-axis passes through the center of the arch. Since the bridge has a span of 60 feet, the length of the major axis is 60, or $2a = 60$ or $a = 30$. The maximum height of the bridge is 20 feet, so $b = 20$. The equation is: $\frac{x^2}{900} + \frac{y^2}{400} = 1$.

The height 5 feet from the center:
$$\frac{5^2}{900} + \frac{y^2}{400} = 1 \rightarrow \frac{y^2}{400} = 1 - \frac{25}{900} \rightarrow y^2 = 400 \cdot \frac{875}{900} \rightarrow y \approx 19.72 \text{ feet}$$
The height 10 feet from the center:
$$\frac{10^2}{900} + \frac{y^2}{400} = 1 \rightarrow \frac{y^2}{400} = 1 - \frac{100}{900} \rightarrow y^2 = 400 \cdot \frac{800}{900} \rightarrow y \approx 18.86 \text{ feet}$$
The height 20 feet from the center:
$$\frac{20^2}{900} + \frac{y^2}{400} = 1 \rightarrow \frac{y^2}{400} = 1 - \frac{400}{900} \rightarrow y^2 = 400 \cdot \frac{500}{900} \rightarrow y \approx 14.91 \text{ feet}$$

77. (a) Set up a coordinate system so that the two stations lie on the x-axis and the origin is midway between them. The ship lies on a hyperbola whose foci are the locations of the two stations. Since the time difference is 0.00032 seconds and the speed of the signal is 186,000 miles per second, the difference in the distances of the ships from each station is:

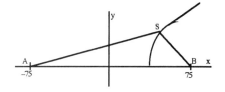

distance $= (186,000)(0.00032) = 59.52$ miles
The difference of the distances from the ship to each station, 59.52, equals $2a$, so $a = 29.76$ and the vertex of the corresponding hyperbola is at $(29.76, 0)$. Since the focus is at $(75, 0)$, following this hyperbola, the ship would reach shore 45.24 miles from the master station.

(b) The ship should follow a hyperbola with a vertex at $(60, 0)$. For this hyperbola, $a = 60$, so the constant difference of the distances from the ship to each station is 120. The time difference the ship should look for is:

$$\text{time} = \frac{120}{186,000} = 0.000645 \text{ seconds}$$

(c) Find the equation of the hyperbola with vertex at $(60, 0)$ and a focus at $(75, 0)$. The form of the equation of the hyperbola is:

$$\frac{x^2}{a^2} - \frac{y^2}{b^2} = 1 \quad \text{where } a = 60.$$

Since $c = 75$ and $b^2 = c^2 - a^2 \rightarrow b^2 = 75^2 - 60^2 = 2025$.

The equation of the hyperbola is: $\dfrac{x^2}{3600} - \dfrac{y^2}{2025} = 1$.

Since the ship is 20 miles off shore, we have $y = 20$. Solve the equation for x:

$$\frac{x^2}{3600} - \frac{20^2}{2025} = 1 \;\rightarrow\; \frac{x^2}{3600} = 1 + \frac{400}{2025} = \frac{97}{81} \;\rightarrow\; x^2 = 3600 \cdot \frac{97}{81}$$

$$x \approx 66 \text{ miles}$$

The ship's location is $(66, 20)$.

79. (a) Use equation (2):

$$x = (100\cos35°)t$$

$$y = -\frac{1}{2}(32)t^2 + (100\sin35°)t + 6$$

(b) The ball is in the air until $y = 0$. Solve:

$$-16t^2 + (100\sin35°)t + 6 = 0$$

$$t = \frac{-100\sin35° \pm \sqrt{(100\sin35°)^2 - 4(-16)(6)}}{2(-16)}$$

$$= \frac{-57.36 \pm \sqrt{3673.9}}{-32} \approx -0.10 \text{ or } 3.69$$

The ball is in the air for about 3.69 seconds. (The negative solution is extraneous.)

(c) The maximum height occurs at the vertex of the quadratic function.

$$t = \frac{-b}{2a} = \frac{-100\sin35°}{2(-16)} \approx 1.79 \text{ seconds}$$

Evaluate the function to find the maximum height:

$$-16(1.79)^2 + 100\sin35°(1.79) + 6 = 57.4 \text{ feet}$$

(d) Find the horizontal displacement:

$$x = (100\cos35°)(3.69) \approx 302 \text{ feet}$$

(e)

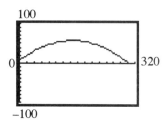

Chapter 11

Systems of Equations and Inequalities

11.1 Systems of Linear Equations: Two Equations Containing Two Unknowns

1. Substituting the values of the variables:
    $$\begin{cases} 2x - y = 5 & \to \quad 2(2) - (-1) = 4 + 1 = 5 \\ 5x + 2y = 8 & \to \quad 5(2) + 2(-1) = 10 - 2 = 8 \end{cases}$$
    Each equation is satisfied, so $x = 2$, $y = -1$ is a solution to the system of equations.

3. Substituting the values of the variables:
    $$\begin{cases} 3x - 4y = 4 & \to \quad 3(2) - 4\left(\frac{1}{2}\right) = 6 - 2 = 4 \\ \frac{1}{2}x - 3y = \frac{-1}{2} & \to \quad \frac{1}{2}(2) - 3\left(\frac{1}{2}\right) = 1 - \frac{3}{2} = \frac{-1}{2} \end{cases}$$
    Each equation is satisfied, so $x = 2$, $y = \frac{1}{2}$ is a solution to the system of equations.

5. Substituting the values of the variables:
    $$\begin{cases} x - y = 3 & \to \quad 4 - 1 = 3 \\ \frac{1}{2}x + y = 3 & \to \quad \frac{1}{2}(4) + 1 = 2 + 1 = 3 \end{cases}$$
    Each equation is satisfied, so $x = 4$, $y = 1$ is a solution to the system of equations.

7. Substituting the values of the variables:
    $$\begin{cases} 3x + 3y + 2z = 4 & \to \quad 3(1) + 3(-1) + 2(2) = 3 - 3 + 4 = 4 \\ x - y - z = 0 & \to \quad 1 - (-1) - 2 = 1 + 1 - 2 = 0 \\ 2y - 3z = -8 & \to \quad 2(-1) - 3(2) = -2 - 6 = -8 \end{cases}$$
    Each equation is satisfied, so $x = 1$, $y = -1$, $z = 2$ is a solution to the system.

9. Solve the first equation for y, substitute into the second equation and solve:
    $$\begin{cases} x + y = 8 & \to \quad y = 8 - x \\ x - y = 4 \end{cases}$$
    $$x - (8 - x) = 4$$
    $$x - 8 + x = 4$$
    $$2x = 12$$
    $$x = 6$$
    Since $x = 6$, $y = 8 - 6 = 2$
    The solution of the system is $x = 6$, $y = 2$.

11. Multiply each side of the first equation by 3 and add the equations:

$$\begin{cases} 5x - y = 13 \\ 2x + 3y = 12 \end{cases} \xrightarrow{\;3\;} \begin{array}{r} 15x - 3y = 39 \\ 2x + 3y = 12 \\ \hline 17x \quad\; = 51 \\ x \quad\; = 3 \end{array}$$

Substitute and solve for y:

$$5(3) - y = 13$$
$$15 - y = 13$$
$$-y = -2$$
$$y = 2$$

The solution of the system is $x = 3,\ y = 2$.

13. Solve the first equation for x and substitute into the second equation:

$$\begin{cases} 3x \quad\;\; = 24 \\ x + 2y = \;\; 0 \end{cases} \rightarrow \quad x = 8$$
$$8 + 2y = 0$$
$$2y = -8$$
$$y = -4$$

The solution of the system is $x = 8,\ y = -4$.

15. Multiply each side of the first equation by 2 and each side of the second equation by 3 to eliminate y:

$$\begin{cases} 3x - 6y = 2 \\ 5x + 4y = 1 \end{cases} \begin{array}{l} \xrightarrow{\;2\;} \\ \xrightarrow{\;3\;} \end{array} \begin{array}{r} 6x - 12y = 4 \\ 15x + 12y = 3 \\ \hline 21x \quad\quad = 7 \\ x \quad\quad = \tfrac{1}{3} \end{array}$$

Substitute and solve for y:

$$3\left(\tfrac{1}{3}\right) - 6y = 2$$
$$1 - 6y = 2$$
$$-6y = 1$$
$$y = \tfrac{-1}{6}$$

The solution of the system is $x = \tfrac{1}{3},\ y = -\tfrac{1}{6}$.

17. Solve the first equation for y, substitute into the second equation and solve:

$$\begin{cases} 2x + y = 1 \quad \rightarrow \quad y = 1 - 2x \\ 4x + 2y = 3 \end{cases}$$
$$4x + 2(1 - 2x) = 3$$
$$4x + 2 - 4x = 3$$
$$0x = 1$$

This has no solution, so the system is inconsistent.

19. Solve the first equation for y, substitute into the second equation and solve:

$$\begin{cases} 2x - y = 0 & \rightarrow \quad 2x = y \\ 3x + 2y = 7 \end{cases}$$

$$3x + 2(2x) = 7$$
$$3x + 4x = 7$$
$$7x = 7$$
$$x = 1$$

Since $x = 1$, $y = 2(1) = 2$

The solution of the system is $x = 1$, $y = 2$.

21. Solve the first equation for x, substitute into the second equation and solve:

$$\begin{cases} x + 2y = 4 & \rightarrow \quad x = 4 - 2y \\ 2x + 4y = 8 \end{cases}$$

$$2(4 - 2y) + 4y = 8$$
$$8 - 4y + 4y = 8$$
$$0y = 0$$

These equations are dependent. Any real number is a solution for y.

The solution of the system is $x = 4 - 2y$, where y is any real number.

23. Multiply each side of the first equation by –5, and add the equations to eliminate x:

$$\begin{cases} 2x - 3y = -1 & \xrightarrow{\;-5\;} & -10x + 15y = 5 \\ 10x + y = 11 & \longrightarrow & \underline{10x + \;\; y = 11} \\ & & 16y = 16 \\ & & y = 1 \end{cases}$$

Substitute and solve for x:

$$2x - 3(1) = -1$$
$$2x - 3 = -1$$
$$2x = 2$$
$$x = 1$$

The solution of the system is $x = 1$, $y = 1$.

25. Solve the second equation for x, substitute into the first equation and solve:

$$\begin{cases} 2x + 3y = 6 \\ x - y = \frac{1}{2} & \rightarrow \quad x = y + \frac{1}{2} \end{cases}$$

$$2\left(y + \frac{1}{2}\right) + 3y = 6$$
$$2y + 1 + 3y = 6$$
$$5y = 5$$
$$y = 1$$

Since $y = 1$, $x = 1 + \frac{1}{2} = \frac{3}{2}$

The solution of the system is $x = \frac{3}{2}$, $y = 1$.

27. Multiply each side of the first equation by –6 and each side of the second equation by 12 to eliminate x:

$$\begin{cases} \frac{1}{2}x + \frac{1}{3}y = 3 \\ \frac{1}{4}x - \frac{2}{3}y = -1 \end{cases} \xrightarrow{\ -6\ } \begin{array}{r} -3x - 2y = -18 \\ \underline{3x - 8y = -12} \\ -10y = -30 \\ y = 3 \end{array}$$

Substitute and solve for x:

$$\frac{1}{2}x + \frac{1}{3}(3) = 3$$
$$\frac{1}{2}x + 1 = 3$$
$$\frac{1}{2}x = 2$$
$$x = 4$$

The solution of the system is $x = 4, \; y = 3$.

29. Add the equations to eliminate y and solve for x:

$$\begin{cases} 3x - 5y = 3 \\ 15x + 5y = 21 \end{cases}$$
$$\overline{18x \qquad = 24}$$
$$x \qquad = \frac{4}{3}$$

Substitute and solve for y:

$$3\left(\frac{4}{3}\right) - 5y = 3$$
$$4 - 5y = 3$$
$$-5y = -1$$
$$y = \frac{1}{5}$$

The solution of the system is $x = \frac{4}{3}, \; y = \frac{1}{5}$.

31. Rewrite letting $a = \frac{1}{x}, \; b = \frac{1}{y}$:

$$\begin{cases} \dfrac{1}{x} + \dfrac{1}{y} = 8 \\ \dfrac{3}{x} - \dfrac{5}{y} = 0 \end{cases} \begin{array}{l} \longrightarrow \quad a + b = 8 \\ \longrightarrow \quad 3a - 5b = 0 \end{array}$$

Solve the first equation for a, substitute into the second equation and solve:

$$\begin{cases} a + b = 8 \quad \rightarrow \quad a = 8 - b \\ 3a - 5b = 0 \end{cases}$$
$$3(8 - b) - 5b = 0$$
$$24 - 3b - 5b = 0$$
$$-8b = -24$$
$$b = 3$$

Since $b = 3, \; a = 8 - 3 = 5$

Thus, $x = \dfrac{1}{a} = \dfrac{1}{5}, \; y = \dfrac{1}{b} = \dfrac{1}{3}$

The solution of the system is $x = \frac{1}{5}, \; y = \frac{1}{3}$.

33. Graph the two equations as y_1 and y_2, and use INTERSECT to solve:
$$\begin{cases} y_1 = \sqrt{2}x - 20\sqrt{7} \\ y_2 = -0.1x + 20 \end{cases}$$

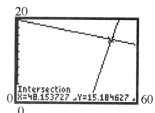

The solution of the system is $x = 48.15$, $y = 15.18$.

35. Solve for y in each equation, graph the two equations as y_1 and y_2, and use INTERSECT to solve:
$$\begin{cases} \sqrt{2}x + \sqrt{3}y + \sqrt{6} = 0 \\ \sqrt{3}x - \sqrt{2}y + 60 = 0 \end{cases}$$
$$\begin{cases} y_1 = \dfrac{-\sqrt{2}x - \sqrt{6}}{\sqrt{3}} \\[4mm] y_2 = \dfrac{\sqrt{3}x + 60}{\sqrt{2}} \end{cases}$$

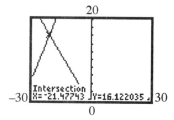

The solution of the system is $x = -21.48$, $y = 16.12$.

37. Solve for y in each equation, graph the two equations as y_1 and y_2, and use INTERSECT to solve:
$$\begin{cases} \sqrt{3}x + \sqrt{2}y = \sqrt{0.3} \\ 100x - 95y = 20 \end{cases}$$
$$\begin{cases} y_1 = \dfrac{-\sqrt{3}x + \sqrt{0.3}}{\sqrt{2}} \\[4mm] y_2 = \dfrac{100x - 20}{95} \end{cases}$$

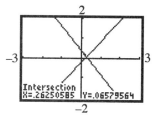

The solution of the system is $x = 0.26$, $y = 0.07$.

39. Solve the system by substitution:
$$Q_s = Q_d$$
$$-200 + 50p = 1000 - 25p$$
$$75p = 1200$$
$$p = 16$$
Therefore, $Q_s = -200 + 50(16) = -200 + 800 = 600$
The equilibrium price is \$16 and the equilibrium quantity is 600 T-shirts.

41. Let l be the length of the rectangle and w be the width of the rectangle. Then:
$$2l + 2w = 90$$
$$l = 2w$$
Solve by substitution:
$$2(2w) + 2w = 90$$
$$4w + 2w = 90$$
$$6w = 90$$
$$w = 15 \text{ feet}$$
$$l = 2(15) = 30 \text{ feet}$$
The dimensions of the floor are 15 feet by 30 feet.

43. Let x = the cost of one cheeseburger and y = the cost of one shake. Then:
$$4x + 2y = 790$$
$$2y = x + 15$$
Solve by substitution:
$$4x + x + 15 = 790$$
$$5x = 775$$
$$x = 155$$
$$2y = 155 + 15$$
$$2y = 170$$
$$y = 85$$
A cheeseburger cost $1.55 and a shake costs $0.85.

45. Let x = the number of pounds of cashews.
Then $x + 30$ is the number of pounds in the mixture.
The value of the cashews is $5x$.
The value of the peanuts is $1.50(30) = 45$.
The value of the mixture is $3(x + 30)$.
Setting up a value equation:
$$5x + 45 = 3(x + 30)$$
$$5x + 45 = 3x + 90$$
$$2x = 45$$
$$x = 22.5$$
22.5 pounds of cashews should be used in the mixture.

47. Let x = the plane's airspeed and y = the wind speed.

	Rate	Time	Distance
With Wind	$x + y$	3	600
Against	$x - y$	4	600

$$(x + y)(3) = 600 \quad \rightarrow \quad x + y = 200$$
$$(x - y)(4) = 600 \quad \rightarrow \quad x - y = 150$$
Solving by elimination:
$$2x = 350$$
$$x = 175$$
$$y = 200 - x = 200 - 175 = 25$$
The airspeed of the plane is 175 mph, and the wind speed is 25 mph.

49. Let x = the number of one design.
    Let y = the number of the second design.
    Then $x + y$ = the total number of sets of dishes.
    $25x + 45y$ = the cost of the dishes.
    Setting up the equations and solving by substitution:
    $$\begin{cases} x + y = 200 \;\rightarrow\; y = 200 - x \\ 25x + 45y = 7400 \end{cases}$$
    $$25x + 45(200 - x) = 7400$$
    $$25x + 9000 - 45x = 7400$$
    $$-20x = -1600$$
    $$x = 80$$
    $$y = 200 - 80 = 120$$
    80 sets of the \$25 dishes and 120 sets of the \$45 dishes should be ordered.

51. Let x = the cost per package of bacon.
    Let y = the cost of a carton of eggs.
    Set up a system of equations for the problem:
    $$\begin{cases} 3x + 2y = 7.45 \\ 2x + 3y = 6.45 \end{cases}$$
    Multiply each side of the first equation by 3 and each side of the second equation by −2 and solve by elimination:
    $$\begin{cases} 3x + 2y = 7.45 \;\xrightarrow{\;3\;}\; 9x + 6y = 22.35 \\ 2x + 3y = 6.45 \;\xrightarrow{\;-2\;}\; \underline{-4x - 6y = -12.90} \end{cases}$$
    $$5x \quad = \quad 9.45$$
    $$x \quad = \quad 1.89$$
    Substitute and solve for y:
    $$3(1.89) + 2y = 7.45$$
    $$5.67 + 2y = 7.45$$
    $$2y = 1.78$$
    $$y = 0.89$$
    A package of bacon costs \$1.89 and a carton of eggs cost \$0.89.
    The refund for 2 packages of bacon and 2 cartons of eggs will be \$5.56.

53. Solve the system by substitution:
    $$R = C$$
    $$8x = 4.5x + 17500$$
    $$3.5x = 17500$$
    $$x = 5000$$
    5000 units must be produced and sold for the firm to break-even.

55. $y = x^2 + bx + c$
     At $(1, 2)$ the equation becomes:
     $$2 = 1^2 + b(1) + c$$
     $$2 = 1 + b + c$$
     $$b + c = 1$$
     At $(-1, 3)$ the equation becomes:
     $$3 = (-1)^2 + b(-1) + c$$
     $$3 = 1 - b + c$$
     $$-b + c = 2$$
     Solve the system by substitution:
     $$\begin{cases} b + c = 1 \\ -b + c = 2 \end{cases} \rightarrow \quad b = 1 - c$$
     $$-(1 - c) + c = 2$$
     $$-1 + c + c = 2$$
     $$2c = 3$$
     $$c = \tfrac{3}{2}$$
     $$b = 1 - \tfrac{3}{2} = \tfrac{-1}{2}$$
     The solution is $b = -\tfrac{1}{2}, \ c = \tfrac{3}{2}$.

11.2 Systems of Linear Equations: Three Equations Containing Three Unknowns

1. Multiply each side of the first equation by –2 and add to the second equation to eliminate x:
    $$\begin{cases} x - y = 6 \\ 2x - 3z = 16 \\ 2y + z = 4 \end{cases} \xrightarrow{-2} \begin{array}{r} -2x + 2y = -12 \\ 2x \quad - 3z = 16 \\ \hline 2y - 3z = 4 \end{array}$$
    Multiply each side of the result by –1 and add to the original third equation to eliminate y:
    $$\begin{array}{r} 2y - 3z = 4 \xrightarrow{-1} -2y + 3z = -4 \\ 2y + z = 4 \longrightarrow \quad 2y + z = 4 \\ \hline 4z = 0 \\ z = 0 \end{array}$$
    Substituting and solving for the other variables:
    $$\begin{array}{ll} 2y + 0 = 4 & 2x - 3(0) = 16 \\ 2y = 4 & 2x = 16 \\ y = 2 & x = 8 \end{array}$$
    The solution is $x = 8, \ y = 2, \ z = 0$.

3.   Multiply each side of the first equation by –2 and add to the second equation to eliminate x; and multiply each side of the first equation by 3 and add to the third equation to eliminate x:

$$\begin{cases} x - 2y + 3z = 7 \\ 2x + y + z = 4 \\ -3x + 2y - 2z = -10 \end{cases}$$

$$\xrightarrow{-2} \quad \begin{array}{r} -2x + 4y - 6z = -14 \\ 2x + y + z = 4 \\ \hline 5y - 5z = -10 \end{array} \xrightarrow{1/5} y - z = -2$$

$$\xrightarrow{3} \quad \begin{array}{r} 3x - 6y + 9z = 21 \\ -3x + 2y - 2z = -10 \\ \hline -4y + 7z = 11 \end{array}$$

Multiply each side of the first result by 4 and add to the second result to eliminate y:

$$\begin{array}{r} y - z = -2 \xrightarrow{4} \quad 4y - 4z = -8 \\ -4y + 7z = 11 \longrightarrow \quad -4y + 7z = 11 \\ \hline 3z = 3 \\ z = 1 \end{array}$$

Substituting and solving for the other variables:

$$\begin{array}{ll} y - 1 = -2 & x - 2(-1) + 3(1) = 7 \\ y = -1 & x + 2 + 3 = 7 \\ & x = 2 \end{array}$$

The solution is $x = 2$, $y = -1$, $z = 1$.

5.   Add the first and second equations to eliminate z:

$$\begin{cases} x - y - z = 1 \\ 2x + 3y + z = 2 \\ 3x + 2y = 0 \end{cases} \longrightarrow \begin{array}{r} x - y - z = 1 \\ 2x + 3y + z = 2 \\ \hline 3x + 2y = 3 \end{array}$$

Multiply each side of the result by –1 and add to the original third equation to eliminate y:

$$\begin{array}{r} 3x + 2y = 3 \xrightarrow{-1} \quad -3x - 2y = -3 \\ 3x + 2y = 0 \longrightarrow \quad 3x + 2y = 0 \\ \hline 0 = -3 \end{array}$$

This result has no solution, so the system is inconsistent.

7.   Add the first and second equations to eliminate x; and multiply the first equation by –3 and add to the third equation to eliminate x:

$$\begin{cases} x - y - z = 1 \\ -x + 2y - 3z = -4 \\ 3x - 2y - 7z = 0 \end{cases}$$

$$\longrightarrow \begin{array}{r} x - y - z = 1 \\ -x + 2y - 3z = -4 \\ \hline y - 4z = -3 \end{array}$$

$$\xrightarrow{-3} \begin{array}{r} -3x + 3y + 3z = -3 \\ 3x - 2y - 7z = 0 \\ \hline y - 4z = -3 \end{array}$$

Multiply each side of the first result by -1 and add to the second result to eliminate y:

$$y - 4z = -3 \xrightarrow{\ -1\ } -y + 4z = 3$$
$$y - 4z = -3 \xrightarrow{\hspace{1cm}} \underline{y - 4z = -3}$$
$$0 = 0$$

The system is dependent. If z is any real number, then $y = 4z - 3$.
Solving for x in terms of z in the first equation:

$$x - (4z - 3) - z = 1$$
$$x - 4z + 3 - z = 1$$
$$x - 5z + 3 = 1$$
$$x = 5z - 2$$

The solution is $x = 5z - 2$, $y = 4z - 3$, z is any real number.

9. Multiply the first equation by -2 and add to the second equation to eliminate x; and add the first and third equations to eliminate x:

$$\begin{cases} 2x - 2y + 3z = 6 \\ 4x - 3y + 2z = 0 \\ -2x + 3y - 7z = 1 \end{cases}$$

$$2x - 2y + 3z = 6 \xrightarrow{\ -2\ } -4x + 4y - 6z = -12$$
$$4x - 3y + 2z = 0 \xrightarrow{\hspace{1cm}} \underline{4x - 3y + 2z = \ \ 0}$$
$$y - 4z = -12$$
$$\xrightarrow{\hspace{1cm}} 2x - 2y + 3z = 6$$
$$\xrightarrow{\hspace{1cm}} \underline{-2x + 3y - 7z = 1}$$
$$y - 4z = 7$$

Multiply each side of the first result by -1 and add to the second result to eliminate y:

$$y - 4z = -12 \xrightarrow{\ -1\ } -y + 4z = 12$$
$$y - 4z = \ \ 7 \xrightarrow{\hspace{1cm}} \underline{y - 4z = 7}$$
$$0 = 19$$

This result has no solution, so the system is inconsistent.

11. Add the first and second equations to eliminate z; and multiply the second equation by 2 and add to the third equation to eliminate z:

$$\begin{cases} x + y - z = 6 \\ 3x - 2y + z = -5 \\ x + 3y - 2z = 14 \end{cases}$$

$$x + y - z = 6 \xrightarrow{\hspace{1cm}} x + y - z = 6$$
$$3x - 2y + z = -5 \xrightarrow{\hspace{1cm}} \underline{3x - 2y + z = -5}$$
$$4x - y = 1$$
$$\xrightarrow{\ 2\ } 6x - 4y + 2z = -10$$
$$\xrightarrow{\hspace{1cm}} \underline{x + 3y - 2z = \ \ 14}$$
$$7x - y = 4$$

Multiply each side of the first result by -1 and add to the second result to eliminate y:

$$4x - y = 1 \xrightarrow{\ -1\ } -4x + y = -1$$
$$7x - y = 4 \xrightarrow{\hspace{1cm}} \underline{7x - y = \ \ 4}$$
$$3x \ \ = 3$$
$$x \ \ = 1$$

Substituting and solving for the other variables:

$$4(1) - y = 1 \qquad\qquad 3(1) - 2(3) + z = -5$$
$$-y = -3 \qquad\qquad 3 - 6 + z = -5$$
$$y = 3 \qquad\qquad z = -2$$

The solution is $x = 1,\ y = 3,\ z = -2$.

13. Add the first and second equations to eliminate z; and multiply the second equation by 3 and add to the third equation to eliminate z:

$$\begin{cases} x + 2y - z = -3 \\ 2x - 4y + z = -7 \\ -2x + 2y - 3z = 4 \end{cases} \longrightarrow \begin{array}{l} x + 2y - z = -3 \\ \underline{2x - 4y + z = -7} \\ 3x - 2y \quad\;\; = -10 \end{array}$$

$$\xrightarrow{\;3\;} \quad 6x - 12y + 3z = -21$$
$$\longrightarrow \quad \underline{-2x + 2y - 3z = \quad 4}$$
$$4x - 10y \quad\;\; = -17$$

Multiply each side of the first result by –5 and add to the second result to eliminate y:

$$3x - 2y = -10 \xrightarrow{\;-5\;} -15x + 10y = 50$$
$$4x - 10y = -17 \longrightarrow \underline{\quad 4x - 10y = -17}$$
$$-11x \quad\;\; = 33$$
$$x \quad\;\; = -3$$

Substituting and solving for the other variables:

$$3(-3) - 2y = -10 \qquad\qquad -3 + 2\left(\tfrac{1}{2}\right) - z = -3$$
$$-9 - 2y = -10 \qquad\qquad -3 + 1 - z = -3$$
$$-2y = -1 \qquad\qquad -z = -1$$
$$y = \tfrac{1}{2} \qquad\qquad z = 1$$

The solution is $x = -3,\ y = \tfrac{1}{2},\ z = 1$.

15. $y = ax^2 + bx + c$

At $(-1, 4)$ the equation becomes:
$$4 = a(-1)^2 + b(-1) + c$$
$$4 = a - b + c$$
$$a - b + c = 4$$

At $(2, 3)$ the equation becomes:
$$3 = a(2)^2 + b(2) + c$$
$$3 = 4a + 2b + c$$
$$4a + 2b + c = 3$$

At $(0, 1)$ the equation becomes:
$$1 = a(0)^2 + b(0) + c$$
$$c = 1$$

The system of equations is:
$$\begin{cases} a - b + c = 4 \\ 4a + 2b + c = 3 \\ \qquad\qquad c = 1 \end{cases}$$

Substitute $c = 1$ into the first and second equations and simplify:

$$\begin{cases} a - b + 1 = 4 & \rightarrow & a - b = 3 & \rightarrow & a = b + 3 \\ 4a + 2b + 1 = 3 & \rightarrow & 4a + 2b = 2 \end{cases}$$

Solve the first equation for a, substitute into the second equation and solve:

$$4(b + 3) + 2b = 2$$
$$4b + 12 + 2b = 2$$
$$6b = -10$$
$$b = -\tfrac{5}{3}$$
$$a = -\tfrac{5}{3} + 3 = \tfrac{4}{3}$$

The solution is $a = \tfrac{4}{3}$, $b = -\tfrac{5}{3}$, $c = 1$.

17. Substitute the expression for I_2 into the second and third equations and simplify:

$$\begin{cases} I_2 = I_1 + I_3 \\ 5 - 3I_1 - 5I_2 = 0 & \rightarrow & 5 - 3I_1 - 5(I_1 + I_3) = 0 & \rightarrow & -8I_1 - 5I_3 = -5 \\ 10 - 5I_2 - 7I_3 = 0 & \rightarrow & 10 - 5(I_1 + I_3) - 7I_3 = 0 & \rightarrow & -5I_1 - 12I_3 = -10 \end{cases}$$

Multiply both sides of the second equation by 5 and multiply both sides of the third equation by -8 to eliminate I_1:

$$\begin{array}{rcl} -8I_1 - 5I_3 = -5 & \xrightarrow{5} & -40I_1 - 25I_3 = -25 \\ -5I_1 - 12I_3 = -10 & \xrightarrow{-8} & \underline{40I_1 + 96I_3 = 80} \\ & & 71I_3 = 55 \\ & & I_3 = \dfrac{55}{71} \end{array}$$

Substituting and solving for the other variables:

$$-8I_1 - 5\left(\frac{55}{71}\right) = -5 \qquad\qquad I_2 = \frac{10}{71} + \frac{55}{71}$$

$$-8I_1 - \frac{275}{71} = -5 \qquad\qquad I_2 = \frac{65}{71}$$

$$-8I_1 = -\frac{80}{71}$$

$$I_1 = \frac{10}{71}$$

The solution is $I_1 = \dfrac{10}{71}$, $I_2 = \dfrac{65}{71}$, $I_3 = \dfrac{55}{71}$.

19. Let x = the number of orchestra seats.
    Let y = the number of main seats.
    Let z = the number of balcony seats.
    Since the total number of seats is 500, $x + y + z = 500$.
    Since the total revenue is \$17,100 if all seats are sold, $50x + 35y + 25z = 17{,}100$.
    If only half of the orchestra seats are sold, the revenue is \$14,600. So,
    $$50\left(\tfrac{1}{2}x\right) + 35y + 25z = 14{,}600.$$
    Multiply each side of the first equation by –25 and add to the second equation to eliminate z; and multiply each side of the third equation by –1 and add to the second equation to eliminate z:

    $$\begin{cases} x + \;\;y + \;\;z = \;\;\;500 \\ 50x + 35y + 25z = 17100 \\ 25x + 35y + 25z = 14600 \end{cases} \xrightarrow{\;-25\;} \begin{aligned} -25x - 25y - 25z &= -12500 \\ 50x + 35y + 25z &= \;\;17100 \\ \hline 25x + 10y \;\;\;\;\;\; &= \;\;\;4600 \end{aligned}$$

    $$\begin{aligned} \xrightarrow{} \quad & 50x + 35y + 25z = \;\;\;17100 \\ \xrightarrow{\;-1\;} \quad & -25x - 35y - 25z = -14600 \\ \hline & 25x \;\;\;\;\;\;\;\;\;\;\;\;\;\;\;\; = \;\;\;2500 \\ & \;\;\;\;x = \;\;100 \end{aligned}$$

    Substituting and solving for the other variables:

    $$\begin{aligned} 25(100) + 10y &= 4600 & 100 + 210 + z &= 500 \\ 2500 + 10y &= 4600 & 310 + z &= 500 \\ 10y &= 2100 & z &= 190 \\ y &= 210 \end{aligned}$$

    There are 100 orchestra seats, 210 main seats, and 190 balcony seats.

21. Let x = the number of servings of chicken.
    Let y = the number of servings of corn.
    Let z = the number of servings of 2% milk.
    Protein equation: $30x + 3y + 9z = 66$
    Carbohydrate equation: $35x + 16y + 13z = 94.5$
    Calcium equation: $200x + 10y + 300z = 910$
    Multiply each side of the first equation by –16 and multiply each side of the second equation by 3 and add them to eliminate y; and multiply each side of the second equation by –5 and multiply each side of the third equation by 8 and add to eliminate y:

    $$\begin{cases} 30x + 3y + \;\;9z = \;\;66 \\ 35x + 16y + \;\;13z = \;\;94.5 \\ 200x + 10y + 300z = 910 \end{cases} \xrightarrow{\;-16\;} \begin{aligned} -480x - 48y - 144z &= -1056 \\ 105x + 48y + \;\;39z &= \;\;283.5 \\ \hline -375x \;\;\;\;\;\;\;\;\; - 105z &= -772.5 \end{aligned}$$

    $$\begin{aligned} \xrightarrow{\;-5\;} \quad & -175x - 80y - \;\;65z = -472.5 \\ \xrightarrow{\;8\;} \quad & 1600x + 80y + 2400z = \;\;7280 \\ \hline & 1425x \;\;\;\;\;\;\;\;\;\; + 2335z = 6807.5 \end{aligned}$$

Multiply each side of the first result by 19 and multiply each side of the second result by 5 to eliminate x:

$$-375x - 105z = -772.5 \xrightarrow{\ 19\ } -7125x - 1995z = -14677.5$$
$$1425x + 2335z = 6807.5 \xrightarrow{\ 5\ } \underline{\ 7125x + 11675z = 34037.5\ }$$
$$9680z = 19360$$
$$z = 2$$

Substituting and solving for the other variables:

$$-375x - 105(2) = -772.5 \qquad 30(1.5) + 3y + 9(2) = 66$$
$$-375x - 210 = -772.5 \qquad 45 + 3y + 18 = 66$$
$$-375x = -562.5 \qquad 3y = 3$$
$$x = 1.5 \qquad y = 1$$

The dietitian should serve 1.5 servings of chicken, 1 serving of corn, and 2 servings of 2% milk.

11.3 Systems of Linear Equations: Matrices

1. Writing the augmented matrix for the system of equations:
$$\begin{cases} x - 5y = 5 \\ 4x + 3y = 6 \end{cases} \rightarrow \begin{bmatrix} 1 & -5 & | & 5 \\ 4 & 3 & | & 6 \end{bmatrix}$$

3. Writing the augmented matrix for the system of equations:
$$\begin{cases} 2x + 3y - 6 = 0 \\ 4x - 6y + 2 = 0 \end{cases} \rightarrow \begin{cases} 2x + 3y = 6 \\ 4x - 6y = -2 \end{cases} \rightarrow \begin{bmatrix} 2 & 3 & | & 6 \\ 4 & -6 & | & -2 \end{bmatrix}$$

5. Writing the augmented matrix for the system of equations:
$$\begin{cases} 0.01x - 0.03y = 0.06 \\ 0.13x + 0.10y = 0.20 \end{cases} \rightarrow \begin{bmatrix} 0.01 & -0.03 & | & 0.06 \\ 0.13 & 0.10 & | & 0.20 \end{bmatrix}$$

7. Writing the augmented matrix for the system of equations:
$$\begin{cases} x - y + z = 10 \\ 3x + 2y = 5 \\ x + y + 2z = 2 \end{cases} \rightarrow \begin{bmatrix} 1 & -1 & 1 & | & 10 \\ 3 & 2 & 0 & | & 5 \\ 1 & 1 & 2 & | & 2 \end{bmatrix}$$

9. Writing the augmented matrix for the system of equations:
$$\begin{cases} x + y - z = 2 \\ 3x - 2y = 2 \\ 5x + 3y - z = 1 \end{cases} \rightarrow \begin{bmatrix} 1 & 1 & -1 & | & 2 \\ 3 & -2 & 0 & | & 2 \\ 5 & 3 & -1 & | & 1 \end{bmatrix}$$

11.
$$\begin{bmatrix} 1 & -3 & -5 & | & -2 \\ 2 & -5 & -4 & | & 5 \\ -3 & 5 & 4 & | & 6 \end{bmatrix} \rightarrow \begin{bmatrix} 1 & -3 & -5 & | & -2 \\ 0 & 1 & 6 & | & 9 \\ -3 & 5 & 4 & | & 6 \end{bmatrix} \rightarrow \begin{bmatrix} 1 & -3 & -5 & | & -2 \\ 0 & 1 & 6 & | & 9 \\ 0 & -4 & -11 & | & 0 \end{bmatrix} \rightarrow \begin{bmatrix} 1 & -3 & -5 & | & -2 \\ 0 & 1 & 6 & | & 9 \\ 0 & 0 & 13 & | & 36 \end{bmatrix}$$

(a) $R_2 = -2r_1 + r_2$ (b) $R_3 = 3r_1 + r_3$ (c) $R_3 = 4r_2 + r_3$

13.
$$\begin{bmatrix} 1 & -3 & 4 & | & 3 \\ 2 & -5 & 6 & | & 6 \\ -3 & 3 & 4 & | & 6 \end{bmatrix} \rightarrow \begin{bmatrix} 1 & -3 & 4 & | & 3 \\ 0 & 1 & -2 & | & 0 \\ -3 & 3 & 4 & | & 6 \end{bmatrix} \rightarrow \begin{bmatrix} 1 & -3 & 4 & | & 3 \\ 0 & 1 & -2 & | & 0 \\ 0 & -6 & 16 & | & 15 \end{bmatrix} \rightarrow \begin{bmatrix} 1 & -3 & 4 & | & 3 \\ 0 & 1 & -2 & | & 0 \\ 0 & 0 & 4 & | & 15 \end{bmatrix}$$

(a) $R_2 = -2r_1 + r_2$ (b) $R_3 = 3r_1 + r_3$ (c) $R_3 = 6r_2 + r_3$

15.
$$\begin{bmatrix} 1 & -3 & 2 & | & -6 \\ 2 & -5 & 3 & | & -4 \\ -3 & -6 & 4 & | & 6 \end{bmatrix} \rightarrow \begin{bmatrix} 1 & -3 & 2 & | & -6 \\ 0 & 1 & -1 & | & 8 \\ -3 & -6 & 4 & | & 6 \end{bmatrix} \rightarrow \begin{bmatrix} 1 & -3 & 2 & | & -6 \\ 0 & 1 & -1 & | & 8 \\ 0 & -15 & 10 & | & -12 \end{bmatrix} \rightarrow \begin{bmatrix} 1 & -3 & 2 & | & -6 \\ 0 & 1 & -1 & | & 8 \\ 0 & 0 & -5 & | & 108 \end{bmatrix}$$

(a) $R_2 = -2r_1 + r_2$ (b) $R_3 = 3r_1 + r_3$ (c) $R_3 = 15r_2 + r_3$

17.
$$\begin{bmatrix} 1 & -3 & 1 & | & -2 \\ 2 & -5 & 6 & | & -2 \\ -3 & 1 & 4 & | & 6 \end{bmatrix} \rightarrow \begin{bmatrix} 1 & -3 & 1 & | & -2 \\ 0 & 1 & 4 & | & 2 \\ -3 & 1 & 4 & | & 6 \end{bmatrix} \rightarrow \begin{bmatrix} 1 & -3 & 1 & | & -2 \\ 0 & 1 & 4 & | & 2 \\ 0 & -8 & 7 & | & 0 \end{bmatrix} \rightarrow \begin{bmatrix} 1 & -3 & 1 & | & -2 \\ 0 & 1 & 4 & | & 2 \\ 0 & 0 & 39 & | & 16 \end{bmatrix}$$

(a) $R_2 = -2r_1 + r_2$ (b) $R_3 = 3r_1 + r_3$ (c) $R_3 = 8r_2 + r_3$

19.
$$\begin{bmatrix} 1 & -3 & -2 & | & 3 \\ 2 & -5 & 2 & | & -1 \\ -3 & -2 & 4 & | & 6 \end{bmatrix} \rightarrow \begin{bmatrix} 1 & -3 & -2 & | & 3 \\ 0 & 1 & 6 & | & -7 \\ -3 & -2 & 4 & | & 6 \end{bmatrix} \rightarrow \begin{bmatrix} 1 & -3 & -2 & | & 3 \\ 0 & 1 & 6 & | & -7 \\ 0 & -11 & -2 & | & 15 \end{bmatrix} \rightarrow \begin{bmatrix} 1 & -3 & -2 & | & 3 \\ 0 & 1 & 6 & | & -7 \\ 0 & 0 & 64 & | & -62 \end{bmatrix}$$

(a) $R_2 = -2r_1 + r_2$ (b) $R_3 = 3r_1 + r_3$ (c) $R_3 = 11r_2 + r_3$

21. $\begin{cases} x = 5 \\ y = -1 \end{cases}$ consistent $x = 5, y = -1$

23. $\begin{cases} x = 1 \\ y = 2 \\ 0 = 3 \end{cases}$ inconsistent

25. $\begin{cases} x + 2z = -1 \\ y - 4z = -2 \\ \quad\ 0 = 0 \end{cases}$ consistent $x = -1 - 2z, y = -2 + 4z, z$ is any real number

27. $\begin{cases} x + y = 8 \\ x - y = 4 \end{cases}$ can be written as: $\begin{bmatrix} 1 & 1 & | & 8 \\ 1 & -1 & | & 4 \end{bmatrix}$

$\rightarrow \begin{bmatrix} 1 & 1 & | & 8 \\ 0 & -2 & | & -4 \end{bmatrix} \rightarrow \begin{bmatrix} 1 & 1 & | & 8 \\ 0 & 1 & | & 2 \end{bmatrix} \rightarrow \begin{bmatrix} 1 & 0 & | & 6 \\ 0 & 1 & | & 2 \end{bmatrix}$

$\quad R_2 = -r_1 + r_2 \quad R_2 = -\frac{1}{2}r_2 \quad R_1 = -r_2 + r_1$

The solution is $x = 6$, $y = 2$.

29. $\begin{cases} x - 5y = -13 \\ 3x + 2y = 12 \end{cases}$ can be written as: $\begin{bmatrix} 1 & -5 & | & -13 \\ 3 & 2 & | & 12 \end{bmatrix}$

$\rightarrow \begin{bmatrix} 1 & -5 & | & -13 \\ 0 & 17 & | & 51 \end{bmatrix} \rightarrow \begin{bmatrix} 1 & -5 & | & -13 \\ 0 & 1 & | & 3 \end{bmatrix} \rightarrow \begin{bmatrix} 1 & 0 & | & 2 \\ 0 & 1 & | & 3 \end{bmatrix}$

$\quad R_2 = -3r_1 + r_2 \quad R_2 = \frac{1}{17}r_2 \quad R_1 = 5r_2 + r_1$

The solution is $x = 2$, $y = 3$.

31. $\begin{cases} 3x - 6y = 24 \\ 5x + 4y = 12 \end{cases}$ can be written as: $\begin{bmatrix} 3 & -6 & | & 24 \\ 5 & 4 & | & 12 \end{bmatrix}$

$\rightarrow \begin{bmatrix} 1 & -2 & | & 8 \\ 5 & 4 & | & 12 \end{bmatrix} \rightarrow \begin{bmatrix} 1 & -2 & | & 8 \\ 0 & 14 & | & -28 \end{bmatrix} \rightarrow \begin{bmatrix} 1 & -2 & | & 8 \\ 0 & 1 & | & -2 \end{bmatrix} \rightarrow \begin{bmatrix} 1 & 0 & | & 4 \\ 0 & 1 & | & -2 \end{bmatrix}$

$\quad R_1 = \frac{1}{3}r_1 \quad R_2 = -5r_1 + r_2 \quad R_2 = \frac{1}{14}r_2 \quad R_1 = 2r_2 + r_1$

The solution is $x = 4$, $y = -2$.

33. $\begin{cases} 2x + y = 1 \\ 4x + 2y = 6 \end{cases}$ can be written as: $\begin{bmatrix} 2 & 1 & | & 1 \\ 4 & 2 & | & 6 \end{bmatrix}$

$\rightarrow \begin{bmatrix} 1 & \frac{1}{2} & | & \frac{1}{2} \\ 4 & 2 & | & 6 \end{bmatrix} \rightarrow \begin{bmatrix} 1 & \frac{1}{2} & | & \frac{1}{2} \\ 0 & 0 & | & 4 \end{bmatrix}$

$\quad R_1 = \frac{1}{2}r_1 \quad R_2 = -4r_1 + r_2$

The system of equations is inconsistent.

35. $\begin{cases} 2x - 4y = -2 \\ 3x + 2y = 3 \end{cases}$ can be written as: $\begin{bmatrix} 2 & -4 & | & -2 \\ 3 & 2 & | & 3 \end{bmatrix}$

$\rightarrow \begin{bmatrix} 1 & -2 & | & -1 \\ 3 & 2 & | & 3 \end{bmatrix} \rightarrow \begin{bmatrix} 1 & -2 & | & -1 \\ 0 & 8 & | & 6 \end{bmatrix} \rightarrow \begin{bmatrix} 1 & -2 & | & -1 \\ 0 & 1 & | & \frac{3}{4} \end{bmatrix} \rightarrow \begin{bmatrix} 1 & 0 & | & \frac{1}{2} \\ 0 & 1 & | & \frac{3}{4} \end{bmatrix}$

$\quad R_1 = \frac{1}{2}r_1 \quad R_2 = -3r_1 + r_2 \quad R_2 = \frac{1}{8}r_2 \quad R_1 = 2r_2 + r_1$

The solution is $x = \frac{1}{2}$, $y = \frac{3}{4}$.

37. $\begin{cases} x + 2y = 4 \\ 2x + 4y = 8 \end{cases}$ can be written as: $\begin{bmatrix} 1 & 2 & | & 4 \\ 2 & 4 & | & 8 \end{bmatrix}$

$\rightarrow \begin{bmatrix} 1 & 2 & | & 4 \\ 0 & 0 & | & 0 \end{bmatrix}$

$\quad R_2 = -2r_1 + r_2$

This is a dependent system and the solution is $x = -2y + 4$, y is any real number .

39. $\begin{cases} 2x + 3y = 6 \\ x - y = \frac{1}{2} \end{cases}$ can be written as: $\begin{bmatrix} 2 & 3 & | & 6 \\ 1 & -1 & | & \frac{1}{2} \end{bmatrix}$

$$\rightarrow \begin{bmatrix} 1 & \frac{3}{2} & | & 3 \\ 1 & -1 & | & \frac{1}{2} \end{bmatrix} \rightarrow \begin{bmatrix} 1 & \frac{3}{2} & | & 3 \\ 0 & -\frac{5}{2} & | & -\frac{5}{2} \end{bmatrix} \rightarrow \begin{bmatrix} 1 & \frac{3}{2} & | & 3 \\ 0 & 1 & | & 1 \end{bmatrix} \rightarrow \begin{bmatrix} 1 & 0 & | & \frac{3}{2} \\ 0 & 1 & | & 1 \end{bmatrix}$$

$\quad R_1 = \frac{1}{2} r_1 \qquad R_2 = -r_1 + r_2 \qquad R_2 = -\frac{2}{5} r_2 \qquad R_1 = -\frac{3}{2} r_2 + r_1$

The solution is $x = \frac{3}{2}$, $y = 1$.

41. $\begin{cases} 3x - 5y = 3 \\ 15x + 5y = 21 \end{cases}$ can be written as: $\begin{bmatrix} 3 & -5 & | & 3 \\ 15 & 5 & | & 21 \end{bmatrix}$

$$\rightarrow \begin{bmatrix} 1 & -\frac{5}{3} & | & 1 \\ 15 & 5 & | & 21 \end{bmatrix} \rightarrow \begin{bmatrix} 1 & -\frac{5}{3} & | & 1 \\ 0 & 30 & | & 6 \end{bmatrix} \rightarrow \begin{bmatrix} 1 & -\frac{5}{3} & | & 1 \\ 0 & 1 & | & \frac{1}{5} \end{bmatrix} \rightarrow \begin{bmatrix} 1 & 0 & | & \frac{4}{3} \\ 0 & 1 & | & \frac{1}{5} \end{bmatrix}$$

$\quad R_1 = \frac{1}{3} r_1 \qquad R_2 = -15 r_1 + r_2 \qquad R_2 = \frac{1}{30} r_2 \qquad R_1 = \frac{5}{3} r_2 + r_1$

The solution is $x = \frac{4}{3}$, $y = \frac{1}{5}$.

43. $\begin{cases} x - y = 6 \\ 2x - 3z = 16 \\ 2y + z = 4 \end{cases}$ can be written as: $\begin{bmatrix} 1 & -1 & 0 & | & 6 \\ 2 & 0 & -3 & | & 16 \\ 0 & 2 & 1 & | & 4 \end{bmatrix}$

$$\rightarrow \begin{bmatrix} 1 & -1 & 0 & | & 6 \\ 0 & 2 & -3 & | & 4 \\ 0 & 2 & 1 & | & 4 \end{bmatrix} \rightarrow \begin{bmatrix} 1 & -1 & 0 & | & 6 \\ 0 & 1 & -\frac{3}{2} & | & 2 \\ 0 & 2 & 1 & | & 4 \end{bmatrix} \rightarrow \begin{bmatrix} 1 & 0 & -\frac{3}{2} & | & 8 \\ 0 & 1 & -\frac{3}{2} & | & 2 \\ 0 & 0 & 4 & | & 0 \end{bmatrix} \rightarrow \begin{bmatrix} 1 & 0 & -\frac{3}{2} & | & 8 \\ 0 & 1 & -\frac{3}{2} & | & 2 \\ 0 & 0 & 1 & | & 0 \end{bmatrix}$$

$\quad R_2 = -2r_1 + r_2 \qquad R_2 = \frac{1}{2} r_2 \qquad \qquad R_1 = r_2 + r_1 \qquad R_3 = \frac{1}{4} r_3$

$\qquad \qquad \qquad \qquad \qquad \qquad \qquad \qquad \qquad \quad R_3 = -2r_2 + r_3$

$$\rightarrow \begin{bmatrix} 1 & 0 & 0 & | & 8 \\ 0 & 1 & 0 & | & 2 \\ 0 & 0 & 1 & | & 0 \end{bmatrix}$$

$\quad R_1 = \frac{3}{2} r_3 + r_1$

$\quad R_2 = \frac{3}{2} r_3 + r_2$

The solution is $x = 8$, $y = 2$, $z = 0$.

45. $\begin{cases} x - 2y + 3z = 7 \\ 2x + y + z = 4 \\ -3x + 2y - 2z = -10 \end{cases}$ can be written as: $\begin{bmatrix} 1 & -2 & 3 & | & 7 \\ 2 & 1 & 1 & | & 4 \\ -3 & 2 & -2 & | & -10 \end{bmatrix}$

$\rightarrow \begin{bmatrix} 1 & -2 & 3 & | & 7 \\ 0 & 5 & -5 & | & -10 \\ 0 & -4 & 7 & | & 11 \end{bmatrix} \rightarrow \begin{bmatrix} 1 & -2 & 3 & | & 7 \\ 0 & 1 & -1 & | & -2 \\ 0 & -4 & 7 & | & 11 \end{bmatrix} \rightarrow \begin{bmatrix} 1 & 0 & 1 & | & 3 \\ 0 & 1 & -1 & | & -2 \\ 0 & 0 & 3 & | & 3 \end{bmatrix}$

$\begin{array}{lll} R_2 = -2r_1 + r_2 & R_2 = \frac{1}{5}r_2 & R_1 = 2r_2 + r_1 \\ R_3 = 3r_1 + r_3 & & R_3 = 4r_2 + r_3 \end{array}$

$\rightarrow \begin{bmatrix} 1 & 0 & 1 & | & 3 \\ 0 & 1 & -1 & | & -2 \\ 0 & 0 & 1 & | & 1 \end{bmatrix} \rightarrow \begin{bmatrix} 1 & 0 & 0 & | & 2 \\ 0 & 1 & 0 & | & -1 \\ 0 & 0 & 1 & | & 1 \end{bmatrix}$

$\begin{array}{ll} R_3 = \frac{1}{3}r_3 & R_1 = -r_3 + r_1 \\ & R_2 = r_3 + r_2 \end{array}$

The solution is $x = 2$, $y = -1$, $z = 1$.

47. $\begin{cases} 2x - 2y - 2z = 2 \\ 2x + 3y + z = 2 \\ 3x + 2y = 0 \end{cases}$ can be written as: $\begin{bmatrix} 2 & -2 & -2 & | & 2 \\ 2 & 3 & 1 & | & 2 \\ 3 & 2 & 0 & | & 0 \end{bmatrix}$

$\rightarrow \begin{bmatrix} 1 & -1 & -1 & | & 1 \\ 2 & 3 & 1 & | & 2 \\ 3 & 2 & 0 & | & 0 \end{bmatrix} \rightarrow \begin{bmatrix} 1 & -1 & -1 & | & 1 \\ 0 & 5 & 3 & | & 0 \\ 0 & 5 & 3 & | & -3 \end{bmatrix} \rightarrow \begin{bmatrix} 1 & -1 & -1 & | & 1 \\ 0 & 5 & 3 & | & 0 \\ 0 & 0 & 0 & | & -3 \end{bmatrix}$

$\begin{array}{lll} R_1 = \frac{1}{2}r_1 & R_2 = -2r_1 + r_2 & R_3 = -r_2 + r_3 \\ & R_3 = -3r_1 + r_3 & \end{array}$

There is no solution. The system is inconsistent.

49. $\begin{cases} -x + y + z = -1 \\ -x + 2y - 3z = -4 \\ 3x - 2y - 7z = 0 \end{cases}$ can be written as: $\begin{bmatrix} -1 & 1 & 1 & | & -1 \\ -1 & 2 & -3 & | & -4 \\ 3 & -2 & -7 & | & 0 \end{bmatrix}$

$\rightarrow \begin{bmatrix} 1 & -1 & -1 & | & 1 \\ -1 & 2 & -3 & | & -4 \\ 3 & -2 & -7 & | & 0 \end{bmatrix} \rightarrow \begin{bmatrix} 1 & -1 & -1 & | & 1 \\ 0 & 1 & -4 & | & -3 \\ 0 & 1 & -4 & | & -3 \end{bmatrix} \rightarrow \begin{bmatrix} 1 & 0 & -5 & | & -2 \\ 0 & 1 & -4 & | & -3 \\ 0 & 0 & 0 & | & 0 \end{bmatrix} \rightarrow \begin{array}{l} x - 5z = -2 \\ y - 4z = -3 \end{array}$

$\begin{array}{lll} R_1 = -r_1 & R_2 = r_1 + r_2 & R_1 = r_2 + r_1 \\ & R_3 = -3r_1 + r_3 & R_3 = -r_2 + r_3 \end{array}$

The solution is $x = 5z - 2$, $y = 4z - 3$, z is any real number .

51. $\begin{cases} 2x - 2y + 3z = 6 \\ 4x - 3y + 2z = 0 \\ -2x + 3y - 7z = 1 \end{cases}$ can be written as: $\begin{bmatrix} 2 & -2 & 3 & | & 6 \\ 4 & -3 & 2 & | & 0 \\ -2 & 3 & -7 & | & 1 \end{bmatrix}$

$$\rightarrow \begin{bmatrix} 1 & -1 & \frac{3}{2} & | & 3 \\ 4 & -3 & 2 & | & 0 \\ -2 & 3 & -7 & | & 1 \end{bmatrix} \rightarrow \begin{bmatrix} 1 & -1 & \frac{3}{2} & | & 3 \\ 0 & 1 & -4 & | & -12 \\ 0 & 1 & -4 & | & 7 \end{bmatrix} \rightarrow \begin{bmatrix} 1 & 0 & -\frac{5}{2} & | & -9 \\ 0 & 1 & -4 & | & -12 \\ 0 & 0 & 0 & | & 19 \end{bmatrix}$$

$R_1 = \frac{1}{2} r_1$ $R_2 = -4r_1 + r_2$ $R_1 = r_2 + r_1$
                              $R_3 = 2r_1 + r_3$ $R_3 = -r_2 + r_3$

There is no solution. The system is inconsistent.

53. $\begin{cases} x + y - z = 6 \\ 3x - 2y + z = -5 \\ x + 3y - 2z = 14 \end{cases}$ can be written as: $\begin{bmatrix} 1 & 1 & -1 & | & 6 \\ 3 & -2 & 1 & | & -5 \\ 1 & 3 & -2 & | & 14 \end{bmatrix}$

$$\rightarrow \begin{bmatrix} 1 & 1 & -1 & | & 6 \\ 0 & -5 & 4 & | & -23 \\ 0 & 2 & -1 & | & 8 \end{bmatrix} \rightarrow \begin{bmatrix} 1 & 1 & -1 & | & 6 \\ 0 & 1 & -\frac{4}{5} & | & \frac{23}{5} \\ 0 & 2 & -1 & | & 8 \end{bmatrix} \rightarrow \begin{bmatrix} 1 & 0 & -\frac{1}{5} & | & \frac{7}{5} \\ 0 & 1 & -\frac{4}{5} & | & \frac{23}{5} \\ 0 & 0 & \frac{3}{5} & | & -\frac{6}{5} \end{bmatrix}$$

$R_2 = -3r_1 + r_2$ $R_2 = -\frac{1}{5} r_2$ $R_1 = -r_2 + r_1$
$R_3 = -r_1 + r_3$ $R_3 = -2r_2 + r_3$

$$\rightarrow \begin{bmatrix} 1 & 0 & -\frac{1}{5} & | & \frac{7}{5} \\ 0 & 1 & -\frac{4}{5} & | & \frac{23}{5} \\ 0 & 0 & 1 & | & -2 \end{bmatrix} \rightarrow \begin{bmatrix} 1 & 0 & 0 & | & 1 \\ 0 & 1 & 0 & | & 3 \\ 0 & 0 & 1 & | & -2 \end{bmatrix}$$

$R_3 = \frac{5}{3} r_3$ $R_1 = \frac{1}{5} r_3 + r_1$
                        $R_2 = \frac{4}{5} r_3 + r_2$

The solution is $x = 1$, $y = 3$, $z = -2$.

55. $\begin{cases} x + 2y - z = -3 \\ 2x - 4y + z = -7 \\ -2x + 2y - 3z = 4 \end{cases}$ can be written as: $\begin{bmatrix} 1 & 2 & -1 & | & -3 \\ 2 & -4 & 1 & | & -7 \\ -2 & 2 & -3 & | & 4 \end{bmatrix}$

$$\rightarrow \begin{bmatrix} 1 & 2 & -1 & | & -3 \\ 0 & -8 & 3 & | & -1 \\ 0 & 6 & -5 & | & -2 \end{bmatrix} \rightarrow \begin{bmatrix} 1 & 2 & -1 & | & -3 \\ 0 & 1 & -\frac{3}{8} & | & \frac{1}{8} \\ 0 & 6 & -5 & | & -2 \end{bmatrix} \rightarrow \begin{bmatrix} 1 & 0 & -\frac{1}{4} & | & -\frac{13}{4} \\ 0 & 1 & -\frac{3}{8} & | & \frac{1}{8} \\ 0 & 0 & -\frac{11}{4} & | & -\frac{11}{4} \end{bmatrix}$$

$R_2 = -2r_1 + r_2$ $R_2 = -\frac{1}{8} r_2$ $R_1 = -2r_2 + r_1$
$R_3 = 2r_1 + r_3$ $R_3 = -6r_2 + r_3$

$$\rightarrow \begin{bmatrix} 1 & 0 & -\frac{1}{4} & -\frac{13}{4} \\ 0 & 1 & -\frac{3}{8} & \frac{1}{8} \\ 0 & 0 & 1 & 1 \end{bmatrix} \rightarrow \begin{bmatrix} 1 & 0 & 0 & -3 \\ 0 & 1 & 0 & \frac{1}{2} \\ 0 & 0 & 1 & 1 \end{bmatrix}$$

$$R_3 = -\frac{4}{11}r_3 \qquad R_1 = \frac{1}{4}r_3 + r_1$$
$$R_2 = \frac{3}{8}r_3 + r_2$$

The solution is $x = -3$, $y = \frac{1}{2}$, $z = 1$.

57. $\begin{cases} 3x + y - z = \frac{2}{3} \\ 2x - y + z = 1 \\ 4x + 2y = \frac{8}{3} \end{cases}$ can be written as: $\begin{bmatrix} 3 & 1 & -1 & \frac{2}{3} \\ 2 & -1 & 1 & 1 \\ 4 & 2 & 0 & \frac{8}{3} \end{bmatrix}$

$$\rightarrow \begin{bmatrix} 1 & \frac{1}{3} & -\frac{1}{3} & \frac{2}{9} \\ 2 & -1 & 1 & 1 \\ 4 & 2 & 0 & \frac{8}{3} \end{bmatrix} \rightarrow \begin{bmatrix} 1 & \frac{1}{3} & -\frac{1}{3} & \frac{2}{9} \\ 0 & -\frac{5}{3} & \frac{5}{3} & \frac{5}{9} \\ 0 & \frac{2}{3} & \frac{4}{3} & \frac{16}{9} \end{bmatrix} \rightarrow \begin{bmatrix} 1 & \frac{1}{3} & -\frac{1}{3} & \frac{2}{9} \\ 0 & 1 & -1 & -\frac{1}{3} \\ 0 & \frac{2}{3} & \frac{4}{3} & \frac{16}{9} \end{bmatrix} \rightarrow \begin{bmatrix} 1 & 0 & 0 & \frac{1}{3} \\ 0 & 1 & -1 & -\frac{1}{3} \\ 0 & 0 & 2 & 2 \end{bmatrix}$$

$$R_1 = \frac{1}{3}r_1 \qquad\qquad R_2 = -2r_1 + r_2 \qquad R_2 = -\frac{3}{5}r_2 \qquad\qquad R_1 = -\frac{1}{3}r_2 + r_1$$
$$R_3 = -4r_1 + r_3 \qquad\qquad\qquad\qquad R_3 = -\frac{2}{3}r_2 + r_3$$

$$\rightarrow \begin{bmatrix} 1 & 0 & 0 & \frac{1}{3} \\ 0 & 1 & -1 & -\frac{1}{3} \\ 0 & 0 & 1 & 1 \end{bmatrix} \rightarrow \begin{bmatrix} 1 & 0 & 0 & \frac{1}{3} \\ 0 & 1 & 0 & \frac{2}{3} \\ 0 & 0 & 1 & 1 \end{bmatrix}$$

$$R_3 = \frac{1}{2}r_3 \qquad R_2 = r_3 + r_2$$

The solution is $x = \frac{1}{3}$, $y = \frac{2}{3}$, $z = 1$.

59. $\begin{cases} x + 2y + z = 1 \\ 2x - y + 2z = 2 \\ 3x + y + 3z = 3 \end{cases}$ can be written as: $\begin{bmatrix} 1 & 2 & 1 & 1 \\ 2 & -1 & 2 & 2 \\ 3 & 1 & 3 & 3 \end{bmatrix}$

$$\rightarrow \begin{bmatrix} 1 & 2 & 1 & 1 \\ 0 & -5 & 0 & 0 \\ 0 & -5 & 0 & 0 \end{bmatrix} \rightarrow \begin{bmatrix} 1 & 2 & 1 & 1 \\ 0 & -5 & 0 & 0 \\ 0 & 0 & 0 & 0 \end{bmatrix} \rightarrow \begin{array}{l} x + 2y + z = 1 \\ \quad -5y = 0 \end{array}$$

$$R_2 = -2r_1 + r_2 \quad R_3 = -r_2 + r_3$$
$$R_3 = -3r_1 + r_3$$

Substitute and solve:
$$y = 0$$
$$x + 2(0) + z = 1$$
$$x + z = 1$$
$$x = 1 - z$$

The solution is $y = 0$, $x = 1 - z$, z is any real number .

61. $\begin{cases} x + y + z + w = 4 \\ 2x - y + z = 0 \\ 3x + 2y + z - w = 6 \\ x - 2y - 2z + 2w = -1 \end{cases}$ can be written as: $\begin{bmatrix} 1 & 1 & 1 & 1 & 4 \\ 2 & -1 & 1 & 0 & 0 \\ 3 & 2 & 1 & -1 & 6 \\ 1 & -2 & -2 & 2 & -1 \end{bmatrix}$

$\rightarrow \begin{bmatrix} 1 & 1 & 1 & 1 & 4 \\ 0 & -3 & -1 & -2 & -8 \\ 0 & -1 & -2 & -4 & -6 \\ 0 & -3 & -3 & 1 & -5 \end{bmatrix} \rightarrow \begin{bmatrix} 1 & 1 & 1 & 1 & 4 \\ 0 & -1 & -2 & -4 & -6 \\ 0 & -3 & -1 & -2 & -8 \\ 0 & -3 & -3 & 1 & -5 \end{bmatrix} \rightarrow \begin{bmatrix} 1 & 1 & 1 & 1 & 4 \\ 0 & 1 & 2 & 4 & 6 \\ 0 & -3 & -1 & -2 & -8 \\ 0 & -3 & -3 & 1 & -5 \end{bmatrix}$

$R_2 = -2r_1 + r_2$ Interchange r_2 and r_3 $R_2 = -r_2$
$R_3 = -3r_1 + r_3$
$R_4 = -r_1 + r_4$

$\rightarrow \begin{bmatrix} 1 & 0 & -1 & -3 & -2 \\ 0 & 1 & 2 & 4 & 6 \\ 0 & 0 & 5 & 10 & 10 \\ 0 & 0 & 3 & 13 & 13 \end{bmatrix} \rightarrow \begin{bmatrix} 1 & 0 & -1 & -3 & -2 \\ 0 & 1 & 2 & 4 & 6 \\ 0 & 0 & 1 & 2 & 2 \\ 0 & 0 & 3 & 13 & 13 \end{bmatrix} \rightarrow \begin{bmatrix} 1 & 0 & 0 & -1 & 0 \\ 0 & 1 & 0 & 0 & 2 \\ 0 & 0 & 1 & 2 & 2 \\ 0 & 0 & 0 & 7 & 7 \end{bmatrix}$

$R_1 = -r_2 + r_1$ $R_3 = \frac{1}{5}r_3$ $R_1 = r_3 + r_1$
$R_3 = 3r_2 + r_3$ $R_2 = -2r_3 + r_2$
$R_4 = 3r_2 + r_4$ $R_4 = -3r_3 + r_4$

$\rightarrow \begin{bmatrix} 1 & 0 & 0 & -1 & 0 \\ 0 & 1 & 0 & 0 & 2 \\ 0 & 0 & 1 & 2 & 2 \\ 0 & 0 & 0 & 1 & 1 \end{bmatrix} \rightarrow \begin{bmatrix} 1 & 0 & 0 & 0 & 1 \\ 0 & 1 & 0 & 0 & 2 \\ 0 & 0 & 1 & 0 & 0 \\ 0 & 0 & 0 & 1 & 1 \end{bmatrix}$

$R_4 = \frac{1}{7}r_4$ $R_1 = r_4 + r_1$
               $R_3 = -2r_4 + r_3$

The solution is $x = 1$, $y = 2$, $z = 0$, $w = 1$.

63. Each of the points must satisfy the equation $y = ax^2 + bx + c$.

     $(1,2)$: $2 = a + b + c$
     $(-2,-7)$: $-7 = 4a - 2b + c$
     $(2,-3)$: $-3 = 4a + 2b + c$

Set up a matrix and solve:

$\begin{bmatrix} 1 & 1 & 1 & 2 \\ 4 & -2 & 1 & -7 \\ 4 & 2 & 1 & -3 \end{bmatrix} \rightarrow \begin{bmatrix} 1 & 1 & 1 & 2 \\ 0 & -6 & -3 & -15 \\ 0 & -2 & -3 & -11 \end{bmatrix} \rightarrow \begin{bmatrix} 1 & 1 & 1 & 2 \\ 0 & 1 & \frac{1}{2} & \frac{5}{2} \\ 0 & -2 & -3 & -11 \end{bmatrix} \rightarrow \begin{bmatrix} 1 & 0 & \frac{1}{2} & -\frac{1}{2} \\ 0 & 1 & \frac{1}{2} & \frac{5}{2} \\ 0 & 0 & -2 & -6 \end{bmatrix}$

      $R_2 = -4r_1 + r_2$ $R_2 = -\frac{1}{6}r_2$ $R_1 = -r_2 + r_1$
      $R_3 = -4r_1 + r_3$ $R_3 = 2r_2 + r_3$

$$\rightarrow \begin{bmatrix} 1 & 0 & \frac{1}{2} & | & -\frac{1}{2} \\ 0 & 1 & \frac{1}{2} & | & \frac{5}{2} \\ 0 & 0 & 1 & | & 3 \end{bmatrix} \rightarrow \begin{bmatrix} 1 & 0 & 0 & | & -2 \\ 0 & 1 & 0 & | & 1 \\ 0 & 0 & 1 & | & 3 \end{bmatrix}$$

$$R_3 = -\tfrac{1}{2}r_3 \qquad R_1 = -\tfrac{1}{2}r_3 + r_1$$
$$R_2 = -\tfrac{1}{2}r_3 + r_2$$

The solution is $a = -2,\, b = 1,\, c = 3$; so the equation is $y = -2x^2 + x + 3$.

65. Each of the points must satisfy the equation $f(x) = ax^3 + bx^2 + cx + d$.

$$\begin{array}{ll} f(-3) = -112: & -27a + 9b - 3c + d = -112 \\ f(-1) = -2: & -a + b - c + d = -2 \\ f(1) = 4: & a + b + c + d = 4 \\ f(2) = 13: & 8a + 4b + 2c + d = 13 \end{array}$$

Set up a matrix and solve:

$$\begin{bmatrix} -27 & 9 & -3 & 1 & | & -112 \\ -1 & 1 & -1 & 1 & | & -2 \\ 1 & 1 & 1 & 1 & | & 4 \\ 8 & 4 & 2 & 1 & | & 13 \end{bmatrix} \rightarrow \begin{bmatrix} 1 & 1 & 1 & 1 & | & 4 \\ -1 & 1 & -1 & 1 & | & -2 \\ -27 & 9 & -3 & 1 & | & -112 \\ 8 & 4 & 2 & 1 & | & 13 \end{bmatrix} \rightarrow \begin{bmatrix} 1 & 1 & 1 & 1 & | & 4 \\ 0 & 2 & 0 & 2 & | & 2 \\ 0 & 36 & 24 & 28 & | & -4 \\ 0 & -4 & -6 & -7 & | & -19 \end{bmatrix}$$

Interchange r_3 and r_1 $R_2 = r_1 + r_2$

$$R_3 = 27r_1 + r_3$$
$$R_4 = -8r_1 + r_4$$

$$\rightarrow \begin{bmatrix} 1 & 1 & 1 & 1 & | & 4 \\ 0 & 1 & 0 & 1 & | & 1 \\ 0 & 36 & 24 & 28 & | & -4 \\ 0 & -4 & -6 & -7 & | & -19 \end{bmatrix} \rightarrow \begin{bmatrix} 1 & 0 & 1 & 0 & | & 3 \\ 0 & 1 & 0 & 1 & | & 1 \\ 0 & 0 & 24 & -8 & | & -40 \\ 0 & 0 & -6 & -3 & | & -15 \end{bmatrix} \rightarrow \begin{bmatrix} 1 & 0 & 1 & 0 & | & 3 \\ 0 & 1 & 0 & 1 & | & 1 \\ 0 & 0 & 1 & -\frac{1}{3} & | & -\frac{5}{3} \\ 0 & 0 & -6 & -3 & | & -15 \end{bmatrix}$$

$$R_2 = \tfrac{1}{2}r_2 \qquad R_1 = -r_2 + r_1 \qquad R_3 = \tfrac{1}{24}r_3$$
$$R_3 = -36r_2 + r_3$$
$$R_4 = 4r_2 + r_4$$

$$\rightarrow \begin{bmatrix} 1 & 0 & 0 & \frac{1}{3} & | & \frac{14}{3} \\ 0 & 1 & 0 & 1 & | & 1 \\ 0 & 0 & 1 & -\frac{1}{3} & | & -\frac{5}{3} \\ 0 & 0 & 0 & -5 & | & -25 \end{bmatrix} \rightarrow \begin{bmatrix} 1 & 0 & 0 & \frac{1}{3} & | & \frac{14}{3} \\ 0 & 1 & 0 & 1 & | & 1 \\ 0 & 0 & 1 & -\frac{1}{3} & | & -\frac{5}{3} \\ 0 & 0 & 0 & 1 & | & 5 \end{bmatrix} \rightarrow \begin{bmatrix} 1 & 0 & 0 & 0 & | & 3 \\ 0 & 1 & 0 & 0 & | & -4 \\ 0 & 0 & 1 & 0 & | & 0 \\ 0 & 0 & 0 & 1 & | & 5 \end{bmatrix}$$

$$R_1 = -r_3 + r_1 \qquad R_4 = -\tfrac{1}{5}r_4 \qquad R_1 = -\tfrac{1}{3}r_4 + r_1$$
$$R_4 = 6r_3 + r_4 \qquad\qquad\qquad R_2 = -r_4 + r_2$$
$$R_3 = \tfrac{1}{3}r_4 + r_3$$

The solution is $a = 3,\, b = -4,\, c = 0,\, d = 5$; so the equation is $f(x) = 3x^3 - 4x^2 + 5$.

67. Let x = the number of servings of salmon steak.
    Let y = the number of servings of baked eggs.
    Let z = the number of servings of acorn squash.
    Protein equation: $\qquad 30x + 15y + 3z = 78$
    Carbohydrate equation: $\quad 20x + 2y + 25z = 59$
    Vitamin A equation: $\qquad 2x + 20y + 32z = 75$
    Set up a matrix and solve:

$$\begin{bmatrix} 30 & 15 & 3 & | & 78 \\ 20 & 2 & 25 & | & 59 \\ 2 & 20 & 32 & | & 75 \end{bmatrix} \rightarrow \begin{bmatrix} 2 & 20 & 32 & | & 75 \\ 20 & 2 & 25 & | & 59 \\ 30 & 15 & 3 & | & 78 \end{bmatrix} \rightarrow \begin{bmatrix} 1 & 10 & 16 & | & 37.5 \\ 20 & 2 & 25 & | & 59 \\ 30 & 15 & 3 & | & 78 \end{bmatrix}$$

$$\text{Interchange } r_3 \text{ and } r_1 \qquad R_1 = \tfrac{1}{2} r_1$$

$$\rightarrow \begin{bmatrix} 1 & 10 & 16 & | & 37.5 \\ 0 & -198 & -295 & | & -691 \\ 0 & -285 & -477 & | & -1047 \end{bmatrix} \rightarrow \begin{bmatrix} 1 & 10 & 16 & | & 37.5 \\ 0 & -198 & -295 & | & -691 \\ 0 & 0 & -\frac{3457}{66} & | & -\frac{3457}{66} \end{bmatrix}$$

$$R_2 = -20r_1 + r_2 \qquad\qquad R_3 = -\tfrac{95}{66} r_2 + r_3$$
$$R_3 = -30r_1 + r_3$$

$$\rightarrow \begin{bmatrix} 1 & 10 & 16 & | & 37.5 \\ 0 & -198 & -295 & | & -691 \\ 0 & 0 & 1 & | & 1 \end{bmatrix}$$

$$R_3 = -\tfrac{66}{3457} r_3$$

Substitute $z = 1$ and solve:

$$-198y - 295(1) = -691 \qquad\qquad x + 10(2) + 16(1) = 37.5$$
$$-198y = -396 \qquad\qquad\qquad x + 36 = 37.5$$
$$y = 2 \qquad\qquad\qquad\qquad x = 1.5$$

The dietitian should serve 1.5 servings of salmon steak, 2 servings of baked eggs, and 1 serving of acorn squash.

69. Let x = the amount invested in Treasury bills.
    Let y = the amount invested in Treasury bonds.
    Let z = the amount invested in corporate bonds.
    Total investment equation: $\qquad x + y + z = 10000$
    Annual income equation: $\qquad 0.06x + 0.07y + 0.08z = 680$
    Condition on investment equation: $\quad z = \tfrac{1}{2} x$

Set up a matrix and solve:

$$\begin{bmatrix} 1 & 1 & 1 & | & 10000 \\ 0.06 & 0.07 & 0.08 & | & 680 \\ 1 & 0 & -2 & | & 0 \end{bmatrix} \rightarrow \begin{bmatrix} 1 & 1 & 1 & | & 10000 \\ 0 & 0.01 & 0.02 & | & 80 \\ 0 & -1 & -3 & | & -10000 \end{bmatrix} \rightarrow \begin{bmatrix} 1 & 1 & 1 & | & 10000 \\ 0 & 1 & 2 & | & 8000 \\ 0 & -1 & -3 & | & -10000 \end{bmatrix}$$

$$R_2 = -0.06r_1 + r_2 \qquad\qquad R_2 = 100r_2$$
$$R_3 = -r_1 + r_3$$

$$\rightarrow \begin{bmatrix} 1 & 0 & -1 & | & 2000 \\ 0 & 1 & 2 & | & 8000 \\ 0 & 0 & -1 & | & -2000 \end{bmatrix} \rightarrow \begin{bmatrix} 1 & 0 & -1 & | & 2000 \\ 0 & 1 & 2 & | & 8000 \\ 0 & 0 & 1 & | & 2000 \end{bmatrix} \rightarrow \begin{bmatrix} 1 & 0 & 0 & | & 4000 \\ 0 & 1 & 0 & | & 4000 \\ 0 & 0 & 1 & | & 2000 \end{bmatrix}$$

$$R_1 = -r_2 + r_1 \qquad R_3 = -r_3 \qquad R_1 = r_3 + r_1$$
$$R_3 = r_2 + r_3 \qquad\qquad\qquad\qquad R_2 = -2r_3 + r_2$$

Carletta should invest $4000 in Treasury bills, $4000 in Treasury bonds, and $2000 in corporate bonds.

71. Let x = the number of Deltas produced.
Let y = the number of Betas produced.
Let z = the number of Sigmas produced.
Painting equation: $10x + 16y + 8z = 240$
Drying equation: $3x + 5y + 2z = 69$
Polishing equation: $2x + 3y + z = 41$
Set up a matrix and solve:

$$\begin{bmatrix} 10 & 16 & 8 & | & 240 \\ 3 & 5 & 2 & | & 69 \\ 2 & 3 & 1 & | & 41 \end{bmatrix} \rightarrow \begin{bmatrix} 1 & 1 & 2 & | & 33 \\ 3 & 5 & 2 & | & 69 \\ 2 & 3 & 1 & | & 41 \end{bmatrix} \rightarrow \begin{bmatrix} 1 & 1 & 2 & | & 33 \\ 0 & 2 & -4 & | & -30 \\ 0 & 1 & -3 & | & -25 \end{bmatrix} \rightarrow \begin{bmatrix} 1 & 1 & 2 & | & 33 \\ 0 & 1 & -2 & | & -15 \\ 0 & 1 & -3 & | & -25 \end{bmatrix}$$

$$R_1 = -3r_2 + r_1 \qquad R_2 = -3r_1 + r_2 \qquad R_2 = \tfrac{1}{2}r_2$$
$$R_3 = -2r_1 + r_3$$

$$\rightarrow \begin{bmatrix} 1 & 0 & 4 & | & 48 \\ 0 & 1 & -2 & | & -15 \\ 0 & 0 & -1 & | & -10 \end{bmatrix} \rightarrow \begin{bmatrix} 1 & 0 & 4 & | & 48 \\ 0 & 1 & -2 & | & -15 \\ 0 & 0 & 1 & | & 10 \end{bmatrix} \rightarrow \begin{bmatrix} 1 & 0 & 0 & | & 8 \\ 0 & 1 & 0 & | & 5 \\ 0 & 0 & 1 & | & 10 \end{bmatrix}$$

$$R_1 = -r_2 + r_1 \qquad R_3 = -r_3 \qquad R_1 = -4r_3 + r_1$$
$$R_3 = -r_2 + r_3 \qquad\qquad\qquad\qquad R_2 = 2r_3 + r_2$$

The company should produce 8 Deltas, 5 Betas, and 10 Sigmas.

73. Rewrite the system as set up and solve the matrix:

$$\begin{cases} -4+8-2I_2 = 0 \\ 8 = 5I_4 + I_1 \\ 4 = 3I_3 + I_1 \\ I_3 + I_4 = I_1 \end{cases} \rightarrow \begin{cases} 2I_2 = 4 \\ I_1 + 5I_4 = 8 \\ I_1 + 3I_3 = 4 \\ I_1 - I_3 - I_4 = 0 \end{cases}$$

$$\begin{bmatrix} 0 & 2 & 0 & 0 & | & 4 \\ 1 & 0 & 0 & 5 & | & 8 \\ 1 & 0 & 3 & 0 & | & 4 \\ 1 & 0 & -1 & -1 & | & 0 \end{bmatrix} \rightarrow \begin{bmatrix} 1 & 0 & 0 & 5 & | & 8 \\ 0 & 2 & 0 & 0 & | & 4 \\ 1 & 0 & 3 & 0 & | & 4 \\ 1 & 0 & -1 & -1 & | & 0 \end{bmatrix} \rightarrow \begin{bmatrix} 1 & 0 & 0 & 5 & | & 8 \\ 0 & 1 & 0 & 0 & | & 2 \\ 0 & 0 & 3 & -5 & | & -4 \\ 0 & 0 & -1 & -6 & | & -8 \end{bmatrix}$$

$$\text{Interchange } r_2 \text{ and } r_1 \quad R_2 = \tfrac{1}{2}r_2$$
$$R_3 = -r_1 + r_3$$
$$R_4 = -r_1 + r_4$$

$$\rightarrow \begin{bmatrix} 1 & 0 & 0 & 5 & | & 8 \\ 0 & 1 & 0 & 0 & | & 2 \\ 0 & 0 & -1 & -6 & | & -8 \\ 0 & 0 & 3 & -5 & | & -4 \end{bmatrix} \rightarrow \begin{bmatrix} 1 & 0 & 0 & 5 & | & 8 \\ 0 & 1 & 0 & 0 & | & 2 \\ 0 & 0 & 1 & 6 & | & 8 \\ 0 & 0 & 0 & -23 & | & -28 \end{bmatrix} \rightarrow \begin{bmatrix} 1 & 0 & 0 & 5 & | & 8 \\ 0 & 1 & 0 & 0 & | & 2 \\ 0 & 0 & 1 & 6 & | & 8 \\ 0 & 0 & 0 & 1 & | & \frac{28}{23} \end{bmatrix}$$

$$\text{Interchange } r_3 \text{ and } r_4 \quad R_3 = -r_3 \quad\quad\quad\quad R_4 = -\tfrac{1}{23}r_4$$
$$R_4 = -3r_3 + r_4$$

$$\rightarrow \begin{bmatrix} 1 & 0 & 0 & 0 & | & \frac{44}{23} \\ 0 & 1 & 0 & 0 & | & 2 \\ 0 & 0 & 1 & 0 & | & \frac{16}{23} \\ 0 & 0 & 0 & 1 & | & \frac{28}{23} \end{bmatrix}$$

$$R_1 = -5r_4 + r_1$$
$$R_3 = -6r_4 + r_3$$

The solution is $I_1 = \frac{44}{23}$, $I_2 = 2$, $I_3 = \frac{16}{23}$, $I_4 = \frac{28}{23}$.

11.4 Systems of Linear Equations: Determinants

For problems 1-10, to answer (b) enter matrix A, then evaluate the determinant of A.

1. (a) Evaluating the determinant:
$$\begin{vmatrix} 3 & 1 \\ 4 & 2 \end{vmatrix} = 3(2) - 4(1) = 6 - 4 = 2$$

3. (a) Evaluating the determinant:
$$\begin{vmatrix} 6 & 4 \\ -1 & 3 \end{vmatrix} = 6(3) - (-1)(4) = 18 + 4 = 22$$

5. (a) Evaluating the determinant:
$$\begin{vmatrix} -3 & -1 \\ 4 & 2 \end{vmatrix} = -3(2) - 4(-1) = -6 + 4 = -2$$

7.   (a)   Evaluating the determinant:

$$\begin{vmatrix} 3 & 4 & 2 \\ 1 & -1 & 5 \\ 1 & 2 & -2 \end{vmatrix} = 3 \begin{vmatrix} -1 & 5 \\ 2 & -2 \end{vmatrix} - 4 \begin{vmatrix} 1 & 5 \\ 1 & -2 \end{vmatrix} + 2 \begin{vmatrix} 1 & -1 \\ 1 & 2 \end{vmatrix}$$

$$= 3\left[-1(-2) - 2(5)\right] - 4\left[1(-2) - 1(5)\right] + 2\left[1(2) - 1(-1)\right]$$
$$= 3(2 - 10) - 4(-2 - 5) + 2(2 + 1)$$
$$= 3(-8) - 4(-7) + 2(3)$$
$$= -24 + 28 + 6$$
$$= 10$$

9.   (a)   Evaluating the determinant:

$$\begin{vmatrix} 4 & -1 & 2 \\ 6 & -1 & 0 \\ 1 & -3 & 4 \end{vmatrix} = 4 \begin{vmatrix} -1 & 0 \\ -3 & 4 \end{vmatrix} - (-1) \begin{vmatrix} 6 & 0 \\ 1 & 4 \end{vmatrix} + 2 \begin{vmatrix} 6 & -1 \\ 1 & -3 \end{vmatrix}$$

$$= 4\left[-1(4) - 0(-3)\right] + 1\left[6(4) - 1(0)\right] + 2\left[6(-3) - 1(-1)\right]$$
$$= 4(-4) + 1(24) + 2(-17)$$
$$= -16 + 24 - 34$$
$$= -26$$

For Problems 11-40, to answer (b), enter D as matrix A, enter D_X as matrix B, enter D_y as matrix C, and enter D_Z as matrix D. Then $x = \det(B)/\det(A)$, $y = \det(C)/\det(A)$, and $z = \det(D)/\det(A)$.

11.   (a)   Set up and evaluate the determinants to use Cramer's Rule:

$$\begin{cases} x + y = 8 \\ x - y = 4 \end{cases}$$

$$D = \begin{vmatrix} 1 & 1 \\ 1 & -1 \end{vmatrix} = 1(-1) - 1(1) = -1 - 1 = -2$$

$$D_x = \begin{vmatrix} 8 & 1 \\ 4 & -1 \end{vmatrix} = 8(-1) - 4(1) = -8 - 4 = -12$$

$$D_y = \begin{vmatrix} 1 & 8 \\ 1 & 4 \end{vmatrix} = 1(4) - 1(8) = 4 - 8 = -4$$

Find the solutions by Cramer's Rule:

$$x = \frac{D_x}{D} = \frac{-12}{-2} = 6 \qquad y = \frac{D_y}{D} = \frac{-4}{-2} = 2$$

(b)   Enter matrix $A = \begin{bmatrix} 1 & 1 \\ 1 & -1 \end{bmatrix}$, matrix $B = \begin{bmatrix} 8 & 1 \\ 4 & -1 \end{bmatrix}$, and matrix $C = \begin{bmatrix} 1 & 8 \\ 1 & 4 \end{bmatrix}$

Then $x = \dfrac{\det(B)}{\det(A)} = 6$ and $y = \dfrac{\det(C)}{\det(A)} = 2$

13. (a)   Set up and evaluate the determinants to use Cramer's Rule:

$$\begin{cases} 5x - y = 13 \\ 2x + 3y = 12 \end{cases}$$

$$D = \begin{vmatrix} 5 & -1 \\ 2 & 3 \end{vmatrix} = 5(3) - 2(-1) = 15 + 2 = 17$$

$$D_x = \begin{vmatrix} 13 & -1 \\ 12 & 3 \end{vmatrix} = 13(3) - 12(-1) = 39 + 12 = 51$$

$$D_y = \begin{vmatrix} 5 & 13 \\ 2 & 12 \end{vmatrix} = 5(12) - 2(13) = 60 - 26 = 34$$

Find the solutions by Cramer's Rule:

$$x = \frac{D_x}{D} = \frac{51}{17} = 3 \qquad y = \frac{D_y}{D} = \frac{34}{17} = 2$$

(b)   Enter matrix $A = \begin{bmatrix} 5 & -1 \\ 2 & 3 \end{bmatrix}$, matrix $B = \begin{bmatrix} 13 & -1 \\ 12 & 3 \end{bmatrix}$, and matrix $C = \begin{bmatrix} 5 & 13 \\ 2 & 12 \end{bmatrix}$

Then $x = \dfrac{\det(B)}{\det(A)} = 3$ and $y = \dfrac{\det(C)}{\det(A)} = 2$

15. (a)   Set up and evaluate the determinants to use Cramer's Rule:

$$\begin{cases} 3x = 24 \\ x + 2y = 0 \end{cases}$$

$$D = \begin{vmatrix} 3 & 0 \\ 1 & 2 \end{vmatrix} = 6 - 0 = 6$$

$$D_x = \begin{vmatrix} 24 & 0 \\ 0 & 2 \end{vmatrix} = 48 - 0 = 48$$

$$D_y = \begin{vmatrix} 3 & 24 \\ 1 & 0 \end{vmatrix} = 0 - 24 = -24$$

Find the solutions by Cramer's Rule:

$$x = \frac{D_x}{D} = \frac{48}{6} = 8 \qquad y = \frac{D_y}{D} = \frac{-24}{6} = -4$$

(b)   Enter matrix $A = \begin{bmatrix} 3 & 0 \\ 1 & 2 \end{bmatrix}$, matrix $B = \begin{bmatrix} 24 & 0 \\ 0 & 2 \end{bmatrix}$, and matrix $C = \begin{bmatrix} 3 & 24 \\ 1 & 0 \end{bmatrix}$

Then $x = \dfrac{\det(B)}{\det(A)} = 8$ and $y = \dfrac{\det(C)}{\det(A)} = -4$

17. (a)   Set up and evaluate the determinants to use Cramer's Rule:

$$\begin{cases} 3x - 6y = 24 \\ 5x + 4y = 12 \end{cases}$$

$$D = \begin{vmatrix} 3 & -6 \\ 5 & 4 \end{vmatrix} = 12 - (-30) = 42$$

$$D_x = \begin{vmatrix} 24 & -6 \\ 12 & 4 \end{vmatrix} = 96 - (-72) = 168$$

$$D_y = \begin{vmatrix} 3 & 24 \\ 5 & 12 \end{vmatrix} = 36 - 120 = -84$$

Find the solutions by Cramer's Rule:

$$x = \frac{D_x}{D} = \frac{168}{42} = 4 \qquad y = \frac{D_y}{D} = \frac{-84}{42} = -2$$

(b)   Enter matrix $A = \begin{bmatrix} 3 & -6 \\ 5 & 4 \end{bmatrix}$, matrix $B = \begin{bmatrix} 24 & -6 \\ 12 & 4 \end{bmatrix}$, and matrix $C = \begin{bmatrix} 3 & 24 \\ 5 & 12 \end{bmatrix}$

Then $x = \dfrac{\det(B)}{\det(A)} = 4$ and $y = \dfrac{\det(C)}{\det(A)} = -2$

19. (a)   Set up and evaluate the determinants to use Cramer's Rule:

$$\begin{cases} 3x - 2y = 4 \\ 6x - 4y = 0 \end{cases}$$

$$D = \begin{vmatrix} 3 & -2 \\ 6 & -4 \end{vmatrix} = -12 - (-12) = 0$$

Since $D = 0$, Cramer's Rule does not apply.

(b)   Enter matrix $A = \begin{bmatrix} 3 & -2 \\ 6 & -4 \end{bmatrix}$. Since $\det(A) = 0$, Cramer's Rule does not apply.

21. (a)   Set up and evaluate the determinants to use Cramer's Rule:

$$\begin{cases} 2x - 4y = -2 \\ 3x + 2y = 3 \end{cases}$$

$$D = \begin{vmatrix} 2 & -4 \\ 3 & 2 \end{vmatrix} = 4 - (-12) = 16$$

$$D_x = \begin{vmatrix} -2 & -4 \\ 3 & 2 \end{vmatrix} = -4 - (-12) = 8$$

$$D_y = \begin{vmatrix} 2 & -2 \\ 3 & 3 \end{vmatrix} = 6 - (-6) = 12$$

Find the solutions by Cramer's Rule:

$$x = \frac{D_x}{D} = \frac{8}{16} = \frac{1}{2} \qquad y = \frac{D_y}{D} = \frac{12}{16} = \frac{3}{4}$$

(b) Enter matrix $A = \begin{bmatrix} 2 & -4 \\ 3 & 2 \end{bmatrix}$, matrix $B = \begin{bmatrix} -2 & -4 \\ 3 & 2 \end{bmatrix}$, and matrix $C = \begin{bmatrix} 2 & -2 \\ 3 & 3 \end{bmatrix}$

Then $x = \dfrac{\det(B)}{\det(A)} = 0.5$ and $y = \dfrac{\det(C)}{\det(A)} = 0.75$

23. (a) Set up and evaluate the determinants to use Cramer's Rule:

$$\begin{cases} 2x - 3y = -1 \\ 10x + 10y = 5 \end{cases}$$

$$D = \begin{vmatrix} 2 & -3 \\ 10 & 10 \end{vmatrix} = 20 - (-30) = 50$$

$$D_x = \begin{vmatrix} -1 & -3 \\ 5 & 10 \end{vmatrix} = -10 - (-15) = 5$$

$$D_y = \begin{vmatrix} 2 & -1 \\ 10 & 5 \end{vmatrix} = 10 - (-10) = 20$$

Find the solutions by Cramer's Rule:

$$x = \frac{D_x}{D} = \frac{5}{50} = \frac{1}{10} \qquad y = \frac{D_y}{D} = \frac{20}{50} = \frac{2}{5}$$

(b) Enter matrix $A = \begin{bmatrix} 2 & -3 \\ 10 & 10 \end{bmatrix}$, matrix $B = \begin{bmatrix} -1 & -3 \\ 5 & 10 \end{bmatrix}$, and matrix $C = \begin{bmatrix} 2 & -1 \\ 10 & 5 \end{bmatrix}$

Then $x = \dfrac{\det(B)}{\det(A)} = 0.1$ and $y = \dfrac{\det(C)}{\det(A)} = 0.4$

25. (a) Set up and evaluate the determinants to use Cramer's Rule:

$$\begin{cases} 2x + 3y = 6 \\ x - y = \tfrac{1}{2} \end{cases}$$

$$D = \begin{vmatrix} 2 & 3 \\ 1 & -1 \end{vmatrix} = -2 - 3 = -5$$

$$D_x = \begin{vmatrix} 6 & 3 \\ \tfrac{1}{2} & -1 \end{vmatrix} = -6 - \tfrac{3}{2} = -\tfrac{15}{2}$$

$$D_y = \begin{vmatrix} 2 & 6 \\ 1 & \tfrac{1}{2} \end{vmatrix} = 1 - 6 = -5$$

Find the solutions by Cramer's Rule:

$$x = \frac{D_x}{D} = \frac{-\tfrac{15}{2}}{-5} = \frac{3}{2} \qquad y = \frac{D_y}{D} = \frac{-5}{-5} = 1$$

(b) Enter matrix $A = \begin{bmatrix} 2 & 3 \\ 1 & -1 \end{bmatrix}$, matrix $B = \begin{bmatrix} 6 & 3 \\ \tfrac{1}{2} & -1 \end{bmatrix}$, and matrix $C = \begin{bmatrix} 2 & 6 \\ 1 & \tfrac{1}{2} \end{bmatrix}$

Then $x = \dfrac{\det(B)}{\det(A)} = 1.5$ and $y = \dfrac{\det(C)}{\det(A)} = 1$

27. (a)   Set up and evaluate the determinants to use Cramer's Rule:

$$\begin{cases} 3x - 5y = 3 \\ 15x + 5y = 21 \end{cases}$$

$$D = \begin{vmatrix} 3 & -5 \\ 15 & 5 \end{vmatrix} = 15 - (-75) = 90$$

$$D_x = \begin{vmatrix} 3 & -5 \\ 21 & 5 \end{vmatrix} = 15 - (-105) = 120$$

$$D_y = \begin{vmatrix} 3 & 3 \\ 15 & 21 \end{vmatrix} = 63 - 45 = 18$$

Find the solutions by Cramer's Rule:

$$x = \frac{D_x}{D} = \frac{120}{90} = \frac{4}{3} \qquad y = \frac{D_y}{D} = \frac{18}{90} = \frac{1}{5}$$

(b)   Enter matrix $A = \begin{bmatrix} 3 & -5 \\ 15 & 5 \end{bmatrix}$, matrix $B = \begin{bmatrix} 3 & -5 \\ 21 & 5 \end{bmatrix}$, and matrix $C = \begin{bmatrix} 3 & 3 \\ 15 & 21 \end{bmatrix}$

Then $x = \dfrac{\det(B)}{\det(A)} = 1.33$ and $y = \dfrac{\det(C)}{\det(A)} = 0.2$

29. (a)   Set up and evaluate the determinants to use Cramer's Rule:

$$\begin{cases} x + y - z = 6 \\ 3x - 2y + z = -5 \\ x + 3y - 2z = 14 \end{cases}$$

$$D = \begin{vmatrix} 1 & 1 & -1 \\ 3 & -2 & 1 \\ 1 & 3 & -2 \end{vmatrix} = 1 \begin{vmatrix} -2 & 1 \\ 3 & -2 \end{vmatrix} - 1 \begin{vmatrix} 3 & 1 \\ 1 & -2 \end{vmatrix} + (-1) \begin{vmatrix} 3 & -2 \\ 1 & 3 \end{vmatrix}$$

$$= 1(4 - 3) - 1(-6 - 1) - 1(9 + 2) = 1 + 7 - 11 = -3$$

$$D_x = \begin{vmatrix} 6 & 1 & -1 \\ -5 & -2 & 1 \\ 14 & 3 & -2 \end{vmatrix} = 6 \begin{vmatrix} -2 & 1 \\ 3 & -2 \end{vmatrix} - 1 \begin{vmatrix} -5 & 1 \\ 14 & -2 \end{vmatrix} + (-1) \begin{vmatrix} -5 & -2 \\ 14 & 3 \end{vmatrix}$$

$$= 6(4 - 3) - 1(10 - 14) - 1(-15 + 28) = 6 + 4 - 13 = -3$$

$$D_y = \begin{vmatrix} 1 & 6 & -1 \\ 3 & -5 & 1 \\ 1 & 14 & -2 \end{vmatrix} = 1 \begin{vmatrix} -5 & 1 \\ 14 & -2 \end{vmatrix} - 6 \begin{vmatrix} 3 & 1 \\ 1 & -2 \end{vmatrix} + (-1) \begin{vmatrix} 3 & -5 \\ 1 & 14 \end{vmatrix}$$

$$= 1(10 - 14) - 6(-6 - 1) - 1(42 + 5) = -4 + 42 - 47 = -9$$

$$D_z = \begin{vmatrix} 1 & 1 & 6 \\ 3 & -2 & -5 \\ 1 & 3 & 14 \end{vmatrix} = 1 \begin{vmatrix} -2 & -5 \\ 3 & 14 \end{vmatrix} - 1 \begin{vmatrix} 3 & -5 \\ 1 & 14 \end{vmatrix} + 6 \begin{vmatrix} 3 & -2 \\ 1 & 3 \end{vmatrix}$$

$$= 1(-28 + 15) - 1(42 + 5) + 6(9 + 2) = -13 - 47 + 66 = 6$$

Find the solutions by Cramer's Rule:

$$x = \frac{D_x}{D} = \frac{-3}{-3} = 1 \qquad y = \frac{D_y}{D} = \frac{-9}{-3} = 3 \qquad z = \frac{D_z}{D} = \frac{6}{-3} = -2$$

(b)   Enter matrix $A = \begin{bmatrix} 1 & 1 & -1 \\ 3 & -2 & 1 \\ 1 & 3 & -2 \end{bmatrix}$, matrix $B = \begin{bmatrix} 6 & 1 & -1 \\ -5 & -2 & 1 \\ 14 & 3 & -2 \end{bmatrix}$,

matrix $C = \begin{bmatrix} 1 & 6 & -1 \\ 3 & -5 & 1 \\ 1 & 14 & -2 \end{bmatrix}$, matrix $D = \begin{bmatrix} 1 & 1 & 6 \\ 3 & -2 & -5 \\ 1 & 3 & 14 \end{bmatrix}$

Then $x = \dfrac{\det(B)}{\det(A)} = 1$, $y = \dfrac{\det(C)}{\det(A)} = 3$, $z = \dfrac{\det(D)}{\det(A)} = -2$

31. (a)   Set up and evaluate the determinants to use Cramer's Rule:

$$\begin{cases} x + 2y - z = -3 \\ 2x - 4y + z = -7 \\ -2x + 2y - 3z = 4 \end{cases}$$

$$D = \begin{vmatrix} 1 & 2 & -1 \\ 2 & -4 & 1 \\ -2 & 2 & -3 \end{vmatrix} = 1 \begin{vmatrix} -4 & 1 \\ 2 & -3 \end{vmatrix} - 2 \begin{vmatrix} 2 & 1 \\ -2 & -3 \end{vmatrix} + (-1) \begin{vmatrix} 2 & -4 \\ -2 & 2 \end{vmatrix}$$

$$= 1(12 - 2) - 2(-6 + 2) - 1(4 - 8) = 10 + 8 + 4 = 22$$

$$D_x = \begin{vmatrix} -3 & 2 & -1 \\ -7 & -4 & 1 \\ 4 & 2 & -3 \end{vmatrix} = -3 \begin{vmatrix} -4 & 1 \\ 2 & -3 \end{vmatrix} - 2 \begin{vmatrix} -7 & 1 \\ 4 & -3 \end{vmatrix} + (-1) \begin{vmatrix} -7 & -4 \\ 4 & 2 \end{vmatrix}$$

$$= -3(12 - 2) - 2(21 - 4) - 1(-14 + 16) = -30 - 34 - 2 = -66$$

$$D_y = \begin{vmatrix} 1 & -3 & -1 \\ 2 & -7 & 1 \\ -2 & 4 & -3 \end{vmatrix} = 1 \begin{vmatrix} -7 & 1 \\ 4 & -3 \end{vmatrix} - (-3) \begin{vmatrix} 2 & 1 \\ -2 & -3 \end{vmatrix} + (-1) \begin{vmatrix} 2 & -7 \\ -2 & 4 \end{vmatrix}$$

$$= 1(21 - 4) + 3(-6 + 2) - 1(8 - 14) = 17 - 12 + 6 = 11$$

$$D_z = \begin{vmatrix} 1 & 2 & -3 \\ 2 & -4 & -7 \\ -2 & 2 & 4 \end{vmatrix} = 1 \begin{vmatrix} -4 & -7 \\ 2 & 4 \end{vmatrix} - 2 \begin{vmatrix} 2 & -7 \\ -2 & 4 \end{vmatrix} + (-3) \begin{vmatrix} 2 & -4 \\ -2 & 2 \end{vmatrix}$$

$$= 1(-16 + 14) - 2(8 - 14) - 3(4 - 8) = -2 + 12 + 12 = 22$$

Find the solutions by Cramer's Rule:

$$x = \frac{D_x}{D} = \frac{-66}{22} = -3 \qquad y = \frac{D_y}{D} = \frac{11}{22} = \frac{1}{2} \qquad z = \frac{D_z}{D} = \frac{22}{22} = 1$$

(b)   Enter matrix $A = \begin{bmatrix} 1 & 2 & -1 \\ 2 & -4 & 1 \\ -2 & 2 & -3 \end{bmatrix}$, matrix $B = \begin{bmatrix} -3 & 2 & -1 \\ -7 & -4 & 1 \\ 4 & 2 & -3 \end{bmatrix}$,

matrix $C = \begin{bmatrix} 1 & -3 & -1 \\ 2 & -7 & 1 \\ -2 & 4 & -3 \end{bmatrix}$, matrix $D = \begin{bmatrix} 1 & 2 & -3 \\ 2 & -4 & -7 \\ -2 & 2 & 4 \end{bmatrix}$

Then $x = \dfrac{\det(B)}{\det(A)} = -3$, $y = \dfrac{\det(C)}{\det(A)} = 0.5$, $z = \dfrac{\det(D)}{\det(A)} = 1$

33. (a) Set up and evaluate the determinants to use Cramer's Rule:

$$\begin{cases} x - 2y + 3z = 1 \\ 3x + y - 2z = 0 \\ 2x - 4y + 6z = 2 \end{cases}$$

$$D = \begin{vmatrix} 1 & -2 & 3 \\ 3 & 1 & -2 \\ 2 & -4 & 6 \end{vmatrix} = 1 \begin{vmatrix} 1 & -2 \\ -4 & 6 \end{vmatrix} - (-2) \begin{vmatrix} 3 & -2 \\ 2 & 6 \end{vmatrix} + 3 \begin{vmatrix} 3 & 1 \\ 2 & -4 \end{vmatrix}$$

$$= 1(6 - 8) + 2(18 + 4) + 3(-12 - 2) = -2 + 44 - 42 = 0$$

Since $D = 0$, Cramer's Rule does not apply.

(b) Enter matrix $A = \begin{bmatrix} 1 & -2 & 3 \\ 3 & 1 & -2 \\ 2 & -4 & 6 \end{bmatrix}$ Since $\det(A) = 0$, Cramer's Rule does not

apply.

35. (a) Set up and evaluate the determinants to use Cramer's Rule:

$$\begin{cases} x + 2y - z = 0 \\ 2x - 4y + z = 0 \\ -2x + 2y - 3z = 0 \end{cases}$$

$$D = \begin{vmatrix} 1 & 2 & -1 \\ 2 & -4 & 1 \\ -2 & 2 & -3 \end{vmatrix} = 1 \begin{vmatrix} -4 & 1 \\ 2 & -3 \end{vmatrix} - 2 \begin{vmatrix} 2 & 1 \\ -2 & -3 \end{vmatrix} + (-1) \begin{vmatrix} 2 & -4 \\ -2 & 2 \end{vmatrix}$$

$$= 1(12 - 2) - 2(-6 + 2) - 1(4 - 8) = 10 + 8 + 4 = 22$$

$$D_x = \begin{vmatrix} 0 & 2 & -1 \\ 0 & -4 & 1 \\ 0 & 2 & -3 \end{vmatrix} = 0 \quad \text{(By Theorem 12)}$$

$$D_y = \begin{vmatrix} 1 & 0 & -1 \\ 2 & 0 & 1 \\ -2 & 0 & -3 \end{vmatrix} = 0 \quad \text{(By Theorem 12)}$$

$$D_z = \begin{vmatrix} 1 & 2 & 0 \\ 2 & -4 & 0 \\ -2 & 2 & 0 \end{vmatrix} = 0 \quad \text{(By Theorem 12)}$$

Find the solutions by Cramer's Rule:

$$x = \frac{D_x}{D} = \frac{0}{22} = 0 \qquad y = \frac{D_y}{D} = \frac{0}{22} = 0 \qquad z = \frac{D_z}{D} = \frac{0}{22} = 0$$

(b) Enter matrix $A = \begin{bmatrix} 1 & 2 & -1 \\ 2 & -4 & 1 \\ -2 & 2 & -3 \end{bmatrix}$, matrix $B = \begin{bmatrix} 0 & 2 & -1 \\ 0 & -4 & 1 \\ 0 & 2 & -3 \end{bmatrix}$,

matrix $C = \begin{bmatrix} 1 & 0 & -1 \\ 2 & 0 & 1 \\ -2 & 0 & -3 \end{bmatrix}$, matrix $D = \begin{bmatrix} 1 & 2 & 0 \\ 2 & -4 & 0 \\ -2 & 2 & 0 \end{bmatrix}$

Then $x = \dfrac{\det(B)}{\det(A)} = 0$, $y = \dfrac{\det(C)}{\det(A)} = 0$, $z = \dfrac{\det(D)}{\det(A)} = 0$

37. (a) Set up and evaluate the determinants to use Cramer's Rule:

$$\begin{cases} x - 2y + 3z = 0 \\ 3x + y - 2z = 0 \\ 2x - 4y + 6z = 0 \end{cases}$$

$$D = \begin{vmatrix} 1 & -2 & 3 \\ 3 & 1 & -2 \\ 2 & -4 & 6 \end{vmatrix} = 1\begin{vmatrix} 1 & -2 \\ -4 & 6 \end{vmatrix} - (-2)\begin{vmatrix} 3 & -2 \\ 2 & 6 \end{vmatrix} + 3\begin{vmatrix} 3 & 1 \\ 2 & -4 \end{vmatrix}$$

$$= 1(6 - 8) + 2(18 + 4) + 3(-12 - 2) = -2 + 44 - 42 = 0$$

Since $D = 0$, Cramer's Rule does not apply.

(b) Enter matrix $A = \begin{bmatrix} 1 & -2 & 3 \\ 3 & 1 & -2 \\ 2 & -4 & 6 \end{bmatrix}$ Since $\det(A) = 0$, Cramer's Rule does not

apply.

39. (a) Rewrite the system letting $u = \dfrac{1}{x}$ and $v = \dfrac{1}{y}$:

$$\begin{cases} \dfrac{1}{x} + \dfrac{1}{y} = 8 \\ \dfrac{3}{x} - \dfrac{5}{y} = 0 \end{cases} \rightarrow \begin{cases} u + v = 8 \\ 3u - 5v = 0 \end{cases}$$

Set up and evaluate the determinants to use Cramer's Rule:

$$D = \begin{vmatrix} 1 & 1 \\ 3 & -5 \end{vmatrix} = -5 - 3 = -8$$

$$D_u = \begin{vmatrix} 8 & 1 \\ 0 & -5 \end{vmatrix} = -40 - 0 = -40$$

$$D_v = \begin{vmatrix} 1 & 8 \\ 3 & 0 \end{vmatrix} = 0 - 24 = -24$$

Find the solutions by Cramer's Rule:

$$u = \frac{D_u}{D} = \frac{-40}{-8} = 5 \qquad v = \frac{D_v}{D} = \frac{-24}{-8} = 3$$

The solutions are $x = \frac{1}{5}$, $y = \frac{1}{3}$

(b)   Enter matrix $A = \begin{bmatrix} 1 & 1 \\ 3 & -5 \end{bmatrix}$, matrix $B = \begin{bmatrix} 8 & 1 \\ 0 & -5 \end{bmatrix}$, and matrix $C = \begin{bmatrix} 1 & 8 \\ 3 & 0 \end{bmatrix}$

Then $u = \dfrac{\det(B)}{\det(A)} = 5$ and $v = \dfrac{\det(C)}{\det(A)} = 3$.

So, $x = \dfrac{1}{5}$ and $y = \dfrac{1}{3}$.

41. Solve for x:

$$\begin{vmatrix} x & x \\ 4 & 3 \end{vmatrix} = 3x - 4x = -x$$

$$-x = 5$$

$$x = -5$$

43. Solve for x:

$$\begin{vmatrix} x & 1 & 1 \\ 4 & 3 & 2 \\ -1 & 2 & 5 \end{vmatrix} = x \begin{vmatrix} 3 & 2 \\ 2 & 5 \end{vmatrix} - 1 \begin{vmatrix} 4 & 2 \\ -1 & 5 \end{vmatrix} + 1 \begin{vmatrix} 4 & 3 \\ -1 & 2 \end{vmatrix}$$

$$= x(15 - 4) - (20 + 2) + (8 + 3) = 11x - 22 + 11 = 11x - 11$$

So, $11x - 11 = 2$

$$11x = 13$$

$$x = \frac{13}{11}$$

45. Solve for x:

$$\begin{vmatrix} x & 2 & 3 \\ 1 & x & 0 \\ 6 & 1 & -2 \end{vmatrix} = x \begin{vmatrix} x & 0 \\ 1 & -2 \end{vmatrix} - 2 \begin{vmatrix} 1 & 0 \\ 6 & -2 \end{vmatrix} + 3 \begin{vmatrix} 1 & x \\ 6 & 1 \end{vmatrix}$$

$$= x(-2x - 0) - 2(-2 - 0) + 3(1 - 6x)$$

$$= -2x^2 + 4 + 3 - 18x$$

$$= -2x^2 - 18x + 7$$

So, $-2x^2 - 18x + 7 = 7$

$$-2x^2 - 18x = 0$$

$$-2x(x + 9) = 0$$

$$x = 0 \text{ or } x = -9$$

47. Let $\begin{vmatrix} x & y & z \\ u & v & w \\ 1 & 2 & 3 \end{vmatrix} = 4$

Then $\begin{vmatrix} 1 & 2 & 3 \\ u & v & w \\ x & y & z \end{vmatrix} = -4$ by Theorem 11 --

The value of the determinant changes sign when two rows are interchanged.

Problems 49 – 53 use the Laws for Determinants in reverse order.

49. Let $\begin{vmatrix} x & y & z \\ u & v & w \\ 1 & 2 & 3 \end{vmatrix} = 4$

$$\begin{vmatrix} x & y & z \\ -3 & -6 & -9 \\ u & v & w \end{vmatrix} = -3\begin{vmatrix} x & y & z \\ 1 & 2 & 3 \\ u & v & w \end{vmatrix} = -3(-1)\begin{vmatrix} x & y & z \\ u & v & w \\ 1 & 2 & 3 \end{vmatrix} = 3(4) = 12$$

$\qquad\qquad\qquad$ Theorem 14 $\qquad\qquad$ Theorem 11

51. Let $\begin{vmatrix} x & y & z \\ u & v & w \\ 1 & 2 & 3 \end{vmatrix} = 4$

$$\begin{vmatrix} 1 & 2 & 3 \\ x-3 & y-6 & z-9 \\ 2u & 2v & 2w \end{vmatrix} = 2\begin{vmatrix} 1 & 2 & 3 \\ x-3 & y-6 & z-9 \\ u & v & w \end{vmatrix} = 2(-1)\begin{vmatrix} x-3 & y-6 & z-9 \\ 1 & 2 & 3 \\ u & v & w \end{vmatrix}$$

$\qquad\qquad\qquad$ Theorem 14 $\qquad\qquad\qquad$ Theorem 11

$$= 2(-1)(-1)\begin{vmatrix} x-3 & y-6 & z-9 \\ u & v & w \\ 1 & 2 & 3 \end{vmatrix} = 2(-1)(-1)\begin{vmatrix} x & y & z \\ u & v & w \\ 1 & 2 & 3 \end{vmatrix} = 2(-1)(-1)(4) = 8$$

$\qquad\qquad$ Theorem 11 $\qquad\qquad\qquad\qquad$ Theorem 15 $\;(R_1 = -3r_3 + r_1)$

53. Let $\begin{vmatrix} x & y & z \\ u & v & w \\ 1 & 2 & 3 \end{vmatrix} = 4$

$$\begin{vmatrix} 1 & 2 & 3 \\ 2x & 2y & 2z \\ u-1 & v-2 & w-3 \end{vmatrix} = 2\begin{vmatrix} 1 & 2 & 3 \\ x & y & z \\ u-1 & v-2 & w-3 \end{vmatrix} = 2(-1)\begin{vmatrix} x & y & z \\ 1 & 2 & 3 \\ u-1 & v-2 & w-3 \end{vmatrix}$$

$\qquad\qquad\qquad$ Theorem 14 $\qquad\qquad\qquad$ Theorem 11

$$= 2(-1)(-1)\begin{vmatrix} x & y & z \\ u-1 & v-2 & w-3 \\ 1 & 2 & 3 \end{vmatrix} = 2(-1)(-1)\begin{vmatrix} x & y & z \\ u & v & w \\ 1 & 2 & 3 \end{vmatrix} = 2(-1)(-1)(4) = 8$$

$\qquad\qquad$ Theorem 11 $\qquad\qquad\qquad\qquad$ Theorem 15 $\;(R_2 = -r_3 + r_2)$

55. Expanding the determinant:

$$\begin{vmatrix} x & y & 1 \\ x_1 & y_1 & 1 \\ x_2 & y_2 & 1 \end{vmatrix} = x\begin{vmatrix} y_1 & 1 \\ y_2 & 1 \end{vmatrix} - y\begin{vmatrix} x_1 & 1 \\ x_2 & 1 \end{vmatrix} + 1\begin{vmatrix} x_1 & y_1 \\ x_2 & y_2 \end{vmatrix}$$

$$= x(y_1 - y_2) - y(x_1 - x_2) + (x_1 y_2 - x_2 y_1) = 0$$

$$x(y_1 - y_2) + y(x_2 - x_1) = x_2 y_1 - x_1 y_2$$

$$y(x_2 - x_1) = x_2 y_1 - x_1 y_2 + x(y_2 - y_1)$$

$$y(x_2 - x_1) - y_1(x_2 - x_1) = x_2 y_1 - x_1 y_2 + x(y_2 - y_1) - y_1(x_2 - x_1)$$

$$(x_2 - x_1)(y - y_1) = x(y_2 - y_1) + x_2 y_1 - x_1 y_2 - y_1 x_2 + y_1 x_1$$

$$(x_2 - x_1)(y - y_1) = (y_2 - y_1)x - (y_2 - y_1)x_1$$

$$(x_2 - x_1)(y - y_1) = (y_2 - y_1)(x - x_1)$$

$$(y - y_1) = \frac{(y_2 - y_1)}{(x_2 - x_1)}(x - x_1)$$

57. Expanding the determinant:

$$\begin{vmatrix} x^2 & x & 1 \\ y^2 & y & 1 \\ z^2 & z & 1 \end{vmatrix} = x^2\begin{vmatrix} y & 1 \\ z & 1 \end{vmatrix} - x\begin{vmatrix} y^2 & 1 \\ z^2 & 1 \end{vmatrix} + 1\begin{vmatrix} y^2 & y \\ z^2 & z \end{vmatrix}$$

$$= x^2(y - z) - x(y^2 - z^2) + 1(y^2 z - z^2 y)$$

$$= x^2(y - z) - x(y - z)(y + z) + yz(y - z)$$

$$= (y - z)\left[x^2 - xy - xz + yz\right]$$

$$= (y - z)\left[x(x - y) - z(x - y)\right]$$

$$= (y - z)(x - y)(x - z)$$

59. Evaluating the determinant to show the relationship:

$$\begin{vmatrix} a_{13} & a_{12} & a_{11} \\ a_{23} & a_{22} & a_{21} \\ a_{33} & a_{32} & a_{31} \end{vmatrix} = a_{13}\begin{vmatrix} a_{22} & a_{21} \\ a_{32} & a_{31} \end{vmatrix} - a_{12}\begin{vmatrix} a_{23} & a_{21} \\ a_{33} & a_{31} \end{vmatrix} + a_{11}\begin{vmatrix} a_{23} & a_{22} \\ a_{33} & a_{32} \end{vmatrix}$$

$$= a_{13}(a_{22}a_{31} - a_{21}a_{32}) - a_{12}(a_{23}a_{31} - a_{21}a_{33}) + a_{11}(a_{23}a_{32} - a_{22}a_{33})$$

$$= a_{13}a_{22}a_{31} - a_{13}a_{21}a_{32} - a_{12}a_{23}a_{31} + a_{12}a_{21}a_{33} + a_{11}a_{23}a_{32} - a_{11}a_{22}a_{33}$$

$$= -a_{11}a_{22}a_{33} + a_{11}a_{23}a_{32} + a_{12}a_{21}a_{33} - a_{12}a_{23}a_{31} - a_{13}a_{21}a_{32} + a_{13}a_{22}a_{31}$$

$$= -a_{11}(a_{22}a_{33} - a_{23}a_{32}) + a_{12}(a_{21}a_{33} - a_{23}a_{31}) - a_{13}(a_{21}a_{32} - a_{22}a_{31})$$

$$= -a_{11}\begin{vmatrix} a_{22} & a_{23} \\ a_{32} & a_{33} \end{vmatrix} + a_{12}\begin{vmatrix} a_{21} & a_{23} \\ a_{31} & a_{33} \end{vmatrix} - a_{13}\begin{vmatrix} a_{21} & a_{22} \\ a_{31} & a_{32} \end{vmatrix}$$

$$= -\left[a_{11}\begin{vmatrix} a_{22} & a_{23} \\ a_{32} & a_{33} \end{vmatrix} - a_{12}\begin{vmatrix} a_{21} & a_{23} \\ a_{31} & a_{33} \end{vmatrix} + a_{13}\begin{vmatrix} a_{21} & a_{22} \\ a_{31} & a_{32} \end{vmatrix}\right]$$

$$= -\begin{vmatrix} a_{11} & a_{12} & a_{13} \\ a_{21} & a_{22} & a_{23} \\ a_{31} & a_{32} & a_{33} \end{vmatrix}$$

61. Set up a 3 by 3 determinant in which the first column and third column are the same and evaluate:

$$\begin{vmatrix} a & b & a \\ c & d & c \\ e & f & e \end{vmatrix} = -b\begin{vmatrix} c & c \\ e & e \end{vmatrix} + d\begin{vmatrix} a & a \\ e & e \end{vmatrix} - f\begin{vmatrix} a & a \\ c & c \end{vmatrix}$$

$$= -b(ce - ce) + d(ae - ae) - f(ac - ac)$$

$$= -b(0) + d(0) - f(0)$$

$$= 0$$

11.5 Matrix Algebra

Problems 1-16: To answer (b), enter the matrices into your graphing utility and perform the operation. The answer should agree with (a).

1. (a) $A + B = \begin{bmatrix} 0 & 3 & -5 \\ 1 & 2 & 6 \end{bmatrix} + \begin{bmatrix} 4 & 1 & 0 \\ -2 & 3 & -2 \end{bmatrix} = \begin{bmatrix} 0+4 & 3+1 & -5+0 \\ 1+(-2) & 2+3 & 6+(-2) \end{bmatrix} = \begin{bmatrix} 4 & 4 & -5 \\ -1 & 5 & 4 \end{bmatrix}$

3. (a) $4A = 4\begin{bmatrix} 0 & 3 & -5 \\ 1 & 2 & 6 \end{bmatrix} = \begin{bmatrix} 4\cdot0 & 4\cdot3 & 4(-5) \\ 4\cdot1 & 4\cdot2 & 4\cdot6 \end{bmatrix} = \begin{bmatrix} 0 & 12 & -20 \\ 4 & 8 & 24 \end{bmatrix}$

5. (a) $3A - 2B = 3\begin{bmatrix} 0 & 3 & -5 \\ 1 & 2 & 6 \end{bmatrix} - 2\begin{bmatrix} 4 & 1 & 0 \\ -2 & 3 & -2 \end{bmatrix}$

$$= \begin{bmatrix} 0 & 9 & -15 \\ 3 & 6 & 18 \end{bmatrix} - \begin{bmatrix} 8 & 2 & 0 \\ -4 & 6 & -4 \end{bmatrix} = \begin{bmatrix} -8 & 7 & -15 \\ 7 & 0 & 22 \end{bmatrix}$$

7. (a) $AC = \begin{bmatrix} 0 & 3 & -5 \\ 1 & 2 & 6 \end{bmatrix} \cdot \begin{bmatrix} 4 & 1 \\ 6 & 2 \\ -2 & 3 \end{bmatrix}$

$$= \begin{bmatrix} 0(4)+3(6)+(-5)(-2) & 0(1)+3(2)+(-5)(3) \\ 1(4)+2(6)+6(-2) & 1(1)+2(2)+6(3) \end{bmatrix} = \begin{bmatrix} 28 & -9 \\ 4 & 23 \end{bmatrix}$$

9. (a) $CA = \begin{bmatrix} 4 & 1 \\ 6 & 2 \\ -2 & 3 \end{bmatrix} \cdot \begin{bmatrix} 0 & 3 & -5 \\ 1 & 2 & 6 \end{bmatrix}$

$$= \begin{bmatrix} 4(0)+1(1) & 4(3)+1(2) & 4(-5)+1(6) \\ 6(0)+2(1) & 6(3)+2(2) & 6(-5)+2(6) \\ -2(0)+3(1) & -2(3)+3(2) & -2(-5)+3(6) \end{bmatrix} = \begin{bmatrix} 1 & 14 & -14 \\ 2 & 22 & -18 \\ 3 & 0 & 28 \end{bmatrix}$$

11. (a) $C(A+B) = \begin{bmatrix} 4 & 1 \\ 6 & 2 \\ -2 & 3 \end{bmatrix} \left(\begin{bmatrix} 0 & 3 & -5 \\ 1 & 2 & 6 \end{bmatrix} + \begin{bmatrix} 4 & 1 & 0 \\ -2 & 3 & -2 \end{bmatrix} \right)$

$= \begin{bmatrix} 4 & 1 \\ 6 & 2 \\ -2 & 3 \end{bmatrix} \cdot \begin{bmatrix} 4 & 4 & -5 \\ -1 & 5 & 4 \end{bmatrix} = \begin{bmatrix} 15 & 21 & -16 \\ 22 & 34 & -22 \\ -11 & 7 & 22 \end{bmatrix}$

13. (a) $AC - 3I_2 = \begin{bmatrix} 0 & 3 & -5 \\ 1 & 2 & 6 \end{bmatrix} \cdot \begin{bmatrix} 4 & 1 \\ 6 & 2 \\ -2 & 3 \end{bmatrix} - 3\begin{bmatrix} 1 & 0 \\ 0 & 1 \end{bmatrix} = \begin{bmatrix} 28 & -9 \\ 4 & 23 \end{bmatrix} - \begin{bmatrix} 3 & 0 \\ 0 & 3 \end{bmatrix} = \begin{bmatrix} 25 & -9 \\ 4 & 20 \end{bmatrix}$

15. (a) $CA - CB = \begin{bmatrix} 4 & 1 \\ 6 & 2 \\ -2 & 3 \end{bmatrix} \cdot \begin{bmatrix} 0 & 3 & -5 \\ 1 & 2 & 6 \end{bmatrix} - \begin{bmatrix} 4 & 1 \\ 6 & 2 \\ -2 & 3 \end{bmatrix} \cdot \begin{bmatrix} 4 & 1 & 0 \\ -2 & 3 & -2 \end{bmatrix}$

$= \begin{bmatrix} 1 & 14 & -14 \\ 2 & 22 & -18 \\ 3 & 0 & 28 \end{bmatrix} - \begin{bmatrix} 14 & 7 & -2 \\ 20 & 12 & -4 \\ -14 & 7 & -6 \end{bmatrix} = \begin{bmatrix} -13 & 7 & -12 \\ -18 & 10 & -14 \\ 17 & -7 & 34 \end{bmatrix}$

17. (a) $\begin{bmatrix} 2 & -2 \\ 1 & 0 \end{bmatrix} \begin{bmatrix} 2 & 1 & 4 & 6 \\ 3 & -1 & 3 & 2 \end{bmatrix}$

$= \begin{bmatrix} 2(2) + (-2)(3) & 2(1) + (-2)(-1) & 2(4) + (-2)(3) & 2(6) + (-2)(2) \\ 1(2) + 0(3) & 1(1) + 0(-1) & 1(4) + 0(3) & 1(6) + 0(2) \end{bmatrix}$

$= \begin{bmatrix} -2 & 4 & 2 & 8 \\ 2 & 1 & 4 & 6 \end{bmatrix}$

(b)   Enter matrix $A = \begin{bmatrix} 2 & -2 \\ 1 & 0 \end{bmatrix}$ and matrix $B = \begin{bmatrix} 2 & 1 & 4 & 6 \\ 3 & -1 & 3 & 2 \end{bmatrix}$.

Then enter $[A][B]$ to obtain the answer in (a).

19. (a) $\begin{bmatrix} 1 & 0 & 1 \\ 2 & 4 & 1 \\ 3 & 6 & 1 \end{bmatrix} \begin{bmatrix} 1 & 3 \\ 6 & 2 \\ 8 & -1 \end{bmatrix} = \begin{bmatrix} 1(1) + 0(6) + 1(8) & 1(3) + 0(2) + 1(-1) \\ 2(1) + 4(6) + 1(8) & 2(3) + 4(2) + 1(-1) \\ 3(1) + 6(6) + 1(8) & 3(3) + 6(2) + 1(-1) \end{bmatrix} = \begin{bmatrix} 9 & 2 \\ 34 & 13 \\ 47 & 20 \end{bmatrix}$

(b)   Enter matrix $A = \begin{bmatrix} 1 & 0 & 1 \\ 2 & 4 & 1 \\ 3 & 6 & 1 \end{bmatrix}$ and matrix $B = \begin{bmatrix} 1 & 3 \\ 6 & 2 \\ 8 & -1 \end{bmatrix}$.

Then enter $[A][B]$ to obtain the answer in (a).

21.   Augment the matrix with the identity and use row operations to find the inverse:

$A = \begin{bmatrix} 2 & 1 \\ 1 & 1 \end{bmatrix} \rightarrow \begin{bmatrix} 2 & 1 & | & 1 & 0 \\ 1 & 1 & | & 0 & 1 \end{bmatrix}$

$\rightarrow \begin{bmatrix} 1 & 1 & | & 0 & 1 \\ 2 & 1 & | & 1 & 0 \end{bmatrix} \rightarrow \begin{bmatrix} 1 & 1 & | & 0 & 1 \\ 0 & -1 & | & 1 & -2 \end{bmatrix} \rightarrow \begin{bmatrix} 1 & 1 & | & 0 & 1 \\ 0 & 1 & | & -1 & 2 \end{bmatrix} \rightarrow \begin{bmatrix} 1 & 0 & | & 1 & -1 \\ 0 & 1 & | & -1 & 2 \end{bmatrix}$

Interchange $R_2 = -2r_1 + r_2$ $R_2 = -r_2$ $R_1 = -r_2 + r_1$
r_1 and r_2

$$A^{-1} = \begin{bmatrix} 1 & -1 \\ -1 & 2 \end{bmatrix}$$

23. Augment the matrix with the identity and use row operations to find the inverse:

$$A = \begin{bmatrix} 6 & 5 \\ 2 & 2 \end{bmatrix} \rightarrow \begin{bmatrix} 6 & 5 & | & 1 & 0 \\ 2 & 2 & | & 0 & 1 \end{bmatrix}$$

$$\rightarrow \begin{bmatrix} 2 & 2 & | & 0 & 1 \\ 6 & 5 & | & 1 & 0 \end{bmatrix} \rightarrow \begin{bmatrix} 2 & 2 & | & 0 & 1 \\ 0 & -1 & | & 1 & -3 \end{bmatrix} \rightarrow \begin{bmatrix} 1 & 1 & | & 0 & \frac{1}{2} \\ 0 & 1 & | & -1 & 3 \end{bmatrix} \rightarrow \begin{bmatrix} 1 & 0 & | & 1 & -\frac{5}{2} \\ 0 & 1 & | & -1 & 3 \end{bmatrix}$$

Interchange $R_2 = -3r_1 + r_2$ $R_1 = \frac{1}{2}r_1$ $R_1 = -r_2 + r_1$

r_1 and r_2 $R_2 = -r_2$

$$A^{-1} = \begin{bmatrix} 1 & -\frac{5}{2} \\ -1 & 3 \end{bmatrix}$$

25. Augment the matrix with the identity and use row operations to find the inverse:

$$A = \begin{bmatrix} 2 & 1 \\ a & a \end{bmatrix} \rightarrow \begin{bmatrix} 2 & 1 & | & 1 & 0 \\ a & a & | & 0 & 1 \end{bmatrix} \text{ where } a \neq 0.$$

$$\rightarrow \begin{bmatrix} 1 & \frac{1}{2} & | & \frac{1}{2} & 0 \\ a & a & | & 0 & 1 \end{bmatrix} \rightarrow \begin{bmatrix} 1 & \frac{1}{2} & | & \frac{1}{2} & 0 \\ 0 & \frac{1}{2}a & | & -\frac{1}{2}a & 1 \end{bmatrix} \rightarrow \begin{bmatrix} 1 & \frac{1}{2} & | & \frac{1}{2} & 0 \\ 0 & 1 & | & -1 & \frac{2}{a} \end{bmatrix} \rightarrow \begin{bmatrix} 1 & 0 & | & 1 & -\frac{1}{a} \\ 0 & 1 & | & -1 & \frac{2}{a} \end{bmatrix}$$

$R_1 = \frac{1}{2}r_1$ $R_2 = -ar_1 + r_2$ $R_2 = \left(\frac{2}{a}\right)r_2$ $R_1 = -\frac{1}{2}r_2 + r_1$

$$A^{-1} = \begin{bmatrix} 1 & -\frac{1}{a} \\ -1 & \frac{2}{a} \end{bmatrix}$$

27. Augment the matrix with the identity and use row operations to find the inverse:

$$A = \begin{bmatrix} 1 & -1 & 1 \\ 0 & -2 & 1 \\ -2 & -3 & 0 \end{bmatrix} \rightarrow \begin{bmatrix} 1 & -1 & 1 & | & 1 & 0 & 0 \\ 0 & -2 & 1 & | & 0 & 1 & 0 \\ -2 & -3 & 0 & | & 0 & 0 & 1 \end{bmatrix}$$

$$\rightarrow \begin{bmatrix} 1 & -1 & 1 & | & 1 & 0 & 0 \\ 0 & -2 & 1 & | & 0 & 1 & 0 \\ 0 & -5 & 2 & | & 2 & 0 & 1 \end{bmatrix} \rightarrow \begin{bmatrix} 1 & -1 & 1 & | & 1 & 0 & 0 \\ 0 & 1 & -\frac{1}{2} & | & 0 & -\frac{1}{2} & 0 \\ 0 & -5 & 2 & | & 2 & 0 & 1 \end{bmatrix} \rightarrow \begin{bmatrix} 1 & 0 & \frac{1}{2} & | & 1 & -\frac{1}{2} & 0 \\ 0 & 1 & -\frac{1}{2} & | & 0 & -\frac{1}{2} & 0 \\ 0 & 0 & -\frac{1}{2} & | & 2 & -\frac{5}{2} & 1 \end{bmatrix}$$

$R_3 = 2r_1 + r_3$ $R_2 = -\frac{1}{2}r_2$ $R_1 = r_2 + r_1$

                                                         $R_3 = 5r_2 + r_3$

$$\rightarrow \begin{bmatrix} 1 & 0 & \frac{1}{2} & | & 1 & -\frac{1}{2} & 0 \\ 0 & 1 & -\frac{1}{2} & | & 0 & -\frac{1}{2} & 0 \\ 0 & 0 & 1 & | & -4 & 5 & -2 \end{bmatrix} \rightarrow \begin{bmatrix} 1 & 0 & 0 & | & 3 & -3 & 1 \\ 0 & 1 & 0 & | & -2 & 2 & -1 \\ 0 & 0 & 1 & | & -4 & 5 & -2 \end{bmatrix}$$

$R_3 = -2r_3$ $R_1 = -\frac{1}{2}r_3 + r_1$

                             $R_2 = \frac{1}{2}r_3 + r_2$

$$A^{-1} = \begin{bmatrix} 3 & -3 & 1 \\ -2 & 2 & -1 \\ -4 & 5 & -2 \end{bmatrix}$$

29. Augment the matrix with the identity and use row operations to find the inverse:

$$A = \begin{bmatrix} 1 & 1 & 1 \\ 3 & 2 & -1 \\ 3 & 1 & 2 \end{bmatrix} \rightarrow \begin{bmatrix} 1 & 1 & 1 & | & 1 & 0 & 0 \\ 3 & 2 & -1 & | & 0 & 1 & 0 \\ 3 & 1 & 2 & | & 0 & 0 & 1 \end{bmatrix}$$

$$\rightarrow \begin{bmatrix} 1 & 1 & 1 & | & 1 & 0 & 0 \\ 0 & -1 & -4 & | & -3 & 1 & 0 \\ 0 & -2 & -1 & | & -3 & 0 & 1 \end{bmatrix} \rightarrow \begin{bmatrix} 1 & 1 & 1 & | & 1 & 0 & 0 \\ 0 & 1 & 4 & | & 3 & -1 & 0 \\ 0 & -2 & -1 & | & -3 & 0 & 1 \end{bmatrix} \rightarrow \begin{bmatrix} 1 & 0 & -3 & | & -2 & 1 & 0 \\ 0 & 1 & 4 & | & 3 & -1 & 0 \\ 0 & 0 & 7 & | & 3 & -2 & 1 \end{bmatrix}$$

$R_2 = -3r_1 + r_2$ $\qquad$ $R_2 = -r_2$ $\qquad$ $R_1 = -r_2 + r_1$

$R_3 = -3r_1 + r_3$ $\qquad\qquad\qquad\qquad\qquad$ $R_3 = 2r_2 + r_3$

$$\rightarrow \begin{bmatrix} 1 & 0 & -3 & | & -2 & 1 & 0 \\ 0 & 1 & 4 & | & 3 & -1 & 0 \\ 0 & 0 & 1 & | & \frac{3}{7} & -\frac{2}{7} & \frac{1}{7} \end{bmatrix} \rightarrow \begin{bmatrix} 1 & 0 & 0 & | & -\frac{5}{7} & \frac{1}{7} & \frac{3}{7} \\ 0 & 1 & 0 & | & \frac{9}{7} & \frac{1}{7} & -\frac{4}{7} \\ 0 & 0 & 1 & | & \frac{3}{7} & -\frac{2}{7} & \frac{1}{7} \end{bmatrix}$$

$R_3 = \frac{1}{7}r_3$ $\qquad\qquad$ $R_1 = 3r_3 + r_1$

$\qquad\qquad\qquad\qquad$ $R_2 = -4r_3 + r_2$

$$A^{-1} = \begin{bmatrix} -\frac{5}{7} & \frac{1}{7} & \frac{3}{7} \\ \frac{9}{7} & \frac{1}{7} & -\frac{4}{7} \\ \frac{3}{7} & -\frac{2}{7} & \frac{1}{7} \end{bmatrix}$$

31. Rewrite the system of equations in matrix form:

$$\begin{cases} 2x + y = 8 \\ x + y = 5 \end{cases} \qquad A = \begin{bmatrix} 2 & 1 \\ 1 & 1 \end{bmatrix}, \quad X = \begin{bmatrix} x \\ y \end{bmatrix}, \quad B = \begin{bmatrix} 8 \\ 5 \end{bmatrix}$$

Find the inverse of A and solve $X = A^{-1}B$:

From Problem 21, $A^{-1} = \begin{bmatrix} 1 & -1 \\ -1 & 2 \end{bmatrix}$ and $X = A^{-1}B = \begin{bmatrix} 1 & -1 \\ -1 & 2 \end{bmatrix}\begin{bmatrix} 8 \\ 5 \end{bmatrix} = \begin{bmatrix} 3 \\ 2 \end{bmatrix}$.

The solution is $x = 3, y = 2$.

33. Rewrite the system of equations in matrix form:

$$\begin{cases} 2x + y = 0 \\ x + y = 5 \end{cases} \qquad A = \begin{bmatrix} 2 & 1 \\ 1 & 1 \end{bmatrix}, \quad X = \begin{bmatrix} x \\ y \end{bmatrix}, \quad B = \begin{bmatrix} 0 \\ 5 \end{bmatrix}$$

Find the inverse of A and solve $X = A^{-1}B$:

From Problem 21, $A^{-1} = \begin{bmatrix} 1 & -1 \\ -1 & 2 \end{bmatrix}$ and $X = A^{-1}B = \begin{bmatrix} 1 & -1 \\ -1 & 2 \end{bmatrix}\begin{bmatrix} 0 \\ 5 \end{bmatrix} = \begin{bmatrix} -5 \\ 10 \end{bmatrix}$.

The solution is $x = -5, y = 10$.

35. Rewrite the system of equations in matrix form:

$$\begin{cases} 6x + 5y = 7 \\ 2x + 2y = 2 \end{cases} \qquad A = \begin{bmatrix} 6 & 5 \\ 2 & 2 \end{bmatrix}, \quad X = \begin{bmatrix} x \\ y \end{bmatrix}, \quad B = \begin{bmatrix} 7 \\ 2 \end{bmatrix}$$

Find the inverse of A and solve $X = A^{-1}B$:

From Problem 23, $A^{-1} = \begin{bmatrix} 1 & -\frac{5}{2} \\ -1 & 3 \end{bmatrix}$ and $X = A^{-1}B = \begin{bmatrix} 1 & -\frac{5}{2} \\ -1 & 3 \end{bmatrix}\begin{bmatrix} 7 \\ 2 \end{bmatrix} = \begin{bmatrix} 2 \\ -1 \end{bmatrix}$.

The solution is $x = 2, y = -1$.

37. Rewrite the system of equations in matrix form:

$$\begin{cases} 6x + 5y = 13 \\ 2x + 2y = 5 \end{cases} \qquad A = \begin{bmatrix} 6 & 5 \\ 2 & 2 \end{bmatrix}, \quad X = \begin{bmatrix} x \\ y \end{bmatrix}, \quad B = \begin{bmatrix} 13 \\ 5 \end{bmatrix}$$

Find the inverse of A and solve $X = A^{-1}B$:

From Problem 23, $A^{-1} = \begin{bmatrix} 1 & -\frac{5}{2} \\ -1 & 3 \end{bmatrix}$ and $X = A^{-1}B = \begin{bmatrix} 1 & -\frac{5}{2} \\ -1 & 3 \end{bmatrix}\begin{bmatrix} 13 \\ 5 \end{bmatrix} = \begin{bmatrix} \frac{1}{2} \\ 2 \end{bmatrix}$.

The solution is $x = \frac{1}{2}$, $y = 2$.

39. Rewrite the system of equations in matrix form:

$$\begin{cases} 2x + y = -3 \\ ax + ay = -a \end{cases} \quad a \neq 0 \qquad A = \begin{bmatrix} 2 & 1 \\ a & a \end{bmatrix}, \quad X = \begin{bmatrix} x \\ y \end{bmatrix}, \quad B = \begin{bmatrix} -3 \\ -a \end{bmatrix}$$

Find the inverse of A and solve $X = A^{-1}B$:

From Problem 25, $A^{-1} = \begin{bmatrix} 1 & -\frac{1}{a} \\ -1 & \frac{2}{a} \end{bmatrix}$ and $X = A^{-1}B = \begin{bmatrix} 1 & -\frac{1}{a} \\ -1 & \frac{2}{a} \end{bmatrix}\begin{bmatrix} -3 \\ -a \end{bmatrix} = \begin{bmatrix} -2 \\ 1 \end{bmatrix}$.

The solution is $x = -2$, $y = 1$.

41. Rewrite the system of equations in matrix form:

$$\begin{cases} 2x + y = \frac{7}{a} \\ ax + ay = 5 \end{cases} \quad a \neq 0 \qquad A = \begin{bmatrix} 2 & 1 \\ a & a \end{bmatrix}, \quad X = \begin{bmatrix} x \\ y \end{bmatrix}, \quad B = \begin{bmatrix} \frac{7}{a} \\ 5 \end{bmatrix}$$

Find the inverse of A and solve $X = A^{-1}B$:

From Problem 25, $A^{-1} = \begin{bmatrix} 1 & -\frac{1}{a} \\ -1 & \frac{2}{a} \end{bmatrix}$ and $X = A^{-1}B = \begin{bmatrix} 1 & -\frac{1}{a} \\ -1 & \frac{2}{a} \end{bmatrix}\begin{bmatrix} \frac{7}{a} \\ 5 \end{bmatrix} = \begin{bmatrix} \frac{2}{a} \\ \frac{3}{a} \end{bmatrix}$.

The solution is $x = \frac{2}{a}$, $y = \frac{3}{a}$.

43. Rewrite the system of equations in matrix form:

$$\begin{cases} x - y + z = 0 \\ -2y + z = -1 \\ -2x - 3y = -5 \end{cases} \qquad A = \begin{bmatrix} 1 & -1 & 1 \\ 0 & -2 & 1 \\ -2 & -3 & 0 \end{bmatrix}, \quad X = \begin{bmatrix} x \\ y \\ z \end{bmatrix}, \quad B = \begin{bmatrix} 0 \\ -1 \\ -5 \end{bmatrix}$$

Find the inverse of A and solve $X = A^{-1}B$:

From Problem 27,

$$A^{-1} = \begin{bmatrix} 3 & -3 & 1 \\ -2 & 2 & -1 \\ -4 & 5 & -2 \end{bmatrix} \text{ and } X = A^{-1}B = \begin{bmatrix} 3 & -3 & 1 \\ -2 & 2 & -1 \\ -4 & 5 & -2 \end{bmatrix}\begin{bmatrix} 0 \\ -1 \\ -5 \end{bmatrix} = \begin{bmatrix} -2 \\ 3 \\ 5 \end{bmatrix}.$$

The solution is $x = -2$, $y = 3$, $z = 5$.

45. Rewrite the system of equations in matrix form:

$$\begin{cases} x - y + z = 2 \\ -2y + z = 2 \\ -2x - 3y = \frac{1}{2} \end{cases} \qquad A = \begin{bmatrix} 1 & -1 & 1 \\ 0 & -2 & 1 \\ -2 & -3 & 0 \end{bmatrix}, \quad X = \begin{bmatrix} x \\ y \\ z \end{bmatrix}, \quad B = \begin{bmatrix} 2 \\ 2 \\ \frac{1}{2} \end{bmatrix}$$

Find the inverse of A and solve $X = A^{-1}B$:
From Problem 27,

$$A^{-1} = \begin{bmatrix} 3 & -3 & 1 \\ -2 & 2 & -1 \\ -4 & 5 & -2 \end{bmatrix} \text{ and } X = A^{-1}B = \begin{bmatrix} 3 & -3 & 1 \\ -2 & 2 & -1 \\ -4 & 5 & -2 \end{bmatrix}\begin{bmatrix} 2 \\ 2 \\ \frac{1}{2} \end{bmatrix} = \begin{bmatrix} \frac{1}{2} \\ -\frac{1}{2} \\ 1 \end{bmatrix}.$$

The solution is $x = \frac{1}{2}$, $y = -\frac{1}{2}$, $z = 1$.

47. Rewrite the system of equations in matrix form:

$$\begin{cases} x + y + z = 9 \\ 3x + 2y - z = 8 \\ 3x + y + 2z = 1 \end{cases} \quad A = \begin{bmatrix} 1 & 1 & 1 \\ 3 & 2 & -1 \\ 3 & 1 & 2 \end{bmatrix}, \quad X = \begin{bmatrix} x \\ y \\ z \end{bmatrix}, \quad B = \begin{bmatrix} 9 \\ 8 \\ 1 \end{bmatrix}$$

Find the inverse of A and solve $X = A^{-1}B$:
From Problem 29,

$$A^{-1} = \begin{bmatrix} -\frac{5}{7} & \frac{1}{7} & \frac{3}{7} \\ \frac{9}{7} & \frac{1}{7} & -\frac{4}{7} \\ \frac{3}{7} & -\frac{2}{7} & \frac{1}{7} \end{bmatrix} \text{ and } X = A^{-1}B = \begin{bmatrix} -\frac{5}{7} & \frac{1}{7} & \frac{3}{7} \\ \frac{9}{7} & \frac{1}{7} & -\frac{4}{7} \\ \frac{3}{7} & -\frac{2}{7} & \frac{1}{7} \end{bmatrix}\begin{bmatrix} 9 \\ 8 \\ 1 \end{bmatrix} = \begin{bmatrix} -\frac{34}{7} \\ \frac{85}{7} \\ \frac{12}{7} \end{bmatrix}.$$

The solution is $x = -\frac{34}{7}$, $y = \frac{85}{7}$, $z = \frac{12}{7}$.

49. Rewrite the system of equations in matrix form:

$$\begin{cases} x + y + z = 2 \\ 3x + 2y - z = \frac{7}{3} \\ 3x + y + 2z = \frac{10}{3} \end{cases} \quad A = \begin{bmatrix} 1 & 1 & 1 \\ 3 & 2 & -1 \\ 3 & 1 & 2 \end{bmatrix}, \quad X = \begin{bmatrix} x \\ y \\ z \end{bmatrix}, \quad B = \begin{bmatrix} 2 \\ \frac{7}{3} \\ \frac{10}{3} \end{bmatrix}$$

Find the inverse of A and solve $X = A^{-1}B$:
From Problem 29,

$$A^{-1} = \begin{bmatrix} -\frac{5}{7} & \frac{1}{7} & \frac{3}{7} \\ \frac{9}{7} & \frac{1}{7} & -\frac{4}{7} \\ \frac{3}{7} & -\frac{2}{7} & \frac{1}{7} \end{bmatrix} \text{ and } X = A^{-1}B = \begin{bmatrix} -\frac{5}{7} & \frac{1}{7} & \frac{3}{7} \\ \frac{9}{7} & \frac{1}{7} & -\frac{4}{7} \\ \frac{3}{7} & -\frac{2}{7} & \frac{1}{7} \end{bmatrix}\begin{bmatrix} 2 \\ \frac{7}{3} \\ \frac{10}{3} \end{bmatrix} = \begin{bmatrix} \frac{1}{3} \\ 1 \\ \frac{2}{3} \end{bmatrix}.$$

The solution is $x = \frac{1}{3}$, $y = 1$, $z = \frac{2}{3}$.

51. Augment the matrix with the identity and use row operations to find the inverse:

$$A = \begin{bmatrix} 4 & 2 \\ 2 & 1 \end{bmatrix} \rightarrow \begin{bmatrix} 4 & 2 & | & 1 & 0 \\ 2 & 1 & | & 0 & 1 \end{bmatrix}$$

$$\rightarrow \begin{bmatrix} 4 & 2 & | & 1 & 0 \\ 0 & 0 & | & -\frac{1}{2} & 1 \end{bmatrix} \rightarrow \begin{bmatrix} 1 & \frac{1}{2} & | & \frac{1}{4} & 0 \\ 0 & 0 & | & -\frac{1}{2} & 1 \end{bmatrix}$$

$$R_2 = -\frac{1}{2}r_1 + r_2 \qquad R_1 = \frac{1}{4}r_1$$

There is no way to obtain the identity matrix on the left; thus, there is no inverse.

53. Augment the matrix with the identity and use row operations to find the inverse:

$$A = \begin{bmatrix} 15 & 3 \\ 10 & 2 \end{bmatrix} \rightarrow \begin{bmatrix} 15 & 3 & 1 & 0 \\ 10 & 2 & 0 & 1 \end{bmatrix}$$

$$\rightarrow \begin{bmatrix} 15 & 3 & 1 & 0 \\ 0 & 0 & -\frac{2}{3} & 1 \end{bmatrix} \rightarrow \begin{bmatrix} 1 & \frac{1}{5} & \frac{1}{15} & 0 \\ 0 & 0 & -\frac{2}{3} & 1 \end{bmatrix}$$

$$R_2 = -\frac{2}{3}r_1 + r_2 \qquad R_1 = \frac{1}{15}r_1$$

There is no way to obtain the identity matrix on the left; thus, there is no inverse.

55. Augment the matrix with the identity and use row operations to find the inverse:

$$A = \begin{bmatrix} -3 & 1 & -1 \\ 1 & -4 & -7 \\ 1 & 2 & 5 \end{bmatrix} \rightarrow \begin{bmatrix} -3 & 1 & -1 & 1 & 0 & 0 \\ 1 & -4 & -7 & 0 & 1 & 0 \\ 1 & 2 & 5 & 0 & 0 & 1 \end{bmatrix}$$

$$\rightarrow \begin{bmatrix} 1 & 2 & 5 & 0 & 0 & 1 \\ 1 & -4 & -7 & 0 & 1 & 0 \\ -3 & 1 & -1 & 1 & 0 & 0 \end{bmatrix} \rightarrow \begin{bmatrix} 1 & 2 & 5 & 0 & 0 & 1 \\ 0 & -6 & -12 & 0 & 1 & -1 \\ 0 & 7 & 14 & 1 & 0 & 3 \end{bmatrix} \rightarrow \begin{bmatrix} 1 & 2 & 5 & 0 & 0 & 1 \\ 0 & 1 & 2 & 0 & -\frac{1}{6} & \frac{1}{6} \\ 0 & 7 & 14 & 1 & 0 & 3 \end{bmatrix}$$

Interchange r_1 and r_3 $\qquad R_2 = -r_1 + r_2 \qquad\qquad R_2 = -\frac{1}{6}r_2$
$$R_3 = 3r_1 + r_3$$

$$\rightarrow \begin{bmatrix} 1 & 0 & 1 & 0 & \frac{1}{3} & \frac{2}{3} \\ 0 & 1 & 2 & 0 & -\frac{1}{6} & \frac{1}{6} \\ 0 & 0 & 0 & 1 & \frac{7}{6} & \frac{11}{6} \end{bmatrix}$$

$$R_1 = -2r_2 + r_1$$
$$R_3 = -7r_2 + r_3$$

There is no way to obtain the identity matrix on the left; thus, there is no inverse.

57. $$\begin{bmatrix} 0.01 & 0.05 & -0.01 \\ 0.01 & -0.02 & 0.01 \\ -0.02 & 0.01 & 0.03 \end{bmatrix}$$

59. $$\begin{bmatrix} 0.02 & -0.04 & -0.01 & 0.01 \\ -0.02 & 0.05 & 0.03 & -0.03 \\ 0.02 & 0.01 & -0.04 & 0.00 \\ -0.02 & 0.06 & 0.07 & 0.06 \end{bmatrix}$$

61. $x = 4.57$, $y = -6.44$, $z = -24.07$ 63. $x = -1.19$, $y = 2.46$, $z = 8.27$

65. (a) The rows of the 2 by 3 matrix represent stainless steel and aluminum. The columns represent 10-gallon, 5-gallon, and 1-gallon.

The 2 by 3 matrix is: The 3 by 2 matrix is:

$$\begin{bmatrix} 500 & 350 & 400 \\ 700 & 500 & 850 \end{bmatrix} \qquad \begin{bmatrix} 500 & 700 \\ 350 & 500 \\ 400 & 850 \end{bmatrix}$$

(b) The 3 by 1 matrix representing the amount of material is:

$$\begin{bmatrix} 15 \\ 8 \\ 3 \end{bmatrix}$$

(c)   The days usage of materials is:

$$\begin{bmatrix} 500 & 350 & 400 \\ 700 & 500 & 850 \end{bmatrix} \cdot \begin{bmatrix} 15 \\ 8 \\ 3 \end{bmatrix} = \begin{bmatrix} 11,500 \\ 17,050 \end{bmatrix}$$

11,500 pounds of stainless steel and 17,050 pounds of aluminum are used each day.

(d)   The 1 by 2 matrix representing cost is:

$$\begin{bmatrix} 0.10 & 0.05 \end{bmatrix}$$

(e)   The total cost of the days production was:

$$\begin{bmatrix} 0.10 & 0.05 \end{bmatrix} \cdot \begin{bmatrix} 11,500 \\ 17,050 \end{bmatrix} = \begin{bmatrix} 2002.50 \end{bmatrix}$$

The total cost of the days production was $2,002.50.

11.6 Partial Fraction Decomposition

1.   The rational expression $\dfrac{x}{x^2 - 1}$ is proper, since the degree of the numerator is less than the degree of the denominator.

3.   The rational expression $\dfrac{x^2 + 5}{x^2 - 4}$ is improper, so perform the division:

$$x^2 - 4 \overline{)\begin{array}{l} 1 \\ x^2 + 5 \end{array}}$$
$$\underline{x^2 - 4}$$
$$9$$

The proper rational expression is:

$$\frac{x^2 + 5}{x^2 - 4} = 1 + \frac{9}{x^2 - 4}$$

5.   The rational expression $\dfrac{5x^3 + 2x - 1}{x^2 - 4}$ is improper, so perform the division:

$$x^2 - 4 \overline{)\begin{array}{l} 5x \\ 5x^3 + 0x^2 + 2x - 1 \end{array}}$$
$$\underline{5x^3 - 20x}$$
$$22x - 1$$

The proper rational expression is:

$$\frac{5x^3 + 2x - 1}{x^2 - 4} = 5x + \frac{22x - 1}{x^2 - 4}$$

7.   The rational expression $\dfrac{x(x - 1)}{(x + 4)(x - 3)} = \dfrac{x^2 - x}{x^2 + x - 12}$ is improper, so perform the division:

$$x^2 + x - 12 \overline{)\begin{array}{l} 1 \\ x^2 - x + 0 \end{array}}$$
$$\underline{x^2 + x - 12}$$
$$-2x + 12$$

The proper rational expression is:

$$\frac{x(x - 1)}{(x + 4)(x - 3)} = 1 + \frac{-2x + 12}{x^2 + x - 12}$$

9.   Find the partial fraction decomposition:

$$\frac{4}{x(x-1)} = \frac{A}{x} + \frac{B}{x-1}$$

   $4 = A(x-1) + Bx$ (Multiply both sides by $x(x-1)$.)

   Let $x = 1$:   then $4 = A(0) + B$, or $B = 4$

   Let $x = 0$:   then $4 = A(-1) + B(0)$, or $A = -4$

$$\frac{4}{x(x-1)} = \frac{-4}{x} + \frac{4}{x-1}$$

11.   Find the partial fraction decomposition:

$$\frac{1}{x(x^2+1)} = \frac{A}{x} + \frac{Bx+C}{x^2+1}$$

   $1 = A(x^2+1) + (Bx+C)x$ (Multiply both sides by $x(x^2+1)$.)

   Let $x = 0$:   then $1 = A(1) + (B(0)+C)(0)$, or $A = 1$

   Let $x = 1$:   then $1 = A(1+1) + (B(1)+C)(1)$, or $1 = 2A+B+C$

   or $1 = 2(1) + B + C$

   or $-1 = B + C$

   Let $x = -1$: then $1 = A(1+1) + (B(-1)+C)(-1)$, or $1 = 2A+B-C$

   or $1 = 2(1) + B - C$

   or $-1 = B - C$

   Solve the system of equations:

   $B + C = -1$

   $\underline{B - C = -1}$

   $2B \quad\quad = -2$ $-1 + C = -1$

   $B \quad\quad = -1$ $C = 0$

$$\frac{1}{x(x^2+1)} = \frac{1}{x} + \frac{-x}{x^2+1}$$

13.   Find the partial fraction decomposition:

$$\frac{x}{(x-1)(x-2)} = \frac{A}{x-1} + \frac{B}{x-2}$$

   $x = A(x-2) + B(x-1)$ (Multiply both sides by $(x-1)(x-2)$.)

   Let $x = 1$:   then $1 = A(1-2) + B(1-1)$ $\rightarrow$ $1 = -A$ $\rightarrow$ $A = -1$

   Let $x = 2$:   then $2 = A(2-2) + B(2-1)$ $\rightarrow$ $2 = B$ $\rightarrow$ $B = 2$

$$\frac{x}{(x-1)(x-2)} = \frac{-1}{x-1} + \frac{2}{x-2}$$

15. Find the partial fraction decomposition:

$$\frac{x^2}{(x-1)^2(x+1)} = \frac{A}{x-1} + \frac{B}{(x-1)^2} + \frac{C}{x+1}$$

(Multiply both sides by $(x-1)^2(x+1)$.)

$$x^2 = A(x-1)(x+1) + B(x+1) + C(x-1)^2$$

Let $x = 1$:   then $1^2 = A(1-1)(1+1) + B(1+1) + C(1-1)^2$

$$\rightarrow 1 = 2B \rightarrow B = \tfrac{1}{2}$$

Let $x = -1$:   then $(-1)^2 = A(-1-1)(-1+1) + B(-1+1) + C(-1-1)^2$

$$\rightarrow 1 = 4C \rightarrow C = \tfrac{1}{4}$$

Let $x = 0$:   then $0^2 = A(0-1)(0+1) + B(0+1) + C(0-1)^2$

$$\rightarrow 0 = -A + B + C \rightarrow A = \tfrac{1}{2} + \tfrac{1}{4} = \tfrac{3}{4}$$

$$\frac{x^2}{(x-1)^2(x+1)} = \frac{\tfrac{3}{4}}{x-1} + \frac{\tfrac{1}{2}}{(x-1)^2} + \frac{\tfrac{1}{4}}{x+1}$$

17. Find the partial fraction decomposition:

$$\frac{1}{x^3-8} = \frac{1}{(x-2)(x^2+2x+4)} = \frac{A}{x-2} + \frac{Bx+C}{x^2+2x+4}$$

(Multiply both sides by $(x-2)(x^2+2x+4)$.)

$$1 = A(x^2+2x+4) + (Bx+C)(x-2)$$

Let $x = 2$:   then $1 = A\left(2^2+2(2)+4\right) + (B(2)+C)(2-2)$

$$\rightarrow 1 = 12A \rightarrow A = \tfrac{1}{12}$$

Let $x = 0$:   then $1 = A\left(0^2+2(0)+4\right) + (B(0)+C)(0-2)$

$$\rightarrow 1 = 4A - 2C \rightarrow 1 = 4\left(\tfrac{1}{12}\right) - 2C$$

$$\rightarrow -2C = \tfrac{2}{3} \rightarrow C = -\tfrac{1}{3}$$

Let $x = 1$:   then $1 = A\left(1^2+2(1)+4\right) + (B(1)+C)(1-2)$

$$\rightarrow 1 = 7A - B - C \rightarrow 1 = 7\left(\tfrac{1}{12}\right) - B + \tfrac{1}{3} \rightarrow B = -\tfrac{1}{12}$$

$$\frac{1}{x^3-8} = \frac{\tfrac{1}{12}}{x-2} + \frac{-\tfrac{1}{12}x - \tfrac{1}{3}}{x^2+2x+4}$$

19. Find the partial fraction decomposition:

$$\frac{x^2}{(x-1)^2(x+1)^2} = \frac{A}{x-1} + \frac{B}{(x-1)^2} + \frac{C}{x+1} + \frac{D}{(x+1)^2}$$

(Multiply both sides by $(x-1)^2(x+1)^2$.)

$$x^2 = A(x-1)(x+1)^2 + B(x+1)^2 + C(x-1)^2(x+1) + D(x-1)^2$$

Let $x = 1$:   then $1^2 = A(1-1)(1+1)^2 + B(1+1)^2 + C(1-1)^2(1+1) + D(1-1)^2$

$$\rightarrow 1 = 4B \rightarrow B = \tfrac{1}{4}$$

Let $x = -1$:   then

$$(-1)^2 = A(-1-1)(-1+1)^2 + B(-1+1)^2 + C(-1-1)^2(-1+1) + D(-1-1)^2$$

$$\rightarrow 1 = 4D \rightarrow D = \tfrac{1}{4}$$

Let $x = 0$: then
$$0^2 = A(0-1)(0+1)^2 + B(0+1)^2 + C(0-1)^2(0+1) + D(0-1)^2$$
$$\rightarrow \quad 0 = -A + B + C + D \quad \rightarrow \quad A - C = \tfrac{1}{4} + \tfrac{1}{4} = \tfrac{1}{2}$$

Let $x = 2$: then
$$2^2 = A(2-1)(2+1)^2 + B(2+1)^2 + C(2-1)^2(2+1) + D(2-1)^2$$
$$\rightarrow \quad 4 = 9A + 9B + 3C + D \quad \rightarrow \quad 9A + 3C = 4 - \tfrac{9}{4} - \tfrac{1}{4} = \tfrac{3}{2}$$
$$\rightarrow \quad 3A + C = \tfrac{1}{2}$$

Solve the system of equations:
$$A - C = \tfrac{1}{2}$$
$$\underline{3A + C = \tfrac{1}{2}}$$
$$4A \quad\;\; = 1$$
$$A \quad\;\; = \tfrac{1}{4} \quad \rightarrow \quad \tfrac{3}{4} + C = \tfrac{1}{2} \quad \rightarrow \quad C = -\tfrac{1}{4}$$
$$\frac{x^2}{(x-1)^2(x+1)^2} = \frac{\tfrac{1}{4}}{x-1} + \frac{\tfrac{1}{4}}{(x-1)^2} + \frac{-\tfrac{1}{4}}{x+1} + \frac{\tfrac{1}{4}}{(x+1)^2}$$

21. Find the partial fraction decomposition:
$$\frac{x-3}{(x+2)(x+1)^2} = \frac{A}{x+2} + \frac{B}{x+1} + \frac{C}{(x+1)^2}$$

(Multiply both sides by $(x+2)(x+1)^2$.)
$$x - 3 = A(x+1)^2 + B(x+2)(x+1) + C(x+2)$$

Let $x = -2$: then $-2 - 3 = A(-2+1)^2 + B(-2+2)(-2+1) + C(-2+2)$
$$\rightarrow \quad -5 = A \quad \rightarrow \quad A = -5$$

Let $x = -1$: then $-1 - 3 = A(-1+1)^2 + B(-1+2)(-1+1) + C(-1+2)$
$$\rightarrow \quad -4 = C \quad \rightarrow \quad C = -4$$

Let $x = 0$: then
$$0 - 3 = A(0+1)^2 + B(0+2)(0+1) + C(0+2) \quad \rightarrow \quad -3 = A + 2B + 2C$$
$$\rightarrow \quad -3 = -5 + 2B + 2(-4) \quad \rightarrow \quad 2B = 10 \quad \rightarrow \quad B = 5$$
$$\frac{x-3}{(x+2)(x+1)^2} = \frac{-5}{x+2} + \frac{5}{x+1} + \frac{-4}{(x+1)^2}$$

23. Find the partial fraction decomposition:
$$\frac{x+4}{x^2(x^2+4)} = \frac{A}{x} + \frac{B}{x^2} + \frac{Cx+D}{x^2+4}$$

(Multiply both sides by $x^2(x^2+4)$.)
$$x + 4 = Ax(x^2+4) + B(x^2+4) + (Cx+D)x^2$$

Let $x = 0$: then $0 + 4 = A(0)(0^2+4) + B(0^2+4) + (C0+D)(0)^2$
$$\rightarrow \quad 4 = 4B \quad \rightarrow \quad B = 1$$

Let $x = 1$: then $1 + 4 = A(1)(1^2+4) + B(1^2+4) + (C(1)+D)(1)^2$
$$\rightarrow \quad 5 = 5A + 5B + C + D \quad \rightarrow \quad 5 = 5A + 5 + C + D$$
$$\rightarrow \quad 5A + C + D = 0$$

Let $x = -1$: then

$$-1 + 4 = A(-1)((-1)^2 + 4) + B((-1)^2 + 4) + (C(-1) + D)(-1)^2$$
$$\rightarrow 3 = -5A + 5B - C + D \rightarrow 3 = -5A + 5 - C + D$$
$$\rightarrow -5A - C + D = -2$$

Let $x = 2$: then $2 + 4 = A(2)(2^2 + 4) + B(2^2 + 4) + (C(2) + D)(2)^2$
$$\rightarrow 6 = 16A + 8B + 8C + 4D \rightarrow 6 = 16A + 8 + 8C + 4D$$
$$\rightarrow 16A + 8C + 4D = -2$$

Solve the system of equations:

$$\begin{aligned} 5A + C + D &= 0 \\ \underline{-5A - C + D} &= -2 \\ 2D &= -2 \qquad\qquad 5A + C - 1 = 0 \\ D &= -1 \qquad\qquad C = 1 - 5A \\ 16A + 8(1 - 5A) + 4(-1) &= -2 \qquad\qquad C = 1 - 5\left(\tfrac{1}{4}\right) \\ 16A + 8 - 40A - 4 &= -2 \qquad\qquad C = 1 - \tfrac{5}{4} \\ -24A &= -6 \qquad\qquad C = -\tfrac{1}{4} \\ A &= \tfrac{1}{4} \end{aligned}$$

$$\frac{x + 4}{x^2(x^2 + 4)} = \frac{\tfrac{1}{4}}{x} + \frac{1}{x^2} + \frac{-\tfrac{1}{4}x - 1}{x^2 + 4}$$

25. Find the partial fraction decomposition:

$$\frac{x^2 + 2x + 3}{(x + 1)(x^2 + 2x + 4)} = \frac{A}{x + 1} + \frac{Bx + C}{x^2 + 2x + 4}$$

(Multiply both sides by $(x + 1)(x^2 + 2x + 4)$.)

$$x^2 + 2x + 3 = A(x^2 + 2x + 4) + (Bx + C)(x + 1)$$

Let $x = -1$: then

$$(-1)^2 + 2(-1) + 3 = A((-1)^2 + 2(-1) + 4) + (B(-1) + C)(-1 + 1)$$
$$\rightarrow 2 = 3A \rightarrow A = \tfrac{2}{3}$$

Let $x = 0$: then $0^2 + 2(0) + 3 = A(0^2 + 2(0) + 4) + (B(0) + C)(0 + 1)$
$$\rightarrow 3 = 4A + C \rightarrow 3 = 4\left(\tfrac{2}{3}\right) + C \rightarrow C = \tfrac{1}{3}$$

Let $x = 1$: then

$$1^2 + 2(1) + 3 = A(1^2 + 2(1) + 4) + (B(1) + C)(1 + 1)$$
$$\rightarrow 6 = 7A + 2B + 2C \rightarrow 6 = 7\left(\tfrac{2}{3}\right) + 2B + 2\left(\tfrac{1}{3}\right)$$
$$\rightarrow 2B = 6 - \tfrac{14}{3} - \tfrac{2}{3} \rightarrow 2B = \tfrac{2}{3} \rightarrow B = \tfrac{1}{3}$$

$$\frac{x^2 + 2x + 3}{(x + 1)(x^2 + 2x + 4)} = \frac{\tfrac{2}{3}}{x + 1} + \frac{\tfrac{1}{3}x + \tfrac{1}{3}}{x^2 + 2x + 4}$$

27. Find the partial fraction decomposition:

$$\frac{x}{(3x-2)(2x+1)} = \frac{A}{3x-2} + \frac{B}{2x+1}$$

(Multiply both sides by $(3x-2)(2x+1)$.)

$$x = A(2x+1) + B(3x-2)$$

Let $x = -\frac{1}{2}$: then $-\frac{1}{2} = A\left(2\left(-\frac{1}{2}\right)+1\right) + B\left(3\left(-\frac{1}{2}\right)-2\right)$

$$\rightarrow \; -\frac{1}{2} = -\frac{7}{2}B \; \rightarrow \; B = \frac{1}{7}$$

Let $x = \frac{2}{3}$: then $\frac{2}{3} = A\left(2\left(\frac{2}{3}\right)+1\right) + B\left(3\left(\frac{2}{3}\right)-2\right) \; \rightarrow \; \frac{2}{3} = \frac{7}{3}A \; \rightarrow \; A = \frac{2}{7}$

$$\frac{x}{(3x-2)(2x+1)} = \frac{\frac{2}{7}}{3x-2} + \frac{\frac{1}{7}}{2x+1}$$

29. Find the partial fraction decomposition:

$$\frac{x}{x^2+2x-3} = \frac{x}{(x+3)(x-1)} = \frac{A}{x+3} + \frac{B}{x-1}$$

(Multiply both sides by $(x+3)(x-1)$.)

$$x = A(x-1) + B(x+3)$$

Let $x = 1$: then $1 = A(1-1) + B(1+3) \; \rightarrow \; 1 = 4B \; \rightarrow \; B = \frac{1}{4}$

Let $x = -3$: then $-3 = A(-3-1) + B(-3+3) \; \rightarrow \; -3 = -4A \; \rightarrow \; A = \frac{3}{4}$

$$\frac{x}{x^2+2x-3} = \frac{\frac{3}{4}}{x+3} + \frac{\frac{1}{4}}{x-1}$$

31. Find the partial fraction decomposition:

$$\frac{x^2+2x+3}{(x^2+4)^2} = \frac{Ax+B}{x^2+4} + \frac{Cx+D}{(x^2+4)^2}$$

(Multiply both sides by $(x^2+4)^2$.)

$$x^2+2x+3 = (Ax+B)(x^2+4) + Cx+D$$

$$x^2+2x+3 = Ax^3 + Bx^2 + 4Ax + 4B + Cx + D$$

$$x^2+2x+3 = Ax^3 + Bx^2 + (4A+C)x + 4B + D$$

$$A = 0$$

$$B = 1$$

$$4A+C = 2 \; \rightarrow \; 4(0)+C = 2 \; \rightarrow \; C = 2$$

$$4B+D = 3 \; \rightarrow \; 4(1)+D = 3 \; \rightarrow \; D = -1$$

$$\frac{x^2+2x+3}{(x^2+4)^2} = \frac{1}{x^2+4} + \frac{2x-1}{(x^2+4)^2}$$

33. Find the partial fraction decomposition:

$$\frac{7x+3}{x^3-2x^2-3x} = \frac{7x+3}{x(x-3)(x+1)} = \frac{A}{x} + \frac{B}{x-3} + \frac{C}{x+1}$$

(Multiply both sides by $x(x-3)(x+1)$.)

$$7x+3 = A(x-3)(x+1) + Bx(x+1) + Cx(x-3)$$

Let $x = 0$:   then $7(0)+3 = A(0-3)(0+1) + B(0)(0+1) + C(0)(0-3)$

$$\rightarrow 3 = -3A \rightarrow A = -1$$

Let $x = 3$:   then $7(3)+3 = A(3-3)(3+1) + B(3)(3+1) + C(3)(3-3)$

$$\rightarrow 24 = 12B \rightarrow B = 2$$

Let $x = -1$:   then

$$7(-1)+3 = A(-1-3)(-1+1) + B(-1)(-1+1) + C(-1)(-1-3)$$

$$\rightarrow -4 = 4C \rightarrow C = -1$$

$$\frac{7x+3}{x^3-2x^2-3x} = \frac{-1}{x} + \frac{2}{x-3} + \frac{-1}{x+1}$$

35. Perform synthetic division to find a factor:

$$2\overline{)\,1 \quad -4 \quad 5 \quad -2\,}$$
$$\,2 \quad -4 \quad 2$$
$$\overline{\,1 \quad -2 \quad 1 \quad 0}$$

$$x^3 - 4x^2 + 5x - 2 = (x-2)(x^2-2x+1) = (x-2)(x-1)^2$$

Find the partial fraction decomposition:

$$\frac{x^2}{x^3-4x^2+5x-2} = \frac{x^2}{(x-2)(x-1)^2} = \frac{A}{x-2} + \frac{B}{x-1} + \frac{C}{(x-1)^2}$$

(Multiply both sides by $(x-2)(x-1)^2$.)

$$x^2 = A(x-1)^2 + B(x-2)(x-1) + C(x-2)$$

Let $x = 2$:   then $2^2 = A(2-1)^2 + B(2-2)(2-1) + C(2-2)$

$$\rightarrow 4 = A \rightarrow A = 4$$

Let $x = 1$:   then $1^2 = A(1-1)^2 + B(1-2)(1-1) + C(1-2)$

$$\rightarrow 1 = -C \rightarrow C = -1$$

Let $x = 0$:   then $0^2 = A(0-1)^2 + B(0-2)(0-1) + C(0-2)$

$$\rightarrow 0 = A + 2B - 2C \rightarrow 0 = 4 + 2B - 2(-1)$$

$$\rightarrow 2B = -6 \rightarrow B = -3$$

$$\frac{x^2}{x^3-4x^2+5x-2} = \frac{4}{x-2} + \frac{-3}{x-1} + \frac{-1}{(x-1)^2}$$

37. Find the partial fraction decomposition:

$$\frac{x^3}{(x^2+16)^3} = \frac{Ax+B}{x^2+16} + \frac{Cx+D}{(x^2+16)^2} + \frac{Ex+F}{(x^2+16)^3}$$

(Multiply both sides by $(x^2+16)^3$.)

$$x^3 = (Ax+B)(x^2+16)^2 + (Cx+D)(x^2+16) + Ex+F$$
$$x^3 = (Ax+B)(x^4+32x^2+256) + Cx^3 + Dx^2 + 16Cx + 16D + Ex + F$$
$$x^3 = Ax^5 + Bx^4 + 32Ax^3 + 32Bx^2 + 256Ax + 256B + Cx^3 + Dx^2$$
$$+ 16Cx + 16D + Ex + F$$
$$x^3 = Ax^5 + Bx^4 + (32A+C)x^3 + (32B+D)x^2 + (256A+16C+E)x$$
$$+ (256B+16D+F)$$

$$A = 0$$
$$B = 0$$
$$32A + C = 1 \;\rightarrow\; 32(0) + C = 1 \;\rightarrow\; C = 1$$
$$32B + D = 0 \;\rightarrow\; 32(0) + D = 0 \;\rightarrow\; D = 0$$
$$256A + 16C + E = 0 \;\rightarrow\; 256(0) + 16(1) + E = 0 \;\rightarrow\; E = -16$$
$$256B + 16D + F = 0 \;\rightarrow\; 256(0) + 16(0) + F = 0 \;\rightarrow\; F = 0$$
$$\frac{x^3}{(x^2+16)^3} = + \frac{x}{(x^2+16)^2} + \frac{-16x}{(x^2+16)^3}$$

39. Find the partial fraction decomposition:

$$\frac{4}{2x^2-5x-3} = \frac{4}{(x-3)(2x+1)} = \frac{A}{x-3} + \frac{B}{2x+1}$$

(Multiply both sides by $(x-3)(2x+1)$.)

$$4 = A(2x+1) + B(x-3)$$

Let $x = -\frac{1}{2}$:   then $4 = A\left(2\left(-\frac{1}{2}\right)+1\right) + B\left(-\frac{1}{2}-3\right)$

$$\rightarrow\; 4 = -\frac{7}{2}B \;\rightarrow\; B = -\frac{8}{7}$$

Let $x = 3$:   then $4 = A(2(3)+1) + B(3-3) \;\rightarrow\; 4 = 7A \;\rightarrow\; A = \frac{4}{7}$

$$\frac{4}{2x^2-5x-3} = \frac{4}{(x-3)(2x+1)} = \frac{\frac{4}{7}}{x-3} + \frac{-\frac{8}{7}}{2x+1}$$

41. Find the partial fraction decomposition:

$$\frac{2x+3}{x^4-9x^2} = \frac{2x+3}{x^2(x-3)(x+3)} = \frac{A}{x} + \frac{B}{x^2} + \frac{C}{x-3} + \frac{D}{x+3}$$

(Multiply both sides by $x^2(x-3)(x+3)$.)

$$2x+3 = Ax(x-3)(x+3) + B(x-3)(x+3) + Cx^2(x+3) + Dx^2(x-3)$$

Let $x = 0$: then

$$2\cdot0+3 = A\cdot0(0-3)(0+3) + B(0-3)(0+3) + C\cdot0^2(0+3) + D\cdot0^2(0-3)$$

$$\rightarrow \ 3 = -9B \ \rightarrow \ B = -\tfrac{1}{3}$$

Let $x = 3$: then

$$2\cdot3+3 = A\cdot3(3-3)(3+3) + B(3-3)(3+3) + C\cdot3^2(3+3) + D\cdot3^2(3-3)$$

$$\rightarrow \ 9 = 54C \ \rightarrow \ C = \tfrac{1}{6}$$

Let $x = -3$: then

$$2(-3)+3 = A(-3)(-3-3)(-3+3) + B(-3-3)(-3+3) + C(-3)^2(-3+3)$$
$$+ D(-3)^2(-3-3)$$

$$\rightarrow \ -3 = -54D \ \rightarrow \ D = \tfrac{1}{18}$$

Let $x = 1$: then

$$2\cdot1+3 = A\cdot1(1-3)(1+3) + B(1-3)(1+3) + C\cdot1^2(1+3) + D\cdot1^2(1-3)$$

$$\rightarrow \ 5 = -8A - 8B + 4C - 2D$$

$$\rightarrow \ 5 = -8A - 8\left(-\tfrac{1}{3}\right) + 4\left(\tfrac{1}{6}\right) - 2\left(\tfrac{1}{18}\right)$$

$$\rightarrow \ 5 = -8A + \tfrac{8}{3} + \tfrac{2}{3} - \tfrac{1}{9} \ \rightarrow \ -8A = \tfrac{16}{9} \ \rightarrow \ A = -\tfrac{2}{9}$$

$$\frac{2x+3}{x^4-9x^2} = \frac{2x+3}{x^2(x-3)(x+3)} = \frac{-\tfrac{2}{9}}{x} + \frac{-\tfrac{1}{3}}{x^2} + \frac{\tfrac{1}{6}}{x-3} + \frac{\tfrac{1}{18}}{x+3}$$

11.7 Systems of Nonlinear Equations

1. $\begin{cases} y = x^2 + 1 \\ y = x + 1 \end{cases}$

Graph: $y_1 = x^2 + 1$; $y_2 = x + 1$

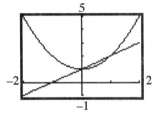

(0, 1) and (1, 2) are the intersection points.

Solve by substitution:
$$x^2 + 1 = x + 1$$
$$x^2 - x = 0$$
$$x(x-1) = 0$$
$$x = 0 \ \text{or} \ x = 1$$
$$y = 1 \qquad y = 2$$
Solutions: (0, 1) and (1, 2)

3. $\begin{cases} y = \sqrt{36 - x^2} \\ y = 8 - x \end{cases}$

Graph: $y_1 = \sqrt{36 - x^2}$; $y_2 = 8 - x$

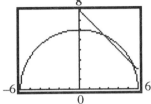

(2.59, 5.41) and (5.41, 2.59) are the intersection points.

Solve by substitution:

$$\sqrt{36 - x^2} = 8 - x$$
$$36 - x^2 = 64 - 16x + x^2$$
$$2x^2 - 16x + 28 = 0$$
$$x^2 - 8x + 14 = 0$$
$$x = \frac{8 \pm \sqrt{64 - 56}}{2}$$
$$x = \frac{8 \pm 2\sqrt{2}}{2}$$
$$x = 4 \pm \sqrt{2}$$

If $x = 4 + \sqrt{2}$, $y = 8 - \left(4 + \sqrt{2}\right) = 4 - \sqrt{2}$

If $x = 4 - \sqrt{2}$, $y = 8 - \left(4 - \sqrt{2}\right) = 4 + \sqrt{2}$

Solutions:

$$\left(4 + \sqrt{2}, 4 - \sqrt{2}\right) \text{ and } \left(4 - \sqrt{2}, 4 + \sqrt{2}\right)$$

5. $\begin{cases} y = \sqrt{x} \\ y = 2 - x \end{cases}$

Graph: $y_1 = \sqrt{x}$; $y_2 = 2 - x$

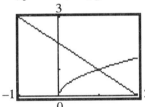

(1, 1) is the intersection point.

Solve by substitution:

$$\sqrt{x} = 2 - x$$
$$x = 4 - 4x + x^2$$
$$x^2 - 5x + 4 = 0$$
$$(x - 4)(x - 1) = 0$$
$$x = 4 \quad \text{or} \quad x = 1$$
$$y = -2 \quad \text{or} \quad y = 1$$

Eliminate (4, –2); it does not check.

Solution: (1, 1)

7. $\begin{cases} x = 2y \\ x = y^2 - 2y \end{cases}$

Solve each equation for y in order to enter it into the graphing utility:

$$y^2 - 2y + 1 = x + 1$$
$$(y - 1)^2 = x + 1$$
$$y - 1 = \pm\sqrt{x + 1}$$
$$y = 1 \pm \sqrt{x + 1}$$

Graph: $y_1 = \dfrac{x}{2}$; $y_2 = 1 + \sqrt{x+1}$;

$y_3 = 1 - \sqrt{x+1}$

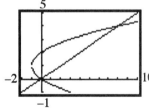

$(0, 0)$ and $(8, 4)$ are the intersection points.

Solve by substitution:

$$2y = y^2 - 2y$$
$$y^2 - 4y = 0$$
$$y(y-4) = 0$$
$$y = 0 \ \text{ or } \ y = 4$$
$$x = 0 \ \text{ or } \ x = 8$$

Solutions: $(0, 0)$ and $(8, 4)$

9. $\begin{cases} x^2 + y^2 = 4 \\ x^2 + 2x + y^2 = 0 \end{cases}$

Graph: $y_1 = \sqrt{4 - x^2}$; $y_2 = -\sqrt{4 - x^2}$; $y_3 = \sqrt{-x^2 - 2x}$; $y_4 = -\sqrt{-x^2 - 2x}$

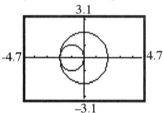

$(-2, 0)$ is the intersection point. (Note: This intersection point is impossible to find on your graphing utility unless you have just the right window and make an excellent guess.)

Substitute 4 for $x^2 + y^2$ in the second equation:

$$2x + 4 = 0$$
$$2x = -4$$
$$x = -2$$
$$y = \sqrt{4 - (-2)^2} = 0$$

Solution: $(-2, 0)$

11. $\begin{cases} y = 3x - 5 \\ x^2 + y^2 = 5 \end{cases}$

Graph: $y_1 = 3x - 5$; $y_2 = \sqrt{5 - x^2}$;

$y_3 = -\sqrt{5 - x^2}$

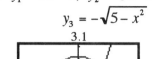
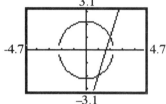

$(1, -2)$ and $(2, 1)$ are the intersection points.

Solve by substitution:

$$x^2 + (3x - 5)^2 = 5$$
$$x^2 + 9x^2 - 30x + 25 = 5$$
$$10x^2 - 30x + 20 = 0$$
$$x^2 - 3x + 2 = 0$$
$$(x - 1)(x - 2) = 0$$
$$x = 1 \qquad \text{or } x = 2$$
$$y = 3(1) - 5 \qquad y = 3(2) - 5$$
$$y = -2 \qquad\qquad y = 1$$

Solutions: $(1, -2)$ and $(2, 1)$

13. $\begin{cases} x^2 + y^2 = 4 \\ y^2 - x = 4 \end{cases}$

Graph: $y_1 = \sqrt{4 - x^2}$; $y_2 = -\sqrt{4 - x^2}$; $y_3 = \sqrt{x + 4}$; $y_4 = -\sqrt{x + 4}$

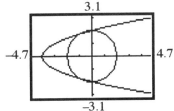

$(-1, 1.73)$, $(-1, -1.73)$, $(0, 2)$, and $(0, -2)$ are the intersection points.

Substitute $x + 4$ for y^2 in the first equation :

$x^2 + x + 4 = 4$

$x^2 + x = 0$

$x(x + 1) = 0$

$x = 0$ or $x = -1$

$y^2 = 4$ $y^2 = 3$

$y = \pm 2$ $y^2 = \pm\sqrt{3}$

Solutions:

$(0, -2), (0, 2), \left(-1, \sqrt{3}\right), \left(-1, -\sqrt{3}\right)$

15. $\begin{cases} xy = 4 \\ x^2 + y^2 = 8 \end{cases}$

Graph: $y_1 = \dfrac{4}{x}$; $y_2 = \sqrt{8 - x^2}$;

$y_3 = -\sqrt{8 - x^2}$

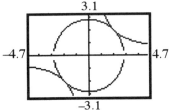

$(-2, -2)$ and $(2, 2)$ are the intersection points.

Solve by substitution:

$x^2 + \left(\dfrac{4}{x}\right)^2 = 8$

$x^2 + \dfrac{16}{x^2} = 8$

$x^4 + 16 = 8x^2$

$x^4 - 8x^2 + 16 = 0$

$\left(x^2 - 4\right)^2 = 0$

$x^2 - 4 = 0$

$x^2 = 4$

$x = 2$ or $x = -2$

$y = 2$ $y = -2$

Solutions: $(-2, -2)$ and $(2, 2)$

17. $\begin{cases} x^2 + y^2 = 4 \\ y = x^2 - 9 \end{cases}$

Graph: $y_1 = x^2 - 9$; $y_2 = \sqrt{4 - x^2}$;

$y_3 = -\sqrt{4 - x^2}$

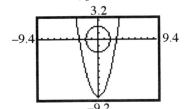

No solution; Inconsistent.

Solve by substitution:

$x^2 + (x^2 - 9)^2 = 4$

$x^2 + x^4 - 18x^2 + 81 = 4$

$x^4 - 17x^2 + 77 = 0$

$x^2 = \dfrac{17 \pm \sqrt{289 - 4(77)}}{2}$

$x^2 = \dfrac{17 \pm \sqrt{-19}}{2}$

There are no real solutions to this expression; Inconsistent.

19. $\begin{cases} y = x^2 - 4 \\ y = 6x - 13 \end{cases}$

    Graph: $y_1 = x^2 - 4;$ $y_2 = 6x - 13$

    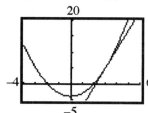

    (3,5) is the intersection point.

    Solve by substitution:
    $$x^2 - 4 = 6x - 13$$
    $$x^2 - 6x + 9 = 0$$
    $$(x - 3)^2 = 0$$
    $$x - 3 = 0$$
    $$x = 3$$
    $$y = 6(3) - 13 = 5$$
    Solutions: (3,5)

21. Solve the second equation for y, substitute into the first equation and solve:
    $$\begin{cases} 2x^2 + y^2 = 18 \\ \quad xy = 4 \ \rightarrow \ y = \dfrac{4}{x} \end{cases}$$

    $$2x^2 + \left(\frac{4}{x}\right)^2 = 18$$
    $$2x^2 + \frac{16}{x^2} = 18$$
    $$2x^4 + 16 = 18x^2$$
    $$2x^4 - 18x^2 + 16 = 0$$
    $$x^4 - 9x^2 + 8 = 0$$
    $$\left(x^2 - 8\right)\left(x^2 - 1\right) = 0$$
    $$x^2 = 8 \qquad \text{or} \ \ x^2 = 1$$
    $$x = \pm\sqrt{8} = \pm 2\sqrt{2} \ \text{ or } \quad x = \pm 1$$

    If $x = 2\sqrt{2}$: $y = \dfrac{4}{2\sqrt{2}} = \sqrt{2}$

    If $x = -2\sqrt{2}$: $y = \dfrac{4}{-2\sqrt{2}} = -\sqrt{2}$

    If $x = 1$: $y = \dfrac{4}{1} = 4$

    If $x = -1$: $y = \dfrac{4}{-1} = -4$

    Solutions: $\left(2\sqrt{2}, \sqrt{2}\right), \left(-2\sqrt{2}, -\sqrt{2}\right), (1, 4), (-1, -4)$

23. Substitute the first equation into the second equation and solve:

$$\begin{cases} y = 2x + 1 \\ 2x^2 + y^2 = 1 \end{cases}$$

$$2x^2 + (2x + 1)^2 = 1$$
$$2x^2 + 4x^2 + 4x + 1 = 1$$
$$6x^2 + 4x = 0$$
$$2x(3x + 2) = 0$$
$$2x = 0 \quad \text{or} \quad 3x + 2 = 0$$
$$x = 0 \quad \text{or} \qquad x = -\tfrac{2}{3}$$

If $x = 0$: $y = 2(0) + 1 = 1$

If $x = -\tfrac{2}{3}$: $y = 2\left(-\tfrac{2}{3}\right) + 1 = -\tfrac{4}{3} + 1 = -\tfrac{1}{3}$

Solutions: $(0, 1), \left(-\tfrac{2}{3}, -\tfrac{1}{3}\right)$

25. Solve the first equation for y, substitute into the second equation and solve:

$$\begin{cases} x + y + 1 = 0 \quad \rightarrow \quad y = -x - 1 \\ x^2 + y^2 + 6y - x = -5 \end{cases}$$

$$x^2 + (-x - 1)^2 + 6(-x - 1) - x = -5$$
$$x^2 + x^2 + 2x + 1 - 6x - 6 - x = -5$$
$$2x^2 - 5x = 0$$
$$x(2x - 5) = 0$$
$$x = 0 \quad \text{or} \quad x = \tfrac{5}{2}$$

If $x = 0$: $y = -(0) - 1 = -1$

If $x = \tfrac{5}{2}$: $y = -\tfrac{5}{2} - 1 = -\tfrac{7}{2}$

Solutions: $(0, -1), \left(\tfrac{5}{2}, -\tfrac{7}{2}\right)$

27. Solve the second equation for y, substitute into the first equation and solve:

$$\begin{cases} 4x^2 - 3xy + 9y^2 = 15 \\ 2x + 3y = 5 \quad \rightarrow \quad y = -\tfrac{2}{3}x + \tfrac{5}{3} \end{cases}$$

$$4x^2 - 3x\left(-\tfrac{2}{3}x + \tfrac{5}{3}\right) + 9\left(-\tfrac{2}{3}x + \tfrac{5}{3}\right)^2 = 15$$
$$4x^2 + 2x^2 - 5x + 4x^2 - 20x + 25 = 15$$
$$10x^2 - 25x + 10 = 0$$
$$2x^2 - 5x + 2 = 0$$
$$(2x - 1)(x - 2) = 0$$
$$x = \tfrac{1}{2} \quad \text{or} \quad x = 2$$

If $x = \tfrac{1}{2}$: $y = -\tfrac{2}{3}\left(\tfrac{1}{2}\right) + \tfrac{5}{3} = \tfrac{4}{3}$

If $x = 2$: $y = -\tfrac{2}{3}(2) + \tfrac{5}{3} = \tfrac{1}{3}$

Solutions: $\left(\tfrac{1}{2}, \tfrac{4}{3}\right), \left(2, \tfrac{1}{3}\right)$

29. Multiply each side of the second equation by 4 and add the equations to eliminate y:

$$\begin{cases} x^2 - 4y^2 = -7 \longrightarrow & x^2 - 4y^2 = -7 \\ 3x^2 + \ y^2 = 31 \ \overset{4}{\longrightarrow} & \underline{12x^2 + 4y^2 = 124} \end{cases}$$

$$\begin{aligned} 13x^2 \quad\quad &= 117 \\ x^2 &= 9 \\ x &= \pm 3 \end{aligned}$$

If $x = 3$: $3(3)^2 + y^2 = 31 \ \rightarrow \ y^2 = 4 \ \rightarrow \ y = \pm 2$

If $x = -3$: $3(-3)^2 + y^2 = 31 \ \rightarrow \ y^2 = 4 \ \rightarrow \ y = \pm 2$

Solutions: $(3, 2), (3, -2), (-3, 2), (-3, -2)$

31. Multiply each side of the first equation by 5 and each side of the second equation by 3 to eliminate y:

$$\begin{cases} 7x^2 - 3y^2 = -5 \ \overset{5}{\longrightarrow} & 35x^2 - 15y^2 = -25 \\ 3x^2 + 5y^2 = 12 \ \overset{3}{\longrightarrow} & \underline{9x^2 + 15y^2 = \quad 36} \end{cases}$$

$$\begin{aligned} 44x^2 \quad\quad &= \quad 11 \\ x^2 &= \ \tfrac{1}{4} \\ x &= \pm \tfrac{1}{2} \end{aligned}$$

If $x = \dfrac{1}{2}$: $3\left(\dfrac{1}{2}\right)^2 + 5y^2 = 12 \ \rightarrow \ 5y^2 = \dfrac{45}{4} \ \rightarrow \ y^2 = \dfrac{9}{4} \ \rightarrow \ y = \pm\dfrac{3}{2}$

If $x = -\dfrac{1}{2}$: $3\left(-\dfrac{1}{2}\right)^2 + 5y^2 = 12 \ \rightarrow \ 5y^2 = \dfrac{45}{4} \ \rightarrow \ y^2 = \dfrac{9}{4} \ \rightarrow \ y = \pm\dfrac{3}{2}$

Solutions: $\left(\tfrac{1}{2}, \tfrac{3}{2}\right), \left(\tfrac{1}{2}, -\tfrac{3}{2}\right), \left(-\tfrac{1}{2}, \tfrac{3}{2}\right), \left(-\tfrac{1}{2}, -\tfrac{3}{2}\right)$

33. Multiply each side of the second equation by 2 and add to eliminate xy:

$$\begin{cases} x^2 + 2xy = 10 \longrightarrow & x^2 + 2xy = 10 \\ 3x^2 - \ xy = \ 2 \ \overset{2}{\longrightarrow} & \underline{6x^2 - 2xy = \ 4} \end{cases}$$

$$\begin{aligned} 7x^2 \quad\quad &= 14 \\ x^2 &= 2 \\ x &= \pm\sqrt{2} \end{aligned}$$

If $x = \sqrt{2}$: $3\left(\sqrt{2}\right)^2 - \sqrt{2} \cdot y = 2 \ \rightarrow \ -\sqrt{2} \cdot y = -4 \ \rightarrow \ y = \dfrac{4}{\sqrt{2}} \ \rightarrow \ y = 2\sqrt{2}$

If $x = -\sqrt{2}$: $3\left(-\sqrt{2}\right)^2 - \left(-\sqrt{2}\right)y = 2 \ \rightarrow \ \sqrt{2} \cdot y = -4 \ \rightarrow \ y = \dfrac{-4}{\sqrt{2}} \ \rightarrow \ y = -2\sqrt{2}$

Solutions: $\left(\sqrt{2}, 2\sqrt{2}\right), \left(-\sqrt{2}, -2\sqrt{2}\right)$

35. Multiply each side of the first equation by 2 and add the equations to eliminate y:

$$\begin{cases} 2x^2 + y^2 = 2 \xrightarrow{\quad 2 \quad} 4x^2 + 2y^2 = 4 \\ x^2 - 2y^2 = -8 \xrightarrow{\quad\quad} \underline{x^2 - 2y^2 = -8} \end{cases}$$

$$5x^2 = -4$$

$$x^2 = \frac{-4}{5}$$

No solution. The system is inconsistent.

37. Multiply each side of the second equation by 2 and add the equations to eliminate y:

$$\begin{cases} x^2 + 2y^2 = 16 \xrightarrow{\quad\quad} x^2 + 2y^2 = 16 \\ 4x^2 - y^2 = 24 \xrightarrow{\quad 2 \quad} \underline{8x^2 - 2y^2 = 48} \end{cases}$$

$$9x^2 = 64$$

$$x^2 = \frac{64}{9}$$

$$x = \pm\frac{8}{3}$$

If $x = \frac{8}{3}$: $\left(\frac{8}{3}\right)^2 + 2y^2 = 16 \rightarrow 2y^2 = \frac{80}{9} \rightarrow y^2 = \frac{40}{9} \rightarrow y = \pm\frac{2\sqrt{10}}{3}$

If $x = -\frac{8}{3}$: $\left(-\frac{8}{3}\right)^2 + 2y^2 = 16 \rightarrow 2y^2 = \frac{80}{9} \rightarrow y^2 = \frac{40}{9} \rightarrow y = \pm\frac{2\sqrt{10}}{3}$

Solutions: $\left(\frac{8}{3}, \frac{2\sqrt{10}}{3}\right), \left(\frac{8}{3}, \frac{-2\sqrt{10}}{3}\right), \left(-\frac{8}{3}, \frac{2\sqrt{10}}{3}\right), \left(-\frac{8}{3}, \frac{-2\sqrt{10}}{3}\right)$

39. Multiply each side of the second equation by 2 and add the equations to eliminate y:

$$\begin{cases} \dfrac{5}{x^2} - \dfrac{2}{y^2} = -3 \xrightarrow{\quad\quad} \dfrac{5}{x^2} - \dfrac{2}{y^2} = -3 \\ \dfrac{3}{x^2} + \dfrac{1}{y^2} = 7 \xrightarrow{\quad 2 \quad} \underline{\dfrac{6}{x^2} + \dfrac{2}{y^2} = 14} \end{cases}$$

$$\dfrac{11}{x^2} \phantom{+\dfrac{2}{y^2}} = 11$$

$$11 = 11x^2$$

$$x^2 = 1$$

$$x = \pm1$$

If $x = 1$: $\dfrac{3}{(1)^2} + \dfrac{1}{y^2} = 7 \rightarrow \dfrac{1}{y^2} = 4 \rightarrow y^2 = \dfrac{1}{4} \rightarrow y = \pm\dfrac{1}{2}$

If $x = -1$: $\dfrac{3}{(-1)^2} + \dfrac{1}{y^2} = 7 \rightarrow \dfrac{1}{y^2} = 4 \rightarrow y^2 = \dfrac{1}{4} \rightarrow y = \pm\dfrac{1}{2}$

Solutions: $\left(1, \dfrac{1}{2}\right), \left(1, -\dfrac{1}{2}\right), \left(-1, \dfrac{1}{2}\right), \left(-1, -\dfrac{1}{2}\right)$

41. Multiply each side of the first equation by –2 and add the equations to eliminate x:

$$\begin{cases} \dfrac{1}{x^4} + \dfrac{6}{y^4} = 6 \\[2mm] \dfrac{2}{x^4} - \dfrac{2}{y^4} = 19 \end{cases} \xrightarrow{\ -2\ } \begin{array}{l} \dfrac{-2}{x^4} - \dfrac{12}{y^4} = -12 \\[2mm] \dfrac{2}{x^4} - \dfrac{2}{y^4} = 19 \\[1mm] \hline \dfrac{-14}{y^4} = 7 \end{array}$$

$$-14 = 7y^4$$
$$y^4 = -2$$

There are no real solutions. The system is inconsistent.

43. Factor the first equation, solve for x, substitute into the second equation and solve:

$$\begin{cases} x^2 - 3xy + 2y^2 = 0 \\ x^2 + xy = 6 \end{cases} \rightarrow (x - 2y)(x - y) = 0 \rightarrow x = 2y \text{ or } x = y$$

Substitute $x = 2y$ and solve: Substitute $x = y$ and solve:

$$\begin{array}{cc}
x^2 + xy = 6 & x^2 + xy = 6 \\
(2y)^2 + (2y)y = 6 & y^2 + y \cdot y = 6 \\
4y^2 + 2y^2 = 6 & y^2 + y^2 = 6 \\
6y^2 = 6 & 2y^2 = 6 \\
y^2 = 1 & y^2 = 3 \\
y = \pm 1 & y = \pm\sqrt{3}
\end{array}$$

If $y = 1$: $x = 2 \cdot 1 = 2$ If $y = \sqrt{3}$: $x = \sqrt{3}$
If $y = -1$: $x = 2(-1) = -2$ If $y = -\sqrt{3}$: $x = -\sqrt{3}$

Solutions: $(2, 1), (-2, -1), \left(\sqrt{3}, \sqrt{3}\right), \left(-\sqrt{3}, -\sqrt{3}\right)$

45. Solve the first equation for y, substitute into the second equation and solve:

$$\begin{cases} xy - x^2 = -3 \\ 3xy - 4y^2 = 2 \end{cases} \rightarrow y = \frac{x^2 - 3}{x}$$

$$3x\left(\frac{x^2 - 3}{x}\right) - 4\left(\frac{x^2 - 3}{x}\right)^2 = 2$$

$$3x^2 - 9 - 4\left(\frac{x^4 - 6x^2 + 9}{x^2}\right) = 2$$

$$3x^4 - 9x^2 - 4x^4 + 24x^2 - 36 = 2x^2$$
$$-x^4 + 13x^2 - 36 = 0$$
$$x^4 - 13x^2 + 36 = 0$$
$$(x^2 - 9)(x^2 - 4) = 0$$
$$x^2 = 9 \text{ or } x^2 = 4$$
$$x = \pm 3 \text{ or } x = \pm 2$$

If $x = 3$: $y = \dfrac{(3)^2 - 3}{3} = \dfrac{6}{3} = 2$

If $x = -3$: $y = \dfrac{(-3)^2 - 3}{-3} = \dfrac{6}{-3} = -2$

If $x = 2$: $y = \dfrac{(2)^2 - 3}{2} = \dfrac{1}{2}$

If $x = -2$: $y = \dfrac{(-2)^2 - 3}{-2} = \dfrac{1}{-2} = -\dfrac{1}{2}$

Solutions: $(3, 2), (-3, -2), \left(2, \frac{1}{2}\right), \left(-2, -\frac{1}{2}\right)$

47. Solve the second equation for x, substitute into the first equation and solve:

$$\begin{cases} x^3 - y^3 = 26 \\ \ \ x - y = \ \ 2 \end{cases} \rightarrow \ \ x = y + 2$$

$$(y + 2)^3 - y^3 = 26$$
$$y^3 + 6y^2 + 12y + 8 - y^3 = 26$$
$$6y^2 + 12y - 18 = 0$$
$$y^2 + 2y - 3 = 0$$
$$(y + 3)(y - 1) = 0$$
$$y = -3 \ \text{ or } \ y = 1$$

If $y = -3$: $x = -3 + 2 = -1$
If $y = 1$: $x = 1 + 2 = 3$

Solutions: $(-1, -3), (3, 1)$

49. Multiply each side of the second equation by $-y$ and add the equations to eliminate y:

$$\begin{cases} y^2 + y + x^2 - x - 2 = 0 \ \longrightarrow \ \ \ y^2 + y + x^2 - x - 2 = 0 \\ \ \ \ \ \ y + 1 + \dfrac{x - 2}{y} = 0 \ \xrightarrow{-y} \ \underline{-y^2 - y \ \ \ \ \ \ - x + 2 = 0} \end{cases}$$

$$x^2 - 2x \ \ \ \ \ = 0$$
$$x(x - 2) = 0$$
$$x = 0 \ \text{ or } \ x = 2$$

If $x = 0$: $y^2 + y + 0^2 - 0 - 2 = 0 \rightarrow y^2 + y - 2 = 0 \rightarrow (y + 2)(y - 1) = 0$
$\rightarrow \ y = -2 \ \text{ or } \ y = 1$

If $x = 2$: $y^2 + y + 2^2 - 2 - 2 = 0 \rightarrow y^2 + y = 0 \rightarrow y(y + 1) = 0$
$\rightarrow \ y = 0 \ \text{ or } \ y = -1$

Solutions: $(0, -2), (0, 1), (2, 0), (2, -1)$

51. Rewrite each equation in exponential form:
$$\begin{cases} \log_x y = 3 \;\rightarrow\; y = x^3 \\ \log_x(4y) = 5 \;\rightarrow\; 4y = x^5 \end{cases}$$
Substitute the first equation into the second and solve:
$$4x^3 = x^5$$
$$x^5 - 4x^3 = 0$$
$$x^3(x^2 - 4) = 0$$
$$x^3 = 0 \;\text{ or }\; x^2 = 4$$
$$x = 0 \;\text{ or }\;\; x = \pm 2$$
The base of a logarithm must be positive, thus $x \neq 0$ and $x \neq -2$.

If $x = 2$: $y = 2^3 = 8$

Solution: (2, 8)

53. Graph: $y_1 = x \wedge (2/3);\quad y_2 = e \wedge (-x)$
Use INTERSECT to solve:

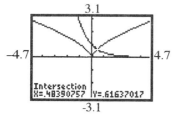

Solution: (0.48, 0.62)

55. Graph: $y_1 = \sqrt[3]{(2 - x^2)};\quad y_2 = 4/x^3$
Use INTERSECT to solve:

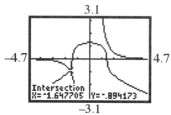

Solution: (−1.65, −0.89)

57. Graph: $y_1 = \sqrt[4]{(12 - x^4)};\quad y_2 = -\sqrt[4]{(12 - x^4)};\quad y_3 = \sqrt{2/x};\quad y_4 = -\sqrt{2/x}$
Use INTERSECT to solve:

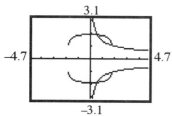

Solutions: (0.58, 1.86), (1.81, 1.05), (1.81, −1.05), (0.58, −1.86)

59. Graph: $y_1 = 2 / x$; $y_2 = \ln x$
     Use INTERSECT to solve:

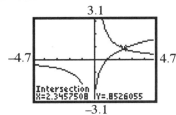

Solution: (2.35, 0.85)

61. Let x and y be the two numbers. The system of equations is:
     $$\begin{cases} x - y = 2 \\ x^2 + y^2 = 10 \end{cases}$$
     Solve the first equation for x, substitute into the second equation and solve:
     $$(y+2)^2 + y^2 = 10$$
     $$y^2 + 4y + 4 + y^2 = 10$$
     $$2y^2 + 4y - 6 = 0$$
     $$y^2 + 2y - 3 = 0$$
     $$(y+3)(y-1) = 0$$
     $$y = -3 \text{ or } y = 1$$
     If $y = -3$: $x = -3 + 2 = -1$
     If $y = 1$: $x = 1 + 2 = 3$
     The two numbers are 1 and 3 or –1 and –3.

63. Let x and y be the two numbers. The system of equations is:
     $$\begin{cases} xy = 4 \\ x^2 + y^2 = 8 \end{cases}$$
     Solve the first equation for x, substitute into the second equation and solve:
     $$\left(\frac{4}{y}\right)^2 + y^2 = 8$$
     $$\frac{16}{y^2} + y^2 = 8$$
     $$16 + y^4 = 8y^2$$
     $$y^4 - 8y^2 + 16 = 0$$
     $$(y^2 - 4)^2 = 0$$
     $$y^2 - 4 = 0$$
     $$y^2 = 4$$
     $$y = \pm 2$$
     If $y = 2$: $x = \dfrac{4}{2} = 2$
     If $y = -2$: $x = \dfrac{4}{-2} = -2$
     The two numbers are 2 and 2 or –2 and –2.

65. Let x and y be the two numbers. The system of equations is:

$$\begin{cases} x - y = xy \\ \dfrac{1}{x} + \dfrac{1}{y} = 5 \end{cases}$$

Solve the first equation for x, substitute into the second equation and solve:

$$x - xy = y$$

$$x(1 - y) = y$$

$$x = \frac{y}{1 - y}$$

$$\frac{1}{\dfrac{y}{1-y}} + \frac{1}{y} = 5$$

$$\frac{1 - y}{y} + \frac{1}{y} = 5$$

$$\frac{2 - y}{y} = 5$$

$$2 - y = 5y$$

$$6y = 2$$

$$y = \frac{1}{3}$$

If $y = \dfrac{1}{3}$: $\qquad x = \dfrac{\frac{1}{3}}{1 - \frac{1}{3}} = \dfrac{\frac{1}{3}}{\frac{2}{3}} = \dfrac{1}{2}$

The two numbers are $\frac{1}{2}$ and $\frac{1}{3}$.

67. $\begin{cases} \dfrac{a}{b} = \dfrac{2}{3} \\ a + b = 10 \end{cases}$

Solve the second equation for a, substitute into the first equation and solve:

$$\frac{10 - b}{b} = \frac{2}{3}$$

$$3(10 - b) = 2b$$

$$30 - 3b = 2b$$

$$30 = 5b$$

$$b = 6$$

$$a = 4$$

$$a + b = 10; \quad b - a = 2$$

The ratio of $a + b$ to $b - a$ is $\dfrac{10}{2} = 5.$

69. Solve the first equation for x, substitute into the second equation and solve:

$$\begin{cases} x + 2y = 0 \;\;\rightarrow\;\; x = -2y \\ (x-1)^2 + (y-1)^2 = 5 \end{cases}$$

$$(-2y-1)^2 + (y-1)^2 = 5$$

$$4y^2 + 4y + 1 + y^2 - 2y + 1 = 5$$

$$5y^2 + 2y - 3 = 0$$

$$(5y-3)(y+1) = 0$$

$$y = \frac{3}{5} \;\; \text{or} \;\; y = -1$$

$$x = -\frac{6}{5} \;\; \text{or} \;\; x = 2$$

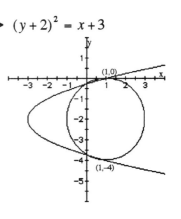

The points of intersection are $\left(-\frac{6}{5}, \frac{3}{5}\right), (2, -1)$.

71. Complete the square on the second equation, substitute into the first equation and solve:

$$\begin{cases} (x-1)^2 + (y+2)^2 = 4 \\ y^2 + 4y - x + 1 = 0 \;\;\rightarrow\;\; y^2 + 4y + 4 = x - 1 + 4 \rightarrow (y+2)^2 = x+3 \end{cases}$$

$$(x-1)^2 + x + 3 = 4$$

$$x^2 - 2x + 1 + x + 3 = 4$$

$$x^2 - x = 0$$

$$x(x-1) = 0$$

$$x = 0 \;\; \text{or} \;\; x = 1$$

If $x = 0$: $(y+2)^2 = 0 + 3 \;\;\rightarrow\;\; y + 2 = \pm\sqrt{3}$

$$\rightarrow \;\; y = -2 \pm \sqrt{3}$$

If $x = 1$: $(y+2)^2 = 1 + 3 \;\;\rightarrow\;\; y + 2 = \pm 2$

$$\rightarrow \;\; y = -2 \pm 2$$

The points of intersection are:

$$\left(0, -2 - \sqrt{3}\right), \left(0, -2 + \sqrt{3}\right), (1, -4), (1, 0).$$

73. Solve the first equation for x, substitute into the second equation and solve:

$$\begin{cases} y = \dfrac{4}{x-3} \;\rightarrow\; x - 3 = \dfrac{4}{y} \;\rightarrow\; x = \dfrac{4}{y} + 3 \\ x^2 - 6x + y^2 + 1 = 0 \end{cases}$$

$$\left(\frac{4}{y}+3\right)^2 - 6\left(\frac{4}{y}+3\right) + y^2 + 1 = 0$$

$$\frac{16}{y^2} + \frac{24}{y} + 9 - \frac{24}{y} - 18 + y^2 + 1 = 0$$

$$\frac{16}{y^2} + y^2 - 8 = 0$$

$$16 + y^4 - 8y^2 = 0$$

$$y^4 - 8y^2 + 16 = 0$$

$$(y^2 - 4)^2 = 0$$

$$y^2 - 4 = 0$$

$$y^2 = 4$$

$$y = \pm 2$$

If $y = 2$: $\quad x = \dfrac{4}{2} + 3 = 5$

If $y = -2$: $\quad x = \dfrac{4}{-2} + 3 = 1$

The points of intersection are: $(1, -2), (5, 2)$.

75. Let x = the width of the rectangle.
    Let y = the length of the rectangle.

$$\begin{cases} 2x + 2y = 16 \\ \quad\;\; xy = 15 \end{cases}$$

Solve the first equation for y, substitute into the second equation and solve:

$$2x + 2y = 16 \qquad\qquad x(8 - x) = 15$$
$$2y = 16 - 2x \qquad\qquad 8x - x^2 = 15$$
$$y = 8 - x \qquad\qquad x^2 - 8x + 15 = 0$$
$$(x - 5)(x - 3) = 0$$
$$x = 5 \text{ or } x = 3$$
$$y = 3 \qquad y = 5$$

The dimensions of the rectangle are 3 inches by 5 inches.

77. Let x = the radius of the first circle.
     Let y = the radius of the second circle.
$$\begin{cases} 2\pi x + 2\pi y = 12\pi \\ \pi x^2 + \pi y^2 = 20\pi \end{cases}$$
Solve the first equation for y, substitute into the second equation and solve:

$$2\pi x + 2\pi y = 12\pi$$
$$x + y = 6$$
$$y = 6 - x$$

$$\pi x^2 + \pi y^2 = 20\pi$$
$$x^2 + y^2 = 20$$
$$x^2 + (6 - x)^2 = 20$$
$$x^2 + 36 - 12x + x^2 = 20$$
$$2x^2 - 12x + 16 = 0$$
$$x^2 - 6x + 8 = 0$$
$$(x - 4)(x - 2) = 0$$
$$x = 4 \text{ or } x = 2$$
$$y = 2 \qquad y = 4$$

The radii of the circles are 2 centimeters and 4 centimeters.

79. The tortoise takes $9 + 3 = 12$ minutes or $\frac{1}{5}$ hour longer to complete the race than the hare.
     Let r = the rate of the hare.
     Let t = the time for the hare to complete the race.
     Then $t + \frac{1}{5}$ = the time for the tortoise.
     $r - 0.5$ = the rate for the tortoise.
     Since the length of the race is 21 meters, the distance equations are:
$$\begin{cases} rt = 21 \\ (r - 0.5)\left(t + \frac{1}{5}\right) = 21 \end{cases}$$
Solve the first equation for r, substitute into the second equation and solve:

$$\left(\frac{21}{t} - \frac{1}{2}\right)\left(t + \frac{1}{5}\right) = 21$$
$$21 + \frac{21}{5t} - \frac{t}{2} - \frac{1}{10} = 21$$
$$210t + 42 - 5t^2 - t = 210t$$
$$5t^2 + t - 42 = 0$$
$$(5t - 14)(t + 3) = 0$$
$$t = \frac{14}{5} \text{ or } t = -3$$

$t = -3$ makes no sense, since time cannot be negative.
Solve for r:
$$r = \frac{21}{\frac{14}{5}} = 21 \cdot \frac{5}{14} = \frac{15}{2} = 7.5$$

The average speed of the hare is 7.5 meters per hour, and the average speed for the tortoise is 7 meters per hour.

81. Let x = the width of the cardboard.
    Let y = the length of the cardboard.
    The width of the box will be $x - 4$, the length of the box will be $y - 4$, and the height is 2. The volume is $V = (x-4)(y-4)(2)$.
    Solve the system of equations:
    $$\begin{cases} xy = 216 \\ 2(x-4)(y-4) = 224 \end{cases}$$
    Solve the first equation for y, substitute into the second equation and solve:
    $$(2x-8)\left(\frac{216}{x} - 4\right) = 224$$
    $$432 - 8x - \frac{1728}{x} + 32 = 224$$
    $$432x - 8x^2 - 1728 + 32x = 224x$$
    $$-8x^2 + 240x - 1728 = 0$$
    $$x^2 - 30x + 216 = 0$$
    $$(x-12)(x-18) = 0$$
    $$x = 12 \quad \text{or} \quad x = 18$$
    $$y = 18 \qquad y = 12$$
    The cardboard should be 12 centimeters by 18 centimeters.

83. Find equations relating area and perimeter:
    $$\begin{cases} x^2 + y^2 = 4500 \\ 3x + 3y + (x - y) = 300 \end{cases}$$
    Solve the second equation for y, substitute into the first equation and solve:
    $$4x + 2y = 300 \qquad\qquad x^2 + (150 - 2x)^2 = 4500$$
    $$2y = 300 - 4x \qquad x^2 + 22500 - 600x + 4x^2 = 4500$$
    $$y = 150 - 2x \qquad\qquad 5x^2 - 600x + 18000 = 0$$
    $$x^2 - 120x + 3600 = 0$$
    $$(x - 60)^2 = 0$$
    $$x - 60 = 0$$
    $$x = 60$$
    $$y = 150 - 2(60) = 30$$
    The sides of the squares are 30 feet and 60 feet.

85. Solve the system for l and w:

$$\begin{cases} 2l + 2w = P \\ lw = A \end{cases}$$

Solve the first equation for l, substitute into the second equation and solve:

$$2l = P - 2w \;\;\rightarrow\;\; l = \frac{P}{2} - w$$

$$\left(\frac{P}{2} - w\right)w = A$$

$$\frac{P}{2}w - w^2 = A$$

$$w^2 - \frac{P}{2}w + A = 0$$

$$w = \frac{\frac{P}{2} \pm \sqrt{\frac{P^2}{4} - 4A}}{2} = \frac{\frac{P}{2} \pm \sqrt{\frac{P^2 - 16A}{4}}}{2} = \frac{\frac{P}{2} \pm \frac{\sqrt{P^2 - 16A}}{2}}{2}$$

$$w = \frac{P \pm \sqrt{P^2 - 16A}}{4}$$

If $w = \dfrac{P + \sqrt{P^2 - 16A}}{4}$ then $l = \dfrac{P}{2} - \dfrac{P + \sqrt{P^2 - 16A}}{4} = \dfrac{P - \sqrt{P^2 - 16A}}{4}$

If $w = \dfrac{P - \sqrt{P^2 - 16A}}{4}$ then $l = \dfrac{P}{2} - \dfrac{P - \sqrt{P^2 - 16A}}{4} = \dfrac{P + \sqrt{P^2 - 16A}}{4}$

If it is required that length be greater than width, then the solution is:

$$w = \frac{P - \sqrt{P^2 - 16A}}{4} \text{ and } l = \frac{P + \sqrt{P^2 - 16A}}{4}$$

87. Solve the equation:

$$m^2 - 4(2m - 4) = 0$$

$$m^2 - 8m + 16 = 0$$

$$(m - 4)^2 = 0$$

$$m - 4 = 0$$

$$m = 4$$

Use the point-slope equation with slope 4 and the point (2, 4) to obtain the equation of the tangent line:

$$y - 4 = 4(x - 2)$$

$$y - 4 = 4x - 8$$

$$y = 4x - 4$$

89. Solve the system:
$$\begin{cases} y = x^2 + 2 \\ y = mx + b \end{cases}$$
Solve the system by substitution:
$$x^2 + 2 = mx + b$$
$$x^2 - mx + 2 - b = 0$$
Note that the tangent line passes through $(1, 3)$. Find the relation between m and b:
$$3 = m(1) + b$$
$$b = 3 - m$$
Substitute into the quadratic to eliminate b:
$$x^2 - mx + 2 - (3 - m) = 0$$
$$x^2 - mx + (m - 1) = 0$$
Find when the discriminant is 0:
$$(-m)^2 - 4(1)(m - 1) = 0$$
$$m^2 - 4m + 4 = 0$$
$$(m - 2)^2 = 0$$
$$m - 2 = 0$$
$$m = 2 \quad b = 3 - 2 = 1$$
The equation of the tangent line is $y = 2x + 1$.

91. Solve the system:
$$\begin{cases} 2x^2 + 3y^2 = 14 \\ \qquad y = mx + b \end{cases}$$
Solve the system by substitution:
$$2x^2 + 3(mx + b)^2 = 14$$
$$2x^2 + 3m^2x^2 + 6mbx + 3b^2 = 14$$
$$(3m^2 + 2)x^2 + 6mbx + 3b^2 - 14 = 0$$
Note that the tangent line passes through $(1, 2)$. Find the relation between m and b:
$$2 = m(1) + b$$
$$b = 2 - m$$
Substitute into the quadratic to eliminate b:
$$(3m^2 + 2)x^2 + 6m(2 - m)x + 3(2 - m)^2 - 14 = 0$$
$$(3m^2 + 2)x^2 + (12m - 6m^2)x + 12 - 12m + 3m^2 - 14 = 0$$
$$(3m^2 + 2)x^2 + (12m - 6m^2)x + (3m^2 - 12m - 2) = 0$$

Find when the discriminant is 0:

$$(12m - 6m^2)^2 - 4(3m^2 + 2)(3m^2 - 12m - 2) = 0$$
$$144m^2 - 144m^3 + 36m^4 - 4(9m^4 - 36m^3 - 24m - 4) = 0$$
$$144m^2 - 144m^3 + 36m^4 - 36m^4 + 144m^3 + 96m + 16 = 0$$
$$144m^2 + 96m + 16 = 0$$
$$9m^2 + 6m + 1 = 0$$
$$(3m + 1)^2 = 0$$
$$3m + 1 = 0$$
$$m = -\frac{1}{3} \qquad b = 2 - \left(-\frac{1}{3}\right) = \frac{7}{3}$$

The equation of the tangent line is $y = -\frac{1}{3}x + \frac{7}{3}$.

93. Solve the system:
$$\begin{cases} x^2 - y^2 = 3 \\ \quad y = mx + b \end{cases}$$
Solve the system by substitution:
$$x^2 - (mx + b)^2 = 3$$
$$x^2 - m^2 x^2 - 2mbx - b^2 = 3$$
$$(1 - m^2)x^2 - 2mbx - b^2 - 3 = 0$$
Note that the tangent line passes through (2, 1). Find the relation between m and b:
$$1 = m(2) + b$$
$$b = 1 - 2m$$
Substitute into the quadratic to eliminate b:
$$(1 - m^2)x^2 - 2m(1 - 2m)x - (1 - 2m)^2 - 3 = 0$$
$$(1 - m^2)x^2 + (-2m + 4m^2)x - 1 + 4m - 4m^2 - 3 = 0$$
$$(1 - m^2)x^2 + (-2m + 4m^2)x + (-4m^2 + 4m - 4) = 0$$
Find when the discriminant is 0:
$$(-2m + 4m^2)^2 - 4(1 - m^2)(-4m^2 + 4m - 4) = 0$$
$$4m^2 - 16m^3 + 16m^4 - 4(4m^4 - 4m^3 + 4m - 4) = 0$$
$$4m^2 - 16m^3 + 16m^4 - 16m^4 + 16m^3 - 16m + 16 = 0$$
$$4m^2 - 16m + 16 = 0$$
$$m^2 - 4m + 4 = 0$$
$$(m - 2)^2 = 0$$
$$m - 2 = 0$$
$$m = 2 \qquad b = 1 - 2(2) = -3$$

The equation of the tangent line is $y = 2x - 3$.

95. Solve for r_1 and r_2:

$$\begin{cases} r_1 + r_2 = -\dfrac{b}{a} \\ r_1 r_2 = \dfrac{c}{a} \end{cases}$$

Substitute and solve:

$$r_1 = -r_2 - \frac{b}{a}$$

$$\left(-r_2 - \frac{b}{a}\right) r_2 = \frac{c}{a}$$

$$-r_2^2 - \frac{b}{a} r_2 - \frac{c}{a} = 0$$

$$a r_2^2 + b r_2 + c = 0$$

$$r_2 = \frac{-b \pm \sqrt{b^2 - 4ac}}{2a}$$

$$r_1 = -r_2 - \frac{b}{a} = -\left(\frac{-b \pm \sqrt{b^2 - 4ac}}{2a}\right) - \frac{2b}{2a} = \frac{-b \mp \sqrt{b^2 - 4ac}}{2a}$$

The solutions are: $\dfrac{-b + \sqrt{b^2 - 4ac}}{2a}$ and $\dfrac{-b - \sqrt{b^2 - 4ac}}{2a}$.

11.8 Systems of Linear Inequalities; Linear Programming

1. $x \geq 0$

Graph the line $x = 0$. Use a solid line since
  the inequality uses $\geq$.
Choose a test point not on the line, such as
  $(2, 0)$. Since $2 \geq 0$ is true, shade the side
  of the line containing $(2, 0)$.

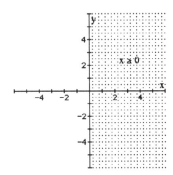

3. $x \geq 4$

Graph the line $x = 4$. Use a solid line since
  the inequality uses $\geq$.
Choose a test point not on the line, such as
  $(5, 0)$. Since $5 \geq 0$ is true, shade the side
  of the line containing $(5, 0)$.

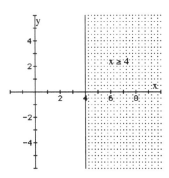

5. $x + y > 1$

Graph the line $x + y = 1$. Use a dashed line
since the inequality uses $>$.
Choose a test point not on the line, such as
$(0, 0)$. Since $0 + 0 > 1$ is false, shade the
opposite side of the line from $(0, 0)$.

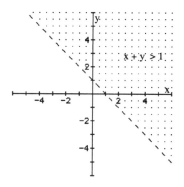

7. $2x + y \geq 6$

Graph the line $2x + y = 6$. Use a solid line
since the inequality uses $\geq$.
Choose a test point not on the line, such as
$(0, 0)$. Since $2(0) + 0 \geq 6$ is false, shade
the opposite side of the line from $(0, 0)$.

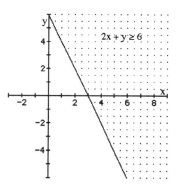

9. $\begin{cases} x + y \leq 2 \\ 2x + y \geq 4 \end{cases}$

(a)   Graph the line $x + y = 2$. Use a solid line since
the inequality uses $\leq$.
Choose a test point not on the line, such as
$(0, 0)$. Since $0 + 0 \leq 2$ is true, shade the
side of the line containing $(0, 0)$.

(b)   Graph the line $2x + y = 4$. Use a solid line
since the inequality uses $\geq$.
Choose a test point not on the line, such as
$(0, 0)$. Since $2(0) + 0 \geq 4$ is false, shade
the opposite side of the line from $(0, 0)$.

(c)   The overlapping region is the solution

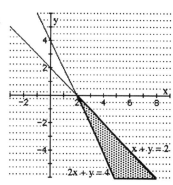

11. $\begin{cases} 2x - y \le 4 \\ 3x + 2y \ge -6 \end{cases}$

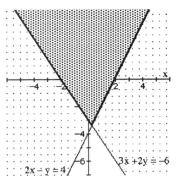

(a) Graph the line $2x - y = 4$. Use a solid line since the inequality uses $\le$.
Choose a test point not on the line, such as $(0, 0)$. Since $2(0) - 0 \le 4$ is true, shade the side of the line containing $(0, 0)$.

(b) Graph the line $3x + 2y = -6$. Use a solid line since the inequality uses $\ge$.
Choose a test point not on the line, such as $(0, 0)$. Since $3(0) + 2(0) \ge -6$ is true, shade the side of the line containing $(0, 0)$.

(c) The overlapping region is the solution

13. $\begin{cases} 2x - 3y \le 0 \\ 3x + 2y \le 6 \end{cases}$

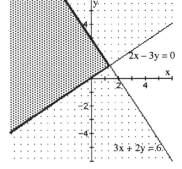

(a) Graph the line $2x - 3y = 0$. Use a solid line since the inequality uses $\le$.
Choose a test point not on the line, such as $(0, 3)$. Since $2(0) - 3(3) \le 0$ is true, shade the side of the line containing $(0, 3)$.

(b) Graph the line $3x + 2y = 6$. Use a solid line since the inequality uses $\le$.
Choose a test point not on the line, such as $(0, 0)$. Since $3(0) + 2(0) \le 6$ is true, shade the side of the line containing $(0, 0)$.

(c) The overlapping region is the solution

15. $\begin{cases} x - 2y \le 6 \\ 2x - 4y \ge 0 \end{cases}$

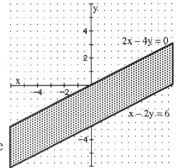

(a) Graph the line $x - 2y = 6$. Use a solid line since the inequality uses $\le$.
Choose a test point not on the line, such as $(0, 0)$. Since $0 - 2(0) \le 6$ is true, shade the side of the line containing $(0, 0)$.

(b) Graph the line $2x - 4y = 0$. Use a solid line since the inequality uses $\ge$.
Choose a test point not on the line, such as $(0, 2)$. Since $2(0) - 4(2) \ge 0$ is false, shade the opposite side of the line from $(0, 2)$.

(c) The overlapping region is the solution

17. $\begin{cases} 2x + y \geq -2 \\ 2x + y \geq 2 \end{cases}$

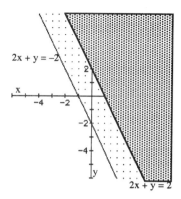

  (a)   Graph the line $2x + y = -2$. Use a solid line
          since the inequality uses $\geq$.
          Choose a test point not on the line, such as
          $(0, 0)$. Since $2(0) + 0 \geq -2$ is true, shade
          the side of the line containing $(0, 0)$.
  (b)   Graph the line $2x + y = 2$. Use a solid line
          since the inequality uses $\geq$.
          Choose a test point not on the line, such as
          $(0, 0)$. Since $2(0) + 0 \geq 2$ is false, shade
          the opposite side of the line from $(0, 0)$.
  (c)   The overlapping region is the solution.

19. $\begin{cases} 2x + 3y \geq 6 \\ 2x + 3y \leq 0 \end{cases}$

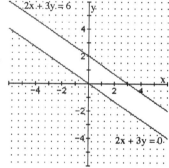

  (a)   Graph the line $2x + 3y = 6$. Use a solid line
          since the inequality uses $\geq$.
          Choose a test point not on the line, such as
          $(0, 0)$. Since $2(0) + 3(0) \geq 6$ is false, shade
          the opposite side of the line from $(0, 0)$.
  (b)   Graph the line $2x + 3y = 0$. Use a solid line
          since the inequality uses $\leq$.
          Choose a test point not on the line, such as
          $(0, 2)$. Since $2(0) + 3(2) \leq 0$ is false, shade
          the opposite side of the line from $(0, 2)$.
  (c)   Since the regions do not overlap, the solution is
          an empty set.

21.   Graph the system of linear inequalities:

$$\begin{cases} x \geq 0 \\ y \geq 0 \\ 2x + y \leq 6 \\ x + 2y \leq 6 \end{cases}$$

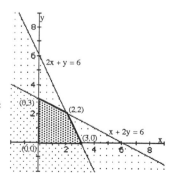

  (a)   Graph $x \geq 0; y \geq 0$. Shaded region is the first
          quadrant.
  (b)   Graph the line $2x + y = 6$. Use a solid line
          since the inequality uses $\leq$.
          Choose a test point not on the line, such as
          $(0, 0)$. Since $2(0) + 0 \leq 6$ is true, shade the
          side of the line containing $(0, 0)$.
  (c)   Graph the line $x + 2y = 6$. Use a solid line
          since the inequality uses $\leq$.
          Choose a test point not on the line, such as
          $(0, 0)$. Since $0 + 2(0) \leq 6$ is true, shade the
          side of the line containing $(0, 0)$.

(d)   The overlapping region is the solution.
(e)   The graph is bounded.
(f)   Find the vertices:
The x-axis and y-axis intersect at $(0, 0)$.
The intersection of $x + 2y = 6$ and the y-axis is $(0, 3)$.
The intersection of $2x + y = 6$ and the x-axis is $(3, 0)$.
To find the intersection of $x + 2y = 6$ and $2x + y = 6$, solve the system:
$$\begin{cases} x + 2y = 6 \quad \rightarrow \quad x = 6 - 2y \\ 2x + y = 6 \end{cases}$$
Substitute and solve:
$$2(6 - 2y) + y = 6$$
$$12 - 4y + y = 6$$
$$-3y = -6$$
$$y = 2$$
$$x = 6 - 2(2) = 6 - 4 = 2$$
The point of intersection is $(2, 2)$.
The four corner points are $(0, 0)$, $(0, 3)$, $(3, 0)$, and $(2, 2)$.

23.   Graph the system of linear inequalities:
$$\begin{cases} x \geq 0 \\ y \geq 0 \\ x + y \geq 2 \\ 2x + y \geq 4 \end{cases}$$

(a)   Graph $x \geq 0$; $y \geq 0$. Shaded region is the first quadrant.

(b)   Graph the line $x + y = 2$. Use a solid line since the inequality uses $\geq$.
Choose a test point not on the line, such as $(0, 0)$. Since $0 + 0 \geq 2$ is false, shade the opposite side of the line from $(0, 0)$.

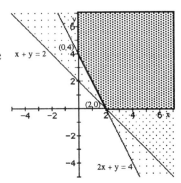

(c)   Graph the line $2x + y = 4$. Use a solid line since the inequality uses $\geq$.
Choose a test point not on the line, such as $(0, 0)$. Since $2(0) + 0 \geq 4$ is false, shade the opposite side of the line from $(0, 0)$.

(d)   The overlapping region is the solution.
(e)   The graph is unbounded.
(f)   Find the vertices:
The intersection of $x + y = 2$ and the x-axis is $(2, 0)$.
The intersection of $2x + y = 4$ and the y-axis is $(0, 4)$.
The two corner points are $(2, 0)$, and $(0, 4)$.

25. Graph the system of linear inequalities:

$$\begin{cases} x \geq 0 \\ y \geq 0 \\ x + y \geq 2 \\ 2x + 3y \leq 12 \\ 3x + y \leq 12 \end{cases}$$

(a) Graph $x \geq 0$; $y \geq 0$. Shaded region is the first quadrant.

(b) Graph the line $x + y = 2$. Use a solid line since the inequality uses $\geq$.
     Choose a test point not on the line, such as $(0, 0)$. Since $0 + 0 \geq 2$ is false, shade the opposite side of the line from $(0, 0)$.

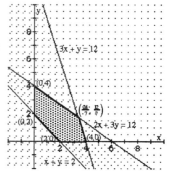

(c) Graph the line $2x + 3y = 12$. Use a solid line since the inequality uses $\leq$.
     Choose a test point not on the line, such as $(0, 0)$. Since $2(0) + 3(0) \leq 12$ is true, shade the side of the line containing $(0, 0)$.

(d) Graph the line $3x + y = 12$. Use a solid line since the inequality uses $\leq$.
     Choose a test point not on the line, such as $(0, 0)$. Since $3(0) + 0 \leq 12$ is true, shade the side of the line containing $(0, 0)$.

(e) The overlapping region is the solution.
(f) The graph is bounded.
(g) Find the vertices:
     The intersection of $x + y = 2$ and the y-axis is $(0, 2)$.
     The intersection of $x + y = 2$ and the x-axis is $(2, 0)$.
     The intersection of $2x + 3y = 12$ and the y-axis is $(0, 4)$.
     The intersection of $3x + y = 12$ and the x-axis is $(4, 0)$.
     To find the intersection of $2x + 3y = 12$ and $3x + y = 12$, solve the system:

$$\begin{cases} 2x + 3y = 12 \\ 3x + y = 12 \end{cases} \rightarrow \quad y = 12 - 3x$$

Substitute and solve:

$$2x + 3(12 - 3x) = 12$$
$$2x + 36 - 9x = 12$$
$$-7x = -24$$
$$x = \tfrac{24}{7}$$
$$y = 12 - 3\left(\tfrac{24}{7}\right) = 12 - \tfrac{72}{7} = \tfrac{12}{7}$$

The point of intersection is $\left(\tfrac{24}{7}, \tfrac{12}{7}\right)$.

The five corner points are $(0, 2)$, $(0, 4)$, $(2, 0)$, $(4, 0)$, and $\left(\tfrac{24}{7}, \tfrac{12}{7}\right)$.

27.   Graph the system of linear inequalities:

$$\begin{cases} x \ge 0 \\ y \ge 0 \\ x + y \ge 2 \\ x + y \le 8 \\ 2x + y \le 10 \end{cases}$$

(a)   Graph $x \ge 0$; $y \ge 0$. Shaded region is the first quadrant.

(b)   Graph the line $x + y = 2$. Use a solid line since the inequality uses $\ge$.

Choose a test point not on the line, such as $(0, 0)$. Since $0 + 0 \ge 2$ is false, shade the opposite side of the line from $(0, 0)$.

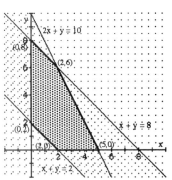

(c)   Graph the line $x + y = 8$. Use a solid line since the inequality uses $\le$.

Choose a test point not on the line, such as $(0, 0)$. Since $0 + 0 \le 8$ is true, shade the side of the line containing $(0, 0)$.

(d)   Graph the line $2x + y = 10$. Use a solid line since the inequality uses $\le$.

Choose a test point not on the line, such as $(0, 0)$. Since $2(0) + 0 \le 10$ is true, shade the side of the line containing $(0, 0)$.

(e)   The overlapping region is the solution.

(f)   The graph is bounded.

(g)   Find the vertices:

The intersection of $x + y = 2$ and the y-axis is $(0, 2)$.

The intersection of $x + y = 2$ and the x-axis is $(2, 0)$.

The intersection of $x + y = 8$ and the y-axis is $(0, 8)$.

The intersection of $2x + y = 10$ and the x-axis is $(5, 0)$.

To find the intersection of $x + y = 8$ and $2x + y = 10$, solve the system:

$$\begin{cases} x + y = 8 \quad \rightarrow \quad y = 8 - x \\ 2x + y = 10 \end{cases}$$

Substitute and solve:

$$2x + 8 - x = 10$$
$$x = 2$$
$$y = 8 - 2 = 6$$

The point of intersection is $(2, 6)$.

The five corner points are $(0, 2)$, $(0, 8)$, $(2, 0)$, $(5, 0)$, and $(2, 6)$.

29. Graph the system of linear inequalities:

$$\begin{cases} x \geq 0 \\ y \geq 0 \\ x + 2y \geq 1 \\ x + 2y \leq 10 \end{cases}$$

(a) Graph $x \geq 0$; $y \geq 0$. Shaded region is the first quadrant.

(b) Graph the line $x + 2y = 1$. Use a solid line since the inequality uses $\geq$.
Choose a test point not on the line, such as $(0, 0)$. Since $0 + 2(0) \geq 1$ is false, shade the opposite side of the line from $(0, 0)$.

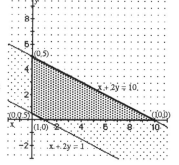

(c) Graph the line $x + 2y = 10$. Use a solid line since the inequality uses $\leq$.
Choose a test point not on the line, such as $(0, 0)$. Since $0 + 2(0) \leq 10$ is true, shade the side of the line containing $(0, 0)$.

(d) The overlapping region is the solution.
(e) The graph is bounded.
(f) Find the vertices:
The intersection of $x + 2y = 1$ and the y-axis is $(0, 0.5)$.
The intersection of $x + 2y = 1$ and the x-axis is $(1, 0)$.
The intersection of $x + 2y = 10$ and the y-axis is $(0, 5)$.
The intersection of $x + 2y = 10$ and the x-axis is $(10, 0)$.
The four corner points are $(0, 0.5)$, $(0, 5)$, $(1, 0)$, and $(10, 0)$.

31. The system of linear inequalities is:

$$\begin{cases} x \geq 0 \\ y \geq 0 \\ x \leq 4 \\ x + y \leq 6 \end{cases}$$

33. The system of linear inequalities is:

$$\begin{cases} x \geq 0 \\ y \geq 15 \\ x \leq 20 \\ x + y \leq 50 \\ x - y \leq 0 \end{cases}$$

35. $z = x + y$

Vertex	Value of $z = x + y$
$(0, 3)$	$z = 0 + 3 = 3$
$(0, 6)$	$z = 0 + 6 = 6$
$(5, 6)$	$z = 5 + 6 = 11$
$(5, 2)$	$z = 5 + 2 = 7$
$(4, 0)$	$z = 4 + 0 = 4$

The maximum value is 11 at $(5, 6)$, and the minimum value is 3 at $(0, 3)$.

37. $z = x + 10y$

Vertex	Value of $z = x + 10y$
(0, 3)	$z = 0 + 10(3) = 30$
(0, 6)	$z = 0 + 10(6) = 60$
(5, 6)	$z = 5 + 10(6) = 65$
(5, 2)	$z = 5 + 10(2) = 25$
(4, 0)	$z = 4 + 10(0) = 4$

The maximum value is 65 at (5, 6), and the minimum value is 4 at (4, 0).

39. $z = 5x + 7y$

Vertex	Value of $z = 5x + 7y$
(0, 3)	$z = 5(0) + 7(3) = 21$
(0, 6)	$z = 5(0) + 7(6) = 42$
(5, 6)	$z = 5(5) + 7(6) = 67$
(5, 2)	$z = 5(5) + 7(2) = 39$
(4, 0)	$z = 5(4) + 7(0) = 20$

The maximum value is 67 at (5, 6), and the minimum value is 20 at (4, 0).

41. Maximize $z = 2x + y$

Subject to $x \geq 0$, $y \geq 0$, $x + y \leq 6$, $x + y \geq 1$
Graph the constraints.

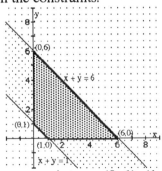

The corner points are (0, 1), (1, 0), (0, 6), (6, 0).
Evaluate the objective function:

Vertex	Value of $z = 2x + y$
(0, 1)	$z = 2(0) + 1 = 1$
(0, 6)	$z = 2(0) + 6 = 6$
(1, 0)	$z = 2(1) + 0 = 2$
(6, 0)	$z = 2(6) + 0 = 12$

The maximum value is 12 at (6, 0).

43.   Minimize $z = 2x + 5y$

Subject to $x \geq 0, \quad y \geq 0, \quad x + y \geq 2, \quad x \leq 5, \quad y \leq 3$
Graph the constraints.

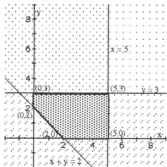

The corner points are $(0, 2)$, $(2, 0)$, $(0, 3)$, $(5, 0)$, $(5, 3)$.
Evaluate the objective function:

Vertex	Value of $z = 2x + 5y$
$(0, 2)$	$z = 2(0) + 5(2) = 10$
$(0, 3)$	$z = 2(0) + 5(3) = 15$
$(2, 0)$	$z = 2(2) + 5(0) = 4$
$(5, 0)$	$z = 2(5) + 5(0) = 10$
$(5, 3)$	$z = 2(5) + 5(3) = 25$

The minimum value is 4 at $(2, 0)$.

45.   Maximize $z = 3x + 5y$

Subject to $x \geq 0, \quad y \geq 0, \quad x + y \geq 2, \quad 2x + 3y \leq 12, \quad 3x + 2y \leq 12$
Graph the constraints.

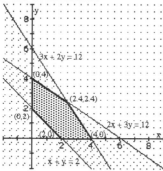

To find the intersection of $2x + 3y = 12$ and $3x + 2y = 12$, solve the system:

$$\begin{cases} 2x + 3y = 12 \\ 3x + 2y = 12 \end{cases} \rightarrow \quad y = 6 - \tfrac{3}{2}x$$

Substitute and solve:

$$2x + 3\left(6 - \tfrac{3}{2}x\right) = 12$$

$$2x + 18 - \tfrac{9}{2}x = 12$$

$$-\tfrac{5}{2}x = -6$$

$$x = \tfrac{12}{5} \qquad y = 6 - \tfrac{3}{2}\left(\tfrac{12}{5}\right) = 6 - \tfrac{18}{5} = \tfrac{12}{5}$$

The point of intersection is $\left(\tfrac{12}{5}, \tfrac{12}{5}\right)$.

The corner points are (0, 2), (2, 0), (0, 4), (4, 0), (2.4, 2.4).
Evaluate the objective function:

Vertex	Value of $z = 3x + 5y$
(0, 2)	$z = 3(0) + 5(2) = 10$
(0, 4)	$z = 3(0) + 5(4) = 20$
(2, 0)	$z = 3(2) + 5(0) = 6$
(4, 0)	$z = 3(4) + 5(0) = 12$
(2.4, 2.4)	$z = 3(2.4) + 5(2.4) = 19.2$

The maximum value is 20 at (0, 4).

47.   Minimize $z = 5x + 4y$

Subject to $x \geq 0,\ y \geq 0,\ x + y \geq 2,\ 2x + 3y \leq 12,\ 3x + y \leq 12$
Graph the constraints.

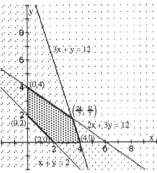

To find the intersection of $2x + 3y = 12$ and $3x + y = 12$, solve the system:

$$\begin{cases} 2x + 3y = 12 \\ 3x + y = 12 \end{cases} \rightarrow \quad y = 12 - 3x$$

Substitute and solve:

$$2x + 3(12 - 3x) = 12$$
$$2x + 36 - 9x = 12$$
$$-7x = -24$$
$$x = \tfrac{24}{7}$$
$$y = 12 - 3\left(\tfrac{24}{7}\right) = 12 - \tfrac{72}{7} = \tfrac{12}{7}$$

The point of intersection is $\left(\tfrac{24}{7}, \tfrac{12}{7}\right)$.

The corner points are (0, 2), (2, 0), (0, 4), (4, 0), $\left(\tfrac{24}{7}, \tfrac{12}{7}\right)$.
Evaluate the objective function:

Vertex	Value of $z = 5x + 4y$
(0, 2)	$z = 5(0) + 4(2) = 8$
(0, 4)	$z = 5(0) + 4(4) = 16$
(2, 0)	$z = 5(2) + 4(0) = 10$
(4, 0)	$z = 5(4) + 4(0) = 20$
$\left(\tfrac{24}{7}, \tfrac{12}{7}\right)$	$z = 5\left(\tfrac{24}{7}\right) + 4\left(\tfrac{12}{7}\right) = \tfrac{120}{7} + \tfrac{48}{7} = \tfrac{168}{7} = 24$

The minimum value is 8 at (0, 2).

49. Maximize $z = 5x + 2y$

Subject to $x \geq 0$, $y \geq 0$, $x + y \leq 10$, $2x + y \geq 10$, $x + 2y \geq 10$

Graph the constraints.

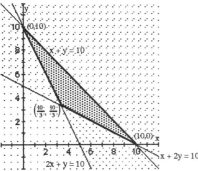

To find the intersection of $2x + y = 10$ and $x + 2y = 10$, solve the system:

$$\begin{cases} 2x + y = 10 & \rightarrow \quad y = 10 - 2x \\ x + 2y = 10 \end{cases}$$

Substitute and solve:

$$x + 2(10 - 2x) = 10$$
$$x + 20 - 4x = 10$$
$$-3x = -10$$
$$x = \tfrac{10}{3}$$
$$y = 10 - 2\left(\tfrac{10}{3}\right) = 10 - \tfrac{20}{3} = \tfrac{10}{3}$$

The point of intersection is $\left(\tfrac{10}{3}, \tfrac{10}{3}\right)$.

The corner points are $(0, 10)$, $(10, 0)$, $\left(\tfrac{10}{3}, \tfrac{10}{3}\right)$.

Evaluate the objective function:

Vertex	Value of $z = 5x + 2y$
$(0, 10)$	$z = 5(0) + 2(10) = 20$
$(10, 0)$	$z = 5(10) + 2(0) = 50$
$\left(\tfrac{10}{3}, \tfrac{10}{3}\right)$	$z = 5\left(\tfrac{10}{3}\right) + 2\left(\tfrac{10}{3}\right) = \tfrac{50}{3} + \tfrac{20}{3} = \tfrac{70}{3} = 23\tfrac{1}{3}$

The maximum value is 50 at $(10, 0)$.

51. Let x = the number of downhill skis produced.

Let y = the number of cross-country skis produced.

The total profit is: $P = 70x + 50y$. Profit is to be maximized; thus, this is the objective function.

The constraints are:

$x \geq 0$, $y \geq 0$ A positive number of skis must be produced.

$2x + y \leq 40$ Only 40 hours of manufacturing time is available.

$x + y \leq 32$ Only 32 hours of finishing time is available.

Graph the constraints.

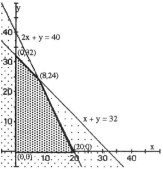

To find the intersection of $x + y = 32$ and $2x + y = 40$, solve the system:

$$\begin{cases} x + y = 32 & \rightarrow \quad y = 32 - x \\ 2x + y = 40 \end{cases}$$

Substitute and solve:

$$2x + 32 - x = 40$$

$$x = 8$$

$$y = 32 - 8 = 24$$

The point of intersection is (8, 24).

The corner points are (0, 0), (0, 32), (20, 0), (8, 24).

Evaluate the objective function:

Vertex	Value of $P = 70x + 50y$
(0, 0)	$P = 70(0) + 50(0) = 0$
(0, 32)	$P = 70(0) + 50(32) = 1600$
(20, 0)	$P = 70(20) + 50(0) = 1400$
(8, 24)	$P = 70(8) + 50(24) = 1760$

The maximum profit is $1760, when 8 downhill skis and 24 cross-country skis are produced.

With the increase of the manufacturing time to 48 hours, we do the following:

The constraints are:

$x \geq 0, \quad y \geq 0$ A positive number of skis must be produced.

$2x + y \leq 48$ Only 48 hours of manufacturing time is available.

$x + y \leq 32$ Only 32 hours of finishing time is available.

Graph the constraints.

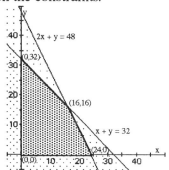

To find the intersection of $x + y = 32$ and $2x + y = 48$, solve the system:

$$\begin{cases} x + y = 32 & \rightarrow \quad y = 32 - x \\ 2x + y = 48 \end{cases}$$

Substitute and solve:

$$2x + 32 - x = 48$$
$$x = 16$$
$$y = 32 - 16 = 16$$

The point of intersection is (16, 16).
The corner points are (0, 0), (0, 32), (24, 0), (16, 16).
Evaluate the objective function:

Vertex	Value of $P = 70x + 50y$
(0, 0)	$P = 70(0) + 50(0) = 0$
(0, 32)	$P = 70(0) + 50(32) = 1600$
(24, 0)	$P = 70(24) + 50(0) = 1680$
(16, 16)	$P = 70(16) + 50(16) = 1920$

The maximum profit is $1920, when 16 downhill skis and 16 cross-country skis are produced.

53. Let x = the number of acres of corn that should be planted.
Let y = the number of acres of soybeans that should be planted.

The total profit is: $P = 250x + 200y$. Profit is to be maximized; thus, this is the objective function.

The constraints are:

$x \geq 0, \ y \geq 0$ A positive number of acres must be planted.

$x + y \leq 100$ Acres available for planting.

$60x + 40y \leq 1800$ Money available for cultivation.

$60x + 60y \leq 2400$ Money available for labor.

Graph the constraints.

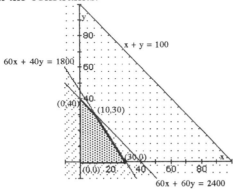

To find the intersection of $60x + 40y = 1800$ and $60x + 60y = 2400$, solve the system:

$$\begin{cases} 60x + 40y = 1800 & \rightarrow \quad 60x = 1800 - 40y \\ 60x + 60y = 2400 \end{cases}$$

Substitute and solve:

$$1800 - 40y + 60y = 2400$$
$$20y = 600$$
$$y = 30$$
$$60x = 1800 - 40(30)$$
$$60x = 600$$
$$x = 10$$

The point of intersection is $(10, 30)$.

The corner points are $(0, 0)$, $(0, 40)$, $(30, 0)$, $(10, 30)$.

Evaluate the objective function:

Vertex	Value of $P = 250x + 200y$
(0, 0)	P = 250(0) + 200(0) = 0
(0, 40)	P = 250(0) + 200(40) = 8000
(30, 0)	P = 250(30) + 200(0) = 7500
(10, 30)	P = 250(10) + 200(30) = 8500

The maximum profit is $8500, when 10 acres of corn and 30 acres of soybeans are planted.

55. Let x = the number of hours that machine 1 is operated.

Let y = the number of hours that machine 2 is operated.

The total cost is: $C = 50x + 30y$. Cost is to be minimized; thus, this is the objective function.

The constraints are:

$x \geq 0$, $y \geq 0$ A positive number of hours must be used.

$x \leq 10$ 10 hours available on machine 1.

$y \leq 10$ 10 hours available on machine 2.

$60x + 40y \geq 240$ At least 240 8-inch plyers must be produced.

$70x + 20y \geq 140$ At least 140 6-inch plyers must be produced.

Graph the constraints.

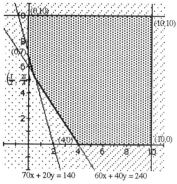

To find the intersection of $60x + 40y = 240$ and $70x + 20y = 140$, solve the system:

$$\begin{cases} 60x + 40y = 240 \\ 70x + 20y = 140 \quad \rightarrow \quad 20y = 140 - 70x \end{cases}$$

Substitute and solve:

$$60x + 2(140 - 70x) = 240$$
$$60x + 280 - 140x = 240$$
$$-80x = -40$$
$$x = \tfrac{1}{2} = 0.5$$
$$20y = 140 - 70\left(\tfrac{1}{2}\right)$$
$$20y = 105$$
$$y = \tfrac{21}{4} = 5.25$$

The point of intersection is $\left(\tfrac{1}{2}, \tfrac{21}{4}\right)$.

The corner points are $(0, 7)$, $(0, 10)$, $(4, 0)$, $(10, 0)$, $(10, 10)$, $\left(\tfrac{1}{2}, \tfrac{21}{4}\right)$.

Evaluate the objective function:

Vertex	Value of $C = 50x + 30y$
$(0, 7)$	$C = 50(0) + 30(7) = 210$
$(0, 10)$	$C = 50(0) + 30(10) = 300$
$(4, 0)$	$C = 50(4) + 30(0) = 200$
$(10, 0)$	$C = 50(10) + 30(0) = 500$
$(10, 10)$	$C = 50(10) + 30(10) = 800$
$\left(\tfrac{1}{2}, \tfrac{21}{4}\right)$	$C = 50\left(\tfrac{1}{2}\right) + 30\left(\tfrac{21}{4}\right) = 182.50$

The minimum cost is \$182.50, when machine 1 is used for 0.5 hours and machine 2 is used for 5.25 hours.

57. Let x = the number of pounds of ground beef.
Let y = the number of pounds of ground pork.

The total cost is: $C = 0.75x + 0.45y$. Cost is to be minimized; thus, this is the objective function.

The constraints are:

$x \geq 0, \; y \geq 0$ A positive number of pounds must be used.

$x \leq 200$ Only 200 pounds of ground beef are available.

$y \geq 50$ At least 50 pounds of ground pork must be used.

$0.75x + 0.60y \geq 0.70(x + y) \rightarrow 0.05x \geq 0.10y$ Leanness condition to be met.

Graph the constraints.

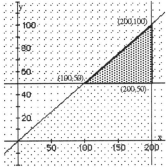

The corner points are $(100, 50)$, $(200, 50)$, $(200, 100)$.

Evaluate the objective function:

Vertex	Value of $C = 0.75x + 0.45y$
(100, 50)	$C = 0.75(100) + 0.45(50) = 97.50$
(200, 50)	$C = 0.75(200) + 0.45(50) = 172.50$
(200, 100)	$C = 0.75(200) + 0.45(100) = 195.00$

The minimum cost is $97.50, when 100 pounds of ground beef and 50 pounds of ground pork are used.

59. Let x = the number of racing skates manufactured.
Let y = the number of figure skates manufactured.
The total profit is: $P = 10x + 12y$. Profit is to be maximized; thus, this is the objective function.
The constraints are:

$x \geq 0,\ \ y \geq 0$ A positive number of skates must be manufactured.

$6x + 4y \leq 120$ Only 120 hours are available for fabrication.

$x + 2y \leq 40$ Only 40 hours are available for finishing.

Graph the constraints.

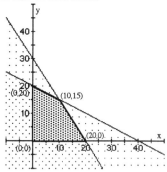

To find the intersection of $6x + 4y = 120$ and $x + 2y = 40$, solve the system:

$$\begin{cases} 6x + 4y = 120 \\ \ \ x + 2y = \ \ 40 \end{cases} \rightarrow \ \ x = 40 - 2y$$

Substitute and solve:

$$6(40 - 2y) + 4y = 120$$
$$240 - 12y + 4y = 120$$
$$-8y = -120$$
$$y = 15$$
$$x = 40 - 2(15) = 10$$

The point of intersection is (10, 15).
The corner points are (0, 0), (0, 20), (20, 0), (10, 15).
Evaluate the objective function:

Vertex	Value of $P = 10x + 12y$
(0, 0)	$P = 10(0) + 12(0) = 0$
(0, 20)	$P = 10(0) + 12(20) = 240$
(20, 0)	$P = 10(20) + 12(0) = 200$
(10, 15)	$P = 10(10) + 12(15) = 280$

The maximum profit is $280, when 10 racing skates and 15 figure skates are produced.

61. Let x = the number of metal fasteners.
     Let y = the number of plastic fasteners.
     The total cost is: $C = 9x + 4y$. Cost is to be minimized; thus, this is the objective
         function.
     The constraints are:

     $x \geq 2, \; y \geq 2$ At least 2 of each fastener must be made.

     $x + y \geq 6$ At least 6 fasteners are needed.

     $4x + 2y \leq 24$ Only 24 hours are available.

     Graph the constraints.

     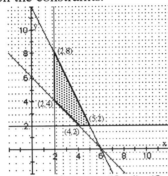

     The corner points are $(2, 4)$, $(2, 8)$, $(4, 2)$, $(5, 2)$.
     Evaluate the objective function:

Vertex	Value of $C = 9x + 4y$
$(2, 4)$	$C = 9(2) + 4(4) = 34$
$(2, 8)$	$C = 9(2) + 4(8) = 50$
$(4, 2)$	$C = 9(4) + 4(2) = 44$
$(5, 2)$	$C = 9(5) + 4(2) = 53$

     The minimum cost is $34, when 2 metal fasteners and 4 plastic fasteners are ordered.

63. Let x = the number of first-class seats.
     Let y = the number of coach seats.
     The constraints are:

     $8 \leq x \leq 16$ Restriction on first-class seats.

     $80 \leq y \leq 120$ Restriction on coach seats.

     (a) $\dfrac{x}{y} \leq \dfrac{1}{12}$ Ratio of seats.

     If $y = 120$, then $\dfrac{x}{120} \leq \dfrac{1}{12}$

     $12x \leq 120$

     $x \leq 10$

     The maximum revenue will be obtained with 120 coach seats and 10 first-class
     seats. (Note that the first-class seats meet their constraint.)

(b) $\dfrac{x}{y} \le \dfrac{1}{8}$ Ratio of seats.

$$\text{If } y = 120, \text{ then } \dfrac{x}{120} \le \dfrac{1}{8}$$
$$8x \le 120$$
$$x \le 15$$

The maximum revenue will be obtained with 120 coach seats and 15 first-class seats. (Note that the first-class seats meet their constraint.)

11   Chapter Review

1.   Solve the first equation for y, substitute into the second equation and solve:
$$\begin{cases} 2x - y = 5 \quad\to\quad y = 2x - 5 \\ 5x + 2y = 8 \end{cases}$$
$$5x + 2(2x - 5) = 8$$
$$5x + 4x - 10 = 8$$
$$9x = 18$$
$$x = 2$$
$$y = 2(2) - 5 = 4 - 5 = -1$$
The solution is $x = 2, \; y = -1$.

3.   Solve the second equation for x, substitute into the first equation and solve:
$$\begin{cases} 3x - 4y = 4 \\ x - 3y = \tfrac{1}{2} \quad\to\quad x = 3y + \tfrac{1}{2} \end{cases}$$
$$3\left(3y + \tfrac{1}{2}\right) - 4y = 4$$
$$9y + \tfrac{3}{2} - 4y = 4$$
$$5y = \tfrac{5}{2}$$
$$y = \tfrac{1}{2}$$
$$x = 3\left(\tfrac{1}{2}\right) + \tfrac{1}{2} = 2$$
The solution is $x = 2, \; y = \tfrac{1}{2}$.

5.   Solve the first equation for x, substitute into the second equation and solve:
$$\begin{cases} x - 2y - 4 = 0 \quad\to\quad x = 2y + 4 \\ 3x + 2y - 4 = 0 \end{cases}$$
$$3(2y + 4) + 2y - 4 = 0$$
$$6y + 12 + 2y - 4 = 0$$
$$8y = -8$$
$$y = -1$$
$$x = 2(-1) + 4 = 2$$
The solution is $x = 2, \; y = -1$.

7. Substitute the first equation into the second equation and solve:

$$\begin{cases} y = 2x - 5 \\ x = 3y + 4 \end{cases}$$

$$x = 3(2x - 5) + 4$$
$$x = 6x - 15 + 4$$
$$-5x = -11$$
$$x = \tfrac{11}{5}$$
$$y = 2\left(\tfrac{11}{5}\right) - 5 = -\tfrac{3}{5}$$

The solution is $x = \tfrac{11}{5}$, $y = -\tfrac{3}{5}$.

9. Multiply each side of the first equation by 5 and each side of the second equation by 30 and add to eliminate y:

$$\begin{cases} x - y + 4 = 0 \quad \xrightarrow{\;5\;} \quad 5x - 5y + 20 = 0 \\ \tfrac{1}{2}x + \tfrac{1}{6}y + \tfrac{2}{5} = 0 \quad \xrightarrow{\;30\;} \quad 15x + 5y + 12 = 0 \end{cases}$$
$$\underline{}$$
$$20x \qquad + 32 = 0$$
$$20x = -32$$
$$x = -\tfrac{8}{5}$$

Substitute and solve for y:

$$-\tfrac{8}{5} - y + 4 = 0 \quad \rightarrow \quad y = \tfrac{12}{5}$$

The solution of the system is $x = -\tfrac{8}{5}$, $y = \tfrac{12}{5}$.

11. Rewrite each equation and add to eliminate y:

$$\begin{cases} x - 2y - 8 = 0 \quad \longrightarrow \quad x - 2y = 8 \\ 2x + 2y - 10 = 0 \quad \longrightarrow \quad 2x + 2y = 10 \end{cases}$$
$$\underline{}$$
$$3x \qquad = 18$$
$$x = 6$$

Substitute and solve for y:

$$6 - 2y = 8$$
$$-2y = 2$$
$$y = -1$$

The solution of the system is $x = 6$, $y = -1$.

13. Solve the first equation for y, substitute into the second equation and solve:

$$\begin{cases} y - 2x = 11 \quad \rightarrow \quad y = 2x + 11 \\ 2y - 3x = 18 \end{cases}$$
$$2(2x + 11) - 3x = 18$$
$$4x + 22 - 3x = 18$$
$$x = -4$$
$$y = 2(-4) + 11 = 3$$

The solution is $x = -4$, $y = 3$.

15. Multiply each side of the first equation by 2 and each side of the second equation by 3 and add to eliminate y:

$$\begin{cases} 2x + 3y - 13 = 0 \\ 3x - 2y \quad\; = 0 \end{cases} \xrightarrow{\;2\;} \begin{array}{l} 4x + 6y - 26 = 0 \\ 9x - 6y \qquad = 0 \end{array}$$

$$\begin{array}{r} 13x \quad - 26 = 0 \\ 13x = 26 \\ x = 2 \end{array}$$

Substitute and solve for y:

$$3(2) - 2y = 0$$
$$-2y = -6$$
$$y = 3$$

The solution of the system is $x = 2, \; y = 3$.

17. Multiply each side of the second equation by –3 and add to eliminate x:

$$\begin{cases} 3x - 2y = 8 \\ x - \frac{2}{3}y = 12 \end{cases} \xrightarrow{\;\;\;\;\;} \begin{array}{l} 3x - 2y = \;\;8 \\ -3x + 2y = -36 \end{array}$$

$$\xrightarrow{-3} \qquad \qquad \qquad 0 = -28$$

The system has no solution, so the system is inconsistent.

19. Multiply each side of the first equation by –2 and add to the second equation to eliminate x; and multiply each side of the first equation by –3 and add to the third equation to eliminate x:

$$\begin{cases} x + 2y - z = \;\;\;6 \\ 2x - \;\;y + 3z = -13 \\ 3x - 2y + 3z = -16 \end{cases} \xrightarrow{-2} \begin{array}{l} -2x - 4y + 2z = -12 \\ 2x - \;\;y + 3z = -13 \end{array}$$

$$-5y + 5z = -25 \xrightarrow{-1/5} y - z = 5$$

$$\xrightarrow{-3} \begin{array}{l} -3x - 6y + 3z = -18 \\ 3x - 2y + 3z = -16 \end{array}$$

$$-8y + 6z = -34$$

Multiply each side of the first result by 8 and add to the second result to eliminate y:

$$\begin{array}{l} y - z = \;\;\;5 \\ -8y + 6z = -34 \end{array} \xrightarrow{\;8\;} \begin{array}{l} 8y - 8z = \;\;40 \\ -8y + 6z = -34 \end{array}$$

$$\begin{array}{r} -2z = \;\;\;6 \\ z = -3 \end{array}$$

Substituting and solving for the other variables:

$$\begin{array}{ll} y - (-3) = 5 & x + 2(2) - (-3) = 6 \\ y = 2 & x + 4 + 3 = 6 \\ & x = -1 \end{array}$$

The solution is $x = -1, \; y = 2, \; z = -3$.

21. $A + C = \begin{bmatrix} 1 & 0 \\ 2 & 4 \\ -1 & 2 \end{bmatrix} + \begin{bmatrix} 3 & -4 \\ 1 & 5 \\ 5 & -2 \end{bmatrix} = \begin{bmatrix} 4 & -4 \\ 3 & 9 \\ 4 & 0 \end{bmatrix}$

23. $6A = 6 \cdot \begin{bmatrix} 1 & 0 \\ 2 & 4 \\ -1 & 2 \end{bmatrix} = \begin{bmatrix} 6 & 0 \\ 12 & 24 \\ -6 & 12 \end{bmatrix}$

25. $AB = \begin{bmatrix} 1 & 0 \\ 2 & 4 \\ -1 & 2 \end{bmatrix} \cdot \begin{bmatrix} 4 & -3 & 0 \\ 1 & 1 & -2 \end{bmatrix} = \begin{bmatrix} 4 & -3 & 0 \\ 12 & -2 & -8 \\ -2 & 5 & -4 \end{bmatrix}$

27. $CB = \begin{bmatrix} 3 & -4 \\ 1 & 5 \\ 5 & -2 \end{bmatrix} \cdot \begin{bmatrix} 4 & -3 & 0 \\ 1 & 1 & -2 \end{bmatrix} = \begin{bmatrix} 8 & -13 & 8 \\ 9 & 2 & -10 \\ 18 & -17 & 4 \end{bmatrix}$

29. Augment the matrix with the identity and use row operations to find the inverse:

$A = \begin{bmatrix} 4 & 6 \\ 1 & 3 \end{bmatrix} \rightarrow \begin{bmatrix} 4 & 6 & | & 1 & 0 \\ 1 & 3 & | & 0 & 1 \end{bmatrix}$

$\rightarrow \begin{bmatrix} 1 & 3 & | & 0 & 1 \\ 4 & 6 & | & 1 & 0 \end{bmatrix} \rightarrow \begin{bmatrix} 1 & 3 & | & 0 & 1 \\ 0 & -6 & | & 1 & -4 \end{bmatrix} \rightarrow \begin{bmatrix} 1 & 3 & | & 0 & 1 \\ 0 & 1 & | & -\frac{1}{6} & \frac{2}{3} \end{bmatrix} \rightarrow \begin{bmatrix} 1 & 0 & | & \frac{1}{2} & -1 \\ 0 & 1 & | & -\frac{1}{6} & \frac{2}{3} \end{bmatrix}$

$\quad$ Interchange $\quad R_2 = -4r_1 + r_2 \quad R_2 = -\frac{1}{6}r_2 \quad R_1 = -3r_2 + r_1$
$\quad r_1$ and r_2

$A^{-1} = \begin{bmatrix} \frac{1}{2} & -1 \\ -\frac{1}{6} & \frac{2}{3} \end{bmatrix}$

31. Augment the matrix with the identity and use row operations to find the inverse:

$A = \begin{bmatrix} 1 & 3 & 3 \\ 1 & 2 & 1 \\ 1 & -1 & 2 \end{bmatrix} \rightarrow \begin{bmatrix} 1 & 3 & 3 & | & 1 & 0 & 0 \\ 1 & 2 & 1 & | & 0 & 1 & 0 \\ 1 & -1 & 2 & | & 0 & 0 & 1 \end{bmatrix}$

$\rightarrow \begin{bmatrix} 1 & 3 & 3 & | & 1 & 0 & 0 \\ 0 & -1 & -2 & | & -1 & 1 & 0 \\ 0 & -4 & -1 & | & -1 & 0 & 1 \end{bmatrix} \rightarrow \begin{bmatrix} 1 & 3 & 3 & | & 1 & 0 & 0 \\ 0 & 1 & 2 & | & 1 & -1 & 0 \\ 0 & -4 & -1 & | & -1 & 0 & 1 \end{bmatrix} \rightarrow \begin{bmatrix} 1 & 0 & -3 & | & -2 & 3 & 0 \\ 0 & 1 & 2 & | & 1 & -1 & 0 \\ 0 & 0 & 7 & | & 3 & -4 & 1 \end{bmatrix}$

$\quad R_2 = -r_1 + r_2 \qquad\qquad R_2 = -r_2 \qquad\qquad R_1 = -3r_2 + r_1$
$\quad R_3 = -r_1 + r_3 \qquad\qquad\qquad\qquad\qquad\qquad R_3 = 4r_2 + r_3$

$\rightarrow \begin{bmatrix} 1 & 0 & -3 & | & -2 & 3 & 0 \\ 0 & 1 & 2 & | & 1 & -1 & 0 \\ 0 & 0 & 1 & | & \frac{3}{7} & -\frac{4}{7} & \frac{1}{7} \end{bmatrix} \rightarrow \begin{bmatrix} 1 & 0 & 0 & | & -\frac{5}{7} & \frac{9}{7} & \frac{3}{7} \\ 0 & 1 & 0 & | & \frac{1}{7} & \frac{1}{7} & -\frac{2}{7} \\ 0 & 0 & 1 & | & \frac{3}{7} & -\frac{4}{7} & \frac{1}{7} \end{bmatrix}$

$\quad\quad R_3 = \frac{1}{7}r_3 \qquad\qquad\qquad R_1 = 3r_3 + r_1$
$\qquad\qquad\qquad\qquad\qquad R_2 = -2r_3 + r_2$

$A^{-1} = \begin{bmatrix} -\frac{5}{7} & \frac{9}{7} & \frac{3}{7} \\ \frac{1}{7} & \frac{1}{7} & -\frac{2}{7} \\ \frac{3}{7} & -\frac{4}{7} & \frac{1}{7} \end{bmatrix}$

33. Augment the matrix with the identity and use row operations to find the inverse:

$$A = \begin{bmatrix} 4 & -8 \\ -1 & 2 \end{bmatrix} \rightarrow \begin{bmatrix} 4 & -8 & | & 1 & 0 \\ -1 & 2 & | & 0 & 1 \end{bmatrix}$$

$$\rightarrow \begin{bmatrix} -1 & 2 & | & 0 & 1 \\ 4 & -8 & | & 1 & 0 \end{bmatrix} \rightarrow \begin{bmatrix} -1 & 2 & | & 0 & 1 \\ 0 & 0 & | & 1 & 4 \end{bmatrix} \rightarrow \begin{bmatrix} 1 & -2 & | & 0 & -1 \\ 0 & 0 & | & 1 & 4 \end{bmatrix}$$

Interchange $\qquad R_2 = 4r_1 + r_2 \qquad R_1 = -r_1$

r_1 and r_2

There is no inverse because there is no way to obtain the identity on the left side. The matrix is singular.

35. $\begin{cases} 3x - 2y = 1 \\ 10x + 10y = 5 \end{cases}$ can be written as : $\begin{bmatrix} 3 & -2 & | & 1 \\ 10 & 10 & | & 5 \end{bmatrix}$

$$\rightarrow \begin{bmatrix} 3 & -2 & | & 1 \\ 1 & 16 & | & 2 \end{bmatrix} \rightarrow \begin{bmatrix} 1 & 16 & | & 2 \\ 3 & -2 & | & 1 \end{bmatrix} \rightarrow \begin{bmatrix} 1 & 16 & | & 2 \\ 0 & -50 & | & -5 \end{bmatrix} \rightarrow \begin{bmatrix} 1 & 16 & | & 2 \\ 0 & 1 & | & \frac{1}{10} \end{bmatrix} \rightarrow \begin{bmatrix} 1 & 0 & | & \frac{2}{5} \\ 0 & 1 & | & \frac{1}{10} \end{bmatrix}$$

$R_2 = -3r_1 + r_2$ Interchange $\quad R_2 = -3r_1 + r_2 \quad R_2 = -\frac{1}{50}r_2 \quad R_1 = -16r_2 + r_1$

r_1 and r_2

The solution is $x = \frac{2}{5}$, $y = \frac{1}{10}$.

37. $\begin{cases} 5x + 6y - 3z = 6 \\ 4x - 7y - 2z = -3 \\ 3x + y - 7z = 1 \end{cases}$ can be written as : $\begin{bmatrix} 5 & 6 & -3 & | & 6 \\ 4 & -7 & -2 & | & -3 \\ 3 & 1 & -7 & | & 1 \end{bmatrix}$

$$\rightarrow \begin{bmatrix} 1 & 13 & -1 & | & 9 \\ 4 & -7 & -2 & | & -3 \\ 3 & 1 & -7 & | & 1 \end{bmatrix} \rightarrow \begin{bmatrix} 1 & 13 & -1 & | & 9 \\ 0 & -59 & 2 & | & -39 \\ 0 & -38 & -4 & | & -26 \end{bmatrix} \rightarrow \begin{bmatrix} 1 & 13 & -1 & | & 9 \\ 0 & 1 & -\frac{2}{59} & | & \frac{39}{59} \\ 0 & -38 & -4 & | & -26 \end{bmatrix}$$

$R_1 = -r_2 + r_1 \qquad R_2 = -4r_1 + r_2 \qquad R_2 = -\frac{1}{59}r_2$

$R_3 = -3r_1 + r_3$

$$\rightarrow \begin{bmatrix} 1 & 0 & -\frac{33}{59} & | & \frac{24}{59} \\ 0 & 1 & -\frac{2}{59} & | & \frac{39}{59} \\ 0 & 0 & -\frac{312}{59} & | & -\frac{52}{59} \end{bmatrix} \rightarrow \begin{bmatrix} 1 & 0 & -\frac{33}{59} & | & \frac{24}{59} \\ 0 & 1 & -\frac{2}{59} & | & \frac{39}{59} \\ 0 & 0 & 1 & | & \frac{1}{6} \end{bmatrix} \rightarrow \begin{bmatrix} 1 & 0 & 0 & | & \frac{1}{2} \\ 0 & 1 & 0 & | & \frac{2}{3} \\ 0 & 0 & 1 & | & \frac{1}{6} \end{bmatrix}$$

$R_1 = -13r_2 + r_1 \qquad R_3 = -\frac{59}{312}r_3 \qquad R_1 = \frac{33}{59}r_3 + r_1$

$R_3 = 38r_2 + r_3 \qquad\qquad\qquad\qquad R_2 = \frac{2}{59}r_3 + r_2$

The solution is $x = \frac{1}{2}$, $y = \frac{2}{3}$, $z = \frac{1}{6}$.

39. $\begin{cases} x \quad\ \ -2z = 1 \\ 2x + 3y \quad\quad = -3 \\ 4x - 3y - 4z = 3 \end{cases}$ can be written as: $\begin{bmatrix} 1 & 0 & -2 & | & 1 \\ 2 & 3 & 0 & | & -3 \\ 4 & -3 & -4 & | & 3 \end{bmatrix}$

$\rightarrow \begin{bmatrix} 1 & 0 & -2 & | & 1 \\ 0 & 3 & 4 & | & -5 \\ 0 & -3 & 4 & | & -1 \end{bmatrix} \rightarrow \begin{bmatrix} 1 & 0 & -2 & | & 1 \\ 0 & 1 & \frac{4}{3} & | & -\frac{5}{3} \\ 0 & -3 & 4 & | & -1 \end{bmatrix} \rightarrow \begin{bmatrix} 1 & 0 & -2 & | & 1 \\ 0 & 1 & \frac{4}{3} & | & -\frac{5}{3} \\ 0 & 0 & 8 & | & -6 \end{bmatrix}$

$R_2 = -2r_1 + r_2 \qquad R_2 = \frac{1}{3}r_2 \qquad\qquad R_3 = 3r_2 + r_3$
$R_3 = -4r_1 + r_3$

$\rightarrow \begin{bmatrix} 1 & 0 & -2 & | & 1 \\ 0 & 1 & \frac{4}{3} & | & -\frac{5}{3} \\ 0 & 0 & 1 & | & -\frac{3}{4} \end{bmatrix} \rightarrow \begin{bmatrix} 1 & 0 & 0 & | & -\frac{1}{2} \\ 0 & 1 & 0 & | & -\frac{2}{3} \\ 0 & 0 & 1 & | & -\frac{3}{4} \end{bmatrix}$

$R_3 = \frac{1}{8}r_3 \qquad\qquad R_1 = 2r_3 + r_1$
$\qquad\qquad\qquad\qquad R_2 = -\frac{4}{3}r_3 + r_2$

The solution is $x = -\frac{1}{2}$, $y = -\frac{2}{3}$, $z = -\frac{3}{4}$.

41. $\begin{cases} x - y + z = 0 \\ x - y - 5z = 6 \\ 2x - 2y + z = 1 \end{cases}$ can be written as : $\begin{bmatrix} 1 & -1 & 1 & | & 0 \\ 1 & -1 & -5 & | & 6 \\ 2 & -2 & 1 & | & 1 \end{bmatrix}$

$\rightarrow \begin{bmatrix} 1 & -1 & 1 & | & 0 \\ 0 & 0 & -6 & | & 6 \\ 0 & 0 & -1 & | & 1 \end{bmatrix} \rightarrow \begin{bmatrix} 1 & -1 & 1 & | & 0 \\ 0 & 0 & 1 & | & -1 \\ 0 & 0 & -1 & | & 1 \end{bmatrix} \rightarrow \begin{bmatrix} 1 & -1 & 0 & | & 1 \\ 0 & 0 & 1 & | & -1 \\ 0 & 0 & 0 & | & 0 \end{bmatrix} \rightarrow \begin{cases} x = y + 1 \\ z = -1 \end{cases}$

$R_2 = -r_1 + r_2 \qquad R_2 = -\frac{1}{6}r_2 \qquad R_1 = -r_2 + r_1$
$R_3 = -2r_1 + r_3 \qquad\qquad\qquad\quad R_3 = r_2 + r_3$

The solution is $x = y + 1$, $z = -1$, y is any real number..

43. $\begin{cases} x - y - z - t = 1 \\ 2x + y + z + 2t = 3 \\ x - 2y - 2z - 3t = 0 \\ 3x - 4y + z + 5t = -3 \end{cases}$ can be written as: $\begin{bmatrix} 1 & -1 & -1 & -1 & | & 1 \\ 2 & 1 & 1 & 2 & | & 3 \\ 1 & -2 & -2 & -3 & | & 0 \\ 3 & -4 & 1 & 5 & | & -3 \end{bmatrix}$

$\rightarrow \begin{bmatrix} 1 & -1 & -1 & -1 & | & 1 \\ 0 & 3 & 3 & 4 & | & 1 \\ 0 & -1 & -1 & -2 & | & -1 \\ 0 & -1 & 4 & 8 & | & -6 \end{bmatrix} \rightarrow \begin{bmatrix} 1 & -1 & -1 & -1 & | & 1 \\ 0 & -1 & -1 & -2 & | & -1 \\ 0 & 3 & 3 & 4 & | & 1 \\ 0 & -1 & 4 & 8 & | & -6 \end{bmatrix} \rightarrow \begin{bmatrix} 1 & -1 & -1 & -1 & | & 1 \\ 0 & 1 & 1 & 2 & | & 1 \\ 0 & 3 & 3 & 4 & | & 1 \\ 0 & -1 & 4 & 8 & | & -6 \end{bmatrix}$

$R_2 = -2r_1 + r_2 \qquad\qquad$ Interchange r_2 and $r_3 \qquad R_2 = -r_2$
$R_3 = -r_1 + r_3$
$R_4 = -3r_1 + r_4$

$$\rightarrow \begin{bmatrix} 1 & 0 & 0 & 1 & | & 2 \\ 0 & 1 & 1 & 2 & | & 1 \\ 0 & 0 & 0 & -2 & | & -2 \\ 0 & 0 & 5 & 10 & | & -5 \end{bmatrix} \rightarrow \begin{bmatrix} 1 & 0 & 0 & 1 & | & 2 \\ 0 & 1 & 1 & 2 & | & 1 \\ 0 & 0 & 0 & 1 & | & 1 \\ 0 & 0 & 1 & 2 & | & -1 \end{bmatrix} \rightarrow \begin{bmatrix} 1 & 0 & 0 & 1 & | & 2 \\ 0 & 1 & 1 & 2 & | & 1 \\ 0 & 0 & 1 & 2 & | & -1 \\ 0 & 0 & 0 & 1 & | & 1 \end{bmatrix}$$

$$R_1 = r_2 + r_1 \qquad R_3 = -\tfrac{1}{2}r_3 \qquad \text{Interchange } r_3 \text{ and } r_4$$
$$R_3 = -3r_2 + r_3 \qquad R_4 = \tfrac{1}{5}r_4$$
$$R_4 = r_2 + r_4$$

$$\rightarrow \begin{bmatrix} 1 & 0 & 0 & 1 & | & 2 \\ 0 & 1 & 0 & 0 & | & 2 \\ 0 & 0 & 1 & 2 & | & -1 \\ 0 & 0 & 0 & 1 & | & 1 \end{bmatrix} \rightarrow \begin{bmatrix} 1 & 0 & 0 & 0 & | & 1 \\ 0 & 1 & 0 & 0 & | & 2 \\ 0 & 0 & 1 & 0 & | & -3 \\ 0 & 0 & 0 & 1 & | & 1 \end{bmatrix}$$

$$R_2 = -r_3 + r_2 \qquad R_1 = -r_4 + r_1$$
$$R_3 = -2r_4 + r_3$$

The solution is $x = 1$, $y = 2$, $z = -3$, $t = 1$.

45. Evaluating the determinant:

$$\begin{vmatrix} 3 & 4 \\ 1 & 3 \end{vmatrix} = 3(3) - 4(1) = 9 - 4 = 5$$

47. Evaluating the determinant:

$$\begin{vmatrix} 1 & 4 & 0 \\ -1 & 2 & 6 \\ 4 & 1 & 3 \end{vmatrix} = 1\begin{vmatrix} 2 & 6 \\ 1 & 3 \end{vmatrix} - 4\begin{vmatrix} -1 & 6 \\ 4 & 3 \end{vmatrix} + 0\begin{vmatrix} -1 & 2 \\ 4 & 1 \end{vmatrix}$$

$$= 1[2(3) - 6(1)] - 4[-1(3) - 6(4)] + 0[-1(1) - 2(4)]$$
$$= 1(6 - 6) - 4(-3 - 24) + 0(-1 - 8)$$
$$= 1(0) - 4(-27) + 0(-9)$$
$$= 0 + 108 + 0$$
$$= 108$$

49. Evaluating the determinant:

$$\begin{vmatrix} 2 & 1 & -3 \\ 5 & 0 & 1 \\ 2 & 6 & 0 \end{vmatrix} = 2\begin{vmatrix} 0 & 1 \\ 6 & 0 \end{vmatrix} - 1\begin{vmatrix} 5 & 1 \\ 2 & 0 \end{vmatrix} + (-3)\begin{vmatrix} 5 & 0 \\ 2 & 6 \end{vmatrix}$$

$$= 2(0 - 6) - 1(0 - 2) + (-3)(30 - 0)$$
$$= -12 + 2 - 90$$
$$= -100$$

51. Set up and evaluate the determinants to use Cramer's Rule:

$$\begin{cases} x - 2y = 4 \\ 3x + 2y = 4 \end{cases}$$

$$D = \begin{vmatrix} 1 & -2 \\ 3 & 2 \end{vmatrix} = 1(2) - 3(-2) = 2 + 6 = 8$$

$$D_x = \begin{vmatrix} 4 & -2 \\ 4 & 2 \end{vmatrix} = 4(2) - 4(-2) = 8 + 8 = 16$$

$$D_y = \begin{vmatrix} 1 & 4 \\ 3 & 4 \end{vmatrix} = 1(4) - 4(3) = 4 - 12 = -8$$

Find the solutions by Cramer's Rule:

$$x = \frac{D_x}{D} = \frac{16}{8} = 2 \qquad y = \frac{D_y}{D} = \frac{-8}{8} = -1$$

53. Set up and evaluate the determinants to use Cramer's Rule:

$$\begin{cases} 2x + 3y = 13 \\ 3x - 2y = 0 \end{cases}$$

$$D = \begin{vmatrix} 2 & 3 \\ 3 & -2 \end{vmatrix} = -4 - 9 = -13$$

$$D_x = \begin{vmatrix} 13 & 3 \\ 0 & -2 \end{vmatrix} = -26 - 0 = -26$$

$$D_y = \begin{vmatrix} 2 & 13 \\ 3 & 0 \end{vmatrix} = 0 - 39 = -39$$

Find the solutions by Cramer's Rule:

$$x = \frac{D_x}{D} = \frac{-26}{-13} = 2 \qquad y = \frac{D_y}{D} = \frac{-39}{-13} = 3$$

55. Set up and evaluate the determinants to use Cramer's Rule:

$$\begin{cases} x + 2y - z = 6 \\ 2x - y + 3z = -13 \\ 3x - 2y + 3z = -16 \end{cases}$$

$$D = \begin{vmatrix} 1 & 2 & -1 \\ 2 & -1 & 3 \\ 3 & -2 & 3 \end{vmatrix} = 1 \begin{vmatrix} -1 & 3 \\ -2 & 3 \end{vmatrix} - 2 \begin{vmatrix} 2 & 3 \\ 3 & 3 \end{vmatrix} + (-1) \begin{vmatrix} 2 & -1 \\ 3 & -2 \end{vmatrix}$$

$$= 1(-3 + 6) - 2(6 - 9) - 1(-4 + 3) = 3 + 6 + 1 = 10$$

$$D_x = \begin{vmatrix} 6 & 2 & -1 \\ -13 & -1 & 3 \\ -16 & -2 & 3 \end{vmatrix} = 6 \begin{vmatrix} -1 & 3 \\ -2 & 3 \end{vmatrix} - 2 \begin{vmatrix} -13 & 3 \\ -16 & 3 \end{vmatrix} + (-1) \begin{vmatrix} -13 & -1 \\ -16 & -2 \end{vmatrix}$$

$$= 6(-3 + 6) - 2(-39 + 48) - 1(26 - 16) = 18 - 18 - 10 = -10$$

$$D_y = \begin{vmatrix} 1 & 6 & -1 \\ 2 & -13 & 3 \\ 3 & -16 & 3 \end{vmatrix} = 1 \begin{vmatrix} -13 & 3 \\ -16 & 3 \end{vmatrix} - 6 \begin{vmatrix} 2 & 3 \\ 3 & 3 \end{vmatrix} + (-1) \begin{vmatrix} 2 & -13 \\ 3 & -16 \end{vmatrix}$$

$$= 1(-39 + 48) - 6(6 - 9) - 1(-32 + 39) = 9 + 18 - 7 = 20$$

$$D_z = \begin{vmatrix} 1 & 2 & 6 \\ 2 & -1 & -13 \\ 3 & -2 & -16 \end{vmatrix} = 1\begin{vmatrix} -1 & -13 \\ -2 & -16 \end{vmatrix} - 2\begin{vmatrix} 2 & -13 \\ 3 & -16 \end{vmatrix} + 6\begin{vmatrix} 2 & -1 \\ 3 & -2 \end{vmatrix}$$

$$= 1(16 - 26) - 2(-32 + 39) + 6(-4 + 3) = -10 - 14 - 6 = -30$$

Find the solutions by Cramer's Rule:

$$x = \frac{D_x}{D} = \frac{-10}{10} = -1 \qquad y = \frac{D_y}{D} = \frac{20}{10} = 2 \qquad z = \frac{D_z}{D} = \frac{-30}{10} = -3$$

57. Find the partial fraction decomposition:

$$\frac{6}{x(x - 4)} = \frac{A}{x} + \frac{B}{x - 4}$$

$$6 = A(x - 4) + Bx \qquad \text{(Multiply both sides by } x(x - 4).)$$

Let $x = 4$: then $6 = A(4 - 4) + B(4) \;\rightarrow\; 4B = 6 \;\rightarrow\; B = \frac{3}{2}$

Let $x = 0$: then $6 = A(0 - 4) + B(0) \;\rightarrow\; -4A = 6 \;\rightarrow\; A = -\frac{3}{2}$

$$\frac{6}{x(x - 4)} = \frac{-\frac{3}{2}}{x} + \frac{\frac{3}{2}}{x - 4}$$

59. Find the partial fraction decomposition:

$$\frac{x - 4}{x^2(x - 1)} = \frac{A}{x} + \frac{B}{x^2} + \frac{C}{x - 1}$$

$$\text{(Multiply both sides by } x^2(x - 1).)$$

$$x - 4 = Ax(x - 1) + B(x - 1) + Cx^2$$

Let $x = 1$: then $1 - 4 = A(1)(1 - 1) + B(1 - 1) + C(1)^2$

$$\rightarrow \; -3 = C \;\rightarrow\; C = -3$$

Let $x = 0$: then $0 - 4 = A(0)(0 - 1) + B(0 - 1) + C(0)^2$

$$\rightarrow \; -4 = -B \;\rightarrow\; B = 4$$

Let $x = 2$: then $2 - 4 = A(2)(2 - 1) + B(2 - 1) + C(2)^2 \;\rightarrow\; -2 = 2A + B + 4C$

$$\rightarrow \; 2A = -2 - 4 - 4(-3) \;\rightarrow\; 2A = 6 \;\rightarrow\; A = 3$$

$$\frac{x - 4}{x^2(x - 1)} = \frac{3}{x} + \frac{4}{x^2} + \frac{-3}{x - 1}$$

61. Find the partial fraction decomposition:

$$\frac{x}{(x^2 + 9)(x + 1)} = \frac{A}{x + 1} + \frac{Bx + C}{x^2 + 9}$$

$$\text{(Multiply both sides by } (x + 1)(x^2 + 9).)$$

$$x = A(x^2 + 9) + (Bx + C)(x + 1)$$

Let $x = -1$: then $-1 = A((-1)^2 + 9) + (B(-1) + C)(-1 + 1)$

$$\rightarrow \; -1 = 10A \;\rightarrow\; A = -\frac{1}{10}$$

Let $x = 1$: then $1 = A(1^2 + 9) + (B(1) + C)(1 + 1) \;\rightarrow\; 1 = 10A + 2B + 2C$

$$\rightarrow \; 1 = 10\left(-\frac{1}{10}\right) + 2B + 2C \;\rightarrow\; 2 = 2B + 2C \;\rightarrow\; B + C = 1$$

Let $x = 0$: then $0 = A(0^2 + 9) + (B(0) + C)(0 + 1) \rightarrow 0 = 9A + C$

$\rightarrow 0 = 9\left(-\frac{1}{10}\right) + C \rightarrow C = \frac{9}{10}$

$B = 1 - C \rightarrow B = 1 - \frac{9}{10} \rightarrow B = \frac{1}{10}$

$$\frac{x}{(x^2 + 9)(x + 1)} = \frac{-\frac{1}{10}}{x + 1} + \frac{\frac{1}{10}x + \frac{9}{10}}{x^2 + 9}$$

63. Find the partial fraction decomposition:

$$\frac{x^3}{(x^2 + 4)^2} = \frac{Ax + B}{x^2 + 4} + \frac{Cx + D}{(x^2 + 4)^2}$$

(Multiply both sides by $(x^2 + 4)^2$.)

$x^3 = (Ax + B)(x^2 + 4) + Cx + D$

$x^3 = Ax^3 + Bx^2 + 4Ax + 4B + Cx + D$

$x^3 = Ax^3 + Bx^2 + (4A + C)x + 4B + D$

$A = 1$

$B = 0$

$4A + C = 0 \rightarrow 4(1) + C = 0 \rightarrow C = -4$

$4B + D = 0 \rightarrow 4(0) + D = 0 \rightarrow D = 0$

$$\frac{x^3}{(x^2 + 4)^2} = \frac{x}{x^2 + 4} + \frac{-4x}{(x^2 + 4)^2}$$

65. Find the partial fraction decomposition:

$$\frac{x^2}{(x^2 + 1)(x^2 - 1)} = \frac{x^2}{(x^2 + 1)(x - 1)(x + 1)} = \frac{A}{x - 1} + \frac{B}{x + 1} + \frac{Cx + D}{x^2 + 1}$$

(Multiply both sides by $(x - 1)(x + 1)(x^2 + 1)$.)

$x^2 = A(x + 1)(x^2 + 1) + B(x - 1)(x^2 + 1) + (Cx + D)(x - 1)(x + 1)$

Let $x = 1$: then $1^2 = A(1 + 1)(1^2 + 1) + B(1 - 1)(1^2 + 1) + (C(1) + D)(1 - 1)(1 + 1)$

$\rightarrow 1 = 4A \rightarrow A = \frac{1}{4}$

Let $x = -1$: then

$(-1)^2 = A(-1 + 1)((-1)^2 + 1) + B(-1 - 1)((-1)^2 + 1) + (C(-1)$
$\qquad\qquad\qquad\qquad\qquad\qquad\qquad\qquad + D)(-1 - 1)(-1 + 1)$

$\rightarrow 1 = -4B \rightarrow B = -\frac{1}{4}$

Let $x = 0$: then

$0^2 = A(0 + 1)(0^2 + 1) + B(0 - 1)(0^2 + 1) + (C(0) + D)(0 - 1)(0 + 1)$

$\rightarrow 0 = A - B - D \rightarrow 0 = \frac{1}{4} - \left(-\frac{1}{4}\right) - D \rightarrow D = \frac{1}{2}$

Let $x = 2$: then

$2^2 = A(2 + 1)(2^2 + 1) + B(2 - 1)(2^2 + 1) + (C(2) + D)(2 - 1)(2 + 1)$

$\rightarrow 4 = 15A + 5B + 6C + 3D \rightarrow 4 = 15\left(\frac{1}{4}\right) + 5\left(-\frac{1}{4}\right) + 6C + 3\left(\frac{1}{2}\right)$

$\rightarrow 6C = 4 - \frac{15}{4} + \frac{5}{4} - \frac{3}{2} \rightarrow 6C = 0 \rightarrow C = 0$

$$\frac{x^2}{(x^2 + 1)(x^2 - 1)} = \frac{x^2}{(x^2 + 1)(x - 1)(x + 1)} = \frac{\frac{1}{4}}{x - 1} + \frac{-\frac{1}{4}}{x + 1} + \frac{\frac{1}{2}}{x^2 + 1}$$

67. Solve the first equation for y, substitute into the second equation and solve:

$$\begin{cases} 2x + y + 3 = 0 & \rightarrow \ y = -2x - 3 \\ \quad x^2 + y^2 = 5 \end{cases}$$

$$x^2 + (-2x - 3)^2 = 5$$
$$x^2 + 4x^2 + 12x + 9 = 5$$
$$5x^2 + 12x + 4 = 0$$
$$(5x + 2)(x + 2) = 0$$
$$x = -\frac{2}{5} \quad \text{or} \ x = -2$$
$$y = -\frac{11}{5} \qquad y = 1$$

Solutions: $\left(-\frac{2}{5}, -\frac{11}{5}\right), (-2, 1)$.

69. Multiply each side of the second equation by 2 and add the equations to eliminate xy:

$$\begin{cases} 2xy + \ y^2 = 10 \ \longrightarrow \qquad 2xy + \ y^2 = 10 \\ -xy + 3y^2 = \ 2 \ \xrightarrow{\ 2\ } \ \underline{-2xy + 6y^2 = \ \ 4} \\ \qquad\qquad\qquad\qquad\qquad 7y^2 = 14 \\ \qquad\qquad\qquad\qquad\qquad \ \ y^2 = 2 \\ \qquad\qquad\qquad\qquad\qquad \ \ y = \pm\sqrt{2} \end{cases}$$

If $y = \sqrt{2}$: $\quad 2x\left(\sqrt{2}\right) + \left(\sqrt{2}\right)^2 = 10 \ \rightarrow \ 2\sqrt{2}x = 8 \ \rightarrow \ x = \dfrac{8}{2\sqrt{2}} = 2\sqrt{2}$

If $y = -\sqrt{2}$: $\quad 2x\left(-\sqrt{2}\right) + \left(-\sqrt{2}\right)^2 = 10 \ \rightarrow \ -2\sqrt{2}x = 8 \ \rightarrow \ x = \dfrac{8}{-2\sqrt{2}} = -2\sqrt{2}$

Solutions: $\left(2\sqrt{2}, \sqrt{2}\right), \left(-2\sqrt{2}, -\sqrt{2}\right)$

71. Substitute into the second equation into the first equation and solve:

$$\begin{cases} x^2 + y^2 = 6y \\ \quad x^2 = 3y \end{cases}$$

$$3y + y^2 = 6y$$
$$y^2 - 3y = 0$$
$$y(y - 3) = 0$$
$$y = 0 \ \text{ or } \ y = 3$$

If $y = 0$: $\qquad x^2 = 3(0) \ \rightarrow \ x^2 = 0 \ \rightarrow \ x = 0$

If $y = 3$: $\qquad x^2 = 3(3) \ \rightarrow \ x^2 = 9 \ \rightarrow \ x = \pm3$

Solutions: $(0, 0), (-3, 3), (3, 3)$

73. Factor the second equation, solve for x, substitute into the first equation and solve:
$$\begin{cases} 3x^2 + 4xy + 5y^2 = 8 \\ x^2 + 3xy + 2y^2 = 0 \end{cases} \rightarrow (x+2y)(x+y) = 0 \rightarrow x = -2y \text{ or } x = -y$$

Substitute $x = -2y$ and solve:

$$3x^2 + 4xy + 5y^2 = 8$$
$$3(-2y)^2 + 4(-2y)y + 5y^2 = 8$$
$$12y^2 - 8y^2 + 5y^2 = 8$$
$$9y^2 = 8$$
$$y^2 = \frac{8}{9}$$
$$y = \pm\frac{2\sqrt{2}}{3}$$

If $y = \frac{2\sqrt{2}}{3}$: $x = -2\left(\frac{2\sqrt{2}}{3}\right) = \frac{-4\sqrt{2}}{3}$

If $y = \frac{-2\sqrt{2}}{3}$: $x = -2\left(\frac{-2\sqrt{2}}{3}\right) = \frac{4\sqrt{2}}{3}$

Substitute $x = -y$ and solve:

$$3x^2 + 4xy + 5y^2 = 8$$
$$3(-y)^2 + 4(-y)y + 5y^2 = 8$$
$$3y^2 - 4y^2 + 5y^2 = 8$$
$$4y^2 = 8$$
$$y^2 = 2$$
$$y = \pm\sqrt{2}$$

If $y = \sqrt{2}$: $x = -\sqrt{2}$
If $y = -\sqrt{2}$: $x = \sqrt{2}$

Solutions: $\left(\frac{-4\sqrt{2}}{3}, \frac{2\sqrt{2}}{3}\right), \left(\frac{4\sqrt{2}}{3}, \frac{-2\sqrt{2}}{3}\right), \left(-\sqrt{2}, \sqrt{2}\right), \left(\sqrt{2}, -\sqrt{2}\right)$

75. Multiply each side of the second equation by $-y$ and add the equations to eliminate y:
$$\begin{cases} x^2 - 3x + y^2 + y = -2 \longrightarrow \quad x^2 - 3x + y^2 + y = -2 \\ \dfrac{x^2 - x}{y} + y + 1 = 0 \xrightarrow{-y} \quad -x^2 + x - y^2 - y = 0 \end{cases}$$
$$\begin{aligned} -2x \quad\quad\quad &= -2 \\ x &= 1 \end{aligned}$$

If $x = 1$: $1^2 - 3(1) + y^2 + y = -2 \rightarrow y^2 + y = 0 \rightarrow y(y+1) = 0$
$\rightarrow y = 0$ or $y = -1$

Note that $y \neq 0$ because that would cause division by zero in the original equation.
Solution: $(1, -1)$

77. Graph the system of linear inequalities:
$$\begin{cases} -2x + y \le 2 \\ \quad x + y \ge 2 \end{cases}$$

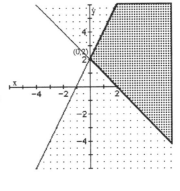

(a) Graph the line $-2x + y = 2$. Use a solid line since the inequality uses $\le$.
Choose a test point not on the line, such as $(0, 0)$. Since $-2(0) + 0 \le 2$ is true, shade the side of the line containing $(0, 0)$.

(b) Graph the line $x + y = 2$. Use a solid line since the inequality uses $\ge$.
Choose a test point not on the line, such as $(0, 0)$. Since $0 + 0 \ge 2$ is false, shade the opposite side of the line from $(0, 0)$.

(c) The overlapping region is the solution.

(d) The graph is unbounded.

(e) Find the vertices:
To find the intersection of $x + y = 2$ and $-2x + y = 2$, solve the system:
$$\begin{cases} \quad x + y = 2 \quad \rightarrow \quad x = 2 - y \\ -2x + y = 2 \end{cases}$$
Substitute and solve:
$$-2(2 - y) + y = 2$$
$$-4 + 2y + y = 2$$
$$3y = 6$$
$$y = 2$$
$$x = 2 - 2 = 0$$
The point of intersection is $(0, 2)$.
The corner point is $(0, 2)$.

79. Graph the system of linear inequalities:
$$\begin{cases} \quad\quad x \ge 0 \\ \quad\quad y \ge 0 \\ \quad x + \;\; y \le 4 \\ 2x + 3y \le 6 \end{cases}$$

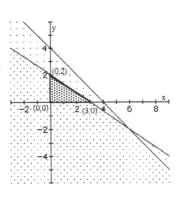

(a) Graph $x \ge 0$; $y \ge 0$. Shaded region is the first quadrant.

(b) Graph the line $x + y = 4$. Use a solid line since the inequality uses $\le$.
Choose a test point not on the line, such as $(0, 0)$. Since $0 + 0 \le 4$ is true, shade the side of the line containing $(0, 0)$.

(c) Graph the line $2x + 3y = 6$. Use a solid line since the inequality uses $\le$.
Choose a test point not on the line, such as $(0, 0)$. Since $2(0) + 3(0) \le 6$ is true, shade the side of the line containing $(0, 0)$.

(d) The overlapping region is the solution.

(e)   The graph is bounded.
(f)   Find the vertices:
       The x-axis and y-axis intersect at (0, 0).
       The intersection of $2x + 3y = 6$ and the y-axis is (0, 2).
       The intersection of $2x + 3y = 6$ and the x-axis is (3, 0).
       The three corner points are (0, 0), (0, 2), and (3, 0).

81.   Graph the system of linear inequalities:
$$\begin{cases} x \geq 0 \\ y \geq 0 \\ 2x + y \leq 8 \\ x + 2y \geq 2 \end{cases}$$

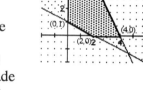

(a)   Graph $x \geq 0; y \geq 0$. Shaded region is the first
       quadrant.
(b)   Graph the line $2x + y = 8$. Use a solid line
       since the inequality uses $\leq$.
       Choose a test point not on the line, such as
       (0, 0). Since $2(0) + 0 \leq 8$ is true, shade the
       side of the line containing (0, 0).
(c)   Graph the line $x + 2y = 2$. Use a solid line
       since the inequality uses $\geq$.
       Choose a test point not on the line, such as
       (0, 0). Since $0 + 2(0) \geq 2$ is false, shade
       the opposite side of the line from (0, 0).
(d)   The overlapping region is the solution.
(e)   The graph is bounded.
(f)   Find the vertices:
       The intersection of $x + 2y = 2$ and the y-axis is (0, 1).
       The intersection of $x + 2y = 2$ and the x-axis is (2, 0).
       The intersection of $2x + y = 8$ and the y-axis is (0, 8).
       The intersection of $2x + y = 8$ and the x-axis is (4, 0).
       The four corner points are (0, 1), (0, 8), (2, 0), and (4, 0).

83.   Maximize $z = 3x + 4y$ Subject to $x \geq 0, \ y \geq 0, \ 3x + 2y \geq 6, \ x + y \leq 8$
      Graph the constraints.

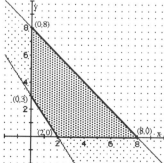

      The corner points are (0, 3), (2, 0), (0, 8), (8, 0).

Evaluate the objective function:

Vertex	Value of $z = 3x + 4y$
(0, 3)	$z = 3(0) + 4(3) = 12$
(0, 8)	$z = 3(0) + 4(8) = 32$
(2, 0)	$z = 3(2) + 4(0) = 6$
(8, 0)	$z = 3(8) + 4(0) = 24$

The maximum value is 32 at (0, 8).

85. Minimize $z = 3x + 5y$

Subject to $x \geq 0$, $y \geq 0$, $x + y \geq 1$, $3x + 2y \leq 12$, $x + 3y \leq 12$
Graph the constraints.

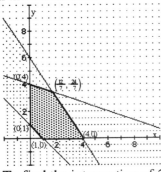

To find the intersection of $3x + 2y = 12$ and $x + 3y = 12$, solve the system:

$$\begin{cases} 3x + 2y = 12 \\ x + 3y = 12 \end{cases} \rightarrow \quad x = 12 - 3y$$

Substitute and solve:

$$3(12 - 3y) + 2y = 12$$
$$36 - 9y + 2y = 12$$
$$-7y = -24$$
$$y = \tfrac{24}{7}$$
$$x = 12 - 3\left(\tfrac{24}{7}\right) = 12 - \tfrac{72}{7} = \tfrac{12}{7}$$

The point of intersection is $\left(\tfrac{12}{7}, \tfrac{24}{7}\right)$.

The corner points are (0, 1), (1, 0), (0, 4), (4, 0), $\left(\tfrac{12}{7}, \tfrac{24}{7}\right)$.
Evaluate the objective function:

Vertex	Value of $z = 3x + 5y$
(0, 1)	$z = 3(0) + 5(1) = 5$
(0, 4)	$z = 3(0) + 5(4) = 20$
(1, 0)	$z = 3(1) + 5(0) = 3$
(4, 0)	$z = 3(4) + 5(0) = 12$
$\left(\tfrac{12}{7}, \tfrac{24}{7}\right)$	$z = 3\left(\tfrac{12}{7}\right) + 5\left(\tfrac{24}{7}\right) = \tfrac{36}{7} + \tfrac{120}{7} = \tfrac{156}{7} \approx 22.3$

The minimum value is 3 at (1, 0).

87. Maximize $z = 5x + 4y$ Subject to $x \geq 0, \ y \geq 0, \ x + 2y \geq 2, \ 3x + 4y \leq 12, \ y \geq x$
Graph the constraints.

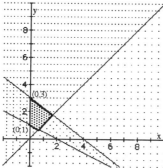

To find the intersection of $x + 2y = 2$ and $y = x$, substitute and solve:

$$x + 2x = 2$$
$$3x = 2$$
$$x = \tfrac{2}{3}$$
$$y = \tfrac{2}{3}$$

The point of intersection is $\left(\tfrac{2}{3}, \tfrac{2}{3}\right)$.

To find the intersection of $y = x$ and $3x + 4y = 12$, substitute and solve:

$$3x + 4x = 12$$
$$7x = 12$$
$$x = \tfrac{12}{7}$$
$$y = \tfrac{12}{7}$$

The point of intersection is $\left(\tfrac{12}{7}, \tfrac{12}{7}\right)$.

The corner points are $(0, 1), (0, 3), \left(\tfrac{2}{3}, \tfrac{2}{3}\right), \left(\tfrac{12}{7}, \tfrac{12}{7}\right)$.
Evaluate the objective function:

Vertex	Value of $z = 5x + 4y$
$(0, 1)$	$z = 5(0) + 4(1) = 4$
$(0, 3)$	$z = 5(0) + 4(3) = 12$
$\left(\tfrac{2}{3}, \tfrac{2}{3}\right)$	$z = 5\left(\tfrac{2}{3}\right) + 4\left(\tfrac{2}{3}\right) = \tfrac{18}{3} = 6$
$\left(\tfrac{12}{7}, \tfrac{12}{7}\right)$	$z = 5\left(\tfrac{12}{7}\right) + 4\left(\tfrac{12}{7}\right) = \tfrac{108}{7} \approx 15.43$

The maximum value is $\tfrac{108}{7}$ at $\left(\tfrac{12}{7}, \tfrac{12}{7}\right)$.

89. Multiply each side of the first equation by -2 and eliminate x:

$$\begin{cases} 2x + \ 5y = 5 \ \ \xrightarrow{\ -2\ } \ \ -4x - 10y = -10 \\ 4x + 10y = A \ \ \xrightarrow{\quad} \quad \underline{\ \ 4x + 10y = \quad A\ } \\ \qquad\qquad\qquad\qquad\qquad\qquad 0 = A - 10 \end{cases}$$

If there are to be infinitely many solutions, the sum in elimination should be $0 = 0$.
Therefore, $A - 10 = 0$ or $A = 10$.

91. $y = ax^2 + bx + c$
At $(0, 1)$ the equation becomes:
$$1 = a(0)^2 + b(0) + c$$
$$c = 1$$
At $(1, 0)$ the equation becomes:
$$0 = a(1)^2 + b(1) + c$$
$$0 = a + b + c$$
$$a + b + c = 0$$
At $(-2, 1)$ the equation becomes:
$$1 = a(-2)^2 + b(-2) + c$$
$$1 = 4a - 2b + c$$
$$4a - 2b + c = 1$$
The system of equations is:
$$\begin{cases} a + b + c = 0 \\ 4a - 2b + c = 1 \\ \qquad\qquad c = 1 \end{cases}$$
Substitute $c = 1$ into the first and second equations and simplify:
$$\begin{cases} a + b + 1 = 0 \\ 4a - 2b + 1 = 1 \end{cases} \begin{array}{l} \rightarrow \\ \rightarrow \end{array} \begin{array}{l} a + b = -1 \\ 4a - 2b = 0 \end{array} \rightarrow a = -b - 1$$
Solve the first equation for a, substitute into the second equation and solve:
$$4(-b - 1) - 2b = 0$$
$$-4b - 4 - 2b = 0$$
$$-6b = 4$$
$$b = -\tfrac{2}{3}$$
$$a = \tfrac{2}{3} - 1 = -\tfrac{1}{3}$$
The quadratic function is $y = -\tfrac{1}{3}x^2 - \tfrac{2}{3}x + 1$.

93. Let x = the number of pounds of coffee that costs \$3.00 per pound.
Let y = the number of pounds of coffee that costs \$6.00 per pound.
Then $x + y = 100$ represents the total amount of coffee in the blend.
The value of the blend will be represented by the equation: $3x + 6y = 3.90(100)$.
Solve the system of equations:
$$\begin{cases} x + y = 100 \\ 3x + 6y = 390 \end{cases} \rightarrow y = 100 - x$$
Solve by substitution:
$$3x + 6(100 - x) = 390$$
$$3x + 600 - 6x = 390$$
$$-3x = -210$$
$$x = 70$$
$$y = 100 - 70 = 30$$
The blend is made up of 70 pounds of the \$3 per pound coffee and 30 pounds of the \$6 per pound coffee.

95. Let x = the number of small boxes.
Let y = the number of medium boxes.
Let z = the number of large boxes.

Oatmeal raisin equation: $x + 2y + 2z = 15$
Chocolate chip equation: $x + y + 2z = 10$
Shortbread equation: $y + 3z = 11$

Multiply each side of the second equation by −1 and add to the first equation to eliminate x:

$$\begin{cases} x + 2y + 2z = 15 \\ x + y + 2z = 10 \\ y + 3z = 11 \end{cases} \xrightarrow{} \begin{array}{r} x + 2y + 2z = 15 \\ \xrightarrow{-1} \quad -x - y - 2z = -10 \\ \hline y = 5 \end{array}$$

Substituting and solving for the other variables:

$$\begin{array}{ll} 5 + 3z = 11 & x + 5 + 2(2) = 10 \\ 3z = 6 & x + 9 = 10 \\ z = 2 & x = 1 \end{array}$$

1 small box, 5 medium boxes, and 2 large boxes of cookies should be purchased.

97. Let x = the length of the lot.
Let y = the width of the lot.

Perimeter equation: $2x + 2y = 68$
Diagonal equation: $x^2 + y^2 = 26^2$

Solve the system of equations:

$$\begin{cases} 2x + 2y = 68 \quad \rightarrow \quad y = 34 - x \\ x^2 + y^2 = 676 \end{cases}$$

Solve by substitution:

$$x^2 + (34 - x)^2 = 676$$
$$x^2 + 1156 - 68x + x^2 = 676$$
$$2x^2 - 68x + 480 = 0$$
$$x^2 - 34x + 240 = 0$$
$$(x - 24)(x - 10) = 0$$
$$x = 24 \ \text{ or } \ x = 10$$
$$y = 10 \ \text{ or } \ y = 24$$

The dimensions of the lot are 24 feet by 10 feet.

99. Let x = the length of one leg.
    Let y = the length of the other leg.
    Perimeter equation: $\qquad x + y + 6 = 14$
    Pythagorean equation: $\quad x^2 + y^2 = 6^2$
    Solve the system of equations:
    $$\begin{cases} x + y = 8 \quad \rightarrow \quad y = 8 - x \\ x^2 + y^2 = 36 \end{cases}$$
    Solve by substitution:
    $$x^2 + (8 - x)^2 = 36$$
    $$x^2 + 64 - 16x + x^2 = 36$$
    $$2x^2 - 16x + 28 = 0$$
    $$x^2 - 8x + 14 = 0$$
    $$x = \frac{8 \pm \sqrt{64 - 56}}{2} = \frac{8 \pm \sqrt{8}}{2} = \frac{8 \pm 2\sqrt{2}}{2} = 4 \pm \sqrt{2}$$
    If $x = 4 - \sqrt{2}$, then $y = 8 - 4 + \sqrt{2} = 4 + \sqrt{2}$
    If $x = 4 + \sqrt{2}$, then $y = 8 - 4 - \sqrt{2} = 4 - \sqrt{2}$
    The legs are $4 + \sqrt{2}$ and $4 - \sqrt{2}$.

101. Let x = the length of the side of the smaller square.
    Then $2x$ = the length of the side of the larger square.
    The needed fencing is $4x + 8x = 12x$.
    Solve the area equation:
    $$x^2 + (2x)^2 = 5000$$
    $$5x^2 = 5000$$
    $$x^2 = 1000$$
    $$x = 10\sqrt{10}$$
    $$12x = 120\sqrt{10} \approx 379.5 \text{ feet of fence are needed}.$$

103. Let x = the amount Katy receives.
    Let y = the amount Mike receives.
    Let z = the amount Danny receives.
    Let w = the amount that Colleen receives.
    Conditions:
    $$x + y + z + w = 45$$
    $$y = 2x$$
    $$w = x$$
    $$z = \tfrac{1}{2}x$$
    Solve by substitution:
    $$x + y + z + w = 45$$
    $$x + 2x + x + \tfrac{1}{2}x = 45$$
    $$\tfrac{9}{2}x = 45$$
    $$x = 10$$
    Katy receives $10, Mike receives $20, Danny receives $5, and Colleen receives $10.

105. Let x = the number of hours for Bruce to do the job alone.
Let y = the number of hours for Bryce to do the job alone.
Let z = the number of hours for Marty to do the job alone.

Then $\dfrac{1}{x}$ represents the fraction of the job that Bruce does in one hour.

$\dfrac{1}{y}$ represents the fraction of the job that Bryce does in one hour.

$\dfrac{1}{z}$ represents the fraction of the job that Marty does in one hour.

The equation representing Bruce and Bryce working together is:
$$\frac{1}{x} + \frac{1}{y} = \frac{1}{\frac{4}{3}} = \frac{3}{4}$$

The equation representing Bryce and Marty working together is:
$$\frac{1}{y} + \frac{1}{z} = \frac{1}{\frac{8}{5}} = \frac{5}{8}$$

The equation representing Bruce and Marty working together is:
$$\frac{1}{x} + \frac{1}{z} = \frac{1}{\frac{8}{3}} = \frac{3}{8}$$

Solve the system of equations:
$$\begin{cases} \dfrac{1}{x} + \dfrac{1}{y} = \dfrac{3}{4} \\[2mm] \dfrac{1}{y} + \dfrac{1}{z} = \dfrac{5}{8} \\[2mm] \dfrac{1}{x} + \dfrac{1}{z} = \dfrac{3}{8} \end{cases}$$

Let $u = \dfrac{1}{x}, \ v = \dfrac{1}{y}, \ w = \dfrac{1}{z}$:
$$\begin{cases} u + v = \frac{3}{4} & \rightarrow & u = \frac{3}{4} - v \\[1mm] v + w = \frac{5}{8} & \rightarrow & w = \frac{5}{8} - v \\[1mm] u + w = \frac{3}{8} \end{cases}$$

Substitute into the third equation and solve:
$$\frac{3}{4} - v + \frac{5}{8} - v = \frac{3}{8}$$
$$-2v = -1$$
$$v = \frac{1}{2}$$
$$u = \frac{3}{4} - \frac{1}{2} = \frac{1}{4}$$
$$w = \frac{5}{8} - \frac{1}{2} = \frac{1}{8}$$

Solve for x, y, and z:
$$x = 4, \ y = 2, \ z = 8 \ (\text{reciprocals})$$
Bruce can do the job in 4 hours, Bryce in 2 hours, and Marty in 8 hours.

107. Let x = the number of gasoline engines produced each week.
Let y = the number of diesel engines produced each week.

The total cost is: $C = 450x + 550y$. Cost is to be minimized; thus, this is the objective function.

The constraints are:

$20 \leq x \leq 60$ number of gasoline engines needed and capacity each week.

$15 \leq y \leq 40$ number of diesel engines needed and capacity each week.

$x + y \geq 50$ number of engines produced to prevent layoffs.

Graph the constraints.

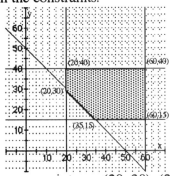

The corner points are (20, 30), (20, 40), (35, 15), (60, 15), (60, 40).
Evaluate the objective function:

Vertex	Value of $C = 450x + 550y$
(20, 30)	$C = 450(20) + 550(30) = 25,500$
(20, 40)	$C = 450(20) + 550(40) = 31,000$
(35, 15)	$C = 450(35) + 550(15) = 24,000$
(60, 15)	$C = 450(60) + 550(15) = 35,250$
(60, 40)	$C = 450(60) + 550(40) = 49,000$

The minimum cost is $24,000, when 35 gasoline engines and 15 diesel engines are produced.

The excess capacity is 15 gasoline engines, since only 20 gasoline engines had to be delivered.

Chapter 12

Sequences; Induction; The Binomial Theorem

12.1 Sequences

1. $a_1 = 1, \ a_2 = 2, \ a_3 = 3, \ a_4 = 4, \ a_5 = 5$

3. $a_1 = \dfrac{1}{1+2} = \dfrac{1}{3}, \ a_2 = \dfrac{2}{2+2} = \dfrac{2}{4} = \dfrac{1}{2}, \ a_3 = \dfrac{3}{3+2} = \dfrac{3}{5}, \ a_4 = \dfrac{4}{4+2} = \dfrac{4}{6} = \dfrac{2}{3},$
$a_5 = \dfrac{5}{5+2} = \dfrac{5}{7}$

5. $a_1 = (-1)^{1+1}(1^2) = 1, \ a_2 = (-1)^{2+1}(2^2) = -4, \ a_3 = (-1)^{3+1}(3^2) = 9,$
$a_4 = (-1)^{4+1}(4^2) = -16, \ a_5 = (-1)^{5+1}(5^2) = 25$

7. $a_1 = \dfrac{2^1}{3^1+1} = \dfrac{2}{4} = \dfrac{1}{2}, \ a_2 = \dfrac{2^2}{3^2+1} = \dfrac{4}{10} = \dfrac{2}{5}, \ a_3 = \dfrac{2^3}{3^3+1} = \dfrac{8}{28} = \dfrac{2}{7},$
$a_4 = \dfrac{2^4}{3^4+1} = \dfrac{16}{82} = \dfrac{8}{41}, \ a_5 = \dfrac{2^5}{3^5+1} = \dfrac{32}{244} = \dfrac{8}{61}$

9. $a_1 = \dfrac{(-1)^1}{(1+1)(1+2)} = \dfrac{-1}{2\cdot3} = \dfrac{-1}{6}, \ a_2 = \dfrac{(-1)^2}{(2+1)(2+2)} = \dfrac{1}{3\cdot4} = \dfrac{1}{12},$
$a_3 = \dfrac{(-1)^3}{(3+1)(3+2)} = \dfrac{-1}{4\cdot5} = \dfrac{-1}{20}, \ a_4 = \dfrac{(-1)^4}{(4+1)(4+2)} = \dfrac{1}{5\cdot6} = \dfrac{1}{30},$
$a_5 = \dfrac{(-1)^5}{(5+1)(5+2)} = \dfrac{-1}{6\cdot7} = \dfrac{-1}{42}$

11. $a_1 = \dfrac{1}{e^1} = \dfrac{1}{e}, \ a_2 = \dfrac{2}{e^2}, \ a_3 = \dfrac{3}{e^3}, \ a_4 = \dfrac{4}{e^4}, \ a_5 = \dfrac{5}{e^5}$

13. $\dfrac{n}{n+1}$ 15. $\dfrac{1}{2^{n-1}}$ 17. $(-1)^{n+1}$

19. $(-1)^{n+1}n$

21. $a_1 = 2,\ a_2 = 3 + 2 = 5,\ a_3 = 3 + 5 = 8,\ a_4 = 3 + 8 = 11,\ a_5 = 3 + 11 = 14$

23. $a_1 = -2,\ a_2 = 2 + (-2) = 0,\ a_3 = 3 + 0 = 3,\ a_4 = 4 + 3 = 7,\ a_5 = 5 + 7 = 12$

25. $a_1 = 5,\ a_2 = 2 \cdot 5 = 10,\ a_3 = 2 \cdot 10 = 20,\ a_4 = 2 \cdot 20 = 40,\ a_5 = 2 \cdot 40 = 80$

27. $a_1 = 3,\ a_2 = \dfrac{3}{2},\ a_3 = \dfrac{\frac{3}{2}}{3} = \dfrac{1}{2},\ a_4 = \dfrac{\frac{1}{2}}{4} = \dfrac{1}{8},\ a_5 = \dfrac{\frac{1}{8}}{5} = \dfrac{1}{40}$

29. $a_1 = 1,\ a_2 = 2,\ a_3 = 2 \cdot 1 = 2,\ a_4 = 2 \cdot 2 = 4,\ a_5 = 4 \cdot 2 = 8$

31. $a_1 = A,\ a_2 = A + d,\ a_3 = (A + d) + d = A + 2d,\ a_4 = (A + 2d) + d = A + 3d,$
$$a_5 = (A + 3d) + d = A + 4d$$

33. $a_1 = \sqrt{2},\ a_2 = \sqrt{2 + \sqrt{2}},\ a_3 = \sqrt{2 + \sqrt{2 + \sqrt{2}}},\ a_4 = \sqrt{2 + \sqrt{2 + \sqrt{2 + \sqrt{2}}}},$
$$a_5 = \sqrt{2 + \sqrt{2 + \sqrt{2 + \sqrt{2 + \sqrt{2}}}}}$$

35. $\displaystyle\sum_{k=1}^{10} 5 = \underbrace{5 + 5 + 5 + \ldots + 5}_{10\ \text{times}} = 50$

37. $\displaystyle\sum_{k=1}^{6} k = 1 + 2 + 3 + 4 + 5 + 6 = 21$

39. $\displaystyle\sum_{k=1}^{5} (5k + 3) = 8 + 13 + 18 + 23 + 28 = 90$

41. $\displaystyle\sum_{k=1}^{3} (k^2 + 4) = 5 + 8 + 13 = 26$

43. $\displaystyle\sum_{k=1}^{6} (-1)^k 2^k = (-1)^1 \cdot 2^1 + (-1)^2 \cdot 2^2 + (-1)^3 \cdot 2^3 + (-1)^4 \cdot 2^4 + (-1)^5 \cdot 2^5 + (-1)^6 \cdot 2^6$
$$= -2 + 4 - 8 + 16 - 32 + 64 = 42$$

45. $\displaystyle\sum_{k=1}^{4} (k^3 - 1) = 0 + 7 + 26 + 63 = 96$

47. $\displaystyle\sum_{k=1}^{n} (k + 2) = 3 + 4 + 5 + 6 + \cdots + (n + 2)$

49. $\displaystyle\sum_{k=1}^{n} \frac{k^2}{2} = \frac{1}{2} + 2 + \frac{9}{2} + 8 + \frac{25}{2} + \cdots + \frac{n^2}{2}$

51. $\displaystyle\sum_{k=0}^{n} \frac{1}{3^k} = 1 + \frac{1}{3} + \frac{1}{9} + \frac{1}{27} + \cdots + \frac{1}{3^n}$

53. $\displaystyle\sum_{k=0}^{n-1} \frac{1}{3^{k+1}} = \frac{1}{3} + \frac{1}{9} + \frac{1}{27} + \cdots + \frac{1}{3^n}$

55. $\displaystyle\sum_{k=2}^{n} (-1)^k \ln k = \ln 2 - \ln 3 + \ln 4 - \ln 5 + \cdots + (-1)^n \ln n$

57. $\displaystyle 1 + 2 + 3 + \cdots + 20 = \sum_{k=1}^{20} k$

59. $\displaystyle \frac{1}{2} + \frac{2}{3} + \frac{3}{4} + \cdots + \frac{13}{13+1} = \sum_{k=1}^{13} \frac{k}{k+1}$

61. $\displaystyle 1 - \frac{1}{3} + \frac{1}{9} - \frac{1}{27} + \cdots + (-1)^6 \left(\frac{1}{3^6}\right) = \sum_{k=0}^{6} (-1)^k \left(\frac{1}{3^k}\right)$

63. $\displaystyle 3 + \frac{3^2}{2} + \frac{3^3}{3} + \cdots + \frac{3^n}{n} = \sum_{k=1}^{n} \frac{3^k}{k}$

65. $\displaystyle a + (a+d) + (a+2d) + \cdots + (a+nd) = \sum_{k=0}^{n} (a+kd)$

67. (a) $B_2 = 1.01(3000) - 100 = \$2930$
    (b)   Put the graphing utility in SEQuence mode. Enter Y= as follows, then examine
          the TABLE:

From the table we see that the balance is below $2000 after 14 payments have
been made. The balance then is $1953.70.

(c) Scrolling down the table, we find that balance is paid off in the 36th month. The last payment is $83.78. There are 35 payments of $100 and the last payment of $83.78. The total amount paid is: $35(100) + 83.78 = \$3583.78$.

(d) The interest expense is: $3583.78 - 3000.00 = \$583.78$

69. (a) $p_1 = 1.03(2000) + 20 = 2080;$ $p_2 = 1.03(2080) + 20 = 2162.4$

(b) Scrolling down the table, we find the trout population exceeds 5000 at the beginning of the 26th month when the population is 5084.

```
Plot1 Plot2 Plot3
nMin=0
\u(n)⊟1.03(u(n-1
))+20
u(nMin)⊟{2000}
\v(n)=
v(nMin)=
\w(n)=
```

n	u(n)
22	4442.9
23	4596.2
24	4754.1
25	4916.7
26	5084.2
27	5256.8
28	5434.5

n=28

71. Since the fund returns 12% compound annually, this is equivalent to a return of 1% monthly. Defining a recursive sequence, we have:

$$a_1 = 5000, \quad a_n = 1.01a_{n-1} + 100$$

Insert the formulas in your graphing utility and find the value at the end of 30 years or 360 months (the value at $n = 361$):

```
Plot1 Plot2 Plot3
nMin=1
\u(n)⊟(u(n-1))*1
.01+100
u(nMin)⊟{5000}
\v(n)=
v(nMin)=
\w(n)=
```

n	u(n)
355	497993
356	503073
357	508203
358	513386
359	518619
360	523906
361	529245

n=361

The value of the account will be approximately $529,245.

73. (a) Since the fund returns 8% compound annually, this is equivalent to a return of 2% each quarter. Defining a recursive sequence, we have:

$$a_1 = 0, \quad a_n = 1.02a_{n-1} + 500$$

(b) Insert the formulas in your graphing utility and use the table feature to find when the value of the account will exceed $100,000:

```
Plot1 Plot2 Plot3
nMin=1
\u(n)⊟(u(n-1))*1
.02+500
u(nMin)⊟{0}
\v(n)=
v(nMin)=
\w(n)=
```

n	u(n)
78	89856
79	92153
80	94496
81	96886
82	99324
83	101810
84	104346

n=84

In the 83rd quarter (during the 21st year) the value of the account will exceed $100,000 with a value of $101,810.

(c) Find the value of the account in 25 years or 100 quarters:

n	$u(n)$
96	139042
97	142323
98	145670
99	149083
100	152565
101	156116
102	159738

$n=102$

The value of the account will be $152,565.

75. (a) Since the interest rate is 6% per annum compounded monthly, this is equivalent to a rate of 0.5% each month. Defining a recursive sequence, we have:
$$a_1 = 150,000, \qquad a_n = 1.005a_{n-1} - 899.33$$

(b) $1.005(150,000) - 899.33 = \$149,850.67$

(c) Enter the recursive formula in Y= and create the table:

```
Plot1 Plot2 Plot3
 nMin=1
·u(n)⊟(u(n-1))*1
.005-899.33
 u(nMin)⊟(15000...
·v(n)=
 v(nMin)=
·w(n)=
```

n	$u(n)$
1	150000
2	149851
3	149701
4	149550
5	149398
6	149246
7	149093

$n=7$

(d) Scroll through the table:

n	$u(n)$
54	140963
55	140769
56	140573
57	140377
58	140180
59	139981
60	139782

$n=60$

At the beginning of the 59th month, or after 58 payments have been made, the balance is below $140,000. The balance is $139,981.

(e) Scroll through the table:

n	$u(n)$
355	5298.7
356	4425.9
357	3548.7
358	2667.1
359	1781.1
360	890.65
361	-4.231

$n=355$

The loan will be paid off at the end of 360 months or 30 years.

(f) The total interest expense is the difference of the total of the payments and the original loan:
$$899.33(359) + 890.65 - 150,000 = \$173,750.12$$

(g) (a) Since the interest rate is 6% per annum compounded monthly, this is equivalent to a rate of 0.5% each month. Defining a recursive sequence, we have:
$$a_1 = 150,000, \qquad a_n = 1.005a_{n-1} - 999.33$$

(b) $1.005(150,000) - 999.33 = \$149,750.67$

(c)   Enter the recursive formula in Y= and create the table:

```
Plot1 Plot2 Plot3
 nMin=1
·u(n)▤(u(n-1))*1
.005-999.33
 u(nMin)▤(15000...
·v(n)=
 v(nMin)=
·w(n)=
```

n	u(n)
1	150000
2	149751
3	149500
4	149248
5	148995
6	148741
7	148485
n=7	

(d)   Scroll through the table:

n	u(n)
34	141079
35	140785
36	140489
37	140192
38	139894
39	139594
40	139293
n=40	

At the beginning of the 38th month, or after 37 payments have been made, the balance is below $140,000. The balance is $139,894.

(e)   Scroll through the table:

n	u(n)
275	4294.6
276	3316.7
277	2333.9
278	1346.3
279	353.69
280	-643.9
281	-1646
n=281	

The loan will be paid off at the end of 279 months or 23 years and 3 months.

(f)   The total interest expense is the difference of the total of the payments and the original loan:
$$999.33(278) + 353.69 - 150,000 = \$128,167.43$$

77. $a_1 = 1$, $a_2 = 1$, $a_3 = 2$, $a_4 = 3$, $a_5 = 5$, $a_6 = 8$, $a_7 = 13$, $a_8 = 21$, $a_n = a_{n-1} + a_{n-2}$
$a_8 = a_7 + a_6 = 13 + 8 = 21$
After 7 months there are 21 mature pairs of rabbits.

79.   1, 1, 2, 3, 5, 8, 13   This is the Fibonacci sequence.

12.2 Arithmetic Sequences

1. $d = a_{n+1} - a_n = (n + 1 + 4) - (n + 4) = n + 5 - n - 4 = 1$
$a_1 = 1 + 4 = 5$, $a_2 = 2 + 4 = 6$, $a_3 = 3 + 4 = 7$, $a_4 = 4 + 4 = 8$

3. $d = a_{n+1} - a_n = (2(n + 1) - 5) - (2n - 5) = 2n + 2 - 5 - 2n + 5 = 2$
$a_1 = 2 \cdot 1 - 5 = -3$, $a_2 = 2 \cdot 2 - 5 = -1$, $a_3 = 2 \cdot 3 - 5 = 1$, $a_4 = 2 \cdot 4 - 5 = 3$

5. $d = a_{n+1} - a_n = (6 - 2(n + 1)) - (6 - 2n) = 6 - 2n - 2 - 6 + 2n = -2$
$a_1 = 6 - 2 \cdot 1 = 4$, $a_2 = 6 - 2 \cdot 2 = 2$, $a_3 = 6 - 2 \cdot 3 = 0$, $a_4 = 6 - 2 \cdot 4 = -2$

7. $d = a_{n+1} - a_n = \left(\frac{1}{2} - \frac{1}{3}(n+1)\right) - \left(\frac{1}{2} - \frac{1}{3}n\right) = \frac{1}{2} - \frac{1}{3}n - \frac{1}{3} - \frac{1}{2} + \frac{1}{3}n = -\frac{1}{3}$

    $a_1 = \frac{1}{2} - \frac{1}{3}\cdot 1 = \frac{1}{6}$, $a_2 = \frac{1}{2} - \frac{1}{3}\cdot 2 = -\frac{1}{6}$, $a_3 = \frac{1}{2} - \frac{1}{3}\cdot 3 = -\frac{1}{2}$, $a_4 = \frac{1}{2} - \frac{1}{3}\cdot 4 = -\frac{5}{6}$

9. $d = a_{n+1} - a_n = \ln 3^{n+1} - \ln 3^n = (n+1)\ln 3 - n\ln 3 = \ln 3(n+1-n) = \ln 3$

    $a_1 = \ln 3^1 = \ln 3$, $a_2 = \ln 3^2 = 2\ln 3$, $a_3 = \ln 3^3 = 3\ln 3$, $a_4 = \ln 3^4 = 4\ln 3$

11. $a_n = a + (n-1)d = 2 + (n-1)3 = 2 + 3n - 3 = 3n - 1$

    $a_5 = 3\cdot 5 - 1 = 14$

13. $a_n = a + (n-1)d = 5 + (n-1)(-3) = 5 - 3n + 3 = 8 - 3n$

    $a_5 = 8 - 3\cdot 5 = -7$

15. $a_n = a + (n-1)d = 0 + (n-1)\frac{1}{2} = \frac{1}{2}n - \frac{1}{2}$

    $a_5 = \frac{1}{2}\cdot 5 - \frac{1}{2} = 2$

17. $a_n = a + (n-1)d = \sqrt{2} + (n-1)\sqrt{2} = \sqrt{2} + \sqrt{2}n - \sqrt{2} = \sqrt{2}n$

    $a_5 = 5\sqrt{2}$

19. $a_1 = 2$, $d = 2$, $a_n = a + (n-1)d$

    $a_{12} = 2 + (12-1)2 = 2 + 11(2) = 2 + 22 = 24$

21. $a_1 = 1$, $d = -2 - 1 = -3$, $a_n = a + (n-1)d$

    $a_{10} = 1 + (10-1)(-3) = 1 + 9(-3) = 1 - 27 = -26$

23. $a_1 = a$, $d = (a+b) - a = b$, $a_n = a + (n-1)d$

    $a_8 = a + (8-1)b = a + 7b$

25. $a_8 = a + 7d = 8$ $a_{20} = a + 19d = 44$

    Solve the system of equations:

    $8 - 7d + 19d = 44$

    $12d = 36$

    $d = 3$

    $a = 8 - 7(3) = 8 - 21 = -13$

    Recursive formula: $a_1 = -13$ $a_n = a_{n-1} + 3$

27. $a_9 = a + 8d = -5$ $a_{15} = a + 14d = 31$

    Solve the system of equations:

    $-5 - 8d + 14d = 31$

    $6d = 36$

    $d = 6$

    $a = -5 - 8(6) = -5 - 48 = -53$

    Recursive formula: $a_1 = -53$ $a_n = a_{n-1} + 6$

29. $a_{15} = a + 14d = 0$ $a_{40} = a + 39d = -50$
     Solve the system of equations:
$$-14d + 39d = -50$$
$$25d = -50$$
$$d = -2$$
$$a = -14(-2) = 28$$
     Recursive formula: $a_1 = 28$ $a_n = a_{n-1} - 2$

31. $a_{14} = a + 13d = -1$ $a_{18} = a + 17d = -9$
     Solve the system of equations:
$$-1 - 13d + 17d = -9$$
$$4d = -8$$
$$d = -2$$
$$a = -1 - 13(-2) = -1 + 26 = 25$$
     Recursive formula: $a_1 = 25$ $a_n = a_{n-1} - 2$

33. $S_n = \dfrac{n}{2}(a + a_n) = \dfrac{n}{2}(1 + (2n - 1)) = \dfrac{n}{2}(2n) = n^2$

35. $S_n = \dfrac{n}{2}(a + a_n) = \dfrac{n}{2}(7 + (2 + 5n)) = \dfrac{n}{2}(9 + 5n) = \dfrac{9}{2}n + \dfrac{5}{2}n^2$

37. $a_1 = 2,\ d = 4 - 2 = 2,\ a_n = a + (n - 1)d$
$$70 = 2 + (n - 1)2$$
$$70 = 2 + 2n - 2$$
$$70 = 2n$$
$$n = 35$$
$$S_n = \dfrac{n}{2}(a + a_n) = \dfrac{35}{2}(2 + 70) = \dfrac{35}{2}(72) = 35(36) = 1260$$

39. $a_1 = 5,\ d = 9 - 5 = 4,\ a_n = a + (n - 1)d$
$$49 = 5 + (n - 1)4$$
$$49 = 5 + 4n - 4$$
$$48 = 4n$$
$$n = 12$$
$$S_n = \dfrac{n}{2}(a + a_n) = \dfrac{12}{2}(5 + 49) = 6(54) = 324$$

41. Using the sum of the sequence feature:

```
sum(seq(3.45N+4.
12,N,1,20,1))
          806.9
```

43. $d = 5.2 - 2.8 = 2.4$

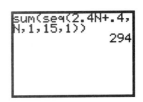

$a = 2.8$
$36.4 = 2.8 + (n-1)2.4$
$36.4 = 2.8 + 2.4n - 2.4$
$36 = 2.4n$
$n = 15$
$a_n = 2.8 + (n-1)2.4 = 2.8 + 2.4n - 2.4 = 2.4n + 0.4$

45. $d = 7.48 - 4.9 = 2.58$
$a = 4.9$
$66.82 = 4.9 + (n-1)2.58$
$66.82 = 4.9 + 2.58n - 2.58$
$64.5 = 2.58n$
$n = 25$

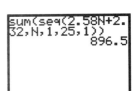

$a_n = 4.9 + (n-1)2.58 = 4.9 + 2.58n - 2.58$
$a_n = 2.58n + 2.32$

47.   Find the common difference of the terms and solve the system of equations:
$$(2x+1) - (x+3) = d \;\rightarrow\; x - 2 = d$$
$$(5x+2) - (2x+1) = d \;\rightarrow\; 3x + 1 = d$$
$$3x + 1 = x - 2$$
$$2x = -3$$
$$x = -\tfrac{3}{2}$$

49.   The total number of seats is: $S = 25 + 26 + 27 + \cdots$
This is the sum of an arithmetic sequence with $d = 1,\; a = 25,$ and $n = 30$.
Find the sum of the sequence:
$$S_{30} = \frac{30}{2}[2(25) + (30-1)(1)] = 15(50 + 29) = 15(79) = 1185$$
There are 1185 seats in the theater.

51.   The lighter colored tiles have 20 tiles in the bottom row and 1 tile in the top row. The number decreases by 1 as we move up the triangle. This is an arithmetic sequence with $a_1 = 20,\; d = -1,$ and $n = 20$. Find the sum:
$$S = \frac{20}{2}[2(20) + (20-1)(-1)] = 10(40 - 19) = 10(21) = 210 \text{ lighter tiles.}$$
The darker colored tiles have 19 tiles in the bottom row and 1 tile in the top row. The number decreases by 1 as we move up the triangle. This is an arithmetic sequence with $a_1 = 19,\; d = -1,$ and $n = 19$. Find the sum:
$$S = \frac{19}{2}[2(19) + (19-1)(-1)] = \frac{19}{2}(38 - 18) = \frac{19}{2}(20) = 190 \text{ darker tiles.}$$

53. Find n in an arithmetic sequence with $a_1 = 10, \ d = 4, \ s_n = 2040$.

$$s_n = \frac{n}{2}\big[2a_1 + (n-1)d\big]$$
$$2040 = \frac{n}{2}\big[2(10) + (n-1)4\big]$$
$$4080 = n\big[20 + 4n - 4\big]$$
$$4080 = n(4n + 16)$$
$$4080 = 4n^2 + 16n$$
$$1020 = n^2 + 4n$$
$$n^2 + 4n - 1020 = 0$$
$$(n + 34)(n - 30) = 0$$
$$n = -34 \text{ or } n = 30$$

There are 30 rows in the corner section of the stadium.

12.3 Geometric Sequences; Geometric Series

1. $r = \dfrac{3^{n+1}}{3^n} = 3^{n+1-n} = 3$

   $a_1 = 3^1 = 3, \ a_2 = 3^2 = 9, \ a_3 = 3^3 = 27, \ a_4 = 3^4 = 81$

3. $r = \dfrac{-3\left(\frac{1}{2}\right)^{n+1}}{-3\left(\frac{1}{2}\right)^{n}} = \left(\frac{1}{2}\right)^{n+1-n} = \frac{1}{2}$

   $a_1 = -3\left(\frac{1}{2}\right)^1 = -\frac{3}{2}, \ a_2 = -3\left(\frac{1}{2}\right)^2 = -\frac{3}{4}, \ a_3 = -3\left(\frac{1}{2}\right)^3 = -\frac{3}{8}, \ a_4 = -3\left(\frac{1}{2}\right)^4 = -\frac{3}{16}$

5. $r = \dfrac{\frac{2^{n+1-1}}{4}}{\frac{2^{n-1}}{4}} = \dfrac{2^n}{2^{n-1}} = 2^{n-(n-1)} = 2$

   $a_1 = \dfrac{2^{1-1}}{4} = \dfrac{2^0}{2^2} = 2^{-2} = \dfrac{1}{4}, \ a_2 = \dfrac{2^{2-1}}{4} = \dfrac{2^1}{2^2} = 2^{-1} = \dfrac{1}{2}, \ a_3 = \dfrac{2^{3-1}}{4} = \dfrac{2^2}{2^2} = 1,$

   $a_4 = \dfrac{2^{4-1}}{4} = \dfrac{2^3}{2^2} = 2$

7. $r = \dfrac{2^{\frac{n+1}{3}}}{2^{\frac{n}{3}}} = 2^{\frac{n+1}{3} - \frac{n}{3}} = 2^{\frac{1}{3}}$

   $a_1 = 2^{\frac{1}{3}}, \ a_2 = 2^{\frac{2}{3}}, \ a_3 = 2^{\frac{3}{3}} = 2, \ a_4 = 2^{\frac{4}{3}}$

9. $r = \dfrac{\dfrac{3^{n+1-1}}{2^{n+1}}}{\dfrac{3^{n-1}}{2^n}} = \dfrac{3^n}{3^{n-1}} \cdot \dfrac{2^n}{2^{n+1}} = 3^{n-(n-1)} \cdot 2^{n-(n+1)} = 3 \cdot 2^{-1} = \dfrac{3}{2}$

$a_1 = \dfrac{3^{1-1}}{2^1} = \dfrac{3^0}{2} = \dfrac{1}{2}, \; a_2 = \dfrac{3^{2-1}}{2^2} = \dfrac{3^1}{2^2} = \dfrac{3}{4}, \; a_3 = \dfrac{3^{3-1}}{2^3} = \dfrac{3^2}{2^3} = \dfrac{9}{8},$

$a_4 = \dfrac{3^{4-1}}{2^4} = \dfrac{3^3}{2^4} = \dfrac{27}{16}$

11. $\{n+2\}$ Arithmetic
$d = (n+1+2) - (n+2) = n+3-n-2 = 1$

13. $\{4n^2\}$ Examine the terms of the sequence: 4, 16, 36, 64, 100, ...
There is no common difference; there is no common ratio; neither.

15. $\left\{3 - \frac{2}{3}n\right\}$ Arithmetic
$d = \left(3 - \frac{2}{3}(n+1)\right) - \left(3 - \frac{2}{3}n\right) = 3 - \frac{2}{3}n - \frac{2}{3} - 3 + \frac{2}{3}n = -\frac{2}{3}$

17. 1, 3, 6, 10, ...   Neither
There is no common difference or common ratio.

19. $\left\{\left(\frac{2}{3}\right)^n\right\}$ Geometric $r = \dfrac{\left(\frac{2}{3}\right)^{n+1}}{\left(\frac{2}{3}\right)^n} = \left(\frac{2}{3}\right)^{n+1-n} = \frac{2}{3}$

21. –1, –2, –4, –8, ...   Geometric $r = \dfrac{-2}{-1} = \dfrac{-4}{-2} = \dfrac{-8}{-4} = 2$

23. $\left\{3^{\frac{n}{2}}\right\}$ Geometric $r = \dfrac{3^{\frac{n+1}{2}}}{3^{\frac{n}{2}}} = 3^{\frac{n+1}{2} - \frac{n}{2}} = 3^{\frac{1}{2}}$

25. $a_5 = 2 \cdot 3^{5-1} = 2 \cdot 3^4 = 2 \cdot 81 = 162$ $a_n = 2 \cdot 3^{n-1}$

27. $a_5 = 5(-1)^{5-1} = 5(-1)^4 = 5 \cdot 1 = 5$ $a_n = 5 \cdot (-1)^{n-1}$

29. $a_5 = 0 \cdot \left(\frac{1}{2}\right)^{5-1} = 0 \cdot \left(\frac{1}{2}\right)^4 = 0$ $a_n = 0 \cdot \left(\frac{1}{2}\right)^{n-1} = 0$

31. $a_5 = \sqrt{2} \cdot \sqrt{2}^{\,5-1} = \sqrt{2} \cdot \sqrt{2}^{\,4} = \sqrt{2} \cdot 4 = 4\sqrt{2}$ $a_n = \sqrt{2} \cdot \sqrt{2}^{\,n-1} = \sqrt{2}^{\,n}$

33. $a = 1, \; r = \frac{1}{2}, \; n = 7$ $a_7 = 1 \cdot \left(\frac{1}{2}\right)^{7-1} = \left(\frac{1}{2}\right)^6 = \frac{1}{64}$

35. $a = 1, \; r = -1, \; n = 9$ $\qquad a_7 = 1 \cdot (-1)^{9-1} = (-1)^8 = 1$

37. $a = 0.4, \; r = 0.1, \; n = 8$ $\qquad a_7 = 0.4 \cdot (0.1)^{8-1} = 0.4(0.1)^7 = 0.00000004$

39. $a = \dfrac{1}{4}, \; r = 2$ $\qquad S_n = a\left(\dfrac{1-r^n}{1-r}\right) = \dfrac{1}{4}\left(\dfrac{1-2^n}{1-2}\right) = -\dfrac{1}{4}(1-2^n)$

41. $a = \dfrac{2}{3}, \; r = \dfrac{2}{3}$ $\qquad S_n = a\left(\dfrac{1-r^n}{1-r}\right) = \dfrac{2}{3}\left(\dfrac{1-\left(\frac{2}{3}\right)^n}{1-\frac{2}{3}}\right) = \dfrac{2}{3}\left(\dfrac{1-\left(\frac{2}{3}\right)^n}{\frac{1}{3}}\right) = 2\left(1-\left(\tfrac{2}{3}\right)^n\right)$

43. $a = -1, \; r = 2$ $\qquad S_n = a\left(\dfrac{1-r^n}{1-r}\right) = -1\left(\dfrac{1-2^n}{1-2}\right) = 1-2^n$

45. Using the sum of the sequence feature:

```
sum(seq(2^N/4,N,
0,14,1))
            8191.75
```

47. Using the sum of the sequence feature:

```
sum(seq((2/3)^N,
N,1,15,1))
        1.995432683
```

49. Using the sum of the sequence feature:

```
sum(seq(-1*2^N,N
,0,14,1))
            -32767
```

51. $a = 1, \; r = \frac{1}{3}$ $\qquad$ Since $|r| < 1, \;\; S_n = \dfrac{a}{1-r} = \dfrac{1}{1-\frac{1}{3}} = \dfrac{1}{\frac{2}{3}} = \dfrac{3}{2}$

53. $a = 8, \; r = \frac{1}{2}$ $\qquad$ Since $|r| < 1, \;\; S_n = \dfrac{a}{1-r} = \dfrac{8}{1-\frac{1}{2}} = \dfrac{8}{\frac{1}{2}} = 16$

55. $a = 2, \; r = -\frac{1}{4}$ $\qquad$ Since $|r| < 1, \;\; S_n = \dfrac{a}{1-r} = \dfrac{2}{1-\left(-\frac{1}{4}\right)} = \dfrac{2}{\frac{5}{4}} = \dfrac{8}{5}$

57. $a = 5, \; r = \frac{1}{4}$ Since $|r| < 1, \;\; S_n = \frac{a}{1-r} = \frac{5}{1-\frac{1}{4}} = \frac{5}{\frac{3}{4}} = \frac{20}{3}$

59. $a = 6, \; r = -\frac{2}{3}$ Since $|r| < 1, \;\; S_n = \frac{a}{1-r} = \frac{6}{1-\left(-\frac{2}{3}\right)} = \frac{6}{\frac{5}{3}} = \frac{18}{5}$

61. Find the common ratio of the terms and solve the system of equations:

$$\frac{x+2}{x} = r$$
$$\frac{x+3}{x+2} = r$$
$$\frac{x+2}{x} = \frac{x+3}{x+2}$$
$$x^2 + 4x + 4 = x^2 + 3x$$
$$x = -4$$

63. The common ratio, $r = 0.90 < 1$. The sum is: $S = \frac{1}{1-0.9} = \frac{1}{0.10} = 10.$
The multiplier is 10.

65. This is an infinite geometric series with $a = 4$, and $r = \frac{1.03}{1.09}$.

Find the sum: Price $= \dfrac{4}{1 - \frac{1.03}{1.09}} = \$72.67$.

67. (a) Find the 10th term of the geometric sequence:
$$a = 2, \; r = 0.9, \; n = 10 \qquad a_{10} = 2(0.9)^{10-1} = 2(0.9)^9 = 0.775 \text{ feet}$$

(b) Find n when $a_n < 1$:
$$2(0.9)^{n-1} < 1$$
$$0.9^{n-1} < 0.5$$
$$(n-1)\log 0.9 < \log 0.5$$
$$n - 1 > \frac{\log 0.5}{\log 0.9}$$
$$n > \frac{\log 0.5}{\log 0.9} + 1 = 7.58$$

On the 8th swing the arc is less than 1 foot.

(c) Find the sum of the first 15 swings:
$$S_{15} = 2\left(\frac{1-(0.9)^{15}}{1-0.9}\right) = 2\left(\frac{1-0.9^{15}}{0.1}\right) = 20\left(1-0.9^{15}\right) = 15.88 \text{ feet}$$

(d) Find the infinite sum of the geometric series:
$$S = \frac{2}{1-0.9} = \frac{2}{0.1} = 20 \text{ feet}$$

69. This is a geometric series with $a = \$18,000, \quad r = 1.05, \quad n = 5$. Find the 5th term:
$$a_5 = 18000(1.05)^{5-1} = 18000(1.05)^4 = \$21,879.11$$

71. Option 1: Total Salary $= \$2,000,000(7) + \$100,000(7) = \$14,700,000$
Option 2: Geometric series with: $a = \$2,000,000, \quad r = 1.045, \quad n = 7$
Find the sum of the geometric series:
$$S = 2,000,000\left(\frac{1-1.045^7}{1-1.045}\right) = \$16,038,304$$
Option 3: Arithmetic series with: $a = \$2,000,000, \quad d = \$95,000, \quad n = 7$
Find the sum of the arithmetic series:
$$S_7 = \frac{7}{2}(2(2,000,000) + (7-1)(95,000)) = \$15,995,000$$
Option 2 provides the most money; Option 1 provides the least money.

73. This is a geometric sequence with $a = 1, \quad r = 2, \quad n = 64$.
Find the sum of the geometric series:
$$S_{64} = 1\left(\frac{1-2^{64}}{1-2}\right) = \frac{1-2^{64}}{-1} = 2^{64} - 1 = 1.845 \times 10^{19} \text{ grains}$$

77. Use the **sum(seq(....))** operation on your graphing utility to evaluate the sum of the powers of x. You will need to guess the correct number of terms until you find the appropriate accuracy. You will end up with the following screens eventually:

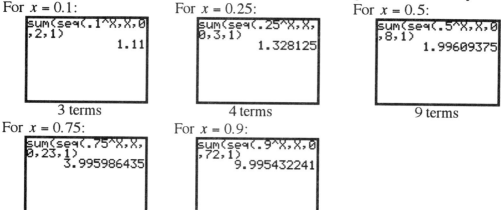

For $x = 0.1$:   For $x = 0.25$:   For $x = 0.5$:

3 terms   4 terms   9 terms

For $x = 0.75$:   For $x = 0.9$:

24 terms   73 terms

Remember you are counting from 0 to n, so there are $n + 1$ terms.
Create the table:

x	$\dfrac{1}{1-x}$	n
0.1	1.11	3
0.25	1.33	4
0.5	2	9
0.75	4	24
0.9	10	73

79. Both options are geometric sequences:

Option A: $a = \$20,000$; $r = 1.06$; $n = 5$

$$a_5 = 20,000(1.06)^{5-1} = 20,000(1.06)^4 = \$25,250$$

$$S_5 = 20000\left(\frac{1 - 1.06^5}{1 - 1.06}\right) = \$112,742$$

Option B: $a = \$22,000$; $r = 1.03$; $n = 5$

$$a_5 = 22,000(1.03)^{5-1} = 22,000(1.03)^4 = \$24,761$$

$$S_5 = 22000\left(\frac{1 - 1.03^5}{1 - 1.03}\right) = \$116,801$$

Option A provides more money in the 5th year, while Option B provides the greatest total amount of money over the 5 year period.

12.4 Mathematical Induction

1. I: $n = 1$: $2 \cdot 1 = 2$ and $1(1 + 1) = 2$

   II: If $2 + 4 + 6 + \cdots + 2k = k(k + 1)$

   then $2 + 4 + 6 + \cdots + 2k + 2(k + 1)$

   $$= \left[2 + 4 + 6 + \cdots + 2k\right] + 2(k + 1) = k(k + 1) + 2(k + 1)$$
   $$= (k + 1)(k + 2)$$

   Conditions I and II are satisfied; the statement is true.

3. I: $n = 1$: $1 + 2 = 3$ and $\frac{1}{2} \cdot 1(1 + 5) = 3$

   II: If $3 + 4 + 5 + \cdots + (k + 2) = \frac{1}{2}k(k + 5)$

   then $3 + 4 + 5 + \cdots + (k + 2) + [(k + 1) + 2]$

   $$= \left[3 + 4 + 5 + \cdots + (k + 2)\right] + (k + 3) = \frac{1}{2}k(k + 5) + (k + 3)$$
   $$= \frac{1}{2}k^2 + \frac{5}{2}k + k + 3 = \frac{1}{2}k^2 + \frac{7}{2}k + 3 = \frac{1}{2}\left(k^2 + 7k + 6\right)$$
   $$= \frac{1}{2}(k + 1)(k + 6)$$

   Conditions I and II are satisfied; the statement is true.

5. I: $n = 1$: $3 \cdot 1 - 1 = 2$ and $\frac{1}{2} \cdot 1(3 \cdot 1 + 1) = 2$

   II: If $2 + 5 + 8 + \cdots + (3k - 1) = \frac{1}{2}k(3k + 1)$

   then $2 + 5 + 8 + \cdots + (3k - 1) + [3(k + 1) - 1]$

   $$= \left[2 + 5 + 8 + \cdots + (3k - 1)\right] + (3k + 2) = \frac{1}{2}k(3k + 1) + (3k + 2)$$
   $$= \frac{3}{2}k^2 + \frac{1}{2}k + 3k + 2 = \frac{3}{2}k^2 + \frac{7}{2}k + 2 = \frac{1}{2}\left(3k^2 + 7k + 4\right)$$
   $$= \frac{1}{2}(k + 1)(3k + 4)$$

   Conditions I and II are satisfied; the statement is true.

7. I: $n = 1$: $2^{1-1} = 1$ and $2^1 - 1 = 1$

    II: If $1 + 2 + 2^2 + \cdots + 2^{k-1} = 2^k - 1$

    then $1 + 2 + 2^2 + \cdots + 2^{k-1} + 2^{k+1-1}$

    $$= \left[1 + 2 + 2^2 + \cdots + 2^{k-1} \right] + 2^k = 2^k - 1 + 2^k$$

    $$= 2 \cdot 2^k - 1 = 2^{k+1} - 1$$

    Conditions I and II are satisfied; the statement is true.

9. I: $n = 1$: $4^{1-1} = 1$ and $\frac{1}{3}\left(4^1 - 1 \right) = 1$

    II: If $1 + 4 + 4^2 + \cdots + 4^{k-1} = \frac{1}{3}\left(4^k - 1 \right)$

    then $1 + 4 + 4^2 + \cdots + 4^{k-1} + 4^{k+1-1}$

    $$= \left[1 + 4 + 4^2 + \cdots + 4^{k-1} \right] + 4^k = \frac{1}{3}\left(4^k - 1 \right) + 4^k$$

    $$= \frac{1}{3} \cdot 4^k - \frac{1}{3} + 4^k = \frac{4}{3} \cdot 4^k - \frac{1}{3} = \frac{1}{3}\left(4 \cdot 4^k - 1 \right) = \frac{1}{3}\left(4^{k+1} - 1 \right)$$

    Conditions I and II are satisfied; the statement is true.

11. I: $n = 1$: $\dfrac{1}{1(1+1)} = \dfrac{1}{2}$ and $\dfrac{1}{1+1} = \dfrac{1}{2}$

    II: If $\dfrac{1}{1 \cdot 2} + \dfrac{1}{2 \cdot 3} + \dfrac{1}{3 \cdot 4} + \cdots + \dfrac{1}{k(k+1)} = \dfrac{k}{k+1}$

    then $\dfrac{1}{1 \cdot 2} + \dfrac{1}{2 \cdot 3} + \dfrac{1}{3 \cdot 4} + \cdots + \dfrac{1}{k(k+1)} + \dfrac{1}{(k+1)(k+1+1)}$

    $$= \left[\dfrac{1}{1 \cdot 2} + \dfrac{1}{2 \cdot 3} + \dfrac{1}{3 \cdot 4} + \cdots + \dfrac{1}{k(k+1)} \right] + \dfrac{1}{(k+1)(k+2)}$$

    $$= \dfrac{k}{k+1} + \dfrac{1}{(k+1)(k+2)} = \dfrac{k}{k+1} \cdot \dfrac{k+2}{k+2} + \dfrac{1}{(k+1)(k+2)}$$

    $$= \dfrac{k^2 + 2k + 1}{(k+1)(k+2)} = \dfrac{(k+1)(k+1)}{(k+1)(k+2)} = \dfrac{k+1}{k+2}$$

    Conditions I and II are satisfied; the statement is true.

13. I: $n = 1$: $1^2 = 1$ and $\frac{1}{6} \cdot 1(1+1)(2 \cdot 1 + 1) = 1$

    II: If $1^2 + 2^2 + 3^2 + \cdots + k^2 = \frac{1}{6} k(k+1)(2k+1)$

    then $1^2 + 2^2 + 3^2 + \cdots + k^2 + (k+1)^2$

    $$= \left[1^2 + 2^2 + 3^2 + \cdots + k^2 \right] + (k+1)^2 = \frac{1}{6} k(k+1)(2k+1) + (k+1)^2$$

    $$= (k+1)\left[\frac{1}{6} k(2k+1) + k + 1 \right] = (k+1)\left[\frac{1}{3} k^2 + \frac{1}{6} k + k + 1 \right]$$

    $$= (k+1)\left[\frac{1}{3} k^2 + \frac{7}{6} k + 1 \right] = \frac{1}{6}(k+1)\left[2k^2 + 7k + 6 \right]$$

    $$= \frac{1}{6}(k+1)(k+2)(2k+3)$$

    Conditions I and II are satisfied; the statement is true.

15. **I:** $n = 1$: $5 - 1 = 4$ and $\frac{1}{2} \cdot 1(9 - 1) = 4$

    II: If $4 + 3 + 2 + \cdots + (5 - k) = \frac{1}{2}k(9 - k)$

        then $4 + 3 + 2 + \cdots + (5 - k) + (5 - (k + 1))$

$$= \left[4 + 3 + 2 + \cdots + (5 - k)\right] + (4 - k) = \frac{1}{2}k(9 - k) + (4 - k)$$

$$= \frac{9}{2}k - \frac{1}{2}k^2 + 4 - k = -\frac{1}{2}k^2 + \frac{7}{2}k + 4 = -\frac{1}{2}\left[k^2 - 7k - 8\right]$$

$$= -\frac{1}{2}(k + 1)(k - 8) = \frac{1}{2}(k + 1)(8 - k) = \frac{1}{2}(k + 1)\left[9 - (k + 1)\right]$$

Conditions I and II are satisfied; the statement is true.

17. **I:** $n = 1$: $1(1 + 1) = 2$ and $\frac{1}{3} \cdot 1(1 + 1)(1 + 2) = 2$

    II: If $1 \cdot 2 + 2 \cdot 3 + 3 \cdot 4 + \cdots + k(k + 1) = \frac{1}{3}k(k + 1)(k + 2)$

        then $1 \cdot 2 + 2 \cdot 3 + 3 \cdot 4 + \cdots + k(k + 1) + (k + 1)(k + 1 + 1)$

$$= \left[1 \cdot 2 + 2 \cdot 3 + 3 \cdot 4 + \cdots + k(k + 1)\right] + (k + 1)(k + 2)$$

$$= \frac{1}{3}k(k + 1)(k + 2) + (k + 1)(k + 2) = (k + 1)(k + 2)\left[\frac{1}{3}k + 1\right]$$

$$= \frac{1}{3}(k + 1)(k + 2)(k + 3)$$

Conditions I and II are satisfied; the statement is true.

19. **I:** $n = 1$: $1^2 + 1 = 2$ is divisible by 2

    II: If $k^2 + k$ is divisible by 2

        then $(k + 1)^2 + (k + 1) = k^2 + 2k + 1 + k + 1 = (k^2 + k) + (2k + 2)$

    Since $k^2 + k$ is divisible by 2 and $2k + 2$ is divisible by 2, then $(k + 1)^2 + (k + 1)$

    is divisible by 2.

Conditions I and II are satisfied; the statement is true.

21. **I:** $n = 1$: $1^2 - 1 + 2 = 2$ is divisible by 2

    II: If $k^2 - k + 2$ is divisible by 2

        then $(k + 1)^2 - (k + 1) + 2 = k^2 + 2k + 1 - k - 1 + 2 = (k^2 - k + 2) + (2k)$

    Since $k^2 - k + 2$ is divisible by 2 and $2k$ is divisible by 2, then

    $(k + 1)^2 - (k + 1) + 2$ is divisible by 2.

Conditions I and II are satisfied; the statement is true.

23. **I:** $n = 1$:   If $x > 1$ then $x^1 = x > 1$.

    II: Assume, for any natural number k, that if $x > 1$, then $x^k > 1$.

    Show that if $x^k > 1$, then $x^{k+1} > 1$:

$$x^{k+1} = x^k \cdot x > 1 \cdot x = x > 1$$

$$\uparrow$$

$$(x^k > 1)$$

Conditions I and II are satisfied; the statement is true.

25.   I: $n = 1$: $a - b$ is a factor of $a^1 - b^1 = a - b$.

     II: If $a - b$ is a factor of $a^k - b^k$

        Show that $a - b$ is a factor of $a^{k+1} - b^{k+1} = a \cdot a^k - b \cdot b^k$

$$= a \cdot a^k - a \cdot b^k + a \cdot b^k - b \cdot b^k = a\left(a^k - b^k\right) + b^k(a - b)$$

        Since $a - b$ is a factor of $a^k - b^k$ and $a - b$ is a factor of $a - b$, then

        $a - b$ is a factor of $a^{k+1} - b^{k+1}$.

    Conditions I and II are satisfied; the statement is true.

27. $n = 1$: $1^2 - 1 + 41 = 41$ is a prime number.

      $n = 41$: $41^2 - 41 + 41 = 41^2$ is not a prime number.

29.   I: $n = 1$: $ar^{1-1} = a$ and $a\left(\dfrac{1 - r^1}{1 - r}\right) = a$

     II: If $a + ar + ar^2 + \cdots + ar^{k-1} = a\left(\dfrac{1 - r^k}{1 - r}\right)$

        then $a + ar + ar^2 + \cdots + ar^{k-1} + ar^{k+1-1}$

$$= \left[a + ar + ar^2 + \cdots + ar^{k-1}\right] + ar^k = a\left(\dfrac{1 - r^k}{1 - r}\right) + ar^k$$

$$= \dfrac{a(1 - r^k) + ar^k(1 - r)}{1 - r} = \dfrac{a - ar^k + ar^k - ar^{k+1}}{1 - r} = a\left(\dfrac{1 - r^{k+1}}{1 - r}\right)$$

    Conditions I and II are satisfied; the statement is true.

31.   I: $n = 3$: $(3 - 2) \cdot 180° = 180°$ which is the sum of the angles of a triangle .

     II: Assume that for any integer k the sum of the angles of a convex polygon with

        k sides is $(k - 2) \cdot 180°$. A convex polygon with $k + 1$ sides consists of a

        convex polygon with k sides plus a triangle . Thus the sum of the angles is

        $(k - 2) \cdot 180° + 180° = (k - 1) \cdot 180°$.

    Conditions I and II are satisfied; the statement is true.

12.5 The Binomial Theorem

1. $\dbinom{5}{3} = \dfrac{5!}{3! \, 2!} = \dfrac{5 \cdot 4 \cdot 3 \cdot 2 \cdot 1}{3 \cdot 2 \cdot 1 \cdot 2 \cdot 1} = \dfrac{5 \cdot 4}{2 \cdot 1} = 10$

3. $\dbinom{7}{5} = \dfrac{7!}{5! \, 2!} = \dfrac{7 \cdot 6 \cdot 5 \cdot 4 \cdot 3 \cdot 2 \cdot 1}{5 \cdot 4 \cdot 3 \cdot 2 \cdot 1 \cdot 2 \cdot 1} = \dfrac{7 \cdot 6}{2 \cdot 1} = 21$

5. $\dbinom{50}{49} = \dfrac{50!}{49! \, 1!} = \dfrac{50 \cdot 49!}{49! \cdot 1} = \dfrac{50}{1} = 50$

7. $\dbinom{1000}{1000} = \dfrac{1000!}{1000!\,0!} = \dfrac{1}{1} = 1$

9. $\dbinom{55}{23} = \dfrac{55!}{23!\,32!} = 1.866442159 \times 10^{15}$

11. $\dbinom{47}{25} = \dfrac{47!}{25!\,22!} = 1.483389769 \times 10^{13}$

13. $(x+1)^5 = \dbinom{5}{0}x^5 + \dbinom{5}{1}x^4 + \dbinom{5}{2}x^3 + \dbinom{5}{3}x^2 + \dbinom{5}{4}x^1 + \dbinom{5}{5}x^0$

$$= x^5 + 5x^4 + 10x^3 + 10x^2 + 5x + 1$$

15. $(x-2)^6 = \dbinom{6}{0}x^6 + \dbinom{6}{1}x^5(-2) + \dbinom{6}{2}x^4(-2)^2 + \dbinom{6}{3}x^3(-2)^3 + \dbinom{6}{4}x^2(-2)^4$

$$+ \dbinom{6}{5}x(-2)^5 + \dbinom{6}{6}(-2)^6$$

$$= x^6 + 6x^5(-2) + 15x^4 \cdot 4 + 20x^3(-8) + 15x^2 \cdot 16 + 6x \cdot (-32) + 64$$

$$= x^6 - 12x^5 + 60x^4 - 160x^3 + 240x - 192x + 64$$

17. $(3x+1)^4 = \dbinom{4}{0}(3x)^4 + \dbinom{4}{1}(3x)^3 + \dbinom{4}{2}(3x)^2 + \dbinom{4}{3}(3x) + \dbinom{4}{4}$

$$= 81x^4 + 4 \cdot 27x^3 + 6 \cdot 9x^2 + 4 \cdot 3x + 1 = 81x^4 + 108x^3 + 54x^2 + 12x + 1$$

19. $\left(x^2 + y^2\right)^5 = \dbinom{5}{0}\left(x^2\right)^5 + \dbinom{5}{1}\left(x^2\right)^4 y^2 + \dbinom{5}{2}\left(x^2\right)^3\left(y^2\right)^2 + \dbinom{5}{3}\left(x^2\right)^2\left(y^2\right)^3$

$$+ \dbinom{5}{4}x^2\left(y^2\right)^4 + \dbinom{5}{5}\left(y^2\right)^5$$

$$= x^{10} + 5x^8 y^2 + 10x^6 y^4 + 10x^4 y^6 + 5x^2 y^8 + y^{10}$$

21. $\left(\sqrt{x} + \sqrt{2}\right)^6 = \dbinom{6}{0}\left(\sqrt{x}\right)^6 + \dbinom{6}{1}\left(\sqrt{x}\right)^5\left(\sqrt{2}\right)^1 + \dbinom{6}{2}\left(\sqrt{x}\right)^4\left(\sqrt{2}\right)^2 + \dbinom{6}{3}\left(\sqrt{x}\right)^3\left(\sqrt{2}\right)^3$

$$\dbinom{6}{4}\left(\sqrt{x}\right)^2\left(\sqrt{2}\right)^4 + \dbinom{6}{5}\left(\sqrt{x}\right)\left(\sqrt{2}\right)^5 + \dbinom{6}{6}\left(\sqrt{2}\right)^6$$

$$= x^3 + 6\sqrt{2}x^{\frac{5}{2}} + 15 \cdot 2x^2 + 20 \cdot 2\sqrt{2}x^{\frac{3}{2}} + 15 \cdot 4x + 6 \cdot 4\sqrt{2}x^{\frac{1}{2}} + 8$$

$$= x^3 + 6\sqrt{2}x^{\frac{5}{2}} + 30x^2 + 40\sqrt{2}x^{\frac{3}{2}} + 60x + 24\sqrt{2}x^{\frac{1}{2}} + 8$$

23. $(ax + by)^5 = \binom{5}{0}(ax)^5 + \binom{5}{1}(ax)^4 \cdot by + \binom{5}{2}(ax)^3(by)^2 + \binom{5}{3}(ax)^2(by)^3$

$$+ \binom{5}{4}ax(by)^4 + \binom{5}{5}(by)^5$$

$$= a^5x^5 + 5a^4x^4by + 10a^3x^3b^2y^2 + 10a^2x^2b^3y^3 + 5axb^4y^4 + b^5y^5$$

25. $n = 10, \ j = 4, \ x = x, \ a = 3$

$$\binom{10}{4}x^6 \cdot 3^4 = \frac{10!}{4!\,6!} \cdot 81x^6 = \frac{10 \cdot 9 \cdot 8 \cdot 7}{4 \cdot 3 \cdot 2 \cdot 1} \cdot 81x^6 = 17{,}010x^6$$

The coefficient of x^6 is 17,010.

27. $n = 12, \ j = 5, \ x = 2x, \ a = -1$

$$\binom{12}{5}(2x)^7 \cdot (-1)^5 = \frac{12!}{5!\,7!} \cdot 128x^7(-1) = \frac{12 \cdot 11 \cdot 10 \cdot 9 \cdot 8}{5 \cdot 4 \cdot 3 \cdot 2 \cdot 1} \cdot (-128)x^7 = -101{,}376x^7$$

The coefficient of x^7 is $-101{,}376$.

29. $n = 9, \ j = 2, \ x = 2x, \ a = 3$

$$\binom{9}{2}(2x)^7 \cdot 3^2 = \frac{9!}{2!\,7!} \cdot 128x^7(9) = \frac{9 \cdot 8}{2 \cdot 1} \cdot 128x^7 \cdot 9 = 41{,}472x^7$$

The coefficient of x^7 is 41,472.

31. $n = 7, \ j = 4, \ x = x, \ a = 3$

$$\binom{7}{4}x^3 \cdot 3^4 = \frac{7!}{4!\,3!} \cdot 81x^3 = \frac{7 \cdot 6 \cdot 5}{3 \cdot 2 \cdot 1} \cdot 81x^3 = 2835x^3$$

33. $n = 9, \ j = 2, \ x = 3x, \ a = -2$

$$\binom{9}{2}(3x)^7 \cdot (-2)^2 = \frac{9!}{2!\,7!} \cdot 2187x^7 \cdot 4 = \frac{9 \cdot 8}{2 \cdot 1} \cdot 8748x^7 = 314{,}928x^7$$

35. The constant term in $\binom{12}{j}(x^2)^{12-j}\left(\frac{1}{x}\right)^j$ occurs when:

$$2(12 - j) = j \ \rightarrow \ 24 - 2j = j \ \rightarrow \ 3j = 24 \ \rightarrow \ j = 8.$$
Evaluate the 9th term:

$$\binom{12}{8}(x^2)^4 \cdot \left(\frac{1}{x}\right)^8 = \frac{12!}{8!\,4!}x^8 \cdot \frac{1}{x^8} = \frac{12 \cdot 11 \cdot 10 \cdot 9}{4 \cdot 3 \cdot 2 \cdot 1}x^0 = 495$$

37. The x^4 term in $\binom{10}{j}(x)^{10-j}\left(\dfrac{-2}{\sqrt{x}}\right)^{j}$ occurs when:

$$10 - j - \tfrac{1}{2}j = 4 \;\rightarrow\; -\tfrac{3}{2}j = -6 \;\rightarrow\; j = 4.$$

Evaluate the 5th term:

$$\binom{10}{4}(x)^6\cdot\left(\dfrac{-2}{\sqrt{x}}\right)^4 = \dfrac{10!}{6!\,4!}\,x^6\cdot\dfrac{16}{x^2} = \dfrac{10\cdot9\cdot8\cdot7}{4\cdot3\cdot2\cdot1}\cdot16x^4 = 3360x^4$$

The coefficient is $3360.$

39.
$$(1.001)^5 = \left(1+10^{-3}\right)^5 = \binom{5}{0}\cdot1^5 + \binom{5}{1}\cdot1^4\cdot10^{-3} + \binom{5}{2}\cdot1^3\cdot\left(10^{-3}\right)^2 + \binom{5}{3}\cdot1^2\cdot\left(10^{-3}\right)^3 + \ldots$$
$$= 1 + 5(0.001) + 10(0.000001) + 10(0.000000001) + \ldots$$
$$= 1 + 0.005 + 0.000010 + 0.000000010 + \ldots$$
$$= 1.00501 \quad (\text{correct to } 5 \text{ decimal places})$$

41.
$$\binom{n}{n} = \dfrac{n!}{n!(n-n)!} = \dfrac{n!}{n!\,0!} = \dfrac{n!}{n!\cdot1} = \dfrac{n!}{n!} = 1$$

43.
$$\binom{n}{0} + \binom{n}{1} + \ldots + \binom{n}{n} = 2^n$$
$$= (1+1)^n$$
$$= \binom{n}{0}\cdot1^n + \binom{n}{1}\cdot1^{n-1}\cdot1 + \binom{n}{2}\cdot1^{n-2}\cdot1^2 + \ldots + \binom{n}{n}\cdot1^{n-n}\cdot1^n$$
$$= \binom{n}{0} + \binom{n}{1} + \ldots + \binom{n}{n}$$

45.
$$\binom{5}{0}\left(\tfrac{1}{4}\right)^5 + \binom{5}{1}\left(\tfrac{1}{4}\right)^4\left(\tfrac{3}{4}\right) + \binom{5}{2}\left(\tfrac{1}{4}\right)^3\left(\tfrac{3}{4}\right)^2 + \binom{5}{3}\left(\tfrac{1}{4}\right)^2\left(\tfrac{3}{4}\right)^3$$
$$+ \binom{5}{4}\left(\tfrac{1}{4}\right)\left(\tfrac{3}{4}\right)^4 + \binom{5}{5}\left(\tfrac{3}{4}\right)^5 = \left(\tfrac{1}{4}+\tfrac{3}{4}\right)^5 = (1)^5 = 1$$

12   Chapter Review

1. $a_1 = (-1)^1\dfrac{1+3}{1+2} = -\dfrac{4}{3}, \quad a_2 = (-1)^2\dfrac{2+3}{2+2} = \dfrac{5}{4}, \quad a_3 = (-1)^3\dfrac{3+3}{3+2} = -\dfrac{6}{5},$

$$a_4 = (-1)^4\dfrac{4+3}{4+2} = \dfrac{7}{6}, \quad a_5 = (-1)^5\dfrac{5+3}{5+2} = -\dfrac{8}{7}$$

3. $a_1 = \dfrac{2^1}{1^2} = \dfrac{2}{1} = 2, \quad a_2 = \dfrac{2^2}{2^2} = \dfrac{4}{4} = 1, \quad a_3 = \dfrac{2^3}{3^2} = \dfrac{8}{9}, \quad a_4 = \dfrac{2^4}{4^2} = \dfrac{16}{16} = 1, \quad a_5 = \dfrac{2^5}{5^2} = \dfrac{32}{25}$

5. $a_1 = 3, \quad a_2 = \tfrac{2}{3}\cdot3 = 2, \quad a_3 = \tfrac{2}{3}\cdot2 = \tfrac{4}{3}, \quad a_4 = \tfrac{2}{3}\cdot\tfrac{4}{3} = \tfrac{8}{9}, \quad a_5 = \tfrac{2}{3}\cdot\tfrac{8}{9} = \tfrac{16}{27}$

7. $a_1 = 2, \; a_2 = 2 - 2 = 0, \; a_3 = 2 - 0 = 2, \; a_4 = 2 - 2 = 0, \; a_5 = 2 - 0 = 2$

9. $\{n + 5\}$ Arithmetic
$$d = (n + 1 + 5) - (n + 5) = n + 6 - n - 5 = 1$$
$$S_n = \frac{n}{2}[6 + n + 5] = \frac{n}{2}(n + 11)$$

11. $\{2n^3\}$ Examine the terms of the sequence: 2, 16, 54, 128, 250, ...
There is no common difference; there is no common ratio; neither.

13. $\{2^{3n}\}$ Geometric $r = \dfrac{2^{3(n+1)}}{2^{3n}} = \dfrac{2^{3n+3}}{2^{3n}} = 2^{3n+3-3n} = 2^3 = 8$
$$S_n = 8\left(\frac{1 - 8^n}{1 - 8}\right) = 8\left(\frac{1 - 8^n}{-7}\right) = \frac{8}{7}(8^n - 1)$$

15.   0, 4, 8, 12, ...   Arithmetic $d = 4 - 0 = 4$
$$S_n = \frac{n}{2}(2(0) + (n - 1)4) = \frac{n}{2}(4(n - 1)) = 2n(n - 1)$$

17. $3, \frac{3}{2}, \frac{3}{4}, \frac{3}{8}, \frac{3}{16}, ...$ Geometric $r = \dfrac{\frac{3}{2}}{3} = \dfrac{3}{2} \cdot \dfrac{1}{3} = \dfrac{1}{2}$ $S_n = \dfrac{3}{1 - \frac{1}{2}} = \dfrac{3}{\frac{1}{2}} = 6$

19.   Neither.   Insufficient information to determine the type of sequence.

21. $\displaystyle\sum_{k=1}^{5}(k^2 + 12) = 13 + 16 + 21 + 28 + 37 = 115$

23. $\displaystyle\sum_{k=1}^{10}(3k - 9) = \sum_{k=1}^{10} 3k - \sum_{k=1}^{10} 9 = 3\sum_{k=1}^{10} k - \sum_{k=1}^{10} 9 = 3\left(\frac{10(10 + 1)}{2}\right) - 10(9) = 165 - 90 = 75$

25. $\displaystyle\sum_{k=1}^{7}\left(\tfrac{1}{3}\right)^k = \frac{1}{3}\left(\frac{1 - \left(\frac{1}{3}\right)^7}{1 - \frac{1}{3}}\right) = \frac{1}{3}\left(\frac{1 - \left(\frac{1}{3}\right)^7}{\frac{2}{3}}\right) = \frac{1}{2}\left(1 - \frac{1}{2187}\right) = \frac{1}{2} \cdot \frac{2186}{2187} = \frac{1093}{2187}$

27.   Arithmetic $a_1 = 3, \; d = 4, \; a_n = a + (n - 1)d$
$a_9 = 3 + (9 - 1)4 = 3 + 8(4) = 3 + 32 = 35$

29.   Geometric $a = 1, \; r = \frac{1}{10}, \; n = 11$ $a_{11} = 1 \cdot \left(\frac{1}{10}\right)^{11-1} = \left(\frac{1}{10}\right)^{10} = \frac{1}{10,000,000,000}$

31.   Arithmetic $a_1 = \sqrt{2}, \; d = \sqrt{2}, \; n = 9, \; a_n = a + (n - 1)d$
$a_9 = \sqrt{2} + (9 - 1)\sqrt{2} = \sqrt{2} + 8\sqrt{2} = 9\sqrt{2}$

33. $a_7 = a + 6d = 31 \qquad a_{20} = a + 19d = 96$

Solve the system of equations:
$$31 - 6d + 19d = 96$$
$$13d = 65$$
$$d = 5$$
$$a = 31 - 6(5) = 31 - 30 = 1$$
General formula: $\{5n - 4\}$

35. $a_{10} = a + 9d = 0 \qquad a_{18} = a + 17d = 8$

Solve the system of equations:
$$-9d + 17d = 8$$
$$8d = 8$$
$$d = 1$$
$$a = -9(1) = -9$$
General formula: $\{n - 10\}$

37. $a = 3, \; r = \frac{1}{3}$ Since $|r| < 1$, $S_n = \dfrac{a}{1-r} = \dfrac{3}{1-\frac{1}{3}} = \dfrac{3}{\frac{2}{3}} = \dfrac{9}{2}$

39. $a = 2, \; r = -\frac{1}{2}$ Since $|r| < 1$, $S_n = \dfrac{a}{1-r} = \dfrac{2}{1-\left(-\frac{1}{2}\right)} = \dfrac{2}{\frac{3}{2}} = \dfrac{4}{3}$

41. $a = 4, \; r = \frac{1}{2}$ Since $|r| < 1$, $S_n = \dfrac{a}{1-r} = \dfrac{4}{1-\frac{1}{2}} = \dfrac{4}{\frac{1}{2}} = 8$

43. I: $n = 1$: $3 \cdot 1 = 3$ and $\dfrac{3 \cdot 1}{2}(1 + 1) = 3$

II: If $3 + 6 + 9 + \cdots + 3k = \dfrac{3k}{2}(k + 1)$

then $3 + 6 + 9 + \cdots + 3k + 3(k + 1)$

$$= \left[3 + 6 + 9 + \cdots + 3k\right] + 3(k + 1) = \frac{3k}{2}(k + 1) + 3(k + 1)$$

$$= (k + 1)\left(\frac{3k}{2} + 3\right) = \frac{3}{2}(k + 1)(k + 2)$$

Conditions I and II are satisfied; the statement is true.

45. I: $n = 1$: $2 \cdot 3^{1-1} = 2$ and $3^1 - 1 = 2$

II: If $2 + 6 + 18 + \cdots + 2 \cdot 3^{k-1} = 3^k - 1$

then $2 + 6 + 18 + \cdots + 2 \cdot 3^{k-1} + 2 \cdot 3^{k+1-1}$

$$= \left[2 + 6 + 18 + \cdots + 2 \cdot 3^{k-1}\right] + 2 \cdot 3^k = 3^k - 1 + 2 \cdot 3^k$$

$$= 3 \cdot 3^k - 1 = 3^{k+1} - 1$$

Conditions I and II are satisfied; the statement is true.

47. I: $n = 1$: $(3 \cdot 1 - 2)^2 = 1$ and $\frac{1}{2} \cdot 1(6 \cdot 1^2 - 3 \cdot 1 - 1) = 1$

  II: If $1^2 + 4^2 + 7^2 + \cdots + (3k - 2)^2 = \frac{1}{2} k \left(6k^2 - 3k - 1\right)$

  then $1^2 + 4^2 + 7^2 + \cdots + (3k - 2)^2 + (3(k + 1) - 2)^2$

$$= \left[1^2 + 4^2 + 7^2 + \cdots + (3k - 2)^2\right] + (3k + 1)^2$$

$$= \frac{1}{2} k \left(6k^2 - 3k - 1\right) + (3k + 1)^2$$

$$= \frac{1}{2}\left[6k^3 - 3k^2 - k + 18k^2 + 12k + 2\right] = \frac{1}{2}\left[6k^3 + 15k^2 + 11k + 2\right]$$

$$= \frac{1}{2}(k + 1)\left[6k^2 + 9k + 2\right] = \frac{1}{2}(k + 1)\left[6k^2 + 12k + 6 - 3k - 3 - 1\right]$$

$$= \frac{1}{2}(k + 1)\left[6(k^2 + 2k + 1) - 3(k + 1) - 1\right]$$

$$= \frac{1}{2}(k + 1)\left[6(k + 1)^2 - 3(k + 1) - 1\right]$$

Conditions I and II are satisfied; the statement is true.

49. $(x + 2)^5 = \binom{5}{0}x^5 + \binom{5}{1}x^4 \cdot 2 + \binom{5}{2}x^3 \cdot 2^2 + \binom{5}{3}x^2 \cdot 2^3 + \binom{5}{4}x^1 \cdot 2^4 + \binom{5}{5} \cdot 2^5$

$$= x^5 + 5 \cdot 2x^4 + 10 \cdot 4x^3 + 10 \cdot 8x^2 + 5 \cdot 16x + 1 \cdot 32$$

$$= x^5 + 10x^4 + 40x^3 + 80x^2 + 80x + 32$$

51. $(2x + 3)^5 = \binom{5}{0}(2x)^5 + \binom{5}{1}(2x)^4 \cdot 3 + \binom{5}{2}(2x)^3 \cdot 3^2 + \binom{5}{3}(2x)^2 \cdot 3^3$

$$+ \binom{5}{4}(2x)^1 \cdot 3^4 + \binom{5}{5} \cdot 3^5$$

$$= 32x^5 + 5 \cdot 16x^4 \cdot 3 + 10 \cdot 8x^3 \cdot 9 + 10 \cdot 4x^2 \cdot 27 + 5 \cdot 2x \cdot 81 + 1 \cdot 243$$

$$= 32x^5 + 240x^4 + 720x^3 + 1080x^2 + 810x + 243$$

53. $n = 9$, $j = 2$, $x = x$, $a = 2$

$$\binom{9}{2}x^7 \cdot 2^2 = \frac{9!}{2! \, 7!} \cdot 4x^7 = \frac{9 \cdot 8}{2 \cdot 1} \cdot 4x^7 = 144x^7$$

The coefficient of x^7 is 144.

55. $n = 7$, $j = 5$, $x = 2x$, $a = 1$

$$\binom{7}{5}(2x)^2 \cdot 1^5 = \frac{7!}{5! \, 2!} \cdot 4x^2(1) = \frac{7 \cdot 6}{2 \cdot 1} \cdot 4x^2 = 84x^2$$

The coefficient of x^2 is 84.

57. This is an arithmetic sequence with $a = 80$, $d = -3$, $n = 25$

  (a) $a_{25} = 80 + (25 - 1)(-3) = 80 - 72 = 8$ bricks

  (b) $S_{25} = \frac{25}{2}(80 + 8) = 25(44) = 1100$ bricks

  1100 bricks are needed to build the steps.

59.   Since the fund returns 10% compound annually, this is equivalent to a return of $\frac{5}{6}\% = 0.00833$ monthly. Defining a recursive sequence, we have:

$$a_1 = 200, \quad a_n = 1.00833a_{n-1} + 200$$

Insert the formulas in your graphing utility and find the value at the end of 20 years or 240 months:

 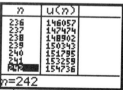

The value of the account will be approximately $151,795.

61.   This is a geometric sequence with $a = 20$, $r = \frac{3}{4}$.

(a)   After striking the ground the third time, the height is $20\left(\frac{3}{4}\right)^3 = \frac{135}{16} \approx 8.44$ feet .

(b)   After striking the ground the nth time, the height is $20\left(\frac{3}{4}\right)^n$ feet .

(c)   If the height is less than 6 inches or 0.5 feet, then:

$$0.5 = 20\left(\tfrac{3}{4}\right)^n$$
$$0.025 = \left(\tfrac{3}{4}\right)^n$$
$$\log 0.025 = n \log\left(\tfrac{3}{4}\right)$$
$$n = \frac{\log 0.025}{\log\left(\tfrac{3}{4}\right)} = 12.82$$

The height is less than 6 inches after the 13th strike.

(d)   Since this is a geometric sequence with $|r| < 1$, the distance is the sum of the two infinite geometric series - the distances going down plus the distances going up.

Distance going down:

$$S_{down} = \frac{20}{1 - \frac{3}{4}} = \frac{20}{\frac{1}{4}} = 80 \text{ feet.}$$

Distance going up:

$$S_{down} = \frac{15}{1 - \frac{3}{4}} = \frac{15}{\frac{1}{4}} = 60 \text{ feet.}$$

The total distance traveled is 140 feet.

63.   a)   Since the interest rate is 6.75% per annum compounded monthly, this is equivalent to a rate of 0.005625% each month. Defining a recursive sequence, we have:

$$a_1 = 190,000, \quad a_n = 1.005625a_{n-1} - 1232.34$$

(b) $1.005625(190,000) - 1232.34 = \$189,836.41$

(c) Enter the recursive formula in Y= and create the table:

```
Plot1 Plot2 Plot3
nMin=1
\u(n)◼(u(n-1))*1
.005625-1232.34
u(nMin)◼(19000...
\v(n)=
v(nMin)=
\w(n)=
```

n	u(n)
1	190000
2	189836
3	189672
4	189506
5	189340
6	189173
7	189005

n=1

(d) Scroll through the table:

n	u(n)
250	101535
251	100874
252	100209
253	99540
254	98868
255	98192
256	97512

n=250

At the beginning of the 253rd month, after 252 payments have been made, the balance is below $100,000. The balance is $99,540.

(e) Scroll through the table:

n	u(n)
355	7246.6
356	6055
357	4856.7
358	3651.7
359	2439.9
360	1221.3
361	■■■

u(n)=-4.20046497

The loan will be paid off at the end of 360 months or 30 years.

(f) The total interest expense is the difference of the total of the payments and the original loan:
$$1232.34(359) + 1221.27 - 190,000 = \$253,631.33$$

(g) a) Since the interest rate is 6.75% per annum compounded monthly, this is equivalent to a rate of 0.005625% each month. Defining a recursive sequence, we have:
$$a_1 = 190,000, \qquad a_n = 1.005625a_{n-1} - 1332.34$$

(b) $1.005625(190,000) - 1332.34 = \$189,736.41$

(c) Enter the recursive formula in Y= and create the table:

```
Plot1 Plot2 Plot3
nMin=1
\u(n)◼(u(n-1))*1
.005625-1332.34
u(nMin)◼(19000...
\v(n)=
v(nMin)=
\w(n)=
```

n	u(n)
1	190000
2	189736
3	189471
4	189205
5	188937
6	188667
7	188396

n=1

(d) Scroll through the table:

n	u(n)
190	101584
191	100823
192	100058
193	99288
194	98515
195	97736
196	96954

n=190

At the beginning of the 193rd month, after 192 payments have been made, the balance is below $100,000. The balance is $99,288.

(e)   Scroll through the table:

n	$u(n)$	
285	6372.7	
286	5076.2	
287	3772.4	
288	2461.3	
289	1142.8	
290	-183.1	
291	-1516	

$n=285$

The loan will be paid off at the end of 289 months.

(f)   The total interest expense is the difference of the total of the payments and the original loan:

$$1332.34(288) + 1142.80 - 190,000 = \$194,856.72$$

Chapter 13

Counting and Probability

13.1 Sets and Counting

1. $A \cup B = \{1, 3, 5, 7, 9\} \cup \{1, 5, 6, 7\} = \{1, 3, 5, 6, 7, 9\}$

3. $A \cap B = \{1, 3, 5, 7, 9\} \cap \{1, 5, 6, 7\} = \{1, 5, 7\}$

5. $(A \cup B) \cap C = (\{1, 3, 5, 7, 9\} \cup \{1, 5, 6, 7\}) \cap \{1, 2, 4, 6, 8, 9\}$
   $= \{1, 3, 5, 6, 7, 9\} \cap \{1, 2, 4, 6, 8, 9\}$
   $= \{1, 6, 9\}$

7. $(A \cap B) \cup C = (\{1, 3, 5, 7, 9\} \cap \{1, 5, 6, 7\}) \cup \{1, 2, 4, 6, 8, 9\}$
   $= \{1, 5, 7\} \cup \{1, 2, 4, 6, 8, 9\}$
   $= \{1, 2, 4, 5, 6, 7, 8, 9\}$

9. $(A \cup C) \cap (B \cup C)$
   $= (\{1, 3, 5, 7, 9\} \cup \{1, 2, 4, 6, 8, 9\}) \cap (\{1, 5, 6, 7\} \cup \{1, 2, 4, 6, 8, 9\})$
   $= \{1, 2, 3, 4, 5, 6, 7, 8, 9\} \cap \{1, 2, 4, 5, 6, 7, 8, 9\}$
   $= \{1, 2, 4, 5, 6, 7, 8, 9\}$

11. $\overline{A} = \{0, 2, 6, 7, 8\}$

13. $\overline{A \cap B} = \overline{\{1, 3, 4, 5, 9\} \cap \{2, 4, 6, 7, 8\}} = \overline{\{4\}} = \{0, 1, 2, 3, 5, 6, 7, 8, 9\}$

15. $\overline{A} \cup \overline{B} = \{0, 2, 6, 7, 8\} \cup \{0, 1, 3, 5, 9\} = \{0, 1, 2, 3, 5, 6, 7, 8, 9\}$

17. $\overline{A \cap \overline{C}} = \overline{\{1, 3, 4, 5, 9\} \cap \{0, 2, 5, 7, 8, 9\}} = \overline{\{5, 9\}} = \{0, 1, 2, 3, 4, 6, 7, 8\}$

19. $\overline{A \cup B \cup C} = \overline{\{1, 3, 4, 5, 9\} \cup \{2, 4, 6, 7, 8\} \cup \{1, 3, 4, 6\}}$
    $= \overline{\{1, 2, 3, 4, 5, 6, 7, 8, 9\}} = \{0\}$

21. $\{a\}, \{b\}, \{c\}, \{d\}, \{a, b\}, \{a, c\}, \{a, d\}, \{b, c\}, \{b, d\}, \{c, d\}, \{a, b, c\}, \{a, b, d\},$
     $\{a, c, d\}, \{b, c, d\}, \{a, b, c, d\}, \phi$

23. $n(A) = 15, n(B) = 20, n(A \cap B) = 10$
     $\quad n(A \cup B) = n(A) + n(B) - n(A \cap B) = 15 + 20 - 10 = 25$

25. $n(A \cup B) = 50, n(A \cap B) = 10, n(B) = 20$
     $\quad n(A \cup B) = n(A) + n(B) - n(A \cap B)$
     $\quad\quad\quad 50 = n(A) + 20 - 10$
     $\quad\quad\quad 40 = n(A)$

27. From the figure:
     $\quad n(A) = 15 + 3 + 5 + 2 = 25$

29. From the figure:
     $\quad n(A \text{ or } B) = n(A \cup B) = n(A) + n(B) - n(A \cap B) = 25 + 20 - 8 = 37$

31. From the figure:
     $\quad n(A \text{ but not } C) = n(A) - n(A \cap C) = 25 - 7 = 18$

33. From the figure:
     $\quad n(A \text{ and } B \text{ and } C) = n(A \cap B \cap C) = 5$

35. Let $A = \{$those who will purchase a major appliance $\}$
     $\quad B = \{$those who will buy a car $\}$
     $\quad n(U) = 500, \; n(A) = 200, \; n(B) = 150, \; n(A \cap B) = 25$
     $\quad n(A \cup B) = n(A) + n(B) - n(A \cap B) = 200 + 150 - 25 = 325$
     $\quad n(\text{purchase neither}) = 500 - 325 = 175$
     $\quad n(\text{purchase only a car}) = 150 - 25 = 125$

37. Construct a Venn diagram:

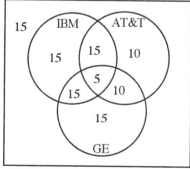

     (a) 15
     (b) 15
     (c) 15
     (d) 25
     (e) 40

39. (a) $n(\text{married}) = n(\text{married, spouse present}) + n(\text{married, spouse absent})$
$$= 54{,}654 + 3{,}232 = 57{,}886 \text{ thousand}$$

 (b) $n(\text{widowed or divorced}) = n(\text{widowed}) + n(\text{divorced})$
$$= 2{,}686 + 8{,}208 = 10{,}894 \text{ thousand}$$

 (c) $n(\text{married, spouse absent or widowed or divorced})$
$$= n(\text{married, spouse absent}) + n(\text{widowed}) + n(\text{divorced})$$
$$= 3{,}232 + 2{,}686 + 8{,}208 = 14{,}126 \text{ thousand}$$

13.2 Permutations and Combinations

1. $P(6, 2) = \dfrac{6!}{(6-2)!} = \dfrac{6!}{4!} = \dfrac{6 \cdot 5 \cdot 4!}{4!} = 30$

3. $P(4, 4) = \dfrac{4!}{(4-4)!} = \dfrac{4!}{0!} = \dfrac{4 \cdot 3 \cdot 2 \cdot 1}{1} = 24$

5. $P(7, 0) = \dfrac{7!}{(7-0)!} = \dfrac{7!}{7!} = 1$

7. $P(8, 4) = \dfrac{8!}{(8-4)!} = \dfrac{8!}{4!} = \dfrac{8 \cdot 7 \cdot 6 \cdot 5 \cdot 4!}{4!} = 1680$

9. $C(8, 2) = \dfrac{8!}{(8-2)!\,2!} = \dfrac{8!}{6!\,2!} = \dfrac{8 \cdot 7 \cdot 6!}{6! \cdot 2 \cdot 1} = 28$

11. $C(7, 4) = \dfrac{7!}{(7-4)!\,4!} = \dfrac{7!}{3!\,4!} = \dfrac{7 \cdot 6 \cdot 5 \cdot 4!}{4! \cdot 3 \cdot 2 \cdot 1} = 35$

13. $C(15, 15) = \dfrac{15!}{(15-15)!\,15!} = \dfrac{15!}{0!\,15!} = \dfrac{15!}{15! \cdot 1} = 1$

15. $C(26, 13) = \dfrac{26!}{(26-13)!\,13!} = \dfrac{26!}{13!\,13!} = 10{,}400{,}600$

17. {*abc, abd, abe, acb, acd, ace, adb, adc, ade, aeb, aec, aed, bac, bad, bae, bca, bcd, bce, bda, bdc, bde, bea, bec, bed, cab, cad, cae, cba, cbd, cbe, cda, cdb, cde, cea, ceb, ced, dab, dac, dae, dba, dbc, dbe, dca, dcb, dce, dea, deb, dec, eab, eac, ead, eba, ebc, ebd, eca, ecb, ecd, eda, edb, edc*}
$$P(5,3) = \dfrac{5!}{(5-3)!} = \dfrac{5!}{2!} = \dfrac{5 \cdot 4 \cdot 3 \cdot 2!}{2!} = 60$$

19. {123, 124, 132, 134, 142, 143, 213, 214, 231, 234, 241, 243, 312, 314, 321, 324, 341, 342, 412, 413, 421, 423, 431, 432}

$$P(4,3) = \frac{4!}{(4-3)!} = \frac{4!}{1!} = \frac{4 \cdot 3 \cdot 2 \cdot 1}{1} = 24$$

21. {abc, abd, abe, acd, ace, ade, bcd, bce, bde, cde}

$$C(5,3) = \frac{5!}{(5-3)!\,3!} = \frac{5 \cdot 4 \cdot 3!}{2 \cdot 1 \cdot 3!} = 10$$

23. {123, 124, 134, 234} $C(4,3) = \dfrac{4!}{(4-3)!\,3!} = \dfrac{4 \cdot 3!}{1!\,3!} = 4$

25. There are 5 choices of shirts and 3 choices of ties; there are (5)(3) = 15 combinations.

27. There are 4 choices for the first letter in the code and 4 choices for the second letter in the code; there are (4)(4) = 16 possible two-letter codes.

29. There are two choices for each of three positions; there are (2)(2)(2) = 8 possible three-digit numbers.

31. To line up the four people, there are 4 choices for the first position, 3 choices for the second position, 2 choices for the third position, and 1 choice for the fourth position. Thus there are (4)(3)(2)(1) = 24 possible ways four people can be lined up.

33. Since no letter can be repeated, there are 5 choices for the first letter, 4 choices for the second letter, and 3 choices for the 3 letter. Thus, there are (5)(4)(3) = 60 possible three-letter codes.

35. There are 26 possible one-letter names. There are (26)(26) = 676 possible two-letter names. There are (26)(26)(26) = 17576 possible three-letter names. Thus, there are 26 + 676 + 17576 = 18,278 possible companies that can be listed on the New York Stock Exchange.

37. A committee of 4 from a total of 7 students is given by:

$$C(7,4) = \frac{7!}{(7-4)!\,4!} = \frac{7!}{3!\,4!} = \frac{7 \cdot 6 \cdot 5 \cdot 4!}{3 \cdot 2 \cdot 1 \cdot 4!} = 35$$

35 committees are possible.

39. There are 2 possible answers for each question. Therefore, there are $2^{10} = 1024$ different possible arrangements of the answers.

41. There are 9 choices for the first digit, and 10 choices for each of the other three digits. Thus, there are (9)(10)(10)(10) = 9000 possible four-digit numbers.

43. There are 5 choices for the first position, 4 choices for the second position, 3 choices for the third position, 2 choices for the fourth position, and 1 choice for the fifth position. Thus, there are (5)(4)(3)(2)(1) = 120 possible arrangements of the books.

45. There are 8 choices for the DOW stocks, 15 choices for the NASDAQ stocks, and 4 choices for the global stocks. Thus, there are $(8)(15)(4) = 480$ different portfolios.

47. The first person can have any of 365 days, the second person can have any of the remaining 364 days, and the third person can have any of the remaining 363 days. Thus, there are $(365)(364)(363) = 48,228,180$ possible ways three people can have different birthdays.

49. Choosing 2 boys from the 4 boys can be done $C(4,2)$ ways, and choosing 3 girls from the 8 girls can be done in $C(8,3)$ ways. Thus, there are a total of:

$$C(4,2) \cdot C(8,3) = \frac{4!}{(4-2)!\,2!} \cdot \frac{8!}{(8-3)!\,3!} = \frac{4!}{2!\,2!} \cdot \frac{8!}{5!\,3!}$$

$$= \frac{4 \cdot 3!}{2 \cdot 1 \cdot 2 \cdot 1} \cdot \frac{8 \cdot 7 \cdot 6 \cdot 5!}{5!\,3!} = 336$$

51. The committee is made up of 2 of 4 administrators, 3 of 8 faculty, and 5 of 20 students. The number of possible committees is:

$$C(4,2) \cdot C(8,3) \cdot C(20,5) = \frac{4!}{(4-2)!\,2!} \cdot \frac{8!}{(8-3)!\,3!} \cdot \frac{20!}{(20-5)!\,5!}$$

$$= \frac{4!}{2!\,2!} \cdot \frac{8!}{5!\,3!} \cdot \frac{20!}{15!\,5!} = \frac{4!}{2 \cdot 1 \cdot 2 \cdot 1} \cdot \frac{8 \cdot 7 \cdot 6 \cdot 5!}{5 \cdot 4!\,3!} \cdot \frac{20 \cdot 19 \cdot 18 \cdot 17 \cdot 16 \cdot 15!}{15!\,5!}$$

$$= 5,209,344 \text{ possible committees}$$

53. There are 9 choices for the first position, 8 choices for the second position, 7 for the third position, etc. There are $9 \cdot 8 \cdot 7 \cdot 6 \cdot 5 \cdot 4 \cdot 3 \cdot 2 \cdot 1 = 9! = 362,880$ possible batting orders.

55. This is a permutation with repetition. There are $\dfrac{9!}{2!\,2!} = 90,720$ different words.

57. There are $C(100, 22)$ ways to form the first committee. There are 78 senators left, so there are $C(78, 13)$ ways to form the second committee. There are $C(65, 10)$ ways to form the third committee. There are $C(55, 5)$ ways to form the fourth committee. There are $C(50, 16)$ ways to form the fifth committee. There are $C(34, 17)$ ways to form the sixth committee. There are $C(17, 17)$ ways to form the seventh committee. The total number of committees =

$$= C(100,22) \cdot C(78,13) \cdot C(65,10) \cdot C(55,5) \cdot C(50,16) \cdot C(34,17) \cdot C(17,17)$$

$$= 1.1568 \times 10^{76}$$

59. Choose 2 players from a group of 6 players. Therefore, there are $C(6,2) = 15$ different teams possible.

61. (a) $C(7,2) \cdot C(3,1) = 21 \cdot 3 = 63$

    (b) $C(7,3) = 35$

    (c) $C(3,3) = 1$

13.3 Probability

1. Probabilities must be between 0 and 1, inclusive. Thus, 0, 0.01, 0.35, and 1 are probabilities.

3. All the probabilities are between 0 and 1.
   The sum of the probabilities is $0.2 + 0.3 + 0.1 + 0.4 = 1$.
   This is a probability model.

5. All the probabilities are between 0 and 1.
   The sum of the probabilities is $0.3 + 0.2 + 0.1 + 0.3 = 0.9$.
   This is not a probability model.

7. The sample space is: $S = \{HH, HT, TH, TT\}$.

   Each outcome is equally likely to occur; so $P(E) = \dfrac{n(E)}{n(S)}$.

   The probabilities are: $P(HH) = \dfrac{1}{4}$, $P(HT) = \dfrac{1}{4}$, $P(TH) = \dfrac{1}{4}$, $P(TT) = \dfrac{1}{4}$.

9. The sample space of tossing two fair coins and a fair die is:
   $$S = \{HH1, HH2, HH3, HH4, HH5, HH6, HT1, HT2, HT3, HT4, HT5,$$
   $$HT6, TH1, TH2, TH3, TH4, TH5, TH6, TT1, TT2, TT3, TT4, TT5, TT6\}$$

   There are 24 equally likely outcomes and the probability of each is $\dfrac{1}{24}$.

11. The sample space for tossing three fair coins is:
    $$S = \{HHH, HHT, HTH, THH, HTT, THT, TTH, TTT\}$$

    There are 8 equally likely outcomes and the probability of each is $\dfrac{1}{8}$.

13. The sample space is:
    $$S = \{1 \text{ Yellow}, 1 \text{ Red}, 1 \text{ Green}, 2 \text{ Yellow}, 2 \text{ Red}, 2 \text{ Green}, 3 \text{ Yellow}, 3 \text{ Red},$$
    $$3 \text{ Green}, 4 \text{ Yellow}, 4 \text{ Red}, 4 \text{ Green}\}$$

    There are 12 equally likely events and the probability of each is $\dfrac{1}{12}$. The probability

    of getting a 2 or 4 followed by a Red is $P(2 \text{ Red}) + P(4 \text{ Red}) = \dfrac{1}{12} + \dfrac{1}{12} = \dfrac{1}{6}$.

15. The sample space is:
S = {1 Yellow Forward, 1 Yellow Backward, 1 Red Forward, 1 Red Backward,
1 Green Forward, 1 Green Backward, 2 Yellow Forward, 2 Yellow Backward,
2 Red Forward, 2 Red Backward, 2 Green Forward, 2 Green Backward,
3 Yellow Forward, 3 Yellow Backward, 3 Red Forward, 3 Red Backward,
3 Green Forward, 3 Green Backward, 4 Yellow Forward, 4 Yellow Backward,
4 Red Forward, 4 Red Backward, 4 Green Forward, 4 Green Backward}

There are 24 equally likely events and the probability of each is $\frac{1}{24}$. The probability
of getting a 1, followed by a Red or Green, followed by a Backward is

$$P(1 \text{ Red Backward}) + P(1 \text{ Green Backward}) = \frac{1}{24} + \frac{1}{24} = \frac{1}{12}.$$

17. The sample space is:
S = {1 1 Yellow, 1 1 Red, 1 1 Green, 1 2 Yellow, 1 2 Red, 1 2 Green, 1 3 Yellow,
1 3 Red, 1 3 Green, 1 4 Yellow, 1 4 Red, 1 4 Green, 2 1 Yellow, 2 1 Red,
2 1 Green, 2 2 Yellow, 2 2 Red, 2 2 Green, 2 3 Yellow, 2 3 Red, 2 3 Green,
2 4 Yellow, 2 4 Red, 2 4 Green, 3 1 Yellow, 3 1 Red, 3 1 Green, 3 2 Yellow,
3 2 Red, 3 2 Green, 3 3 Yellow, 3 3 Red, 3 3 Green, 3 4 Yellow, 3 4 Red,
3 4 Green, 4 1 Yellow, 4 1 Red, 4 1 Green, 4 2 Yellow, 4 2 Red, 4 2 Green,
4 3 Yellow, 4 3 Red, 4 3 Green, 4 4 Yellow, 4 4 Red, 4 4 Green}

There are 48 equally likely events and the probability of each is $\frac{1}{48}$. The probability of
getting a 2, followed by a 2 or 4, followed by a Red or Green is

$$P(2\ 2 \text{ Red}) + P(2\ 4 \text{ Red}) + P(2\ 2 \text{ Green}) + P(2\ 4 \text{ Green}) = \frac{1}{48} + \frac{1}{48} + \frac{1}{48} + \frac{1}{48} = \frac{1}{12}$$

19. A, B, C, F

21. B

23. Let $P(\text{tails}) = x$, then $P(\text{heads}) = 4x$
$$\begin{aligned} x + 4x &= 1 \\ 5x &= 1 \\ x &= \tfrac{1}{5} \qquad P(\text{tails}) = \tfrac{1}{5}, \quad P(\text{heads}) = \tfrac{4}{5} \end{aligned}$$

25. $P(2) = P(4) = P(6) = x \qquad P(1) = P(3) = P(5) = 2x$
$$\begin{aligned} P(1) + P(2) + P(3) + P(4) + P(5) + P(6) &= 1 \\ 2x + x + 2x + x + 2x + x &= 1 \\ 9x &= 1 \\ x &= \tfrac{1}{9} \end{aligned}$$
$P(2) = P(4) = P(6) = \tfrac{1}{9} \qquad P(1) = P(3) = P(5) = \tfrac{2}{9}$

27. $P(E) = \dfrac{n(E)}{n(S)} = \dfrac{n\{1,2,3\}}{10} = \dfrac{3}{10}$

29. $P(E) = \dfrac{n(E)}{n(S)} = \dfrac{n\{2,4,6,8,10\}}{10} = \dfrac{5}{10} = \dfrac{1}{2}$

31. $P(\text{white}) = \dfrac{n(\text{white})}{n(S)} = \dfrac{5}{5+10+8+7} = \dfrac{5}{30} = \dfrac{1}{6}$

33.   The sample space is: S = {BBB, BBG, BGB, GBB, BGG, GBG, GGB, GGG}

$P(3 \text{ boys}) = \dfrac{n(3 \text{ boys})}{n(S)} = \dfrac{1}{8}$

35.   The sample space is:
S = {BBBB, BBBG, BBGB, BGBB, GBBB, BBGG, BGBG, GBBG, BGGB, GBGB, GGBB, BGGG, GBGG, GGBG, GGGB, GGGG}

$P(1 \text{ girl}, 3 \text{ boys}) = \dfrac{n(1 \text{ girl}, 3 \text{ boys})}{n(S)} = \dfrac{4}{16} = \dfrac{1}{4}$

37. $P(\text{sum of two die is } 7) = \dfrac{n(\text{sum of two die is } 7)}{n(S)}$

$= \dfrac{n\{1,6 \text{ or } 2,5 \text{ or } 3,4 \text{ or } 4,3 \text{ or } 5,2 \text{ or } 6,1\}}{n(S)} = \dfrac{6}{36} = \dfrac{1}{6}$

39. $P(\text{sum of two die is } 3) = \dfrac{n(\text{sum of two die is } 3)}{n(S)} = \dfrac{n\{1,2 \text{ or } 2,1\}}{n(S)} = \dfrac{2}{36} = \dfrac{1}{18}$

41. $P(A \cup B) = P(A) + P(B) - P(A \cap B) = 0.25 + 0.45 - 0.15 = 0.55$

43. $P(A \cup B) = P(A) + P(B) = 0.25 + 0.45 = 0.70$

45. $P(A \cup B) = P(A) + P(B) - P(A \cap B)$
$0.85 = 0.60 + P(B) - 0.05$
$P(B) = 0.85 - 0.60 + 0.05 = 0.30$

47. $P(\text{not victim}) = 1 - P(\text{victim}) = 1 - 0.253 = 0.747$

49. $P(\text{not in 70's}) = 1 - P(\text{in 70's}) = 1 - 0.3 = 0.7$

51. $P(\text{white or green}) = P(\text{white}) + P(\text{green}) = \dfrac{n(\text{white}) + n(\text{green})}{n(S)} = \dfrac{9+8}{9+8+3} = \dfrac{17}{20}$

53. $P(\text{not white}) = 1 - P(\text{white}) = 1 - \dfrac{n(\text{white})}{n(S)} = 1 - \dfrac{9}{20} = \dfrac{11}{20}$

55. $P(\text{strike or one}) = P(\text{strike}) + P(\text{one}) = \dfrac{n(\text{strike}) + n(\text{one})}{n(S)} = \dfrac{3+1}{8} = \dfrac{4}{8} = \dfrac{1}{2}$

57.   There are 30 households out of 100 with an income of $30,000 or more.

$P(E) = \dfrac{n(E)}{n(S)} = \dfrac{n(30,000 \text{ or more})}{n(\text{total households})} = \dfrac{30}{100} = \dfrac{3}{10}$

59. There are 40 households out of 100 with an income of less than $20,000.

$$P(E) = \frac{n(E)}{n(S)} = \frac{n(\text{less than } \$20{,}000)}{n(\text{total households})} = \frac{40}{100} = \frac{2}{5}$$

61. (a) $P(1 \text{ or } 2) = P(1) + P(2) = 0.24 + 0.33 = 0.57$
     (b) $P(1 \text{ or more}) = P(1) + P(2) + P(3) + P(4 \text{ or more})$
           $\qquad\qquad = 0.24 + 0.33 + 0.21 + 0.17 = 0.95$
     (c) $P(3 \text{ or fewer}) = P(0) + P(1) + P(2) + P(3) = 0.05 + 0.24 + 0.33 + 0.21 = 0.83$
     (d) $P(3 \text{ or more}) = P(3) + P(4 \text{ or more}) = 0.21 + 0.17 = 0.38$
     (e) $P(\text{less than } 2) = P(0) + P(1) = 0.05 + 0.24 = 0.29$
     (f) $P(\text{less than } 1) = P(0) = 0.05$
     (g) $P(1, 2, \text{ or } 3) = P(1) + P(2) + P(3) = 0.24 + 0.33 + 0.21 = 0.78$
     (h) $P(2 \text{ or more}) = P(2) + P(3) + P(4 \text{ or more}) = 0.33 + 0.21 + 0.17 = 0.71$

63. (a) $P(\text{freshman or female}) = P(\text{freshman}) + P(\text{female}) - P(\text{freshman and female})$
$$= \frac{n(\text{freshman}) + n(\text{female}) - n(\text{freshman and female})}{n(S)}$$
$$= \frac{18 + 15 - 8}{33} = \frac{25}{33}$$
     (b) $P(\text{sophomore or male}) = P(\text{sophomore}) + P(\text{male}) - P(\text{sophomore and male})$
$$= \frac{n(\text{sophomore}) + n(\text{male}) - n(\text{sophomore and male})}{n(S)}$$
$$= \frac{15 + 18 - 8}{33} = \frac{25}{33}$$

65. $P(\text{at least 2 with same birthday}) = 1 - P(\text{none with same birthday})$
$$= 1 - \frac{n(\text{different birthdays})}{n(S)}$$
$$= 1 - \frac{365 \cdot 364 \cdot 363 \cdot 362 \cdot 361 \cdot 360 \cdot \ldots \cdot 354}{365^{12}}$$
$$= 1 - 0.833$$
$$= 0.167$$

67. The sample space for picking 5 out of 10 numbers in a particular order contains
$$P(10,5) = \frac{10!}{(10-5)!} = \frac{10!}{5!} = 30{,}240 \text{ possible outcomes.}$$
     One of these is the desired outcome. Thus, the probability of winning is:
$$P(E) = \frac{n(E)}{n(S)} = \frac{n(\text{winning})}{n(\text{total possible outcomes})} = \frac{1}{30240}$$

69. (a) $P(3 \text{ heads}) = \dfrac{C(5,3)}{2^5} = \dfrac{10}{32} = \dfrac{5}{16}$
     (b) $P(0 \text{ heads}) = \dfrac{C(5,0)}{2^5} = \dfrac{1}{32}$

71. (a) $P(\text{sum} = 7 \text{ three times}) = P(\text{sum} = 7) \cdot P(\text{sum} = 7) \cdot P(\text{sum} = 7)$

$$= \frac{1}{6} \cdot \frac{1}{6} \cdot \frac{1}{6} = \frac{1}{216}$$

(b) $P(\text{sum} = 7 \text{ or } 11 \text{ at least twice})$

$$= P(\text{sum} = 7 \text{ or } 11) \cdot P(\text{sum} = 7 \text{ or } 11) \cdot P(\text{sum} \neq 7 \text{ or } 11) +$$
$$P(\text{sum} = 7 \text{ or } 11) \cdot P(\text{sum} = 7 \text{ or } 11) \cdot P(\text{sum} = 7 \text{ or } 11)$$

$$= \frac{8}{36} \cdot \frac{8}{36} \cdot \frac{28}{36} + \frac{8}{36} \cdot \frac{8}{36} \cdot \frac{8}{36} = 0.049$$

73. $P(\text{all } 5 \text{ defective}) = \dfrac{n(5 \text{ defective})}{n(S)} = \dfrac{1}{C(30,5)} = 7.02 \times 10^{-6}$

$P(\text{at least } 2 \text{ defective}) = 1 - (P(\text{none defective}) + P(\text{one defective}))$

$$= 1 - \left(\frac{C(5,0) \cdot C(25,5)}{C(30,5)} + \frac{C(5,1) \cdot C(25,4)}{C(30,5)} \right) = 1 - 0.817 = 0.183$$

75. $P(\text{one of 5 coins is valued at more than } \$10,000) = \dfrac{C(49,4) \cdot C(1,1)}{C(50,5)} = 0.1$

13.4 Analyzing Univariate Data; Probabilities from Data

1. (a)

Cause of Death	Probability
Accidents and adverse effects	0.4031
Homicide and legal intervention	0.2034
Suicide	0.1427
Malignant neoplasms	0.0476
Diseases of heart	0.0287
Human immunodeficiency virus infection	0.0192
Congenital anomalies	0.0127
Chronic obstructive pulmonary diseases	0.0066
Pneumonia and influenza	0.0057
Cerebrovascular diseases	0.0049
All other causes	0.1254

(b) $P(\text{suicide}) = 0.1427 = 14.27\%$

(c) $P(\text{suicide or malignant neoplasms}) = P(\text{suicide}) + P(\text{malignant neoplasms})$
$$= 0.1427 + 0.0476 = 0.1903 = 19.03\%$$

(d) $P(\text{not suicide nor malignant neoplasms}) = 1 - P(\text{suicide or malignant neoplasms})$
$$= 1 - 0.1903 = 0.8097 = 80.97\%$$

3. (a)

Age	Probability
20-24	0.0964
25-29	0.1044
30-34	0.1195
35-39	0.1224
40-44	0.1155
45-49	0.1039
50-54	0.0815
55-59	0.0651
60-64	0.0569
65-69	0.0507
70-74	0.0419
75-79	0.0279
80-84	0.0140

(b) $P(45-49) = 0.1039 = 10.39\%$

(c) $P(50-54 \text{ or } 55-59) = P(50-54) + P(55-59)$
$$= 0.0815 + 0.0651 = 0.1466 = 14.66\%$$

(d) $P(\text{not } 45-49) = 1 - P(45-49) = 1 - 0.1039 = 0.8961 = 89.61\%$

5. (a)

Tuition	Probability
0-999	0.0094
1000-1999	0.0065
2000-2999	0.0421
3000-3999	0.0617
4000-4999	0.0786
5000-5999	0.0786
6000-6999	0.0907
7000-7999	0.1104
8000-8999	0.1291
9000-9999	0.1029
10000-10999	0.0973
11000-11999	0.0767
12000-12999	0.0571
13000-13999	0.0318
14000-14999	0.0271

(b) $P(11000-11999) = 0.0767 = 7.67\%$

(c) $P(8000-10999) = P(8000-8999) + P(9000-9999) + P(10000-10999)$
$$= 0.1291 + 0.1029 + 0.0973 = 0.3293 = 32.93\%$$

(d) $P(\text{not } 11000-11999) = 1 - 0.0767 = 0.9233 = 92.33\%$

7. (a)

Marital Status	Probability
Married, spouse present	0.5805
Married, spouse absent	0.0343
Widowed	0.0285
Divorced	0.0872
Never married	0.2695

(b) $P(\text{married, spouse present}) = 0.5805 = 58.05\%$

(c) $P(\text{never married}) = 0.2695 = 26.95\%$

(d) $P(\text{married, spouse present or married, spouse absent})$
$$= P(\text{married, spouse present}) + P(\text{married, spouse absent})$$
$$= 0.5805 + 0.0343 = 0.6148 = 61.48\%$$

(e) $P(\text{not married, spouse present}) = 1 - 0.5805 = 0.4195 = 41.95\%$

9. $P(\text{Titleist}) = \dfrac{n(\text{Titleist})}{n(S)} = \dfrac{4}{10} = \dfrac{2}{5}$

11. $P(2\text{ boys, 1 girl}) = \dfrac{n(2\text{ boys, 1 girl})}{n(S)} = \dfrac{20}{50} = \dfrac{2}{5}$

13. $P(\text{tornado, 5-6p,m,}) = \dfrac{n(\text{tornado, 5-6p,m,})}{n(S)} = \dfrac{3262}{28538} = 0.1143 = 11.43\%$

$P(\text{tornado, not 5-6p,m,}) = 1 - P(\text{tornado, 5-6p,m,})$
$$= 1 - \dfrac{3262}{28538} = 1 - 0.1143 = 0.8857 = 88.57\%$$

15. $P(\text{cloudy}) = \dfrac{n(\text{cloudy})}{n(S)} = \dfrac{11.2}{30} = 0.3733 = 37.33\%$

$P(\text{clear}) = \dfrac{n(\text{clear})}{n(S)} = \dfrac{7.3}{30} = 0.2433 = 24.33\%$

$P(\text{not clear}) = 1 - P(\text{clear}) = 1 - 0.2433 = 0.7567 = 75.67\%$

17. $P(\text{thunderstorm day}) = \dfrac{n(\text{thunderstorm day})}{n(S)} = \dfrac{6.4}{30} = 0.2133 = 21.33\%$

$P(\text{not thunderstorm day}) = 1 - P(\text{thunderstorm day})$
$$= 1 - 0.2133 = 0.7867 = 78.67\%$$

19. (a)

Field of Home Run	Probability
Left	0.4194
Left center	0.3387
Center	0.1935
Right center	0.0484
Right	0

(b) $P(\text{left field}) = 0.4194 = 41.94\%$

(c) $P(\text{center field}) = 0.1935 = 19.35\%$

(d) $P(\text{right field}) = 0 = 0\%$

(e) Answers will vary.

13 Chapter Review

1. $A \cup B = \{1, 3, 5, 7\} \cup \{3, 5, 6, 7, 8\} = \{1, 3, 5, 6, 7, 8\}$

3. $A \cap C = \{1, 3, 5, 7\} \cap \{2, 3, 7, 8, 9\} = \{3, 7\}$

5. $\overline{A} \cup \overline{B} = \overline{\{1, 3, 5, 7\}} \cup \overline{\{3, 5, 6, 7, 8\}} = \{2, 4, 6, 8, 9\} \cup \{1, 2, 4, 9\} = \{1, 2, 4, 6, 8, 9\}$

7. $\overline{B \cap C} = \overline{\{3, 5, 6, 7, 8\} \cap \{2, 3, 7, 8, 9\}} = \overline{\{3, 7, 8\}} = \{1, 2, 4, 5, 6, 9\}$

9. $n(A) = 8, \ n(B) = 12, \ n(A \cap B) = 3$
   $n(A \cup B) = n(A) + n(B) - n(A \cap B) = 8 + 12 - 3 = 17$

11. From the figure:
    $n(A) = 20 + 2 + 6 + 1 = 29$

13. From the figure:
    $n(A \text{ and } C) = n(A \cap C) = 1 + 6 = 7$

15. From the figure:
    $n(\text{neither in } A \text{ nor in } C) = n(\overline{A \cup C}) = 20 + 5 = 25$

17. $5! = 5 \cdot 4 \cdot 3 \cdot 2 \cdot 1 = 120$

19. $P(8,3) = \dfrac{8!}{(8-3)!} = \dfrac{8!}{5!} = \dfrac{8 \cdot 7 \cdot 6 \cdot 5!}{5!} = 336$

21. $C(8,3) = \dfrac{8!}{(8-3)! \, 3!} = \dfrac{8!}{5! \, 3!} = \dfrac{8 \cdot 7 \cdot 6 \cdot 5!}{5! \cdot 3 \cdot 2 \cdot 1} = 56$

23. There are 2 choices of material, 3 choices of color, and 10 choices of size. The complete assortment would have: $2 \cdot 3 \cdot 10 = 60$ suits.

25. There are two possible outcomes for each game or
$2 \cdot 2 \cdot 2 \cdot 2 \cdot 2 \cdot 2 \cdot 2 = 2^7 = 128$ outcomes for 7 games.

27. Since order is significant, this is a permutation.
$$P(9,4) = \frac{9!}{(9-4)!} = \frac{9!}{5!} = \frac{9 \cdot 8 \cdot 7 \cdot 6 \cdot 5!}{5!} = 3024 \text{ ways to seat 4 people in 9 seats.}$$

29. Choose 4 runners - order is not significant:
$$C(8,4) = \frac{8!}{(8-4)!\,4!} = \frac{8!}{4!\,4!} = \frac{8 \cdot 7 \cdot 6 \cdot 5 \cdot 4!}{4 \cdot 3 \cdot 2 \cdot 1 \cdot 4!} = 70 \text{ ways a squad can be chosen.}$$

31. Choose 14 teams 2 at a time:
$$C(14,2) = \frac{14!}{(14-2)!\,2!} = \frac{14!}{12!\,2!} = \frac{14 \cdot 13 \cdot 12!}{12! \cdot 2 \cdot 1} = 91 \text{ ways to pair 14 teams.}$$

33. There are $8 \cdot 10 \cdot 10 \cdot 10 \cdot 10 \cdot 10 \cdot 2 = 1,600,000$ possible phone numbers.

35. There are $24 \cdot 9 \cdot 10 \cdot 10 \cdot 10 = 216,000$ possible license plates.

37. Since there are repeated letters:
$$\frac{7!}{2! \cdot 2!} = \frac{7 \cdot 6 \cdot 5 \cdot 4 \cdot 3 \cdot 2 \cdot 1}{2 \cdot 1 \cdot 2 \cdot 1} = 1260 \text{ different words can be formed.}$$

39. (a) $C(9,4) \cdot C(9,3) \cdot C(9,2) = 126 \cdot 84 \cdot 36 = 381,024$ committees can be formed.
    (b) $C(9,4) \cdot C(5,3) \cdot C(2,2) = 126 \cdot 10 \cdot 1 = 1260$ committees can be formed.

41. (a)

Ethnic Group	Probability
Asian, Pacific Islander	0.3522
Black	0.0166
American Indian, Eskimo, Aleut	0.0011
Mexican American, Puerto Rican, or other Hispanic	0.0399
White (non-Hispanic)	0.5836
Unknown	0.0066

(b) $P(\text{Asian, Pacific Islander}) = 0.3522 = 35.22\%$

(c) $P(\text{Asian, Pacific Islander or Mexican American, Puerto Rican})$
$= P(\text{Asian, Pacific Islander}) + P(\text{Mexican American, Puerto Rican})$
$= 0.3522 + 0.0399 = 0.3921 = 39.21\%$

(d) $P(\text{not Asian, Pacific Islander}) = 1 - P(\text{Asian, Pacific Islander})$
$= 1 - 0.3522 = 0.6478 = 64.78\%$

43. (a)

Age	Probability
25-29	0.2001
30-34	0.1599
35-39	0.1480
40-44	0.1403
45-49	0.1244
50-54	0.1019
55-59	0.0756
60-64	0.0498

(b) $P(35 - 39) = 0.1480 = 14.80\%$

(c) $P(35 - 39 \text{ or } 55 - 59) = P(35 - 39) + P(55 - 59)$

$\qquad\qquad = 0.1480 + 0.0756 = 0.2236 = 22.36\%$

(d) $P(\text{not } 35 - 39) = 1 - P(35 - 39) = 1 - 0.1480 = 0.8520 = 85.20\%$

45. (a) $365 \cdot 364 \cdot 363 \cdot 362 \cdot \ldots \cdot 348 = 8.634628387 \times 10^{45}$

(b) $P(\text{no one has same birthday}) = \dfrac{365 \cdot 364 \cdot 363 \cdot 362 \cdot \ldots \cdot 348}{365^{18}} = 0.6531 = 65.31\%$

(c) $P(\text{at least 2 have same birthday}) = 1 - P(\text{no one has same birthday})$

$\qquad\qquad = 1 - 0.6531 = 0.3469 = 34.69\%$

47. (a) $P(\text{unemployed}) = 0.054 = 5.4\%$

(b) $P(\text{not unemployed}) = 1 - P(\text{unemployed}) = 1 - 0.054 = 0.946 = 94.6\%$

49. $P(\$1 \text{ bill}) = \dfrac{n(\$1 \text{ bill})}{n(S)} = \dfrac{4}{9}$

51. Let S be all possible selections, let D be a card that is divisible by 5, and let PN be a 1 or a prime number.

$\qquad n(S) = 100$

$\qquad n(D) = 20 \qquad$ (There are 20 number divisible by 5 between 1 and 100.)

$\qquad n(PN) = 26 \qquad$ (There are 26 prime numbers or 1 less than or equal to 100.)

$\qquad P(D) = \dfrac{n(D)}{n(S)} = \dfrac{20}{100} = \dfrac{1}{5} = 0.2$

$\qquad P(PN) = \dfrac{n(PN)}{n(S)} = \dfrac{26}{100} = \dfrac{13}{50} = 0.26$

53. (a) $P(5 \text{ heads}) = \dfrac{n(5 \text{ heads})}{n(S)} = \dfrac{C(10,5)}{2^{10}} = \dfrac{\frac{10!}{5!\,5!}}{1024} = \dfrac{252}{1024} = 0.2461$

(b) $P(\text{all heads}) = \dfrac{n(\text{all heads})}{n(S)} = \dfrac{1}{2^{10}} = \dfrac{1}{1024} = 0.00098$

Chapter 14

A Preview of Calculus: The Limit, Derivative, and Integral of a Function

14.1 Finding Limits Using Tables and Graphs

1. $\lim\limits_{x\to 2}\left(4x^3\right)$

X	Y1
1.99	31.522
1.999	31.952
1.9999	31.995
2.0001	32.005
2.001	32.048
2.01	32.482

Y1▤4X^3

$\lim\limits_{x\to 2}\left(4x^3\right) = 32$

3. $\lim\limits_{x\to 0}\dfrac{x+1}{x^2+1}$

X	Y1
-.01	.9899
-.001	.999
-1E-4	.9999
1E-4	1.0001
.001	1.001
.01	1.0099

Y1▤(X+1)/(X²+1)

$\lim\limits_{x\to 0}\dfrac{x+1}{x^2+1} = 1$

5. $\lim\limits_{x\to 4}\dfrac{x^2-4x}{x-4}$

X	Y1
3.99	3.99
3.999	3.999
3.9999	3.9999
4.0001	4.0001
4.001	4.001
4.01	4.01

Y1▤(X²-4X)/(X-4)

$\lim\limits_{x\to 4}\dfrac{x^2-4x}{x-4} = 4$

7. $\lim\limits_{x\to 0}\left(e^x+1\right)$

X	Y1
-.01	1.99
-.001	1.999
-1E-4	1.9999
1E-4	2.0001
.001	2.001
.01	2.0101

Y1▤e^(X)+1

$\lim\limits_{x\to 0}\left(e^x+1\right) = 2$

9. $\lim\limits_{x\to 0}\dfrac{\cos x-1}{x}$

X	Y1
-.01	.005
-.001	5E-4
-1E-4	5E-5
1E-4	-5E-5
.001	-5E-4
.01	-.005

Y1▤(cos(X)-1)/X

$\lim\limits_{x\to 0}\dfrac{\cos x-1}{x} = 0$

11. $\lim\limits_{x\to 2} f(x) = 3$

13. $\lim\limits_{x\to 2} f(x) = 4$

15. $\lim\limits_{x\to 3} f(x)$ does not exist because as x gets closer to 3, but is less than 3, $f(x)$ gets closer to 3. However, as x gets closer to 3, but is greater than 3, $f(x)$ gets closer to 6.

17. $f(x) = 3x + 1$

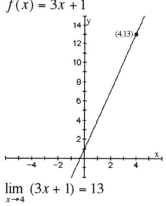

$$\lim_{x \to 4} (3x + 1) = 13$$

19. $f(x) = 1 - x^2$

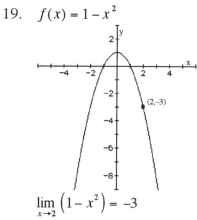

$$\lim_{x \to 2} \left(1 - x^2\right) = -3$$

21. $f(x) = |2x|$

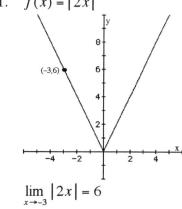

$$\lim_{x \to -3} |2x| = 6$$

23. $f(x) = \sin x$

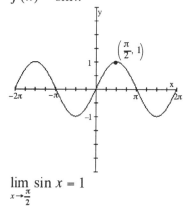

$$\lim_{x \to \frac{\pi}{2}} \sin x = 1$$

25. $f(x) = e^x$

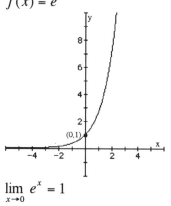

$$\lim_{x \to 0} e^x = 1$$

27. $f(x) = \dfrac{1}{x}$

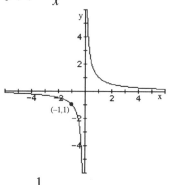

$$\lim_{x \to -1} \frac{1}{x} = -1$$

29. $f(x) = \begin{cases} x^2 & x \geq 0 \\ 2x & x < 0 \end{cases}$

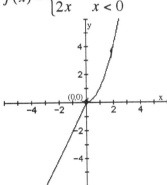

$\lim\limits_{x \to 0} f(x) = 0$

since $\lim\limits_{x \to 0^-} f(x) = \lim\limits_{x \to 0^+} f(x) = 0$

31. $f(x) = \begin{cases} 3x & x \leq 1 \\ x+1 & x > 1 \end{cases}$

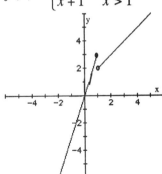

$\lim\limits_{x \to 1} f(x)$ does not exist

since $\lim\limits_{x \to 1^-} f(x) \neq \lim\limits_{x \to 1^+} f(x)$

33. $f(x) = \begin{cases} x & x < 0 \\ 1 & x = 0 \\ 3x & x > 0 \end{cases}$

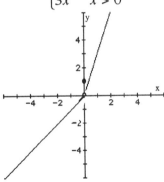

$\lim\limits_{x \to 0} f(x) = 0$

since $\lim\limits_{x \to 0^-} f(x) = \lim\limits_{x \to 0^+} f(x) = 0$

35. $f(x) = \begin{cases} \sin x & x \leq 0 \\ x^2 & x > 0 \end{cases}$

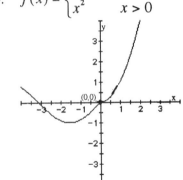

$\lim\limits_{x \to 0} f(x) = 0$

since $\lim\limits_{x \to 0^-} f(x) = \lim\limits_{x \to 0^+} f(x) = 0$

37. $\lim\limits_{x \to 1} \dfrac{x^3 - x^2 + x - 1}{x^4 - x^3 + 2x - 2}$

X	Y1
.99	.66663
.999	.66667
.9999	.66667
1	ERROR
1.0001	.66667
1.001	.66667
1.01	.66663

Y1⊟(X^3-X²+X-1)...

$\lim\limits_{x \to 1} \dfrac{x^3 - x^2 + x - 1}{x^4 - x^3 + 2x - 2} = 0.67$

39. $\lim\limits_{x \to 2} \dfrac{x^3 - 2x^2 + 4x - 8}{x^2 + x - 6}$

X	Y1
1.99	1.5952
1.999	1.5995
1.9999	1.6
2	ERROR
2.0001	1.6
2.001	1.6005
2.01	1.6048

Y1⊟(X^3-2X²+4X-...

$\lim\limits_{x \to 2} \dfrac{x^3 - 2x^2 + 4x - 8}{x^2 + x - 6} = 1.60$

41. $\lim\limits_{x \to -1} \dfrac{x^3 + 2x^2 + x}{x^4 + x^3 + 2x + 2}$

$\lim\limits_{x \to -1} \dfrac{x^3 + 2x^2 + x}{x^4 + x^3 + 2x + 2} = 0$

14.2 Algebraic Techniques for Finding Limits

1. $\lim\limits_{x \to 1} 5 = 5$

3. $\lim\limits_{x \to 4} x = 4$

5. $\lim\limits_{x \to 2} (3x + 2) = 3(2) + 2 = 8$

7. $\lim\limits_{x \to -1} (3x^2 - 5x) = 3(-1)^2 - 5(-1) = 8$

9. $\lim\limits_{x \to 1} (5x^4 - 3x^2 + 6x - 9) = 5(1)^4 - 3(1)^2 + 6(1) - 9 = 5 - 3 + 6 - 9 = -1$

11. $\lim\limits_{x \to 1} (x^2 + 1)^3 = \left(\lim\limits_{x \to 1} (x^2 + 1)\right)^3 = \left(1^2 + 1\right)^3 = 2^3 = 8$

13. $\lim\limits_{x \to 1} \sqrt{5x + 4} = \sqrt{\lim\limits_{x \to 1} (5x + 4)} = \sqrt{5(1) + 4} = \sqrt{9} = 3$

15. $\lim\limits_{x \to 0} \dfrac{x^2 - 4}{x^2 + 4} = \dfrac{\lim\limits_{x \to 0}(x^2 - 4)}{\lim\limits_{x \to 0}(x^2 + 4)} = \dfrac{0^2 - 4}{0^2 + 4} = \dfrac{-4}{4} = -1$

17. $\lim\limits_{x \to 2} (3x - 2)^{\frac{5}{2}} = \left(\lim\limits_{x \to 2} (3x - 2)\right)^{\frac{5}{2}} = \left(3(2) - 2\right)^{\frac{5}{2}} = 4^{\frac{5}{2}} = 32$

19. $\lim\limits_{x \to 2} \dfrac{x^2 - 4}{x^2 - 2x} = \lim\limits_{x \to 2} \dfrac{(x - 2)(x + 2)}{x(x - 2)} = \lim\limits_{x \to 2} \dfrac{x + 2}{x} = \dfrac{2 + 2}{2} = \dfrac{4}{2} = 2$

21. $\lim\limits_{x \to -3} \dfrac{x^2 - x - 12}{x^2 - 9} = \lim\limits_{x \to -3} \dfrac{(x - 4)(x + 3)}{(x - 3)(x + 3)} = \lim\limits_{x \to -3} \dfrac{x - 4}{x - 3} = \dfrac{-3 - 4}{-3 - 3} = \dfrac{-7}{-6} = \dfrac{7}{6}$

23. $\lim\limits_{x \to 1} \dfrac{x^3 - 1}{x - 1} = \lim\limits_{x \to 1} \dfrac{(x - 1)(x^2 + x + 1)}{x - 1} = \lim\limits_{x \to 1} x^2 + x + 1 = 1^2 + 1 + 1 = 3$

25. $\lim\limits_{x \to -1} \dfrac{(x + 1)^2}{x^2 - 1} = \lim\limits_{x \to -1} \dfrac{(x + 1)^2}{(x - 1)(x + 1)} = \lim\limits_{x \to -1} \dfrac{x + 1}{x - 1} = \dfrac{-1 + 1}{-1 - 1} = \dfrac{0}{-2} = 0$

27. $\lim\limits_{x\to1}\dfrac{x^3-x^2+x-1}{x^4-x^3+2x-2}=\lim\limits_{x\to1}\dfrac{x^2(x-1)+1(x-1)}{x^3(x-1)+2(x-1)}=\lim\limits_{x\to1}\dfrac{(x-1)(x^2+1)}{(x-1)(x^3+2)}=\lim\limits_{x\to1}\dfrac{x^2+1}{x^3+2}$

$$=\dfrac{1^2+1}{1^3+2}=\dfrac{2}{3}$$

29. $\lim\limits_{x\to2}\dfrac{x^3-2x^2+4x-8}{x^2+x-6}=\lim\limits_{x\to2}\dfrac{x^2(x-2)+4(x-2)}{(x+3)(x-2)}=\lim\limits_{x\to2}\dfrac{(x-2)(x^2+4)}{(x+3)(x-2)}=\lim\limits_{x\to2}\dfrac{x^2+4}{x+3}$

$$=\dfrac{2^2+4}{2+3}=\dfrac{8}{5}$$

31. $\lim\limits_{x\to-1}\dfrac{x^3+2x^2+x}{x^4+x^3+2x+2}=\lim\limits_{x\to-1}\dfrac{x(x^2+2x+1)}{x^3(x+1)+2(x+1)}=\lim\limits_{x\to-1}\dfrac{x(x+1)^2}{(x+1)(x^3+2)}$

$$=\lim\limits_{x\to-1}\dfrac{x(x+1)}{x^3+2}=\dfrac{-1(-1+1)}{(-1)^3+2}=\dfrac{-1(0)}{-1+2}=\dfrac{0}{1}=0$$

33. $\lim\limits_{x\to2}\dfrac{f(x)-f(2)}{x-2}=\lim\limits_{x\to2}\dfrac{(5x-3)-7}{x-2}=\lim\limits_{x\to2}\dfrac{5x-10}{x-2}=\lim\limits_{x\to2}\dfrac{5(x-2)}{x-2}=\lim\limits_{x\to2}5=5$

35. $\lim\limits_{x\to3}\dfrac{f(x)-f(3)}{x-3}=\lim\limits_{x\to3}\dfrac{x^2-9}{x-3}=\lim\limits_{x\to3}\dfrac{(x-3)(x+3)}{x-3}=\lim\limits_{x\to3}x+3=3+3=6$

37. $\lim\limits_{x\to-1}\dfrac{f(x)-f(-1)}{x-(-1)}=\lim\limits_{x\to-1}\dfrac{x^2+2x-(-1)}{x+1}=\lim\limits_{x\to-1}\dfrac{x^2+2x+1}{x+1}=\lim\limits_{x\to-1}\dfrac{(x+1)^2}{x+1}$

$$=\lim\limits_{x\to-1}x+1=-1+1=0$$

39. $\lim\limits_{x\to0}\dfrac{f(x)-f(0)}{x-0}=\lim\limits_{x\to0}\dfrac{3x^3-2x^2+4-4}{x}=\lim\limits_{x\to0}\dfrac{3x^3-2x^2}{x}=\lim\limits_{x\to0}(3x^2-2x)=0$

41. $\lim\limits_{x\to1}\dfrac{f(x)-f(1)}{x-1}=\lim\limits_{x\to1}\dfrac{\dfrac{1}{x}-1}{x-1}=\lim\limits_{x\to1}\dfrac{\dfrac{1-x}{x}}{x-1}=\lim\limits_{x\to1}\dfrac{-1(x-1)}{x(x-1)}=\lim\limits_{x\to1}\dfrac{-1}{x}=\dfrac{-1}{1}=-1$

43. $\lim\limits_{x\to0}\dfrac{\tan x}{x}=\lim\limits_{x\to0}\dfrac{\dfrac{\sin x}{\cos x}}{x}=\lim\limits_{x\to0}\dfrac{\sin x}{x}\cdot\dfrac{1}{\cos x}=\lim\limits_{x\to0}\dfrac{\sin x}{x}\cdot\lim\limits_{x\to0}\dfrac{1}{\cos x}=1\cdot\dfrac{\lim\limits_{x\to0}1}{\lim\limits_{x\to0}\cos x}$

$$=1\cdot\dfrac{1}{1}=1$$

45. $\lim\limits_{x\to0}\dfrac{3\sin x+\cos x-1}{4x}=\lim\limits_{x\to0}\left(\dfrac{3\sin x}{4x}+\dfrac{\cos x-1}{4x}\right)=\lim\limits_{x\to0}\dfrac{3\sin x}{4x}+\lim\limits_{x\to0}\dfrac{\cos x-1}{4x}$

$$=\dfrac{3}{4}\lim\limits_{x\to0}\dfrac{\sin x}{x}+\dfrac{1}{4}\lim\limits_{x\to0}\dfrac{\cos x-1}{x}=\dfrac{3}{4}\cdot1+\dfrac{1}{4}\cdot0=\dfrac{3}{4}$$

14.3 One-Sided Limits; Continuous Functions

1. The domain of $f(x) = 2x + 3$ is all real numbers. Therefore, $f(x)$ is continuous everywhere.

3. The domain of $f(x) = 3x^2 + x$ is all real numbers. Therefore, $f(x)$ is continuous everywhere.

5. The domain of $f(x) = 4\sin x$ is all real numbers. Therefore, $f(x)$ is continuous everywhere.

7. The domain of $f(x) = 2\tan x$ is all real numbers except odd integer multiples of $\frac{\pi}{2}$. Therefore, $f(x)$ is continuous everywhere except where $x = \frac{k\pi}{2}$ where k is an odd integer. $f(x)$ is discontinuous at $x = \frac{k\pi}{2}$ where k is an odd integer.

9. $f(x) = \frac{2x + 5}{x^2 - 4} = \frac{2x + 5}{(x - 2)(x + 2)}$. The domain of $f(x)$ is all real numbers except $x = 2$ and $x = -2$. Therefore, $f(x)$ is continuous everywhere except at $x = 2$ and $x = -2$. $f(x)$ is discontinuous at $x = 2$ and $x = -2$.

11. $f(x) = \frac{x - 3}{\ln x}$. The domain of $f(x)$ is $(0, 1)$ or $(1, \infty)$. Thus, $f(x)$ is continuous on the interval $(0, \infty)$ except at $x = 1$. $f(x)$ is discontinuous at $x = 1$.

13. $\lim\limits_{x \to 1^+} (2x + 3) = 2(1) + 3 = 5$

15. $\lim\limits_{x \to 1^-} \left(2x^3 + 5x\right) = 2(1)^3 + 5(1) = 2 + 5 = 7$

17. $\lim\limits_{x \to \frac{\pi}{2}^+} \sin x = \sin\frac{\pi}{2} = 1$

19. $\lim\limits_{x \to 2^+} \frac{x^2 - 4}{x - 2} = \lim\limits_{x \to 2^+} \frac{(x + 2)(x - 2)}{x - 2} = \lim\limits_{x \to 2^+} (x + 2) = 2 + 2 = 4$

21. $\lim\limits_{x \to -1^-} \frac{x^2 - 1}{x^3 + 1} = \lim\limits_{x \to -1^-} \frac{(x + 1)(x - 1)}{(x + 1)(x^2 - x + 1)} = \lim\limits_{x \to -1^-} \frac{x - 1}{x^2 - x + 1} = \frac{-1 - 1}{(-1)^2 - (-1) + 1} = \frac{-2}{3}$

23. $\lim\limits_{x \to -2^+} \frac{x^2 + x - 2}{x^2 + 2x} = \lim\limits_{x \to -2^+} \frac{(x + 2)(x - 1)}{x(x + 2)} = \lim\limits_{x \to -2^+} \frac{x - 1}{x} = \frac{-2 - 1}{-2} = \frac{-3}{-2} = \frac{3}{2}$

25. $f(x) = x^3 - 3x^2 + 2x - 6$; $c = 2$

    1. $f(2) = 2^3 - 3 \cdot 2^2 + 2 \cdot 2 - 6 = -6$

    2. $\lim\limits_{x \to 2^-} f(x) = 2^3 - 3 \cdot 2^2 + 2 \cdot 2 - 6 = -6$

    3. $\lim\limits_{x \to 2^+} f(x) = 2^3 - 3 \cdot 2^2 + 2 \cdot 2 - 6 = -6$

Thus, $f(x)$ is continuous at $c = 2$.

27. $f(x) = \dfrac{x^2 + 5}{x - 6}$; $c = 3$

    1. $f(3) = \dfrac{3^2 + 5}{3 - 6} = \dfrac{14}{-3} = -\dfrac{14}{3}$

    2. $\lim\limits_{x \to 3^-} f(x) = \dfrac{3^2 + 5}{3 - 6} = \dfrac{14}{-3} = -\dfrac{14}{3}$

    3. $\lim\limits_{x \to 3^+} f(x) = \dfrac{3^2 + 5}{3 - 6} = \dfrac{14}{-3} = -\dfrac{14}{3}$

Thus, $f(x)$ is continuous at $c = 3$.

29. $f(x) = \dfrac{x + 3}{x - 3}$; $c = 3$

Since $f(x)$ is not defined at $c = 3$, the function is not continuous at $c = 3$.

31. $f(x) = \dfrac{x^3 + 3x}{x^2 - 3x}$; $c = 0$

Since $f(x)$ is not defined at $c = 0$, the function is not continuous at $c = 0$.

33. $f(x) = \begin{cases} \dfrac{x^3 + 3x}{x^2 - 3x} & \text{if } x \neq 0 \\ 1 & \text{if } x = 0 \end{cases}$; $c = 0$

    1. $f(0) = 1$

    2. $\lim\limits_{x \to 0^-} f(x) = \lim\limits_{x \to 0^-} \dfrac{x^3 + 3x}{x^2 - 3x} = \lim\limits_{x \to 0^-} \dfrac{x(x^2 + 3)}{x(x - 3)} = \lim\limits_{x \to 0^-} \dfrac{x^2 + 3}{x - 3} = \dfrac{3}{-3} = -1$

Since $\lim\limits_{x \to 0^-} f(x) \neq f(c)$, the function is not continuous at $c = 0$.

35. $f(x) = \begin{cases} \dfrac{x^3 + 3x}{x^2 - 3x} & \text{if } x \neq 0 \\ -1 & \text{if } x = 0 \end{cases}$; $c = 0$

    1. $f(0) = -1$

    2. $\lim\limits_{x \to 0^-} f(x) = \lim\limits_{x \to 0^-} \dfrac{x^3 + 3x}{x^2 - 3x} = \lim\limits_{x \to 0^-} \dfrac{x(x^2 + 3)}{x(x - 3)} = \lim\limits_{x \to 0^-} \dfrac{x^2 + 3}{x - 3} = \dfrac{3}{-3} = -1$

    3. $\lim\limits_{x \to 0^+} f(x) = \lim\limits_{x \to 0^+} \dfrac{x^3 + 3x}{x^2 - 3x} = \lim\limits_{x \to 0^+} \dfrac{x(x^2 + 3)}{x(x - 3)} = \lim\limits_{x \to 0^+} \dfrac{x^2 + 3}{x - 3} = \dfrac{3}{-3} = -1$

The function is continuous at $c = 0$.

37. $f(x) = \begin{cases} \dfrac{x^3 - 1}{x^2 - 1} & \text{if } x < 1 \\ 2 & \text{if } x = 1 \; ; \quad c = 1 \\ \dfrac{3}{x + 1} & \text{if } x > 1 \end{cases}$

     1. $f(1) = 2$

     2. $\displaystyle\lim_{x \to 1^-} f(x) = \lim_{x \to 1^-} \frac{x^3 - 1}{x^2 - 1} = \lim_{x \to 1^-} \frac{(x-1)(x^2 + x + 1)}{(x-1)(x+1)} = \lim_{x \to 1^-} \frac{x^2 + x + 1}{x + 1} = \frac{3}{2}$

Since $\displaystyle\lim_{x \to 1^-} f(x) \neq f(c)$, the function is not continuous at $c = 1$.

39. $f(x) = \begin{cases} 2e^x & \text{if } x < 0 \\ 2 & \text{if } x = 0 \; ; \quad c = 0 \\ \dfrac{x^3 + 2x^2}{x^2} & \text{if } x > 0 \end{cases}$

     1. $f(0) = 2$

     2. $\displaystyle\lim_{x \to 0^-} f(x) = \lim_{x \to 0^-} 2e^x = 2e^0 = 2 \cdot 1 = 2$

     3. $\displaystyle\lim_{x \to 0^+} f(x) = \lim_{x \to 0^+} \frac{x^3 + 2x^2}{x^2} = \lim_{x \to 0^+} \frac{x^2(x + 2)}{x^2} = \lim_{x \to 0^+} (x + 2) = 0 + 2 = 2$

The function is continuous at $c = 0$.

41. Domain: $\left\{ x \mid -8 \le x < -6 \text{ or } -6 < x < 4 \text{ or } 4 < x \le 6 \right\}$

43. x-intercepts: $-8, -5, -3$

45. $f(-8) = 0$; $f(-4) = 2$

47. $\displaystyle\lim_{x \to -6^-} f(x) = +\infty$

49. $\displaystyle\lim_{x \to -4^-} f(x) = 2$

51. $\displaystyle\lim_{x \to 2^-} f(x) = 1$

53. $\displaystyle\lim_{x \to 4} f(x)$ does exist. $\displaystyle\lim_{x \to 4} f(x) = 0$ since $\displaystyle\lim_{x \to 4^-} f(x) = \lim_{x \to 4^+} f(x) = 0$

55. f is not continuous at -6 because $f(-6)$ does not exist.

57. f is continuous at 0 because $f(0) = \displaystyle\lim_{x \to 0^-} f(x) = \lim_{x \to 0^+} f(x) = 3$

59. f is not continuous at 4 because $f(4)$ does not exist.

61. $R(x) = \dfrac{x-1}{x^2-1} = \dfrac{x-1}{(x-1)(x+1)}$. The domain of R is $\left\{ x \mid x \neq -1, x \neq 1 \right\}$. Thus R is

discontinuous at both -1 and 1. $\lim\limits_{x \to -1^-} R(x) = \lim\limits_{x \to -1^-} \dfrac{x-1}{(x-1)(x+1)} = \lim\limits_{x \to -1^-} \dfrac{1}{x+1} = -\infty$

since when $x < -1, \dfrac{1}{x+1} < 0$, and as x approaches $-1, \dfrac{1}{x+1}$ becomes unbounded.

$\lim\limits_{x \to -1^+} \dfrac{1}{x+1} = +\infty$ since when $x > -1, \dfrac{1}{x+1} > 0$, and as x approaches $-1, \dfrac{1}{x+1}$

becomes unbounded. The behavior near $c = 1$ is: $\lim\limits_{x \to 1} R(x) = \lim\limits_{x \to 1} \dfrac{1}{x+1} = \dfrac{1}{2}$. Note

there is a hole in the graph at $\left(1, \dfrac{1}{2}\right)$.

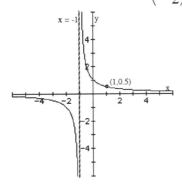

63. $R(x) = \dfrac{x^2+x}{x^2-1} = \dfrac{x(x+1)}{(x-1)(x+1)}$. The domain of R is $\left\{ x \mid x \neq -1, x \neq 1 \right\}$. Thus R is

discontinuous at both -1 and 1. $\lim\limits_{x \to 1^-} R(x) = \lim\limits_{x \to 1^-} \dfrac{x(x+1)}{(x-1)(x+1)} = \lim\limits_{x \to 1^-} \dfrac{x}{x-1} = -\infty$

since when $x < 1, \dfrac{x}{x-1} < 0$, and as x approaches $1, \dfrac{x}{x-1}$ becomes unbounded.

$\lim\limits_{x \to 1^+} \dfrac{x}{x-1} = +\infty$ since when $x > 1, \dfrac{x}{x-1} > 0$, and as x approaches $1, \dfrac{x}{x-1}$ becomes

unbounded. The behavior near $c = -1$ is: $\lim\limits_{x \to -1} R(x) = \lim\limits_{x \to -1} \dfrac{x}{x-1} = \dfrac{-1}{-2} = \dfrac{1}{2}$. Note

there is a hole in the graph at $\left(-1, \dfrac{1}{2}\right)$.

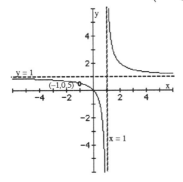

65. $R(x) = \dfrac{x^3 - x^2 + x - 1}{x^4 - x^3 + 2x - 2} = \dfrac{x^2(x-1) + 1(x-1)}{x^3(x-1) + 2(x-1)} = \dfrac{(x-1)(x^2+1)}{(x-1)(x^3+2)} = \dfrac{x^2+1}{x^3+2}, \; x \ne 1$

There is a vertical asymptote where $x^3 + 2 = 0$. $x = -\sqrt[3]{2}$ is a vertical asymptote.
There is a hole in the graph at $x = 1$.

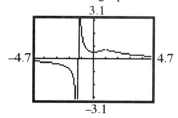

67. $R(x) = \dfrac{x^3 - 2x^2 + 4x - 8}{x^2 + x - 6} = \dfrac{x^2(x-2) + 4(x-2)}{(x+3)(x-2)} = \dfrac{(x-2)(x^2+4)}{(x+3)(x-2)} = \dfrac{x^2+4}{x+3}, \; x \ne 2$

There is a vertical asymptote where $x + 3 = 0$. $x = -3$ is a vertical asymptote. There
is a hole in the graph at $x = 2$.

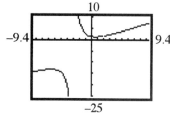

69. $R(x) = \dfrac{x^3 + 2x^2 + x}{x^4 + x^3 + 2x + 2} = \dfrac{x(x^2 + 2x + 1)}{x^3(x+1) + 2(x+1)} = \dfrac{x(x+1)^2}{(x+1)(x^3+2)} = \dfrac{x(x+1)}{x^3+2}, \; x \ne -1$

There is a vertical asymptote where $x^3 + 2 = 0$. $x = -\sqrt[3]{2}$ is a vertical asymptote.
There is a hole in the graph at $x = -1$.

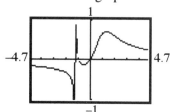

14.4 The Tangent Problem; The Derivative

1. $f(x) = 3x + 5$ at $(1, 8)$

$$m_{tan} = \lim_{x \to 1} \frac{f(x) - f(1)}{x - 1} = \lim_{x \to 1} \frac{3x + 5 - 8}{x - 1}$$

$$= \lim_{x \to 1} \frac{3x - 3}{x - 1} = \lim_{x \to 1} \frac{3(x - 1)}{x - 1} = \lim_{x \to 1} 3 = 3$$

Tangent Line: $y - 8 = 3(x - 1)$
$$y - 8 = 3x - 3$$
$$y = 3x + 5$$

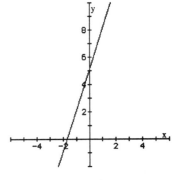

3. $f(x) = x^2 + 2$ at $(-1, 3)$

$$m_{tan} = \lim_{x \to -1} \frac{f(x) - f(-1)}{x + 1} = \lim_{x \to -1} \frac{x^2 + 2 - 3}{x + 1}$$

$$= \lim_{x \to -1} \frac{x^2 - 1}{x + 1} = \lim_{x \to -1} \frac{(x + 1)(x - 1)}{x + 1}$$

$$= \lim_{x \to -1} (x - 1) = -1 - 1 = -2$$

Tangent Line: $y - 3 = -2(x - (-1))$
$$y - 3 = -2x - 2$$
$$y = -2x + 1$$

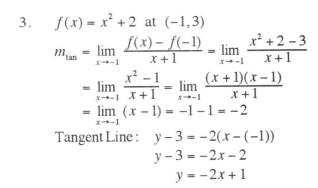
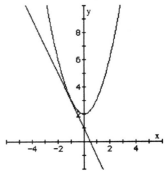

5. $f(x) = 3x^2$ at $(2, 12)$

$$m_{tan} = \lim_{x \to 2} \frac{f(x) - f(2)}{x - 2} = \lim_{x \to 2} \frac{3x^2 - 12}{x - 2}$$

$$= \lim_{x \to 2} \frac{3(x^2 - 4)}{x - 2} = \lim_{x \to 2} \frac{3(x + 2)(x - 2)}{x - 2}$$

$$= \lim_{x \to 2} 3(x + 2) = 3(2 + 2) = 12$$

Tangent Line: $y - 12 = 12(x - 2)$
$$y - 12 = 12x - 24$$
$$y = 12x - 12$$

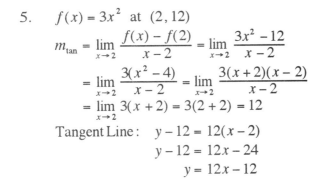
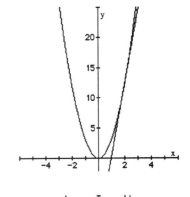

7. $f(x) = 2x^2 + x$ at $(1, 3)$

$$m_{tan} = \lim_{x \to 1} \frac{f(x) - f(1)}{x - 1} = \lim_{x \to 1} \frac{2x^2 + x - 3}{x - 1}$$

$$= \lim_{x \to 1} \frac{(2x + 3)(x - 1)}{x - 1} = \lim_{x \to 1} (2x + 3)$$

$$= 2(1) + 3 = 5$$

Tangent Line: $y - 3 = 5(x - 1)$
$$y - 3 = 5x - 5$$
$$y = 5x - 2$$

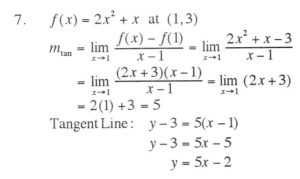
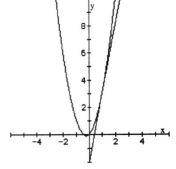

9. $f(x) = x^2 - 2x + 3$ at $(-1, 6)$

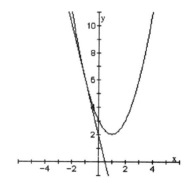

$$m_{\tan} = \lim_{x \to -1} \frac{f(x) - f(-1)}{x + 1} = \lim_{x \to -1} \frac{x^2 - 2x + 3 - 6}{x + 1}$$

$$= \lim_{x \to -1} \frac{x^2 - 2x - 3}{x + 1} = \lim_{x \to -1} \frac{(x + 1)(x - 3)}{x + 1}$$

$$= \lim_{x \to -1} (x - 3) = -1 - 3 = -4$$

Tangent Line : $y - 6 = -4(x - (-1))$
$$y - 6 = -4x - 4$$
$$y = -4x + 2$$

11. $f(x) = x^3 + x$ at $(2, 10)$

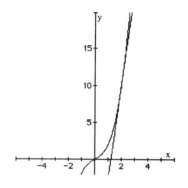

$$m_{\tan} = \lim_{x \to 2} \frac{f(x) - f(2)}{x - 2} = \lim_{x \to 2} \frac{x^3 + x - 10}{x - 2}$$

$$= \lim_{x \to 2} \frac{x^3 - 8 + x - 2}{x - 2}$$

$$= \lim_{x \to 2} \frac{(x - 2)(x^2 + 2x + 4) + x - 2}{x - 2}$$

$$= \lim_{x \to 2} \frac{(x - 2)(x^2 + 2x + 4 + 1)}{x - 2}$$

$$= \lim_{x \to 2} (x^2 + 2x + 5) = 4 + 4 + 5 = 13$$

Tangent Line : $y - 10 = 13(x - 2)$
$$y - 10 = 13x - 26$$
$$y = 13x - 16$$

13. $f(x) = -4x + 5$ at 3

$$f'(3) = \lim_{x \to 3} \frac{f(x) - f(3)}{x - 3} = \lim_{x \to 3} \frac{-4x + 5 - (-7)}{x - 3} = \lim_{x \to 3} \frac{-4x + 12}{x - 3} = \lim_{x \to 3} \frac{-4(x - 3)}{x - 3}$$

$$= \lim_{x \to 3} (-4) = -4$$

15. $f(x) = x^2 - 3$ at 0

$$f'(0) = \lim_{x \to 0} \frac{f(x) - f(0)}{x - 0} = \lim_{x \to 0} \frac{x^2 - 3 - (-3)}{x} = \lim_{x \to 0} \frac{x^2}{x} = \lim_{x \to 0} x = 0$$

17. $f(x) = 2x^2 + 3x$ at 1

$$f'(1) = \lim_{x \to 1} \frac{f(x) - f(1)}{x - 1} = \lim_{x \to 1} \frac{2x^2 + 3x - 5}{x - 1} = \lim_{x \to 1} \frac{(2x + 5)(x - 1)}{x - 1} = \lim_{x \to 1} (2x + 5) = 7$$

19. $f(x) = x^3 + 4x$ at -1

$$f'(-1) = \lim_{x \to -1} \frac{f(x) - f(-1)}{x - (-1)} = \lim_{x \to -1} \frac{x^3 + 4x - (-5)}{x + 1} = \lim_{x \to -1} \frac{x^3 + 1 + 4x + 4}{x + 1}$$

$$= \lim_{x \to -1} \frac{(x + 1)(x^2 - x + 1) + 4(x + 1)}{x + 1} = \lim_{x \to -1} \frac{(x + 1)(x^2 - x + 1 + 4)}{x + 1}$$

$$= \lim_{x \to -1} (x^2 - x + 5) = (-1)^2 - (-1) + 5 = 7$$

21. $f(x) = x^3 + x^2 - 2x$ at 1

$$f'(1) = \lim_{x \to 1} \frac{f(x) - f(1)}{x - 1} = \lim_{x \to 1} \frac{x^3 + x^2 - 2x - 0}{x - 1} = \lim_{x \to 1} \frac{x(x^2 + x - 2)}{x - 1}$$

$$= \lim_{x \to 1} \frac{x(x + 2)(x - 1)}{x - 1} = \lim_{x \to 1} x(x + 2) = 1(1 + 2) = 3$$

23. $f(x) = \sin x$ at 0

$$f'(0) = \lim_{x \to 0} \frac{f(x) - f(0)}{x - 0} = \lim_{x \to 0} \frac{\sin x - 0}{x - 0} = \lim_{x \to 0} \frac{\sin x}{x} = 1$$

25. Use NDeriv:

27. Use NDeriv:

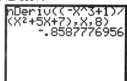

29. Use NDeriv:

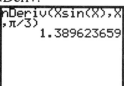

31. Use NDeriv:

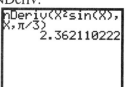

33. Use NDeriv:

35. $V(r) = 3\pi r^2$ at $r = 3$

$$V'(3) = \lim_{r \to 3} \frac{V(r) - V(3)}{r - 3} = \lim_{r \to 3} \frac{3\pi r^2 - 27\pi}{r - 3} = \lim_{r \to 3} \frac{3\pi(r^2 - 9)}{r - 3}$$

$$= \lim_{r \to 3} \frac{3\pi(r - 3)(r + 3)}{r - 3} = \lim_{r \to 3} 3\pi(r + 3) = 3\pi(3 + 3) = 18\pi$$

At the instant $r = 3$ feet, the volume of the cone is increasing at a rate of 18π cubic feet.

37. $V(r) = \frac{4}{3}\pi r^3$ at $r = 2$

$$V'(2) = \lim_{r \to 2} \frac{V(r) - V(2)}{r - 2} = \lim_{r \to 2} \frac{\frac{4}{3}\pi r^3 - \frac{32}{3}\pi}{r - 2} = \lim_{r \to 2} \frac{\frac{4}{3}\pi(r^3 - 8)}{r - 2}$$

$$= \lim_{r \to 2} \frac{\frac{4}{3}\pi(r - 2)(r^2 + 2r + 4)}{r - 2} = \lim_{r \to 2} \frac{4}{3}\pi(r^2 + 2r + 4) = \frac{4}{3}\pi(4 + 4 + 4) = 16\pi$$

At the instant $r = 2$ feet, the volume of the sphere is increasing at a rate of 16π cubic feet.

39. (a) $-16t^2 + 96t = 0$
$-16t(t - 6) = 0$
$t = 0$ or $t = 6$
The ball strikes the ground after 6 seconds.

(b) $\frac{\Delta s}{\Delta t} = \frac{s(2) - s(0)}{2 - 0} = \frac{-16(2)^2 + 96(2) - 0}{2} = \frac{128}{2} = 64$ feet / sec

(c) $s'(t) = \lim_{t \to t_0} \frac{s(t) - s(t_0)}{t - t_0} = \lim_{t \to t_0} \frac{-16t^2 + 96t - \left(-16t_0^2 + 96t_0\right)}{t - t_0}$

$= \lim_{t \to t_0} \frac{-16t^2 + 16t_0^2 + 96t - 96t_0}{t - t_0} = \lim_{t \to t_0} \frac{-16\left(t^2 - t_0^2\right) + 96\left(t - t_0\right)}{t - t_0}$

$= \lim_{t \to t_0} \frac{-16\left(t - t_0\right)\left(t + t_0\right) + 96\left(t - t_0\right)}{t - t_0} = \lim_{t \to t_0} \frac{\left(t - t_0\right)\left(-16\left(t + t_0\right) + 96\right)}{t - t_0}$

$= \lim_{t \to t_0} \left(-16\left(t + t_0\right) + 96\right) = \left(-16\left(t_0 + t_0\right) + 96\right) = -32t_0 + 96$

(d) $s'(2) = -32(2) + 96 = -64 + 96 = 32$ feet / sec

(e) $s'(t) = 0$
$-32t + 96 = 0$
$-32t = -96$
$t = 3$ seconds

(f) $s(3) = -16(3)^2 + 96(3) = -144 + 288 = 144$ feet

(g) $s'(6) = -32(6) + 96 = -192 + 96 = -96$ feet / sec

41. (a) $\frac{\Delta s}{\Delta t} = \frac{s(4) - s(1)}{4 - 1} = \frac{917 - 987}{3} = \frac{-70}{3} = -23\frac{1}{3}$ feet / sec

(b) $\frac{\Delta s}{\Delta t} = \frac{s(3) - s(1)}{3 - 1} = \frac{945 - 987}{2} = \frac{-42}{2} = -21$ feet / sec

(c) $\frac{\Delta s}{\Delta t} = \frac{s(2) - s(1)}{2 - 1} = \frac{969 - 987}{1} = \frac{-18}{1} = -18$ feet / sec

(d) $s(t) = -2.631t^2 - 10.269t + 999.933$

(e) $s'(1) = \lim\limits_{t \to 1} \dfrac{s(t) - s(1)}{t - 1} = \lim\limits_{t \to 1} \dfrac{-2.631t^2 - 10.269t + 999.933 - 987.033}{t - 1}$

$= \lim\limits_{t \to 1} \dfrac{-2.631t^2 - 10.269t + 12.9}{t - 1}$

$= \lim\limits_{t \to 1} \dfrac{-2.631t^2 + 2.631t - 12.9t + 12.9}{t - 1}$

$= \lim\limits_{t \to 1} \dfrac{-2.631t(t - 1) - 12.9(t - 1)}{t - 1} = \lim\limits_{t \to 1} \dfrac{(-2.631t - 12.9)(t - 1)}{t - 1}$

$= \lim\limits_{t \to 1} (-2.631t - 12.9) = -2.631(1) - 12.9 = -15.531 \text{ feet / sec}$

The instant $t = 1$, the instantaneous speed of the ball is -15.531 feet / sec.

14.5 The Area Problem; The Integral

1. $A \approx f(1) \cdot 1 + f(2) \cdot 1 = 1 \cdot 1 + 2 \cdot 1 = 1 + 2 = 3$

3. $A \approx f(0) \cdot 2 + f(2) \cdot 2 + f(4) \cdot 2 + f(6) \cdot 2 = 10 \cdot 2 + 6 \cdot 2 + 7 \cdot 2 + 5 \cdot 2$
$= 20 + 12 + 14 + 10 = 56$

5.   (a)   Graph $f(x) = 3x$:

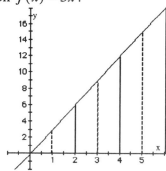

(b) $A \approx f(0)(2) + f(2)(2) + f(4)(2) = 0(2) + 6(2) + 12(2) = 0 + 12 + 24 = 36$

(c) $A \approx f(2)(2) + f(4)(2) + f(6)(2) = 6(2) + 12(2) + 18(2) = 12 + 24 + 36 = 72$

(d) $A \approx f(0)(1) + f(1)(1) + f(2)(1) + f(3)(1) + f(4)(1) + f(5)(1)$
$= 0(1) + 3(1) + 6(1) + 9(1) + 12(1) + 15(1) = 0 + 3 + 6 + 9 + 12 + 15 = 45$

(e) $A \approx f(1)(1) + f(2)(1) + f(3)(1) + f(4)(1) + f(5)(1) + f(6)(1)$
$= 3(1) + 6(1) + 9(1) + 12(1) + 15(1) + 18(1) = 3 + 6 + 9 + 12 + 15 + 18 = 63$

(f)   The actual area is the area of a triangle: $A = \dfrac{1}{2}(6)(18) = 54$

7. (a) Graph $f(x) = -3x + 9$:

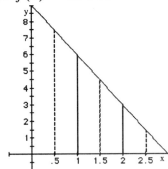

(b) $A \approx f(0)(1) + f(1)(1) + f(2)(1) = 9(1) + 6(1) + 3(1) = 9 + 6 + 3 = 18$

(c) $A \approx f(1)(1) + f(2)(1) + f(3)(1) = 6(1) + 3(1) + 0(1) = 6 + 3 + 0 = 9$

(d) $A \approx f(0)\left(\frac{1}{2}\right) + f\left(\frac{1}{2}\right)\left(\frac{1}{2}\right) + f(1)\left(\frac{1}{2}\right) + f\left(\frac{3}{2}\right)\left(\frac{1}{2}\right) + f(2)\left(\frac{1}{2}\right) + f\left(\frac{5}{2}\right)\left(\frac{1}{2}\right)$

$= 9\left(\frac{1}{2}\right) + \frac{15}{2}\left(\frac{1}{2}\right) + 6\left(\frac{1}{2}\right) + \frac{9}{2}\left(\frac{1}{2}\right) + 3\left(\frac{1}{2}\right) + \frac{3}{2}\left(\frac{1}{2}\right)$

$= \frac{9}{2} + \frac{15}{4} + 3 + \frac{9}{4} + \frac{3}{2} + \frac{3}{4} = \frac{63}{4}$

(e) $A \approx f\left(\frac{1}{2}\right)\left(\frac{1}{2}\right) + f(1)\left(\frac{1}{2}\right) + f\left(\frac{3}{2}\right)\left(\frac{1}{2}\right) + f(2)\left(\frac{1}{2}\right) + f\left(\frac{5}{2}\right)\left(\frac{1}{2}\right) + f(3)\left(\frac{1}{2}\right)$

$= \frac{15}{2}\left(\frac{1}{2}\right) + 6\left(\frac{1}{2}\right) + \frac{9}{2}\left(\frac{1}{2}\right) + 3\left(\frac{1}{2}\right) + \frac{3}{2}\left(\frac{1}{2}\right) + 0\left(\frac{1}{2}\right)$

$= \frac{15}{4} + 3 + \frac{9}{4} + \frac{3}{2} + \frac{3}{4} + 0 = \frac{45}{4}$

(f) The actual area is the area of a triangle: $A = \frac{1}{2}(3)(9) = \frac{27}{2} = 13.5$

9. (a) Graph $f(x) = x^2 + 2$, $[0, 4]$:

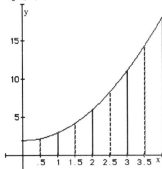

(b) $A \approx f(0)(1) + f(1)(1) + f(2)(1) + f(3)(1) = 2(1) + 3(1) + 6(1) + 11(1)$

$= 2 + 3 + 6 + 11 = 22$

(c) $A \approx f(0)\left(\dfrac{1}{2}\right) + f\left(\dfrac{1}{2}\right)\left(\dfrac{1}{2}\right) + f(1)\left(\dfrac{1}{2}\right) + f\left(\dfrac{3}{2}\right)\left(\dfrac{1}{2}\right) + f(2)\left(\dfrac{1}{2}\right) + f\left(\dfrac{5}{2}\right)\left(\dfrac{1}{2}\right)$
$$+ f(3)\left(\dfrac{1}{2}\right) + f\left(\dfrac{7}{2}\right)\left(\dfrac{1}{2}\right)$$
$$= 2\left(\dfrac{1}{2}\right) + \dfrac{9}{4}\left(\dfrac{1}{2}\right) + 3\left(\dfrac{1}{2}\right) + \dfrac{17}{4}\left(\dfrac{1}{2}\right) + 6\left(\dfrac{1}{2}\right) + \dfrac{33}{4}\left(\dfrac{1}{2}\right) + 11\left(\dfrac{1}{2}\right) + \dfrac{57}{4}\left(\dfrac{1}{2}\right)$$
$$= 1 + \dfrac{9}{8} + \dfrac{3}{2} + \dfrac{17}{8} + 3 + \dfrac{33}{8} + \dfrac{11}{2} + \dfrac{57}{8} = \dfrac{51}{2}$$

(d) $A = \displaystyle\int_0^4 (x^2 + 2)\,dx$

(e) Use fnInt function:

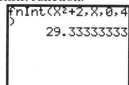

11. (a) Graph $f(x) = x^3$, $[0, 4]$:

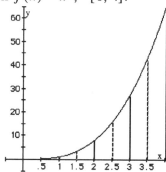

(b) $A \approx f(0)(1) + f(1)(1) + f(2)(1) + f(3)(1) = 0(1) + 1(1) + 8(1) + 27(1)$
$$= 0 + 1 + 8 + 27 = 36$$

(c) $A \approx f(0)\left(\dfrac{1}{2}\right) + f\left(\dfrac{1}{2}\right)\left(\dfrac{1}{2}\right) + f(1)\left(\dfrac{1}{2}\right) + f\left(\dfrac{3}{2}\right)\left(\dfrac{1}{2}\right) + f(2)\left(\dfrac{1}{2}\right) + f\left(\dfrac{5}{2}\right)\left(\dfrac{1}{2}\right)$
$$+ f(3)\left(\dfrac{1}{2}\right) + f\left(\dfrac{7}{2}\right)\left(\dfrac{1}{2}\right)$$
$$= 0\left(\dfrac{1}{2}\right) + \dfrac{1}{8}\left(\dfrac{1}{2}\right) + 1\left(\dfrac{1}{2}\right) + \dfrac{27}{8}\left(\dfrac{1}{2}\right) + 8\left(\dfrac{1}{2}\right) + \dfrac{125}{8}\left(\dfrac{1}{2}\right) + 27\left(\dfrac{1}{2}\right) + \dfrac{343}{8}\left(\dfrac{1}{2}\right)$$
$$= 0 + \dfrac{1}{16} + \dfrac{1}{2} + \dfrac{27}{16} + 4 + \dfrac{125}{16} + \dfrac{27}{2} + \dfrac{343}{16} = 49$$

(d) $A = \displaystyle\int_0^4 x^3\,dx$

(e) Use fnInt function:

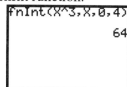

13. (a) Graph $f(x) = \dfrac{1}{x}$, [1, 5]:

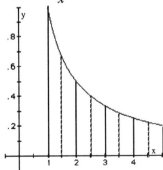

(b) $A \approx f(1)(1) + f(2)(1) + f(3)(1) + f(4)(1) = 1(1) + \dfrac{1}{2}(1) + \dfrac{1}{3}(1) + \dfrac{1}{4}(1)$

$= 1 + \dfrac{1}{2} + \dfrac{1}{3} + \dfrac{1}{4} = \dfrac{25}{12}$

(c) $A \approx f(1)\left(\dfrac{1}{2}\right) + f\left(\dfrac{3}{2}\right)\left(\dfrac{1}{2}\right) + f(2)\left(\dfrac{1}{2}\right) + f\left(\dfrac{5}{2}\right)\left(\dfrac{1}{2}\right) + f(3)\left(\dfrac{1}{2}\right) + f\left(\dfrac{7}{2}\right)\left(\dfrac{1}{2}\right)$

$+ f(4)\left(\dfrac{1}{2}\right) + f\left(\dfrac{9}{2}\right)\left(\dfrac{1}{2}\right)$

$= 1\left(\dfrac{1}{2}\right) + \dfrac{2}{3}\left(\dfrac{1}{2}\right) + \dfrac{1}{2}\left(\dfrac{1}{2}\right) + \dfrac{2}{5}\left(\dfrac{1}{2}\right) + \dfrac{1}{3}\left(\dfrac{1}{2}\right) + \dfrac{2}{7}\left(\dfrac{1}{2}\right) + \dfrac{1}{4}\left(\dfrac{1}{2}\right) + \dfrac{2}{9}\left(\dfrac{1}{2}\right)$

$= \dfrac{1}{2} + \dfrac{1}{3} + \dfrac{1}{4} + \dfrac{1}{5} + \dfrac{1}{6} + \dfrac{1}{7} + \dfrac{1}{8} + \dfrac{1}{9} = \dfrac{4690}{2520} \approx 1.829$

(d) $A = \displaystyle\int_1^5 \dfrac{1}{x}\,dx$

(e) Use fnInt function:

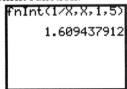

15. (a) Graph $f(x) = e^x$, [−1, 3]:

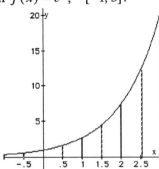

(b) $A \approx \big(f(-1) + f(0) + f(1) + f(2)\big)(1) = (0.3679 + 1 + 2.7183 + 7.3891)(1)$

$= 11.4753$

(c) $A \approx \left(f(-1) + f\left(-\frac{1}{2}\right) + f(0) + f\left(\frac{1}{2}\right) + f(1) + f\left(\frac{3}{2}\right) + f(2) + f\left(\frac{5}{2}\right) \right) \cdot \left(\frac{1}{2}\right)$

$= (0.3679 + 0.6065 + 1 + 1.6487 + 2.7183 + 4.4817 + 7.3891 + 12.1825)(0.5)$

$= 30.3947(0.5) = 15.1974$

(d) $A = \int_{-1}^{3} e^x \, dx$

(e)   Use fnInt function:

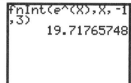

17. (a)   Graph $f(x) = \sin x$, $[0, \pi]$:

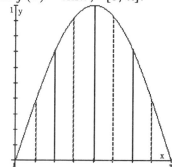

(b) $A \approx \left(f(0) + f\left(\frac{\pi}{4}\right) + f\left(\frac{\pi}{2}\right) + f\left(\frac{3\pi}{4}\right) \right)\left(\frac{\pi}{4}\right) = \left(0 + \frac{\sqrt{2}}{2} + 1 + \frac{\sqrt{2}}{2} \right)\left(\frac{\pi}{4}\right)$

$\left(1 + \sqrt{2} \right)\left(\frac{\pi}{4}\right) \approx 1.8961$

(c) $A \approx \left(f(0) + f\left(\frac{\pi}{8}\right) + f\left(\frac{\pi}{4}\right) + f\left(\frac{3\pi}{8}\right) + f\left(\frac{\pi}{2}\right) + f\left(\frac{5\pi}{8}\right) + f\left(\frac{3\pi}{4}\right) + f\left(\frac{7\pi}{8}\right) \right)\left(\frac{\pi}{8}\right)$

$= (0 + 0.3827 + 0.7071 + 0.9239 + 1 + 0.9239 + 0.7071 + 0.3827)(0.3927)$

$= 5.0274(0.3927) = 1.9743$

(d) $A = \int_{0}^{\pi} \sin x \, dx$

(e)   Use fnInt function:

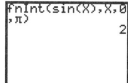

19. (a) The integral represents the area under the graph of $f(x) = 3x + 1$ from 0 to 4.
    (b) 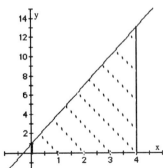 (c)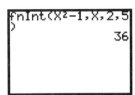

21. (a) The integral represents the area under the graph of $f(x) = x^2 - 1$ from 2 to 5.
    (b) 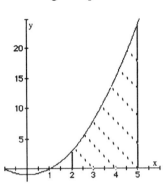 (c)

23. (a) The integral represents the area under the graph of $f(x) = \sin x$ from 0 to $\dfrac{\pi}{2}$.
    (b) 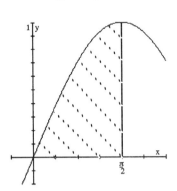 (c)

25. (a) The integral represents the area under the graph of $f(x) = e^x$ from 0 to 2.
    (b)   (c)

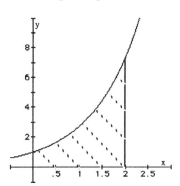

```
fnInt(e^(X),X,0,
2)
            6.389056099
```

14   Chapter Review

1. $\lim\limits_{x \to 2} \left(3x^2 - 2x + 1\right) = 3(2)^2 - 2(2) + 1 = 12 - 4 + 1 = 9$

3. $\lim\limits_{x \to -2} \left(x^2 + 1\right)^2 = \left(\lim\limits_{x \to -2} \left(x^2 + 1\right)\right)^2 = \left((-2)^2 + 1\right)^2 = 5^2 = 25$

5. $\lim\limits_{x \to 3} \sqrt{x^2 + 7} = \sqrt{\lim\limits_{x \to 3} (x^2 + 7)} = \sqrt{3^2 + 7} = \sqrt{16} = 4$

7. $\lim\limits_{x \to 1^-} \sqrt{1 - x^2} = \sqrt{\lim\limits_{x \to 1^-} (1 - x^2)} = \sqrt{1 - 1^2} = \sqrt{0} = 0$

9. $\lim\limits_{x \to 2} (5x + 6)^{\frac{3}{2}} = \left(\lim\limits_{x \to 2} (5x + 6)\right)^{\frac{3}{2}} = (5(2) + 6)^{\frac{3}{2}} = 16^{\frac{3}{2}} = 64$

11. $\lim\limits_{x \to -1} \dfrac{x^2 + x + 2}{x^2 - 9} = \dfrac{\lim\limits_{x \to -1}(x^2 + x + 2)}{\lim\limits_{x \to -1}(x^2 - 9)} = \dfrac{(-1)^2 + (-1) + 2}{(-1)^2 - 9} = \dfrac{2}{-8} = -\dfrac{1}{4}$

13. $\lim\limits_{x \to 1} \dfrac{x - 1}{x^3 - 1} = \lim\limits_{x \to 1} \dfrac{x - 1}{(x - 1)(x^2 + x + 1)} = \lim\limits_{x \to 1} \dfrac{1}{x^2 + x + 1} = \dfrac{1}{1^2 + 1 + 1} = \dfrac{1}{3}$

15. $\lim\limits_{x \to -3} \dfrac{x^2 - 9}{x^2 - x - 12} = \lim\limits_{x \to -3} \dfrac{(x - 3)(x + 3)}{(x - 4)(x + 3)} = \lim\limits_{x \to -3} \dfrac{x - 3}{x - 4} = \dfrac{-3 - 3}{-3 - 4} = \dfrac{-6}{-7} = \dfrac{6}{7}$

17. $\lim\limits_{x \to -1^-} \dfrac{x^2 - 1}{x^3 - 1} = \lim\limits_{x \to -1^-} \dfrac{(x + 1)(x - 1)}{(x - 1)(x^2 + x + 1)} = \lim\limits_{x \to -1^-} \dfrac{x + 1}{x^2 + x + 1} = \dfrac{-1 + 1}{(-1)^2 + (-1) + 1} = \dfrac{0}{1} = 0$

19. $\lim\limits_{x \to 2} \dfrac{x^3 - 8}{x^3 - 2x^2 + 4x - 8} = \lim\limits_{x \to 2} \dfrac{(x-2)(x^2 + 2x + 4)}{x^2(x-2) + 4(x-2)} = \lim\limits_{x \to 2} \dfrac{(x-2)(x^2 + 2x + 4)}{(x-2)(x^2 + 4)}$

$$= \lim\limits_{x \to 2} \dfrac{x^2 + 2x + 4}{x^2 + 4} = \dfrac{2^2 + 2(2) + 4}{2^2 + 4} = \dfrac{12}{8} = \dfrac{3}{2}$$

21. $\lim\limits_{x \to 3} \dfrac{x^4 - 3x^3 + x - 3}{x^3 - 3x^2 + 2x - 6} = \lim\limits_{x \to 3} \dfrac{x^3(x-3) + 1(x-3)}{x^2(x-3) + 2(x-3)} = \lim\limits_{x \to 3} \dfrac{(x-3)(x^3 + 1)}{(x-3)(x^2 + 2)} = \lim\limits_{x \to 3} \dfrac{x^3 + 1}{x^2 + 2}$

$$= \dfrac{3^3 + 1}{3^2 + 2} = \dfrac{28}{11}$$

23. $f(x) = 3x^4 - x^2 + 2; \ c = 5$

     1. $f(5) = 3(5)^4 - 5^2 + 2 = 1852$

     2. $\lim\limits_{x \to 5^-} f(x) = 3(5)^4 - 5^2 + 2 = 1852$

     3. $\lim\limits_{x \to 5^+} f(x) = 3(5)^4 - 5^2 + 2 = 1852$

   Thus, $f(x)$ is continuous at $c = 5$.

25. $f(x) = \dfrac{x^2 - 4}{x + 2}; \ c = -2$

   Since $f(x)$ is not defined at $c = -2$, the function is not continuous at $c = -2$.

27. $f(x) = \begin{cases} \dfrac{x^2 - 4}{x + 2} & \text{if } x \neq -2 \\ 4 & \text{if } x = -2 \end{cases}; \ c = -2$

     1. $f(-2) = 4$

     2. $\lim\limits_{x \to -2^-} f(x) = \lim\limits_{x \to -2^-} \dfrac{x^2 - 4}{x + 2} = \lim\limits_{x \to -2^-} \dfrac{(x-2)(x+2)}{x + 2} = \lim\limits_{x \to -2^-} (x - 2) = -4$

   Since $\lim\limits_{x \to -2^-} f(x) \neq f(c)$, the function is not continuous at $c = -2$.

29. $f(x) = \begin{cases} \dfrac{x^2 - 4}{x + 2} & \text{if } x \neq -2 \\ -4 & \text{if } x = -2 \end{cases}; \ c = -2$

     1. $f(-2) = -4$

     2. $\lim\limits_{x \to -2^-} f(x) = \lim\limits_{x \to -2^-} \dfrac{x^2 - 4}{x + 2} = \lim\limits_{x \to -2^-} \dfrac{(x-2)(x+2)}{x + 2} = \lim\limits_{x \to -2^-} (x - 2) = -4$

     3. $\lim\limits_{x \to -2^+} f(x) = \lim\limits_{x \to -2^+} \dfrac{x^2 - 4}{x + 2} = \lim\limits_{x \to -2^+} \dfrac{(x-2)(x+2)}{x + 2} = \lim\limits_{x \to -2^+} (x - 2) = -4$

   The function is continuous at $c = -2$.

31. Domain: $\left\{ x \mid -6 \leq x < 2 \text{ or } 2 < x < 5 \text{ or } 5 < x \leq 6 \right\}$

33. x-intercepts: $1, 6$

35. $f(-6) = 2;\ f(-4) = 1$

37. $\lim\limits_{x \to -4^-} f(x) = 4$

39. $\lim\limits_{x \to -2^-} f(x) = -2$

41. $\lim\limits_{x \to 2^-} f(x) = -\infty$

43. $\lim\limits_{x \to 0} f(x)$ does not exist. $\lim\limits_{x \to 0^-} f(x) = 4 \neq \lim\limits_{x \to 0^+} f(x) = 1$

45. f is not continuous at -2 because $\lim\limits_{x \to -2^-} f(x) \neq \lim\limits_{x \to -2^+} f(x)$

47. f is not continuous at 0 because $\lim\limits_{x \to 0^-} f(x) \neq \lim\limits_{x \to 0^+} f(x)$

49. f is continuous at 4 because $f(4) = \lim\limits_{x \to 4^-} f(x) = \lim\limits_{x \to 4^+} f(x)$

51. $R(x) = \dfrac{x+4}{x^2-16} = \dfrac{x+4}{(x-4)(x+4)}$. The domain of R is $\left\{ x \,\middle|\, x \neq -4,\ x \neq 4 \right\}$. Thus

R is discontinuous at both -4 and 4. $\lim\limits_{x \to 4^-} R(x) = \lim\limits_{x \to 4^-} \dfrac{x+4}{(x-4)(x+4)}$

$= \lim\limits_{x \to 4^-} \dfrac{1}{x-4} = -\infty$ since when $x < 4$, $\dfrac{1}{x-4} < 0$, and as x approaches 4, $\dfrac{1}{x-4}$

becomes unbounded. $\lim\limits_{x \to 4^+} \dfrac{1}{x-4} = +\infty$ since when $x > 4$, $\dfrac{1}{x-4} > 0$, and as x

approaches 4, $\dfrac{1}{x-4}$ becomes unbounded. The behavior near $c = -4$ is:

$\lim\limits_{x \to -4} R(x) = \lim\limits_{x \to -4} \dfrac{1}{x-4} = \dfrac{1}{-8}$. Note there is a hole in the graph at $\left(-4, -\dfrac{1}{8} \right)$.

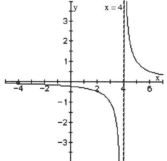

53. $R(x) = \dfrac{x^3 - 2x^2 + 4x - 8}{x^2 - 11x + 18} = \dfrac{x^2(x-2) + 4(x-2)}{(x-9)(x-2)} = \dfrac{(x-2)(x^2+4)}{(x-9)(x-2)} = \dfrac{x^2+4}{x-9}, \; x \ne 2$

There is a vertical asymptote where $x - 9 = 0$. $x = 9$ is a vertical asymptote. There is a hole in the graph at $x = 2$.

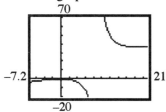

55. $f(x) = 2x^2 + 8x$ at $(1, 10)$

$$m_{\tan} = \lim_{x \to 1} \frac{f(x) - f(1)}{x - 1} = \lim_{x \to 1} \frac{2x^2 + 8x - 10}{x - 1}$$

$$= \lim_{x \to 1} \frac{2(x+5)(x-1)}{x-1} = \lim_{x \to 1} 2(x+5)$$

$$= 2(1+5) = 12$$

Tangent Line: $\quad y - 10 = 12(x - 1)$
$$y - 10 = 12x - 12$$
$$y = 12x - 2$$

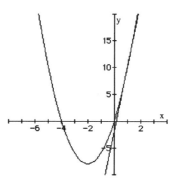

57. $f(x) = x^2 + 2x - 3$ at $(-1, -4)$

$$m_{\tan} = \lim_{x \to -1} \frac{f(x) - f(-1)}{x + 1} = \lim_{x \to -1} \frac{x^2 + 2x - 3 - (-4)}{x + 1}$$

$$= \lim_{x \to -1} \frac{x^2 + 2x + 1}{x + 1} = \lim_{x \to -1} \frac{(x+1)^2}{x+1} = \lim_{x \to -1} (x + 1)$$

$$= -1 + 1 = 0$$

Tangent Line: $\quad y - (-4) = 0(x - (-1))$
$$y + 4 = 0$$
$$y = -4$$

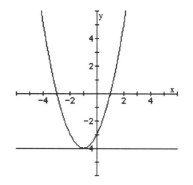

59. $f(x) = x^3 + x^2$ at $(2, 12)$

$$m_{\tan} = \lim_{x \to 2} \frac{f(x) - f(2)}{x - 2} = \lim_{x \to 2} \frac{x^3 + x^2 - 12}{x - 2}$$

$$= \lim_{x \to 2} \frac{x^3 - 2x^2 + 3x^2 - 12}{x - 2}$$

$$= \lim_{x \to 2} \frac{x^2(x-2) + 3(x-2)(x+2)}{x - 2}$$

$$= \lim_{x \to 2} \frac{(x-2)(x^2 + 3x + 6)}{x - 2}$$

$$= \lim_{x \to 2} (x^2 + 3x + 6) = 4 + 6 + 6 = 16$$

Tangent Line: $\quad y - 12 = 16(x - 2) \quad \rightarrow \quad y - 12 = 16x - 32 \quad \rightarrow \quad y = 16x - 20$

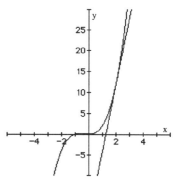

61. $f(x) = -4x^2 + 5$ at 3

$$f'(3) = \lim_{x \to 3} \frac{f(x) - f(3)}{x - 3} = \lim_{x \to 3} \frac{-4x^2 + 5 - (-31)}{x - 3} = \lim_{x \to 3} \frac{-4x^2 + 36}{x - 3}$$

$$= \lim_{x \to 3} \frac{-4(x^2 - 9)}{x - 3} = \lim_{x \to 3} \frac{-4(x - 3)(x + 3)}{x - 3} = \lim_{x \to 3} (-4)(x + 3) = -4(6) = -24$$

63. $f(x) = x^2 - 3x$ at 0

$$f'(0) = \lim_{x \to 0} \frac{f(x) - f(0)}{x - 0} = \lim_{x \to 0} \frac{x^2 - 3x - 0}{x} = \lim_{x \to 0} \frac{x(x - 3)}{x} = \lim_{x \to 0} (x - 3) = -3$$

65. $f(x) = 2x^2 + 3x + 2$ at 1

$$f'(1) = \lim_{x \to 1} \frac{f(x) - f(1)}{x - 1} = \lim_{x \to 1} \frac{2x^2 + 3x + 2 - 7}{x - 1} = \lim_{x \to 1} \frac{2x^2 + 3x - 5}{x - 1}$$

$$= \lim_{x \to 1} \frac{(2x + 5)(x - 1)}{x - 1} = \lim_{x \to 1} (2x + 5) = 7$$

67. Use NDeriv:

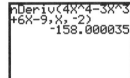

69. Use NDeriv:

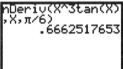

71. (a) $-16t^2 + 96t + 112 = 0$

$-16(t^2 - 6t - 7) = 0$

$-16(t + 1)(t - 7) = 0$

$t = -1$ or $t = 7$

The ball strikes the ground after 7 seconds.

(b) $-16t^2 + 96t + 112 = 112$

$-16t^2 + 96t = 0$

$-16t(t - 6) = 0$

$t = 0$ or $t = 6$

The ball passes the rooftop after 6 seconds.

(c) $\dfrac{\Delta s}{\Delta t} = \dfrac{s(2) - s(0)}{2 - 0} = \dfrac{-16(2)^2 + 96(2) + 112 - 112}{2} = \dfrac{128}{2} = 64$ feet / sec

(d) $s'(t) = \lim_{t \to t_0} \dfrac{s(t) - s(t_0)}{t - t_0} = \lim_{t \to t_0} \dfrac{-16t^2 + 96t + 112 - \left(-16t_0^2 + 96t_0 + 112\right)}{t - t_0}$

$= \lim_{t \to t_0} \dfrac{-16t^2 + 16t_0^2 + 96t - 96t_0}{t - t_0} = \lim_{t \to t_0} \dfrac{-16\left(t^2 - t_0^2\right) + 96(t - t_0)}{t - t_0}$

$= \lim_{t \to t_0} \dfrac{-16(t - t_0)(t + t_0) + 96(t - t_0)}{t - t_0} = \lim_{t \to t_0} \dfrac{(t - t_0)\left(-16(t + t_0) + 96\right)}{t - t_0}$

$= \lim_{t \to t_0} \left(-16(t + t_0) + 96\right) = \left(-16(t_0 + t_0) + 96\right) = -32t_0 + 96$

(e) $s'(2) = -32(2) + 96 = -64 + 96 = 32$ feet / sec

(f) $s'(t) = 0$

$-32t + 96 = 0$

$-32t = -96$

$t = 3$ seconds

(g) $s'(6) = -32(6) + 96 = -192 + 96 = -96$ feet / sec

(h) $s'(7) = -32(7) + 96 = -224 + 96 = -128$ feet / sec

73. (a) $\dfrac{\Delta R}{\Delta x} = \dfrac{8775 - 2340}{130 - 25} = \dfrac{6435}{105} = \$61.29$ / watch

(b) $\dfrac{\Delta R}{\Delta x} = \dfrac{6975 - 2340}{90 - 25} = \dfrac{4635}{65} = \$71.31$ / watch

(c) $\dfrac{\Delta R}{\Delta x} = \dfrac{4375 - 2340}{50 - 25} = \dfrac{2035}{25} = \$81.40$ / watch

(d) $R(x) = -0.25x^2 + 100.01x - 1.24$

(e) $R'(25) = \lim\limits_{x \to 25} \dfrac{R(x) - R(25)}{x - 25} = \lim\limits_{x \to 25} \dfrac{-0.25x^2 + 100.01x - 1.24 - 2342.76}{x - 25}$

$= \lim\limits_{x \to 25} \dfrac{-0.25x^2 + 100.01x - 2344}{x - 25}$ (Divide)

$= \lim\limits_{x \to 25} (-0.25x + 93.76) = -0.25(25) + 93.76 = \$87.51$ / watch

75. (a)   Graph $f(x) = 2x + 3$:

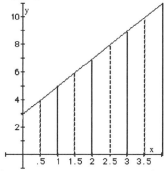

(b) $A \approx \big(f(0) + f(1) + f(2) + f(3)\big)(1) = (3 + 5 + 7 + 9)(1) = 24(1) = 24$

(c) $A \approx \big(f(1) + f(2) + f(3) + f(4)\big)(1) = (5 + 7 + 9 + 11)(1) = 32(1) = 32$

(d) $A \approx \left(f(0) + f\!\left(\dfrac{1}{2}\right) + f(1) + f\!\left(\dfrac{3}{2}\right) + f(2) + f\!\left(\dfrac{5}{2}\right) + f(3) + f\!\left(\dfrac{7}{2}\right) \right) \cdot \left(\dfrac{1}{2}\right)$

$= (3 + 4 + 5 + 6 + 7 + 8 + 9 + 10)\left(\dfrac{1}{2}\right) = 52\left(\dfrac{1}{2}\right) = 26$

(e) $A \approx \left(f\!\left(\dfrac{1}{2}\right) + f(1) + f\!\left(\dfrac{3}{2}\right) + f(2) + f\!\left(\dfrac{5}{2}\right) + f(3) + f\!\left(\dfrac{7}{2}\right) + f(4) \right) \cdot \left(\dfrac{1}{2}\right)$

$= (4 + 5 + 6 + 7 + 8 + 9 + 10 + 11)\left(\dfrac{1}{2}\right) = 60\left(\dfrac{1}{2}\right) = 30$

(f)   The actual area is the area of a trapezoid: $A = \dfrac{1}{2}(3 + 11)(4) = \dfrac{56}{2} = 28$

77. (a) Graph $f(x) = 4 - x^2$, $[-1, 2]$:

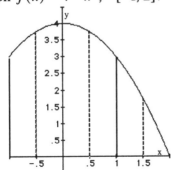

(b) $A \approx \left(f(-1) + f(0) + f(1) \right)(1) = (3 + 4 + 3)(1) = 10(1) = 10$

(c) $A \approx \left(f(-1) + f\left(-\frac{1}{2}\right) + f(0) + f\left(\frac{1}{2}\right) + f(1) + f\left(\frac{3}{2}\right) \right) \cdot \left(\frac{1}{2}\right)$

$= \left(3 + \frac{15}{4} + 4 + \frac{15}{4} + 3 + \frac{7}{4} \right)\left(\frac{1}{2}\right) = \frac{61}{4}\left(\frac{1}{2}\right) = \frac{61}{8} = 7.625$

(d) $A = \int_{-1}^{2} \left(4 - x^2\right) dx$

(e) Use fnInt function:

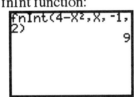

79. (a) Graph $f(x) = \frac{1}{x^2}$, $[1, 4]$:

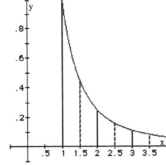

(b) $A \approx \left(f(1) + f(2) + f(3) \right)(1) = \left(1 + \frac{1}{4} + \frac{1}{9} \right)(1) = \frac{49}{36}(1) = \frac{49}{36} = 1.361$

(c) $A \approx \left(f(1) + f\left(\frac{3}{2}\right) + f(2) + f\left(\frac{5}{2}\right) + f(3) + f\left(\frac{7}{2}\right) \right) \cdot \left(\frac{1}{2}\right)$

$= \left(1 + \frac{4}{9} + \frac{1}{4} + \frac{4}{25} + \frac{1}{9} + \frac{4}{49} \right)\left(\frac{1}{2}\right) = 1.024$

(d) $A = \int_{1}^{4} \left(\frac{1}{x^2}\right) dx$

(e)   Use fnInt function:

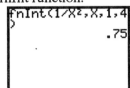

81.   (a)   The integral represents the area under the graph of $f(x) = 9 - x^2$ from -1 to 3.

(b)

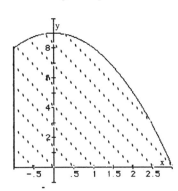

(c)

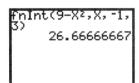

83.   (a)   The integral represents the area under the graph of $f(x) = e^x$ from -1 to 1.

(b)

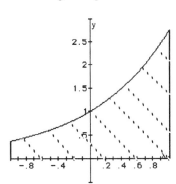

(c)

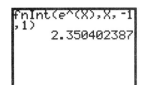

A

Appendix

Section 1 Topics from Algebra

1. (a) $\{2, 5\}$
   (b) $\{-6, 2, 5\}$
   (c) $\{-6, \frac{1}{2}, -1.333\ldots, 2, 5\}$
   (d) $\{\pi\}$
   (e) $\{-6, \frac{1}{2}, -1.333\ldots, \pi, 2, 5\}$

3. (a) $\{1\}$
   (b) $\{0, 1\}$
   (c) $\{0, 1, \frac{1}{2}, \frac{1}{3}, \frac{1}{4}\}$
   (d) None
   (e) $\{0, 1, \frac{1}{2}, \frac{1}{3}, \frac{1}{4}\}$

5. (a) None
   (b) None
   (c) None
   (d) $\{\sqrt{2}, \pi, \sqrt{2}+1, \pi+\frac{1}{2}\}$
   (e) $\{\sqrt{2}, \pi, \sqrt{2}+1, \pi+\frac{1}{2}\}$

7.

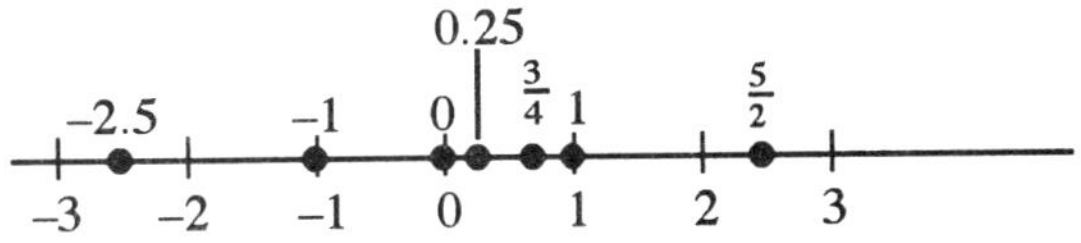

9. $\frac{1}{2} > 0$

11. $-1 > -2$

13. $\pi > 3.14$

15. $\frac{1}{2} = 0.5$

17. $\frac{2}{3} < 0.67$

19. $x > 0$

21. $x < 2$

23. $x \le 1$

25. $2 < x < 5$

27. Graph on the number line:

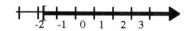

29. Graph on the number line:

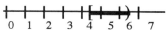

31. Graph on the number line:

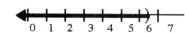

33. Graph on the number line:

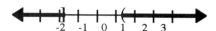

35. $d(D, E) = d(1,3) = |3 - 1| = |2| = 2$

37. $d(A, E) = d(-3,3) = |3 - (-3)| = |6| = 6$

39. $x + 2y = -2 + 2 \cdot 3 = -2 + 6 = 4$

41. $5xy + 2 = 5(-2)(3) + 2 = -30 + 2 = -28$

43. $\dfrac{2x}{x - y} = \dfrac{2(-2)}{-2 - 3} = \dfrac{-4}{-5} = \dfrac{4}{5}$

45. $\dfrac{3x + 2y}{2 + y} = \dfrac{3(-2) + 2(3)}{2 + 3} = \dfrac{-6 + 6}{5} = \dfrac{0}{5} = 0$

47. $|x + y| = |3 + (-2)| = |1| = 1$

49. $|x| + |y| = |3| + |-2| = 3 + 2 = 5$

51. $\dfrac{|x|}{x} = \dfrac{|3|}{3} = \dfrac{3}{3} = 1$

53. $|4x - 5y| = |4(3) - 5(-2)| = |12 + 10| = |22| = 22$

55. $||4x| - |5y|| = ||4(3)| - |5(-2)|| = ||12| - |-10|| = |12 - 10| = |2| = 2$

57. $C = \frac{5}{9}(F - 32) = \frac{5}{9}(32 - 32) = \frac{5}{9}(0) = 0$

59. $C = \frac{5}{9}(F - 32) = \frac{5}{9}(77 - 32) = \frac{5}{9}(45) = 25$

61. (a) If $x = 1000$, $C = 4000 + 2x = 4000 + 2(1000) = 4000 + 2000 = \$6000$
    (b) If $x = 2000$, $C = 4000 + 2x = 4000 + 2(2000) = 4000 + 4000 = \$8000$

63. $|x - 115| \le 5$
    (a) $|x - 115| = |113 - 115| = |-2| = 2 \le 5$ 113 volts is acceptable.
    (b) $|x - 115| = |109 - 115| = |-6| = 6 \not\le 5$ 109 volts is not acceptable.

65. $|x - 3| \le 0.01$
    (a) $|x - 3| = |2.999 - 3| = |-0.001| = 0.001 \le 0.01$ A radius of 2.999 centimeters is acceptable.
    (b) $|x - 3| = |2.89 - 3| = |-0.11| = 0.11 \not\le 0.01$ A radius of 2.89 centimeters is not acceptable.

67. $\frac{1}{3} = 0.333333\ldots > 0.333$ $\frac{1}{3}$ is larger by approximately $0.0003333\ldots$

69. No. No.

71. $0.9999\ldots = 1$

Section 2 Integer Exponents

1. $4^2 = 16$

3. $4^{-2} = \dfrac{1}{4^2} = \dfrac{1}{16}$

5. $-4^{-2} = -\dfrac{1}{4^2} = -\dfrac{1}{16}$

7. $4^0 \cdot 2^{-3} = 1 \cdot \dfrac{1}{2^3} = 1 \cdot \dfrac{1}{8} = \dfrac{1}{8}$

9. $2^{-3} + \left(\dfrac{1}{2}\right)^3 = \dfrac{1}{2^3} + \dfrac{1^3}{2^3} = \dfrac{1}{8} + \dfrac{1}{8} = \dfrac{1}{4}$

11. $3^{-6} \cdot 3^4 = 3^{-6+4} = 3^{-2} = \dfrac{1}{3^2} = \dfrac{1}{9}$

13. $\dfrac{\left(3^2\right)^2}{\left(2^3\right)^2} = \dfrac{3^4}{2^6} = \dfrac{81}{64}$

15. $\left(\dfrac{2}{3}\right)^{-3} = \dfrac{1}{\left(\dfrac{2}{3}\right)^3} = \dfrac{1}{\dfrac{2^3}{3^3}} = \dfrac{3^3}{2^3} = \dfrac{27}{8}$

17. $\dfrac{2^3 \cdot 3^2}{2^4 \cdot 3^{-2}} = \dfrac{2^3}{2^4} \cdot \dfrac{3^2}{3^{-2}} = 2^{3-4} \cdot 3^{2-(-2)} = 2^{-1} \cdot 3^4 = \dfrac{1}{2} \cdot 81 = \dfrac{81}{2}$

19. $\left(\dfrac{9}{2}\right)^{-2} = \dfrac{1}{\left(\dfrac{9}{2}\right)^2} = \dfrac{1}{\dfrac{9^2}{2^2}} = \dfrac{2^2}{9^2} = \dfrac{4}{81}$

21. $\dfrac{2^{-2}}{3} = \dfrac{\dfrac{1}{2^2}}{3} = \dfrac{\dfrac{1}{4}}{3} = \dfrac{1}{4} \cdot \dfrac{1}{3} = \dfrac{1}{12}$

23. $\dfrac{-3^{-1}}{2^{-1}} = \dfrac{-\dfrac{1}{3}}{\dfrac{1}{2}} = -\dfrac{1}{3} \cdot \dfrac{2}{1} = -\dfrac{2}{3}$

25. $x^0 y^2 = 1 \cdot y^2 = y^2$

27. $x y^{-2} = x \cdot \dfrac{1}{y^2} = \dfrac{x}{y^2}$

29. $\left(8x^3\right)^{-2} = \dfrac{1}{\left(8x^3\right)^2} = \dfrac{1}{8^2 \cdot x^6} = \dfrac{1}{64x^6}$

31. $-4x^{-1} = -4 \cdot \dfrac{1}{x} = -\dfrac{4}{x}$ 33. $3x^0 = 3 \cdot 1 = 3$

35. $\dfrac{x^{-2}y^3}{xy^4} = \dfrac{x^{-2}}{x} \cdot \dfrac{y^3}{y^4} = x^{-2-1}y^{3-4} = x^{-3}y^{-1} = \dfrac{1}{x^3} \cdot \dfrac{1}{y} = \dfrac{1}{x^3 y}$

37. $x^{-1}y^{-1} = \dfrac{1}{x} \cdot \dfrac{1}{y} = \dfrac{1}{xy}$

39. $\dfrac{x^{-1}}{y^{-1}} = \dfrac{\frac{1}{x}}{\frac{1}{y}} = \dfrac{1}{x} \cdot \dfrac{y}{1} = \dfrac{y}{x}$

41. $\left(\dfrac{4y}{5x}\right)^{-2} = \dfrac{1}{\left(\frac{4y}{5x}\right)^2} = \dfrac{1}{\frac{(4y)^2}{(5x)^2}} = \dfrac{(5x)^2}{(4y)^2} = \dfrac{5^2 \cdot x^2}{4^2 \cdot y^2} = \dfrac{25x^2}{16y^2}$

43. $x^{-2}y^{-2} = \dfrac{1}{x^2} \cdot \dfrac{1}{y^2} = \dfrac{1}{x^2 y^2}$

45. $\dfrac{x^{-1}y^{-2}z^3}{x^2 y z^3} = \dfrac{x^{-1}}{x^2} \cdot \dfrac{y^{-2}}{y} \cdot \dfrac{z^3}{z^3} = x^{-1-2}y^{-2-1}z^{3-3} = x^{-3}y^{-3}z^0 = \dfrac{1}{x^3} \cdot \dfrac{1}{y^3} \cdot 1 = \dfrac{1}{x^3 y^3}$

47. $\dfrac{(-2)^3 x^4 (yz)^2}{3^2 xy^3 z^4} = \dfrac{-8x^4 y^2 z^2}{9xy^3 z^4} = \dfrac{-8}{9}x^{4-1}y^{2-3}z^{2-4} = \dfrac{-8}{9}x^3 y^{-1}z^{-2} = \dfrac{-8}{9}x^3 \cdot \dfrac{1}{y} \cdot \dfrac{1}{z^2} = \dfrac{-8x^3}{9yz^2}$

49. $\left(\dfrac{3x^{-1}}{4y^{-1}}\right)^{-2} = \dfrac{1}{\left(\frac{3x^{-1}}{4y^{-1}}\right)^2} = \dfrac{1}{\frac{(3x^{-1})^2}{(4y^{-1})^2}} = \dfrac{(4y^{-1})^2}{(3x^{-1})^2} = \dfrac{4^2 y^{-2}}{3^2 x^{-2}} = \dfrac{16 \cdot \frac{1}{y^2}}{9 \cdot \frac{1}{x^2}} = \dfrac{16}{9} \cdot \dfrac{x^2}{y^2} = \dfrac{16x^2}{9y^2}$

51. $\dfrac{(xy^{-1})^{-2}}{xy^3} = \dfrac{x^{-2}y^2}{xy^3} = \dfrac{x^{-2}}{x} \cdot \dfrac{y^2}{y^3} = x^{-2-1}y^{2-3} = x^{-3}y^{-1} = \dfrac{1}{x^3 y}$

53. $\left(\dfrac{x}{y^2}\right)^{-2} \cdot (y^2)^{-1} = \dfrac{x^{-2}}{(y^2)^{-2}} \cdot \dfrac{1}{y^2} = \dfrac{x^{-2}}{y^{-4}y^2} = \dfrac{x^{-2}}{y^{-2}} = \dfrac{\frac{1}{x^2}}{\frac{1}{y^2}} = \dfrac{y^2}{x^2}$

55. If $x = 2$, $2x^3 - 3x^2 + 5x - 4 = 2 \cdot 2^3 - 3 \cdot 2^2 + 5 \cdot 2 - 4 = 16 - 12 + 10 - 4 = 10$
    If $x = 1$, $2x^3 - 3x^2 + 5x - 4 = 2 \cdot 1^3 - 3 \cdot 1^2 + 5 \cdot 1 - 4 = 2 - 3 + 5 - 4 = 0$

57. $(0.3)^4 = \left(\dfrac{3}{10}\right)^4 = \dfrac{3^4}{10^4} = \dfrac{81}{10000} = 0.0081$

59. $\dfrac{(666)^4}{(222)^4} = \left(\dfrac{666}{222}\right)^4 = 3^4 = 81$

61. $(8.2)^6 \approx 304{,}006.671$

63. $(6.1)^{-3} \approx 0.004$

65. $(-2.8)^6 \approx 481.890$

67. $(-8.11)^{-4} \approx 0.000$

Section 3 Polynomials

1. $2x^3$ Monomial; Variable: x; Coefficient: 2; Degree: 3

3. $\dfrac{8}{x}$ Not a monomial.

5. $-2x$ Monomial; Variable: x; Coefficient: –2; Degree: 1

7. $\dfrac{8x}{5}$ Monomial; Variable: x; Coefficient: $\dfrac{8}{5}$; Degree: 1

9. $x^2 + 9$ Not a monomial.

11. $3x^2 - 5$ Polynomial; Degree: 2

13. 5   Polynomial; Degree: 0

15. $3x^2 - \dfrac{5}{x}$ Not a polynomial.

17. $2y^3 - \sqrt{2}$ Polynomial; Degree: 3

19. $\dfrac{x^2 + 5}{x^3 - 1}$ Not a polynomial.

21. $(x^2 + 4x + 5) + (3x - 3) = x^2 + (4x + 3x) + (5 - 3) = x^2 + 7x + 2$

23. $(x^3 - 2x^2 + 5x + 10) - (2x^2 - 4x + 3) = x^3 - 2x^2 + 5x + 10 - 2x^2 + 4x - 3$
$$= x^3 + (-2x^2 - 2x^2) + (5x + 4x) + (10 - 3)$$
$$= x^3 - 4x^2 + 9x + 7$$

25. $(x^2 - 3x + 1) + 2(3x^2 + x - 4) = x^2 - 3x + 1 + 6x^2 + 2x - 8 = 7x^2 - x - 7$

27. $6(x^3 + x^2 - 3) - 4(2x^3 - 3x^2) = 6x^3 + 6x^2 - 18 - 8x^3 + 12x^2 = -2x^3 + 18x^2 - 18$

29. $x(x^2 + x - 4) = x^3 + x^2 - 4x$

31. $(x + 1)(x^2 + 2x - 4) = x(x^2 + 2x - 4) + 1(x^2 + 2x - 4)$
$$= x^3 + 2x^2 - 4x + x^2 + 2x - 4$$
$$= x^3 + 3x^2 - 2x - 4$$

33. $(x + 2)(x + 4) = x^2 + 4x + 2x + 8 = x^2 + 6x + 8$

35. $(2x + 5)(x + 2) = 2x^2 + 4x + 5x + 10 = 2x^2 + 9x + 10$

37. $(x - 4)(x + 2) = x^2 + 2x - 4x - 8 = x^2 - 2x - 8$

39. $(x - 3)(x - 2) = x^2 - 2x - 3x + 6 = x^2 - 5x + 6$

41. $(2x + 3)(x - 2) = 2x^2 - 4x + 3x - 6 = 2x^2 - x - 6$

43. $(-2x + 3)(x - 4) = -2x^2 + 8x + 3x - 12 = -2x^2 + 11x - 12$

45. $(x - 7)(x + 7) = x^2 - 7^2 = x^2 - 49$

47. $(2x + 3)(2x - 3) = (2x)^2 - 3^2 = 4x^2 - 9$

49. $(x + 4)^2 = x^2 + 2 \cdot x \cdot 4 + 4^2 = x^2 + 8x + 16$

51. $(x - 4)^2 = x^2 - 2 \cdot x \cdot 4 + 4^2 = x^2 - 8x + 16$

53. $(2x - 3)^2 = (2x)^2 - 2(2x)(3) + 3^2 = 4x^2 - 12x + 9$

55. $(x - 2)^3 = x^3 - 3 \cdot x^2 \cdot 2 + 3 \cdot x \cdot 2^2 - 2^3 = x^3 - 6x^2 + 12x - 8$

57. $(2x + 1)^3 = (2x)^3 + 3(2x)^2(1) + 3(2x) \cdot 1^2 + 1^3 = 8x^3 + 12x^2 + 6x + 1$

Section 4 Polynomial Division; Synthetic Division

1. Divide:

$$\begin{array}{r} 4x^2 - 11x + 23 \\ x+2\overline{)4x^3 - 3x^2 + x + 1} \\ \underline{4x^3 + 8x^2} \\ -11x^2 + x \\ \underline{-11x^2 - 22x} \\ 23x + 1 \\ \underline{23x + 46} \\ -45 \end{array}$$

Check:

$(x+2)(4x^2 - 11x + 23) + (-45)$

$= 4x^3 - 11x^2 + 23x + 8x^2 - 22x + 46 - 45$

$= 4x^3 - 3x^2 + x + 1$

The quotient is $4x^2 - 11x + 23$; the remainder is –45.

3. Divide:

$$\begin{array}{r} 4x^2 + 13x + 53 \\ x-4\overline{)4x^3 - 3x^2 + x + 1} \\ \underline{4x^3 - 16x^2} \\ 13x^2 + x \\ \underline{13x^2 - 52x} \\ 53x + 1 \\ \underline{53x - 212} \\ 213 \end{array}$$

Check:

$(x - 4)(4x^2 + 13x + 53) + 213$

$= 4x^3 + 13x^2 + 53x - 16x^2 - 52x - 212 + 213$

$= 4x^3 - 3x^2 + x + 1$

The quotient is $4x^2 + 13x + 53$; the remainder is 213.

5. Divide:

$$\begin{array}{r} 4x - 3 \\ x^2+2\overline{)4x^3 - 3x^2 + x + 1} \\ \underline{4x^3 + 8x} \\ -3x^2 - 7x \\ \underline{-3x^2 - 6} \\ -7x + 7 \end{array}$$

Check:

$(x^2 + 2)(4x - 3) + (-7x + 7)$

$= 4x^3 - 3x^2 + 8x - 6 - 7x + 7$

$= 4x^3 - 3x^2 + x + 1$

The quotient is $4x - 3$; the remainder is $-7x + 7$.

7. Divide:

$$\begin{array}{r} 2 \\ 2x^3-1\overline{)4x^3 - 3x^2 + x + 1} \\ \underline{4x^3 - 2} \\ -3x^2 + x + 3 \end{array}$$

Check:

$(2x^3 - 1)(2) + (-3x^2 + x + 3)$

$= 4x^3 - 2 - 3x^2 + x + 3$

$= 4x^3 - 3x^2 + x + 1$

The quotient is 2; the remainder is $-3x^2 + x + 3$.

9.   Divide:

$$\begin{array}{r} 2x - \frac{5}{2} \\ 2x^2 + x + 1 \overline{\smash{\big)}\ 4x^3 - 3x^2 + \ x + 1} \\ \underline{4x^3 + 2x^2 + 2x} \\ -5x^2 - \ x \\ \underline{-5x^2 - \frac{5}{2}x - \frac{5}{2}} \\ \frac{3}{2}x + \frac{7}{2} \end{array}$$

Check:

$(2x^2 + x + 1)(2x - \frac{5}{2}) + (\frac{3}{2}x + \frac{7}{2})$

$= 4x^3 - 5x^2 + 2x^2 - \frac{5}{2}x + 2x - \frac{5}{2} + \frac{3}{2}x + \frac{7}{2}$

$= 4x^3 - 3x^2 + x + 1$

The quotient is $2x - \frac{5}{2}$; the remainder is $\frac{3}{2}x + \frac{7}{2}$.

11.   Divide:

$$\begin{array}{r} x - \frac{3}{4} \\ 4x^2 + 1 \overline{\smash{\big)}\ 4x^3 - 3x^2 + x + 1} \\ \underline{4x^3 \qquad\quad + x} \\ -3x^2 \qquad +1 \\ \underline{-3x^2 \qquad\ -\frac{3}{4}} \\ \frac{7}{4} \end{array}$$

Check:

$(4x^2 + 1)(x - \frac{3}{4}) + \frac{7}{4}$

$= 4x^3 - 3x^2 + x - \frac{3}{4} + \frac{7}{4}$

$= 4x^3 - 3x^2 + x + 1$

The quotient is $x - \frac{3}{4}$; the remainder is $\frac{7}{4}$.

13.   Divide:

$$\begin{array}{r} x^3 + x^2 + x + 1 \\ x - 1 \overline{\smash{\big)}\ x^4 + 0x^3 + 0x^2 + 0x - 1} \\ \underline{x^4 - \ x^3} \\ x^3 \\ \underline{x^3 - \ x^2} \\ x^2 \\ \underline{x^2 - \ x} \\ x - 1 \\ \underline{x - 1} \\ 0 \end{array}$$

Check:

$(x - 1)(x^3 + x^2 + x + 1) + 0$

$= x^4 + x^3 + x^2 + x - x^3 - x^2 - x - 1$

$= x^4 - 1$

The quotient is $x^3 + x^2 + x + 1$; the remainder is 0.

15.   Divide:

$$\begin{array}{r} x^2 + 1 \\ x^2 - 1 \overline{\smash{\big)}\ x^4 + 0x^3 + 0x^2 + 0x - 1} \\ \underline{x^4 \qquad\ - \ x^2} \\ x^2 \\ \underline{x^2 \qquad - 1} \\ 0 \end{array}$$

Check:

$(x^2 - 1)(x^2 + 1) + 0$

$= x^4 + x^2 - x^2 - 1$

$= x^4 - 1$

The quotient is $x^2 + 1$; the remainder is 0.

17. Divide:

$$\begin{array}{r} -4x^2 - 3x - 3 \\ x-1{\overline{\smash{\big)}\,-4x^3 + x^2 + 0x - 4}} \\ \underline{-4x^3 + 4x^2} \\ -3x^2 \\ \underline{-3x^2 + 3x} \\ -3x - 4 \\ \underline{-3x + 3} \\ -7 \end{array}$$

Check:

$$(x-1)(-4x^2 - 3x - 3) + (-7)$$
$$= -4x^3 - 3x^2 - 3x + 4x^2 + 3x + 3 - 7$$
$$= -4x^3 + x^2 - 4$$

The quotient is $-4x^2 - 3x - 3$; the remainder is -7.

19. Divide:

$$\begin{array}{r} x^2 - x - 1 \\ x^2+x+1{\overline{\smash{\big)}\,x^4 + 0x^3 - x^2 + 0x + 1}} \\ \underline{x^4 + x^3 + x^2} \\ -x^3 - 2x^2 \\ \underline{-x^3 - x^2 - x} \\ -x^2 + x + 1 \\ \underline{-x^2 - x - 1} \\ 2x + 2 \end{array}$$

Check:

$$(x^2 + x + 1)(x^2 - x - 1) + 2x + 2$$
$$= x^4 + x^3 + x^2 - x^3 - x^2 - x - x^2 - x - 1 + 2x + 2$$
$$= x^4 - x^2 + 1$$

The quotient is $x^2 - x - 1$; the remainder is $2x + 2$.

21. Divide:

$$\begin{array}{r} -x^2 \\ -x^2+1{\overline{\smash{\big)}\,x^4 + 0x^3 - x^2 + 0x + 1}} \\ \underline{x^4 \qquad - x^2} \\ 1 \end{array}$$

Check:

$$-x^2(-x^2 + 1) + 1$$
$$= x^4 - x^2 + 1$$

The quotient is $-x^2$; the remainder is 1.

23. Divide:

$$\begin{array}{r} x^2 + ax + a^2 \\ x-a{\overline{\smash{\big)}\,x^3 + 0x^2 + 0x - a^3}} \\ \underline{x^3 - ax^2} \\ ax^2 \\ \underline{ax^2 - a^2x} \\ a^2x - a^3 \\ \underline{a^2x - a^3} \\ 0 \end{array}$$

Check:

$$(x - a)(x^2 + ax + a^2) + 0$$
$$= x^3 + ax^2 + a^2x - ax^2 - a^2x - a^3$$
$$= x^3 - a^3$$

The quotient is $x^2 + ax + a^2$; the remainder is 0.

25. Divide:

$$\begin{array}{r} x^3 + ax^2 + a^2x + a^3 \\ x - a\overline{)x^4 + 0x^3 + 0x^2 + 0x - a^4} \end{array}$$

$$\underline{x^4 - ax^3}$$

$$ax^3$$

$$\underline{ax^3 - a^2x^2} \qquad\qquad \text{Check:}$$

$$a^2x^2 \qquad\qquad\qquad (x-a)(x^3 + ax^2 + a^2x + a^3) + 0$$

$$\underline{a^2x^2 - a^3x} \qquad\qquad = x^4 + ax^3 + a^2x^2 + a^3x - ax^3 - a^2x^2 - a^3x - a^4$$

$$a^3x - a^4 \qquad\qquad = x^4 - a^4$$

$$\underline{a^3x - a^4}$$

$$0$$

The quotient is $x^3 + ax^2 + a^2x + a^3$; the remainder is 0.

27. Use synthetic division:

$$\begin{array}{r} 2\overline{)1 \;\; -1 \;\;\; 2 \;\;\;\; 4} \\ \underline{ \;\;\;\; 2 \;\;\; 2 \;\;\;\; 8} \\ 1 \;\;\;\; 1 \;\;\; 4 \;\; 12 \end{array}$$

Quotient: $x^2 + x + 4$ Remainder: 12

29. Use synthetic division:

$$\begin{array}{r} 3\overline{)3 \;\;\; 2 \;\; -1 \;\;\;\; 3} \\ \underline{ \;\;\; 9 \;\; 33 \;\; 96} \\ 3 \;\; 11 \;\; 32 \;\; 99 \end{array}$$

Quotient: $3x^2 + 11x + 32$ Remainder: 99

31. Use synthetic division:

$$\begin{array}{r} -3\overline{)1 \;\;\; 0 \;\; -4 \;\;\;\; 0 \;\;\;\; 1 \;\;\;\;\;\; 0} \\ \underline{ \; -3 \;\;\; 9 \;\; -15 \;\; 45 \;\; -138} \\ 1 \;\; -3 \;\;\; 5 \;\; -15 \;\; 46 \;\; -138 \end{array}$$

Quotient: $x^4 - 3x^3 + 5x^2 - 15x + 46$ Remainder: −138

33. Use synthetic division:

$$\begin{array}{r} 1\overline{)4 \;\;\; 0 \;\; -3 \;\;\; 0 \;\;\; 1 \;\;\; 0 \;\;\; 5} \\ \underline{ \;\;\; 4 \;\;\; 4 \;\;\; 1 \;\;\; 1 \;\;\; 2 \;\;\; 2} \\ 4 \;\;\; 4 \;\;\; 1 \;\;\; 1 \;\;\; 2 \;\;\; 2 \;\;\; 7 \end{array}$$

Quotient: $4x^5 + 4x^4 + x^3 + x^2 + 2x + 2$ Remainder: 7

35. Use synthetic division:

$$\begin{array}{r} -1.1\overline{)0.1 \;\;\;\;\; 0 \;\;\; 0.2 \;\;\;\;\;\;\; 0} \\ \underline{ \; -0.11 \;\;\; 0.121 \;\; -0.3531} \\ 0.1 \;\; -0.11 \;\;\; 0.321 \;\; -0.3531 \end{array}$$

Quotient: $0.1x^2 - 0.11x + 0.321$ Remainder: −0.3531

37. Use synthetic division:

$$\begin{array}{r|rrrrr}
1) & 1 & 0 & 0 & 0 & -1 \\
   &   & 1 & 1 & 1 & 1 & 1 \\
\hline
   & 1 & 1 & 1 & 1 & 1 & 0
\end{array}$$

Quotient: $x^4 + x^3 + x^2 + x + 1$ Remainder: 0

39. Use synthetic division:

$$\begin{array}{r|rrrr}
2) & 4 & -3 & -8 & 4 \\
   &   & 8 & 10 & 4 \\
\hline
   & 4 & 5 & 2 & 8
\end{array}$$

Remainder $= 8 \neq 0$; therefore $x - 2$ is not a factor of $f(x)$.

41. Use synthetic division:

$$\begin{array}{r|rrrrr}
2) & 3 & -6 & 0 & -5 & 10 \\
   &   & 6 & 0 & 0 & -10 \\
\hline
   & 3 & 0 & 0 & -5 & 0
\end{array}$$

Remainder $= 0$; therefore $x - 2$ is a factor of $f(x)$.

43. Use synthetic division:

$$\begin{array}{r|rrrrrrr}
-3) & 3 & 0 & 0 & 82 & 0 & 0 & 27 \\
    &   & -9 & 27 & -81 & -3 & 9 & -27 \\
\hline
    & 3 & -9 & 27 & 1 & -3 & 9 & 0
\end{array}$$

Remainder $= 0$; therefore $x + 3$ is a factor of $f(x)$.

45. Use synthetic division:

$$\begin{array}{r|rrrrrrr}
-4) & 4 & 0 & -64 & 0 & 1 & 0 & -15 \\
    &   & -16 & 64 & 0 & 0 & -4 & 16 \\
\hline
    & 4 & -16 & 0 & 0 & 1 & -4 & 1
\end{array}$$

Remainder $= 1 \neq 0$; therefore $x + 3$ is not a factor of $f(x)$.

47. Use synthetic division:

$$\begin{array}{r|rrrrr}
\frac{1}{2}) & 2 & -1 & 0 & 2 & -1 \\
             &   & 1 & 0 & 0 & 1 \\
\hline
             & 2 & 0 & 0 & 2 & 0
\end{array}$$

Remainder $= 0$; therefore $x - \frac{1}{2}$ is a factor of $f(x)$.

Section 5 Factoring

1. $3x + 6 = 3(x + 2)$

3. $ax^2 + a = a(x^2 + 1)$

5. $x^3 + x^2 + x = x(x^2 + x + 1)$

7. $2x^2 - 2x = 2x(x - 1)$

9. $x^2 - 1 = x^2 - 1^2 = (x - 1)(x + 1)$

11. $4x^2 - 1 = (2x)^2 - 1^2 = (2x - 1)(2x + 1)$

13. $x^2 - 16 = x^2 - 4^2 = (x - 4)(x + 4)$

15. $25x^2 - 4 = (5x - 2)(5x + 2)$

17. $x^2 + 2x + 1 = (x + 1)^2$

19. $x^2 - 10x + 25 = (x - 5)^2$

21. $4x^2 + 4x + 1 = (2x + 1)^2$

23. $16x^2 + 8x + 1 = (4x + 1)^2$

25. $x^3 - 27 = x^3 - 3^3 = (x - 3)(x^2 + 3x + 9)$

27. $x^3 + 27 = x^3 + 3^3 = (x + 3)(x^2 - 3x + 9)$

29. $8x^3 + 27 = (2x)^3 + 3^3 = (2x + 3)(4x^2 - 6x + 9)$

31. $x^2 + 5x + 6 = (x + 2)(x + 3)$

33. $x^2 + 7x + 10 = (x + 2)(x + 5)$

35. $x^2 - 10x + 16 = (x - 2)(x - 8)$

37. $x^2 - 7x - 8 = (x + 1)(x - 8)$

39. $2x^2 + 4x + 3x + 6 = 2x(x + 2) + 3(x + 2) = (x + 2)(2x + 3)$

41. $2x^2 - 4x + x - 2 = 2x(x - 2) + 1(x - 2) = (x - 2)(2x + 1)$

43. $6x^2 + 9x + 4x + 6 = 3x(2x + 3) + 2(2x + 3) = (2x + 3)(3x + 2)$

45. $3x^2 + 4x + 1 = (3x + 1)(x + 1)$

47. $2z^2 + 5z + 3 = (2z + 3)(z + 1)$

49. $3x^2 - 2x - 8 = (3x + 4)(x - 2)$

51. $3x^2 + 10x - 8 = (3x - 2)(x + 4)$

53. $x^2 - 36 = (x - 6)(x + 6)$

55. $1 - 4x^2 = 1^2 - (2x)^2 = (1 - 2x)(1 + 2x)$

57. $x^2 + 7x + 10 = (x + 2)(x + 5)$

59. $x^2 - 2x + 8$ is prime because there are no factors of 8 whose sum is –2.

61. $x^2 + 4x + 16$ is prime because there are no factors of 16 whose sum is 4.

63. $15 + 2x - x^2 = -(x^2 - 2x - 15) = -(x - 5)(x + 3)$

65. $3x^2 - 12x - 36 = 3(x^2 - 4x - 12) = 3(x - 6)(x + 2)$

67. $y^4 + 11y^3 + 30y^2 = y^2(y^2 + 11y + 30) = y^2(y + 5)(y + 6)$

69. $4x^2 + 12x + 9 = (2x + 3)^2$

71. $3x^2 + 4x + 1 = (3x + 1)(x + 1)$

73. $x^4 - 81 = (x^2 - 9)(x^2 + 9) = (x - 3)(x + 3)(x^2 + 9)$

75. $x^6 - 2x^3 + 1 = (x^3 - 1)^2 = \left[(x - 1)(x^2 + x + 1)\right]^2 = (x - 1)^2(x^2 + x + 1)^2$

77. $x^7 - x^5 = x^5(x^2 - 1) = x^5(x - 1)(x + 1)$

79. $5 + 16x - 16x^2 = -(16x^2 - 16x - 5) = -(4x - 5)(4x + 1)$

81. $4y^2 - 16y + 15 = (2y - 5)(2y - 3)$

83. $1 - 8x^2 - 9x^4 = -(9x^4 + 8x^2 - 1) = -(9x^2 - 1)(x^2 + 1) = -(3x - 1)(3x + 1)(x^2 + 1)$

85. $x(x + 3) - 6(x + 3) = (x + 3)(x - 6)$

87. $(x + 2)^2 - 5(x + 2) = (x + 2)[(x + 2) - 5] = (x + 2)(x - 3)$

89. $x^3 + 2x^2 - x - 2 = x^2(x + 2) - (x + 2) = (x + 2)(x^2 - 1) = (x + 2)(x - 1)(x + 1)$

91. $x^4 - x^3 + x - 1 = x^3(x - 1) + (x - 1) = (x - 1)(x^3 + 1) = (x - 1)(x + 1)(x^2 - x + 1)$

93. Factors of 4   1, 4   2, 2   –1, –4   –2, –2
    Sum   5   4   –5   –4
    None of the sums of the factors is 0, so $x^2 + 4$ is prime.

Section 6 Solving Equations

1. Solve:
$$3x + 2 = x$$
$$3x + 2 - 2 = x - 2$$
$$3x = x - 2$$
$$3x - x = x - 2 - x$$
$$2x = -2$$
$$\frac{2x}{2} = \frac{-2}{2}$$
$$x = -1$$

3. Solve:
$$2t - 6 = 3 - t$$
$$2t - 6 + 6 = 3 - t + 6$$
$$2t = 9 - t$$
$$2t + t = 9 - t + t$$
$$3t = 9$$
$$\frac{3t}{3} = \frac{9}{3}$$
$$t = 3$$

5. Solve:
$$6 - x = 2x + 9$$
$$-3x = 3$$
$$x = -1$$

7. Solve:
$$3 + 2n = 5n + 7$$
$$-3n = 4$$
$$n = -\frac{4}{3}$$

9. Solve:
$$2(3 + 2x) = 3(x - 4)$$
$$6 + 4x = 3x - 12$$
$$x = -18$$

11. Solve:
$$8x - (3x + 2) = 3x - 10$$
$$5x - 2 = 3x - 10$$
$$2x = -8$$
$$x = -4$$

13. Solve:
$$\tfrac{3}{2}x + 2 = \tfrac{1}{2} - \tfrac{1}{2}x$$
$$2x = -\tfrac{3}{2}$$
$$x = -\tfrac{3}{2} \cdot \tfrac{1}{2}$$
$$x = -\tfrac{3}{4}$$

15. Solve:
$$\tfrac{1}{2}x - 5 = \tfrac{3}{4}x$$
$$4\left(\tfrac{1}{2}x - 5\right) = 4\left(\tfrac{3}{4}x\right)$$
$$2x - 20 = 3x$$
$$-20 = x$$

17. Solve:
$$\tfrac{2}{3}p = \tfrac{1}{2}p + \tfrac{1}{3}$$
$$6\left(\tfrac{2}{3}p\right) = 6\left(\tfrac{1}{2}p + \tfrac{1}{3}\right)$$
$$4p = 3p + 2$$
$$p = 2$$

19. Solve:
$$x^2 - 7x + 12 = 0$$
$$(x - 3)(x - 4) = 0$$
$$x - 3 = 0 \ \text{ or } \ x - 4 = 0$$
$$x = 3 \ \text{ or } \quad x = 4$$
Solution: $\{3, 4\}$

21. Solve:
$$2x^2 + 5x - 3 = 0$$
$$(2x - 1)(x + 3) = 0$$
$$2x - 1 = 0 \text{ or } x + 3 = 0$$
$$2x = 1 \text{ or } \quad x = -3$$
$$x = \tfrac{1}{2}$$

Solution: $\left\{-3, \tfrac{1}{2}\right\}$

23. Solve:
$$x^3 = 9x$$
$$x^3 - 9x = 0$$
$$x(x^2 - 9) = 0$$
$$x(x - 3)(x + 3) = 0$$
$$x = 0 \text{ or } x - 3 = 0 \text{ or } x + 3 = 0$$
$$x = 0 \text{ or } \quad x = 3 \text{ or } \quad x = -3$$

Solution: $\{-3, 0, 3\}$

25. Solve:
$$x^3 + x^2 - 20x = 0$$
$$x(x^2 + x - 20) = 0$$
$$x(x + 5)(x - 4) = 0$$
$$x = 0 \text{ or } x + 5 = 0 \text{ or } x - 4 = 0$$
$$x = 0 \text{ or } \quad x = -5 \text{ or } \quad x = 4$$

Solution: $\{-5, 0, 4\}$

Section 7 Rational Expressions

1. $\dfrac{x^2 - 1}{x}$ Part (c) must be excluded.

The value $x = 0$ must be excluded from the domain because it causes division by 0.

3. $\dfrac{x}{x^2 - 9} = \dfrac{x}{(x - 3)(x + 3)}$ Part (a) must be excluded.

The values $x = -3$ and $x = 3$ must be excluded from the domain because they cause division by 0.

5. $\dfrac{x^2}{x^2 + 1}$ None of the given values are excluded.

The domain is all real numbers.

7. $\dfrac{x^2 + 5x - 10}{x^3 - x} = \dfrac{x^2 + 5x - 10}{x(x-1)(x+1)}$ Parts (b), (c), and (d) must be excluded.

The values $x = 0$, $x = 1$, and $x = -1$ must be excluded from the domain because they cause division by 0.

9. $\dfrac{3x + 9}{x^2 - 9} = \dfrac{3(x+3)}{(x-3)(x+3)} = \dfrac{3}{x-3}$

11. $\dfrac{x^2 - 2x}{3x - 6} = \dfrac{x(x-2)}{3(x-2)} = \dfrac{x}{3}$

13. $\dfrac{24x^2}{12x^2 - 6x} = \dfrac{24x^2}{6x(2x-1)} = \dfrac{4x}{2x-1}$

15. $\dfrac{x^2 + 4x - 5}{x^2 - 2x + 1} = \dfrac{(x+5)(x-1)}{(x-1)(x-1)} = \dfrac{x+5}{x-1}$

17. $\dfrac{x^2 - 4}{x^2 + 5x + 6} = \dfrac{(x-2)(x+2)}{(x+3)(x+2)} = \dfrac{x-2}{x+3}$

19. $\dfrac{x^2 + 5x - 14}{2 - x} = \dfrac{(x+7)(x-2)}{-1(x-2)} = -(x+7)$

21. $\dfrac{3x + 6}{5x^2} \cdot \dfrac{x}{x^2 - 4} = \dfrac{3(x+2)}{5x^2} \cdot \dfrac{x}{(x-2)(x+2)} = \dfrac{3}{5x(x-2)}$

23. $\dfrac{4x^2}{x^2 - 16} \cdot \dfrac{x-4}{2x} = \dfrac{4x^2}{(x-4)(x+4)} \cdot \dfrac{x-4}{2x} = \dfrac{2x}{x+4}$

25. $\dfrac{x^2 - 3x - 10}{x^2 + 2x - 35} \cdot \dfrac{x^2 + 4x - 21}{x^2 + 9x + 14} = \dfrac{(x-5)(x+2)}{(x+7)(x-5)} \cdot \dfrac{(x+7)(x-3)}{(x+7)(x+2)} = \dfrac{x-3}{x+7}$

27. $\dfrac{\dfrac{6x}{x^2 - 4}}{\dfrac{3x-9}{2x+4}} = \dfrac{6x}{x^2 - 4} \cdot \dfrac{2x+4}{3x-9} = \dfrac{6x}{(x-2)(x+2)} \cdot \dfrac{2(x+2)}{3(x-3)} = \dfrac{4x}{(x-2)(x-3)}$

29. $\dfrac{\dfrac{x^2 + 7x + 12}{x^2 - 7x + 12}}{\dfrac{x^2 + x - 12}{x^2 - x - 12}} = \dfrac{x^2 + 7x + 12}{x^2 - 7x + 12} \cdot \dfrac{x^2 - x - 12}{x^2 + x - 12} = \dfrac{(x+3)(x+4)}{(x-3)(x-4)} \cdot \dfrac{(x-4)(x+3)}{(x+4)(x-3)} = \dfrac{(x+3)^2}{(x-3)^2}$

31. $\dfrac{\dfrac{2x^2 - x - 28}{3x^2 - x - 2}}{\dfrac{4x^2 + 16x + 7}{3x^2 + 11x + 6}} = \dfrac{2x^2 - x - 28}{3x^2 - x - 2} \cdot \dfrac{3x^2 + 11x + 6}{4x^2 + 16x + 7} = \dfrac{(2x + 7)(x - 4)}{(3x + 2)(x - 1)} \cdot \dfrac{(3x + 2)(x + 3)}{(2x + 7)(2x + 1)}$

$$= \dfrac{(x - 4)(x + 3)}{(x - 1)(2x + 1)}$$

33. $\dfrac{x + 1}{x - 3} + \dfrac{2x - 3}{x - 3} = \dfrac{x + 1 + 2x - 3}{x - 3} = \dfrac{3x - 2}{x - 3}$

35. $\dfrac{3x + 5}{2x - 1} - \dfrac{2x - 4}{2x - 1} = \dfrac{(3x + 5) - (2x - 4)}{2x - 1} = \dfrac{3x + 5 - 2x + 4}{2x - 1} = \dfrac{x + 9}{2x - 1}$

37. $\dfrac{4}{x - 2} + \dfrac{x}{2 - x} = \dfrac{4}{x - 2} - \dfrac{x}{x - 2} = \dfrac{4 - x}{x - 2}$

39. $\dfrac{4}{x - 1} - \dfrac{2}{x + 2} = \dfrac{4(x + 2)}{(x - 1)(x + 2)} - \dfrac{2(x - 1)}{(x + 2)(x - 1)} = \dfrac{4x + 8 - 2x + 2}{(x + 2)(x - 1)} = \dfrac{2x + 10}{(x + 2)(x - 1)}$

$$= \dfrac{2(x + 5)}{(x + 2)(x - 1)}$$

41. $\dfrac{x}{x + 1} + \dfrac{2x - 3}{x - 1} = \dfrac{x(x - 1)}{(x + 1)(x - 1)} + \dfrac{(2x - 3)(x + 1)}{(x - 1)(x + 1)} = \dfrac{x^2 - x + 2x^2 - x - 3}{(x - 1)(x + 1)}$

$$= \dfrac{3x^2 - 2x - 3}{(x - 1)(x + 1)}$$

43. $\dfrac{x - 3}{x + 2} - \dfrac{x + 4}{x - 2} = \dfrac{(x - 3)(x - 2)}{(x + 2)(x - 2)} - \dfrac{(x + 4)(x + 2)}{(x - 2)(x + 2)} = \dfrac{x^2 - 5x + 6 - (x^2 + 6x + 8)}{(x + 2)(x - 2)}$

$$= \dfrac{x^2 - 5x + 6 - x^2 - 6x - 8}{(x + 2)(x - 2)} = \dfrac{-11x - 2}{(x + 2)(x - 2)}$$

45. $\dfrac{x^3}{(x - 1)^2} - \dfrac{x^2 + 1}{x} = \dfrac{x^3 \cdot x}{(x - 1)^2 x} - \dfrac{(x^2 + 1)(x - 1)^2}{x(x - 1)^2} = \dfrac{x^4 - (x^2 + 1)(x^2 - 2x + 1)}{x(x - 1)^2}$

$$= \dfrac{x^4 - (x^4 - 2x^3 + 2x^2 - 2x + 1)}{x(x - 1)^2} = \dfrac{2x^3 - 2x^2 + 2x - 1}{x(x - 1)^2}$$

47. $\dfrac{x}{x^2 - 7x + 6} - \dfrac{x}{x^2 - 2x - 24} = \dfrac{x}{(x - 6)(x - 1)} - \dfrac{x}{(x - 6)(x + 4)}$

$$= \dfrac{x(x + 4)}{(x - 6)(x - 1)(x + 4)} - \dfrac{x(x - 1)}{(x - 6)(x + 4)(x - 1)}$$

$$= \dfrac{x^2 + 4x - x^2 + x}{(x - 6)(x + 4)(x - 1)} = \dfrac{5x}{(x - 6)(x + 4)(x - 1)}$$

49.
$$\frac{4x}{x^2-4}-\frac{2}{x^2+x-6}=\frac{4x}{(x-2)(x+2)}-\frac{2}{(x+3)(x-2)}$$
$$=\frac{4x(x+3)}{(x-2)(x+2)(x+3)}-\frac{2(x+2)}{(x+3)(x-2)(x+2)}$$
$$=\frac{4x^2+12x-2x-4}{(x-2)(x+2)(x+3)}=\frac{4x^2+10x-4}{(x-2)(x+2)(x+3)}$$
$$=\frac{2(2x^2+5x-2)}{(x-2)(x+2)(x+3)}$$

51.
$$\frac{x+4}{x^2-x-2}-\frac{2x+3}{x^2+2x-8}=\frac{x+4}{(x-2)(x+1)}-\frac{2x+3}{(x+4)(x-2)}$$
$$=\frac{(x+4)(x+4)}{(x-2)(x+1)(x+4)}-\frac{(2x+3)(x+1)}{(x+4)(x-2)(x+1)}$$
$$=\frac{x^2+8x+16-(2x^2+5x+3)}{(x-2)(x+1)(x+4)}=\frac{-x^2+3x+13}{(x-2)(x+1)(x+4)}$$

53.
$$\frac{1}{h}\left(\frac{1}{x+h}-\frac{1}{x}\right)=\frac{1}{h}\left(\frac{1\cdot x}{(x+h)x}-\frac{1(x+h)}{x(x+h)}\right)=\frac{1}{h}\left(\frac{x-x-h}{x(x+h)}\right)=\frac{-h}{hx(x+h)}=\frac{-1}{x(x+h)}$$

55.
$$\frac{1+\dfrac{1}{x}}{1-\dfrac{1}{x}}=\frac{\dfrac{x}{x}+\dfrac{1}{x}}{\dfrac{x}{x}-\dfrac{1}{x}}=\frac{\dfrac{x+1}{x}}{\dfrac{x-1}{x}}=\frac{x+1}{x}\cdot\frac{x}{x-1}=\frac{x+1}{x-1}$$

57.
$$\frac{x-\dfrac{1}{x}}{x+\dfrac{1}{x}}=\frac{\dfrac{x^2}{x}-\dfrac{1}{x}}{\dfrac{x^2}{x}+\dfrac{1}{x}}=\frac{\dfrac{x^2-1}{x}}{\dfrac{x^2+1}{x}}=\frac{x^2-1}{x}\cdot\frac{x}{x^2+1}=\frac{(x-1)(x+1)}{x^2+1}$$

59.
$$\frac{\dfrac{x+4}{x-2}-\dfrac{x-3}{x+1}}{x+1}=\frac{\dfrac{(x+4)(x+1)}{(x-2)(x+1)}-\dfrac{(x-3)(x-2)}{(x+1)(x-2)}}{x+1}=\frac{\dfrac{x^2+5x+4-(x^2-5x+6)}{(x-2)(x+1)}}{x+1}$$
$$=\frac{10x-2}{(x-2)(x+1)}\cdot\frac{1}{x+1}=\frac{2(5x-1)}{(x-2)(x+1)^2}$$

61.
$$\frac{\dfrac{x-2}{x+2}+\dfrac{x-1}{x+1}}{\dfrac{x}{x+1}-\dfrac{2x-3}{x}}=\frac{\dfrac{(x-2)(x+1)}{(x+2)(x+1)}+\dfrac{(x-1)(x+2)}{(x+1)(x+2)}}{\dfrac{x^2}{(x+1)(x)}-\dfrac{(2x-3)(x+1)}{x(x+1)}}=\frac{\dfrac{x^2-x-2+x^2+x-2}{(x+2)(x+1)}}{\dfrac{x^2-(2x^2-x-3)}{x(x+1)}}$$
$$=\frac{\dfrac{2x^2-4}{(x+2)(x+1)}}{\dfrac{-x^2+x+3}{x(x+1)}}=\frac{2(x^2-2)}{(x+2)(x+1)}\cdot\frac{x(x+1)}{-(x^2-x-3)}$$
$$=\frac{2x(x^2-2)}{-(x+2)(x^2-x-3)}$$

63. $\dfrac{1}{f} = (n-1)\left(\dfrac{1}{R_1} + \dfrac{1}{R_2}\right)$

$\dfrac{R_1 \cdot R_2}{f} = (n-1)\left(\dfrac{1}{R_1} + \dfrac{1}{R_2}\right) R_1 \cdot R_2$

$\dfrac{R_1 \cdot R_2}{f} = (n-1)(R_2 + R_1)$

$\dfrac{f}{R_1 \cdot R_2} = \dfrac{1}{(n-1)(R_2 + R_1)}$

$f = \dfrac{R_1 \cdot R_2}{(n-1)(R_2 + R_1)}$

$f = \dfrac{0.1(0.2)}{(1.5-1)(0.2+0.1)} = \dfrac{0.02}{0.5(0.3)} = \dfrac{0.02}{0.15} = \dfrac{2}{15}$

Section 8 Radicals; Rational Exponents

1. $\sqrt{8} = \sqrt{4\cdot 2} = 2\sqrt{2}$

3. $\sqrt[3]{16x^4} = \sqrt[3]{8\cdot 2\cdot x^3 \cdot x} = 2x\sqrt[3]{2x}$

5. $\sqrt[3]{\sqrt{x^6}} = \sqrt[6]{x^6} = x$

7. $\sqrt{\dfrac{32x^3}{9x}} = \sqrt{\dfrac{32x^2}{9}} = \dfrac{\sqrt{16\cdot 2\,x^2}}{\sqrt{9}} = \dfrac{4x\sqrt{2}}{3}$

9. $\sqrt[4]{x^{12}y^8} = \sqrt[4]{\left(x^3\right)^4\left(y^2\right)^4} = x^3 y^2$

11. $\sqrt[4]{\dfrac{x^9 y^7}{x y^3}} = \sqrt[4]{x^8 y^4} = x^2 y$

13. $\sqrt{36x} = 6\sqrt{x}$

15. $\sqrt{3x^2}\sqrt{12x} = \sqrt{36x^2 \cdot x} = 6x\sqrt{x}$

17. $\left(\sqrt{5}\,\sqrt[3]{9}\right)^2 = 5\sqrt[3]{81} = 5\sqrt[3]{27\cdot 3} = 5\cdot 3\sqrt[3]{3} = 15\sqrt[3]{3}$

19. $\left(3\sqrt{6}\right)\left(2\sqrt{2}\right) = 6\sqrt{12} = 6\sqrt{4\cdot 3} = 6\cdot 2\sqrt{3} = 12\sqrt{3}$

21. $\left(\sqrt{3}+3\right)\left(\sqrt{3}-1\right) = \sqrt{9} - \sqrt{3} + 3\sqrt{3} - 3 = 3 + 2\sqrt{3} - 3 = 2\sqrt{3}$

23. $\left(\sqrt{x}-1\right)^2 = \sqrt{x}^2 - 2\sqrt{x} + 1 = x + 1 - 2\sqrt{x}$

25. $\dfrac{1}{\sqrt{2}} = \dfrac{1}{\sqrt{2}} \cdot \dfrac{\sqrt{2}}{\sqrt{2}} = \dfrac{\sqrt{2}}{\sqrt{4}} = \dfrac{\sqrt{2}}{2}$

27. $\dfrac{-\sqrt{3}}{\sqrt{5}} = \dfrac{-\sqrt{3}}{\sqrt{5}} \cdot \dfrac{\sqrt{5}}{\sqrt{5}} = \dfrac{-\sqrt{15}}{\sqrt{25}} = \dfrac{-\sqrt{15}}{5}$

29. $\dfrac{\sqrt{3}}{5-\sqrt{2}} = \dfrac{\sqrt{3}}{5-\sqrt{2}} \cdot \dfrac{5+\sqrt{2}}{5+\sqrt{2}} = \dfrac{5\sqrt{3}+\sqrt{6}}{25-\sqrt{4}} = \dfrac{5\sqrt{3}+\sqrt{6}}{23}$

31. $\dfrac{2-\sqrt{5}}{2+3\sqrt{5}} = \dfrac{2-\sqrt{5}}{2+3\sqrt{5}} \cdot \dfrac{2-3\sqrt{5}}{2-3\sqrt{5}} = \dfrac{4-6\sqrt{5}-2\sqrt{5}+3\sqrt{25}}{4-9\sqrt{25}} = \dfrac{19-8\sqrt{5}}{-41}$

33. $\dfrac{\sqrt{x+h}-\sqrt{x}}{\sqrt{x+h}+\sqrt{x}} = \dfrac{\sqrt{x+h}-\sqrt{x}}{\sqrt{x+h}+\sqrt{x}} \cdot \dfrac{\sqrt{x+h}-\sqrt{x}}{\sqrt{x+h}-\sqrt{x}} = \dfrac{x+h-2\sqrt{x(x+h)}+x}{x+h-x}$

$\qquad = \dfrac{2x+h-2\sqrt{x(x+h)}}{h}$

35. $8^{\frac{2}{3}} = \left(2^3\right)^{\frac{2}{3}} = 2^2 = 4$

37. $(-27)^{\frac{1}{3}} = \left((-3)^3\right)^{\frac{1}{3}} = -3$

39. $16^{\frac{3}{2}} = \left(4^2\right)^{\frac{3}{2}} = 4^3 = 64$

41. $9^{\frac{-3}{2}} = \left(3^2\right)^{\frac{-3}{2}} = 3^{-3} = \dfrac{1}{3^3} = \dfrac{1}{27}$

43. $\left(\dfrac{9}{8}\right)^{\frac{3}{2}} = \dfrac{9^{\frac{3}{2}}}{8^{\frac{3}{2}}} = \dfrac{\left(3^2\right)^{\frac{3}{2}}}{\left(2^3\right)^{\frac{3}{2}}} = \dfrac{3^3}{2^{\frac{9}{2}}} = \dfrac{27}{2^4 \cdot 2^{\frac{1}{2}}} = \dfrac{27}{16\sqrt{2}} = \dfrac{27}{16\sqrt{2}} \cdot \dfrac{\sqrt{2}}{\sqrt{2}} = \dfrac{27\sqrt{2}}{32}$

45. $\left(\dfrac{8}{9}\right)^{\frac{-3}{2}} = \left(\dfrac{9}{8}\right)^{\frac{3}{2}} = \dfrac{9^{\frac{3}{2}}}{8^{\frac{3}{2}}} = \dfrac{\left(3^2\right)^{\frac{3}{2}}}{\left(2^3\right)^{\frac{3}{2}}} = \dfrac{3^3}{2^{\frac{9}{2}}} = \dfrac{27}{2^4 \cdot 2^{\frac{1}{2}}} = \dfrac{27}{16\sqrt{2}} = \dfrac{27}{16\sqrt{2}} \cdot \dfrac{\sqrt{2}}{\sqrt{2}} = \dfrac{27\sqrt{2}}{32}$

47. $x^{\frac{3}{4}} \, x^{\frac{1}{3}} \, x^{\frac{-1}{2}} = x^{\frac{3}{4}+\frac{1}{3}+\frac{-1}{2}} = x^{\frac{7}{12}}$

49. $\left(x^3 y^6\right)^{\frac{1}{3}} = \left(x^3\right)^{\frac{1}{3}}\left(y^6\right)^{\frac{1}{3}} = x\, y^2$

51. $\left(x^2 y\right)^{\frac{1}{3}}\left(x y^2\right)^{\frac{2}{3}} = x^{\frac{2}{3}} y^{\frac{1}{3}} x^{\frac{2}{3}} y^{\frac{4}{3}} = x^{\frac{4}{3}} y^{\frac{5}{3}}$

53. $\left(16 x^2 y^{\frac{-1}{3}}\right)^{\frac{3}{4}} = \left(2^4 x^2 y^{\frac{-1}{3}}\right)^{\frac{3}{4}} = 2^3 x^{\frac{3}{2}} y^{\frac{-1}{4}} = \dfrac{8 x^{\frac{3}{2}}}{y^{\frac{1}{4}}}$

55. $\dfrac{x}{(1+x)^{\frac{1}{2}}} + 2(1+x)^{\frac{1}{2}} = \dfrac{x + 2(1+x)^{\frac{1}{2}}(1+x)^{\frac{1}{2}}}{(1+x)^{\frac{1}{2}}} = \dfrac{x + 2(1+x)}{(1+x)^{\frac{1}{2}}} = \dfrac{x + 2 + 2x}{(1+x)^{\frac{1}{2}}} = \dfrac{3x + 2}{(1+x)^{\frac{1}{2}}}$

57. $\dfrac{\sqrt{1+x} - x\cdot\dfrac{1}{2\sqrt{1+x}}}{1+x} = \dfrac{\sqrt{1+x} - \dfrac{x}{2\sqrt{1+x}}}{1+x} = \dfrac{\dfrac{2\sqrt{1+x}\sqrt{1+x} - x}{2\sqrt{1+x}}}{1+x}$

$$= \dfrac{2(1+x) - x}{2(1+x)^{\frac{1}{2}}}\cdot\dfrac{1}{1+x} = \dfrac{2 + x}{2(1+x)^{\frac{3}{2}}}$$

59. $(x+1)^{\frac{3}{2}} + x\cdot\frac{3}{2}(x+1)^{\frac{1}{2}} = (x+1)^{\frac{1}{2}}\left(x + 1 + \frac{3}{2}x\right) = (x+1)^{\frac{1}{2}}\left(\frac{5}{2}x + 1\right) = \frac{1}{2}(x+1)^{\frac{1}{2}}(5x + 2)$

61. $6 x^{\frac{1}{2}}\left(x^2 + x\right) - 8 x^{\frac{3}{2}} - 8 x^{\frac{1}{2}} = 2 x^{\frac{1}{2}}\left(3(x^2 + x) - 4x - 4\right) = 2 x^{\frac{1}{2}}\left(3 x^2 - x - 4\right)$

$$= 2 x^{\frac{1}{2}}(3x - 4)(x + 1)$$

Section 9 Geometry Review

1. $a = 5,\ b = 12,\ c^2 = a^2 + b^2 = 5^2 + 12^2 = 25 + 144 = 169 \ \rightarrow\ c = 13$

3. $a = 10,\ b = 24,\ c^2 = a^2 + b^2 = 10^2 + 24^2 = 100 + 576 = 676 \ \rightarrow\ c = 26$

5. $a = 7,\ b = 24,\ c^2 = a^2 + b^2 = 7^2 + 24^2 = 49 + 576 = 625 \ \rightarrow\ c = 25$

7. $5^2 = 3^2 + 4^2 \ \rightarrow\ 25 = 9 + 16 \ \rightarrow\ 25 = 25$
   The given triangle is a right triangle. The hypotenuse is 5.

9. $6^2 = 4^2 + 5^2 \ \rightarrow\ 36 = 16 + 25 \ \rightarrow\ 36 \neq 41$
   The given triangle is not a right triangle.

11. $25^2 = 7^2 + 24^2 \;\rightarrow\; 625 = 49 + 576 \;\rightarrow\; 625 = 625$
   The given triangle is a right triangle. The hypotenuse is 25.

13. $6^2 = 3^2 + 4^2 \;\rightarrow\; 36 = 9 + 16 \;\rightarrow\; 36 \neq 25$
   The given triangle is not a right triangle.

15. $A = l \cdot w = 4 \cdot 2 = 8 \text{ in}^2$

17. $A = \frac{1}{2} b \cdot h = \frac{1}{2}(2)(4) = 4 \text{ in}^2$

19. $A = \pi r^2 = \pi(5)^2 = 25\pi \text{ m}^2 \qquad C = 2\pi r = 2\pi(5) = 10\pi \text{ m}$

21. $V = lwh = 8 \cdot 4 \cdot 7 = 224 \text{ ft}^2$

23. $V = \frac{4}{3}\pi r^3 = \frac{4}{3}\pi \cdot 4^3 = \frac{256}{3}\pi \text{ cm}^3 \qquad S = 4\pi r^2 = 4\pi \cdot 4^2 = 64\pi \text{ cm}^2$

25. $V = \pi r^2 h = \pi(9)^2(8) = 648\pi \text{ in}^3$

27. The diameter of the circle is 2, so its radius is 1. $A = \pi r^2 = \pi(1)^2 = \pi$ square units

29. The diameter of the circle is the length of the diagonal of the square.
   $$d^2 = 2^2 + 2^2 = 4 + 4 = 8 \;\rightarrow\; d = \sqrt{8} = 2\sqrt{2} \qquad r = \sqrt{2}$$
   The area of the circle is: $A = \pi r^2 = \pi\left(\sqrt{2}\right)^2 = 2\pi$ square units

31. The total distance traveled is 4 times the circumference of the wheel.
   Total distance $= 4C = 4(\pi d) = 4\pi \cdot 16 = 64\pi = 201.1$ inches

33. Area of the border = area of EFGH – area of ABCD $= 10^2 - 6^2 = 100 - 36 = 64 \text{ ft}^2$

35. Area of the window = area of the rectangle + area of the semicircle.
   $$A = (6)(4) + \tfrac{1}{2}\pi \cdot 2^2 = 24 + 2\pi = 30.28 \text{ ft}^2$$
   Perimeter of the window = 2 heights + width + one-half the circumference.
   $$P = 2(6) + 4 + \tfrac{1}{2}\pi(4) = 12 + 4 + 2\pi = 16 + 2\pi = 22.28 \text{ feet}$$

37. Convert 20 feet to miles, and solve the Pythagorean theorem to find the distance:
   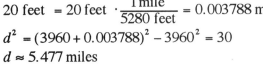
   $$20 \text{ feet} = 20 \text{ feet} \cdot \frac{1 \text{ mile}}{5280 \text{ feet}} = 0.003788 \text{ miles}$$
   $$d^2 = (3960 + 0.003788)^2 - 3960^2 = 30$$
   $$d \approx 5.477 \text{ miles}$$
   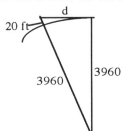

39. Convert 100 feet to miles, and solve the Pythagorean theorem to find the distance:

$$100 \text{ feet } = 100 \text{ feet } \cdot \frac{1 \text{ mile}}{5280 \text{ feet}} = 0.018939 \text{ miles}$$

$$d^2 = (3960 + 0.018939)^2 - 3960^2 = 150$$

$$d \approx 12.247 \text{ miles}$$

Convert 150 feet to miles, and solve the Pythagorean theorem to find the distance:

$$150 \text{ feet } = 150 \text{ feet } \cdot \frac{1 \text{ mile}}{5280 \text{ feet}} = 0.028409 \text{ miles}$$

$$d^2 = (3960 + 0.028409)^2 - 3960^2 = 225$$

$$d \approx 15 \text{ miles}$$

Section 10   Completing the Square; The Quadratic Formula

1. $x^2 - 4x + \underline{\quad}$ 4 should be added to complete the square.

3. $x^2 + \frac{1}{2}x + \underline{\quad}$ $\frac{1}{16}$ should be added to complete the square.

5. $x^2 - \frac{2}{3}x + \underline{\quad}$ $\frac{1}{9}$ should be added to complete the square.

7. $x^2 + y^2 - 4x + 4y - 1 = 0$
$$x^2 - 4x + \underline{\quad} + y^2 + 4y + \underline{\quad} = 1$$
$$x^2 - 4x + 4 + y^2 + 4y + 4 = 1 + 4 + 4$$
$$(x - 2)^2 + (y + 2)^2 = 9$$

9. $x^2 + y^2 + 6x - 2y + 1 = 0$
$$x^2 + 6x + \underline{\quad} + y^2 - 2y + \underline{\quad} = -1$$
$$x^2 + 6x + 9 + y^2 - 2y + 1 = -1 + 9 + 1$$
$$(x + 3)^2 + (y - 1)^2 = 9$$

11. $x^2 + y^2 - x - y - \frac{1}{2} = 0$
$$x^2 - x + \underline{\quad} + y^2 - y + \underline{\quad} = \frac{1}{2}$$
$$x^2 - x + \frac{1}{4} + y^2 - y + \frac{1}{4} = \frac{1}{2} + \frac{1}{4} + \frac{1}{4}$$
$$\left(x - \frac{1}{2}\right)^2 + \left(y - \frac{1}{2}\right)^2 = 1$$

13. $x^2 + 4x - 21 = 0$

$x^2 + 4x = 21$

$x^2 + 4x + 4 = 21 + 4$

$(x + 2)^2 = 25$

$x + 2 = \pm 5$

$x = -2 \pm 5$

$\{-7, 3\}$

15. $x^2 - \frac{1}{2}x = \frac{3}{16}$

$x^2 - \frac{1}{2}x + \frac{1}{16} = \frac{3}{16} + \frac{1}{16}$

$\left(x - \frac{1}{4}\right)^2 = \frac{1}{4}$

$x - \frac{1}{4} = \pm\frac{1}{2}$

$x = \frac{1}{4} \pm \frac{1}{2}$

$\left\{\frac{-1}{4}, \frac{3}{4}\right\}$

17. $3x^2 + x - \frac{1}{2} = 0$

$3x^2 + x = \frac{1}{2}$

$x^2 + \frac{1}{3}x = \frac{1}{6}$

$x^2 + \frac{1}{3}x + \frac{1}{36} = \frac{1}{6} + \frac{1}{36}$

$\left(x + \frac{1}{6}\right)^2 = \frac{7}{36}$

$x + \frac{1}{6} = \pm\frac{\sqrt{7}}{6}$

$x = -\frac{1}{6} \pm \frac{\sqrt{7}}{6}$

$\left\{\frac{-1 - \sqrt{7}}{6}, \frac{-1 + \sqrt{7}}{6}\right\}$